Microbial Biotechnology

Principles and Applications

Third Edition

Microbial Biotechnology

Principles and Applications

Third Edition

Editor

Lee Yuan Kun

National University of Singapore

World Scientific

NEW JERSEY · LONDON · SINGAPORE · BEIJING · SHANGHAI · HONG KONG · TAIPEI · CHENNAI

Published by

World Scientific Publishing Co. Pte. Ltd.

5 Toh Tuck Link, Singapore 596224

USA office: 27 Warren Street, Suite 401-402, Hackensack, NJ 07601

UK office: 57 Shelton Street, Covent Garden, London WC2H 9HE

Library of Congress Cataloging-in-Publication Data
Microbial biotechnology : principles and applications / edited by Yuan Kun Lee,
National University of Singapore, Singapore. -- 3rd edition.
    pages cm
    Includes bibliographical references and index.
    ISBN 978-9814366816 (hardcover : alk. paper) --
    ISBN 978-9814366823 (pbk. : alk. paper)
    1. Microbial biotechnology. I. Lee, Y. K. (Yuan Kun)
    TP248.27.M53M52 2013
    660.6'2--dc23
                                        2012046471

British Library Cataloguing-in-Publication Data
A catalogue record for this book is available from the British Library.

Typeset by Stallion Press
Email: enquiries@stallionpress.com

Printed by FuIsland Offset Printing (S) Pte Ltd Singapore

Preface

Life science research and industry is developing rapidly all over the world. Microbial biotechnology is increasingly being regarded as a core subject in most university and polytechnic life science courses. There are already a number of excellent general textbooks on microbiology and biotechnology in the market that deal with the basic principles of the field. In order to complement these general textbooks, this book aims to focus on the applications of microbial biotechnological principles. A teaching-based format is adopted, whereby working problems, as well as answers to frequently asked questions are included to supplement the main text. This textbook also includes real life examples on how the application of microbial biotechnological principles has achieved breakthroughs in both research and industrial production.

The rapidly expanding molecular biology, genetic, protein and metabolic techniques and approaches have significant impact on the Microbial bio-technology; this leads to the addition of three new chapters in the third edition of this textbook. They are the Chapter 3: Application of -Omics Technologies in Microbial Fermentation, Chapter 5: Microbial Genome Mining for Identifying Antimicrobial Targets, and the Chapter 21: Bacterial Biofilm: Molecular Characterization and Impacts on Water Management". The Chapter 1: Microbial Screening and Chapter 15: Transgenic Plants are re-written as fresh chapters. For completion, Part VI is expanded to include Biomining (Chapter 23: Microbial Biomining). Most of the others chapters are updated in this new edition.

Although this textbook is written for university undergraduates and poly-technic students, it contains sufficient details to be used as a reference book for postgraduate students and lecturers. It may also service as a source book for corporate planners, managers and applied research personnel.

Lee Yuan Kun
Editor

Contents

List of Contributors

Anwar, Azlinda, BSc (Hons), PhD
Program in Infectious Diseases, Duke-NUS Graduate Medical School
8 College Rd, Singapore 169857
Email: azlinda.anwar@duke-nus.edu.sg

Bu, Yun Ping, BSc, PhD
Temasek Life Sciences Laboratory
National University of Singapore
1 Research Link, Singapore 117604
Email: Buyp@tll.org.sg

Candlish, John, PhD, LLM
Faculty of Medicine and Health Sciences, University of Malaysia
Sarawak, Kuching, East Malaysia
Email: jkcandlish@yahoo.com.sg

Chang, Siao Yun, BSc (Hons), PhD
Technology & Water Quality Office, Public Utilities Board (PUB)
82 Toh Guan Road East, WaterHub, Singapore 608576
Email: Siao_Yun_ Chang@pub.gov.sg

Chow, Vincent T. K., MBBS, MSc, Dip Microbiology, FRC Path, MD, PhD
Infectious Diseases Program
Department of Microbiology, Yong Loo Lin School of Medicine
National University Health System
National University of Singapore

Kent Ridge, Singapore
Email: micctk@nus.edu.sg

Chu, Lee Man, BSc (Hons), PhD
Department of Biology, The Chinese University of Hong Kong
Shatin, NT, Hong Kong SAR, China
Email: leemanchu@cuhk.edu.hk

Chua, Kim Lee, BSc (Hons), Dip Ed, PhD
Department of Biochemistry, Yong Loo Lin School of Medicine
National University of Singapore
MD 7, 8 Medical Drive, Singapore 117597
Email: kim_lee_chua@nuhs.edu.sg

Deng, Lih Wen, BSc (Hons), PhD
Department of Biochemistry, Yong Loo Lin School of Medicine
National University of Singapore
MD 7, 8 Medical Drive, Singapore 117597
Email: bchdlw@nus.edu.sg

Gao, Pingping, BSc, PhD
Technology & Water Quality Office, Public Utilities Board (PUB)
82 Toh Guan Road East, WaterHub, Singapore 608576
Email: pingping_ gao@pub.gov.sg

He, Fang
Institute of Molecular Agrobiology
National University of Singapore
1 Research Link, Singapore 117604

Hong, Yan, BSc, PhD
Temasek Life Sciences Laboratory
National University of Singapore
1 Research Link, Singapore 117604
Email: hongy@tll.org.sg

Inglis, Tim J. J., BM, DM, PhD, FRCPath, DTM&H, FRCPA
School of Pathology and Laboratory Medicine
University of Western Australia
Crawley WA 6009, Perth, Australia
Email: tim.inglis@uwa.edu.au

Kwang, Jimmy, DVM, PhD
Institute of Molecular Agrobiology
National University of Singapore
1 Research Link, Singapore 117604
Email: kwang@tll.org.sg

Lee, Dong-Yup, BSc (Hons), PhD
Bioprocessing Technology Institute, A*STAR
20 Biopolis Way, #06–01 Centros, Singapore 138668, and
Department of Chemical and Biomolecular Engineering
National University of Singapore
4 Engineering Drive 4, Blk E5 #02–09, Singapore 117576
Email: cheld@nus.edu.sg

Lee, Yuan Kun, BSc (Hons), PhD
Department of Microbiology, Yong Loo Lin School of Medicine
National University of Singapore
5 Science Drive 2, Singapore 117597
Email: micleeyk@nus.edu.sg

Lim, Allan, BSc (Hons), PhD
Nestlé R&D Center Pte Ltd
15A, Changi Business Park Central 1
#05–02/03 Eightrium@Changi Business Park
Singapore 486035
Email: Allan.Lim@rdsg.nestle.com

Loke, Paxton, BSc (Hons), PhD
Ian Potter Hepatitis Research Laboratory
The Macfarlane Burnet Institute
  for Medical Research and Public Health
Commercial Road, Melbourne, Vic 3004, Australia

O'Toole, Desmond K., BSc, PhD
Department of Biology and Chemistry, City University of Hong Kong
83 Tat Chee Avenue, Kowloon, Hong Kong, China
Email: d.k.otoole@gmail.com

Premkumar, Jayaraman
Biomedical Engineering Research Centre
Nanyang Technological University, Singapore 637553

Sakharkar, Kishore R.
Biomedical Engineering Research Centre
Nanyang Technological University, Singapore 637553

Sakharkar, Meena K.
Graduate School of Life and Environmental Sciences
University of Tsukuba, Ibaraki, Japan 305–8572

Sim, Tiow Suan, BSc (Hons), PhD
Department of Microbiology, Yong Loo Lin School of Medicine
National University of Singapore
5 Science Drive 2, Singapore 117597
Email: tiow_suan_sim@nuhs.edu.sg

Song, Keang Peng, BDS, PhD
School of Science, Monash University
Sunway Campus, Selangor, Malaysia
Email: song.keang.peng@sci.monash.edu.my

Tan, Hai Meng, BSc (Hons), PhD
Kemin Industries (Asia) Pte Ltd
12 Senoko Drive, Singapore 758200
Email: haimeng.tan@kemin.com

Tan, Theresa May Chin, BSc, PhD
Department of Biochemistry, Yong Loo Lin School of Medicine
National University of Singapore
MD 7, 8 Medical Drive, Singapore 117597
Email: bchtant@nus.edu.sg

Teng, Wee Lin BSc (Hons), PhD
Technology & Water Quality Office, Public Utilities Board (PUB)
82 Toh Guan Road East, WaterHub, Singapore 608576
Email: sarah_teng@pub.gov.sg

Too, Heng-Phon, BSc, PhD
Department of Biochemistry, Yong Loo Lin School of Medicine
National University of Singapore
MD 7, 8 Medical Drive, Singapore 117597, and
Chemical and Pharmaceutical Engineering
Singapore-Massachusetts Institute of Technology
E4–04–10, 4 Engineering Drive 3, Singapore 117576
Email: bchtoohp@nus.edu.sg

Wong, Po-Keung, BSc (Hons), PhD
Department of Biology, The Chinese University of Hong Kong
Shatin, NT, Hong Kong SAR, China
Email: pkwong@cuhk.edu.hk

Wong, Victor Vai-Tak, BSc (Hons), PhD
Bioprocessing Technology Institute, A*STAR
20 Biopolis Way, #06–01 Centros, Singapore 138668, and
Department of Chemical and Biomolecular Engineering
National University of Singapore
4 Engineering Drive 4, Blk E5 #02–09, Singapore 117576
Email: victor.wong@lonza.com

Yew, Wen Shan
Department of Biochemistry, Yong Loo Lin School of Medicine
National University of Singapore
MD 7, 8 Medical Drive, Singapore 117597

Yusufi, Faraaz Noor Khan, BSc (Hons), PhD
Bioprocessing Technology Institute, A* STAR
20 Biopolis Way, #06-01 Centros, Singapore 138668
Email: faraaz_yusufi@bti.a-star.edu.sg

Part I

Principles of Microbial Biotechnology

The use of microorganisms for large-scale industrial purposes has a long history, which is long before the realization of the activities of the microorganisms. For centuries, beer, wine, vinegar, soy sauce and other fermented foods were produced through spontaneous fermentation of natural occurring microorganisms or the use of carry-over microbial seeds from the previous batch of production. The quality and productivity of these early products were very often inconsistent. The development of scientific screening and isolation methods allows the selection of desirable natural occurring or mutated microorganisms for specific purposes. These methods, coupled with the advancement of the technical know-how in large-scale sterilization of culture media, in provision of adequate oxygen supply and in mixing homogeneity of the culture systems, enable the exploitation of both anaerobic (yeast and some bacteria) and aerobic microorganisms (fungi and some bacteria). Common examples are: the development of large-scale processes for the production of citric acid, amino acids and antibiotics; in improved biotransformation of steroid hormones, as well as the mass production of many enzymes. The diverse catalytic activities of microorganisms are being used more and more widely to perform specific chemical reactions in the industrial production processes.

Microbial biotechnology was pushed to a new height in the eighties when continuous fermentation and airlift fermentation processes were developed for the production of food and feed grade microbial protein from industrial by-products, such as methanol and alkanes. These processes lead to considerable savings in capital, energy and labor costs.

Modern techniques of gene manipulation, and advanced bioinformatics and biocomputing are powerful tools for genomic, proteomic and metabolomic research. The scientific breakthroughs that ensued have made feasible the industrial manufacturing of non-microbial products, such as human growth hormone, interferon and viral vaccines; and defined approaches in the monitoring and control of fementation processes.

It is the aim of Part I of this book to provide the readers with in-depth and comprehensive scientific knowledge in the areas as listed below, so as to facilitate the understanding of the various applications of micro-organisms and the production of their bioactive molecules in the biotechnological systems:

- Screening for microbial products
- Bioprocess technology
- Enzymology
- Manipulation of genes and metabolic pathways
- Application of bioinformatics and biocomputing.

Chapter 1

Microbial Screening

Allan Lim
Nestlé R&D Center Pte Ltd
15A, Changi Business Park Central 1
#05-02/03 Eightrium@Changi Business Park, Singapore 486035

Tan Hai-Meng
Kemin Industries (Asia) Pte Ltd
12 Senoko Drive, Singapore 758200

1.1. Introduction

Microbes can be harnessed as "mini-factories" to produce primary and secondary metabolites for human applications such as ethanol for breweries, citric acid for food, and antibiotics for disease prophylaxis and treatment (Fig. 1.1). Since most of the metabolic pathways in microorganisms are very similar, the primary objective of screening is to search for the strain(s) that is most efficient and cost effective in producing these metabolites. Other considerations such as safety and sensory profiles are particularly important in the food and beverage industries. For instance, enormous resources are used to isolate special yeast strains (e.g. *Saccharomyces uvarum (carlsbergenesis)* and *Torulaspora delbrueckii*) that confer unique flavors to beers. In contrast, microbes for non-food applications have much less restrictions, and they can even be mutants generated in laboratories. For example, genetically-modified *Escherichia coli* and *Pseudomomas* spp. are now used extensively in bioremediation.

A screen is a systematic assay procedure that allows testing of numerous compounds for a particular activity, phenotype for a defined

3

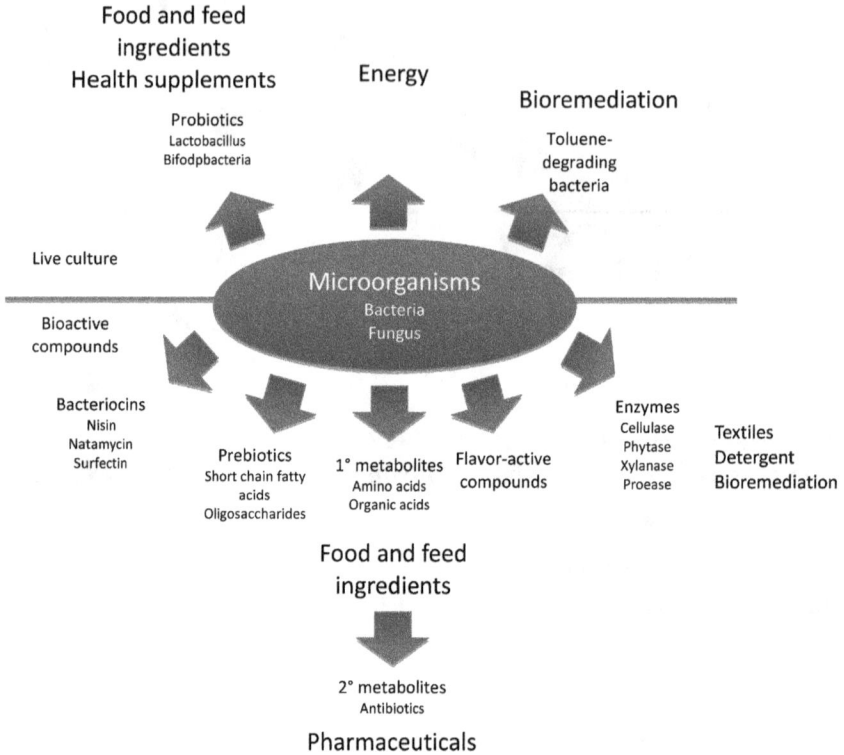

Food and feed
ingredients
Health supplements

Energy

Bioremediation

Probiotics
Lactobacillus
Bifodpbacteria

Toluene-
degrading
bacteria

Live culture

Microorganisms
Bacteria
Fungus

Bioactive
compounds

Bacteriocins
Nisin
Natamycin
Surfectin

Prebiotics
Short chain fatty
acids
Oligosaccharides

1° metabolites
Amino acids
Organic acids

Flavor-active
compounds

Enzymes
Cellulase
Phytase
Xylanase
Proease

Textiles
Detergent
Bioremediation

Food and feed
ingredients

2° metabolites
Antibiotics

Pharmaceuticals

Fig. 1.1. Application of microorganisms in various industries.

application. The better the design of the screen, the higher the chance of finding the strain in the shortest time. Access to unique or previously unculturable microorganisms will also ensure the success of screening. Many companies even patent sources or screening procedures for competitive advantage, such as the process of "bioprospecting" a proprietary genetic bank of microbes from extreme ecosystems (e.g. volcanoes and deep sea hydrothermal vents). Proprietary screening technology will then facilitate rapid discovery and evolution of unique and tailored enzymes from this genetic bank to suit various challenging industrial environments. Companies are also known to partner with governmental research institutions (e.g. BIOTEC, National Centre for Genetic Engineering and Biotechnology, Thailand) to screen thousands of enzymes and microorganisms from diverse environments for possible applications in disease treatment and drug development. It is usual for such collaborations to have the country compensated either in monetary or non-monetary means,

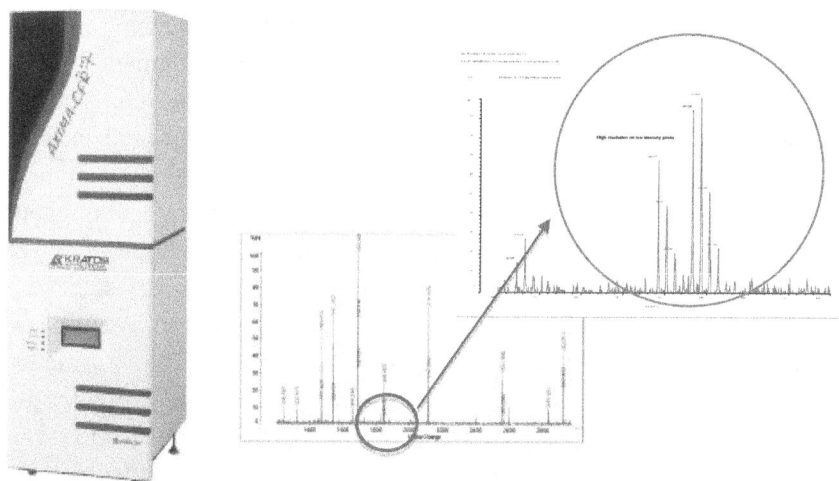

Fig. 1.2. High Performance MALDI-TOF MS designed for the mass analysis of biomolecules. High quality peptide mapping fingerprints (PMF) could be generated easily and resolved at high resolution. (Courtesy of Shimadzu (Asia Pacific) Pte Ltd.)

such as capacity building and technology transfer. As legal owners of these microorganisms and natural compounds, the research institutions will also continue to evaluate them in their home country for applications outside the scope of the commercial collaboration.

The scale and speed of screening has also been dramatically increasing at an unprecedented speed due to the availability of advanced instrumentations, such as the MALDI-TOF/TOF and LC-MALDI for high throughput screening of metabolites (Fig. 1.2).

Although high throughput screening was primarily developed for drug screening, it is increasingly being deployed in screening of microbials for the target metabolites. Today, this is often coupled with high throughput DNA sequencing so that shortlisted microbial strains can be rapidly characterized.

1.2. Screening Process

Several key areas are involved in the screening process of microorganisms and their metabolites. The first is the development and validation of assay(s), which upon further assessment for its reproducibility and compatibility with the physicochemical characteristics of the samples converting it to a primary or pilot screen. The primary screening process makes possible the estimation of hit rates and for large numbers of samples to be screened rapidly to obtain sufficient data to initiate a full screening program. The second area is high throughput screening, which enables large numbers of potential samples to be screened with minimum numbers of steps using semi- or fully automated systems. The high throughput screening system will also perform retesting and confirmation of active metabolites from the initial screen. The third area is secondary screening of the active metabolites or "hits" from the primary screen to eliminate false positive results. At this stage, active fractions are further characterized based on heat stability, chemical classes, molecular weights, pH optimum and enzyme inhibition patterns. Following purification and structure elucidation of active metabolites, the potential molecules are then subjected to further *in vivo* testing.

1.3. Screening Methods

The use of appropriate isolation procedures is an important initial step in the development of preliminary screens for useful secondary metabolites. Basically, this involves the selective isolation of producer-bacteria and a demonstration, usually visual, of their secondary metabolites. Following selection and pre-treatment of the material containing bacterial strains of interest, specific selective laboratory media are used for the growth and isolation of the microorganisms of interest. Pre-treatment procedures used to increase the relative number of producer microorganisms e.g. antibiotic-producing actinomycetes relative to bacteria on isolation plates can include processes of simple drying and dry heating to reduce the number of undesired Gram-negative bacteria. The latter can cause swarming on the isolation plates but the aerial spores of most actinoycetes are relatively resistant to desiccation. Incubation conditions generally include temperatures of 25–30°C and periods of 7–14 days although longer incubation periods may be needed for slower growing strains. In some cases, the incorporation of spore activating and bactericidal agents in isolation media such as sodium doedcyl sulphate, phenol, antibiotics and yeast extract in combination with dry heat can help increase the recovery of actinomycetes while suppressing the counts of bacteria. Antibiotics such as rifampicin, novobiocin and oxytetracylcine are commonly used for the isolation of *Acitnomadura, Micromonospora* and *Streptoverticillium*, respectively. At the same time, antifungal agents such as cycloheximide and nyastatin which do not inhibit actinomycetes are regularly added to eliminate fungal contamination. To help in the isolation of specific taxa of actinomycetes, chromogenic or fluorogenic substrates for genus specific enzymes can be incorporated into the selective media used.

Once the microorganisms of interest have been isolated, several methods can be used to investigate the antimicrobial activities of secondary metabolites produced by these microorganisms. This can involve overlaying the isolation plate with a single test or indicator organism. Another variation, also known as the agar-spot test, has also been used for the detection of antagonistic activity between bacteria. These methods can be quite laborious since only one indicator organism at a time can be applied to each isolation plate. As an alternative, replica plating of selected colonies can be used to study a range of indicator organisms. However, the spread of motile bacteria on the surface of each isolation plate can limit the usefulness of this method.

The sensitivity of an isolation medium can be influenced by its nutrient composition, pH, selective agents and incubation temperature. Typically, nonselective and/or selective enrichments are required to increase the sensitivity of detection of target bacteria. Generally, the stepwise enrichment process begins with pre-enrichment or primary enrichment, in which contaminated samples are incubated in a nutritious nonselective medium to allow for the growth of target bacteria. Pre-enriched samples are then transferred to a secondary selective enrichment medium, where the normal flora is suppressed but the target bacteria is allowed to grow. Finally, the producer bacteria is streaked on selective isolation agar and isolated as a pure culture. Pure producer bacterial culture is then transferred into a broth system to begin the fermentation and optimization process of enhancing the production of active secondary metabolites. Filtration of fermented broth through a membrane filter (0.20–0.45 μm pore size) is routinely used to separate cells from the aqueous phase. The filtrate is then subjected to screening against a spectrum of indicator organisms. The disk-activity assay has been the method of choice used in the comparison of relative activities of active filtrates from various bacterial fermented broths. Sterile analytical paper disks (0.5 inch in diameter) are immersed in sample filtrate and placed onto the surface of solidified agar medium seeded with an indicator organism. Following overnight incubation at a specific temperature, the relative activities between different filtrates can be observed from size of the zone of inhibition on each medium seeded with indicator organisms. Alternatively, a modified agar-well diffusion method can be used to examine the activity of the active filtrates from various bacterial fermented broths. In this method, wells are aseptically created by a hole-borer in a solidified medium seeded with indicator organisms. Similarly, the relative activities between different filtrates are determined based on the sizes of the zone of inhibition on the solidified medium. However, both techniques using the disk and well assays are laborious and time-consuming when screening the active filtrate against a wide range of bacteria. Therefore, to enable rapid screening of active samples, screens may be carried out using semi- or fully automated high throughput screening systems. The major advantage of using the high throughput screening process is in the quick turn-around time to screen numerous samples against a wider range of microorganisms.

This screening method may also be extended to perform numerous bioassays to determine the minimum-inhibitory concentrations (MIC) of the active fractions against a range of indicator-microorganisms. In the minimum-inhibitory concentration assay, an active sample undergoes a two-fold dilution procedure using suitable broth and a portion from each dilution is transferred into the wells of a microtiter plate. Following incubation for 24–48 hours, the optical density of each culture broth in the wells is measured spectrophotometrically with a microtiter plate reader at a stipulated wavelength. Automated high throughput and target directed screening are now used together with structure-based design and combinatorial chemistry in drug discovery. High throughput screening is used to generate lead compounds in the initial part of the drug discovery process. Such lead compounds are then subjected to more vigorous biological and chemical analyses. The application of combinatorial chemistry generates a whole library of new variant compounds that can be tested for the elucidation of the structure-function relationship.

Where the intention is to isolate microorganisms capable of carrying out degradation of environmental contaminants, target microorganisms can be cultivated in the presence of the contaminant, and the growth of the microorganisms followed either spectrophotometrically or by plating. Alternatively, the disappearance or drop in the concentration of the contaminant, or the concomitant appearance of the breakdown product of the contaminant can be analyzed over a time course.

The task of identifying the main active component(s) in bacterial filtrates can be complicated by the presence of both organic acids and antimicrobial peptides or proteins. It has been established that the predominant organic compounds found in a fermented product by *P. freudenreichii* ssp. *shermanii* are diacetyl, acetic, propionic, and lactic acids as well as an antimicrobial peptide. A method to remove organic acids from fermented broth involves a lyophilization process at a vacuum of 150 millitorrs and a temperature of –40°C. Lyophilization under such conditions would cause the acids to volatilize and evaporate from crude fermented extract leaving behind the largely unaffected antimicrobial substances for further analysis. Alternatively, solvents may be used to extract organic acids from fermented products. The samples can be centrifuged to remove cell debris and growth medium components. Bioactive supernatants are then extracted with an equal volume of organic solvent (e.g. ethyl-acetate) for 1 hour. The fractions obtained are allowed to evaporate under vacuum to dryness using a rotary-evaporator set at 70°C to remove the organic solvent. Samples are rehydrated with distilled water and then analyzed for the presence of organic acids by gas-liquid chromatography. A simpler technique with hydroxyl-treated anion-exchanger has been found to be effective in binding organic acids can be used to obtain acid-free active ingredient(s) from crude extract.

The proteinaceous nature of these antimicrobial substances is then determined using proteolytic enzymes. Generally, antimicrobial peptides are not sensitive to lipases, amylases, RNAses or DNAses. The effect of heat, pH, hydrostatic pressure, chemicals and ions may be included for assessment of the antimicrobial peptides present in the active filtrate or supernatant fractions.

The active peptides or proteins can be quantified using various molecular weight cut-off membranes at ultra-high pressure under nitrogen gas. All retentate and filtrate samples can be assayed using the disk-activity or well-assay methods for antimicrobial activity against specific indicator organisms. Active molecular weight cut-off filtrate or retentate is then analyzed for total protein concentration using the Lowry Protein Assay. Isoelectric focusing is a sensitive technique for protein separation based on the principle that protein migration in an electrically charged field will cease at its isoelectric point. The isoelectric point of a protein is defined as the point of electrical neutrality (zero-charge). Individual proteins or peptides are separated into specific bands based on a pH gradient. The individual bands are then eluted and tested for antimicrobial activity against specific indicator-organisms.

Other qualitative methods to determine molecular weights for partially purified active components include gel filtration and SDS-polyacrylamide gel electrophoresis. Further purification of the active compounds may involve multi-steps such as separation and concentration of the molecules by different chromatographic methods, including hydrophobic, ion exchange, and size exclusion chromatography. Purified antimicrobial peptides are then analyzed and sequenced to determine their amino acids composition using mass spectrometry and Edman degradation.

1.4. Screening Microorganisms for Probiotic Functions

The probiotic *Lactobacillus acidophilus* La1 is one of the few probiotic bacteria discovered and commercialized (Bernet *et al.*, 1994). Using *in vitro* assays, the adhesive and protective properties of the probiotic *Lactoacillus acidophilus* La1 were confirmed (Fig. 1.3). In the first assay, four proprietary strains of *Lactobacillus acidophilus* strains were incubated with monolayer cultures of Caco-2 cells and HT-29-MTX on cover slips for one hour at 37°C/10% CO2/90% air. After washing with phosphate buffered saline, the number of adherent bacteria was counted manually. In the second assay, the protective function of *Lactobacillus acidophilus* was demonstrated by challenging the Lactobacilli-bound Caco-2 cells with radiolabeled pathogenic strains of *E. coli*. *Lactobacillus acidophilus* La1 was found to be have the highest adherence to Caco-2 and HT-29-MTX cells, and highest inhibition to the adhesion of pathogenic bacteria (enterotoxigenic and enteropathogenic *E. coli*, and *Yersinia pseudotuberculosis* and *Salmonella tiphymurium*) to Caco-2 cells.

On the other hand, in an attempt to develop an active microbial strain to inhibit *C. perfringens*, the causative agent of necrotic enteritis, in place of antibiotic growth promoters or AGPs, *Bacillus spp.* and *Lactobacillus spp.*

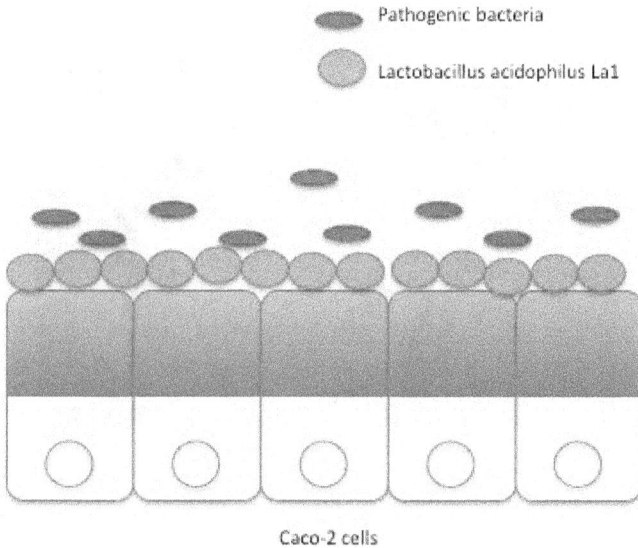

Fig. 1.3. *Lactobacillus acidophilus* La1 was able to prevent adhesion of pathogenic bacteria such as *E. coli, Salmonella sp.* and *Yersina sp.* to Caco-2 cells. (Adapted from Pfeifer *et al.*, 1997.)

were originally cultured from different sections of the gastrointestinal tracts of healthy chickens and screened for anti-clostridial activity against *Clostridium perfringens* ATCC 13124 (Teo and Tan, 2005). Using such an approach, a commercial product CloSTAT™, containing *Bacillus subtilis* PB6 was developed (Fig. 1.4).

This patented proprietary strain, closely associated with the intestinal epithelium, was able to tolerate gastric and bile conditions. The *Bacillus subtilis* PB6 strain is able to form spores and this is a huge advantage in the strain's survival during pellet formation, which is widely used for broiler feed production. *Bacillus subtilis* PB6 is found to secrete a group of peptides that, while inactive towards beneficial gut bacteria such as *Lactobacillus* and *Bidifidobacterium* spp., are inhibitory towards common enteric pathogens such as *Clostridium perfringens*, implicated in outbreaks of necrotic enteritis in broiler chickens and *Campylobacter* spp. in *in vitro* well diffusion assays. Using LC/MS and Q-TOF analyses, the active metabolites have been found to possess a cyclic structure comprising seven amino acids, collectively known as surfactins (Fig. 1.5). The

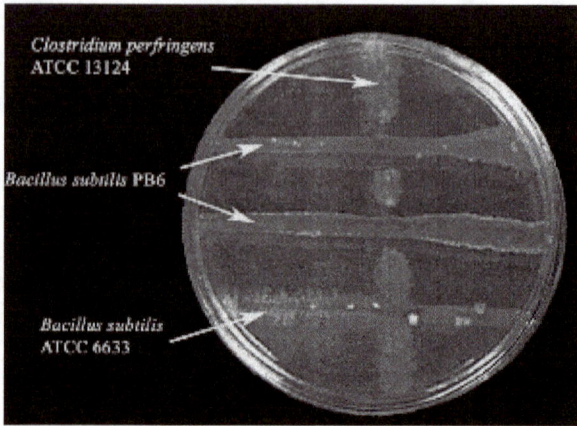

Fig. 1.4. Antagonistic assay showing the zone of clearance of *Clostrium perfringens* ATCC 13124 by *Bacillus subtilis* PB6.

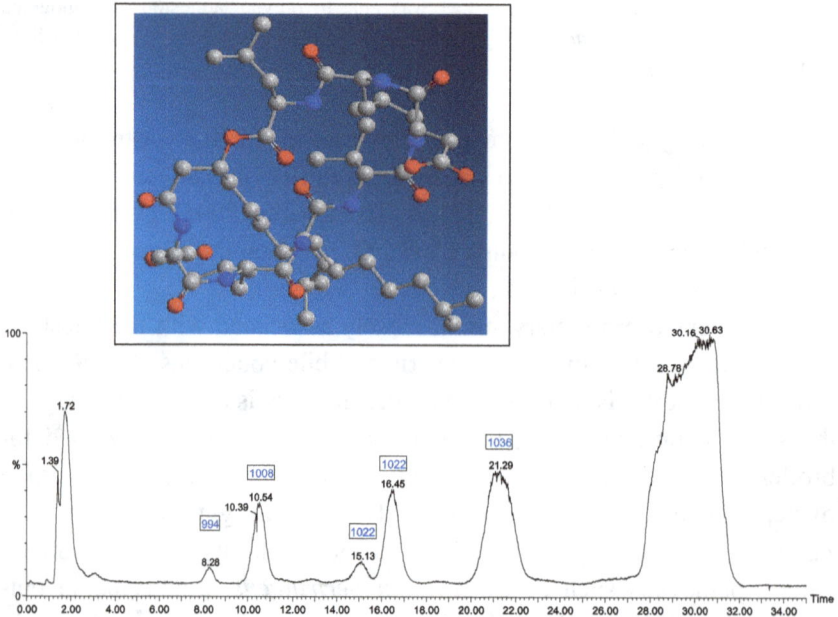

Fig. 1.5. LC-MS chromatogram of the extract of *B. subtilis* PB6 containing surfactins. (Insert: structure of a surfactin.)

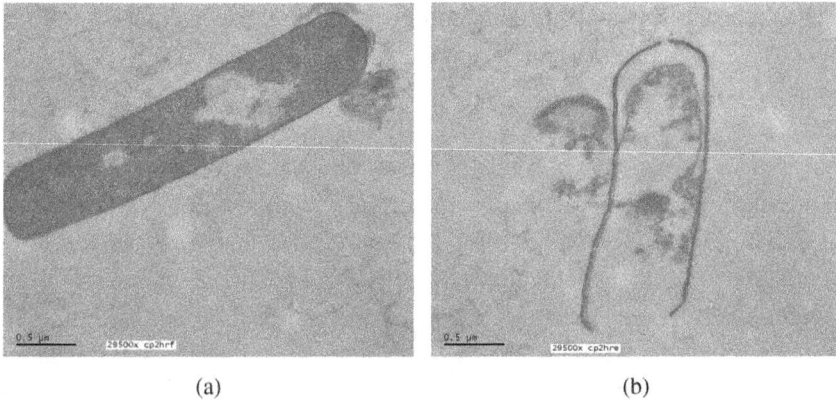

(a)   (b)

Fig. 1.6.   The effect of CloSTAT™ on *Clostridium perfringens* at 37°C. (a) Disruption of cell wall and loss of cytoplasmic contents into the exterior after 1 hour. (b) Rupture and death of cell after 4 hours (transmission electron micrograph magnification: 29000 ×).

mode of killing action is due to the ability of the surfactin molecules to form pores on the cell walls of the bacteria as shown using electron microscopy (Fig. 1.6).

1.5. Screening Microorganisms for Enzymes

Microbial enzymes have been harnessed by various industries for many decades. The use of bacterial protease in commercial detergents in 1960s marked the birth of industrial microbial enzymes. Today, microbial enzymes are used in food processing (α-amylase, pectinase), textiles (cellulase), detergent (protease, lipase), animal feed additive (α-amylase, protease, cellulase, xylanase, phytase). The classical screening procedure relies primarily on colorimetric or viscometric measurements to identify bacterial or fungal strains with the highest activities, and is therefore often the rate-limiting step. Today, this process is replaced by combination of molecular tools and metabolomics (Dalbøge and Lange, 1998).

1.5.1. *Phytase*

Phytase is a unique enzyme that is capable of dephosphorylating myoinositol hexakisphosphate resulting in inositol and inorganic phosphates. It is found in many microorganisms, especially *Aspergillus* species, and some mammalian tissues. This enzyme has gained much attention in the feed industry due to its proven ability to increase the bio-availability of nutrients in the feed. Although phytase was originally dis-covered in *E. coli*, the current industrial phytases are obtained from the fermentation of *Aspergillus niger*. There are 2 *phy*A genes: the 3-phytase (EC 3.1.3.8) gene isolated from *Aspergillus ficuum*, and expressed in *Aspergillus niger*; and the 6-phytase (EC 3.1.3.26) gene derived from *Peniophora lycii* and expressed in *Aspergillus niger* (Haefner *et al.*, 2005).

The screening assay for phytase–producing strains is quite straight forward as phytase expression is stimulated by phytate and wheat bran extract medium. Phytase activity can be easily determined by spectropho-tometry the amount of phosphomolybdate formed by the inorganic phos-phate released into the medium. One unit of phytase is defined as the amount of enzyme hydrolyzing 1 µmol of Pi per minute at 37°C.

1.5.2. *Cellulase*

Cellulases belong to a family of enzymes that catalyze the hydrolysis of cellulose into sugars. Complete hydrolysis of cellulose to glucose requires the action of 3 enzymatic activities: endo-β-glucanase (EC 3.2.1.4), exo-β-glucanase (EC 3.2.1.91) and β-glucosidase (EC 3.2.1.21). Microorganisms with cellulolytic activities play an important role in the biosphere by recycling cellulose. The demand for cellulases has increased rapidly in recent years due to their application in many industries, such as textiles, detergent, food, animal feed, bio-fuel, paper and pulp, pharma-ceutical, and waste management.

Screening for cellulase-producing microorganisms has been widely carried out by flooding carboxymethylcellulose (CMC) plates either with 1% hexa-adecyltrimethyl ammonium bromide or with 0.1% Congo red followed by 1 M NaCl. A more efficient method replaces the stains with Gram's iodine (Kasana *et al.*, 2008). The advantage of this method is that the zone of hydrolysis is much more distinct and the reaction can be short-ened from 30 to 40 minutes to less than 5 minutes. Shortlisted strains can then be further characterized by quantitative assays based on the accumu-lation of monosaccharides or reducing sugars after hydrolysis, the reduction in substrate quantity, or change in the physical properties of substrates. By virtue of the more complex nature of insoluble cellulose such as crystalline or amorphous cellulose, the success of obtaining high-efficiency cellulase from such screens would be higher than those based on soluble cellulose such as CMC.

1.6. Screening Microorganisms for Metabolites

1.6.1. *Primary metabolite: Citric acid*

Citric acid is one of the most important additives for food and averages, accounting for 75% of the world production of 1.6×10^6 tons in 2007. Natural citric acid was first commercially produced in 1917. Several strains of *Apergillus niger* then produced significant amounts of citric acid at pH of 2.5–3.5 and in the presence of high sugar concentrations. Today, *Aspergillus niger* remains as the most widely-used microorganism to produce citric acid by submerged and lately solid-state fermentation.

A large number of fungi and bacteria have been evaluated for citric acid production with varying degrees of success. These include *Arthrobacter paraffinens, Bacillus licheniformis* and *Corynebacterium ssp., Aspergillus niger, A. aculeatus, A. carbonarius, A. awamori, A. foetidus, A. fonsecaeus, A. phoenicis* and *Penicillium janthinellum*; and yeasts such as *Candida tropicalis, C. oleophila, C. guilliermondii, C. citroformans, Hansenula anamola* and *Yarrowia lipolytica*. Unfortunately, most of these microorganisms could not produce commercially acceptable yields of citric acid. More importantly, citric acid is a metabolite of energy metabolism and is therefore unlikely to be accumulated at high amounts under normal circumstances. Today, mutants of *Aspergillus niger* created by γ-radiation or ultraviolet rays are screened for citric acid production.

1.6.2. *Secondary metabolite: Antibiotics*

The main approach in which new antibiotics have been discovered has been by screening of other groups of microorganisms besides *Streptomyces, Penicillium,* and *Bacillus*. In the screening approach, a large number of putative antibiotic-producing microorganisms are obtained from nature and pure isolates tested for antibiotic production by observing for diffusible materials that are inhibitory to the test or indicator bacteria. The classical method for testing potential antibiotic-producing microbial isolates is the cross-streak method, used by Fleming in his studies on penicillin. Isolates that demonstrate evidence of antibiotic production are then subjected to further studies to determine if the antibiotics they produce are new. When an organism producing a new antibiotic is discovered, it is produced in large quantity, purified and tested for cytotoxicity and therapeutic activity in infected animals. Most new antibiotics will fail the *in vivo* testings but a few of these new antibiotics that prove useful medically are then produced commercially. Since antibiotic-producing strains isolated from nature rarely produce the desired antibiotic at sufficiently high concentration, it is necessary to isolate new high yielding strains. In the commercial production of penicillin, the yield of this antibiotic was increased by 50,000-fold using strain selection and appropriate medium development. Strain selection involves mutagenesis of the wild type culture, screening for mutants and testing of these mutants for enhance antibiotic production.

1.7. Screening Microorganisms for Flavor

The use of an electronic nose (SMartNoses system (LDZ), SMartNose SA, Switzerland; Fig. 1.7) to screen lactic acid bacteria with aroma-forming capabilities was reported as far back as 2004 by Marilley *et al.* (2004) Lactic acid bacteria isolated from five different Swiss Gruyere cheese factories were incubated under anaerobic conditions, and suspended the cells in Ultra High Temperature (UHT) milk supplemented with casamino acid solution in sealed glass vials. The volatiles from the vials were then screened for flavor compounds from catabolism of the amino acids after 10 days using Inside Needle Dynamic Extraction (IN-DEx) pre-concentration syringe containing a permethylated cyclodextrin silicon derivative. The results show that the strains classified by volatile compounds were in conformity with the classification obtained with the molecular method. The MS-based electronic nose can thus be an efficient tool in screening for new aroma-producing strains.

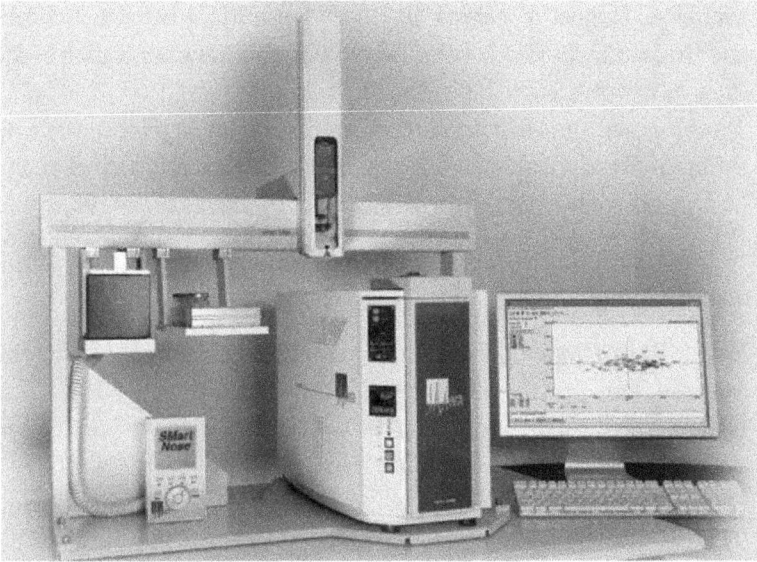

Fig. 1.7. SMartNoses system (LDZ) was used for screening flavor-producing lactobacillus strains by Marriley *et al.* (2007). (Courtesy of SMart Nose SA, Switzerland, http:// smartnose.com.)

References

1. Bernet MF, Brassart D, Neeser JR and Servin AL. *Lactobacillus acidophilus* LA 1 binds to cultured human intestinal cell lines and inhibits cell attachment and cell invasion by enterovirulent bacteria. *Gut* **35**: 483–489, 1994.
2. Dalbøge H and Lange L. Using molecular techniques to identify new microbial biocatalysts. *Trends Biotechnol.* **16**: 265–271, 1998.
3. Haefner S, Knietsch A, Scholten E, Braun J, Lohscheidt M and Zelder O. Biotechnological production and applications of phytases. *Appl. Environ. Microbiol.* **68**: 588–597, 2005.
4. Kasana RC, Salwan RS, Dhar H, Dutt S and Gulati A. A rapid and easy method for the detection of microbial cellulases on agar plates using gram's iodine. *Curr. Micribiol.* **57**: 503–507, 2008.
5. Marilley L, Ampuero S, Zesiger T and Casey M.G.. Screening of aroma-producing lactic acid bacteria with an electronic nose. *Int. Dairy J.* **14**(10): 849–856, 2004.

6.   Pfeifer A, Donnet A, Neeser JR, Link-Amster H, Rochat F, Schiffrin E and Brassart D. Probiotic latic acid bacteria for a new generation of foods. *Aust. J. Pharmacy* **78**: 730–733, 1997.
7.   Teo AY-L and Tan H-M. Inhibition of *Clostridium perfringens* by a novel strain of *Bacillus subtilis* isolated from the gastrointestinal tracts of healthy chickens. *Appl. Environ. Microbiol.* **71**(8): 4185–4190, 2005.

Further Reading

1.   Baltz RH, Davies JE and Demain AL. *Manual of Industrial Microbiology and Biotechnology*, 3rd ed. ASM Press, 2010.
2.   Glazer AN and Nikaido H. *Microbial Biotechnology: Fundamentals of Applied Microbiology*. Cambridge University Press, 2007.
3.   Okafor N. *Modern Industrial Microbiology and Biotechnology*. SBN 978-1-57808-513-2. Science Publishers, 2009.
4.   Saier MH. Beneficial bacteria and bioremediation. *Water Air Soil Pollut.* **184**: 1–3, 2007
5.   Teo AY-L and Tan H-M. Screening for microbial products. In: *Microbial Biotechnology: Principles and Applications*, 2nd ed. YK Lee. World Scientific Publishing Co. Singapore, 2003.

Websites

Todar's Online Textbook of Bacteriology
http://www.textbookofbacteriology.net/

Society for General Microbiology
http://www.microbiologyonline.org.uk/

Questions for Thought…

1.   Why have there been no major discoveries of antibiotics in recent years? Do you think this is due to the limitations of current screening methods?

2. What part of the screening process can be automated to achieve a higher throughput of screening results and discovery of unique products?
3. How might the screening process be modified so as to extend the applications of products discovered in one area e.g. food into other areas e.g. cosmetics?

Chapter 2

Bioprocess Technology

Lee Yuan Kun
Department of Microbiology, Yong Loo Lin School of Medicine
National University of Singapore
5 Science Drive 2, Singapore 117597
Email: micleeyk@nus.edu.sg

Bioprocess technology is the industrial application of biological processes involving living cells or their components to effect desired transformation of substrates. The major advantages of bioprocesses over traditional chemical processes are that they require mild reaction conditions, are more specific and efficient, and produce renewable by-products (biomass). The development of recombinant DNA technology has expanded and extended the potential of bioprocesses.

2.1. Microbial Growth

When all nutrients required for cell growth are present in non-growth limiting quantity, i.e. at sufficiently high concentrations (so that minor changes do not significantly affect the reaction rate) and the culture environment is favorable, most unicellular microorganisms reproduce asexually. The size and biomass of individual cell increase with time, resulting in **biomass growth**. Eventually, the DNA content doubles in quantity and cell division occurs upon complete division of a cell into two progenies of equal genome and approximately identical size, thereby increasing the population number. This increase in the population number of cells in a culture is known as **population growth** (Fig. 2.1).

21

Box 2.1

What is Biomass?

Biomass is defined as the materials essential for the structure and reproduction of a living organism. Thus, storage products such as starch and glycogen are not true biomass and should be subtracted from the weight of the biological material measured, to give the true biomass. The biomass is often expressed as a concentration, e.g. g-biomass/L-culture.

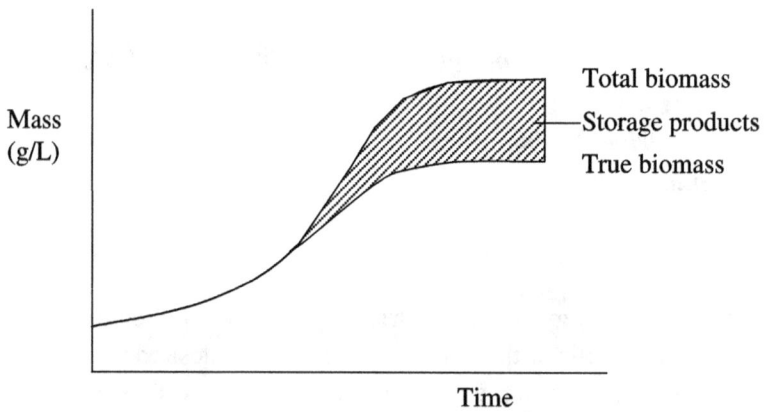

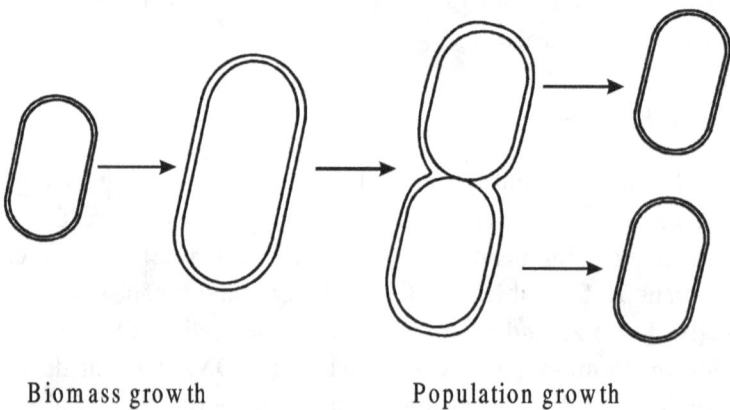

Biomass growth   Population growth

Fig. 2.1. Bacterial cell growth and division.

2.1.1. *Growth kinetics*

Growth kinetics deal with the rate of cell growth and how it is affected by various chemical and physical conditions. During the course of growth, the heterogeneous mixture of young and old cells is continuously changing and adapting itself in the media environment, which is also continuously changing in physical and chemical conditions.

There are distinct growth phases in the growth curve of a microbial culture. A typical growth curve includes the following phases: (1) lag phase, (2) growth phase (accelerating, exponential and decelerating growth phase), and (3) stationary phase and death phase as depicted in Fig. 2.2.

(1) Lag Phase

When an inoculum is introduced to a fresh culture medium, the culture may not grow immediately at the maximum rate, thus giving rise to a lag phase. This is essentially a period of adaptation of cells to a new environment (e.g. change in pH, increase in nutrient supply, reduced growth inhibitors). The cells increase in size and weight rather than increase in numbers. The duration of growth lag is defined in Box 2.4. Lag phase is considered a non-productive period of a fermentation process. Thus, it is often desirable

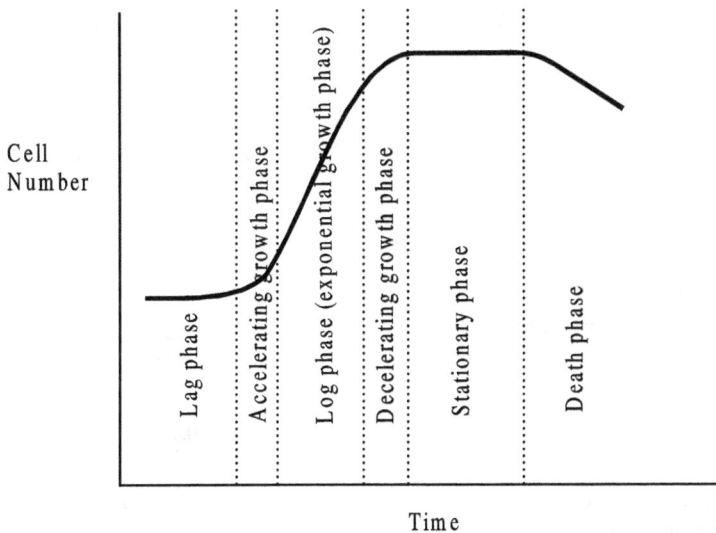

Fig. 2.2. Growth curve of a bacterial culture.

Box 2.2

Why is it Necessary to Understand Microbial Growth Kinetics?

Microbiologists should understand microbial growth kinetics for two basic reasons:

(1) For appropriate interpretation of growth data

Take for example the case of studying the pH profile of a newly isolated bacterium in a batch culture system. The new bacterial culture previously cultured at pH 7.0 was used as the inoculum for the same medium adjusted to pH 5.5 and 7.5. The cultures were incubated at the same temperature for a period of time. Thereafter either the final cell number was enumerated or biomass was measured and the results are as depicted in the following figure.

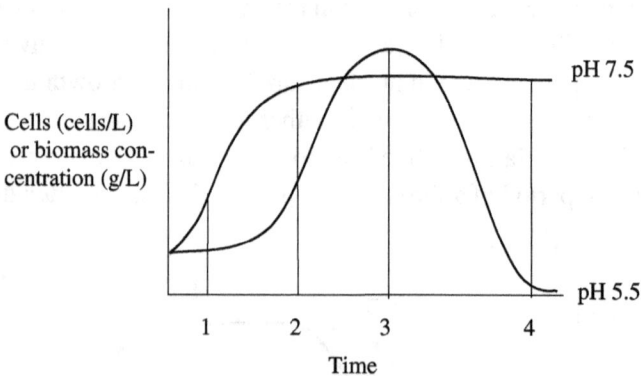

The interpretation of the outcome would depend largely on when the final sample was taken. If the cultures were sampled at time 1, it appears that at pH 5.5 growth of the culture was completely inhibited. In fact, the inoculum used was previously cultured at pH 7.0, was still in the lag phase at time 1, when the cells were adapting to the new pH condition.

   If the cultures were sampled at time 2, it is easy to conclude that the culture could grow at both pH 5.5 and 7.5, and the bacterium grew faster at pH 7.5.

   When the cultures were sampled at time 3, it may appear that pH 5.5 is a favorable pH condition for faster growth. The truth is that the mineral nutrients in the medium are more soluble in this acidic solution, rendering them more available for bacterial growth, hence a higher biomass concentration.

Box 2.2   (*Cont'd*)

If the cultures were sampled at time 4, the vast difference in the cell population of the two pH conditions may lead to the conclusion that the bacterial cells lysed at pH 5.5. However, it is the high maintenance energy requirement at pH 5.5 that caused rapid losses of biomass in the stationary growth phase.

In actual fact, the exponential growth rate of the bacterium was the same at pH 5.5 and 7.5.

(2) For optimization of growth and product formation

Knowledge of microbial growth kinetics is essential to determine when to harvest the culture for different purposes. For a growth-linked product, it is desirable to harvest the culture at the late exponential growth phase. On the other hand, for a non-growth-linked product, it would be desirable to harvest the culture at the stationary growth phase (see Sec. 2.2).

to minimize or control the duration of the lag phase. Base on its mechanism, growth lag could be differentiated into **apparent lag** and **true lag**.

Immediately after inoculation, a portion of the seed culture population may grow at the maximum rate while the rest fail to grow, thus resulted in an apparent **lag phase**. For example, if half of the cells in the inoculum are not viable, and the other half of the cells grow at the maximum rate, it may appear that the culture is growing at half of the maximum rate.

True lag occurs when the culture is not able to grow at its maximum rate initially due to one of the following factors:

(a) Change in nutrient

A change in nutrient supplied in the culture medium, e.g. concentration, component, will probably involve the induction of one or more new enzymes, which could take a few minutes to many hours. The lag is thus due to the induction of these new enzymes and the time required to synthesize the optimal amount of these enzymes. For some microbial cultures, the induction of enzymes to utilize a new carbon and energy substrate could only occur when a small amount of the original carbon and

Box 2.3

Why is the Inoculum Size Usually 10% v/v of the Culture Volume?

Inoculum is often taken from a culture in the late exponential or early stationary growth phase. The desired size is usually 10% of the culture volume. An inoculum size that is too large would dilute the rate-limiting substrate concentration in the medium, resulted in a lower specific growth rate. On the other hand, a low inoculum size would result in a long lag phase. In a culture of alga *Chlorella*, glycolate in the culture supernatant needs to be accumulated to a critical concentration before cell growth is possible (see also Inoculum Effect on True Lag, Sec. 2.1.1(1)).

Box 2.4

When does the Lag Phase in a Growth Curve End?

Let us plot the graph of ln X (X = biomass) versus incubation time, T, as shown in the figure below, and extrapolate the exponential growth line to meet the original biomass concentration. The time when the two lines intercept is defined as the end of the lag phase.

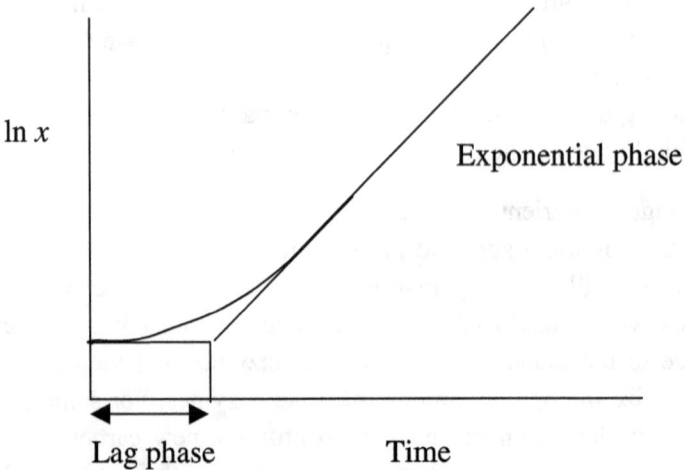

energy substrate is present. It is possible that the cells do not have sufficient carbon and energy storage to support synthesis of the new enzymes. For example, the penicillin producing fungus *Penicillium chrysogenum* could metabolize both glucose and lactose in a mixed substrate fermentation, but could not metabolize lactose if it is added into a glucose starved culture. It is always desirable to adapt an inoculum in the same medium as in the final culture system to minimize the duration of the lag phase.

At times, cells from a nutrient starved inoculum could experience **substrate-accelerated death**, when they are inoculated into a nutritionally rich medium (Fig. 2.3). This could be interpreted as a case where the rapid flux of the growth-limiting substrate (often the carbon-energy substrate) led to unbalanced enzymatic activities and disruption of cellular metabolic control. A rapid decrease in the intracellular level of some essential metabolites, such as cyclic nucleotide, will result in loss in viability. For example, a lactose-limited culture of *Klebsiella aerogenes* shows rapid death when starved in non-nutrient buffer containing lactose.

(b) Change in culture conditions
The change in physical environment of a culture often arises when cells are in the late exponential and stationary phases, where pH and other growth parameters have changed due to cellular activity, are inoculated into a fresh culture medium. The pH of a microbial culture may decrease due to

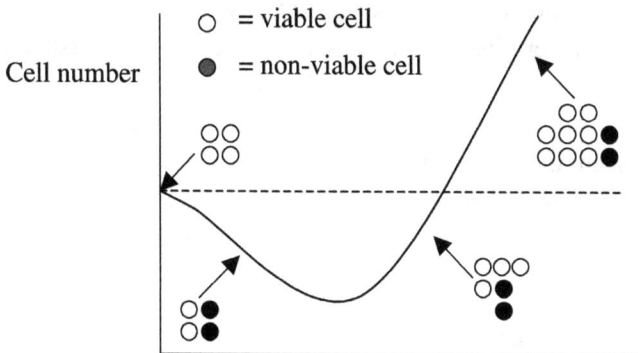

Fig. 2.3. Substrate accelerated cell death results from the transfer of nutritionally starved cells to a nutritionally rich medium.

the consumption of NH_4^+ in the medium or production of organic acids, e.g. lactic acid by *Lactobacillus* culture. The pH of culture medium may increase due to the consumption of NO_3^{++} in the culture medium. In these cases, growth lag represents the adaptation of cells to the new culture conditions.

In the studies of growth parameters in batch culture system, cultures are often inoculated with inoculum of the same source and then incubated at different conditions, e.g. various temperatures. Cells incubated at conditions that deviate from the original culture conditions of the inoculum may experience a growth lag. Some cultures may even fail to grow in the conditions that deviate too much from the original culture condition. Thus, the outcome of the study is cell history dependent and could lead to biased conclusions. It is therefore desirable to preculture the cells in some particular culture conditions before they are used as inoculum to study the effect on cell growth.

(c) Presence of an inhibitor

After the initial inoculation, a period of time may be required to reduce the amount of an inhibitory substrate (e.g. alcohol, phenol and hydrocarbon, or an inhibitory medium component like antibiotic and heavy metal) to a concentration which permits maximum growth rate.

In some cases, the product of inoculum may inhibit growth. For example, members of *Lactobacillaceae* when transferred from an anaerobic to an aerobic medium may accumulate H_2O_2 in the cells. This could happen when a heavy inoculum of *lactobacilli* (low oxygen partial pressure in the original culture) is inoculated into a freshly prepared medium (oxygen partial pressure at air saturation). Growth would only be possible after the synthesis of the enzyme peroxidase and removal of H_2O_2. As such, increasing the inoculum size would prolong the lag period, as the production of H_2O_2 is proportional to the cell concentration. On the contrary, for most of the other causes of growth lag, a bigger inoculum size would shorten the lag period (see below).

(d) Spore germination

Where the inoculum is made up of spores, vegetative growth is only possible after germination of spores.

(e) Inoculum effect

The physiological stage (age and size) of the inoculum is an important factor in determining the length of growth lag. The duration of lag period is prolonged when the inoculum added is at the initial stage of the exponential growth phase. Nevertheless, the lag period could be eliminated by the addition of culture filtrate taken from a culture in the late exponential growth phase, as the intermediate products in the filtrate are readily available by increasing the inoculum size, or by adding intermediates of citric acid cycle, such as asparagine.

Growth lag could represent "priming" of various essential metabolic pathways before substrates are metabolized at the maximum rates for conversion into biomass. Thus, a small inoculum may not be able to grow in a minimal medium containing only essential nutrient. An inoculum of *Klebsiella aerogenes* is not able to grow in a minimal glucose-ammonia medium when the initial cell concentration is less than 10^5/mL. The addition of amino acids such as asparagine, alanine or peptones (partially hydrolyzed proteins) helps to lower the minimal inoculum concentration.

The duration of the lag period is shortened when the growth of the inoculum added to the culture approaches the end of the exponential growth phase, as it consists mainly of young and active cells. The lag period increases when the inoculum is at the stationary growth phase. This could be attributed to the re-organization necessary in the cells to reverse the change caused by cessation of growth.

(2) Growth Phase

There are three stages at the growth phase, namely the accelerating growth phase, the exponential (or logarithmic) growth and the decelerating growth phase.

At the late lag period, the cells have adjusted to the new environment and begin to grow and multiply (**accelerating growth phase**), and eventually enter the **exponential** (or **logarithmic**) **growth phase**, where the cells grow and divide rapidly, at a relatively constant rate, as the exponential function of time. The time required to achieve a doubling of the number of viable cells is termed **doubling time** (t_d). It is also termed **generation time**, as this is the duration the culture takes to grow and produce a generation of cells. The number of cells in an exponentially growing microbial culture could be expressed mathematically as follows:

$$2^0 N_0 \rightarrow 2^1 N_0 \rightarrow 2^2 N_0 \rightarrow 2^3 N_0 \rightarrow 2^n N_0$$

where N_0 = Initial number of cells
            n = Number of doubling (generations)

It can be seen from Table 2.1 that the doubling time of bacterial cells measured at their optimal growth temperature varies widely. Some organisms take slightly longer than others to double their population. In general, the higher the optimal growth temperature, as in the thermophiles, the shorter is the doubling time. Furthermore, small, structurally simple bacteria such as the spherical and rod-shape bacteria double faster than large, structurally complex helical and filamentous bacteria.

The number of doubling (n) at a time interval t, is determined by the relation t/t_d. Thus, the number of cells (N_t) in an exponentially growing culture after incubated for soe time, t, can be estimated as:

$$N_t = N_0 \cdot 2^n$$
$$= N_0 \cdot 2^{t/t_d}$$
$$N_t / N_0 = 2^{t/t_d}$$
$$\ln(N_t / N_0) = (\ln 2) t/t_d \qquad (2.1)$$

Table 2.1. Doubling time of various bacteria measured at their optimal growth temperature.

Organism	Temperature (°C)	Doubling time (min)	Shape
Bacillus stearothermophilus	60	11	Single rod
Escherichia coli	37	20	Single rod
Staphylococcus aureus	37	28	Cluster coccus
Streptococcus faecalis	30	62	Chain coccus
Lactobacillus casei	30	96	Single rod
Nostoc japonicum	25	570	Filament
Anabaena cylindrica	25	840	Chain coccus with heterocyst
Treponema pallidum	37	1980	Helix

During the exponential growth phase, the growth rate of the cells is proportional to the biomass of cells. Since biomass generally can be measured more accurately than the number of cells, the basic microbial growth equations are very often expressed in terms of mass. A biomass component such as protein may be used as an alternative to direct weighing of biomass. Hence, Eq. (2.1) can be modified by assuming the biomass concentration at time 0 (initial) an time t as X_0 and X_t respectively:

$$\ln\left(X_t/X_0\right)/t = 0.693/t_d$$
$$d\left(\ln X\right)/td = 0.693/t_d$$
$$d\left(\ln X\right)/dX \cdot dX/td = 0.693/t_d$$
$$1/X \cdot dX/td = 0.693/t_d$$
$$\mu = 0.693/t_d$$

μ represents the **specific growth rate** (g-biomass/g-biomass/h or h^{-1}) of the culture. It defines the fraction of increase in biomass over a unit time, i.e. an increase of certain g-biomass from every gram of existing biomass per hour. Specific growth rate represents the average growth rate of all cells present in a culture, but not necessary the maximum specific growth rate of the individual cells, as most microbial cultures divide asynchronously.

The expression of the rate of microbial growth as specific growth rate is crucial to avoid the effect of cell concentration. For a microorganism which has a doubling time of 1 h, the **output rate** of a culture with a concentration of 1 g/L is 1g-biomass/L/h, whereas at a biomass concentration of 10 g/L, the output rate is 10 g-biomass/L/h.

After the exponential growth phase is the **decelerating growth phase** where the culture is in a transient state. During this stage there are feed/back mechanisms that regulate the bacterial enzymes involved in key metabolic steps to enable the bacteria to withstand starvation. There is much turnover of protein for the culture to cope with this period of low substrate availability. Examples of the feedback mechanisms involved are Stringent Response (RNA content decreases), Ntr System and Pho System (activity of enzymes changes markedly).

Box 2.5

What is a Stringent Response?

Some bacteria (e.g. Enterobacteriaceae), on experiencing a depletion of nutrition, especially nitrogen source (amino acid starvation), can initiate a stringent response to reduce the rates of protein and other macromolecular synthesis, by decreasing the synthesis of ribosomal RNA. Using this mechanism of transcription control, the cells are able to shut down a number of energy-draining activities as a survival mechanism under poor growth conditions.

Box 2.6

What are Ntr and Pho Systems?

The activity of **Ntr system** is activated in response to nitrogen starvation, when ammonia is the growth limiting substrate. The bacteria are able to turn on transcription of genes for glutamine synthetase. The system enables the bacteria (e.g. Enterobacteriaceae) to assimilate very low levels of ammonia by catalyzing the assimilation of ammonia into glutamine in an ATP-dependent reaction. It also activates additional operons (e.g. *nac* operon) that are involved in the utilization of organic nitrogen sources. This permits the cells to use alternate sources of nitrogen.

A phosphate utilization network **(Pho system)** is activated on inorganic phosphate starvation (in Enterobacteriaceae), resulting in the production of a high concentration of alkaline phosphatase, so that phosphate can be obtained from organic sources.

(3) Stationary and Death Phase (or Decline Phase)

A growing bacterial culture eventually reaches a phase during which there is no further net increase in bacterial cell numbers. This is called the **stationary growth phase**. During the stationary phase, the growth rate is equal to the death rate. A bacterial population may reach stationary growth when a required nutrient is exhausted, or when inhibitory end products are accumulated or when physical conditions changed. The various metabolites formed in this stage are often of great biotechnological interest.

The **death phase** (or **decline phase**) is the result of the inability of the bacteria to carry out further reproduction as condition in the medium become less and less supportive of cell division. Eventually the number of viable bacterial cells begins to decline at an exponential rate. Industrial fermentation is usually interrupted at the end of the exponential growth phase or before the death phase begins.

At the stationary phase, the maximum concentration of biomass that can be achieved in a given medium is influenced by the following factors:

(a) Amount of growth-limiting substrate supplied

The common **growth-limiting substrate** for aerobic culture is the supply of O_2. The solubility of O_2 in water is low, approximately 7 mg/L at 35°C. In the production of microbial protein (or single cell protein, SCP) using *Candida utilis*, assuming that the specific growth rate of the yeast culture is 0.3 h^{-1}, the growth yield for oxygen (Y_{O2}) is 3.2 g-biomass/g-O_2 and the cell concentration is 10 g/L, the amount of oxygen dissolved in the medium will be totally consumed within 0.45 minutes.

$$\text{Rate of } O_2 \text{ consumption} = 10 \times 0.3/3.2$$
$$= 0.94 \text{ g-}O_2/L \cdot h$$
$$\text{Exhaustion of } O_2 \text{ in the culture} = 7 \times 10^{-3}/0.94$$
$$= 7.5 \times 10^{-3} \text{ h}$$
$$= 0.45 \text{ min}$$

Agitation and aeration are necessary to maintain the growth of an aerobic microbial culture. Eventually, growth will cease due to the depletion of an essential substrate included in the culture medium.

(b) Accumulation of inhibitory products

The most frequently occurring toxic products are H^+ and OH^- which can change the pH value of the culture medium. Some microorganisms, e.g. lactic acid bacteria, produce organic acid (lactic acid) as the end product of carbon metabolism. Inclusion of an organic acid substrate such as acetate in a medium often leads to an increase in the pH value due to the excessive uptake and metabolism of anions over cations. The incorporation of NH_3^+ from NH_4Cl in protein synthesis results in the release of H^+, hence lowering the pH value of the culture. Conversely, the inclusion of $NaNO_3$ as the nitrogen source results in the accumulation of

OH^-. Some microorganisms are able to utilize urea as nitrogen as well as carbon substrate; hence, the pH of the medium could remain unchanged or decrease depending on the mechanism of urea uptake.

Under anaerobic condition, yeast metabolizes sugars to ethanol. At a concentration of 6–15% (w/v), which is about the concentration of ethanol in wine, ethanol is inhibitory to cell growth.

(c) Cell death

Cells may die due to errors in cell division or assaulted by environmental factors, such as extreme pH and osmolality. The **Exponential Death Law** states that cell death rate is proportional to the number of viable cells present, i.e.,

$$dX = - kX \cdot dt$$

where k = **specific death rate** (h^{-1}). The value of k is determined by the environmental conditions. Thus,

$$\ln X = \ln X_0 + (\mu - k)t. \tag{2.2}$$

The effective growth rate is determined by ($\mu - k$). There is no net growth when μ equals k.

(d) Maximum packing density

Provided all culture parameters are favorable, the limit to biomass concentration will be the maximum packing density of cells in the biomass. Based on studies in plant and animal tissues, the maximum packing density of cells (number/cm^3) was estimated to be $10^{12}/V$, where V = individual cell volume in μm^3. This maximum possible density corresponding to dry weight is about 10% (w/v), i.e. 100 g/L. In most of the industrial fermentation processes, the biomass concentration ranges between 20 and 50 g/L. It is technically difficult to satisfy the demands for O_2 and other nutrients at high biomass concentrations.

If energy source is in excess, energy reserves may be stored in the storage products such as starch, glycogen, lipid and peptide. After exhaustion of energy source and reserve, part of the cellular substances must be consumed to provide energy (**endogenous metabolism**) for the turnover and maintenance of cell integrity (**maintenance energy**). Hence, loss of dry weight, autolysis and loss of viability may commence, with the resulting onset of decline (death) phase. The onset of the decline phase is dependent on the type of organism and the culture conditions.

2.1.2. Variations in growth kinetics

Deviations from the norm in the growth kinetics described above were reported. These kinetic variations provide useful insights to the nutrient metabolism and physiological status of the cultures, which will be discussed in the following sections.

(1) Conserved Substrate
Some nitrogen and phosphorous substrates, even if they are growth limiting, may conserve in cells with or without simple modification, as the rate of uptake of the substrate is faster than the rate of metabolism of the substrate. In this case, the decelerating growth phase could be prolonged for a number of generations by drawing the substrate from the conserved substrate pool, with apparent depletion of the substrate in the culture medium, before the onset of stationary growth phase.

(2) Synchrony in Cell Division
Synchrony in cell division occurs when all cells divide at the same time. It could be induced by a marked change in the substrate level, culture

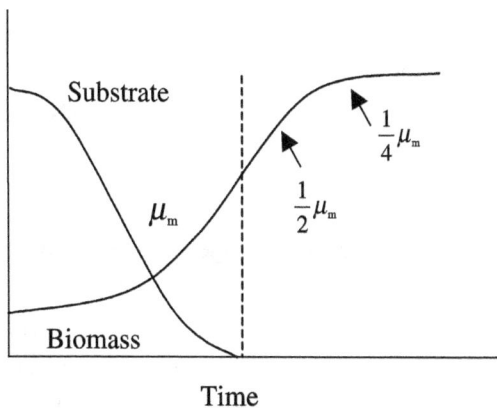

Fig. 2.4. Bacterial growth curve with a rate-limiting substrate that is conserved within the cells.

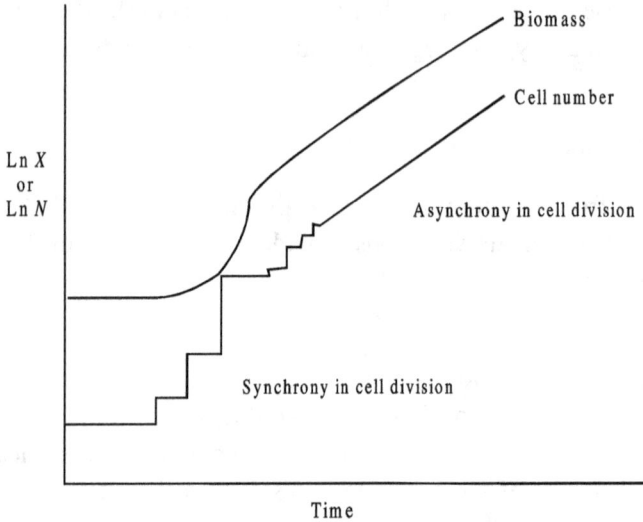

Fig. 2.5. Synchrony in cell division in a bacterial culture. N = cell number; X = biomass concentration.

temperature and light intensity, resulting in an initial rapid increase in the number of cells after the lag period. As the number of organisms increases, nutrients are consumed, metabolic wastes accumulate, living space may become limited and aerobes suffer from oxygen depletion. The culture normally becomes completely asynchronous after 2 to 3 cell divisions, due to the slight differences in the length of cell cycle of individual cells. The cells subsequently grow at a more or less constant exponential growth rate. The synchronous growth phase should not be included in the estimation of the exponential growth rate.

(3) Diauxic Growth

The presence of a preferential carbon substrate may inhibit the synthesis of the enzymatic system involved in the uptake and metabolism of a second carbon substrate. For example, *E. coli* culture may metabolize glucose preferentially in a mixed substrate medium containing both glucose and lactose. The lactose would be utilized as a carbon source only when the glucose is exhausted. Thus, in the growth curve of such a culture, an inflexion in the growth curve or even a decline in biomass occurs after

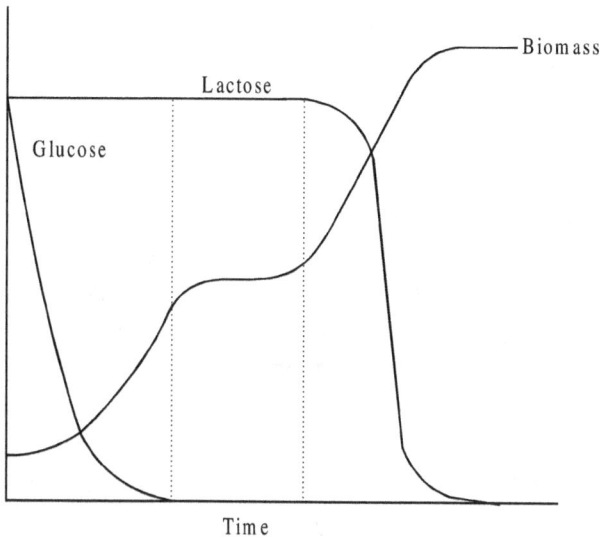

Fig. 2.6. Diauxic growth on two substrates, glucose and lactose in a *Escherichia coli* culture.

glucose is used up while lactose is being processed. This biphasic (two-phase) pattern of growth is known as diauxie or diauxic growth.

2.1.3. *Effect of growth-limiting substrate concentration*

The specific growth rate of a microbial culture is ultimately determined by the growth-limiting substrate concentration. This could be understood by regarding biomass as a component in a biochemical reaction:

$$X + s + \text{other nutrients} \leftrightarrow Xs \rightarrow \text{biomass} + \text{products}$$

where
$$X = \text{free biomass concentration}$$
$$s = \text{concentration of growth-limiting substrate}$$
$$Xs = \text{concentration of biomass-substrate complex}$$

Therefore, the total biomass concentration can be expressed as:

$$X_T = X + Xs \tag{2.3}$$

The equilibrium concentration of the various components in the biochemical reaction described above is determined by the **Saturation Constant** (K_s, g/L):

$$X \cdot s/Xs = K_s \tag{2.4}$$

K_s is defined as the concentration of the rate-limiting substrate to support the specific growth rate at half of the maximum value, i.e. $0.5\,\mu_m$. A higher K_s value would indicate that a higher concentration of the rate-limiting substrate is necessary to achieve a certain amount of biomass-substrate complex Xs in order to produce a certain amount of biomass. Thus, K_s is inversely proportional to the affinity of the cells for the substrate.

Since the **specific rate of substrate assimilation**, q (g-substrate/g-cell/h) is proportional to Xs formed,

$$q = q_m \cdot Xs/X_T \tag{2.5}$$

where q_m is the value of q when substrate is saturated, i.e., the maximum specific rate of substrate assimilation.

By substituting Xs/X_T in Eq. (2.5) with $s/K_s + s$ derived from Eqs. (2.3) and (2.4), we obtain

$$q = q_m \cdot s/K_s + s. \tag{2.6}$$

Assuming that all substrates assimilated are used for biomass production, the specific growth rate is directly proportional to the specific rate of substrate assimilation:

$$q = \mu/Y$$

where the yield constant Y (see Sec. 2.1.4) is the growth yield $= dX/ds$ (g-cell/g-substrate).

Thus Eq. (2.6) can be re-written as:

$$\mu = \mu_m \cdot s/(K_s + s) \tag{2.7}$$

This equation is often called the **Monod Relation**, since Jacques Monod in 1942 first showed empirically that the expression accorded well with the relation of bacterial growth rate to substrate concentration.

Box 2.7

How to Determine the Saturation Constant of a Growth-Limiting Substrate?

The saturation constant could be determined from the plots of specific growth rate, μ versus substrate concentration, s or $1/\mu$ versus $1/s$,

(a) According to Eq. (2.7): $\mu = \mu_m \cdot s/(K_s + s)$

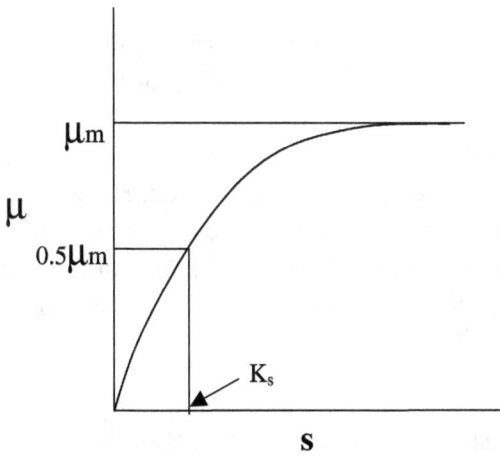

(b) According to the equation: $1/\mu = (K_s + s)/\mu_m \cdot s$
$$= K_s/\mu_m \cdot s + 1/\mu_m$$

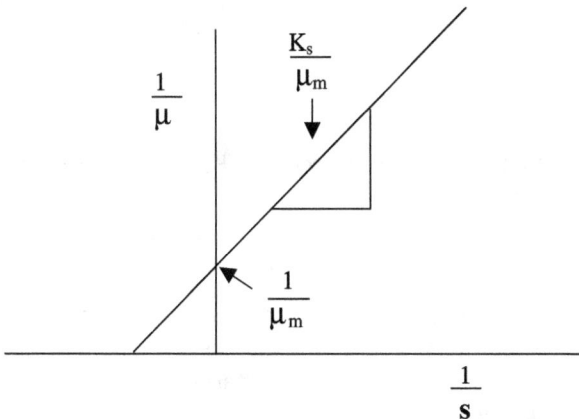

Monod Relation suggests that the specific growth rate of a microbial culture is determined by the highest possible growth rate of the culture (μ_m), the concentration of the growth-limiting substrate (s) and the affinity for the substrate (K_s). The value of μ_m is determined by growth parameters such as temperature, pH and medium osmolality.

Moreover, when

$$\mu = 0.5 \ \mu_m$$
$$s = K_s$$

Hence, K_s could be determined by measuring the value of μ at various values of s (Box 2.7). Sample values of K_s for various carbon substrates and microorganisms are as shown in Table 2.2.

The K_s value for the same substrate (glucose) could be very different for two different microorganisms. In this case, *K. aerogenes* has a higher affinity for glucose than *S. serevisiae*. The microorganism *P. extorquens* has higher affinity for methanol (lower K_s value) than formate.

Knowledge of K_s values is of practical important for:

(1) Designing Culture Medium
In order to achieve the maximum specific growth rate, concentration of all essential substrates needs to be higher than their respective K_s value (see Sec. 2.3 for the design of growth medium).

(2) Predicting Growth Kinetics
Knowledge of K_s would allow us to estimate the specific growth rate of a microbial culture in respond to a certain concentration of the growth-limiting substrate according to Eq. (2.7). The theoretical maximum

Table 2.2. The saturation constant (K_s) of various carbon substrate and microorganisms.

Organism	Substrate	K_s ($\times 10^{-6}$M)
Klebsiella aerogenes	Glucose	20
Saccharomyces serevisiae	Glucose	140
Pseudomonas extorquens	Methanol	20
Pseudomonas extorquens	Formate	228

specific growth rate could be obtained from the plot of $1/\mu$ versus $1/s$ (Box 2.7).

(3) Interpreting Microbial Interactions in an Ecosystem

It is rare that a single species of microorganism is found in a natural habitat. The co-existence of different microorganisms competing for the same growth-limiting substrate could be understood from the following example.

Assuming that two bacteria, A and B, are present in a small volume of water competing for the same growth-limiting substrate. And μ_m of bacterium A is greater than that of bacterium B, while $K_{SA} > K_{SB}$. As such, we would expect a cross over at s_c in the μ versus s plots of the two bacteria (Fig. 2.7). When some amount of the substrate, which exceeds the s_c value, is added into the ecosystem, bacterium A (having a higher μ_m) would outgrow bacterium B in number, and the substrate decreases in concentration. However, when the residual concentration of the substrate drops below the s_c value, the specific growth rate of bacterium B would now be higher than that of bacterium A. Thus, the two bacteria take turns to increase in number, and no one bacterium would replace the other in the ecosystem.

The situation would be different if the concentration of the growth-limiting substrate does not decrease to below s_c or increase to above s_c. In this case, bacteria A and B would have same population size.

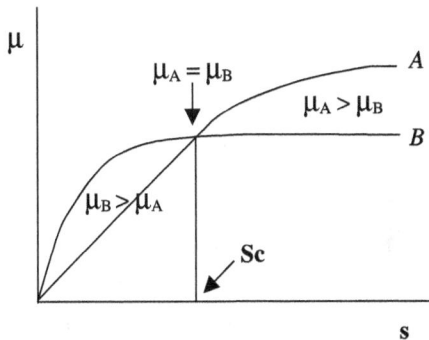

Fig. 2.7. Growth curves of two bacteria, A and B, in a mixed culture competing for the same rate-limiting substrate. μ = specific growth rate; s = rate-limiting substrate concentration.

2.1.4. *Growth yield*

It has been observed frequently that the amount of cell mass produced is proportional to the amount of a limiting substrate consumed.

Growth yield ($Y_{X/S}$, g-cells/g-substrate) is defined as the amount of biomass produced (dX) through the consumption of unit quantity of a substrate (ds), i.e.

$$Y_{X/S} = dX/ds.$$

It is an expression of the conversion efficiency of the substrate to biomass.

Knowledge of growth yield is of practical importance. In many industrial production processes, it is desirable to maintain high growth yield to achieve efficient conversion of the substrates to biomass and products. Biological waste treatment process represents the other extreme where it is desirable to achieve the lowest possible sludge (biomass) formation, with the maximum oxidation of carbon in the effluent to CO_2.

Growth yield also determines the cost of process operation. In the example of Microbial Protein (Single Cell Protein) production, the final product is biomass with a high protein content of about 50%. The yield value determines the biomass output rate as well as the cost of aeration and cooling can be illustrated as below:

(i) The cost of aeration (O_2 supply) in a fermentation process using methanol (a derivative of a by product of petroleum refinery industry) is determined by the O_2 demand of a fermentation system.

The mass balance of microbial growth could be expressed by the following equation:

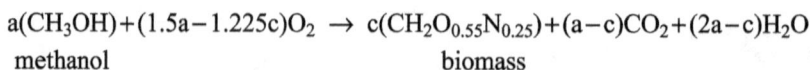

$$a(CH_3OH)+(1.5a-1.225c)O_2 \rightarrow c(CH_2O_{0.55}N_{0.25})+(a-c)CO_2+(2a-c)H_2O$$
methanol   biomass

For any value of growth yield on methanol (c/a) there is a corresponding O_2 requirement ($1.5a - 1.225c$). Thus, a low growth yield of biomass (c) would result in a higher demand ($1.5a - 1.225c$) of O_2 (air supply) for the oxidation of unit quantity of methanol (a) into CO_2. In essence, as it requires 2 moles of oxygen to produce 1 mole of carbon dioxide, while only 0.55 mole of oxygen is necessary to produce 1 C-mole of biomass, there is a higher O_2 demand if more carbon in methanol is oxidized to CO_2. Hence this results in higher cost of production due to the increase in the cost of aeration.

Insufficient O_2 supply would not allow the full oxidation of methanol, leading to even lower level of biomass production.

(ii) The energy input to sustain optimal culture temperature in a fermentation process using methanol.

Energy balance of microbial growth on methanol could be expressed using the following equation:

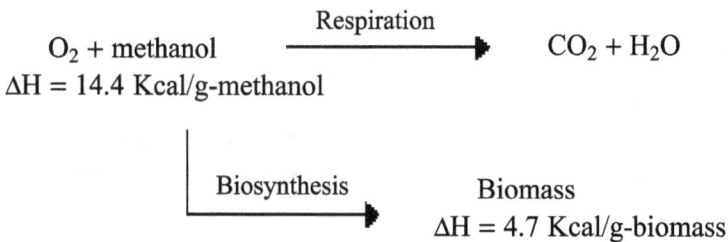

$$O_2 + methanol \xrightarrow{\text{Respiration}} CO_2 + H_2O$$
$$\Delta H = 14.4 \text{ Kcal/g-methanol}$$

Biosynthesis → Biomass
$\Delta H = 4.7$ Kcal/g-biomass

Thus, the lower the growth yield on the energy substrate, the larger amount of the chemical energy stored in the substrate would be released as heat through respiration process. In order to maintain the optimal culture temperature in the fermentation system, energy supply is needed for the circulation of cooling water through the system.

There are several biological factors and culture parameters that determine the growth yield of a microbial culture.

(1) Type of Microorganism
Among the examples cited in Table 2.3, the bacterium *Pseudomonas* has the lowest growth yield on glucose, whereas the fungus *Aspergillus* converted 75% of the carbon in glucose into biomass. The filamentous nature of the fungus facilitates separation of the cells from the culture medium. Therefore, *Aspergillus* would be a desirable candidate for Microbial Protein production. Other factors to be considered are the content of protein and the composition of amino acids. Digestibility of the biomass and safety aspect of the organism as food needs to be evaluated too.

Table 2.3. Growth yield of various microorganisms.

Microorganism	$Y_{X/S}$[a] (C eq[b] biomass/C eq glucose)
Pseudomonas fluorescens	0.47
Penicillium chrysogenum	0.55
Candida utilis	0.68
Aspergillus nidulans	0.75

[a]Growth yield.
[b]Carbon equivalent.

(2) Substrates

(a) Carbon source

The C-substrate is also the energy source for heterotrophic bacterial culture. The energy content of different carbon substrates varies, which are as shown in Table 2.4.

The salt of organic acid oxalic acid has the lowest energy content (2.4 Kcal/g-carbon) among the C-substrates listed, and the maximum growth yield of microbial cultures reported in the literature is also the lowest (0.14 g-biomass/g-carbon). The maximum growth yield seems to increase with increasing energy content of the various C-substrates, down the list from oxalate to glycerol. However, higher energy content in methanol and methane does not support a higher growth yield.

The data imply that the microbial cultures were energy-limited in media containing the first four C-substrates (oxalate to glucose). Thus, higher growth yield was obtained with C-substrate of higher energy content.

In those C-substrates, which have an energy content of higher than 11.0 Kcal/g-carbon, the microbial cultures were probably carbon-limited. There is no difference in the growth yield with increasing energy content of the substrate.

This brings us to the question of whether it is possible to utilize the excess energy from an energy-rich substrate, such as alcohol and sugar alcohol, in the metabolism of the carbon in a carbon-rich substrate, such as organic acid and sugar. Such **co-metabolism** of the two substrates would achieve a growth yield that is higher than any one of the substrate alone.

Table 2.4. Energy content of various carbon substrates.

C-substrate	Energy content (Kcal/g-carbon)	Maximum $Y_{X/S}$[a] (g-biomass/g-carbon)
Oxalate	2.4	0.14
Formate	5.1	0.53
Citrate	6.5	0.80
Glucose	9.2	1.03
Glycerol	11.0	1.42
Methanol	14.6	1.43
Methane	17.4	1.46

[a]Growth yield.

Table 2.5. Growth yield of single substrates and substrate mixtures.

Microorganism	Substrate	Molar ratio	$Y_{X/S}$[a] (g-biomass/g-substrate)
Candida utilis	Sucrose		0.50
	Ethanol		0.68
	Sucrose/ethanol	13.8:1	0.68
Paracoccus denitrificans	Formate		0.07
	Mannitol		0.52
	Formate/mannitol	4:1	0.70

[a]Growth yield.

Table 2.5 presents the growth yield of single substrates and substrate mixtures.

In the case of *Candida utilis*, the energy from one part of ethanol was sufficient to bring up the growth yield from 13.8 parts of sucrose to 0.68 g-biomass/g-sucrose, which represents a 36% increase from 0.50 g-biomass/g-sucrose. When formate and mannitol were mixed in a molar ratio of 4 to 1, the growth yield from the mixture was 10 times that of formate and 1.3 times that of mannitol.

The enhancement of growth yield in **mixed substrate fermentation** can only be observed in microbial cultures where the substrates are metabolized simultaneously. Some C-substrates are assimilated preferentially over other C-substrates, and a diauxic growth kinetic is observed in medium containing more than one C-substrates (see Sec. 2.1.2(3)).

(b) Efficiency of ATP generation

The oxygen content and other culture parameters determine the pathways of which a carbon/energy substrate is metabolized to yield different amounts of biochemically available energy in the form of ATP. As can be seen in Table 2.6, 30 ATP moles are generated from a mole of glucose via oxidative phosphorylation, but only 1–3 moles of ATP are generated in anaerobic condition via the various pathways.

It is generally true that the growth yield of biomass per mole of carbon/ energy substrate is the highest under the condition of carbon/energy limitation. Under the condition of carbon excess, there is often production of overflow metabolites, e.g. acetate, α-ketoglutarate, which reduce yield of biomass on the C-substrate. Overflow metabolites are intermediate metabolites that are produced over and above the base line levels for the formation of biomass, resulting in the accumulation of metabolites within the cells or release of these into the culture medium.

(c) Nitrogen source

Inorganic N sources (NO_3, N_2) need to be reduced to the level of $-NH_2$ for incorporation into protein molecules, thus less reductant is available for

Table 2.6. The number of ATP molecules generated on the metabolism of glucose at different culture conditions and metabolic pathways.

Substrate	Condition	ATP (mole/mole substrate)
Glucose	Aerobic ($P/O^a = 3$)	30
Glucose	Anaerobic (formate:acetate = 1:1)	3[b]
Glucose	Anaerobic (ethanol + CO_2)	2[b]
Glucose	Anaerobic (ethanol + CO_2)	1[c]

[a]Phosphorus oxygen ratio.
[b]Via Embden-Myerhof pathway.
[c]Via Entner-Duodoroff pathway.

energy generation, with an up to 20% resultant decrease in growth yield on the energy substrate.

When organic N-substrates like amino acids are included in culture media, they may serve as pre-fabricated carbon skeleton and give a high apparent growth yield on glucose. If these organic N-substrates are catabolized, energy would be generated as an energy source.

(d) Environment

Oscillation or fluctuation in environmental conditions, such as pH, dissolved O_2 and supply of carbon substrate, may cause **uncoupling** of substrate oxidation from biosynthesis. The growth yield of *Pseudomonas methylotropha* on methanol was found to decrease with increasing interval of addition of methanol into the culture, and corresponding increase in the yield of carbon dioxide.

The phenomenon of uncoupling of substrate oxidation from bio-synthesis has been exploited in the design of bioreactor for wastewater treatment. In ICI's deep shaft process and Union Carbide's Unox process, oscillation in the temperature, oxygen concentration and substrate level are allowed in large volume (tower-type) bioreactors, giving rise to a high rate of substrate oxidation and corresponding low biomass growth yield. Thus, biodegradable carbon in wastewater is transformed into gaseous CO_2, and relatively little solid mass is left.

(e) Maintenance energy requirement

Microbes and cells require energy for both growth and other maintenance purpose, like turnover of cellular materials, maintaining concentration gradient across cell membrane and motility.

Under a particular set of environmental conditions, it is assumed that energy is consumed at a constant rate for maintenance, $(ds/dt)_m = mX$,

---

Box 2.8

Uncoupling

The separation **(uncoupling)** of the two processes: formation of the protonmotive force and its utilization to generate ATP, leading to dissipation of energy without ATP formation.

where m is **maintenance coefficient** (e.g. g-glucose/g-biomass/h), X is the biomass concentration (e.g. g-biomass/L). Hence, energy balance can be expressed as:

Total rate of consumption	=	Rate of consumption for growth	+	Rate of consumption for maintenance
$\mu X/Y_E$	=	$\mu X/Y_G$	+	mX

where Y_E is overall growth yield and substrate is the energy source. Y_G is true growth yield, when $m = 0$. Hence,

$$1/Y_E = 1/Y_G + m/\mu. \tag{2.8}$$

Thus, if m is a constant, the graph of $1/Y_E$ against $1/\mu$ will be a straight line, with slope m, and intercept $1/Y_G$ on the ordinate (Fig. 2.8).
Alternatively

$$q_E \cdot X = \mu X/Y_G + m \cdot X$$

where q_E is specific rate of energy uptake $= \mu/Y_E$.
As shown in Fig. 2.9,

$$q_E = \mu/Y_G + m \tag{2.9}$$

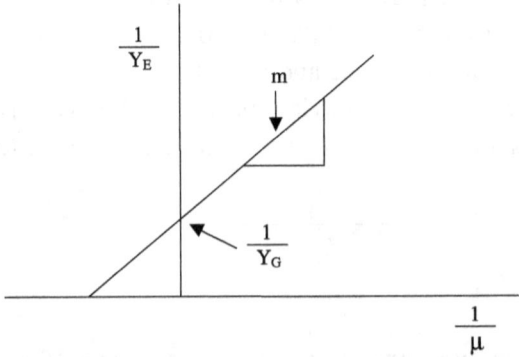

Fig. 2.8. A plot of $1/Y_E$ versus $1/\mu$ according to the equation $1/Y_E = 1/Y_G + m/\mu$, where Y_E is the overall growth yield, Y_G is the true growth yield; m, the maintenance coefficient; μ the specific growth rate.

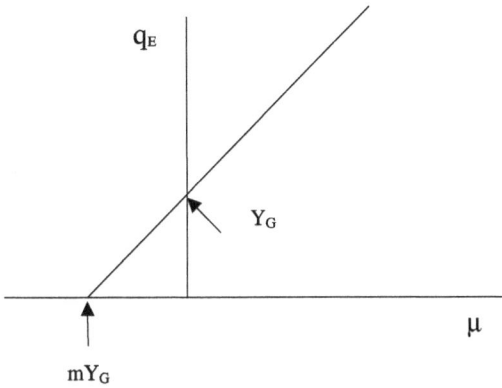

Fig. 2.9. Plot of q_E versus μ according to the equation $q_E = \mu/Y_G + m$, where q_E is the specific rate of energy uptake; μ the specific growth rate; Y_G, the true growth yield and m, the maintenance coefficient.

The fraction of energy that is channeled to maintenance would deprive the culture of the energy for growth, i.e.

$$\mu = \mu_m \cdot s/(s + K_s) - m \cdot Y_G. \qquad (2.10)$$

Thus, if the maintenance energy requirement makes up a significant component in the energy balance equation, the plot of μ versus s would not pass through the origin (Fig. 2.10).

There are several factors that could affect the maintenance energy requirement of a culture.

• Environment
Environmental parameters, which fall outside the optimal range for growth, may have an effect on the maintenance energy requirement of microbial cultures. The energy presumably is consumed for maintaining a favorable intracellular environment for optimal metabolic activities and growth.

It appears that the ionic strength of media has an effect on maintenance energy requirement. For example, in the culture of *Saccharomyces cerevisiae* cited in Table 2.7, the incorporation of 1M salt resulted in a 10-fold increase in maintenance energy for maintaining an osmotic balance across the cell membrane. Whereas N_2-fixing *Azotobacter vinelandii* spent

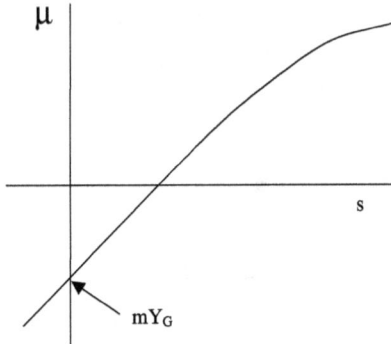

Fig. 2.10. The relationship between specific growth rate, μ and substrate concentration, s in a microbial culture with measurable maintenance energy requirement.

Table 2.7. Maintenance energy requirement of some protists (eukaryotes other than plants, animals or fungi) with glucose as the energy source.

Organism	Growth conditions	Maintenance energy (g-glucose/g-biomass/h)
Saccharomyces cerevisiae	Anaerobic	0.036
	Anaerobic, NaCl (1 M)	0.360
Azotobacter vinelandii	Fixing N_2, dissolved O_2 tension 0.2 atm	1.5
	Fixing N_2, dissolved O_2 tension 0.02 atm	0.15

energy to remove O_2 which could be inhibitory to the highly reduced nitrogenase system.

• Metabolite leakage

Microbial cells are known to release polysaccharides, glycolic acid, amines, amides and vitamins into the culture environment. These extracellular metabolites may play a role in their survival and reproduction, however, as they do not contribute directly to cell growth they are considered maintenance requirements.

- Intermittent energy supply

A decrease in growth yield is often observed if the energy source of a microbial culture is not supplied continuously (see (d) above). This could be due to the uncoupling of substrate oxidation to yield CO_2 and release of latent chemical energy as heat. The released heat may not serve any physiological function, but since it does not contribute to the growth and reproduction of a microbial culture, it is included in the category of maintenance energy.

2.2. Product Formation in Microbial Culture

Microbial cells produce a wide range of industrially useful biomolecules, which can be broadly grouped into three categories, based on their production kinetics (Table 2.8).

(1) Growth-Linked Products

These are products which formation (dp) is directly proportional to the amount of biomass formed (dX). That is:

$$dp = Y_{p/x} \, dX$$

where

$Y_{p/x}$ = product yield referred to biomass formed (g-product/g-biomass)

The rate of product formation could be expressed as follow:

$$dp/dt = Y_{p/x} \cdot dX/dt$$
$$= Y_{p/x} \cdot \mu \cdot X.$$

i.e.

$$dp/dt \cdot 1/X = Y_{p/x} \cdot \mu.$$

Thus, the specific rate of product formation is given by:

$$q_p = Y_{p/x} \cdot \mu. \tag{2.11}$$

$Y_{p/x}$ could be a constant value, e.g. DNA production, or proportional to the specific growth rate, e.g. RNA production.

Table 2.8.   Classification of microbial products.

Production kinetics	Type of microbial products	Examples
Growth-linked	Structural components of cells	Microbial protein, antigens
	Enzymes	
	extracellular	Amylases
	intracellular	β-Galactosidase
	Intermediary metabolites	Amino acids, citric acid
	Transformed substrates	Steroids
Partial growth-linked	End products of energy metabolism	Ethanol, methane, lactic acid
Non growth-linked	Energy storage compounds	Glycogen, starch
	Secondary metabolites	Antibiotics

In order to achieve high productivity ($R = qp \cdot X = Y_{p/x} \cdot \mu \cdot X$, g-product/L/h), the specific growth rate (μ) and the biomass concentration (X) of the culture should be maintained at the highest possible values.

(2) Partial Growth-Linked Product
When product formation is partly growth-linked and partly independent of specific growth rate, the specific rate of product formation could be expressed as follow:

$$q_p = Y_{p/x} \cdot \mu + \beta \tag{2.12}$$

The first term on the right hand side of the equation represents the growth-linked component, whereas β is a constant specific rate of product formation independent of the specific growth rate. Formation of end products of energy metabolism, e.g. lactic acid, follows this relation, where β includes lactic acid formation, which results from either the maintenance energy requirement or uncoupling of ATP.

(3) Non Growth-Linked Product
The product can either be, (i) q_p that is independent of specific growth rate, e.g. penicillin production by *Penicilium*, or (ii) q_p that varies with specific growth rate in a complex way, e.g. spore production by

Bacillus subtilis: $q_p = q_{pmax} - k\mu$, where q_{pmax}, the maximum specific rate of product formation, and k is a constant.

2.3. Culture Medium for Fermentation

Culture medium is an important factor in optimizing cell growth and product formation in fermentation processes. It provides not only the nutrients and culture environment necessary for optimal cell growth, but also induces production of desirable products, suppresses formation of undesirable products and represents a major component of the total cost of production. The following factors are to be considered in the design of a desirable culture medium.

(1) Major Nutrient Requirement
Bulk nutrient components include carbon, hydrogen, nitrogen, oxygen, phosphorus and sulfur that are required for the synthesis of biomass and target products. The chemical composition of fermentation products

Table 2.9. Chemical composition and formula of some fermentation products.

Product	Chemical composition (% dry weight)						Chemical formula
	C	H	N	O	P	S	
Biomass							
Bacteria	53.0	7.3	12.0	19.0			$CH_{1.67}N_{0.20}O_{0.27}$
Yeast	44.7	6.2	8.5	31.2	1.1	0.6	$CH_{1.64}N_{0.16}O_{0.52}P_{0.01}S_{0.005}$
Fungus	43.0	6.9	8.0	35.0			$CH_{1.93}N_{0.16}O_{0.61}$
Antibiotics							
Erythromycin							$C_{37}H_{67}NO_{13}$
Penicillin G							$C_{16}H_{18}N_2O_4S$
Organic acids							
Citric acid							$C_6H_8O_7$
Lactic acid							$C_3H_6O_3$
Alcohol							
Butanol							$C_4H_{10}O$
Ethanol							$C_2H_6O_3$

listed in Table 2.9 provides some ideas on the minimum requirement of major nutrients.

A microbial culture would require more nitrogen for the synthesis of one mole of biomass than one mole of antibiotic. For example, for the production of one mole of bacteria, one mole carbon requires 0.20 mole nitrogen whereas for the production of one mole of erythromycin one mole carbon requires only 1/37 mole of nitrogen. Nutrient requirement is to a large extent determined by the product yield referred to per unit of biomass formed ($Y_{p/x}$, see Sec. 2.2). If the product yield for a particular growth substrate is not known, it is desirable to experimentally determine the requirement by carrying out batch fermentation in which the substrate concerned is supplied in growth-limiting concentrations while all other substrates are present in excess.

From such a study, the requirement for a particular substrate for the production of a unit biomass or product could be obtained from the slope of the plot (growth yield = biomass or product formed/substrate consumed) of cell concentration (or product concentration) versus substrate concentration. Often, in such a plot, the intercept is not at the origin. If there is carry-over of the substrate from the inoculum, the plot will intercept on the Y-axis, i.e. there is an increase in biomass even when no such substrate was added in the medium. An intercept of the plot on the x-axis indicates that a fraction of the substrate was not available for cell growth. The possible explanations are: (i) The energy substrate could have been consumed for maintenance purposes (see Sec. 2.1.4(2e)); (ii) the culture may have a high saturation constant (K_s, see Sec. 2.1.3) for the substrate; (iii) otherwise the substrate could have been "lost" by forming precipitation or complex with other non-cellular materials or destroyed during sterilization (glutamine, thiamine, riboflavin are heat labile) and storage (thiamine could be oxidized readily; riboflavin is light sensitive). It is common that magnesium, urea (or ammonium), sodium (or potassium) and phosphate form complexes and precipitated. For this reason, magnesium salt should be autoclaved separately from phosphate. In addition, trace elements may form less soluble salts with other cations in the medium during autoclaving. Chelating agents (e.g. organic ligands CDTA, EDTA and NTA) are often included in the medium to retain metal ions in solution, however they may compete with cells for the elements.

It should be kept in mind that nutrient requirement may depend on other available substrates or absence of certain substrates in the medium. Cells may metabolize organic substrates like amino acids when simple sugar is not available as energy and carbon sources (for energy substrate requirement see Sec. 2.1.4(2)). Sodium may replace some of the functions of potassium if the later is present in insufficient quantity.

(2) Essential Nutrients

Some strains of microorganisms (autotrophic strains) may have special nutrient requirement for growth and product formation, because of their inability to synthesize an essential amino acid, fatty acid, vitamin or nucleotide. Animal cells and tissue cultures require essential amino acids for growth (e.g. arginine, histidine, isoleucine, leucine, lysine, methionine, phenylalanine, threonine and valine).

(3) Trace Elements

These are elements which are essential for cell growth when present at low concentration, but are toxic at high concentration. The estimated requirements for some of the important trace elements are: Ca 0.100, Fe 0.015, Mn 0.005, Zn 0.005, Cu 0.001, Co 0.001, and Mo 0.001 (in g/100 g dry biomass). Their toxic concentrations are generally considered to be in the order of 10^{-4} M. Trace element deficiency usually slow down the growth rate (longer lag phase) rather than limiting the

Table 2.10. Typical composition of sugar cane molasses.

	% w/v		$\mu g/100g$ dry wt
Dry matter	78–85	Thiamine	830
Sucrose	33	Riboflavin	250
Invert sugars	21	Pyridoxine	650
Other organic matters	20	Niacinamide	2100
Total nitrogen	0.4–1.5	Pantothenic acid	2140
P_2O_5	0.6–2.0	Folic acid	4
CaO	0.1–1.1	Biotin	120
MgO	0.03–0.1		
K_2O	2.6–5.0		
Ash	7–11		

Table 2.11. Typical composition of malt extract (often used for fungus, yeast and actinomycete cultivation).

	% w/w
Maltose	52
Hexoses (glucose, fructose)	19
Sucrose	2
Dextrin	15
Other carbohydrates	4
Total nitrogen	5
Ash	1.5
Water	2

Table 2.12. Typical composition of yeast extract produced by autolysis.

	% w/v		% w/w total amino acid
Dry matter	70	Alanine	3.4
Total nitrogen	9	Amino butyric acid	0.1
Protein	55	Arginine	2.1
NaCl	<1	Asparagine	3.8
		Cystine	0.3
		Glutamic acid	7.2
		Glycine	1.6
	ppm	Histidine	0.9
Thiamine	20–30	Isoleucine	2.0
Riboflavin	50–70	Leucine	2.9
Pyridoxine	25–35	Lysine	3.2
Niacinamide	600	Methionine	0.5
Pantothenic acid	200	Ornithine	0.3
		Phenylalanine	1.6
		Proline	1.6
		Serine	1.9
		Threonine	1.9
		Tyrosine	0.8
		Valine	2.3

Table 2.13. Typical composition of peptones.

	Casein peptone % w/w	Soy peptone % w/w
Total nitrogen	13	9
Amino nitrogen	7	2
Carbohydrate	—	29
NaCl	3	6
Ash	6	14
Water	5	4
Alanine	16.8	3.3
Arginine	30.2	1.1
Aspartic acid	8.7	2.4
Glutamic acid	38.6	5.6
Glycine	4.4	2.3
Histidine	13.3	1.1
Isoleucine	29.9	2.8
Leucine	71.9	9.4
Lysine	61.1	7.9
Methionine	22.7	3.3
Phenylanine	33.0	3.7
Proline	7.4	0.8
Serine	28.7	4.8
Threonine	21.5	1.9
Tryptophan	8.6	1.6
Tyrosine	14.0	0.7
Valine	36.4	3.8

Table 2.14. Typical composition of soy meal (often used in antibiotic fermentations).

	% w/w
Protein	50
Carbohydrate	30
(Sucrose, stachyose, raffinose, arabinoglucan, arabinan, acidic polysaccharides)	
Fat	1
Lecithin	1.8

biomass concentration. In a chemostat culture, one would observe a decreasing steady state biomass concentration with dilution rate, as in the cases of CO_2 (for photosynthetic culture) and O_2 limitation.

(4) Complex Substrates

Complex, sometimes indefinable substrates are often used in industrial fermentation for economic reasons. These are usually by-products of food industry, such as molasses from sugar refinery, malt extract from brewery, starchy waste from starch manufacturing factory, yeast extract (autolysis of baker's yeast at 50–55°C), peptones (acid or enzymatic hydrolysis of meat and animal organs or plant seeds) and soy meal from soybean oil factory (Tables 2.10, 2.11, 2.12, 2.13 and 2.14). Sulfite liquors and cellulose from paper industry are used. Even by-products of petroleum industry such as methane, methanol and alkanes (C_{12} to C_{18}) are suitable carbon substrates for some fermentation processes.

2.4. Microbial Cultivation Methods

In order to realize the production potential of a microbial culture, the provision of a culture system which would allow optimum expression of the desirable genetic traits is necessary. Culture systems could generally be categorized as closed systems and open systems. In a closed culture system, some essential components of the system (e.g. culture and products) cannot be supplied or withdrawn from the reactor during the process. Whereas in the open culture system, further addition of culture medium can be made and the microbial culture and products can be removed from the culture system during cultivation process.

2.4.1. Closed culture system

Based on their culture homogeneity, closed culture systems can be classified as homogeneously mixed batch culture and the plug-flow culture, where localized mixing of the whole culture is allowed.

(1) Batch Culture

A batch-wise culture system where culture medium and inoculum are added into a culture vessel at the beginning of the fermentation process and then incubated at suitable temperature and gaseous environment for an appropriate period of time is called a **batch culture**. The microbial culture in batch culture system goes through a lag phase, accelerating growth phase, exponential growth phase, decelerating growth phase, stationary phase and sometimes the decline phase depends on the end product desired (see Sec. 2.1.1). The substrate concentration in the culture medium and growth parameters, such as pH, changes correspondingly throughout the growth phases. Thus, the physiology of the microorganisms is always in a transient stage, subjected to a continually changing culture conditions. Consequently, product formation is confined to a certain period of cultivation, e.g. antibiotics would only be produced in the decelerating and stationary growth phases.

The batch culture system is still widely used in certain industrial processes, e.g. brewery industry, for its ease of operation; its less stringent sterilization; and its easy management of feed stocks. These advantages allow the use of unskilled labor and low risk of financial loss. Low level of microbial contamination in fermented products is at time tolerable, as long as the microbial contaminants are not pathogenic and do not alter the desired properties of the product, such as taste, color and texture.

(2) Plug-Flow Culture
In a plug-flow system, the culture vessel is usually in the form of a long cylindrical pipe (Fig. 2.11). Substrates and microbial inoculum are added at the inlet of the vessel, and the organisms carry out their biological activity as the liquid passes through the system. If there is minimum mixing in the culture such as in a pipe with large diameter, the culture will go through all the growth phases as in a batch culture, while being pushed along the pipe. That is, the composition of the nutrient, the cell population, gaseous content and products vary at different locations within the tubular reactor. By adjusting the rate that the culture is moving along the pipe and the length of the pipe, specific products for a particular growth phase could be obtained at the outlet of the vessel. Thus, this is a continuous flow culture process with the growth kinetics of a batch culture. Plug-flow culture systems have been used in the industry for continuous production of fermented milk and antibiotic production.

Inoculum

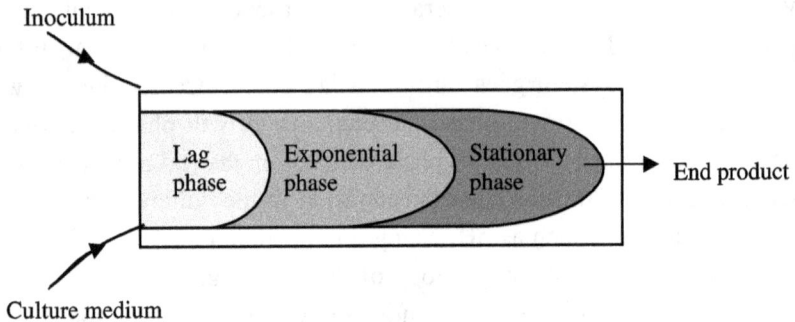

Culture medium

Fig. 2.11. A plug-flow culture system.

2.4.2. Open culture system

To maintain a microbial culture at a steady state, i.e. with its physiology fully adjusted to the culture conditions, concentration of substrates needs to be maintained and end product needs to be removed simultaneously. Thus, a fresh sterile culture medium is added to the bioreactor continuously to replenish the substrates that are consumed, and simultaneously diluting the end products that are accumulating in the system. This is often referred to as the continuous flow culture system.

Three main types of open system can be distinguished:

(1) Chemostat Culture
A chemostat is a special type of continuous flow culture system (Fig. 2.12), which consists of a perfectly mixed homogeneous suspension of microbial cells, in which the medium is fed at a constant rate (F) and culture is harvested at the same rate (F), so that the culture volume remains constant (V). The rate at which the culture is diluted through the addition of fresh medium is defined as:

Dilution rate, $\quad D = F/V.$

Biomass balance is achieved when: Net increase = growth − harvest

$$V \cdot dX = V \cdot \mu X \cdot dt - FX \cdot dt$$
$$dX/dt = (\mu - D) \cdot X$$

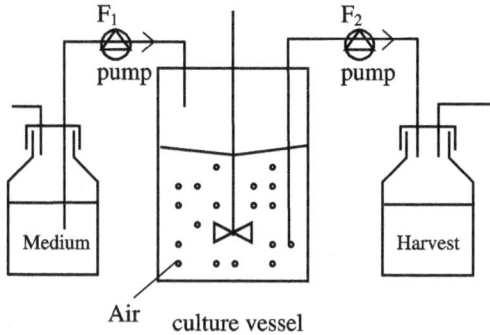

Fig. 2.12. A Chemostat culture system.

If there is a **steady state** whereby the rate of growth equal the rate of cell removal, the biomass concentration remains constant, i.e. $dX/dt = 0$. Then, $(\mu - D) = 0$ or $\mu = D$.

Steady state in chemostat was experimentally established in bacterial, fungal and algal cultures. The theory indicates that it is possible to fix the μ of a culture at any value from zero to the maximum by adjusting the value of D.

A chemostat culture system can be used for the following purposes:

(i) To vary the specific growth rate (or metabolic rate) of a culture without changing the environmental condition, other than the concentration of growth-limiting substrate. This application is particularly useful in the understanding of the relationship between growth rate and product formation or a physiological function of the microbial culture.

(ii) To fix the specific growth rate while varying a culture condition, i.e. converse of (i). The application is useful in the study of the effect of the culture environment without interfering the growth rate.

(iii) To maintain substrate limited growth with a constant specific growth rate. This could not be achieved in a batch culture, as the growth rate is determined by the concentration of the substrate in the culture medium.

(iv) To obtain time-independent balanced growth (in steady state), with phenotype fully adjusted to environment, i.e. the culture is history independent, unlike cells in batch culture which performance is often dictated by the physiological state of the inoculum and culture conditions that the cells were exposed to prior to inoculation.

(v) To maintain maximum output rate of biomass or product. As is shown in Sec. 2.2, the rate of product formation is ultimately determined by the growth rate of the culture.

In the chemostat culture, the growth-limiting substrate balance can be expressed as:

$$\text{Net increase} = \text{Input} - \text{Output} - \text{Substrate used for growth}$$
$$V \cdot ds = F \cdot s_m \cdot dt - F \cdot s_r \cdot dt - V \cdot \mu \cdot X \cdot dt/Y_{X/S}$$

where

s_m = Substrate concentration in the fresh culture medium
s_r = Residual substrate concentration in the culture in steady state

$$ds/dt = F/V \cdot (s_m - s_r) - \mu \cdot X/Y_{X/S}$$
$$= D \cdot (s_m - s_r) - \mu \cdot X/Y_{X/S}$$

In steady state, $D = \mu$, $ds/dt = 0$,

$$X = Y_{X/S} \cdot (s_m - s_r) \qquad\qquad (2.13)$$

Since $\mu = \mu_m \cdot s_r/(s_r + K_s)$,

$$D = D_c \cdot s_r/(s_r + K_s)$$

The critical dilution rate (D_c), which equals the maximum specific growth rate, is the highest possible dilution rate that a steady state could be established. Above which, complete washout (see (2)(i) below) of cells is observed (Fig. 2.13).

The primary effect of varying the dilution rate (D) is to change the residual concentration of the growth-limiting substrate, thereby affecting a change in μ, i.e. μ of a chemostat culture at steady state is determined by s_r only, independent of s_m.

The steady state biomass concentration can be defined and calculated by the following equation:

$$X = Y_{X/S} \cdot (s_m - K_s \cdot D/[\mu_m - D]).$$

Since μ_m, K_s and $Y_{X/S}$ are constants, the steady state concentration of biomass is effectively determined by s_m and D.

Chemostat culture system has been used in Microbial Protein production, e.g. ICI's Pruteen (*Methylophilus methylotrophus*) production from methanol; Rank Hovis McDougall's mycoprotein (*Fusarium graminearum*) production from carbohydrate containing waste from flour milling plant.

Box 2.9

Comparison of the Output Rate of Batch and that of Chemostat Culture

The **biomass output rate** ($R = \mu X$) can be compared quantitatively as follows:

Let X_m = Maximum biomass concentration in a batch culture
X_0 = Initial biomass concentration in a batch culture
μ_m = Specific growth rate in the batch culture until the rate limiting substrate is exhausted
t_c = Duration of the batch culture
t_a = Delay time which includes any lag period and the time required for recycling of the culture vessel between batches
t_d = Doubling time

Thus, in a batch culture:

$$t_c = 1/\mu_m \cdot \ln(X_m/X_0) + t_a$$
$$R_{batch} = Y_{x/s} \cdot s_m/t_c$$
$$= \mu_m \cdot Y_{x/s} \cdot s_m/(\ln[X_m/X_0] + \mu_m \cdot t_a)$$

assuming that in a chemostat, $D = \mu_m$.

Thus, in steady state:

$$R_{chemotat} = \mu_m \cdot Y_{x/s} \cdot s_m$$
$$R_{chemostat}/R_{batch} = \ln(X_m/X_0) + \mu_m \cdot t_a$$
$$= \ln(X_m/X_0) + 0.693t_a/t_d$$

As a general practice, the ratio of $X_m:X_0 = 10:1$, which makes the maximum output rate of chemostat at least 2.3 times that of batch culture (when $t_a = 0$). Frequently, t_a is many times greater than t_d (doubling time of the culture).

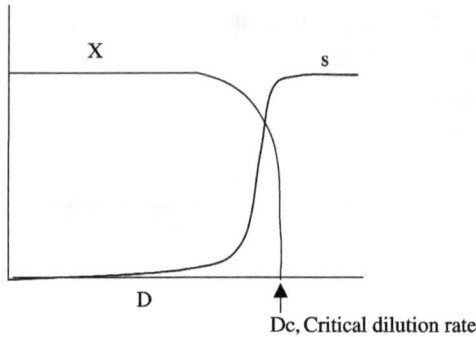

Fig. 2.13. The steady state biomass (X) and rate-limiting substrate concentrations (s) at various dilution rates (D) in a chemostat culture.

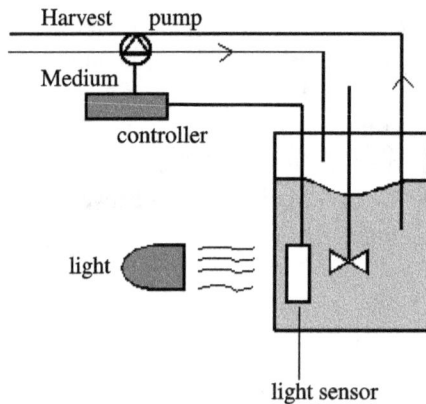

Fig. 2.14. Turbidostat culture system.

(2) Turbidostat

Turbidostat is a continuous flow culture in which the biomass concentration in the culture vessel is maintained at constant by monitoring the turbidity of the culture through the use of a photoelectric cell (an optical sensing device), and regulating the feeding rate of nutrient solution. Thus, the addition of fresh medium to the system is activated when the turbidity of the culture exceeds the set level. The specific growth rate of the culture is therefore allowed to adjust itself in respond to the growth-limiting parameter. Figure 2.14 is the schematic of a turbidostat.

The applications of turbidostat are as follows:

(i) To fix biomass concentration and allow the specific growth rate to adjust itself. This application is particularly useful for slow growing microorganisms and those with complex cell cycles. In the latter case, the μ_m of a culture may be very different at various stages in the cell cycle, resulting in complete **washout** (total removal of cells from the culture system through excessive dilution) of the culture in a fixed dilution rate chemostat.

(ii) To maintain a constant biomass concentration under unstable environment conditions, such as for the treatment of wastewater containing variable concentration of toxic compounds. Temporary suspension of specific growth rate by toxic compounds would result in washout in a simple chemostat with fixed dilution rate.

(iii) To select for fast growing microbial strains, e.g. resistance to an inhibitor. The dilution rate of a turbidostat is determined by the growth rate of the culture, thus evolution of a resistant strain would cause an increase in the dilution rate and washout of the slow growing cells (e.g. those could not tolerate the inhibitor) from the culture system.

(3) Fed-Batch Culture (FBC)

Unlike the conventional batch cultivation method where all the substrates are added at the beginning of the process, fed batch process is a batch culture with continuous feeding of nutrients while the effluent is removed discontinuously. The volume of the culture could be variable, by feeding the complete medium **(Variable Volume FBC)** or constant, by feeding the growth-limiting substrate in the form of solid, concentrated solution, gas or light **(Constant Volume FBC)**. The critical elements are added in small concentrations at the beginning of the fermentation and continue to be added intermittently in small doses during the production.

In the case where at the end of a cultivation cycle, the culture is partially withdrawn and replaced with fresh complete medium, thereby initiating a new cycle. This is termed the **cyclic FBC** (Fig. 2.15).

Listed below are the various applications of FBC:

(i) Like the chemostat, FBC is a mean of achieving substrate-limited growth, but the specific growth rate of the culture is not a constant value at steady state. The specific growth rate of the culture would

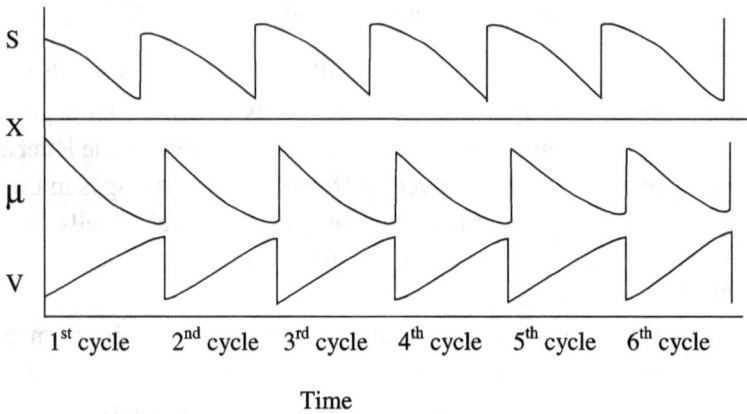

Fig. 2.15. Growth kinetics in a cyclic variable volume fed batch culture. V = volume of the culture, X = biomass concentration, μ = specific growth rate, S = rate-limiting substrate concentration.

decrease with time with increasing culture volume (in a variable volume FBC) or cell concentration (in a constant volume FBC).

(ii)   FBC is also used to achieve periodic shift up and down in the specific growth rate with control over the frequency and magnitude of the shift. Such a periodic shift in metabolic rate may stimulate "overflow metabolism" and secondary metabolism, as in the case of cephalosporin C production.

(iii)   To achieve a steady state in continuous flow culture of filamentous moulds. It is technically difficult to maintain a small volume chemostat of fungal culture, because of the difficulty in maintaining the homogeneity in the harvest stream of the culture system. Beside, FBC system requires only a mechanism to feed the substrate and maintenance of constant volume in the culture is not necessary.

An example of industrial FBC is the penicillin production process. Specific growth rate of *Penicillium chrysogenum* culture is maintained at below maximum values through appropriately adjusting the feed rate of carbohydrate, thus penicillin production rate is always maintained at the optimal level for a prolonged period.

(iv)   The mechanism of FBC simulates some ecosystems, e.g. an infected bladder, is a case of variable volume FBC. The principle of continuous culture depicts that the pathogen in an infected bladder (cystitis) would eventually be washed-out if the dilution rate (i.e. rate of topping up with urine and discharge of urine from bladder) exceeds the specific growth rate of the pathogen.

Further Reading

1. Bull AT and Dalton H, eds. *Comprehensive Biotechnology Vol. 1: The Principles of Biotechnology*. Scientific Fundamentals, Pergamon Press, Oxford, 1985.
2. Crueger W and Crueger A. *Biotechnology: A Textbook of Industrial Microbiology*. Sinauer Associates, Sunderland Massachusetts, 1990.
3. Demain AL and Solomon NA, eds. *Biology of Industrial Microorganisms*. Butterworth, Stoneham Massachusetts, 1985.
4. Glazer AN and Nikaido H. *Microbial Biotechnology: Fundamentals of Applied Microbiology*. WH Freeman, New York, 1995.
5. Harrison DEF. Efficiency of Microbial Growth. In: *Companion to Microbiology*, eds. AT Bull and PM Meadow. Longman, London, 1978, pp. 155–180.
6. Kolter R, Siegele DA and Tormo A. The stationary phase of the bacteria life cycle. *Annu. Rev. Microbiol.* **47**: 855–874, 1993.
7. Mager WH and De Kruijff AJJ. Stress-induced transcriptional activation. *Microbiol. Rev.* **59**: 506–531, 1995.
8. Pirt SJ. *Principles of Microbe and Cell Cultivation*. Blackwell Scientific, Oxford, 1975.
9. Rehm HJ and Reed G, eds. *Biotechnology Vol.1: Microbial Fundamentals*. VCH Publishers, Weinheim, 1995.
10. Russell JB and Cook GM. Energetics of bacterial growth: Balance of anabolic and catabolic reactions. *Microbiol. Rev.* **59**: 48–62, 1995.
11. Slater JH. Microbial growth dynamics. In: *Comprehensive Biotechnology Vol. 1*, eds. AT Bull and H Dalton. Pergamon Press, Oxford, 1985, pp. 189–213.
12. Stouthamer AH and van Verseveld HW. Stoichiometry of microbial growth. In: *Comprehensive Biotechnology Vol. 1*, eds. AT Bull and H Dalton. Pergamon Press, Oxford, 1985, pp. 215–238.

13. Vanek Z and Hostalek Z, eds. *Overproduction of Microbial Metabolites: Strain Improvement and Process Control Strategies.* Butterworth, Stoneham Massachusetts, 1986.

Websites

- About Biotech
  http://outcast.gene.com/ae/AB/
- Access Excellence
  http://www.gene.com.ae
- Biotechnology Industry Organization
  http://www.accessexcellence.com/AB/BC/Bioprcess Technology.html
- Biotechnology Information Center
  http://www.inform.umed.edu:8080/EdRes/Topic/AgrEnv/Biotech
- Encyclopedia of Bioprocess Technology
  http://www.cbs.umn.edu/bpti/MCF/ebt.html
- Genetic Engineering News
  http://www.genengnews.com
- Microbiology Laboratory Simulation Programme
  http://www.leeds.ac.uk/bionet/compend/bnt05pst.html

Questions for Thought...

1. Explain, with examples, how the effects of rate limiting substrate concentration on the growth of microbial culture could be quantified at both the theoretical and practical levels.
2. What are the conditions which allow the establishment of a stable mixed culture system? Use an example to describe the interactions among the microbial population in mixed culture fermentation.
3. How do you improve the productivity of a growth-linked product produced by a genetically engineered bacterium in a batch culture process? Give your reasons.
4. Two culture media each containing the same molar equivalent of carbon in the form of molasses or sucrose were used to culture yeast. What could be the possible reasons that the yeast culture in the

medium containing molasses was found to achieve a higher biomass growth yield?

5. Explain with examples, the importance of growth yield in fermentation processes.

6. Two batch cultures of yeast, containing an initial glucose concentration of 0.5 and 2 g/L respectively, were compared. It was observed that the specific growth rate and final biomass concentration of the culture containing 2 g/L glucose were lower than the one containing 0.5 g/L glucose. The final concentration of alcohol was, however, comparable. Give possible explanations.

7. Compare and contrast closed and open microbial culture methods.

8. If you are working for a dairy company, how would you convince your company that you could achieve a higher output rate for the production of fermented milk in a continuous flow fermentation process than the currently used batch culture system?

Chapter 3

Application of -Omics Technologies in Microbial Fermentation

Victor Wong Vai-Tak[*,†,a], **Faraaz Noor Khan Yusufi**[*,b]
and Lee Dong-Yup [*,†,c]
Bioprocessing Technology Institute, A*STAR
20 Biopolis Way, #06-01 Centros, Singapore 138668
†*Department of Chemical and Biomolecular Engineering*
National University of Singapore
4 Engineering Drive 4, Blk E5 #02-09, Singapore 117576
Emails: [a]*victor.wong@lonza.com*
[b]*faraaz_yusufi@bti.a-star.edu.sg*
[c]*cheld@nus.edu.sg*

3.1. Introduction

The use of microorganisms as "cell factories" offers a rich repository of metabolic reactions to convert raw materials (e.g. glucose and amino acids) to a desired product (e.g. metabolite or protein). The development of recombinant DNA technology has further extended the range of products from those that are naturally occurring in the host organism to those produced through insertion of heterologous pathways. However, very often, the yield and productivity of the desired product is too low and it is necessary to improve the performance of the production host strain to establish an economically viable process. Traditionally, strain improvement has proceeded via multiple rounds of random mutagenesis and high-throughput screening to identify a high-productivity strain. The process is laborious and the strain often accumulates deleterious mutations in the process. With genome sequences becoming readily available for many production organisms, a more gene-targeted, rational approach to redirect substrates through the network of metabolic reactions are being applied to improve the yield, productivity and robustness of the production host. This new approach is termed metabolic engineering and usually involves the overexpression or deletion of selected genes from the production host or the introduction of genes from other organisms into the production host.

Identifying the appropriate genes to target for metabolic engineering is significantly aided by an understanding of the entire complement of genes ("genome"), mRNA transcripts ("transcriptome"), proteins ("proteome") or metabolites ("metabolome") involved or affected by the production of the desired product. The corresponding technology for the analysis of each of these "-omes" is collectively termed "-omics" technologies. Figure 3.1 shows the cascade of -omics technologies and the type of information it reveals about the production host.

Typically, -omics technologies are applied to compare samples from two or more experimental conditions. Table 3.1 lists some examples of comparisons that can be done and their corresponding information that may be obtained. The overall goal is to identify gene targets that should

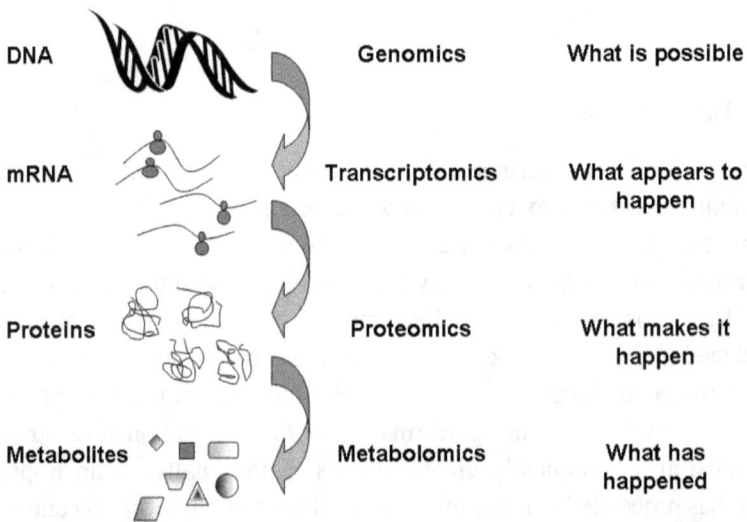

DNA		Genomics	What is possible
mRNA		Transcriptomics	What appears to happen
Proteins		Proteomics	What makes it happen
Metabolites		Metabolomics	What has happened

Fig. 3.1.   The -omics cascade.

Table 3.1.   Examples of experimental conditions compared using -omics technologies.

Comparison	Information obtained
Strains from random mutagenesis screens (e.g. high-productivity vs. low-productivity)	• Identify the genes or pathways associated with the desired phenotypes (e.g. high-productivity, fast growth, robustness) • Identify the genes or pathways associated with deleterious phenotypes (e.g. low-productivity, slow growth, instability) • Identify limiting metabolic substrate • Identify positive and negative regulatory genes/proteins
Before vs. after induction of recombinant protein	• Identify limiting metabolic substrate in production of recombinant protein • Identify positive and negative regulatory genes/proteins
Plasmid-bearing vs. non-plasmid bearing cells	• Identify limiting metabolic substrate in production of recombinant plasmid DNA • Identify positive and negative regulatory genes/proteins
Different stages of growth (e.g. exponential vs. stationary phase)	• Identify positive and negative regulatory genes/proteins • Identify metabolic pathways that are activated or repressed during various stages of growth
Growth in different media or fermentation conditions (e.g. rich vs. minimal media, exposure to heat shock or nutrient limitation)	• Identify metabolic pathways that are activated or repressed under different media or fermentation conditions

be overexpressed, deleted or introduced from other hosts in order to improve the performance of the production host.

3.2. Genomics and Transcriptomics

Genomics refers to the large-scale analysis of the genome of an organism and includes determining the entire DNA sequence of an organism and annotating the function of genome sequences. The genomes of several microorganisms widely used for industrial bioprocesses have already been sequenced, including *Saccharomyces cerevisiae*, *Escherichia coli*, *Bacillus subtilis* and *Pichia pastoris*. With the development of new rapid and cost-effective sequencing technologies, genome sequencing is increasingly being used as a tool to analyze microbial hosts with different phenotypes (e.g. strains from random mutagenesis screens with different productivities). The annotation of genome sequences has evolved into a discipline known as functional genomics, which focuses on developing experimental and theoretical tools to determine the gene functions of the entire genome. The annotation of the genome is a key step to bridge the gap between data-driven approaches (genome sequence collections) and knowledge-based approaches (e.g. application of the genome sequences to guide metabolic engineering).

As illustrated in Fig. 3.1, the genome sequence only provides information about what the microorganism is capable to do (analogous to the complete parts list of the "cell factory"). While this information is useful, it does not reveal if the genes have been expressed. To do that, researchers study the messenger RNA (mRNA) or transcription level of the genes. The analysis of the entire complement of transcripts in the production host is known as transcriptomics or expression profiling. A popular method to study transcript changes on a genome-wide scale is DNA microarray. Figure 3.2 gives an overview of the DNA microarray process.

3.2.1. *DNA microarray construction*

Expression profiling by DNA microarrays compares the relative amount of mRNA expressed in two or more experimental conditions, such as the examples listed in Table 3.1. Researchers first create a DNA library for a production host organism by cloning out sequences specific for each gene using polymerase chain reaction (PCR). For genomes that have been fully sequenced, the library created will be a fuller representation of the transcriptome of the organism compared to those genomes that have only been partially determined. Alternatively, synthetic oligonucleotides designed to match parts of the sequence specific for each gene can be used instead of PCR products. To construct the DNA microarray, the PCR products or oligonucleotides corresponding to each gene are immobilized

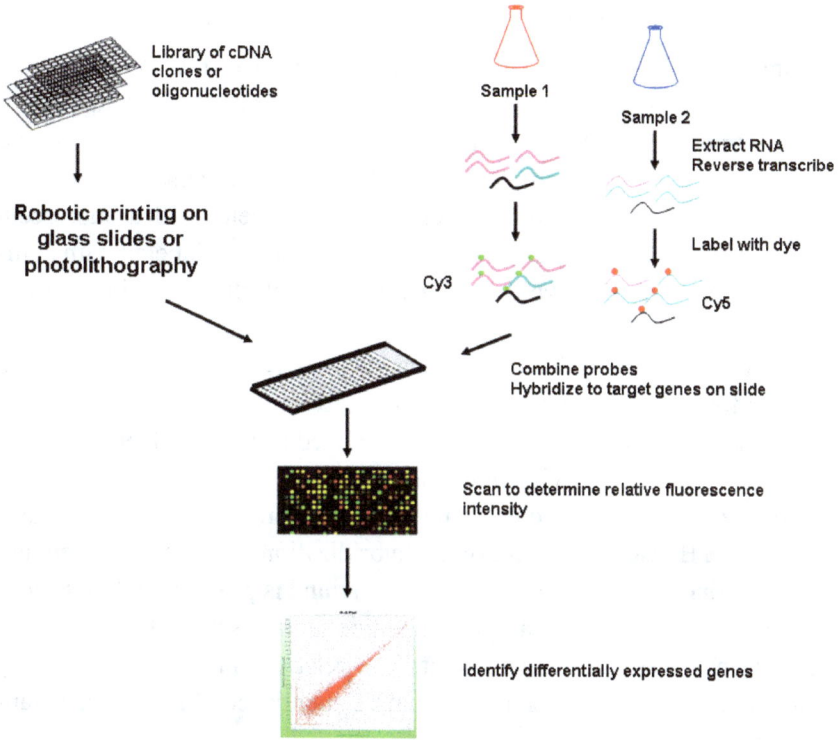

Fig. 3.2.   DNA microarray process.

onto a glass slide in a grid pattern. Earlier forms of microarrays used pins attached to robotic arms to spot the DNA onto the glass surface at densities of 250–1000 spots/cm^2. Later technology (e.g. Affymetrix's GeneChip) used photolithography to synthesize oligonucleotides directly onto silicon substrates at densities of 60000 spots/cm^2. Software programs are used to keep track of the gene identity of each spot in the grid pattern. Each gene is usually spotted more than once and the average reading determined from replicates on each slide. Control spots are also included on each slide. Typically, these are from genome sequences that are unrelated to the production host. Positive control spots randomly spaced across the grids help to assess the labeling efficiency and detect any spatial bias in hybridization. Negative controls give an indication of non-specific hybridization.

3.2.2. Hybridization and scanning

Samples of cells from the reference and test experimental conditions are harvested and resuspended in solutions that limit the degradation of mRNA transcripts. The mRNA from each sample is then extracted and converted into cDNA (complementary DNA) using reverse transcriptase. During reverse transcription, each cDNA is also labeled with a fluorescent dye. In two-color microarrays, each of the 2 samples is labeled with a different fluorescent dye. Typically, Cyanine3 (which fluoresces in the green part of the spectrum) and Cyanine5 (which fluoresces in the red part of the spectrum) are used. The dye is swapped in replicate experiments to account for any labeling bias. Equal concentrations of labeled cDNA from the two samples are then mixed and hybridized against the DNA microarray slides. During hybridization, each strand of labeled cDNA from the sample would bind to the complementary strand immobilized on the DNA microarray. By careful control of the hybridization conditions, the amount of cDNA that binds to its complementary strand is proportional to its concentration present in the sample. The slide is then scanned to detect the fluorescence intensity of each spot. The relative fluorescence intensity would thus be an indication of the relative abundance of a particular transcript in the two samples.

3.2.3. Analysis

Typically, samples from replicate fermentation runs are hybridized against multiple arrays in order to ensure conclusions drawn are statistically and biologically valid. However, slight variation between runs and during hybridization means that the results from various hybridizations need to be normalized before they can be compared. After normalization, the ratio of the fluorescence intensity for each spot is determined. Based on the ratio, statistical methods are used to identify candidate genes showing significant differential expression. Various statistical methods developed for normalization and analysis of results from -omics studies are described in Section 3.5.1. Although DNA microarrays can analyze a large number of genes simultaneously, the technique lacks the quantitative precision of more targeted analytical methods. Hence, DNA microarrays are usually performed first to identify a list of candidate genes. The microarray results are then validated on these selected genes using more quantitative techniques such as quantitative polymerase chain reaction (qPCR) or western blotting.

3.3. Proteomics

While the analysis of transcript levels described above can provide useful insights, there are limitations. Often, the amount of protein translated from a given amount of mRNA depends on the gene and physiological state of the cell. Even if a particular mRNA is abundant, the corresponding protein may be low because the mRNA was degraded rapidly or translated inefficiently. Some proteins may need to be phosphorylated or complexed with other proteins to be active. Furthermore, in higher eukaryotes, a single gene can result in multiple proteins through alternative splicing. Alternative splicing is especially common in mammalian cells, though it is rare in single cell eukaryotes like yeast. Thus, in addition to studying transcriptomic changes, researchers are also interested in analyzing changes in the proteome, or the entire complement of proteins produced by the production host.

The basic proteomic workflow is illustrated in Fig. 3.3.

The first step is the extraction of proteins from the cells. After harvesting the cells by centrifugation, the cell pellet is washed to remove the remaining media and resuspended in a buffered salt solution. Lysis may be carried out by various methods including high pressure homogenization (e.g. French press or Fluidizer), repeated freeze-thaw or sonication. As the genomic DNA that is released can increase the viscosity of the lysate and interfere with recovery of soluble proteins, the DNA may be precipitated with polyethyleneimine. Alternatively, the lysate may be passed through a DEAE-cellulose column to retain the DNA or treated with nucleases.

Following extraction, the protein mixture must be separated before analysis. A number of methods have been developed for large

| Protein extraction | Protein separation | Digestion into peptides | Mass spectrometry |

Fig. 3.3.   Proteomics work flow.

scale separation and analysis of protein samples. A popular method is 2-dimensional gel electrophoresis (2-DE), which is a gel electrophoresis technique to separate thousands of proteins based on two protein properties. The separated proteins are then identified using either mass spectrometry or other methods. Alternatively, a non-gel based method is iTRAQ (isobaric tag for relative and absolute quantitation), which relies on differential labeling of peptides with isobaric tags that cleave into reporter ions of distinct masses after collision induced dissociation in the mass spectrometer.

3.3.1. *Two-dimensional gel electrophoresis*

This technique separates proteins in a gel based on two protein properties, usually the isoelectric point (pI) and the mass. As it is unlikely that two proteins will be similar in two distinct properties, 2-DE allows a more effective separation than 1-DE. As a result, this technique allows the qualitative and quantitative analysis of the expression of thousands of proteins simultaneously.

The first dimension of separation is based on pI and is also known as isoelectric focusing (IEF). This makes use of a strip of gel with a pH gradient, called an IPG (immobilized pH gradient) strip. The protein sample is loaded on IPG strips and an electric potential is applied across the strip. Proteins with a negative charge will migrate towards the positive electrode, while positively charged proteins will move towards the negative electrode. The proteins will accumulate in the gel position corresponding to their pI, where they have neutral net charge.

After IEF, the IPG strips are treated with sodium dodecyl sulfate (SDS), urea and reducing agents like dithiothreitol to denature the proteins so that they are unfolded into long straight molecules. The length of the unfolded protein is approximately proportional to the mass. The SDS binds to the unfolded protein in proportion to the unfolded length and confers a negative charge to the protein. The treated IPG strip is then applied to the top of a SDS-PAGE gel (sodium dodecylsulfate-polyacrylamide gel electrophoresis) and an electric potential applied at 90° to the first field. The proteins will migrate towards the more positive side of the gel according to their mass-to-charge ratio. As the polyacrylamide gel acts like a molecular sieve to retard protein migration,

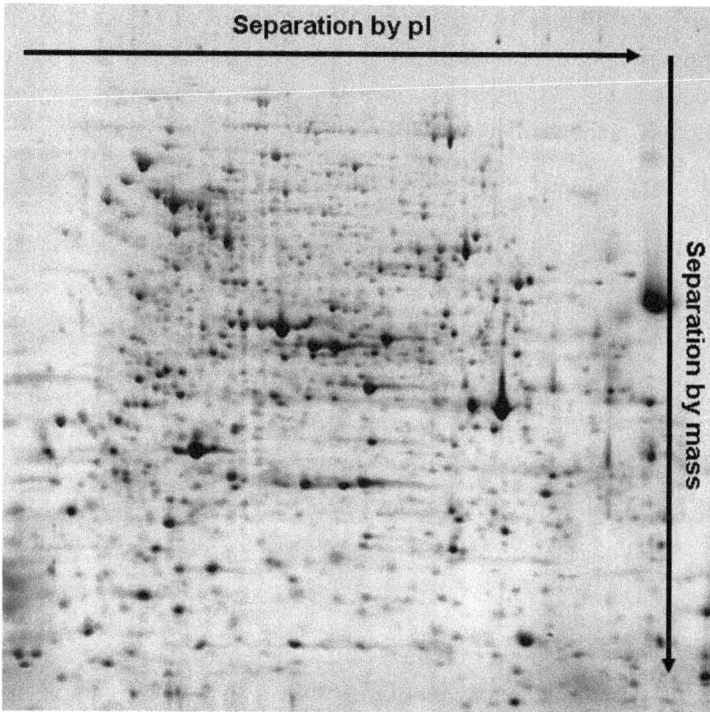

Fig. 3.4.   2-D gel electrophoresis.

the larger proteins will be retained nearer the top of the gel while the smaller proteins will migrate faster towards the bottom of the gel. As a result, the proteins will be spread out on the gel, as shown in Fig. 3.4. To visualize the protein spots, the gel may be stained with Coomassie Brilliant Blue, colloidal silver or a fluorescent dye. The intensity of staining is approximately proportional to the amount of protein, thus enabling comparison of relative protein levels in different gels.

A typical experiment may involve 2-DE separation of proteins extracted from cells under different conditions (e.g. Table 3.1) in various gels to compare the proteomic differences. Analysis of 2-DE gel images may be assisted by software programs, though visual confirmation by a human expert is often required, as automatic software programs are not effective in analyzing incompletely separated or faint spots. Protein gel spots of interest could be those which show significant differential expression between the treated and control sample or are highly expressed (or absent) under specific culture conditions. The gel spots of interest may be directly excised from the 2-DE gel or after blotting to membranes. The protein identity may be determined by immunoblotting, N-terminal sequencing or internal peptide sequencing. A more high-throughput technique is peptide mass fingerprinting. In this technique, peptides are first created from the excised protein spots by endoproteinase (e.g. trypsin) digestion before mass spectrometry to determine the mass profile of the peptides (similar to a "fingerprint" of the protein). These are then searched against a library of theoretical peptide masses generated from protein databases.

While it is a powerful technique, 2-DE has several inherent disadvantages. Despite the assistance of automatic image analysis software, labor intensive manual analysis must still be performed to quantify changes and remove technical artifacts. Low abundance proteins may also not be sufficiently visible on the gel to detect. A protein may migrate in multiple spots due to protein modifications (e.g. oxidation, incomplete trypsin cleavage) so that quantification of total changes for a protein requires locating and identifying all the spots. Protein specific differences may also lead to variable staining that is independent of the protein amount. Limitations in resolution may result in a given spot containing multiple proteins. Finally, some large hydrophobic proteins, low molecular weight proteins and those with extreme pI values cannot be separated and analyzed by 2-DE.

3.3.2. *Isobaric tag for relative and absolute quantitation*

Commonly known as iTRAQ, this technique can overcome some of the limitations of gel-based methods. As shown in Fig. 3.5, each iTRAQ tag consists of a reporter group (based on N-methylpiperazine) and a mass balance group (carbonyl group). The combinations of these 2 groups are such that each intact tag has the same mass (isobaric). A NHS-ester derivative enables the tag to attach to the N-termini of proteolytic peptides from different protein samples. Currently, the iTRAQ tags are available as a collection of 4-plex or 8-plex tags, enabling the labeling of four or eight protein samples respectively. 4-plex tags means the reporter ions have mass/charge ratios of 114, 115, 116 and 118. Similarly, 8-plex tags consists of reporter ions with mass/charge ratios from 113 to 121 (except

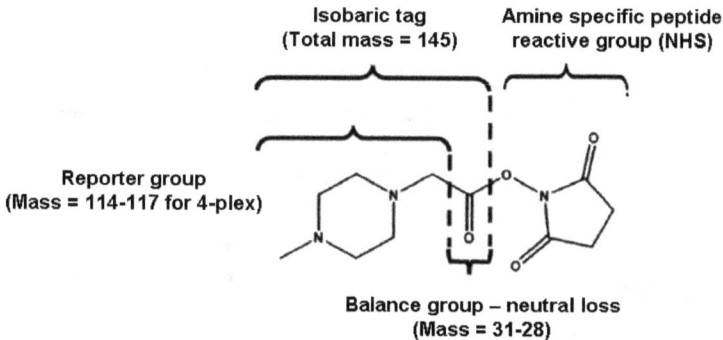

Isobaric tag
(Total mass = 145)

Amine specific peptide
reactive group (NHS)

Reporter group
(Mass = 114-117 for 4-plex)

Balance group – neutral loss
(Mass = 31-28)

Fig. 3.5.   Structure of iTRAQ reagent.

120). The isobaric design means that differentially labeled peptides appear as a single peak in MS scans. Only when subjected to collision induced dissociation in the MS/MS mode will the reporter ion be released from the peptides, allowing relative quantitation of the peptide from the peak area ratios of the reporter ions.

An overview of the iTRAQ method is illustrated in Fig. 3.6. Each protein sample is first reduced, alkylated and digested with trypsin to generate peptides. The peptides from each sample are then labeled with one of the iTRAQ tags. The differentially labeled samples are then combined and subjected to strong cation exchange HPLC (high-pressure liquid chromatography) fractionation. The fractions are further subjected to reverse-phase HPLC before spotting on a plate for MALDI-MS (matrix-assisted laser desorption/ionization — mass spectrometry). Within the MS, peptides that match a preset ion score are subjected to MS/MS fragmentation, which releases the reporter ions. Relative expression ratios for each peptide are calculated from the peak areas of the reporter ions. Eventually, the protein expression ratio is calculated from the average ratio of its constituent peptide peak area ratios.

3.4. Metabolomics

As shown in Fig. 3.1, metabolites are the products or intermediates of the network of biochemical reactions in the cell. A primary metabolite is directly involved in the growth, development and reproduction of the

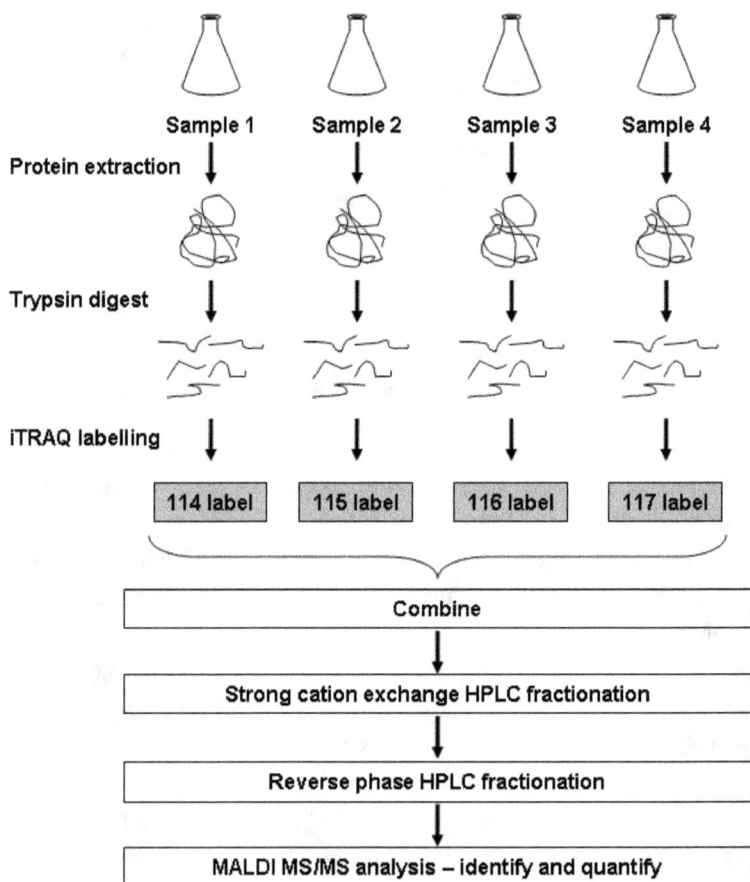

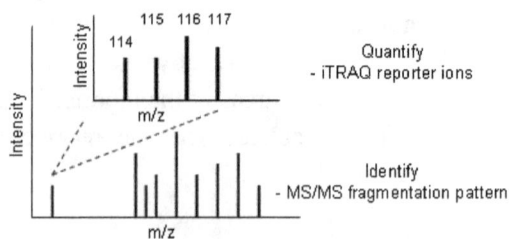

Fig. 3.6.   Overview of iTRAQ process.

organism. In contrast, a secondary metabolite is not directly involved in these processes but still plays an important role in the organism's survival. Many products of industrial bioprocesses using microorganisms are actually secondary metabolites, such as antibiotics and pigments. Metabolomics refers to the identification and quantification of intracellular and extracellular metabolites formed by the microorganism as part of its metabolism. Compared to the other -omics technologies, metabolomics is often more challenging. A majority of metabolites participate in multiple biochemical reactions; hence, their concentration is influenced by the activity of numerous enzymes involved in its synthesis and conversion, resulting in a large range of concentrations and rapid turnover. This necessitates sample preparation techniques that can rapidly quench the biochemical reactions and preserve the integrity of the metabolites. Another limitation is due to the complexity of the metabolome. DNA and RNA are composed of combinations of four nucleotides, while proteins are composed of combinations of 20 amino acids; in contrast, metabolites comprise a wide range of compounds with different structures, functional groups and physicochemical properties. As a result, no single analytical method can simultaneously analyze the complete range of metabolites.

3.4.1. *Types of metabolome analysis*

Different approaches are used in studying the metabolome. Metabolite profiling refers to the analysis of a group of specific metabolites (e.g. amino acids or a group of metabolites involved in a specific pathway) in a semi-quantitative or quantitative way. A restricted quantitative analysis of metabolites participating in a specific part of the metabolism is called metabolite target analysis. For example, researchers may focus on metabolites involved in the glycolytic pathway under different fermentation conditions, or compare the profile between different production strains for metabolites involved in the conversion of a precursor to the final antibiotic product. As the identities of the metabolites are known, a suitable analytical method can be selected to detect and quantify each metabolite species. On the other hand, in metabolite fingerprinting, researchers compare the spectra from NMR (nuclear magnetic resonance) or MS (mass spectrometry) analysis of the intracellular metabolites produced by a cell.

The spectra usually do not provide information about specific metabolites but gives a "fingerprint" of the intracellular metabolites in a given sample. The corresponding analysis of the extracellular metabolites is known as metabolite foot printing. These latter two approaches may be used to classify different samples or as a preliminary diagnostic test for a disease. A truly metabolomic approach, similar to transcriptomics or proteomics, would involve the identification and quantification of all the metabolites produced by a cell. Such an approach is currently not possible due to technical limitations, although researchers are able to analyze a fraction of the metabolome using a combination of analytical methods.

3.4.2. *Sample preparation*

Figure 3.7 shows the general procedure for sample preparation in metabolome analysis. As discussed above, metabolites have rapid turnover rates, a diverse range of physicochemical properties and wide range of concentrations. Hence, researchers need to take all these factors into consideration when deciding a sample preparation procedure.

Primary metabolites have turnover rates in the order of seconds or less as they tend to be involved in multiple reactions. For example, the turnover rate for intracellular glucose is estimated to be 1 mM/s while the rate for ATP is 1.5 mM/s. On the other hand, secondary metabolites usually accumulate in the cells or are secreted into the extracellular medium and tend to have comparatively slower turnover rates. Similarly, extracellular metabolites are usually more diluted and have slower turnover compared to intracellular metabolites. In metabolic engineering, researchers are

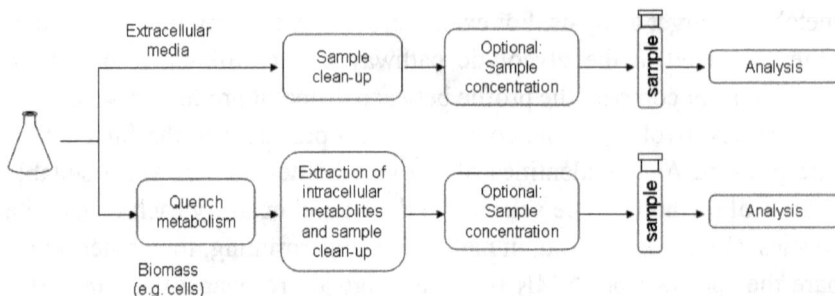

Fig. 3.7.   Sample preparation for metabolome analysis.

usually interested in the impact of environmental conditions or genetic alteration on cell metabolism and product formation, hence, metabolome analysis would have to encompass both primary and secondary metabolites. Quenching of metabolism is the first and extremely important step in sample preparation. The high turnover rate of metabolites necessitates rapid quenching methods. The procedure must also be able to maintain cell membrane integrity to minimize loss of metabolites from the cell. The method used depends on both the cell type and the type of metabolites. For microbial cultures, direct sampling into the quenching solution is often used to minimize changes in metabolite levels. As the levels of intra- and extracellular metabolites in Gram-negative bacteria, like *E. coli*, are in osmotic equilibrium, leakage of intracellular metabolites also needs to be minimized by careful control of the osmolality of the quenching solution. A reported protocol for analysis of intracellular amino acids uses fast filtration under vacuum to separate bacterial cells from the extracellular medium and washing with cold saline solution (0.9%) at $-0.5°C$. Other quenching methods use cold buffered aqueous solutions containing methanol or ethanol. For example, a widely used protocol for yeast involves spraying yeast cultures into 60% (v/v) cold methanol solution at $-40°C$.

After quenching, the intracellular metabolites need to be extracted from the cells. Ideally, the procedure should extract all metabolites equally without bias. However, this is rarely possible in practice and choices have to be made concerning which metabolites should be measured and which can be "sacrificed" in order to get good reproducible measurements of the metabolites of interest. Alternatively, multiple extraction procedures may be needed to analyze as many metabolites as possible. The extraction procedure should also ensure stability of the metabolites and solubility in the extraction solvent. The first step in extraction is to lyse the cells and this may be achieved by physical (e.g. homogenizer, freeze-thaw, sonication) or chemical means (e.g. detergent, acid/alkali treatment, organic solvents). The choice of cell lysis method depends on the type of cell and whether the metabolites of interest can withstand the lysis procedure. The second step is to extract the metabolites into the solvent mixture. As metabolites can range from polar to non-polar compounds, a mixture of aqueous and organic solvent is usually used. For extracellular metabolites, sample clean-up is required to remove compounds that may interfere with subsequent analytical techniques. The major group of interfering compounds are proteins and these are usually removed by precipitating out with cold organic solvents (e.g. methanol or ethanol).

The final step in sample preparation is concentration. This is especially necessary for metabolites that are present at very low concentrations. This is usually done by lyophilization followed by reconstitution in a small volume of buffer. Alternatively, the sample may be concentrated by blow-drying with an inert gas (e.g. nitrogen).

3.4.3. Detection methods

The diverse range of physicochemical properties of metabolites makes it virtually impossible to find one analytical method that can measure all metabolites. Typically, multiple methods are needed for comprehensive coverage of the metabolome. NMR (nuclear magnetic resonance) spectroscopy and MS (mass spectrometry) are currently the most frequently used detection methods for metabolome analysis.

NMR measures the spin and magnetic moment properties of the nuclei in a molecule, which depends on the local chemical environment of the nuclei. The nuclei commonly measured for biological samples are the hydrogen isotope ^1H, the carbon isotope ^{13}C and the phosphor isotope ^{31}P. Software programs help to deconvolute the NMR spectra so that structural information about the molecule in the sample may be obtained. Hence, NMR spectroscopy is very useful for structure characterization of unknown compounds. It is also a relatively fast technique and highly reproducible. However, the sensitivity of NMR is only in the 10^{-6} mole range, which is much lower compared to the 10^{-15} mole range that is achievable by MS. Moreover, NMR equipment needs to be installed in a special magnetically shielded environment, often necessitating a special facility to house the machine. Data analysis for complex mixtures is also more difficult compared to MS.

The most important advantage of MS is its high sensitivity. Very often, MS is used in conjunction with chromatographic separation techniques, such as gas chromatography (GC) or liquid chromatography (LC), which enables the sample mixture to be separated before it enters the mass spectrometer. Hence, the spectra of highly complex biological samples may be easier to deconvolute on GC/LC-MS than with NMR. It can be used to confirm the identity of compounds present in complex samples, and help in the identification of unknown or unexpected compounds. However, chromatographic separation may take up to a few hours, resulting in longer analysis times compared to NMR. MS may also be less reproducible compared to NMR for various reasons, including ion suppression effects, in which metabolites at a higher concentration interferes with the spectrum of metabolites with low concentration, thus increasing variability. Nevertheless, NMR and MS-based methods can be used to complement each other. A proposed strategy may be to use GC/LC-MS for the initial exploration of the metabolome, followed by NMR analysis for structural determination of unknown compounds.

3.5. Bioinformatics

The high-throughput omics technologies described in this chapter generate large amounts of data that need to be preprocessed and normalized to transform the raw data into a suitable format before further analysis. We also have to remove biases that may have been introduced due to biological variation, experimental error and instrument artifacts. Although several approaches have been developed to process data from transcriptomic, proteomic and metabolomic experiments individually, there are several general techniques that can be applied to any type of high throughput data. Figure 3.8 shows an overview of the steps involved in the bioinformatics processing of -omics data.

3.5.1. *Data preprocessing and normalization*

Experimental data is often obtained as raw intensity values, which is heavily skewed with many observations at low intensity levels. Since many statistical tests assume a normal or Gaussian distribution it is preferable to transform the data to a log scale, thereby leading to reduced skewness and more evenly distributed data. If a control observation is available it is possible to divide the observed intensity by the control intensity to obtain a fold change ratio. Scaling fold changes to the log scale results in the data being expressed as log ratios. In addition to log scaling, data can

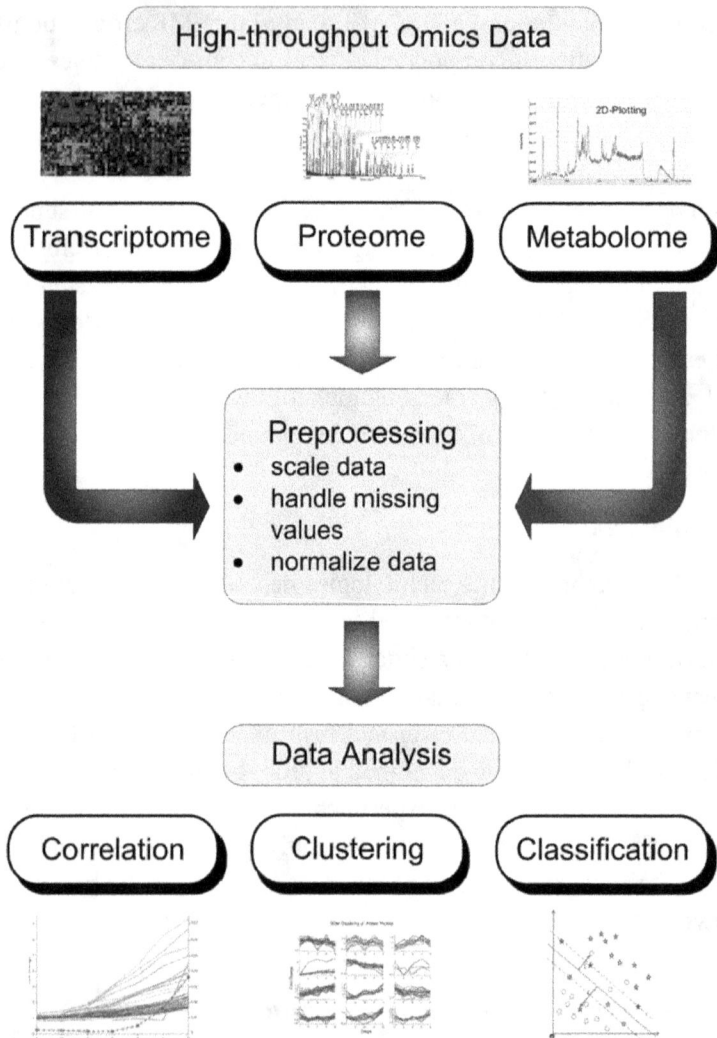

Fig. 3.8.   Overview of the steps involved in bioinformatics processing of -omics data. Raw data from high-throughput experiments needs to be preprocessed before it can be analyzed using various statistical analysis methods.

also be standardized such that its mean is zero and standard deviation is one. Other popular techniques include z-score transformations, or scaling the data in order to force it to fit a Gaussian distribution.

Data generated from -omics experiments often contain missing values. For example, some protein may not be observed in a certain sample, and thus has no value assigned to it. Most statistical methods require a complete dataset, and thus missing values need to be treated appropriately before analysis can be done. Probably, the simplest approach is to discard those rows of data which contain missing values, but a large amount of useful data might be lost in this way. A more useful option is to replace the missing value with the average value observed for that row of data. Alternatively, it is also possible to take the average of the nearest neighbors of the missing observation and substitute in that value instead of the average of the entire row.

There are several more advanced and specialized methods available for preprocessing and normalizing specific types of -omics data. In two color microarrays, a difference between the intensities of different color dyes is known as one of the bias sources and can be reduced by Linear Regression and LOWESS (locally weighted linear regression) that are the most commonly used technique for such dye based microarrays. For oligonucleotide arrays, the MAS5 and RMA algorithms are popular methods for normalization. In proteomic experiments, mass spectrometry (MS) is commonly used. Since the results from the MS are simply a series of mass peaks, additional software must be used to convert the peaks into meaningful peptide sequences. Thus, several tools such as Mascot and Sequest can be employed to perform spectra processing and identify peptides by searching protein sequence databases.

Similar to proteomic experiments, mass spectrometry is also widely used for metabolomic studies. In most cases, LC-MS data needs to be filtered to remove the effects of noise before other processing can be performed. Noise in LC-MS data can either be chemical noise caused by leftover molecules in buffers and solvents, or random noise caused by variations in the detector. Traditional signal processing techniques such as mean filtering, median filtering or filtering based on the Savitzky-Golay method can be used to filter noise from the spectra data. Once the data has been filtered, peak identification algorithms are used to detect all signals caused by true ions while also avoiding detection of false positives. However, different LC-MS experiments can report the same metabolite with slightly different mass and retention time values. Thus, alignment algorithms can be used to identify corresponding peaks from different experiments. For a given peak from one experiment, the algorithm attempts to find the closest matching peak in terms of mass and retention time within a predefined threshold in other experiments.

For more details on the above algorithms, the reader can refer to the "Further Reading" section at the end of this chapter.

3.5.2. *Data analysis*

Once the data from the -omics experiments has been preprocessed, it can be analyzed through statistical procedures to address a variety of biological questions such as the ones described in Table 3.1. Generally speaking, -omics experiments measure the changes in expression of genes, proteins and metabolites across different experimental conditions. Several techniques commonly used for statistical analysis include correlation, principal component analysis, clustering and classification.

Correlation is a measure of similarity or linear dependence between two variables. Pearson's correlation coefficient can be simply calculated to find which variables show similar behavior in an experiment. The value of the correlation coefficient varies between -1 and 1. Positive values of the coefficient indicate a strong similarity with 1 indicating the highest possible similarity. Negative values indicate an inverse relationship between variables while values close to 0 imply that there is no similarity or that the variables are independent of each other.

Other statistical techniques aim to subdivide the observed genes, proteins or metabolites into groups; and can either follow a supervised or an unsupervised approach. In supervised methods the data is already divided into different classes, such as mutant vs. wild-type, and the variable expression profiles then need to be distributed among these known classes. In contrast, the data is not divided into predefined groups in unsupervised methods, and may be grouped into any number of classes.

Principal component analysis is an unsupervised approach which can be used to reduce the dimensionality of a data set. It can also be used to discover major variations in the data by projecting the data onto a new set of coordinates which are the principal components. These principal components are orthogonal to each other and are not correlated, thus the variance described by one component is not captured by any of the others. The components are numbered with the first principal component describing the most variation in the data set, followed by the second component and so on.

One of the goals of such data analysis is to identify groups of genes, proteins, or metabolites which share a common or similar activity pattern or biological behavior. Clustering is a commonly applied unsupervised technique to group observations into a set of mutually exclusive classes. Observations which are grouped into one cluster are highly similar to other observations within their group, but dissimilar to observations classified into other clusters. One popular clustering method is hierarchical clustering. Hierarchical clustering partitions the dataset into a hierarchical series of nested clusters using the following procedure:

1. Calculate the distance between all variables. Find the smallest distance. If several pairs share the same similarity, use a predetermined rule to decide between alternatives.
2. Fuse the two selected clusters to produce a new cluster that now contains at least two objects. Calculate the distance between the new cluster and all other clusters.
3. Repeat steps 1 and 2 until only a single cluster remains.
4. Draw a tree representing the results.

The resulting clusters can be graphically represented by a tree diagram called a dendrogram. Figure 3.9 describes the clustering process described above and the resulting dendrogram. The distance between branches on a dendrogram indicates how similar or dissimilar clusters are, as well as describing the hierarchical structure of the data set. Clustering can either be performed "top-down" or "bottom-up", referring to the order in which clusters are formed. For example, in "bottom-up" clustering of gene expression data, each individual gene starts as its

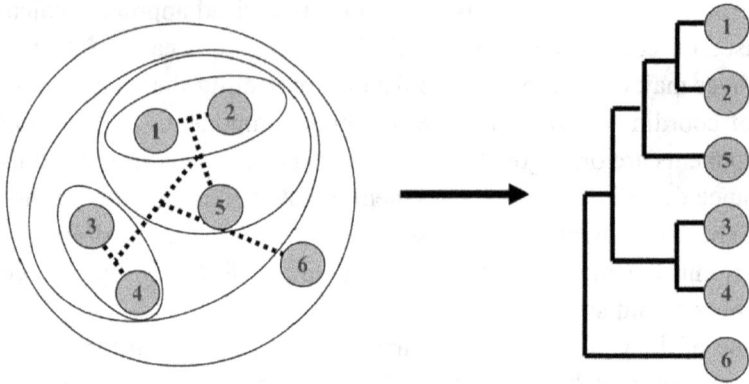

Fig. 3.9.   Graphical representation of hierarchical clustering. Similar variables are joined together into clusters iteratively until the entire dataset fits in one cluster. The resulting dendrogram shows how distant the different variables are from each other.

own cluster, and is then merged with other genes with similar expression profiles. In this way, the final cluster contains all genes and is represented by the root of the dendrogram. The dendrogram can be "cut" at any level to obtain the desired number of clusters. Other popular methods for clustering include K-means clustering and self-organizing maps clustering.

Class prediction, or classification, refers to a family of supervised methods that aim to assign expression profiles into predetermined classes. These methods can either define a function to separate observations, or learn from supplied training data and classify new observations based on their similarity to previously observed data. The principle of supervised classification algorithms is to minimize the intra-class and maximize the inter-class differences. Linear discriminant analysis is one of the popular function based classification methods, while methods such as decision trees, K-nearest neighbor, and support vector machines are learning based methods.

Further Reading

1. Britton RA. DNA microarrays and bacterial gene expression. *Methods Enzymol.* **370**: 264–278, 2003.

2. Choe LH, Aggarwal K, Franck Z and Lee KH. A comparison of the consistency of proteome quantitation using two-dimensional electrophoresis and shotgun isobaric tagging in *Escherichia coli* cells. *Electrophoresis* **26**: 2437–2449, 2005.

3. Liebler DC. *Introduction to Proteomics: Tools for the New Biology.* Humana Press, Totowa, New Jersey, 2002.

4. Lindon JC, Nicholson JK and Holmes E, eds. *Handbook of Metabnomomics and Metabolomics.* Elsevier, Amsterdam, The Netherlands, 2007.

5. Otero JM and Nielsen J. Industrial systems biology. *Biotechnol. Bioeng.* **105**: 439–460, 2010.

6. Picataggio SK, Templeton LJ, Smulski DR and LaRossa RA. Transcript profiling of *Escherichia coli* using high-density DNA microarrays. *Methods Enzymol.* **358**: 177–188, 2002.

7. Ramsay G. DNA chips: State-of-the art. *Nat. Biotechnol.* **16**: 40–44, 1998.

8. Tomita M and Nishioka T, eds. *Metabolomics — The Frontier of Systems Biology.* Springer, Tokyo, 2003.

9. Villas-Bôas SG, Roessner U, Hansen MAE, Smedsgaard J and Nielsen J, eds. *Metabolome Analysis: An Introduction.* John Wiley & Sons, Hoboken, New Jersey, 2007.

10. Warner JR, Patnaik R and Gill RT. Genomics enabled approaches in strain engineering. *Curr. Opin. Microbiol.* **12**: 223–230, 2009.

11. Wilkins MR, Pasquali C, Appel RD, Ou K, Golaz O, Sanchez JC, Yan JX, Gooley AA, Hughes G, Humphery-Smith I, Williams KL and Hochstrasser DF. From proteins to proteomes: Large scale protein identification by two-dimensional electrophoresis and amino acid analysis. *Biotechnol. (N.Y.)* **14**: 61–65, 1996.

12. Ye RW, Wang T, Bedzyk L and Croker KM. Applications of DNA microarrays in microbial systems. *J. Microbiol. Methods* **47**: 257–272, 2001.

13. Zhang A. *Advanced Analysis of Gene Expression Microarray Data.* World Scientific, Singapore, 2006.

14. Zieske LR. A perspective on the use of iTRAQ reagent technology for protein complex and profiling studies. *J. Exp. Bot.* **57**: 1501–1508, 2006.

Questions for Thought...

1. When comparing the expression profiles between strains, why is it necessary to compare samples taken from different growth phases of each strain?

2. Explain why proteins identified to be differentially expressed from proteomic analysis may not be differentially expressed from transcriptomics analysis. Are there also cases where the reverse happens?

Chapter 4

Enzymology

Theresa Tan May Chin[a] and Deng Lih Wen[b]
Department of Biochemistry, Yong Loo Lin School of Medicine
National University of Singapore
MD 7, 8 Medical Drive, Singapore 117597
[a]Emails: bchtant@nus.edu.sg
[b]bchdlw@nus.edu.sg

4.1. Enzymes as Catalysts

Life depends on a series of coordinated chemical reactions. Many of these reactions however proceed too slowly to sustain life and hence proteins with catalytic functions have evolved. An enzyme is a protein that functions as a catalyst. The role of an enzyme (or a catalyst) is to speed up the rate of biochemical reactions without itself undergoing any permanent changes. At the end of the reaction, the enzyme remains unchanged. The rate of a reaction in the presence of an enzyme is typically 10^6 to 10^{14} folds greater than the uncatalyzed reaction. How is this achieved? In any chemical reaction, a reactant must contort into an unfavorable high-energy conformation before it converts into the product. The formation of this unfavorable transition state represents an activation energy barrier that must be overcome before the reaction can occur. Enzymes act to lower the activation energy barrier and therefore facilitate the progress of the reaction (Fig. 4.1). Enzyme-catalyzed reactions can thus proceed at a greater rate.

All enzymes are protein molecules with the exception of a small group of catalytic RNA molecules termed ribozymes. Extreme conditions such as high temperature or extreme pH lead to a denatured protein enzyme and

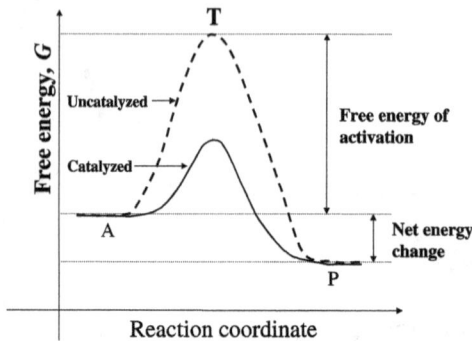

Fig. 4.1.   Transition state diagram of an uncatalyzed and a catalyzed reaction. The net free energy change is the same for both but the free energy of activation is lowered in the presence of a catalyst. Hence, in the presence of a catalyst, it is easier for the reactant, A, to reach the transition state, T and be converted to the product, P.

this in turn destroys its catalytic activity. Most human enzymes have an optimal temperature around 37°C. At low temperatures, the rate of an enzymatic reaction is reduced. At high temperatures, denaturation of the enzyme occurs and loss of activity is observed. For most enzymes, an increase in reaction rate is usually observed as the pH goes from an acidic level toward physiological range and then declines as pH progresses towards the alkaline range. Hence, enzymes, unlike chemical catalysts, catalyze reactions under relatively mild conditions: moderate temperature, atmospheric pressure and nearly neutral pH.

Each enzymatic reaction proceeds with great specificity. How do enzymes achieve specificity? An enzyme binds to the substrate (reactant); forms an enzyme-substrate complex and converts the substrate into the product. This occurs within a cleft or a pocket on the enzyme called the active (or catalytic) site. The specificity of an enzymatic reaction results from the three-dimensional arrangement of specific amino-acids residues in the enzyme that forms the substrate-binding site. Two models have been used to describe enzyme-substrate interaction. In the lock and key model, the enzyme is viewed as the lock and the substrate as the key. The active site of an enzyme molecule is represented as complementary to the shape of the substrate (Fig. 4.2). This model presents the active site as a rigid unchanging structure. In the induced fit model, the enzyme is not presented as a rigid molecule. Rather, the enzyme undergoes conformational change as the

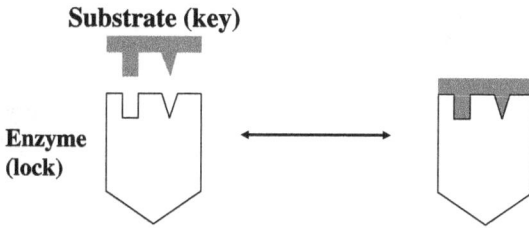

Fig. 4.2. Lock and key model for enzyme-substrate interaction. The enzyme is analogous to a lock while the substrate is viewed as the key that fits specifically into the lock.

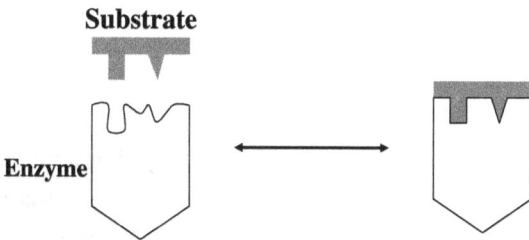

Fig. 4.3. Induced fit model of enzyme-substrate interaction. Substrate binding will alter the conformation of the active site such that the enzyme and substrate fit each other more precisely.

substrate binds and as the reaction proceeds (Fig. 4.3). Such conformational changes have indeed been proven to occur by X-ray crystallographic studies, as in the case of the yeast hexokinase enzyme (Steitz *et al.*, 1981).

4.2. Nomenclature

Before the Enzyme Commission (EC) recommended the present system of nomenclature, enzymes were given trivial names. The trivial names consisted of the suffix -ase added to the substrate of the enzyme (e.g. urease), or implied something about the nature of the reaction (e.g. alcohol dehydrogenase), or simply did not give any clue to what the enzyme does (e.g. trypsin).

The EC was set up by the International Union of Biochemistry and Molecular Biology (IUBMB). Based on the EC recommendation, enzymes are classified and named according to the nature of the chemical reactions they catalyzed (Enzyme Nomenclature, 1992). There are six major classes of enzyme reactions (Table 4.1). The major classes are further subdivided

Table 4.1. Classification of enzyme reactions.

Classification	Type of reaction	Class of enzyme
1	Oxidation-reduction reactions	Oxidoreductases
2	Transfer of functional groups	Transferases
3	Hydrolysis reactions	Hydrolases
4	Addition of groups to double bonds or formation of double bonds	Lyases
5	Isomerization reactions	Isomerases
6	Bond formation coupled with hydrolysis of high energy compounds such as ATP	Ligases

into subclasses and sub-subclasses. Each enzyme has a unique EC number and a systematic name with the suffix -ase. The systematic name of an enzyme consists of the name of its substrate(s) followed by a word ending with "ase" specifying the type of reaction the enzyme catalyzes. The enzyme, glucokinase, catalyzes the transfer of a phosphate from ATP to D-glucose. Its EC number is 2.7.1.2 and its systematic name is ATP:D-glucose 6-phosphotransferase. The number represents the class, subclass, sub-subclass and its arbitrarily assigned serial number in the sub-subclass.

The ENZYME database (http://tw.expasy.org/enzyme/) is a repository of information on the nomenclature of enzymes. It contains the following data for each enzyme: EC number, recommended name, alternative names, catalytic activity, cofactors, links to the SWISS-PROT protein sequence entries that correspond to the enzyme and links to human diseases associated with a deficiency of the enzyme.

4.3. Cofactors

Enzymes have evolved to cope with the necessity of catalyzing biochemical reactions within a living cell. To improve the repertoire of reactions which enzymes are able to catalyze, cofactors are often necessary. Without cofactors, the functional groups present on the side chains of amino acids on enzymes can only facilitate acid-base reactions, form certain types of transient covalent bonds and take part in charge-charge interactions. The presence of cofactors provides additional functional groups and allows enzymes to catalyze other types of reactions (e.g. oxidation-reduction reactions). Thus, enzymes can be divided into two groups: those that rely solely on their protein structure to carry out their catalytic functions and those that require an additional cofactor to carry out the catalytic activity. An enzyme without its cofactor is called an apoenzyme. A catalytically active enzyme-cofactor complex is referred to as a holoenzyme. Some cofactors are transiently associated while others known as prosthetic groups are permanently associated with the enzymes, often by covalent bonds.

Cofactors may be metal ions (such as Zn^{++}, Fe^{++}) or organic molecules. Such organic molecules are referred to as coenzymes and are derived from vitamins. Flavin coenzymes (flavin adenine dinucleotide, FAD and flavin mononucleotide, FMN) are examples of coenzymes and the flavin component is derived from the vitamin riboflavin (also known as vitamin B_2).

4.4. Regulation of Enzyme Activity

Regulation of enzyme activity can be achieved in many different ways. The different modes of regulation include: control of enzyme concentration, synthesis of inactive precursors, covalent modification of enzymes, regulatory proteins and allosteric controls.

1 Control of Enzyme Concentration

The concentration of a given enzyme in a living cell depend on both its synthesis and degradation rates. Bacteria can rapidly adapt to their environment by producing enzymes that can metabolize available nutrients. Bacteria grown in the absence of lactose express little of these enzymes. In the presence of lactose, the synthesis of β-galactosidase and galactose permease is rapidly induced and the bacteria can now utilize lactose as a nutrient source. The enzymes are expressed only when required and the regulation is achieved by controlling gene expression.

2 Inactive Precursors

Regulation of enzyme activity can also be achieved by synthesizing enzymes as inactive precursors. The inactive precursors are known as zymogens or proenzymes. Digestive enzymes are usually synthesized as zymogens and proteolytic cleavage is needed to activate the enzyme. Chymotrypsin is secreted from the pancreas as chymotrypsinogen and it is activated in the digestive tract by the enzyme trypsin. Trypsin cleaves a small peptide from the N-terminal region and converts the inactive chymotrypsinogen to chymotrypsin.

3   Covalent Modification

Protein phosphorylation and dephosphorylation (the attachment and removal of a phosphoryl group) is a major mechanism employed by hormones for intracellular signaling and for regulation of enzymatic activity. The transfer of a phosphate group from ATP to a specific serine, threonine or tyrosine residue on the target protein is mediated by protein kinases and the removal of the phosphate group is mediated by protein phosphatases. Phosphorylation leads to a conformational change and can either activate or inactive the enzyme. Glycogen phosphorylase, which catalyzes the breakdown of glycogen, is regulated by phosphorylation in response to the presence of hormones such as insulin and glucagon.

4   Regulatory Proteins

Regulatory proteins can modulate the activity of an enzyme. These can either activate or inhibit enzyme activity. When the regulatory protein binds to its target enzyme, a conformation change occurs in the enzyme and this affects the function of the enzyme. Calmodulin, a 17 kD protein serves as a calcium sensor in eukaryotic cells. Ca^{++} can bind to multiple sites on calmodulin. The Ca^{++}-calmodulin complex in turn binds to many target enzymes and modulates their activity.

5   Allosteric Control

The enzyme that catalyzes the first step of a biosynthetic pathway is usually regulated by the ultimate product (Fig. 4.4). This form of regulation is termed feedback regulation. Feedback regulation allows for quick adjustment to the cell's requirement for a particular product and prevents excessive synthesis. Feedback regulation makes use of the properties of allosteric enzymes.

Allosteric effectors (allosteric activators and inhibitors) affect the activity of allosteric enzymes. These effectors are molecules other than the substrates and bind at sites separate from the active site of the enzyme. The site where the allosteric effector binds is called the allosteric site

$$A + B \xrightarrow{\text{Enzyme 1}} C$$

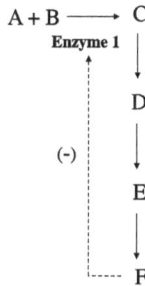

Fig. 4.4.   Regulation of a pathway by feedback inhibition. C, D and E are intermediates in this hypothetical pathway. The final product, F exerts an inhibitory effect on Enzyme 1 which catalyzes the first reaction in the pathway.

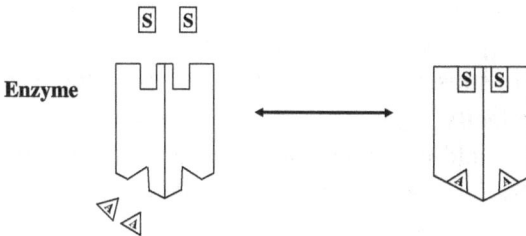

Fig. 4.5.   Binding of substrate (S) molecules and effector (A) molecules to an allosteric enzyme. The allosteric site is distinct from the substrate-binding site.

(Fig. 4.5). Binding of an allosteric effector causes a conformational change of the enzyme so that the affinity for the substrate also changes. Allosteric enzymes contain multiple subunits. They also exhibit positive cooperativity. Binding of the substrate to one subunit facilitates the binding of substrate to one or more of the remaining subunits. An allosteric activator serves either to activate the enzyme or to stabilize the active state of the enzyme thus facilitating substrate binding to its own and other subunits. Conversely, an allosteric inhibitor inactivates the enzyme and prevents substrate binding.

4.5. The Michaelis-Menten Equation

In 1913, Leonor Michaelis and Maud Menten proposed a general theory of enzyme action consistent with observed enzyme kinetics. The enzyme, E associates with its substrate S to form an ES complex with the rate constant of k_1. This association is reversible. The ES complex has two fates.

It can dissociate to E and S with the rate constant of k_{-1} or the reaction can proceed and the product P is formed (with the rate constant of k_2).

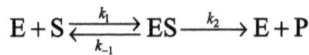

$$E + S \underset{k_{-1}}{\overset{k_1}{\rightleftharpoons}} ES \xrightarrow{k_2} E + P$$

The Michaelis-Menten equation,

$$v_0 = \frac{V_{max} \cdot [S]}{K_m + [S]}$$

relates the initial velocity (v_0) of an enzyme reaction to the concentration of the substrate, [S]. V_{max} is the maximal velocity. K_m is defined as the concentration of substrate at which the velocity of the reaction is at half the maximal velocity. K_m is also equal to $(k_{-1} + k_2) / k_1$. A plot of v_0 versus [S] would thus yield a rectangular hyperbola (Naqui, 1986) (Fig. 4.6). At infinite substrate concentration, the reaction proceeds under saturation conditions and the initial velocity of reaction approaches V_{max}. At low substrate concentration ([S] << K_m), the equation simplifies to $v_0 = V_{max} [S] / K_m$ (Fig. 4.6).

The Michaelis-Menten equation can be algebraically transformed into a linear equation that is more useful in plotting experimental data. One such

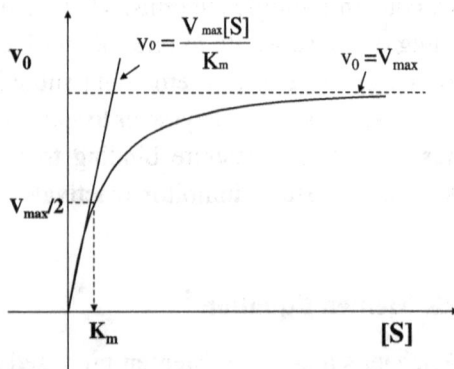

Fig. 4.6.   The Michaelis-Menten plot of initial velocity (v_0) versus substrate concentration, [S]. V_{max} is the maximal velocity while K_m is the concentration of S at $V_{max}/2$.

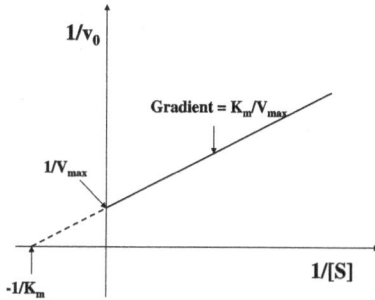

Fig. 4.7.   The Lineweaver-Burk plot of $1/v_0$ versus $1/[S]$. v_0 is the initial velocity while [S] is the substrate concentration.

transformation is the Lineweaver-Burk equation. By taking the reciprocal on both sides, the Michaelis-Menten equation is transformed into

$$\frac{1}{v_0} = \left(\frac{K_m}{V_{max}}\right)\frac{1}{[S]} + \frac{1}{V_{max}}$$

A plot of $1/v_0$ versus $1/[S]$ would yield a straight line and the constants V_{max} and K_m can be easily determined. This line has a slope of K_m/V_{max}, an intercept of $-1/K_m$ on the $1/[S]$ axis and an intercept of $1/V_{max}$ on the $1/v_0$ axis (Fig. 4.7).

The value of K_m is dependent on the assay conditions (pH, temperature, presence of inhibitors or activators) but independent of the amount and purity of the enzyme. K_m is the substrate concentration at which v_0 is half the maximal velocity and is also equivalent to $(k_{-1} + k_2)/k_1$. When $k_2 \ll k_{-1}$, K_m reduces to k_{-1}/k_1, which is defined as the dissociate constant of the ES complex. The K_m value thus provides a means to compare the affinities of enzymes for their respective substrates. A low K_m value would indicate that the enzyme has high affinity for the substrate.

The catalytic constant k_{cat} (also known as turnover number) of an enzyme is a measure of its maximal catalytic activity. It is defined as $k_{cat} = V_{max}/[E]_T$ where $[E]_T$ is the total enzyme concentration. k_{cat} equals to k_2. It is the number of substrate molecules that is converted into the product by an enzyme molecule in a unit time when the enzyme is fully saturated with its substrate. However under physiological conditions, the substrate concentration is seldom at saturation and k_{cat} values cannot be interpreted meaningfully. The constant k_{cat}/K_m is used instead to provide a measure of the enzyme's catalytic efficiency under non-saturating substrate concentration ([S] < K_m, Fig. 4.6), where

$$v_0 = \frac{V_{max}[S]}{K_m} = \frac{k_{cat}}{K_m}[E]_T[S]$$

The above kinetic analysis is for a single substrate reaction. Can it be applied to a multi-substrate reaction? It is beyond the scope here to describe such reactions but it suffices to say that K_m and V_{max} values can still be determined. The usual practice is to determine the velocity of reaction by varying just one substrate while keeping the concentration of all the other cosubstrates constant. In such situations, Michaelis-Menten kinetics will be observed.

Are there situations where a deviation from Michaelis-Menten kinetics is observed? A deviation from Michaelis-Menten behavior is indicative of allosteric regulation (see Sec. 4.4). Hexokinase catalyzes the transfer of a phosphate group from ATP to glucose. Hexokinase exists as a number of different tissue-specific isoenzymes. Isoenzymes are proteins with different amino acid sequences, which catalyze the same reaction. They also exhibit different kinetics and regulatory properties. The hexokinases found in most human tissues follow Michaelis-Menten kinetics and a plot of v_0 versus [S] yield a rectangular hyperbola (Fig. 4.6). However, glucokinase (which is the hexokinase isoenzyme found in the liver) behaves differently and a sigmoidal curve is observed (Fig. 4.8). Such a sigmoidal curve is characteristic of an enzymes exhibiting substrate cooperativity and the value of [S] at which v_0 is half the maximal velocity is designated as $K_{0.5}$ (Fig. 4.8). In the presence of an allosteric activator, the curve becomes more nearly hyperbolic, with a decrease in $K_{0.5}$. Conversely, $K_{0.5}$ increases in the presence of an allosteric inhibitor.

4.6. Inhibition of Enzyme Activity

The activity of an enzyme can be regulated. The various modes of regulation have been described in Sec. 4.4. Besides these regulatory mechanisms,

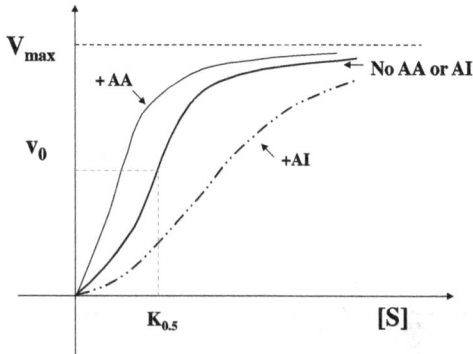

Fig. 4.8. Kinetic profile of an allosteric enzyme in the presence of an allosteric activator (AA) or an allosteric inhibitor (AI). In the absence of an activator or inhibitor, a sigmoidal curve is obtained.

exogenous compounds can also affect the enzymatic activity. A compound that binds to the enzyme and decreases the velocity of the enzymatic reaction is an inhibitor. Irreversible inhibitors are compounds that form covalent bonds or other strong interactions with the enzyme. Examples of irreversible inhibitors with therapeutic uses include the antibiotic penicillin. Penicillin reacts covalently with bacterial glycopeptide transpeptidase to form the penicilloyl-enzyme complex, rendering the enzyme inactive. Glycopeptide transpeptidase is essential for cell wall biosynthesis. Thus, treatment with penicillin disrupts bacterial cell wall synthesis.

Reversible inhibitors on the other hand, do not form covalent bonds with the enzyme and the inhibitor-enzyme complex can dissociate. There are three forms of reversible inhibition: competitive, uncompetitive and mix (or noncompetitive) inhibition. All three types of reversible inhibition can easily be distinguished using the Lineweaver-Burk plot of $1/v_0$ verses $1/[S]$ (Fig. 4.9).

1   Competitive Inhibition

A competitive inhibitor is a structural analog of the substrate. It resembles the structure of the substrate. It will thus compete with the substrate for binding to the active site of the enzyme (Fig. 4.10). Competitive inhibition can be overcome by increasing the substrate concentration. At high substrate concentration, all the active sites will

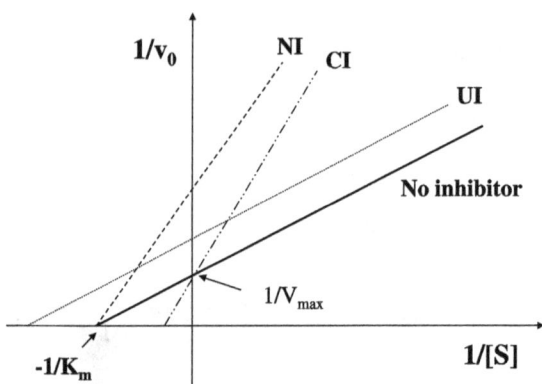

Fig. 4.9.   Lineweaver-Burk plots in the presence of a competitive inhibitor (CI), a non-competitive inhibitor (NI) and an uncompetitive inhibitor (UI).

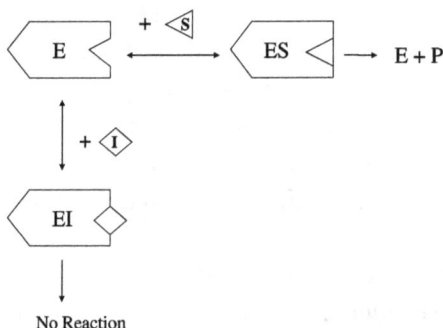

No Reaction

Fig. 4.10.   Competitive inhibition. The inhibitor (I) competes with the substrate (S) for the active site of the enzyme (E).

be occupied by the substrate and the inhibitor will not be able to bind. Hence, competitive inhibition will lead to an increase in the K_m value but the maximal velocity (V_{max}) remains unchanged (Fig 4.9). The mode of action of many therapeutic drugs is based on competitive inhibition. The therapeutic drug methotrexate is a structural analog of dihydrofolate. It competes with dihydrofolate for binding to the active site of the enzyme dihydrofolate reductase.

2   Uncompetitive Inhibition

An uncompetitive inhibitor behaves differently from a competitive inhibitor. It binds at a site distinct from the substrate binding site and

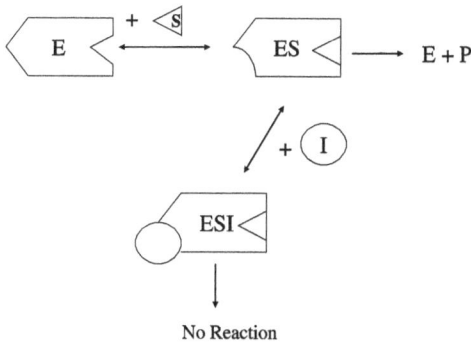

Fig. 4.11.   Uncompetitive inhibition. The inhibitor (I) can only bind to the enzyme-substrate (ES) complex.

it only binds to the enzyme-substrate complex. Binding of the substrate to the enzyme leads to a conformational change in the enzyme and allows the inhibitor to interact with the enzyme-substrate complex (Fig. 4.11). Formation of the substrate-enzyme-inhibitor complex prevents the reaction from proceeding. Uncompetitive inhibition decreases both the K_m and V_{max} values (Fig 4.9).

3   Mixed Inhibition

Mixed inhibition is also known as noncompetitive inhibition. Noncompetitive inhibitors do not compete with the substrate for the active site. Noncompetitive inhibitors bind to the enzyme at a site that is distinct from the active site. Such inhibitors can bind either to the free enzyme or the enzyme-substrate complex (Fig. 4.12). Mixed inhibition may or may not alter the K_m value of a reaction but a decrease in the V_{max} will be observed (Fig. 4.9).

4.7. Industrial Applications of Enzymes

Enzymes are useful catalysts. The advantages of an enzyme compared to other types of catalysts include specificity (lack of side reactions), high efficiency, catalysis under mild conditions (low temperature and atmospheric pressure) and environmentally friendly with low pollution. The catalytic ability of enzymes has been harnessed and used for centuries. The earliest uses include the use of enzymes present in microorganisms for brewing, bread making and cheese making. These had been carried out

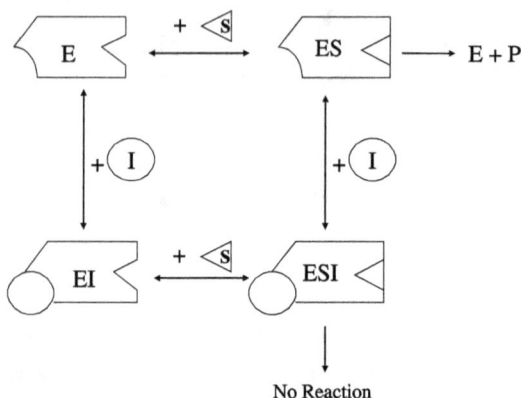

Fig. 4.12. Noncompetitive (mixed) inhibition. The inhibitor can bind to either the enzyme (E) or the enzyme-substrate (ES) complex.

before there was any understanding of the biochemical reactions and the enzymes involved. Today, enzymes are used in the chemical, pharmaceutical, food and biofuel industry. The USA industrial enzyme market was valued at $394.1 M in 2006 and is expected to more than double by the year 2013 to $748.9M (US Industrial Enzyme Markets, 2009). Worldwide, it is expected to exceed US$2.9 billion by 2012.

It is beyond the scope of this section to provide a detailed account of the many and varied uses of enzymes. A selection of examples is used to allow for the appreciation of the industrial and clinical applications of enzymes.

1   Brewing
    An examination of the brewing process will illustrate how enzymes are utilized for the production of beer. Note that for brewing, intact yeast and barley seeds are used and isolation of the required enzymes is not necessary.
    The starting material is starch, which is present in the barley seed. Starch is a polysaccharide and yeast is unable to ferment starch. This can be overcome by allowing the barley seeds to germinate. During germination, endo-1,3-β-D-glucanase, amylases and peptidases are released. Starch granules are found within the cells of the endosperm.

The walls of the endosperm are surrounded by polymers including 1,3-β-D-glucans. The enzyme endo-1,3-β-D-glucanase will degrade the glucans so that the amylases can come into contact with the starch granules. Amylases catalyze the hydrolysis of starch to simple sugars including glucose, maltose, and other oligosaccharides. In addition, peptidases are also released and these enzymes will catalyze the degradation of endogenous proteins within the seeds. The resulting peptides and amino acids are utilized by the yeast for growth. The primary role of yeast in brewing is to provide the necessary enzymes for anaerobic glycolysis. Ethanol is the by-product of anaerobic glycolysis. Hence, for brewing, both the germinating barley seed and the yeast provide the full complement of enzymes for converting the starting raw material starch to the final product ethanol (Pollock, 1987).

2 Biofuel

Today in the current climate of white biotechnology, ethanol and other liquid biofuels are produced in increasing quantities and at lower cost. This third wave of biotechnology is driven by the rising cost and the depletion of oil and fossil fuels as well as global environmental concern with gas and carbon emissions. Bioethanol currently accounts for 94% of all biofuel production. In 2004, the production of ethanol was 10.75 billion gallons and this doubled to 20.37 billion gallons in 2008.

Similar to the process of brewing where ethanol is produced as beverage, the production of bioethanol also involves the use of yeast to provide the full complement of enzymes for the fermentation process. This is followed by the separation and purification the ethanol produced. Simple sugars (such as that obtained from sugarcane, sugar beet or sorghum) or polysaccharides such as starch (from maize, wheat or root crops) as well as lignocellulosic biomass can all serve as sources of sugars for fermentation. When starch is used, enzymatic hydrolysis of starch to glucose by α-amylase and amyloglucosidase is necessary prior to fermentation. Much interest and research is now focused on the use of lignocellulosic biomass. This allows for the utilization of agricultural waste such as corn husks or wood chips. However, such waste contains cellulose and hemicellulose which are resistant to hydrolysis. Thus when lignocellulosic biomass is used, pre-treatment is necessary to prepare the biomass for hydrolysis to glucose. Although acid hydrolysis is the more developed technology, the focus currently is on developing enzymatic hydrolysis technology as this will allow for reduction in cost and better yield of sugars with little degradation (Demirbas, 2011; Mussatto *et al.*, 2010).

3 Antibiotic and Therapeutic products

Microorganisms are also utilized for the production of many useful therapeutic products. These include the production of antibiotics such as erythromycin, streptomycin and penicillin by *Streptomyces spp* and *Penicillium spp* and the production of lovastatin by *Aspergillus terreus*. Antibiotics are secondary metabolites of bacteria and fungi and hence the level of antibiotic production is generally low. Much work has been carried out to enhance the microbial pathways in order to optimize antibiotic production. The classical methods for microbial strain improvements using mutational techniques is now augmented by current technologies such as recombinant DNA technology and metabolic engineering to increase metabolic flux through the desired biosynthetic pathway in the microorganism (Vaishnav and Demain, 2009).

4 Clinical Applications

In clinical applications, enzymes can be used in one of the following manners: as diagnostic tools, as analytical reagents or as therapeutic agents.

Currently, more than 20 different enzymes are assayed routinely in clinical chemistry laboratories. Detection of abnormal levels of enzymes in body fluids can aid diagnosis of diseases. Elevated serum levels of aspartate amino transferase (AST), alanine amino transferease (ALT), alkaline phosphatase (ALP) and γ-glutamyltransferase (GGT) are observed in liver disease. So how are these assayed?

The key feature when assaying for enzyme activity is the need for an end product that can be easily detected and measured. The product usually absorbs light in the UV or visible range. ALP is assayed by measuring the hydrolysis of 4-nitrophenyl phosphate. The products of this reaction are phosphate and 4-nitrophenol. The latter absorbs at 405 nm. However not all reactions produce at least a product which can be easily measured. An example is the assay for ALT activity.

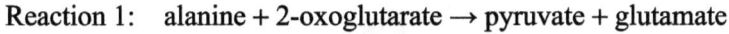

Reaction 1:   alanine + 2-oxoglutarate → pyruvate + glutamate
Reaction 2:   pyruvate + NADH + H^+ → lactate + NAD^+

The products pyruvate and glutamate do not absorb light. Hence it is necessary to couple the ALT reaction (Reaction 1) to a second reaction (Reaction 2). The second reaction is catalyzed by lactate dehydrogenase (LDH). Pyruvate, the product from the ALT reaction, is used as a substrate for the subsequent LDH reaction. In the presence of the coenzyme NADH, LDH converts pyruvate to lactate. NADH absorbs at 340 nm while NAD does not. The change in A_{340} due to oxidation of NADH can be easily measured.

Enzymes are also used as analytical reagents in clinical chemistry applications. In the above example, the enzyme LDH is used as an analytical reagent for the assay for ALT. Another example is the use of the enzyme glucose oxidase to assay blood glucose levels.

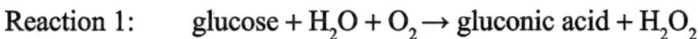

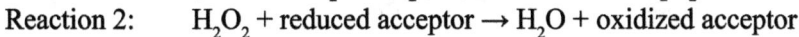

Reaction 1:   glucose + H_2O + O_2 → gluconic acid + H_2O_2
Reaction 2: H_2O_2 + reduced acceptor → H_2O + oxidized acceptor

Glucose oxidase oxidizes glucose to gluconic acid and hydrogen peroxide (Reaction1). The hydrogen peroxide produced then oxidizes an acceptor (Reaction 2). The acceptor commonly used is o-dianisidine. Oxidized o-dianisidine absorbs between 425 and 475 nm.

Enzyme-linked immunosorbant assay (ELISA) also utilizes enzymatic reactions for detection of antigen-antibody interaction (Fig. 4.13). The binding of an antibody to an antigen cannot be easily detected. To facilitate detection, a secondary antibody with an enzyme attached to it is used. Enzymes commonly used for ELISA assays include alkaline phosphatase and horse raddish peroxidase. Both enzymes can react with specific substrates to produce products that are easily detected.

Enzymes are also used for a variety of therapeutic purposes (Table 4.2). Streptokinase, urokinase and human tissue plasminogen activator (tPA) are effective activators of plasminogen. These are used as thrombolytic

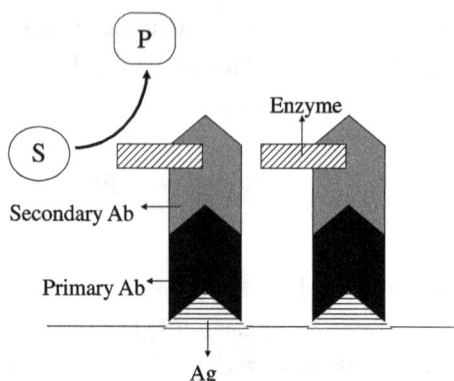

Fig. 4.13. Enzyme-linked immunosorbant (ELISA) assay. The primary antibody (Ab) recognizes the immobilized antigen (Ag) and the secondary Ab specifically interacts with the primary Ab. These interactions can be detected using an appropriate substrate (S) for the enzyme attached to the secondary Ab. The product (P) formed can be easily measured.

Table 4.2. Therapeutic uses of enzymes.

Enzyme	Application
Collagenase	Skin ulcers
Trypsin	Anti-inflammatory agent
Uricase	Gout
Hyaluronidase	Drug administration
Tissue plasminogen activator	Thrombolytic agent
Streptokinase	Thrombolytic agent
Urokinase	Thrombolytic agent
Factor IXa	Hemophilia B

agents and are useful in removing blood clots. Intravenous administration of therapeutic enzyme molecules may elicit allergic, anaphylactic and immunological responses. In addition, the enzyme is usually rapidly removed from circulation and degraded. Much work had been carried out to overcome the antigenicity and short plasma half-lives of therapeutic proteins. An example is tPA. Human tPA has been cloned and the recombinant protein is used for myocardical infarction. tPA has poor solubility, is rapidly cleared by the liver and hence a large dose is necessary. However, the side effects of a large dose include bleeding and non-clotting. Recombinant DNA and protein engineering technology have been used to create an improved version of this protein. The engineered tPA is a smaller protein with increased solubility, better pharmacokinetics and improved specificity and affinity at the clot with little effect on peripheral coagulation (Buckel, 1996).

5   Immobilized Enzymes

The attachment of an enzyme to a support will enable us to reuse the enzyme without a tedious recovery process. Enzymes may be immobilized on an inert support in a number of ways. These include adsorption, covalent linkages, entrapment within a matrix and encapsulation. Immobilized β-galactosidase has been used to hydrolyze lactose to glucose and galactose. Immobilization is achieved by the adsorption of the enzyme onto silica particles. This immobilized enzyme has been used to produce a mixture of glucose and galactose, which is used as a sweetening agent. Immobilized β-galactosidase is also used to remove lactose from milk. Lactose intolerance is a common problem in many parts of the world and this has led to a demand for lactose-free milk.

Analytical reagents may also be immobilized. Glucose dehydrogenase can be immobilized onto nylon tubing and this can be used and reused for measuring blood glucose levels. As little as 0.18 mg of enzyme can be used for more than 1000 glucose determinations. The ability to immobilize enzymes and other analytical reagents has also led to a wide range of home testing kits for urine and blood samples.

The above examples illustrate how enzymes can be utilized for industrial and clinical applications. In some instances, the use of a suitable microorganism with the appropriate complement of enzymes is sufficient (e.g. brewing) but in other instances, purified enzymes are necessary (e.g. clinical applications). In addition, the progress in genetic engineering and recombinant DNA technology, the knowledge of the genomic sequences of microbes and the continual development of new manipulation tools have led to the continual generation of genetically engineered microorganisms to enhance yield of desired products (e.g. antibiotics) and recombinant enzymes which have improved features such as enhanced catalytic activity, increased solubility or improved thermal stability.

References

1. Buckel P. Recombinant proteins for therapy. *TiPS* **17**: 450–456, 1996.
2. Demirbas A. Competitive liquid biofuels from biomass. *Appl. Energy* **88**: 17–28, 2011.
3. Enzyme Nomenclature. *Recommendations of the Nomenclature Committee of the International Union of Biochemistry and Molecular Biology.* Academic Press, London, 1992.
4. Mussatto S, Dragone G, Guimaraes P, Silva J, Carnerio L, Roberto I, Vicente A, Domingues L and Teixeria J. Technological trends, global market, challenges of bio-ethanol production. *Biotechnol. Adv.* **28**: 817–830, 2010.
5. Naqui A. Where are the asymptotes of Michaelis-Menten? *TiBS* **1**: 64–65, 1986.
6. Pollock J. *Brewing Science.* Academic Press, New York, 1987.
7. Steitz TA, Shoham M and Bennett WS Jr. Structural dynamics of yeast hexokinase during catalysis. *Philos. Trans. R. Soc. Lond. B. Biol. Sci.* **293**: 43–52, 1981.
8. The US Industrial Enzyme Markets. *Focus on Catalysis*, 2009, issue 1, p. 2.
9. Vaishnav P and Demain A. *Encyclopedia of Microbiology — Industrial Biotechnology.* Elsevier Inc., 2009, pp. 335–348.

Further Reading

1. Copeland RA. *Enzymes — A Practical Introduction to Structure, Mechanism and Data Analysis*, 2nd ed. Wiley-VCH, Inc, USA, 2002.
2. Price NC and Stevens L. *Fundamentals of Enzymology — The Cell and Molecular Biology of Catalytic Proteins*, 3rd ed. Oxford University Press, 2000.

Website

http://tw.expasy.org/enzyme/

Questions for Thought...

1. The table below shows the initial velocities (v_0) of an enzyme-catalyzed reaction with the corresponding substrate ([S]) concentrations

v_0 (μmole/min)	[S] (x 10^{-5} mole /l)
200	100
180	35
130	12
100	6
50	2

a) Determine the K_m and V_{max} values for the enzyme.

b) In the presence of an inhibitor I, the K_m value increases but the V_{max} remains unchanged.

    i) What is the nature of this inhibition?

    ii) How can this inhibition be overcome?

2. a) Two different enzymes, A and B, catalyze the same reaction. What is the term used to describe such enzymes?

    b) Compound X is a competitive inhibitor of Enzyme A. Another compound, Y interacts with the N-terminal domain of Enzyme A. The N-terminal domain of Enzyme B is different from that of Enzyme A. Kinetic analysis shows that Y is a non-competitive inhibitor for Enzyme A. How will X and Y affect the K_m and V_{max} of Enzyme B?

3. A student S was given two different preparations of enzyme E and told to determine the K_m and V_{max} values. Enzyme E was isolated from rat livers using the same protocol on two different occasions. He noticed that the K_m values were identical but the V_{max} values differ. How can he account for his observations?

4. a) Explain the principle of coupled enzyme assays.
    b) You are interested in determining the concentration of ATP in your sample. With the following reagents

    i)   glucose
    ii) glucokinase
    iii) glucose-6-phosphate dehydrogenase
    iv) NADP⁺

    outline how you would set up an assay to determine the concentration of ATP in your sample.
    c) In place of glucose, you are provided with lactose, galactosidase, ATP and reagents ii, iii and iv. Using these reagents, illustrate how you can determine the activity of galactosidase in your sample.

Chapter 5

Microbial Genome Mining for Identifying Antimicrobial Targets

Kishore R. Sakharkar and Jayaraman Premkumar
Biomedical Engineering Research Centre
Nanyang Technological University, Singapore 637553

Meena K. Sakharkar
Graduate School of Life and Environmental Sciences
University of Tsukuba, Ibaraki, Japan 305-8572

Vincent T. K. Chow
Infectious Diseases Program
Department of Microbiology, Yong Loo Lin School of Medicine
National University Health System
National University of Singapore, Kent Ridge, Singapore

5.1. Introduction

Antibiotics are crucial in the fight against infectious diseases caused by bacteria and other microbes. Antimicrobial chemotherapy has been a major reason for the dramatic rise in average human life expectancy since their discovery by Louis Pasteur, Robert Koch, Alexander Fleming, and other scientists. Some antibiotics have also been discovered by semisynthetic modifications of various natural compounds. However, disease-causing microbes that are resistant to antibiotic therapy pose increasing public health challenges globally. Nowadays, about 70% of the bacteria that cause infections in certain hospitals are resistant to at least one of the drugs most commonly used for treatment. Concurrently, the pace of development of novel

117

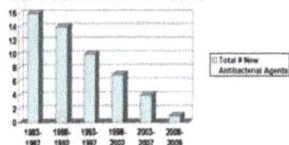

THE PIPELINE OF NEW ANTIBIOTICS IS DRYING UP

IDSA

In spite of the pressing need for new drugs to treat resistant infections, there simply are not enough new antibiotics in the pharmaceutical pipeline to keep pace. Major pharmaceutical companies with the R&D "muscle" to make progress are losing interest in the antibiotics market, even as they increase their overall R&D budgets. Of greatest concern is the dearth of resources being invested i...

A growing number of drug companies appear to be withdrawing from new antibiotic research and development.

1. There is **increasing pressure by health care community to curtail the 'unnecessary use'** of antibacterial agents.
2. Most antibacterial use is for **short courses of therapy**. In general, targeting chronic diseases where patients can look forward to months, even years, of treatment (e.g. for rheumatoid arthritis or depression or 'life-style drugs') is a more cost effective area for drug discovery at a variety of levels.
3. **Anti-infective drugs** are considered 'life-saving' medications and, as such, are often the subject of **aggressive price controls**, especially outside of the United States.
4. The expense is considerable— from $100M to $800M and private producers of antimicrobials are exposed to **liability claims for adverse events**.
5. Finally, **new antibiotics run the risk of rapid obsolescence**, because use leads to greater resistance and the need for new drugs.

Risks out-weigh incentives

INCENTIVES

RISKS

Fig. 5.1.   The pipeline for new antimicrobial agents is drying up as big pharmaceutical companies are moving away from antibacterial drug discovery.

antibiotics is dawdling. The absolute number of new licensed antibiotics has declined, and there is a shortage of new agents against Gram-negative bacteria. The prime reasons for drying up of the antibiotic pipeline, and for big pharmaceutical companies moving away from antibiotic research and development include: (a) antibiotics are under aggressive price control for being "life-saving" drugs, and are also used for short-course therapy; (b) a greater chance of liability claims for adverse reactions; and (c) less lucrative than drugs for life-style and chronic conditions, e.g. drugs for hypertension, diabetes or Alzheimer's disease (Fig. 5.1). Hence, there is an urgent need for novel therapies against microbial pathogens.

Thus, the discovery of new antimicrobial targets and consequently new antimicrobial agents is of immediate concern. Towards this end, the availability of several complete microbial genome sequences is facilitating computational approaches to understand bacterial genomes and DNA structure/function relationships. One of the most interesting applications of the genomic data of pathogenic bacteria and viruses is in the design of new vaccines and antimicrobial molecules. Additionally, techniques to "mine" the data, i.e. to analyze the genomic sequences to identify genes, their functions, and their possible relationships to health and disease processes are gaining momentum. Once genes relevant to specific key disease processes are shortlisted, the focus moves on to the discovery and development of small molecules, antibodies, proteins, or a combination of these, in search of drugs that will target the virulence genes or dysregulated pathways to overcome the therapeutic challenge of infectious diseases. In addition, the genomics approaches facilitate a better understanding of the molecular mechanisms of infectious diseases, with the possibility of exploiting genomic information for the discovery of vaccine candidates leading to a new paradigm in vaccine development. In this chapter, we highlight some gene-mining strategies for the discovery of novel antimicrobial targets and agents that may potentially be explored for antimicrobial drug discovery and vaccine design using *Pseudomonas aeruginosa* as a case study.

5.2. What are Targets?

A target is a defined molecule or structure within the organism, which is linked to a particular disease. For disease intervention, the drug target may either be blocked, inhibited or activated by a drug (small organic molecules, antibodies, therapeutic proteins). Bakheet and Doig analyzed 48 non-redundant antibiotic target proteins from all bacteria, 22 antibiotic target proteins and 4243 non-drug targets from *Escherichia coli* to identify differences in their properties, and used the machine learning approach to predict new potential drug targets.

5.3. Approaches for Genome Mining for Identifying Targets

Global efforts on sequencing the genomes of microbes have focused primarily on pathogens that encompass the majority of all genome projects, and have generated an enormous amount of raw material for *in silico* analysis. These data pose a major challenge in the post-genomic era, i.e. to fully exploit this treasure trove for the identification of novel putative targets for therapeutic intervention. Genomics can be applied to evaluate the suitability of potential targets using two main criteria, i.e. "essentiality" and "selectivity".

5.3.1. *Essential genes*

Identification of essential genes has been one of the primary contributions of genomics to antibiotic target discovery. The target must be essential for the growth, replication, viability or survival of the microorganism of interest, i.e. encoded by genes critical for pathogenic life-stages. Genes that are conserved in different genomes often turn out to be essential. A gene is deemed to be essential if the cell cannot tolerate its inactivation by mutation, and its status is confirmed using conditional lethal mutants. The functions encoded by essential genes are considered to constitute the foundation of life of the organism, and are therefore likely to be common to all cells. Thus, identification and characterization of essential genes for the establishment and/or maintenance of infection may be the basis to elucidate novel and effective antimicrobials against bacteria, especially if these genes are conserved in various bacterial pathogens.

Targets have been identified among intrinsically disordered proteins proteins (IDPs) that have disordered regions. IDPs frequently function by molecular recognition, and usually undergo a binding-induced folding transition upon binding to a suitable partner. They are reported to be interaction hubs in complex protein networks. It is proposed that proteins that are most essential for bacterial cell functioning and viability have larger numbers of interactions, i.e. they are well-connected proteins or hubs in protein interaction networks (PINs). The advantage of using such hub proteins as drug targets lies in their essentiality, non-replaceable position in the PINs, and lower rate of mutation, which can help to counter bacterial resistance. It is reported that highly interconnected proteins play important roles in the central metabolism of the bacterial cell, and are several times more essential than proteins that interact with only a few other neighbors.

In parallel, network models are crucial for understanding complex networks, and help to explain the origin of observed network characteristics. The "load-point" of a metabolite in a metabolic network is defined as the ratio of the number of k-shortest paths passing through the metabolite and its nearest neighbor links. These load-point values provide a global view of the metabolic network, and aid in the analysis of the metabolic pathway reactions. Pathways that are highly connected in cellular metabolism tend to have high load values. Moreover, the lethality of an enzyme depends on the number of connections it has in the whole metabolic network. Enzymes with large numbers of connections are observed to be highly essential, and hence targeting them would result in disruption of the entire metabolic network. On the other hand, "choke-point" enzymes are those that participate in a reaction that consumes a unique specific metabolite (substrate) or uniquely produces a specific metabolite (product) in the metabolic network. These choke-point enzymes are crucial points in the metabolic pathway and inactivation of these important enzymes may lead to the disruption of the metabolic network of the bacterium. Thus, choke-point enzymes are proposed as potential drug targets.

The advent of flux balance analysis (FBA) and related *in silico* approaches contribute novel techniques to predict cellular phenotypes using genome-scale metabolic reconstructions under different environmental conditions. Constraint-based flux analysis is utilized to deduce the metabolic phenotype from the genotype, and this plays a critical role in drug targeting. The widely-used FBA method assumes that the metabolic network will reach a steady state that satisfies certain constraints (e.g. mass balance and flux limitations), and maximizes biomass production. FBA is shown to accurately predict essential genes in yeast and *Escherichia coli*. Thus, it can serve as a useful tool for the rational identification of drug targets in pathogens. A list of essential proteins in *Pseudomonas aeruginosa* metabolism predicted by FBA of the genome-scale metabolic model iMO1056 is available.

The adaptive evolution of benign bacteria into pathogens usually involves the acquisition of foreign genes encoding for specific virulence factors. An alternate or additional evolutionary mechanism employed by pathogenic bacteria is the subtle modification of an existing gene or genes through point mutations that confer a selective advantage in the virulence niche. Virulent factors provide the bacterium the ability to invade the host niche; preferentially colonize a specific host organ or tissue (i.e. strategies for tissue tropism); effectively consume available nutrients; evade host antibacterial defences, and inflict damage to the host. Virulent factors may thus be essential for survival of the bacterium, and contribute towards the elucidation of pathogenic mechanisms in infectious diseases and to the development of novel approaches for disease treatment and prevention.

A list of all currently available essential genes is compiled into the database of essential genes (DEG), that includes the essential genes identified in the genomes of *Mycoplasma genitalium, Haemophilus influenzae, Vibrio cholerae, Staphylococcus aureus, Escherichia coli,* and *Saccharomyces cerevisiae.* Concurrently, the availability of human genome sequences represents a major step in drug discovery. These resources provide a basis for addressing the complexities and conundrums in drug discovery by computational methods.

5.3.2. *Selectivity*

The microbial target for treatment should not have any well-conserved homolog in the host, in order to avoid cytotoxicity issues. This can help to avoid expensive dead-ends when a lead target is identified and investigated in great detail only to discover at a later stage that all its inhibitors are invariably toxic to the host. It is also possible to identify targets that are conserved in different strains of one specific pathogen. On the other hand, drugs designed for specific pathogens may be desired when long-term therapy for a known pathogen is required, to minimize effects on the body's normal flora and the development of drug resistance. Since the human gut and oral flora consist of microbes that are considered to influence the physiology, nutrition, immunity, and development of the host, interfering with their metabolism may have adverse effects. In line with this concept, an interesting approach designated "differential genome display" is proposed for the prediction of potential drug targets. This approach relies on the fact that genomes of parasitic microorganisms are generally much smaller, and encode fewer proteins than the genomes of free-living organisms. The genes that are present in the genome of a parasitic bacterium, but absent in the genome of a closely related free-living bacterium, are therefore likely to be important for pathogenicity and may be considered candidate drug targets. A complementary approach to target identification by bioinformatics is described in concordance analysis of microbial genomes. A simple and efficient computational tool has been developed that can determine concordances of putative gene products, unraveling sets of proteins conserved across one set of user-specified genomes, but that are not present in another set of user-specified genomes. However, there are no guarantees that if there is no homology, toxicity is not observed. An example is the large ribosomal subunit which is targeted by chloremphenicol. Despite the difference in the structure of the prokaryotic and eukaryotic ribosomes, a functionally conserved area is found in the ribosomes of mitochondria and bacteria resulting in adverse effects in certain patients.

5.3.3. *Structural genomics*

Proteins sharing a high degree of sequence identity (>30%) may have similar fold(s) depending on their lengths. Thus, a powerful approach for assigning previously unknown molecular function to a protein is to determine and compare the three-dimensional structure of the protein (rather than the amino acid sequence only) against the protein structure database (PDB). If there are significant structural homologs, the protein under investigation is predicted to have similar molecular properties. These predictions can then be tested experimentally. However, the power of the structural genomics approach will increase with time as larger numbers of protein structures become available. Furthermore, some computational modeling strategies indicate that such an approach offers improved functional predictions compared to alignments of amino acid sequences.

Three-dimensional protein structures aid in the detailed understanding of the molecular basis of protein functions. Sequence information together with three-dimensional structure contribute significant insights for the design of rational experiments such as site-directed mutagenesis, mutational studies, or structure-based design of potential inhibitors. It is therefore likely that structural genomics will continue to help gather substantial knowledge for the increasing number of potential drug targets.

Due to the rapid progress in structural genomics, acquiring the high-resolution structure of a target can provide considerable advantages at several phases of the drug discovery process. However, certain aspects of the discovery process are not currently aided by possession of a high resolution structure of the target. Nonetheless, efficiencies can be realized at many points in the process when employing the structure of a protein target to guide efforts. This is an active field of research, in which the tools for deriving advantage from the structures are emerging, and in which there is still substantial room for improvement of the techniques used.

X-ray crystallography and nuclear magnetic resonance techniques have enabled high-resolution three-dimensional structure or reliable homology modeling of target proteins to be achieved in a relatively short time.

Additional methods will emerge from the increasing numbers of available protein structures and co-structures, and the wealth of high-throughput screening (HTS) information gathered thus far. The information databases linking structure and ligand-binding information also contribute to this effort. Computational models will focus efforts on proteins that are more likely to yield high-affinity ligands for drug discovery. Furthermore, the large numbers of essential bacterial proteins with high-resolution structures provide excellent testing grounds for such methods.

The availability of high-resolution protein structures at the beginning of a drug discovery project permits *in silico* screening techniques to supplement experimental HTS. *In silico* screening employs high-resolution protein structure information to computationally test and identify small molecules more likely to bind the protein. In the process, greater numbers of compounds can be screened *in silico*, and subsequent experimental testing of compounds identified from virtual screens can be undertaken using HTS methods.

5.3.4. *Drug target identification and validation in Pseudomonas aeruginosa*

Identification of potential drug targets is the first step in the process of modern drug discovery and development. An integrative computational approach exploiting the knowledge from various genomic technologies and literature together with sequence comparisons to predict the potential drug targets more efficiently is exemplified by *P. aeruginosa* as a case study. A total of 4408 genes are identified as conserved in several completely sequenced *P. aeruginosa* genomes (Table 5.1).

Table 5.1.   Characteristics of the completely sequenced *P. aeruginosa* genomes.

Strain name	RefSeq	GenBank	Length (Mbp)	GC content	Proteins	RNAs
PA01	NC_002516	AE004091	6.2644	66.0%	5566	98
PA14	NC_008463	CP000438	6.53765	66.0%	5892	72
PA7	NC_009656	CP000744	6.6	66.0%	6286	75
LESB58	NC_011770	FM209186	6.6	66.0%	5925	102

Of these, 2984 genes have no human homologs, while 328 of these are predicted as essential in one or more completely sequenced bacterial genomes and cater to a wide variety of biological processes. Further screening of these 328 putative targets identified 13 of them as virulent factors based on the database of virulence genes of *P. aeruginosa*; 21 of them have disordered regions based on IUPRED software; and 27 of them are at choke-points in biological pathways (Fig. 5.2). Interestingly, none of the genes overlapped in the three lists. Nonetheless, since gene disruption data are not available for all the genes in all the pathogens, this approach allows one to hazard a "first-order guess" on the probability that any untested gene is essential and may act a probable drug target. It has facilitated distilling of genomic data into shortlists of potential therapeutic and vaccine targets for further detailed analysis.

5.4. Genomics and Target-Based Antibacterial Discovery

For many years, the potential of the folic acid biosynthesis pathway as a target for the development of antibiotics has been acknowledged and validated by the clinical use of several drugs. Sulfamethoxazole (a sulfonamide), competitively inhibits the binding of para aminobenzoic acid (pABA) to the enzyme dihydropteroate synthase (DHPS) that catalyzes the formation of dihydrofolate. Trimethoprim inhibits dihydrofolate reductase (DHFR) enzyme which catalyzes the NADPH-dependent reduction of dihydrofolate to tetrahydrofolate, an important cofactor involved in supplying one carbon for the synthesis of purines, pyrimidines, methionine and many other amino acids. Almost all prokaryotes need to synthesize folate compounds *de novo*, starting with GTP and utilizing several different enzymes in a multi-step pathway, whereas eukaryotes (including humans) are able to utilize the dietary folates by uptake via a carrier-mediated transport system. Hence, DHFR and DHPS are good targets for antifolates. Although knowledge of the folate pathway has advanced enormously, there is an urgent need to design novel inhibitors to meet the challenges of multidrug resistance development in many pathogens. Both prokaryotes and eukaryotes require reduced folate cofactors for the biosynthesis of many cellular components. Folates are

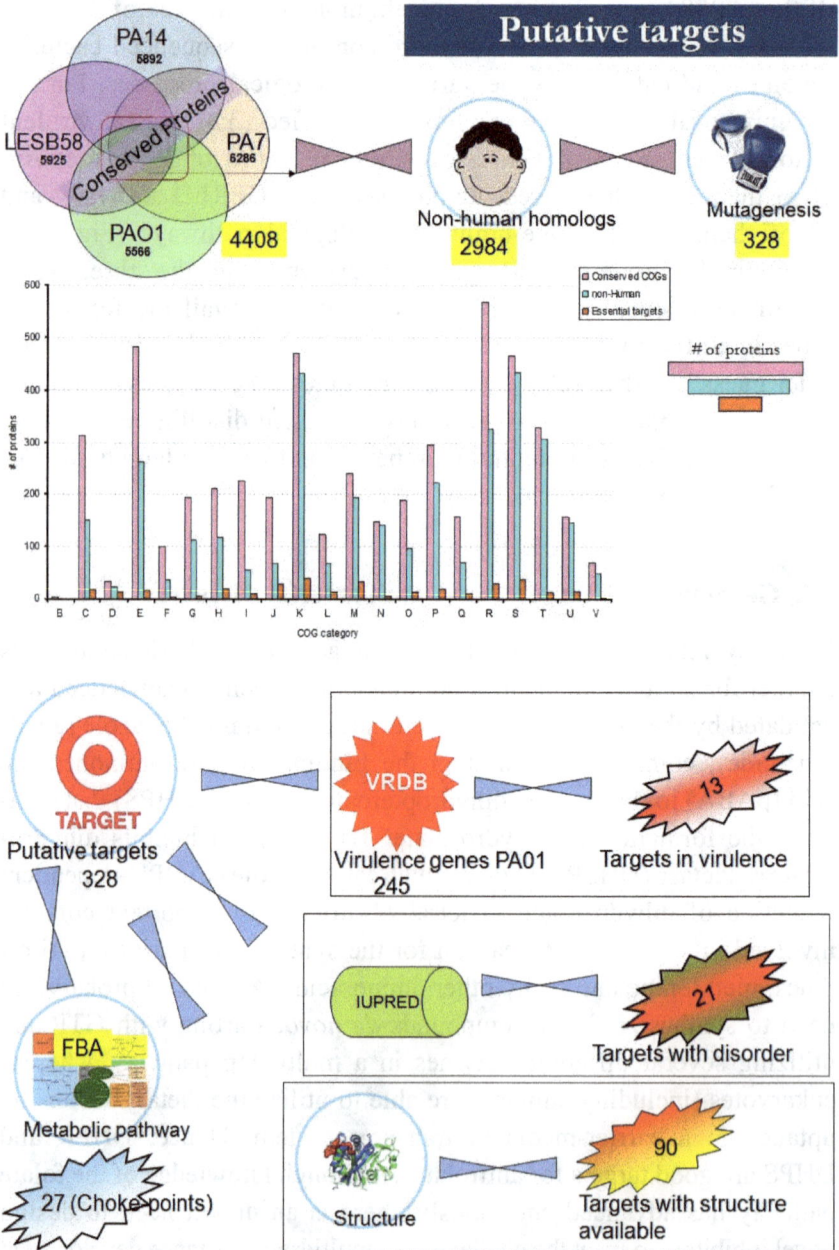

Fig. 5.2.   A schematic representation of the various genome mining methods for the identification of drug targets in *Pseudomonas aeruginosa*.

essential for all types of cells, but the mechanisms for obtaining folates differ between prokaryotes and eukaryotic cells. In plants and micro-organisms, the folates must be synthesized *de novo* through the folate biosynthesis pathway. However, in mammals, the preformed folates are utilized through an active transport system, using the membrane-associated folate transport system for reduced folates and pyrimidines. An ideal drug target should have no homologs in the host cell, since the putative drug may interact with it and lead to adverse effects in the patient (Table 5.2 and Fig. 5.3). Nonetheless, it is noteworthy that tri-methoprim (an antibiotic used to treat urinary tract infections) works by inhibiting dihydrofolate reductase (a folic acid biosynthesis enzyme), despite the presence of a close human homolog of this enzyme. Using a structural genomics approach, it is proposed that certain phytochemicals (protocatechuic acid, gallic acid, quercetin and myricetin) possibly bind to *P. aeruginosa* DHFR (Fig. 5.4), and in combination with sulfamethoxazole may inhibit different steps in the same pathway, thereby leading to synergistic activity. On the other hand, indifferent or additive interaction of trimethoprim with phyto-chemicals may be due to the binding of these phytochemicals into the active site cavity of DHFR, thus resulting in interactions with common target residues of trimethoprim, leading to competitive inhibition and no inhibition of cell growth.

Table 5.2. Characteristics of six genes for the folic acid biosynthesis pathway in *P. aeruginosa*.

Name	Gene	Tag	Cons	Non-human	Ess	Stru	Dis	VF
GTP cyclohydrolase I (H)	folE1	PA3438	√	×	√	√	×	×
Dihydroneopterin aldolase (H)	folB	PA0582	√	√	√	√	×	×
Dihydrofolate reductase (H)	folA	PA0350	√	×	×	√	×	×
2-amino-4-hydroxy-6-hydroxymethyldihydropteridine pyrophosphokinase (H)	folk	PA4728	√	√	√	√	×	×
Dihydropteroate synthase (H)	folP	PA4750	√	√	√	√	×	×
Folypolyglutamate synthetase (H)	folC	PA3111	√	×	×	√	×	×

Cons: Conserved; Ess: Essential; Stru: Structure; Dis: IDR; VF: Virulence factor.

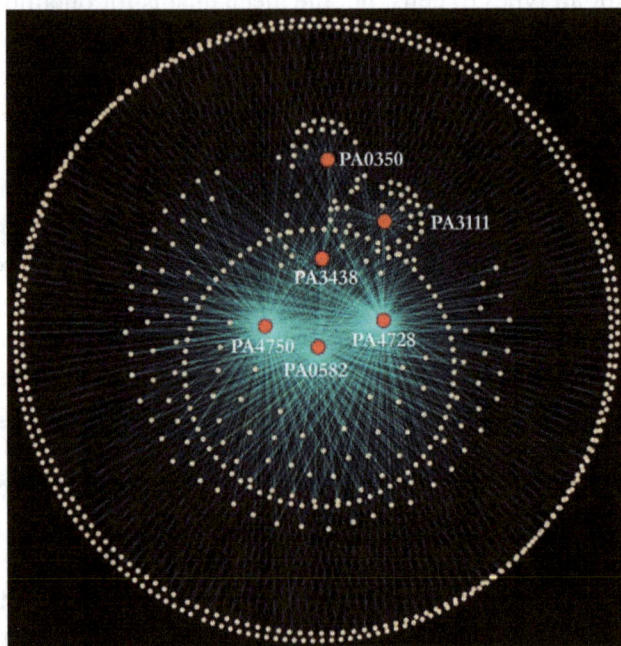

Fig. 5.3.   Protein-protein interaction network analyses for the six enzymes in the folic acid biosynthesis pathway.

5.5. Microbial Genome Mining and Vaccine Design

Vaccines represent one of the most effective ways to prevent infectious diseases and to minimize their impact on the human population. With the recent advancements in computational approaches to understanding genome data, the identification of novel potential vaccine candidates using the combination of *in silico* analysis of genome sequences (a procedure known as "genome mining"), and the knowledge from functional and structural genomics is gaining momentum. A key step in pathogenesis and infection is the interaction between a pathogen's surface-exposed or secreted proteins and host cells. Hence, antigens from pathogen surface proteins or proteins mediating virulence and infection (such as adhesion, invasion, secretion, signaling, abrogating host responses, toxicity, motility and lipoproteins) are of interest as antigens in vaccines. The availability of complete genomic sequences of pathogens, and a plethora of new

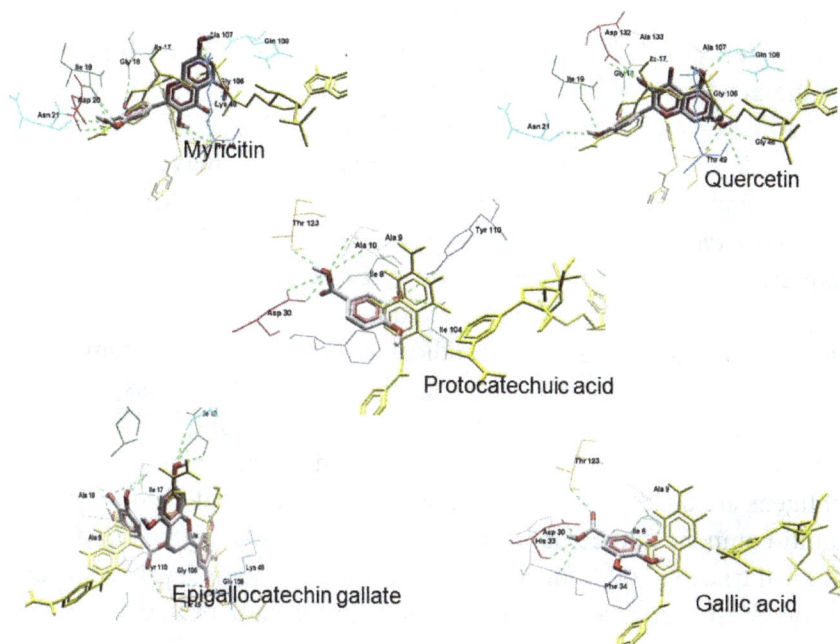

Fig. 5.4. Docking of some phytochemicals in the active site of dihydrofolate reductase.

bioinformatics tools and databases are facilitating such antigen identification. Research is also ongoing to elucidate the structures of surface molecules in order to understand the molecular mechanisms of interactions between the pathogen and the host.

5.5.1. *Reverse vaccinology*

According to this approach using either the genome of a single pathogenic isolate or the pan-genome (the genomic information from several isolates), the pathogen genome(s) is/are sequenced and analyzed *in silico* to predict which of the genes are most likely to encode surface-localized proteins or those with homologies to known bacterial factors involved in pathogenesis and virulence. The selected genes are then cloned, expressed as recombinant proteins, purified, and tested in an animal model for their capacity to confer protection against the pathogen considered.

The *Neisseria meningitidis* serogroup B (MenB) reverse vaccinology project provided the "proof of concept" for this approach. Analysis of the genome sequence of the virulent MenB strain MC58 predicted 2158 open reading frames (ORFs). Of these, 570 ORFs are predicted to encode surface-exposed or secreted proteins that may be accessible to the immune system. Subsequently, the ability of antigens to be expressed in *Escherichia coli* as recombinant proteins, and confirmation on their expression on the cell surface by enzyme immunoassay and flow cytometry are established for 91 candidates. Their capacity for immunogenicity is observed for 28 candidates, as measured by serum bactericidal assays and/or passive protection in infant rat assays. These proteins are further subjected to pan-genome analyses, and a panel of diverse meningococcal isolates are screened to determine whether the antigens are conserved. This approach resulted in the development of a multi-component recombinant MenB vaccine that entered Phase III clinical trials in 2008. Since then, reverse vaccinology has become one of the most preferred and fundamental approaches for vaccine design *in silico*.

5.6. Conclusions and Future Prospects

The major challenges in microbial genomics and mining genome data are to predict the functions of gene products, the behavior of organisms and communities from their sequences, and to exploit such genomic data to develop improved tools for managing infectious diseases. There is considerable genetic diversity among different microbial species and even different strains of the same species. Differences in the sequence and structure of genomes from members of a microbial population reflect the composite effects of mutation, recombination, and selection. Genome mining can facilitate understanding of the history and evolution of microbes, and their relationships with humans with specific reference to pathogenic phenomena, and can elucidate the genetic profile of a given pathogen in the context of its associated disease. This can set the stage for developing new therapeutics, and for expanding our knowledge of currently used antimicrobial agents. Besides this, it facilitates an enhanced understanding of the cultivation-resistant microbes using metagenomics data, and the mechanisms of antimicrobial action. These analyses and data constitute valuable assets of our armamentarium in our unrelenting battle against drug-resistant microbial infections.

Further Reading

1. Achari A, Somers DO, Champness JN, *et al.* Crystal structure of the anti-bacterial sulfonamide drug target dihydropteroate synthase. *Nature Struct. Biol.* **4**: 490–497, 1997.
2. Anurag M and Dash D. Unraveling the potential of intrinsically disordered proteins as drug targets: Application to *Mycobacterium tuberculosis. Mol. BioSyst.* **5**: 1752–1757, 2009.
3. Bagnoli F, Baudner B, Mishra RP, Bartolini E, Fiaschi L, Mariotti P, Nardi-Dei V, Boucher P and Rappuoli R. Designing the next generation of vaccines for global public health. *OMICS* **15**: 545–566, 2011.
4. Bakheet TM and Doig AJ. Properties and identification of antibiotic drug targets. *BMC Bioinformatics* **11**: 195, 2010.
5. Bruccoleri RE, Dougherty TJ and Davison DB. Concordance analysis of microbial genomes. *Nucleic Acids Res.* **26**: 4482–4486, 1998.
6. Covert MW, Knight EM, Reed JL, Herrgard MJ and Palsson BO. Integrating high-throughput and computational data elucidates bacterial networks. *Nature* **429**: 92–96, 2004.
7. Dunker AK, Cortese MS, Romero P, Iakoucheva LM and Uversky VN. Flexible nets. The roles of intrinsic disorder in protein interaction networks. *FEBS J.* **272**: 5129–5148, 2005.
8. Dziuda DM. *Data Mining for Genomics and Proteomics: Analysis of Gene and Protein Expression Data.* Wiley, 2010.
9. Edwards JS, Ibarra RU and Palsson BO. *In silico* predictions of *Escherichia coli* metabolic capabilities are consistent with experimental data. *Nature Biotechnol.* **19**: 125–130, 2001.
10. Frenz S. Better antibiotics through chemistry. *Nature Rev. Drug Discov.* **3**: 900, 2004.
11. Giuliani MM, Adu-Bobie J, Comanducci M, Arico B, Savino S, *et al.* A universal vaccine for serogroup B meningococcus. *Proc. Natl. Acad. Sci. USA* **103**: 10834–10839, 2006.

12. Grigoriev IV and Kim SH. Detection of protein fold similarity based on correlation of amino acid properties. *Proc. Natl. Acad. Sci. USA* **96**: 14318–14323, 1999.

13. Hawser S, Lociuro S and Islam K. Dihydrofolate reductase inhibitors as antibacterial agents. *Biochem. Pharmacol.* **71**: 941–948, 2006.

14. Huovinen P. Resistance to trimethoprim-sulfamethoxazole. *Clin. Infect. Dis.* **32**: 1608–1614, 2001.

15. Huynen M, Diaz-Lazcoz Y and Bork P. Differential genome display. *Trends Genet.* **13**: 389–390, 1997.

16. Itaya M. An estimation of minimal genome size required for life. *FEBS Lett.* **362**: 257–260, 1995.

17. Jain R, Rivera MC and Lake JA. Horizontal transfer among genomes: The complexity hypothesis. *Proc. Natl. Acad. Sci. USA* **181**: 3801–3806, 1999.

18. Jayaraman P, Sakharkar KR, Sing LC, Chow VT and Sakharkar MK. Insights into antifolate activity of phytochemicals against *Pseudomonas aeruginosa*. *J. Drug Target.* **19**: 179–188, 2010.

19. Jayaraman P, Sakharkar MK, Lim CS, Tang TH and Sakharkar KR. Activity and interactions of antibiotic and phytochemical combinations against *Pseudomonas aeruginosa in vitro*. *Int. J. Biol. Sci.* **6**: 556–568, 2010.

20. Jeong H, Mason S, Barabási AL and Oltvai ZN. Lethality and centrality in protein networks. *Nature* **411**: 41–42, 2001.

21. *Journal of Data Mining in Genomics and Proteomics* (http://omicsonline. org/jdmgphome.php).

22. Kobayashi K, Ehrlich SD, Albertini A, Amati G, Andersen KK, Arnaud M, Asai K, Ashikaga S, Aymerich S, Bessieres P, *et al.* Essential *Bacillus subtilis* genes. *Proc. Natl. Acad. Sci. USA* **100**: 4678–4683, 2003.

23. Kolter R, Siegele DA and Tormo A. The stationary phase of the bacteria life cycle. *Annu. Rev. Microbiol.* **47**: 855–874, 1993.

24. Koonin EV, Tatusov RL and Galperin MY. Beyond complete genomes: From sequence to structure and function. *Curr. Opin. Struct. Biol.* **8**: 355–363, 1998.

25. Kuepfer L, Sauer U and Blank LM. Metabolic functions of duplicate genes in *Saccharomyces cerevisiae*. *Genome Res.* **15**: 1421–1430, 2005.

26. Livermore DM. Discovery research: The scientific challenge of finding new antibiotics. *J. Antimicrob. Chemother.* **66**: 1941–1944, 2011.

27. Nagaraj NS and Singh OV. Using genomics to develop novel antibacterial therapeutics. *Crit. Rev. Microbiol.* **36**: 340–348, 2011.

28. Perumal D, Lim CS, Sakharkar KR and Sakharkar MK. "Load points" and "choke points" as nodes for prioritizing drug targets in *Pseudomonas aeruginosa*. *Curr. Bioinformatics* **4**: 48–53, 2009.

29. Perumal D, Samal A, Sakharkar KR and Sakharkar MK. Targeting multiple targets in *Pseudomonas aeruginosa* PAO1 using flux balance analysis of a reconstructed genome-scale metabolic network. *J. Drug Target.* **19**: 1–13, 2011.

30. Price ND, Reed JL and Palsson BO. Genome-scale models of microbial cells: Evaluating the consequences of constraints. *Nat. Rev. Microbiol.* **2**: 886–897, 2004.

31. Rahman SA and Schomburg D. Observing local and global properties of metabolic pathways: "Load points" and "choke points" in the metabolic networks. *Bioinformatics* **15**: 1767–1774, 2006.

32. Rappuoli R, Black S and Lambert PH. Vaccine discovery and translation of new vaccine technology. *Lancet* **378**: 360–368, 2011.

33. Rinaudo CD, Telford JL, Rappuoli R and Seib KL. Vaccinology in the genome era. *J. Clin. Invest.* **119**: 2515–2525, 2009.

34. Sakharkar KR, Sakharkar MK and Chow VT. A novel genomics approach for the identification of drug targets in pathogens, with special reference to *Pseudomonas aeruginosa*. *In Silico Biol.* **4**: 355–360, 2004.

35. Schmid MB. Crystallizing new approaches for antimicrobial drug discovery. *Biochem. Pharmacol.* **71**: 1048–1056, 2006.

36. Schneider G and Fechner U. Computer-based *de novo* design of drug-like molecules. *Nature Rev. Drug Discov.* **4**: 649–663, 2005.

37. Tatusov RL, Koonin EV and Lipman DJ. A genomic perspective on protein families. *Science* **278**: 631–637, 1997.

38. Vivona S, Gardy JL, Ramachandran S, Brinkman FS, Raghava GP, Flower DR and Filippini F. Computer-aided biotechnology: From immuno-informatics to reverse vaccinology. *Trends Biotechnol.* **26**: 190–200, 2008.

39. Yeh I, Hanekamp T, Tsoka S, Karp PD and Altman RB. Computational analysis of *Plasmodium falciparum* metabolism: Organizing genomic information to facilitate drug discovery. *Genome Res.* **14**: 917–924, 2004.

40. Zarembinski TI, Hung LIW, Mueller-Dieckmann HJ, Kim KK, Yokota M, Kim R and Kim SH. Structure-based assignment of the biochemical function of a hypothetical protein: A test case of structural genomics. *Proc. Natl. Acad. Sci. USA* **95**: 15189–15193, 1998.

41. Zhang R, Ou HY and Zhang CT. DEG: A database of essential genes. *Nucleic Acids Res.* **32**: D271–D272, 2004.

Chapter 6

Manipulation of Genes

Azlinda Anwar*, Too Heng-Phon† and Chua Kim Lee†
*Program in Infectious Diseases
Duke-NUS Graduate Medical School
8 College Rd, Singapore 169857
†Department of Biochemistry, Yong Loo Lin School of Medicine
National University of Singapore
MD 7, 8 Medical Drive, Singapore 117597

This chapter focuses on the ways in which molecular genetic techniques can be used to manipulate genes in order to alter the expression and production of microbial products, including the expression of novel recombinant products. "Classical" (*in vivo*) genetic techniques are essentially limited by two factors. Firstly, they can only be applied to the existing genetic complement of an organism, i.e. they are restricted to naturally occurring genes or relatively minor modifications of these genes. It is not possible to get an organism to make a product totally foreign to that organism using these techniques. Secondly, with classical techniques, one can only work on the basis of phenotype and mutants are selected by their effect on the observable characteristics of the organism. This not only limits the changes that can be selected, but also means that it can be difficult to determine the nature of the mutation that has caused the alteration in the phenotype.

The advent of recombinant DNA technology (also referred to as gene cloning or *in vitro* genetic manipulation) has dramatically broadened the spectrum of microbial genetic manipulations. The basis of this technology is in the use of restriction endonucleases and DNA ligases as a means to specifically cut and paste fragments of DNA. In this way, foreign DNA fragments can be introduced into a vector molecule (a plasmid or a bacteriophage), which enables the DNA to be replicated after being introduced into a bacterial cell.

6.1. Construction of Recombinant DNA

Today many areas of biological research require the use of recombinant technology in one form or the other. The ability to modify and clone genes has become an essential tool in research and has accelerated the rate of discovery and the development of bioindustries.

The basic steps in DNA cloning involve the following steps (Fig. 6.1):

1. A fragment of DNA is inserted into a carrier DNA molecule, called a *vector*, to produce a *recombinant DNA*.
2. The recombinant DNA is then introduced into a host cell, where it can multiply and produce numerous copies of itself within the host. The most commonly used host is the bacteria, although other hosts can also be used to propagate the recombinant DNA.
3. Further amplification of the recombinant DNA is achieved when the host cell divides, carrying the recombinant DNA in their progenies, where further vector replication can occur.
4. After a large number of divisions and replications, a colony or clone of identical host cells is produced, carrying one or more copies of the recombinant DNA. The recombinant DNA is now said to be cloned.
5. The colony carrying the recombinant DNA of interest is then identified, isolated and analyzed.

From the schematic diagram, one can appreciate that DNA cloning is a relatively straightforward procedure. Essentially the basic principles in DNA cloning involve cutting a piece of DNA (restriction digestion) and pasting the restricted DNA into a vector (ligation). In order to produce more of this recombinant DNA for easy manipulations, the DNA is then introduced and propagated in host cells (transformation and amplification). Once the clone containing the desired DNA is selected, the recombinant plasmid can finally be isolated and further explored (analysis). Each of

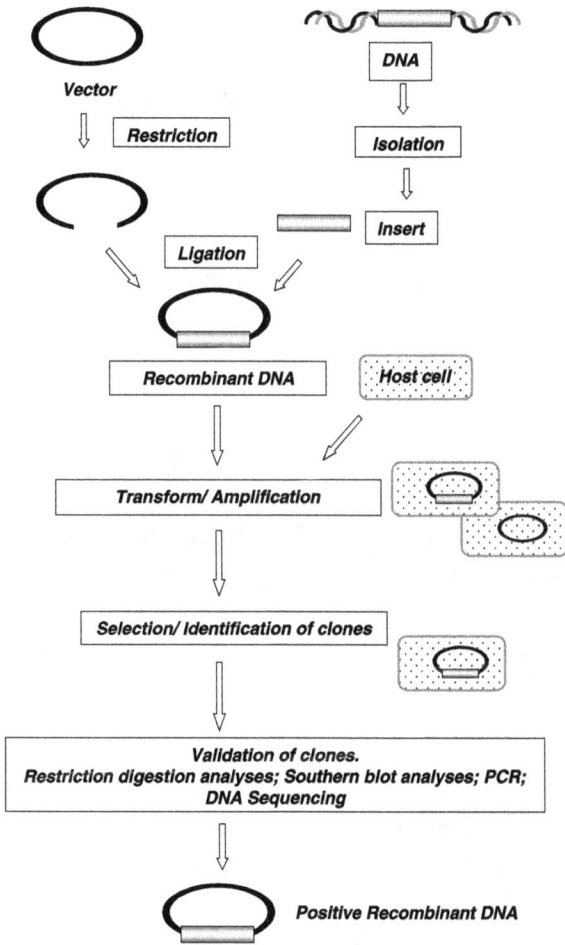

Fig. 6.1. Basic steps in gene cloning.

the steps in the schematic diagram will be described in detail in order to guide the students in the principles of DNA cloning.

6.1.1. *Enzymes used in gene manipulations*

Recombinant DNA technology will not be possible had it not been for the discovery and exploitation of the functions of restriction and modifying enzymes. Within a cell, these enzymes participate in essential processes such as DNA replication and degradation of undesirable DNA. A number of bacterial strains have been identified that can produce these enzymes in high quantities. After purification from the bacterial cell extracts, these enzymes are still able to carry out their natural reactions when the appropriate artificial condition is provided. Currently, a number of these enzymes are produced commercially by recombinant DNA technologies, and are supplied as highly purified and well-characterized enzyme preparations.

The enzymes that are used in gene manipulations can generally be classified into 5 classes based on their functions:

1. **Nucleases**, which cut or degrade DNA molecules.
2. **Polymerases** can copy and make new strands of DNA.
3. **Ligases**, which join pieces of DNA molecules together.
4. **Modifying enzymes**, which modify the DNA by adding or removing chemical groups.
5. **Topoisomerase**, which remove or introduce supercoils from covalently closed-circular DNA.

The functions of these enzymes and how they are used in DNA cloning will be described briefly. Three classes of enzymes that are more widely used in DNA cloning will however be described in detail. A point to note is that although most enzymes can be classified into a particular class of function, a few can display multiple functions. For example, polymerases have both the functions of making new strands of a DNA as well as degrading it.

(1) Nuclease
The function of this class of enzymes is to degrade or digest DNA by breaking the phosphodiester bonds that link one nucleotide to the other in a DNA strand. There are two types of nucleases:

(a) **Exonuclease**, which digests nucleic acids at either ends of the DNA.
(b) **Endonuclease**, which digest nucleic acids from within the DNA.

The main difference within different endonucleases or exonucleases is their ability to digest either single-stranded or double-stranded DNA.

For example, while the enzyme *Exonuclease* III (*Exo* III) is able to cleave double-stranded DNA, the enzyme *Exonuclease* VII (*Exo* VII) is only able to cleave single-stranded DNA. Similar distinctions can be applied to endonucleases. Whereas the endonuclease DNAase I, isolated from bovine pancreas, is able to digest double-stranded DNA, the mung bean nuclease (from the sprouts of mung bean) can only digest single-stranded DNA. Both DNAse I and mung bean nuclease are examples of endonucleases that cleave DNA at non-specific sites. There is yet another group of endonucleases that can only cleave DNA at restricted sites defined by specific DNA sequences. This group of enzyme is known as restriction endonucleases and is the most commonly utilized enzymes in recombinant DNA technology.

Restriction Endonuclease

This group of enzymes was first discovered when it was observed that some strains of bacteria were immune to bacteriophage infection. It is now known that these strains of bacteria had produced endonucleases to restrict the propagation of new phage particles by digesting the foreign DNA before it can replicate. During this process, the bacteria protect their genome by modifying their DNA through methylation to prevent the action of these degradative enzymes.

These restriction endonucleases have been given names consisting of 3 italicized letters (representing the bacteria from which the restriction enzymes isolated from), followed by a letter to identify the strain (if applicable), and end with a Roman numerical to distinguish the different enzymes isolated from the same bacteria. Some examples are given below to illustrate this nomenclature:

*Eco*R I is the first restriction/modification enzyme isolated from *Escherichia coli strain R*.

*Hin*d III refers to the third system isolated from *Haemophilus influenzae strain R_d*.

*Bam*H I was the first restriction/modification system isolated from *Bacillus amyloliquefaciens* H.

To date, more than 2900 restriction enzymes are known, and they have been generally classified into Type I, Type II or Type III specific classes.

Types I and III enzymes are complex, and are not commonly used in genetic engineering. These enzymes cleave the DNA at a substantial distance from their recognition sites. Type II restrictions enzymes, on the other hand, play a central role in recombinant technology.

The central feature of the Type II restriction enzymes is that each enzyme has a specific recognition sequence. The DNA is cleaved either *within* or *very near* to the recognition sequence. It is important to note that these enzymes are very specific and will cut the DNA only when their recognition sequences are present. The recognition sites of Type II enzymes are usually palindromic (two fold symmetry) and are 4, 5, 6 or more base pairs long. Some examples of the most commonly used restriction endonucleases are shown in Table 6.1.

Note that some restriction endonucleases, for example *Bam*H I, cut the DNA asymmetrically (as shown by the slash). This results in each fragment having four unpaired bases (in this case GATC) at both ends, referred to as "sticky" or cohesive ends. These single-stranded regions are complementary and will therefore tend to pair with DNA fragments with compatible sequences. For example, in the case of *Bam*H I-restricted DNA, the DNA can form complement pairing with either a *Bam*H I or *Bgl* II restricted DNA.

Some enzymes, such as *Pst* I, leave a four-base sticky end at the 3' end of the fragment, while others, such as *Sma* I and *Pvu* II, cleave in the center of the recognition sequence, generating blunt-ended fragments.

Table 6.1. Some examples of the frequently used restriction endonucleases. The sequences shown are that of the positive strand of DNA from the 5' to 3' direction.

Enzyme	Recognition site	Blunt or cohesive end
Hind III	A/AGCTT	cohesive
*Bam*H I	G/GATCC	cohesive
Bgl II	A/GATCT	cohesive
Pst I	CTGCA/G	cohesive
Sma I	CCC/GGG	blunt
Pvu II	CAG/CTG	blunt
Alu I	AG/CT	blunt
Not I	GC/GGCCGC	cohesive

(2) Polymerases

Polymerases are enzymes that can synthesize new strands of nucleic acids that are complementary to an existing piece of DNA or RNA. These enzymes can only do so when the template has a double-stranded region that can act as a primer to initiate nucleic acid polymerization.

There are 3 main types of polymerases that are frequently used in recombinant DNA technology (Fig. 6.2). The first, DNA polymerase I, attaches itself to the single-stranded portion of a largely double-stranded DNA and initiates synthesis from the 5' end of the DNA, replacing the

a) Basic reaction

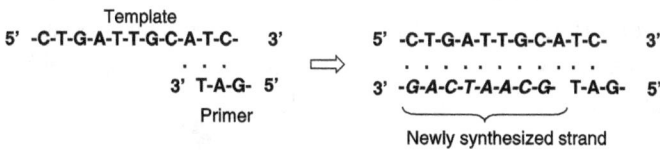

Template
5' -C-T-G-A-T-T-G-C-A-T-C-   3'   5' -C-T-G-A-T-T-G-C-A-T-C-   3'
                                  ⟹
                3' T-A-G- 5'   3' -G-A-C-T-A-A-C-G- T-A-G- 5'
                Primer   Newly synthesized strand

b) DNA polymerase I

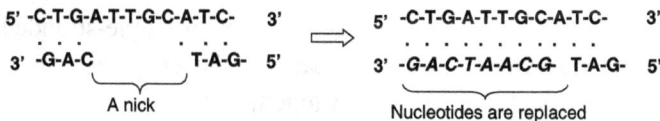

5' -C-T-G-A-T-T-G-C-A-T-C-   3'   5' -C-T-G-A-T-T-G-C-A-T-C-   3'
                                  ⟹
3' -G-A-C   T-A-G- 5'   3' -G-A-C-T-A-A-C-G- T-A-G- 5'
      A nick   Nucleotides are replaced

c) Klenow fragment

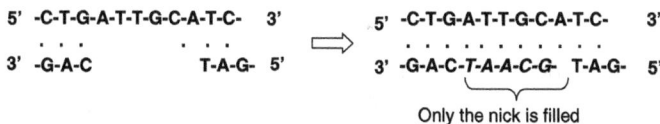

5' -C-T-G-A-T-T-G-C-A-T-C-   3'   5' -C-T-G-A-T-T-G-C-A-T-C-   3'
                                  ⟹
3' -G-A-C   T-A-G- 5'   3' -G-A-C-T-A-A-C-G- T-A-G- 5'
                                          Only the nick is filled

d) Reverse transcriptase

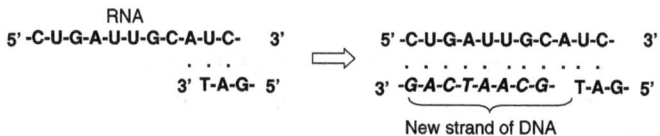

RNA
5' -C-U-G-A-U-U-G-C-A-U-C-   3'   5' -C-U-G-A-U-U-G-C-A-U-C-   3'
                                  ⟹
                3' T-A-G- 5'   3' -G-A-C-T-A-A-C-G- T-A-G- 5'
                                          New strand of DNA

Fig. 6.2. The major polymerases used in DNA cloning.

old strand as it proceeds. One can therefore see that this enzyme has dual functions: DNA synthesis and degradative activities.

The second most commonly used polymerase enzyme is the Klenow fragment, which is essentially derived from DNA polymerase I. The DNA polymerase I enzyme has been modified to remove the DNA degradative function, retaining only the DNA polymerization function. This modified enzyme, called the Klenow fragment, is thus able to synthesize new DNA strands complementary to the single-stranded portion. It cannot, however, replace the preceding double-stranded DNA as it does not have the nuclease activity. The major application of the Klenow fragment is in DNA sequencing.

The third type of polymerase enzyme important in recombinant DNA technology is reverse transcriptase. Unlike the two other polymerases, the reverse transcriptase uses RNA as the starting template. The ability of this enzyme to synthesize a DNA strand complementary to an RNA template is crucial to a technique called cDNA cloning.

(3) Ligases

DNA ligases serve to link DNA strands together by forming phosphodiester bonds between the 5'-phosphate and the 3'-OH termini of the discontinuous strands. In the cell, these enzymes serve to repair single-stranded breaks (nicks) that arise during DNA replication. The enzymes can also join together 2 individual pieces of DNA together. The most commonly used DNA ligase is the T4 DNA ligase that is purified from *E. coli* infected with the T4 bacteriophage. The use of these enzymes in DNA cloning will be described in the later part of the chapter.

(4) Modifying Enzymes

These enzymes modify the DNA by either adding or removing a chemical group. There are numerous modifying enzymes, but the 3 most commonly used ones are:

(a) **Alkaline phosphatase** removes a phosphate group from the 5' end of the DNA. This is usually done to prevent vector religation.

(b) **Polynucleotide kinase** acts in reverse of the alkaline phosphatase by adding a phosphate group to the 5-terminus of a DNA.

(c) **Terminal transferase** adds on one or more nucleotides on the 3' end of a DNA.

(5) Topoisomerases

This set of enzymes is able to change the conformation of a closed circular DNA by adding or removing supercoils. These enzymes are used mainly in study of DNA replication, and are not too widely used in DNA cloning.

6.1.2. Use of restriction endonucleases in recombinant DNA technology

Restriction digestion of DNA is one of the most fundamental techniques in recombinant DNA technology. A typical DNA restriction digestion reaction is illustrated in Fig. 6.3. The amount of DNA to be digested will depend on the nature of the experiment. For example, if a routine cloning experiment is to be performed, typically 1–2 μg of DNA is sufficient. However if one is constructing a genetic library, more starting DNA (~10–20 μg) is needed. The second point to consider is the choice of the restriction endonuclease to be used. There are hundreds of commercially available restriction enzymes, supplied as purified enzymes with known concentrations. By convention, 1 unit of enzyme is defined as the amount required to digest 1 μg of DNA in 1 hour at 37°C. Before adding the enzyme, the DNA solution

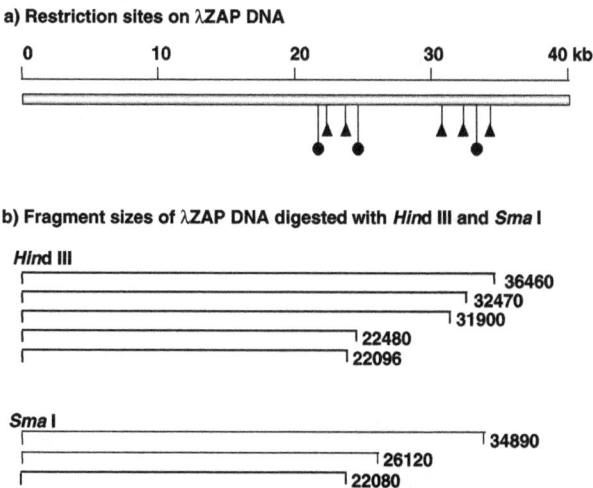

Fig. 6.3. Restriction of the λZAP DNA using *Hin*d III and *Sma* I. The sizes of each fragment are indicated in base pairs.

must be adjusted to provide a suitable reaction condition to ensure optimal enzyme activity. Most enzymes function in pH 7.4, but require different Mg^{2+} concentration and ionic strength (usually provided by NaCl). All commercial enzymes are supplied with their respective buffer in a 10X working concentration. This concentrated buffer will have to be diluted to a 1X concentration in a digestion reaction. The optimal temperature for most restriction endonucleases to function is at 37°C, although there are some enzymes that require a lower temperature to function optimally (e.g. *Sma* I is used at 25°C).

To illustrate this technique, let us consider digesting a sample of λZAP DNA (of concentration 0.5 μg/μL) with *Hin*d III (Fig. 6.4).

We would start with 2 μg of DNA for restriction digestion. Hence 4 μL of the DNA sample is needed for the reaction, which is performed in a typical 1.5 mL microfuge tube. Next, we would need to provide the proper environment for the enzyme to work in. A typical restriction digestion reaction is carried out in a 20 μL volume. Therefore, we would need 2 μL of a 10X *Hin*d III buffer solution. The restriction endonuclease can now be added. A commercial *Hin*d III enzyme is typically supplied at a concentration of 10 U/μL. Since we are using 2 μg of DNA, a 0.5 μL of the enzyme (5 U total) is more than sufficient to completely digest the

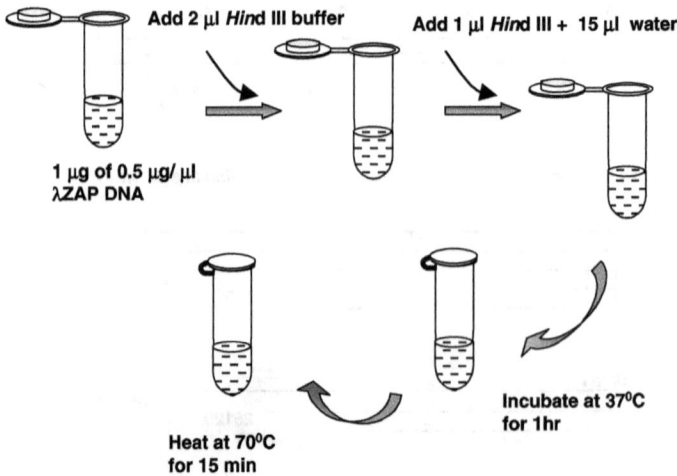

Fig. 6.4. Restriction digest of λZAP DNA with *Hin*d III.

DNA. Since the total reaction volume required is 20 μL, 9.5 μL of water is finally added to the reaction mix. The digestion reaction is then carried out at 37°C for 1 hour to allow for complete digestion of the DNA. At the end of the incubation period, the enzyme must be destroyed to prevent accidental digestion of other DNA molecules that may be added at a later stage. This can be achieved by simply denaturing the enzyme by heating the reaction mix at 37°C for 10 min, or adding EDTA to chelate the Mg^{2+}, thus rendering the enzyme inactive.

6.1.3. *Analysis of the result of restriction endonuclease digestion*

As can be predicted from Fig. 6.3, restriction digestion of the DNA will yield multiple DNA fragments, the sizes of which will depend on the distance between the specific restriction sites for *Hin*d III. Clearly, a method is needed to separate the DNA fragments in order to determine the number and sizes of the different fragments. One of the easiest methods to achieve this is by using the technique of gel electrophoresis. Eventually we would also like to be able to acquire more information about the DNA in terms of the relative positions of different restriction sites within the DNA. This is called a restriction map.

(1) Separation of DNA Molecules by Gel Electrophoresis
Nucleic acids have net negatively charged phosphates with pKa of ~2. Hence in neutral conditions (pH ~7), the molecules will have an overall negative charge. Thus when the DNA molecules are subjected to an electric field in a buffer, the molecules will migrate to the positive end. The DNA molecules, to a great extend, possess a constant charge to size ratio, irrespective of their sizes. Hence nucleic acids of all sizes will migrate at a similar rate in liquids and it is therefore not possible to separate them in liquids. One way to separate the DNA fragments according to their sizes is by carrying out electrophoresis in gels that act as molecular sieves. Gels are semi-solid matrices of polymers entrapping a buffer. Since the DNA molecules have to weave through this matrix mesh to get to the positive end, the sizes of the molecules become a factor; the smaller the DNA fragment, the faster it can migrate through the gel. Two of the most commonly used gel matrices are agarose and polyacrylamide, or a mixture

Table 6.2. Guide to resolution of linear DNA versus gel concentrations.

Agarose (%)	Resolution for linear DNA (kb)
0.5	1.0–40
0.8	0.5–30
1.0	0.3–20
1.5	0.2–5
2.0	0.1–2

of both. In practice, the composition of the gel matrix will determine the sizes of the DNA to be separated. For example, a 1% agarose gel is used to separate DNA fragments of sizes 500–10,000 bp (Table 6.2).

If separation of smaller pieces of DNA fragments is required, differing by just bases, then a polyacrylamide gel can be used. The resolution of double-stranded DNA fragments using nondenaturing polyacrylamide gel is shown in Table 6.3:

Table 6.3. Resolution of DNA in polyacrylamide gels.

Polyacrylamide (%)	Double-stranded DNA (bp)
3.5	100–1000
5.0	100–500
8.0	60–400
12	50–200
20	5–100

(2) Visualization of DNA Molecules in a Gel

The easiest way to visualize DNA fragments in a gel is by using fluorescent dyes. A number of fluorescent dyes can be used for this purpose. One of the most commonly used is ethidium bromide, EtBr. EtBr molecule intercalates (inserts) between the stacked bases of double-stranded nucleic acid or nucleic acid stacked into some secondary structures, as in the case of RNA. The intercalated EtBr will emit a fluorescent light when excited at the UV range (200–400 nm), indicating the position of the DNA fragment. It is important to note that prolonged exposure of the DNA to UV will induce strand breakage within the DNA. Other fluorescent dyes, for example SYBR Gold and SYBR Green, are also available for DNA visualization.

(3) Estimation of the Sizes of DNA Molecules
In gel electrophoresis, the DNA fragments will be separated according to their sizes. The most common way to estimate the sizes of DNA is by including a DNA marker in the electrophoretic run. DNA markers are standard restriction digests, usually of λ DNA, comprising fragments of known sizes. As the sizes of the DNA markers are known, the fragment sizes in the experimental digest can be estimated by comparing the positions of the DNA bands in the two tracks.

There are a number of DNA markers available commercially catering to various ranges of DNA sizes. The choice of markers used in electrophoresis is dependent on the size range of the experimental DNA fragments. For example, if the size of a PCR product, less than a kilobase long, is to be estimated, then it would be more appropriate to use a 100 bp (size range of 100–500 bp) DNA marker rather than a λ DNA/ *Hin*d III marker (size range of 125–3,000 bp).

(4) Mapping the Positions of Different Restriction Sites
The next step in a restriction analysis is to determine the relative position of different restriction sites within the DNA, i.e. to build a restriction map of the DNA. This is necessary if one wants to locate and cut a particular region of the DNA.

To construct a restriction map, a series of restriction digests need to be performed. As a guide in choosing the types of restriction enzymes to use, one should first identify one or more unique restriction sites in the vector (i.e. 5' or 3' to the DNA insert) to anchor a relative map position. A set of digestion experiments is then performed, using one restriction enzyme per tube. The number and sizes of the fragments produced by each restriction endonuclease must then be determined by gel electrophoresis. Next, a series of double digestions, where the DNA is restricted with two restriction endonucleases, needs to be done. If the enzymes share similar conditions required for optimal enzymatic activity, then the double

Enzyme	Number of fragments	Sizes (kb)
EcoR I	3	2.0, 4.0, 5.0
Hind III	2	3.0, 8.0
EcoR I + Hind III	5	1.0, 1.0, 2.0, 3.0, 4.0

Conclusion:
1. The number of restriction sites for each of the enzymes is: EcoR I   3
                                                                Hind III   2
2. The Hind III and EcoR I sites can be mapped as follow:

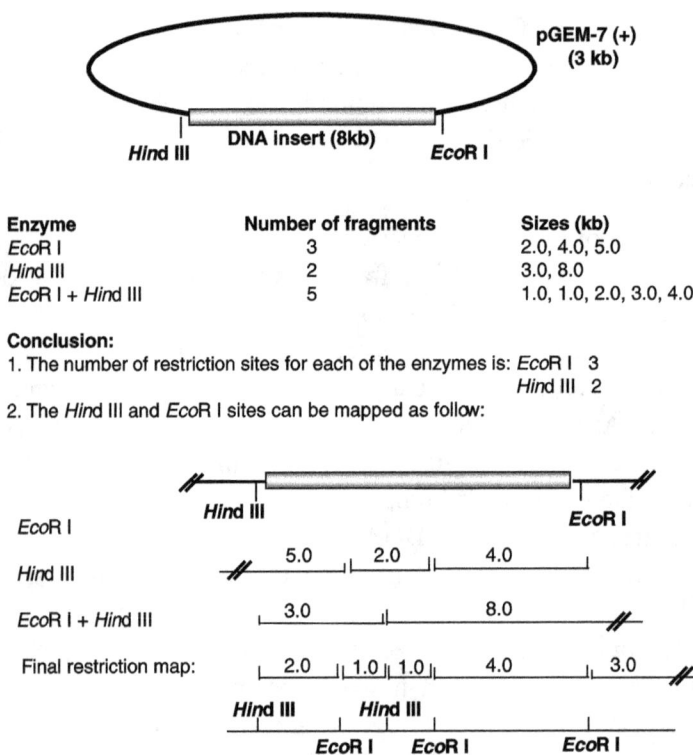

Fig. 6.5. Construction of a restriction map.

digestion can be done in a single-step reaction (i.e. the enzymes are added at the same time). However, if the enzymes do not share similar enzymatic buffer conditions, then the digestion has to be done one after the other. Comparing the results of the two types of digestion will allow many of the restriction sites to be mapped within a piece of DNA (Fig. 6.5).

6.1.4. DNA ligase and its use in recombinant DNA technology

The final step in the construction of a recombinant DNA molecule is the joining of the DNA fragment and the vector in a process known as ligation. As mentioned in Sec. 6.1.1, this reaction is mediated by DNA ligases.

Figure 6.6 shows two examples of a ligation reaction, one between blunt-ended DNA fragments and the other between DNA with cohesive

a) Ligating blunt ends

b) Ligating cohesive ends ligation

c) A typical linker

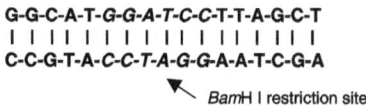

G-G-C-A-T-*G-G-A-T-C-C*-T-T-A-G-C-T
| | | | | | | | | | | | | | | | | |
C-C-G-T-A-*C-C-T-A-G-G*-A-A-T-C-G-A

↖ *Bam*H I restriction site

d) Ligation using linkers

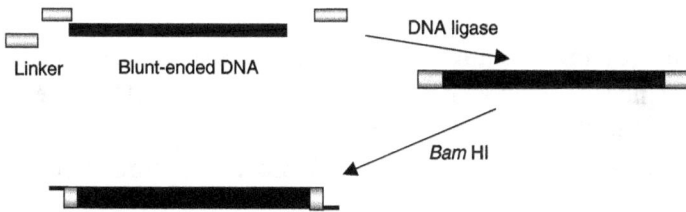

Linker   Blunt-ended DNA

DNA ligase

Bam HI

Fig. 6.6. DNA ligation.

ends. In general, ligation reactions between cohesive ends are more efficient compared to blunt end ligations. This is generally governed by the kinetics of the two ends of the DNA molecules meeting each other before a ligation reaction can proceed. In the case of cohesive end ligations, the single-stranded sequences from both strands of the DNA form a complement with each other, thus forming a stable albeit transient structure for the ligase to work on.

In contrast, without the aid of complementary sequence formation, blunt-ended DNA molecules rely on chance association to bring the two ends together. In order to increase the efficiency of ligation, blunt-end ligations usually have to be done in the presence of higher amounts of DNA molecules. Alternatively, we can modify the blunt ends of DNA fragments to form cohesive ends by adding short nucleic acids in the form of linkers or adaptors.

(1) Linkers

Linkers are short, blunt-ended pieces of double-stranded DNA of known sequences, with one or more restriction sequences inserted within it (Fig. 6.6). The linkers, which can be synthesized in large quantities, are attached to blunt-ended DNA molecules by DNA ligase. Although this is a blunt end ligation, this reaction can be performed with high efficiency as the linkers can be added in high concentrations. It is important to realize that the use of high linker concentrations would inevitably cause multiple linkers to be attached to either side of the DNA molecule. However, restriction digestion with the appropriate enzymes will create cohesive ends within the linkers. This modified DNA-linker molecule will now be ready to be ligated to a cloning vector restricted with the appropriate restriction enzymes.

(2) Adaptors

Adaptors are similar to linkers in that they are also double-stranded small pieces of nucleic acids that are synthesized *in vitro*. They are however different from linkers in that they do not have blunt ends, but are made with cohesive ends instead (Fig. 6.7). To prevent the formation of adaptor

a) An example of a typical adaptor and a 5'-modified adaptor

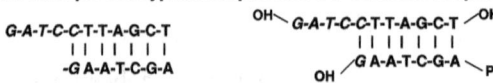

b) Typical adaptors could ligate to one another

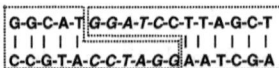

c) Ligation with the modified adaptors

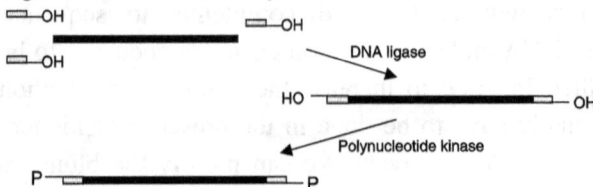

Fig. 6.7. Use of adaptors in the cloning of DNA.

concatamers during ligation, the 5' cohesive ends of the adaptors could be modified to incorporate a 5'-OH group instead. Hence, the enzyme DNA ligase will not be able to make phosphodiester bonds between the adaptors (recall that DNA ligase only makes a phosphodiester bond between a 5'-phosphate terminus and a 3'-OH terminus; see Sec. 6.1.1). Hence, the adaptors can be ligated to the DNA molecules but not to themselves. After the adaptors have been attached, the 5'-OH cohesive ends are then modified to 5'-phosphate by adding a phosphate group using the enzyme polynucleotide kinase (see Sec. 6.1.1). The modified DNA-adaptor molecules are now ready to be inserted into an appropriate cloning vector.

6.2. Host Organisms

The use of living cells as hosts to propagate and amplify recombinant DNA is one of the cornerstones in molecular biology. Among the many hosts used in biotechnology, bacteria have been and still are extensively used to produce genetic vectors as well as in the production of recombinant proteins. Other hosts including yeast, plant, insect and mammalian cells, are also exploited for the production of recombinant proteins for specialized purposes. Every host has its own unique properties that can be harnessed for different uses.

One of the objectives of molecular biotechnology is to produce therapeutic proteins. Today, many of the commonly used bacterial strains have been selected or engineered to make them suitable for specific purposes. As an example, the host restriction or modification systems of many laboratory strains have been rendered deficient in these systems and hence allow the propagation of unmodified DNA (e.g. DNA fragments from ligation or PCR products).

E. coli is the abbreviated name for the bacterium in the Family Enterobacteriaceae called *Escherichia* (Genus) *coli* (Species). *E. coli* represents approximately 0.1% of the total bacteria within an adult intestine. This bacterium is a small rod-shaped organism with a circular genome of about 4500 kb in length. The whole genome has now been sequenced. Its genetic property has been the subject of intensive research over the last 30 years and a wide variety of mutants have been isolated and characterized. With the recent availability of their genomic sequences, it is now possible to carry out targeted mutations to produce novel engineered strains for recombinant DNA studies and protein production. The rapid growth of *E. coli* both in minimal and rich media makes it an ideal host for the propagation of recombinant DNA and proteins *in vivo*. Almost all laboratory strains are derivatives of the wild type isolates K-12 or B strains and generally do not carry *EcoR* I or other type II restriction systems.

6.2.1. *Growth*

There are two common ways to grow *E. coli* in the laboratory: on solid or in liquid media. These media are designed to attain both high biomass and yield of recombinant DNA. This is especially useful for low yield vectors (e.g. BACs) where the number of copies in one cell can be very low in commonly used media, e.g. Luria broth (LB). The development of rich media has been successful for both fed-batch fermentations and small-scale applications. Some of these media go by the labels of "Superbroth", "Terrific Broth" and H15. By obtaining a high yield of vectors, there is therefore no need to use large centrifuges for subsequent manipulations of the host cells. It is also possible to affect the solubility of recombinant proteins by manipulating the growth parameters and media contents.

6.2.2. *Handling, culturing and storage*

It is very important, when working with bacteria, that the cultures are kept pure and properly stored. Note that air-borne bacteria can often confound proper storage. Hence, it is advisable to work in the flow hood. All glassware and media should be sterilized either by autoclaving or filtering through 0.2 or 0.4 μm filters.

Pure cultures can be obtained by propagating bacterial cultures from single, isolated colonies on solid agar plates. This procedure is a powerful and simple method to obtain single, pure species. If the stock is contaminated with another strain and there is a mean to verify the correct strain, then it is possible to plate out and select for the correct strain. For example, XL-1 blue cells (Stratagene) harbor a transposon that confers tetracycline resistance. If this cell culture is contaminated with another strain without this genetic marker, then by simply plating out into individual clones on a selection media (e.g. tetracycline), the contaminant can be eliminated.

The bacterial cultures can be stored for long periods in stab vials, as frozen glycerol cultures or as lyophilized cultures. Stabs should be transferred every 2 years whereas frozen glycerol cultures at −80°C or lower can be stored indefinitely. Storage at −20°C in a common-use freezer will result in lose of viability due to fluctuations in temperature with frequent opening and closing of the freezer.

For frequent laboratory use, bacteria can be streaked out on solid agar plates and stored at 4°C. It is advisable to replate these cultures every 1–2 month to maintain viability.

6.2.3. Introduction of genes into bacterial host cells

The ability to transfer recombinant DNA into a living host is a necessity in molecular biology. In nature, genetic transfers are common occurrences. There are 3 mechanisms of genetic transfer in prokaryotes:

- **Conjugation**: transfer via "conjugative plasmids" requiring cell-cell contact.
- **Transduction**: transfer by bacteriophage. Plasmids (usually smaller than the phage genome) can be encapsidated into phage proheads to produce plasmid-transducing agents. This form of gene transfer is highly efficient.
- **Transformation**: introduction of "naked" DNA into bacteria cells. This is one of the most commonly used methods to introduce recombinant DNA into host cells.

(1) Transformation

This method of transferring genetic material (as free DNA) to recipient cells does not require the contribution from intact donor cell (e.g. need of pili, i.e. the "sex organs" of bacteria). This method is not restricted to plasmids alone but can also be used to transform large DNA molecules. Figure 6.8 illustrates the two general methods for transforming bacteria:

(a) A chemical method utilizing salt and heat shock to promote DNA uptake.
(b) Electroporation utilizing electric shock to facilitate DNA entry.

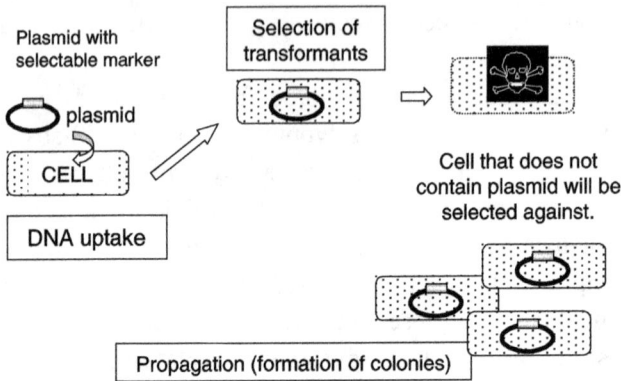

Fig. 6.8. Transformation of bacterial cells using plasmid DNA.

To assess the success of a transformation, there are three related steps that need to be considered: the entry of the DNA (uptake), production of a selectable phenotype (e.g. antibiotic resistance) and the propagation of these transformed cells to form visible colonies.

(2) Competence State

In most host cells, including *E. coli*, competent cells do not occur naturally. For these cells, a competent state has to be induced artificially. Competence is usually a consequence of a nutritional shift-down (decrease of the availability of one or several nutrients, or of available energy) and usually result in decrease in the rate or a blockage of DNA synthesis. The exact mechanism is not well understood.

Several growth media and conditions have been developed empirically to induce competence. The kinetics of competence varies greatly from one strain to another (i.e. some strains may show competence for several hours and drops off with time). A number of metabolites can stimulate the establishment of competence in some strains. Although the biochemical mechanisms are not clearly understood, important modifications of the cell envelopes take place during development of competence.

(a) Chemical induced competence

This method of inducing competence is used extensively in many laboratories. There are a number of available procedures using various chemicals including calcium chloride, cobalt chloride, rubidium chloride and others. The transformation efficiency ranges from 10^6 colony forming units (cfu)/μg to greater than 10^9 cfu/μg of DNA, rivaling electroporation. As with other means of transformation, it is highly dependent on the strains used.

The strategy is as follows: cells are grown overnight to act as the starting inoculum. A small portion of this will be used to seed a fresh batch of media, and are then grown with high aeration. When the cell growth is in the logarithmic phase, the cells are collected and processed. Here, the cells are chemically treated to induce competence. All along the process, the cells are treated gently and in the cold to prevent lysis. It is thought that these chemical treatments facilitate the recruitment of channels where DNA molecules can enter. In the process of DNA uptake, the molecules are incubated with the cells at 4°C for a period of half to one hour. During this time, DNA molecules, being predominantly negatively charged, complex with cations (calcium, rubidium or cobalt) and precipitate onto the cell surface. A brief heat-shock treatment enhances the uptake of these complexes through the channels. There is still no concrete evidence that this actually happens although all the observations are consistent with this model of DNA uptake.

The cells can be prepared and kept frozen in glycerol and stored at −70°C for at least 6 months with little loss in efficiency. Batches of the cells prepared in exactly the same manner will vary. Hence it is essential to determine the transformation efficiency of every batch made. Several factors will affect the transformation efficiency, including the density of the cells at harvest, the absolute requirement for maintaining the cells at 0–4°C throughout the processing and the period of time at 4°C before freezing down for storage.

Competent cells of high efficiency for transformation can be prepared in the laboratory. However, this method is rather involved and tedious, requiring experience in treating the cells. It is no longer necessary to prepare competent cells if the usage is low. There are now commercially available competent cells that come in the form of supercompetent or ultracompetent cell preparations.

Since some portions of the host cells will not have taken up the DNA, it is necessary to be able to select for the recombinant colonies. This selection marker, usually in the form of an antibiotic resistance gene, is carried by the transforming DNA. Plating the transformed bacteria onto selective media would thus enable the transformants to be distinguished from the non-transformants. The next challenge is to distinguish colonies containing the recombinant DNA from the ones that contain self-ligated vectors. With most cloning vectors, insertion of the DNA fragment would destroy the integrity of a marker gene present in the vector. Recombinant colonies can then be identified by the characteristics that are not displayed by the host cells due to the inactivated gene.

(b) Electroporation

In general, this method produces better transformation efficiency and is more successful with many different cell types including plants, yeast, insects and mammalian cells that are resistant to chemical methods. Large molecules are also electroporated with high efficiency. Thus, this method can also be used to transfer drugs or even whole genome into cells of all types. The maximal efficiency occurs when the pulse is strong enough to yield 40–80% cell death. Therefore for electroporation, one needs a higher number of cells to start off in order to obtain a reasonable number of transformants. Like chemical methods, each cell type requires an optimal set of parameters to produce a maximum number of transformants. These are usually determined empirically and many protocols have been published.

Electroporation (electrically induced pore formation) is currently viewed as a bilayer membrane phenomenon, where short (μs to ms) electrical field pulses cause the transmembrane voltage to rise to about 0.5–1.0 V, thus causing pore formation. The membrane is then reversibly permeabilized, thereby enhancing the transport of molecules across the membrane.

The electroporators generally have two components: a generator to provide voltage pulses of electrical fields and a chamber in which cells are subjected to the electrical fields (Fig. 6.9). The transient power required by the generators to deliver the electrical energy to the chambers can easily far exceed the electrical power available in the laboratory outlets. To circumvent this limitation, electrical energy is first stored in capacitors. The accumulated voltage is then discharged, at high power, into the chamber.

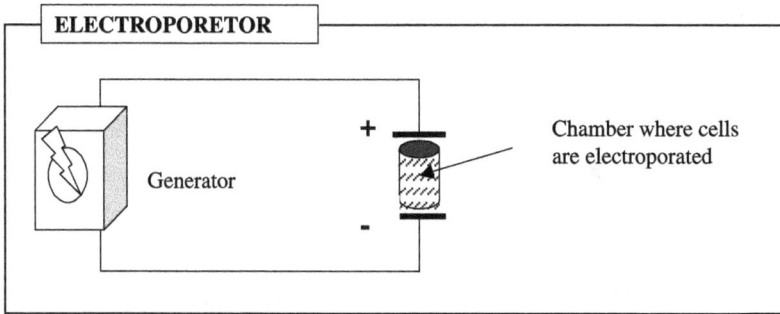

Fig. 6.9. Delivery of DNA into cells by electroporation.

Bacteria require much higher field strengths (6–25 kv) than most eukaryotic cells. For electroporation, the density of cells to be used should be approximately 1–2×10^6 cells/mL. However, this may give rise to a significant proportion of fused cells. This is particularly true with cells that do not possess cell walls (e.g. mammalian cells). This process is called "*electrofusion*".

The electrocompetent cells are prepared in the same way as the cells prepared by chemical competence, except that only water or a 10% glycerol solution is used throughout the preparation. As with chemical methods, the efficiency will vary between different batches of cells, and between different strains. In the preparation of electrocompetent cells, all centrifugations must be gentle and all manipulations must be carried in the cold (0–4°C).

(3) Transfection of Bacterial Cells

There are two ways by which recombinant DNA in a phage vector can be introduced into bacteria: transfection and *in vitro* packaging. Transfection is a process equivalent to transformation, except that the DNA to be transferred is a phage DNA rather than a plasmid. As in the transformation process, the phage DNA is incubated with the competent bacterial cells and the DNA uptake induced by heat-shock. Alternatively the phage recombinant DNA can also be packaged *in vitro*, using the necessary purified phage proteins, to form mature phage particles, which can now infect host cells. This method of genetic transfer is more efficient than transfection with the recombinant phage DNA. When the infected cells are spread onto a solid agarose media, cell lysis caused by the propagated phage particles, can be visualized as plaques on a lawn of bacteria.

6.2.4. *Other hosts*

Beside *E. coli*, the other commonly used hosts include yeast, plant, insect and mammalian cells in culture. These hosts are eukaryotes and are generally used for the expression of specific proteins rather than for propagating recombinant DNA. Protein expression and the controls involved will be discussed in the latter part of the chapter.

Unlike prokaryotes, eukaryotes can carry out a number of post-translational modifications of expressed proteins. These include:

1. Proteolytic processing of precursor proteins.
2. Specific glycosylations.
3. Post-translational modifications of amino acids.
4. Correct disulfide bond linkages.

Precisely how an expressed protein is modified post-translationally is highly dictated by the type of eukaryotic host used. Hence, with certain proteins, the correct choice of a host is essential.

(1) Yeast as Expression Systems
The common baker's yeast (*Saccharomyces cerevisiae*) has been used extensively as a host to express cloned eukaryotic genes. There are a number of reasons for this. This is a single cell and is extremely well known genetically (the whole genome has been completely sequenced) and physiologically. It is easier to manipulate and can be grown in both small culture vessels and large-scale bioreactors. There are several strong promoters and naturally occurring plasmids that have been isolated and well characterized. This organism is capable of carrying out some post-translational modifications. The most important factor is that this strain of yeast has been in use in baking and brewing industries, and has been listed by the U.S. Food and Drug Administration as a "generally recognized as safe" (GRAS) organism. Thus, the use of this organism for the production of human therapeutics (e.g. drugs or pharmaceuticals) does not require as much verification as compared to unapproved host cells. A

number of proteins have been produced in this host including, Hepatitis B and HIV vaccines, insulin, blood coagulation factor XIIIa and others.

Other strains of yeast are also used for the purposes of recombinant protein production. They include *Kluyveromyces lactis* (used commercially to produce lactase), *Schizosaccharomyces pombe*, *Pichia pastoris* and *Hansenula polymorpha*.

(2) Cultured Insect Cells as Expression Systems

Fruit fly (*Drosophila melanogaster*) and baculoviruses are currently used to produce high amounts of recombinant proteins. In the former case, a number of cell-lines (e.g. S2) have been developed and have attained immortalized non-tumorgenic growth. These cells in culture can then be transfected with vectors with the appropriate promoters.

Baculoviruses infect invertebrates, including many insect cells. This virus does not infect humans and hence, is explored as a potential host to produce therapeutic proteins. As this virus harbor a large genome, a number of steps is involved in generating recombinant baculovirus which will then be used to infect insect cells. The infected cells in culture will then produce the protein of interest.

(3) Plants as Expression Systems

There are good reasons to develop plants that express foreign substances. The addition of certain gene(s) often improves the agricultural, horticultural or ornamental value of a crop plant. These "transgenic" plants can act as living bioreactors for the inexpensive production of many economically important metabolites or proteins. There are a number of effective DNA-delivery systems and expression vectors that can be used on a range of plant cells.

(4) Mammalian Cells as Expression Systems

Immortalized mammalian cells grown *in vitro*, has been invaluable in the studies of gene expressions and mechanisms underlying many diseases (e.g. cancers). Many mammalian cells from a variety of body tissues have been successfully grown in plastic wares bathed in specialized liquid media in controlled environment ($37°C$; 5% CO_2/95% O_2), often supplemented with bovine (cow) serum. After a defined number of generations, these cells die in a programmed fashion. Occasionally, some cells escape death and they thrive in the environment and begin to develop immortality. This does not necessary mean that the cells are "cancerous". A cell-line can stay immortal and not be a tumor. Tumorigenic cells can stay immortal in culture and when transplanted into immune deficient mice, develop into full-blown cancers. There are a large number of mammalian cells used for molecular genetic applications. Some of these are also used to produce recombinant proteins.

6.3. Vectors

Plasmids are widely distributed throughout prokaryotes. These are self-replicating entities (replicons), independent of the chromosome. They can be stably inherited in an extrachromosomal state, but are generally dispensable. However, not all extrachromosomal nucleic acids are necessarily plasmids. Some of the heterogeneous circular molecules found in *Bacillus megaterium* and the replicative forms of the filamentous phages (M13, Fd, f1) are not plasmids. However, parts of these replicative forms have been incorporated into the backbones of many vectors.

Most plasmids exist as *double-stranded* circular DNA. The replication of all plasmids is, to an extent, dependent on their host. Plasmids also exhibit selective replicative capabilities in specific hosts. This is referred to as the host-range of the plasmids. Therefore not all hosts can harbor the same plasmids.

The development of recombinant DNA technology has enabled the modifications of existing plasmids to improve their use as cloning vectors. There are many types of cloning vectors including plasmids, fosmids, bacteriophages and others. They all have common features that make them suitable for use as vehicles for genetic engineering. These vectors can also be easily modified to enable them to express heterologous proteins efficiently.

6.3.1. *Some general requirements for a cloning vector*

1. All vectors must be able to replicate within a host cell if these are to be used as "shuttle vectors". Some episomal vectors (vectors that are not integrated into the chromosomes) do not seem to replicate

in mammalian cells (inappropriate origin of replication), but can be effectively used to express recombinant genes. Some mammalian expression vectors, however, do replicate under intense selection pressures.

2. There must be efficient ways of introducing vector DNA into a host cell.
3. There must be ways to isolate the vector DNA from the host DNA.
4. The presence of specific restriction sites in the vector that can be used for cloning an insert DNA.
5. The vector should have at least a selectable marker in the vector to indicate the presence of the vector in host cell. This can be antibiotic or other xenobiotic markers, or merely reporter groups, for e.g. luciferase, green fluorescent protein (GFP) and β-galactosidase.
6. It is desirable to have unique restriction enzyme sites within a selectable or screenable marker so that the presence of a DNA insert in the cloning vector can be selected or screened.

A number of these vectors are available commercially. They can be used for cloning in prokaryotes, yeast, baculovirus, mammalian and

Table 6.4. Some examples of vectors and their maximum insert sizes.

Vectors	Size of inserts (kb)
Plasmids (pBR322)/Phagemids (pGEM, pBSK)	0–10
Bacteriophage λ (insertion vectors; e.g. λgt10, λgt11, Lambda Zap, Unizap XR, Zap-express)	0–10
Bacteriophage λ (replacement vectors; Lambda Fix II, Lambda DASH II, EMBL, Charon)	9–23
Cosmid vectors (SuperCos 1, pWE15, COS)	25–45
Bacteriophage P1	60–100
PAC (P1 artificial chromosomes)	100–150
BAC (bacterial artificial chromosomes)	100–300
YAC (yeast artificial chromosomes)	200–2000
HAECs	
MACS	

other cells (e.g. plants). These vectors come in many different names. They are designed to have specific functions (host specific, expression, recombination) and can encode inserts of different sizes (Table 6.4).

6.3.2. Vector systems

Vectors can be classified as "shuttle vectors" and "expression vectors". Shuttle vectors are used generally for the propagation of DNA. The prokaryotic vectors for carrying DNA inserts, which are of high copy numbers, are regarded as shuttle vectors. By simple manipulations, these vectors can be made to express proteins as well. The distinctions between these two types are gradually eroding, as new vectors that can be used for both propagation and expression have been designed. Some of these vectors are packaged into viruses for high efficiency gene transfer (infection as opposed to transduction or transformation).

There are basic features and principles that all vectors rest on. The vectors available today are rather versatile, with many variations of a common theme. The designs of eukaryotic vectors have evolved from those found in prokaryotic systems, often borrowing and wedding specific functions. The earlier plasmids are derivatives of naturally occurring sequences that are host compatible, e.g. pBR322. Nowadays, many commercial companies, including Promega, Invitrogen, Clontech, Stratagene and others, have spliced together sequences from phage DNA and plasmids, giving rise to a modified type of plasmid called *phagemid*. Phagemids are therefore plas*mid*s that have been synthetically manipulated to contain a small segment of sequences from a filamentous *phage* (M13, fd or f1).

6.3.3. Prokaryotic cloning vectors

A generic prokaryotic cloning vector has some of the following common features:

1. An origin of replication: *ori*.
2. An antibiotic resistance gene: e.g. ampicillin, kanamycin, tetracycline, etc.

3. A selectable marker: e.g. color selection with β-galactosidase (*lacZ*).
4. An available multiple cloning sites (MCS).
5. Other accessories:

    (i) Promoters for *in vitro* syntheses of RNA: SP6/T7/T3.
    (ii) f1 origin to allow production of single-stranded DNA.
    (iii) MCS sequences that allow unidirectional deletion of inserts.

Let us examine the structures of two commonly used vectors in the laboratory that are commercially available, Bluescript (pBSK; Stratagene) and pGEM (Promega) (Fig. 6.10). They are known as phagemids, and are designated as **pBSK(+/−)** and **pGEM-XZf(+/−)**, where **X** denotes vector number, **f** the presence of phage origin of replication, and **(+/−)** the orientation of the single-stranded DNA produced.

(a) **pGEM series**. This series comprises the vectors pGEM-3Z, pGEM-4Z, pGEM-3Zf(+/−), pGEM-5Zf(+/−), pGEM-7Zf(+/−), pGEM-9Zf(−), pGEM-11Zf(+/−) and pGEM-13Zf(+). The parental plasmid for this series is the pUC vector. These vectors contain the pMB1 origin of replication. The early series of pGEM vectors (e.g. pGEM-3Z and -4Z) do not contain an f1 origin for the production of single-stranded DNA. This feature, plus the addition of new MCS sequences, is incorporated into the later generation of pGEM vectors to allow for increased versatility and convenience in gene manipulations.

(b) **pBIISK(+/−)/pBIIKS(+/−)**. Both are the same vectors, except that the orientation of the MCS site has been reversed between the T7 and T3 promoters. These phagemids contain the ColE1 of origin of replication. The pBIISK vectors are derivatives of pUC19 (which is a derivative of pBR322) and, like the pGEM vectors, replicate in high numbers because of the lack of the pBR322 *rom* gene.

(c) Both pGEM and pBIISK/KS vectors use ampicillin resistance as a selection marker. Therefore transformants will acquire resistance to ampillicin.

(d) Both vectors have phage promoters; T7/SP6 in the pGEM series, and T7/T3 in pBIIKS/SK. The promoter sequences flank the insert and contain all the information required for the initiation of RNA transcription by T7, T3 or SP6 RNA polymerase. These polymerases

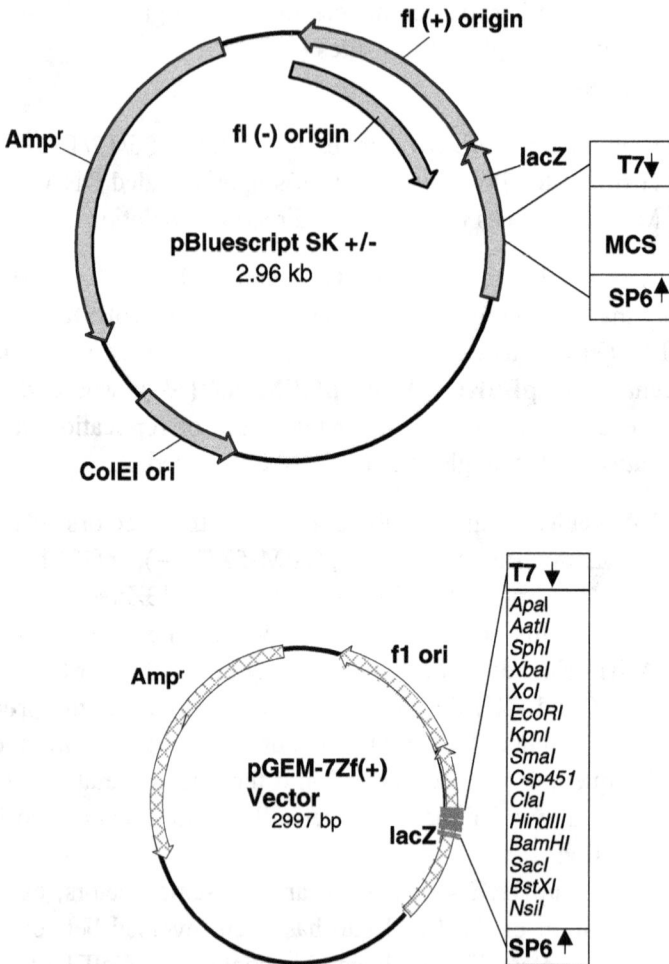

Fig. 6.10. Vector Maps of pBluescript SK (Genbank #X52528 [SK(+)] and #X52330 [SK(−)]) and pGEM-7Zf(+) (Genbank X65310). Adapted from Stratagene and Promega, 2001.

do not cross-initiate and hence, can be used to produce highly specific RNA transcripts. The RNA transcripts produced can be used for a number of studies. For example, the transcripts can be used as riboprobes for nucleic acid hybridizations, as a template for *in vitro* translation of proteins, or in gene expression studies.

(e) Both vectors have two origins of DNA replication, one which is specific for *E. coli* (ColE1 for BSK and pMB1 for pGem), and the other from the single-stranded DNA phage f1. The ColE1 or pMB1 origins are used for plasmid replication. However, following phage f1 infection, the other origin is used and single-stranded DNA is produced. This is useful for single-stranded DNA sequencing and some mutagenesis strategies.

Early cloning studies were rather cumbersome. The most common strategy was to subclone the DNA fragment from a plasmid or lambda vector, into filamentous bacteriophage M13 vector for the production of a single-stranded DNA template that can then be used for sequencing. Once sequencing has been completed, the template had to be recloned into another specialized vector, such as a vector designed for efficient *in vitro* transcription, or an expression vector for protein expression. If site-directed mutagenesis experiments were to be carried out, the mutant forms of interest had to be then subcloned at each step. These rather time-consuming steps have now been eliminated though the development of vectors that possess, in a single construct many features required for specialized DNA manipulations.

(f) **Genetic markers**. Both vectors have convenient ways to select for the presence of DNA insert colorimetrically. This selection method is based on the biochemical properties of the *lac* operon (see Sec. 6.3.4).

(g) **Protein expression**. Foreign DNA encoding protein sequences can be expressed in these two vectors, as long as the sequences are in-frame to the expression of β-galactosidase gene, which is under the control of the *lac* promoter. An adaptation of this strategy is the use of pGEM-T vector where random sequences of a protein can be incorporated and expressed as a fusion protein of β-galactosidase. It is advisable to use a host with the *lac*I^q present in order to increase the yield of proteins produced.

The T7 promoter can also be exploited for protein expression. By inserting the foreign DNA encoding protein sequences under the control of the T7 promoter and adding a ribosome-binding site (RBS) plus an ATG start site, it is possible to keep the expression tightly turned off until a T7 RNA polymerase is expressed in certain host cells (e.g. BL21(DE3)).

(h) **Gene mapping**. The unique *BssH* II restriction sites flanking the
T3/T7 promoters of pBIISK offer an easy and rapid method to map
large and small DNA inserts. To do this, complete *BssH* II digestion
of the clone is first carried out, followed by partial restriction
digestion with the restriction enzymes of choice. The partial fragments
from the restriction digestion will then be separated by agarose gel
electrophoresis, transferred to a nylon membrane, probed with labeled
T3/T7 to determine the lengths of the partial fragments (see Sec. 6.4.1).
The different lengths of the digested fragments identify the location of
the restriction enzyme sites relative to the T3/T7 positions.

(i) **MCS unidirectional deletion of insert**. An alternative to the above
partial restriction map analysis is simply to exploit the use of the MCS.
If the insert DNA has restriction sites similar to that in the MCS,
complete restriction will result in the excision of some vector and
insert sequences. If sequencing is carried out from one direction, it
will then be possible to gain internal sequence information.

(j) **Unidirectional deletions with *Exonuclease* III**. The restrictions sites
have been arranged to allow unidirectional deletions of the cloned
insert with *Exo* III. This enzyme is a 3–5' single-stranded exonuclease
and require double-stranded DNA template. Hence, *Exo* III does not
digest 3' end single-stranded overhangs, but exclusively digest the
3' ends of blunt ends or 5' overhangs. Therefore, double digestion with
a 3', and a 5' or blunt end producing enzyme, followed by *Exo* III
digestion, and the subsequent removal of the remaining single-stranded
DNA by mung bean nuclease will result in unidirectional deletions of
the insert DNA. This method allows the serial deletion of large inserts
for sequencing from one direction.

6.3.4. *Lac operon: A model for the control of gene expression*

A number of cloning and expression vectors utilize the *lacZ* system as
either a selection marker or as a reporter gene. It is therefore important
to have a basic understanding of this system.

The *lac* operon found in *E. coli* is one of the best-known examples
of a bacterial operon (Fig. 6.11). Structurally it consists of 3 genes:
lacZ (coding for β-galactosidase), *lacY* (coding for a permease which is

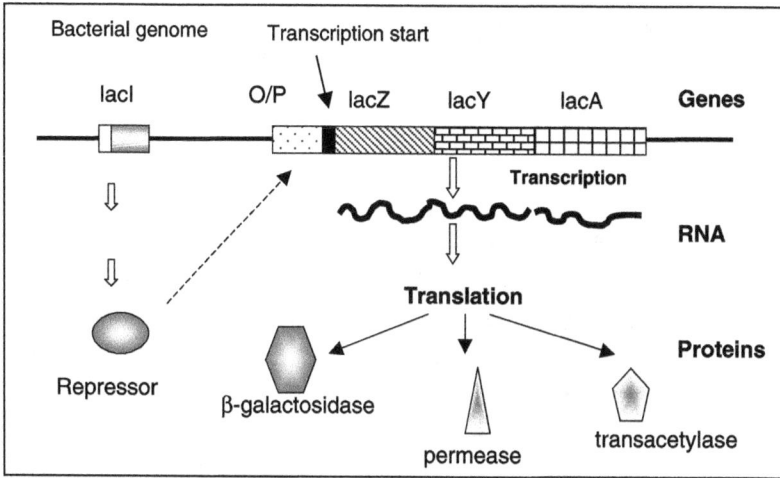

O/P = operator and promoter

Fig. 6.11. The *lac* operon.

needed for uptake of lactose) and *lacA* (coding for an enzyme called thiogalactosidase transacetylase). A regulatory region containing the promoter site is found at the 5' end, overlapping with a sequence known as the operator.

The *lacI*q repressor protein binds to the operator region that physically overlaps with the promoter (which drives the *lac* operon) and the first 20 or so bases to be transcribed, thus preventing the initiation of transcription. The *lacI*q gene is not part of the *lac* operon. The bound *lac* repressor protein has an affinity for a molecule called allolactose, which is a derivative of lactose. This binding to allolactose is allosteric and will prevent the repressor from binding to the operator region. Therefore, in the absence of lactose, the repressor is active and binds to operator, preventing transcription. However in the presence of exogenous lactose, the repressor is inactivated and the operon is then expressed. The cell then produces the enzymes that are necessary for the metabolism of lactose. As lactose will be metabolized as soon as induction begins, thus diminishing the inductive effect, it is not a good inducer of the *lac* operon. In practice, a synthetic analogue of lactose that will not be broken down easily by β-galactosidase is used instead. An analogue that is commonly used in the laboratory is isopropyl-1-thio-β-galactosidase (IPTG). Not all substances that are substrates for breakdown by β-galactosidase, can act as inducers as they are not recognized by the repressor. One example of such a chemical is 5-bromo-4-chloro-3-indolyl-β-D-galactopyranoside, commonly known as X-gal.

Some *E. coli* have a modified *lacZ* gene, lacking a segment referred to as *lacZ'* which codes for the α-peptide portion of β-galactosidase. These bacteria can only synthesize the full active enzyme when they harbor a plasmid that carries the missing *lacZ'* segment. This phenomenon is called α-complementation and is exploited in a technique used to distinguish a recombinant colony within a background of non-recombinants. In this technique, the transformed bacterial cells are plated onto solid media containing X-gal and IPTG. The active β-galactosidase enzyme breaks X-gal down to form a product that is deep blue in color. Since the MCS site resides within the *lacZ'* gene in most plasmids, insertion of a DNA fragment into the vector will inactivate the β-galactosidase enzyme, resulting in the formation of white colonies instead. This selection technique is also applicable to distinguish bacterial cells infected with recombinant bacteriophage DNA that has the *lacZ'* gene in its genome (e.g. λZAPII vector).

6.3.5. λ *bacteriophage vectors*

Two problems hinder the use of plasmids for all genetic manipulations: the severe size limitation of the insert DNA, and the efficiency of gene transfer. Although the efficiency of transformation has dramatically increased by the use of electroporation and specific chemical treatments, it remains a hindrance. As an alternative, infection as a way of introducing DNA into cells show efficiency close to unity (100%). Attention was therefore directed at bacteriophage λ as it possesses a number of desirable qualities required for cloning, a highly efficiency method of gene transfer as it infects the bacterial cells, and the ability to insert large DNA fragments (~20 kb) by manipulating the genome.

To design vectors based on this phage, it is necessary to insert the DNA of interest into the λ replicating sequences (replicon) *in vitro*. The resulting recombinant DNA must be able to infect *E. coli* at high efficiency. The latter requirement is met by developing an *in vitro* packaging system, which mimics the way in which wild-type λ DNA is packaged in a phage coat protein, resulting in high infection efficiency.

Two major strategies are used in generating λ vectors: replacement vectors and insertion vectors. As the name implies, replacement vectors are those which have some segments of its sequences deleted to accommodate the sequences of interest. It is observed that DNA molecules between the sizes of 37 to 52 kb can be stably packaged into the phage head. Exploiting this finding, removing the region of λ genome that is nonessential for growth (central portion where the genes are necessary for integration and recombination) and inserting the DNA of interest to make up the required size, foreign DNA of up to 23 kb in length can be easily incorporated. Hence, these vectors are commonly used for genomic cloning (Fig. 6.12).

For cDNA (complement DNA generated from mRNA) cloning, where the sizes of DNA fragments are generally not as large (< 10 kb), these sequences can be cloned into modified λ vectors (insertion vectors).

With an insertion vector, the non-essential segment has been deleted, and the two arms ligated together. Therefore this vector has at least one restriction site into which a DNA fragment can be inserted. The newer versions of λ insertion vectors (e.g. Lambda ZAP II vector) allow easy *in vivo* excision of insert DNA in a phagemid form. This is enormously

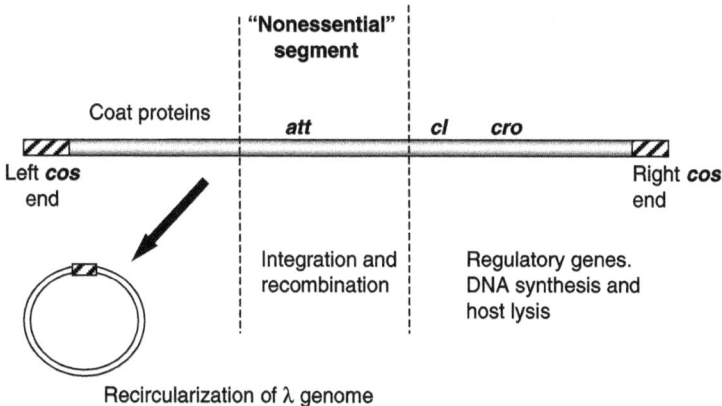

Fig. 6.12. The bacteriophage λ as a cloning vector.

convenient as it is easier to isolate and purify plasmid DNA as opposed to the purification of phage DNA.

The *in vivo* excision depends on the DNA sequences of the phagemid (BSK) and the presence of a variety of proteins, including filamentous phage (M13)-derived proteins. The M13 phage proteins recognize the f1 bacteriophage "origin of replication" found within the pBluescript sequences. This f1 origin of replication can be divided into 2 overlying parts: (a) the site of initiation (I), and (b) the site of termination (T) for DNA synthesis. These two regions have been subcloned separately into the lambda ZAP II vector. The lambda phage vector is made accessible to the M13-derived proteins by simultaneously infecting a particular strain of *E. coli* (the SOLR strain) with the lambda phage vector and the bacteriophage. The "helper" proteins (from M13) then recognize the initiator DNA that is within the lambda vector. One of these proteins nicks one of the two strands of DNA and a new DNA synthesis begins, duplicating the DNA sequences downstream of the nick site. This synthesis continues through the cloned insert until the termination signal is encountered. The single-stranded DNA molecule is then circularized by the gene II product of the M13 phage. This results in a circular DNA molecule containing the DNA sequences between the initiator and terminator sites.

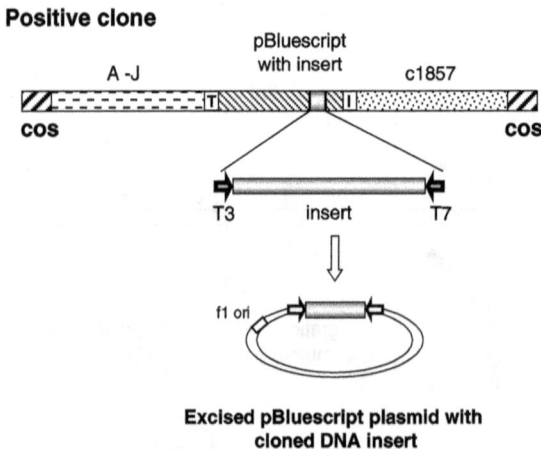

Fig. 6.13. The newer versions of λ insertion vectors (e.g. Lambda ZAP II vector) allow easy *in vivo* excision of insert DNA in a phagemid form.

In this case, this includes all sequences of the pBluescript and the insert DNA (Fig. 6.13).

6.3.6. *Other vectors*

(1) Cosmid Vectors
Effective packaging of the λDNA molecules can only occur if the *cos* sites are present. It is now known that only a small region around the *cos* site is required for recognition by the packaging system. Plasmids have been constructed containing a fragment of the λDNA including the *cos* site. These are called *cosmids* and are used in conjunction with the *in vitro* packaging system. The packaging into the phage head requires a large amount of DNA. If the vector is only 5 kb in length, then to be incorporated into the phage head, an insertion of between 32 to 47 kb of DNA (gene of interest) is required. This large amount of DNA cannot be inserted into any of the known λ vectors. Note that after packaging *in vitro*, the phage can infect the appropriate host. The recombinant cosmid DNA is injected and circularizes like a phage DNA, but replicates as a normal plasmid without expressing phage proteins. Selection of transformants is done on the basis of a vector-drug-resistance marker carried on the plasmid portion.

(2) Bacterial Artificial Chromosome (BAC) Vectors
Most of the commonly used vectors for propagating small inserts have high copy numbers (discussed in other sections). One disadvantage of possessing high copy numbers is that these vectors often show structural instability of inserts (especially for large inserts and those with repeated sequences common in higher eukaryotes). Thus, this limits the use of these high copy number bacterial vectors.

To overcome these limitations, attention has recently been focussed on vectors that exist in low copy number replicons, e.g. *E. coli* fertility plasmid, the F-factor. This plasmid occurs naturally in many strains and is maintained at a copy number of 1–2 per cell. Vectors based on the F-factor system can accept large foreign DNA fragments (>300 kb). Using electroporation, it is possible to introduce these recombinants into *E. coli* efficiently. However, it must be noted that the copy number will be low within the cells. The consequence of this is that, to prepare these vectors, efficient isolation procedures must be used.

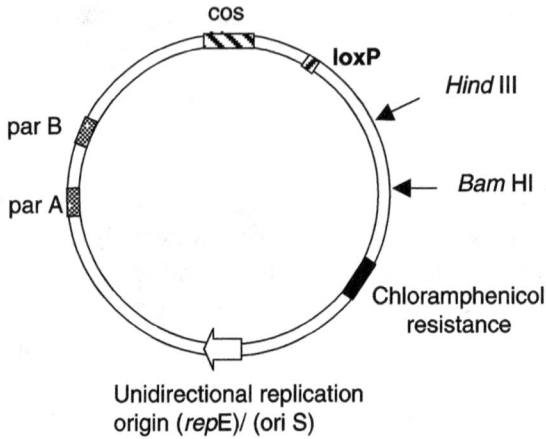

Fig. 6.14. A BAC vector.

An example of a BAC vector derived from a mini-F plasmid is shown in Fig. 6.14.

(3) Yeast Artificial Chromosomes (YACS)
As the upper size limit of cosmid cloning is about 35,045 kb, to cover the whole human genome in a theoretical contiguous chain would require about 70, 000 clones. What is required is the development of vectors that can accommodate insert sizes even bigger than in the cosmids. This will reduce the number of clones to be handled. The minimum structural element for a linear chromosome is the presence of an origin of replication (autonomous replicating sequence, *ars*), telomeres and a centromere.

A linear molecule that behaves like a chromosome in yeast has been constructed by combining an *ars* and a centromere (both from yeast) with telomeres from Tetrahymena (Fig. 6.15). These YACS allow the insertion of large pieces of foreign DNA (~1 Mb) into this linear artificial chromosome.

There are a number of operational problems with YACS. The existence of chimeric DNA segments from different regions of the genome into a single clone is one of such problems. Another problem is that of instability, which tends to delete internal regions in the inserts. Other problems include the difficulty of removing the yeast host chromosome background from the

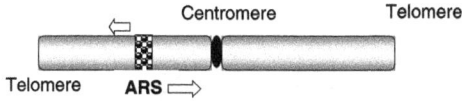

Fig. 6.15. Yeast artificial chromosome as a cloning vector.

YACS and low yield of DNA from YACS. A number of these concerns are currently being addressed.

(4) Shuttle Vectors

Gene cloning and associated techniques are also extremely valuable for investigating all aspects of the behavior of a wide range of bacteria. *E. coli* genes are often expressed satisfactorily in other Gram-negative organisms, but the reverse is usually not true. Genes from *Pseudomonads*, for example, are only poorly expressed in *E. coli* and have to be analyzed in their original hosts. In many cases, the techniques available for these organisms are limited. For example, transformation (or electroporation) frequencies may be relatively low, and the methods for screening gene libraries may be difficult. It is therefore advantageous to carry out the initial cloning of a gene using *E. coli* as the host organism. After the desired DNA construct has been obtained, it can be recovered and then inserted into the target host. This procedure requires a vector, which can replicate in either organism. Certain plasmids are capable of existing in a variety of hosts and can function as shuttle vectors. A shuttle vector carries replication origins for two different plasmids on the same vector (Fig. 6.16). This will facilitate preliminary manipulations in a well-characterized system, such as *E. coli*, and then transfer the clone to the organism of interest. If the clone is to be expressed in both hosts, then it may also be necessary to provide two promoters as well.

(5) Broad Host Range Vectors

Broad host range plasmids can be used as an alternative to dual-origin shuttle vectors. These plasmids belong to two incompatibility groups, P and Q/P4. P group plasmids are usually derived from *Pseudomonads* and the prototype of this group is RP4 (also known as RP1 or RK2). RP4 is a broad host range plasmid and is able to replicate in some Gram-positive

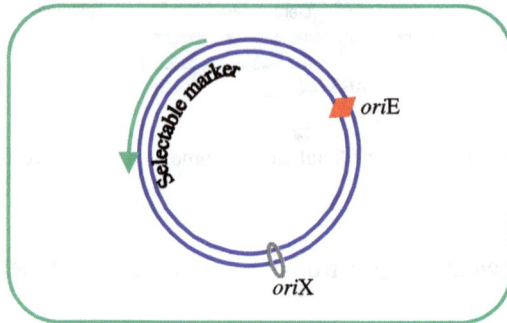

In *E. coli*, plasmid replication is
initiated using *ori*E.

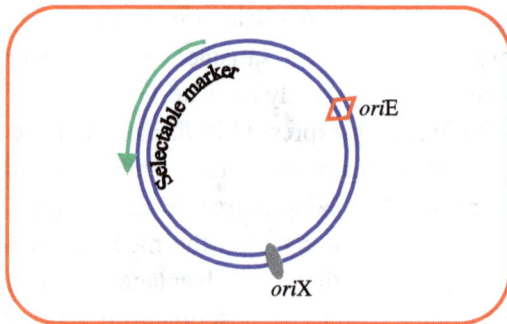

In the alternative host, the plasmid
replication is initiated using the second
origin of replication, *ori*X.

Fig. 6.16. Shuttle vectors with dual origins of replication.

bacteria and in most Gram-negative bacteria. The P group of plasmids
contain replication and transfer functions which are widely dispersed on
the bacterial chromosome, thus making it extremely difficult to construct
lower molecular weight derivatives which could be used as cloning
vehicles. One example is pRK2501, which is 11 kb in length, but lacks all
transfer and mobilization functions. Since this vector is transfer-deficient
and non-mobilizable, it satisfies important requirements for the biological
safety of vectors with broad host-range. A system, which employs two
plasmids, has been developed for cloning in *Rhizobia* (Fig. 6.17). The first
plasmid, pRK290, is the cloning vehicle which is transfer-deficient but still

The donor *E. coli* cells contains plasmid pRK2013 with RK2 transfer functions and ColE1 replicon. Replication of pRK2013 is restricted to *E. coli*.

pRK2013 is transferred into recipient *E. coli* containnng recombinant pRK290 which carries the RK2 replicon and confers broad host specificity.

pRK2013 mobilizes pRK290 and allows it to be transferrred into the target recipient.

Target recipient containing recombinant pRK290 plasmid.

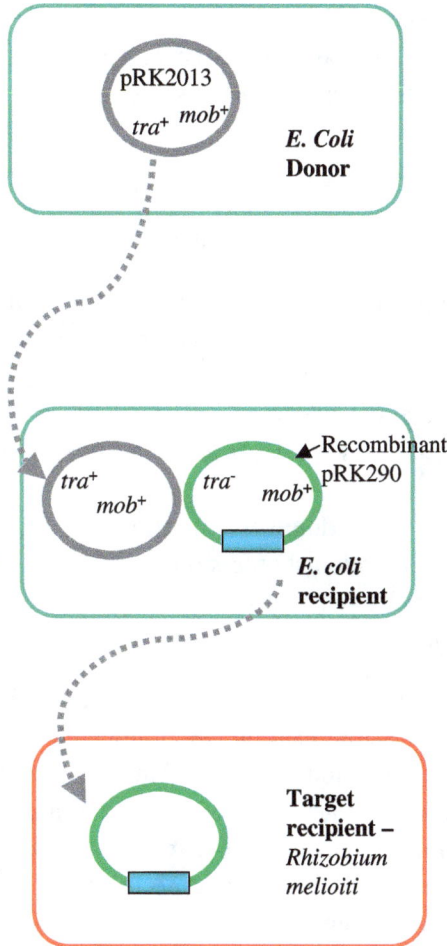

Fig. 6.17. Use of broad host specificity plasmids for gene transfer into other host organisms.

possess all mobilization functions (Tra⁻ Mob⁺) and can replicate equally well in *E. coli* and *Rhizobium* because it contains the RK2 replicon. This vector can be mobilized by a second plasmid residing in the donor cells, and hence can be transferred to suitable recipient cells, e.g. *Rhizobium*. The mobilizing plasmid in the donor is pRK2013, which possesses all transfer functions and also contains the neomycin resistance region of RK2 and the ColE1 replicon, which restricts its replication in *E. coli* but not in

Rhizobium. When this plasmid enters *E. coli* cells, the transfer functions of pRK2013 are expressed and mobilize plasmid pRK290, which can replicate in *Rhizobium* cells.

6.3.7. *Prokaryotic expression system*

In general, expression vectors should have the following properties:

1. **A promoter**, which marks the start of gene transcription. This sequence is recognised by the RNA polymerase.
2. **Inducible elements**, which provides the ability to control the level of protein expression.
3. **Accessory elements for stability** allows for the maintenance of a stable transcript.
4. **Transcriptional termination**, which marks the end of a gene where transcription should stop.
5. **Affinity tag** that allows the protein to be easily isolated and purified (optional).

The promoter is the most important factor of an expression vector. In all expression vectors, the cloning site is located just downstream from a promoter and ribosome binding site (RBS), allowing heterologous regulation of gene expression. The important sequences for *E. coli* RNA polymerase are the TTGACA and TATAAT (Pribnow box) sequences found at position -35 and -10, respectively. The most commonly used promoters are those from the *lac* or *trp* operons, a combination of the two called the *tac* promoter (*trp* and *lac*), and the λ_{PL} promoter from phage λ. These promoters are inducible and, in the case of the *lac* promoter, transcription or derepression of the gene is brought about by the addition of IPTG. The ability to control levels of expression is important: continuous high levels of expression may affect plasmid stability or the protein product may be toxic to the host organism.

Efficient translation of mRNA in prokaryotic cells cannot occur without a ribosomal binding site (RBS). This includes an initiation codon, AUG, together with a consensus sequence, AGGAGGU, which typically lie 3–12 bases upstream of the AUG. This is the Shine-Dalgarno sequence (SD). The sequence of the RBS, together with the secondary structure of the mRNA, has an effect on translation efficiency of mRNA. To obtain expression of foreign genes in *E. coli*, it is necessary to incoporate a RBS into the recombinant DNA construct. Furthermore, the SD sequence must be located at the optimal distance from the translation start codon.

More sophisticated expression vectors, with more efficient expression signals, are used when it is necessary to obtain very high levels of formation of the product of the cloned gene, for example in commercial processes. Extremely high levels of specific protein, representing perhaps 50% of total cell protein, can be produced using such a vector. However, diverting so much of the cell's resources into a product that is useless as far as the cell is concerned will slow down the growth rate (which will also exacerbate plasmid instability). The usual strategy for overcoming this problem is to use a promoter that is subject to repression, so that the bacteria can be grown to high cell density before product formation is induced by altering the growth conditions.

The choice of a promoter will depend on the requirements of the expression. For example, if it is desired that the cloned DNA be regulated to prevent inhibition of host cell growth, the *trp*, *lac* or λ promoters might do well. The appropriate repressor molecule is supplied as a regulator. The *tac* promoter, a combination of the −35 region of the *trp* promoter and the −10 region of the *lac* promoter, also has a high efficiency of expression. Many phage promoters are also very efficient for in protein expression. For example, the phage T5 promoter out-competes all *E. coli* promoters, thus allowing the expression of large amounts of the recombinant protein in *E. coli*.

An example of a prokaryotic expression vector that is designed for high protein expression and easy purification is the pQE-9, which belongs to the pQE series of vector (Fig. 6.18).

This vector uses the phage T5 promoter, which can be recognized by the *E. coli* RNA polymerase. The incorporation of a double *lac* operator repressor system provides a tightly regulated, high-level expression of recombinant proteins in *E. coli*. In addition, the vector contains a 6xHis tag segment that will allow easy purification of the expressed recombinant products *in vitro*.

Phages such as T3, T7 and the *Salmonella* phage SP6, code for their own RNA polymerases. Each of these polymerases specifically recognizes

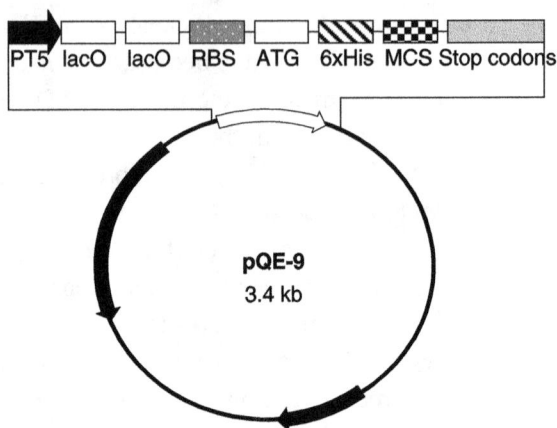

Fig. 6.18. An example of a prokaryotic expression vector that is designed for high protein expression and easy purification. Adapted from Qiagen 2001.

its own particular promoter and ignores those of other phages or the host bacterium. Thus, by using such a phage promoter to drive the expression of a foreign gene, the expression of recombinant protein will be absolutely dependent on the availability of the phage RNA polymerase. By regulating the availability of the phage RNA polymerase using an inducible promoter, such as the λ_{PL} promoter, which is repressed by providing the cI repressor *in trans*, allows the expression of the recombinant protein to be tightly regulated.

Sometimes, a particular mRNA transcript made from cloned DNA is poorly translated in the host cell. This lack of efficiency may be due to an increased susceptibility to degradative enzymes or to the use of rare codons in the mRNA. The only solution for the latter problem is to change the host cells to one that has more suitable pattern of codon usage or else to use site-specific mutagenesis to change the codons themselves. In the case of the former problem, changing the structure of the mRNA molecule itself can often prevent excessive degradation. In particular, mRNA stability can be enhanced by adding sequences such as REP onto the end of the cloned DNA so that a suitable hairpin loop is formed. The loop prevents access of the exonuclease to the end of the mRNA.

6.3.8. *Isolation and purification of plasmids*

An obvious prerequisite for cloning in plasmids is the ability to purify the plasmid DNA. The strategy is to release the plasmid DNA from the cell and then it purify away from the genomic DNA and cell debris. The step that demands the most attention is the initial steps of cell lysis and neutralization of the reaction. The incomplete lysis of the cells will reduce the recovery and purity of the plasmid DNA. It is important to note that cell lysis must be carried out with minimal physical force. After neutralization, the chromosomal DNA of high molecular weight can be removed together with the cell debris, by high-speed centrifugation, to yield a clear lysate where the plasmid resides. There are a number of ways of isolating plasmids and many of these rely on the alkaline lysis procedure that is both simple and effective.

(1) Alkaline Lysis
This is largely based on the observation that there is a narrow range of pH (12–12.5) within which linear DNA, but not supercoiled DNA, denatures (Fig. 6.19). However, prolonged exposure to this pH will eventually denature even the supercoiled DNA. The method is rather simple but highly effective, even with large vectors like BACs. Plasmid containing cells are first lysed in high pH and SDS. The lysate is then neutralized. The chromosomal DNA renatures and aggregates to form an insoluble network. The high concentration of acetate also causes the protein-SDS complexes (and excess dodecylsuphate) to precipitate. Provided the pH of the alkaline lysis step is carefully controlled, super-coiled plasmids will remain in the native state. The insoluble precipitates are then removed by centrifugation. The plasmid can be purified by extracting any remaining bacterial proteins using a mixture of phenol and chloroform (organic phase). The purified plasmid, which is retained in the aqueous phase, is then recovered by precipitation using salt and ethanol. This process is known as ethanol precipitation.

(2) Boiling Method
This is based on the principle that the supercoiled DNA is not denatured as much as the chromosomal DNA. When the extract cools down, the chromosomal DNA will reanneal but due to its complexity will form

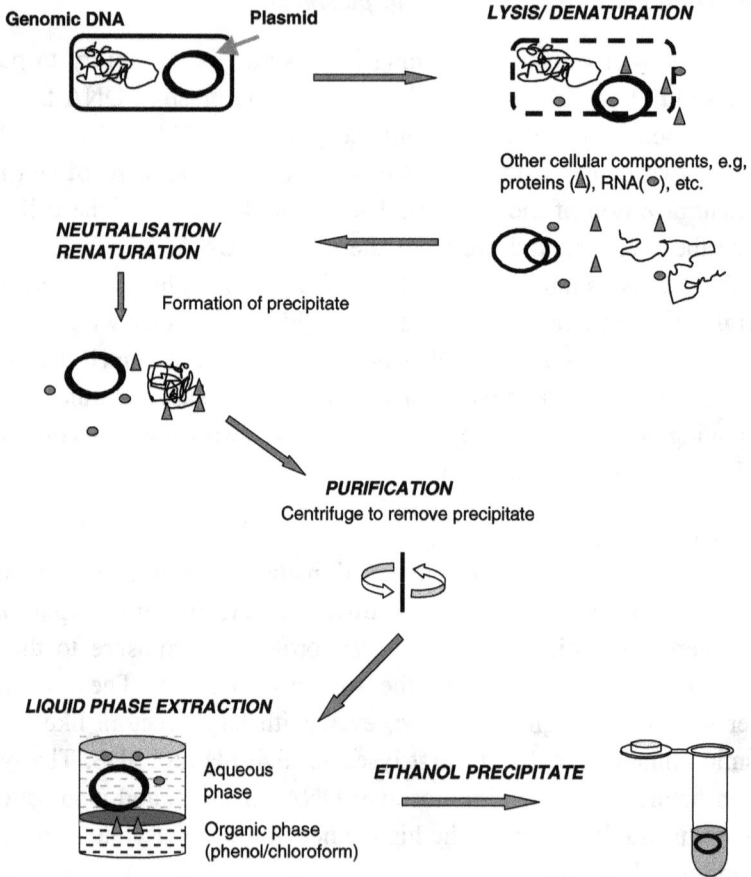

Fig. 6.19. Isolation of plasmid DNA using alkali lysis.

insoluble "tangles" and hence precipitate. As in alkaline lysis, this can be removed by centrifugation. Although this method is similar to the above method, it is not highly reliable and is dependent on host strain.

6.4. Identification of Clones

The success or failure of a cloning experiment relies mainly on the ability to identify and select for the particular clone of interest. This is especially true if a large collection of recombinant DNA representing the entire genome of an organism (a genetic library) has been cloned. Once this hurdle is overcome and a clone of interest is obtained, the scientist is then able to use a wide number of techniques to explore and extract information from the gene.

There are many ways of identifying and analyzing DNA often with the ultimate objective of gaining an insight into its sequence and function. This section describes some of the methods commonly used to isolate, identify and confirm the clone carrying the gene of interest. One of the simplest but powerful methods to identify the correct recombinant DNA directly is by a technique known as **hybridization probing**. This technique forms the basis for methods such as colony and plaque lifts, as well as Southern blot hybridization. The piece of unknown DNA can also be identified by PCR using specific primers. The DNA can eventually be confirmed by a technique called DNA sequencing.

6.4.1. *Hybridization probing: Its use in identifying specific clones*

Hybridization of nucleic acids immobilized on filters occupies a central role in molecular biology. This method has been used extensively in the detection of gene expression and identification of genes. This is based on the properties of strand separations (denaturation), the ability to immobilize nucleic acids on solid matrices and the ability to **probe** for the gene of interest by the formation of a stable duplex between complementary nucleic acid sequences.

In filter hybridization, nucleic acids are bound to a special filter (membrane) in such a way that intramolecular reannealing is prevented, but yet allows the formation of a hybrid with a single-stranded probe added in solution. The probes are incubated with the filter-bound nucleic acids (targets) under conditions favoring the formation of duplexes. Unbound probes are then washed away and the probe-target hybrids are visualized. We shall illustrate the use of this technique in the examples below.

(1) Colony Blotting and Plaque Lift Assays
Nucleic acid hybridization can be used to identify the DNA of interest contained in either a bacterial colony or a phage plaque. This method has been extensively used to screen cDNA and genomic libraries, constructed

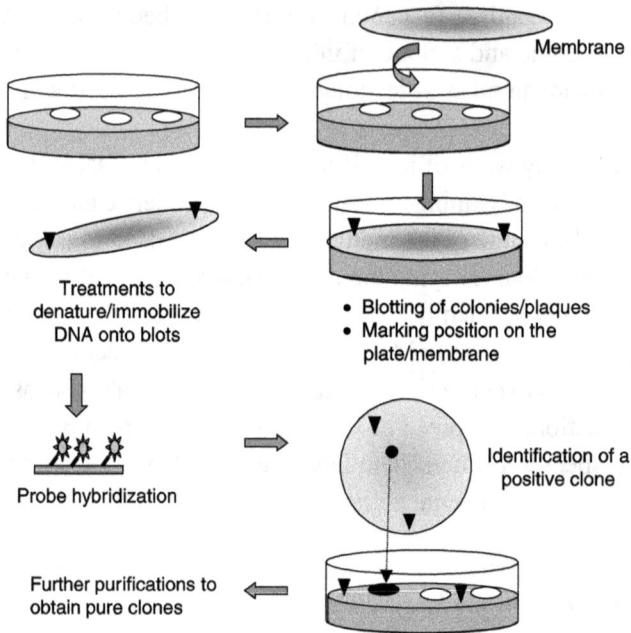

Fig. 6.20. Identification of positive clones by colony transfer onto a membrane and hybridization probing.

in either bacteria (colony blot) or bacteriophages (plaque lifts), for clones carrying the desired genes of interest.

The first step in trying to identify the clones of interest within a library is to transfer the colonies or plaques onto either nitrocellulose or nylon filter membranes (Fig. 6.20). The filter membranes are placed on the bacterial colonies or plaques and marked with Indian ink on the plates for orientation purposes. After a few minutes, colonies/plaques would have been blotted onto the membranes. The membranes are then lifted off the plates, denatured and treated so that the DNA released from the colonies or plaques are immobilized onto the membranes. The single-stranded DNA can then be tightly bound to the membrane by baking it at 80°C if a nitrocellulose membrane is used, or irradiating the membrane with UV if a nylon membrane is used instead.

Next, to locate the DNA of interest within a library of DNA, the membranes are then probed using a labeled piece of complementary DNA, in a solution of chemicals that will promote nucleic acid hybridization. Positive colonies or plaques containing related sequences to the probe used are then identified and subsequently rescued from the agar plates.

In this process, unrelated colonies or plaques in close proximity to the positive clone will inevitably be rescued simultaneously. Further serial dilutions and screenings of the colonies or plaques are therefore necessary to isolate pure clones. These clones are then amplified for the isolation and purification of the insert DNA. Finally the DNA fragments can then subcloned into other vectors of choice for further characterizations, for example restriction mapping and sequencing.

(2) *In vitro* Labeling of Probes

The probes used to identify target sequences can be in the form of double-stranded DNA, single-stranded DNA (synthetic oligonucleotides) or as RNA (see Sec. 6.3.3(iv)). These probes can be labeled either internally or at the terminus end. Two competing methods can be used to generate these probes: radioactive and non-radioactive. The principles underlying both methods are the same except for the differences in the tags and detection methods used.

In general, long probes are usually labeled internally to achieve high specific activities. These probes generally provide the strongest hybridization signals. A major disadvantage of internally radiolabeled probes is that it suffers from radioautolysis. Non-radioactive labeling detection methods are becoming more acceptable and in some cases, can replace radioactive methods. However, this is not without limitations. More steps are generally required for non-radioactive methods and this may introduce more errors. In addition, more optimization of the experimental conditions is required. A major advantage of using non-radioactive methods is that it is not hazardous and the probes generated can be stored indefinitely.

Below are two examples of the commonly used technique for specific probe labeling. The principles are essentially the same for both radioactive and non-radioactive formats. All methods listed here are commercially available as kits.

(a) Nick translation

The term "translation" in this method must not be confused with the translation of mRNA into proteins. Translation here refers to the progressive extension of the newly made strand of DNA by the polymerase.

Single-stranded nicks (internal cleavage of one strand of the duplex DNA) are first generated at random in the double-stranded DNA using low levels of DNAase I. DNA polymerase I is then added. At the nicked sites, the 5' to 3' *exonuclease* activity of DNA polymerase I remove nucleotides from the 5' end of the nick, while the 5' to 3' *polymerase* activity adds dNTP and labeled dNTP to the 3' end (see Sec. 6.1). Hence, the nick is translated downstream, with the labeled nucleotides replacing the nucleotides located between the original and the new position of the nick. This results in a nick-translated probe that is labeled on both strands with interspersed sequences of the newly synthesized labels and the original (unlabeled) DNA. This method works best with probes of 500 base pair or more. The incorporated dNTP can either be radiolabeled nucleotides or chemically modified non-radioactive nucleotides (Fig. 6.21).

(b) Random priming

This uses a large excess of random hexamers (or longer oligomers) as nonspecific primers. Both single-stranded or double-stranded DNA is first denatured. Random hexamers (NNNNNN, where N is any nucleotides) are then incubated with the template where they will hybridize to specific sequences by chance. Next, the Klenow fragment of T7 DNA polymerase will extend the primer-template. The template strand remains unlabeled, whereas the newly synthesized strand(s) are fully labeled by Klenow. As little as 1–5 ng of template DNA is sufficient for synthesis of high-specific

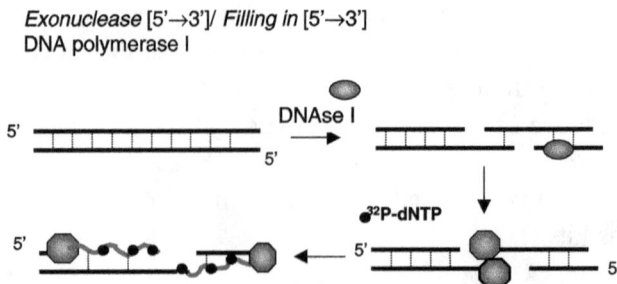

Fig. 6.21. Labeling of probe DNA by nick translation.

RANDOM LABELING METHOD

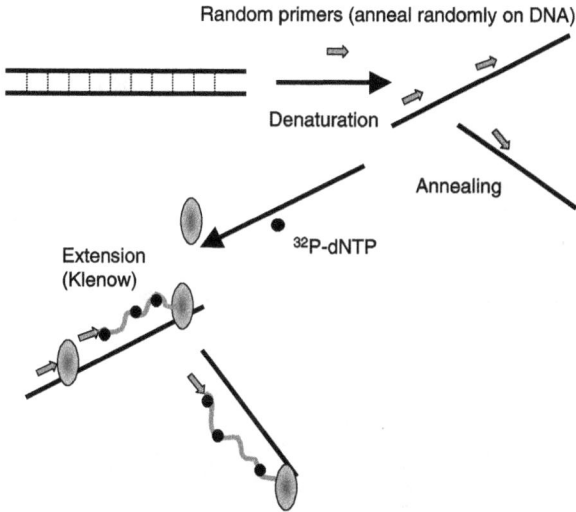

Fig. 6.22. Labeling of probe DNA by random priming.

probes. It appears that probes less than 500 bp are more efficiently labeled than longer probes. There are commercially available Klenow fragments that do not have the 3'–5' proofreading exonuclease domain. Thus, these enzymes have higher processitivity (Fig. 6.22).

(3) Southern Blot: Its Use in Confirming the Identity of a Clone
Once positive colonies have been identified by colony of plaque hybridization, and the DNA isolated and purified (Sec. 6.3.8), it is often a good practice to confirm the identity of a given piece of DNA by restriction map (see Sec. 6.1.3) followed by Southern blot analyzes. Southern blot is a method developed for the transfer of single-stranded and denatured double-stranded DNA from an agarose gel to a membrane. The blotted DNA is immobilized on the membrane but is still able to hybridize to labeled probes.

The method involves is as shown in Fig. 6.23.

The aim of treating the gels before the blotting step is to aid the transfer of large fragments of DNA onto the membrane. The rate of DNA transfer

```
┌─────────────────────────────┐        ┌─────────────────┐
│ Agarose gel electrophoresis │ ─────▶ │  Gel treatment  │
│          of DNA             │        │                 │
└─────────────────────────────┘        └─────────────────┘
                                                 │
                                                 ▼
┌─────────────────────────────┐        ┌─────────────────┐
│   Treatment of membrane     │ ◀───── │ Blotting/ Transfer │
└─────────────────────────────┘        └─────────────────┘
```

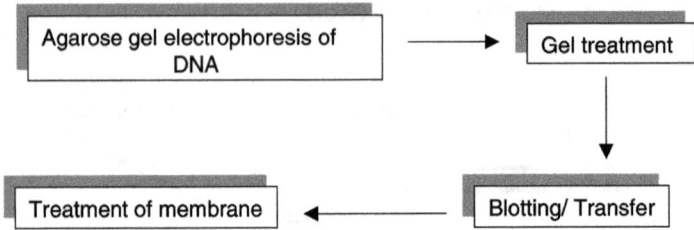

Fig. 6.23. Steps in Southern blotting.

A. Upward transfer of DNA from gel to membrane.

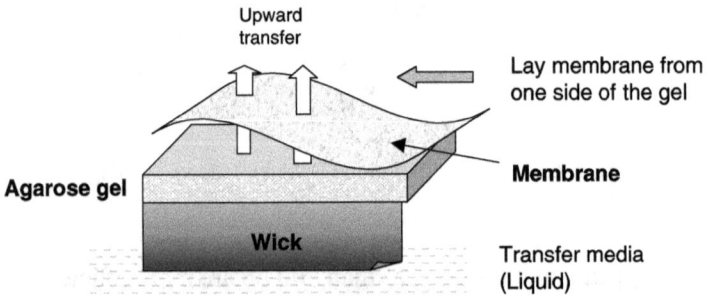

B. Assembly of an "upward" transfer southern blot

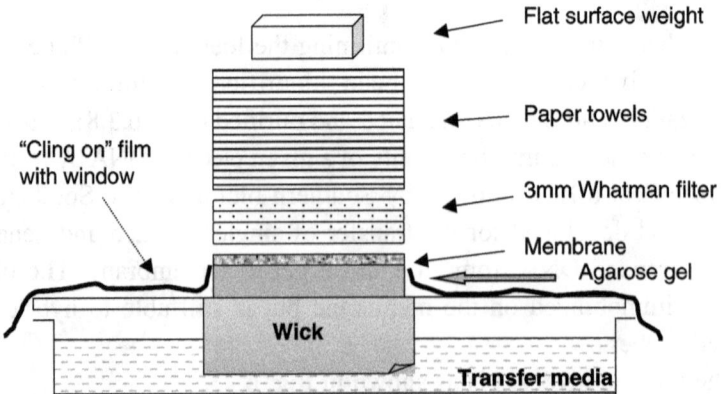

Fig. 6.24. Southern blotting using the "upward transfer" method.

from the gel to the membrane depends on the size of the DNA. The larger the DNA, the less efficient is the transfer to the filter. Using this procedure, the DNA in the gel is depurinated, denatured and neutralized using a series of solutions.

Capillary blotting is the most commonly used transfer method in many laboratories, as it does not require expensive instrumentation. There are two ways to carry this out. Traditionally, Southern blotting is carried out in the "upward transfer" fashion (Fig. 6.24). Quite often, the use of too heavy a weight at the top of the assembly will result in the gel getting squashed. This can result in the smearing of the DNA during transfer. An alternative to this mode of transfer is the "downward transfer". Basically, it shares the same principle, except that the arrangements of the assembly is opposite to that of the "upward transfer". The advantage is that it does not deform the gel during transfer. It is recommended as a rapid method for transfer, using minimal amounts of transfer solution. However, the transfer for a defined short period of time may not result in complete transfer.

After the restricted DNA fragments have been immobilized onto the membrane, the position and identity of the desired piece of DNA can be detected and visualized by nucleic acid hybridization, using its complementary labeled probes.

6.4.2. *Polymerase chain reaction*

As mentioned earlier, a successful cloning experiment relies mainly on the ability to identify and select the particular clone of interest. In this respect, the polymerase chain reaction (PCR) has proven to be an invaluable technique for the amplification of a specific copy of a gene of interest, and thus enhancing its successful cloning and subsequent selection and identification. Using a pair of gene-specific primers (GSPs), it is possible to selectively amplify millions of copies of a gene of interest from the microbial genome. In addition, by the incorporation of restriction endonuclease sites into these GSPs, PCR cloning will also enhance the efficiency of ligation between vector and insert DNA, or to allow in-frame insertion into an expression vector for the production of recombinant proteins.

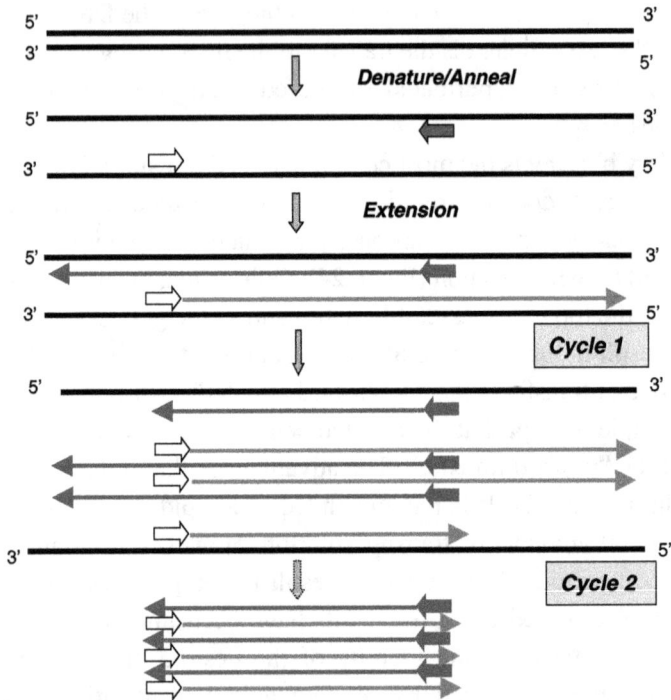

Fig. 6.25. Steps in the polymerase chain reaction.

(1) Basic Principles and Optimizing PCR

In PCR, short oligonucleotides act as specific primers hybridizing to a target DNA molecule. Under optimal conditions, two oligonucleotide primers complementary to the opposite strands of the template, can then be used to initiate DNA synthesis *in vitro* using thermostable DNA polymerises (Fig. 6.25). The hybridization conditions are carefully controlled to limit *in vitro* DNA synthesis due primarily to template sequences located between the two oligonucleotide primers. After annealing, the DNA polymerase will then extend the primers, generating an additional new copy of the template sequences located between the two primers. On completion of the primer extension, the sample is heated to denature the DNA duplex, regenerating approximately twice as much single-stranded template for primer annealing in the next cycle. A series of cycles is then performed, where the DNA template is denatured, oligonucleotide primers annealed specifically, and the primers extended by the thermostable DNA polymerase in each cycle. The repeated cycling results in a geometric accumulation of the products flanked by the two primers.

In slightly more detail, the initial extension products resulting from the original template do not have a distinct length because the polymerase will continue to synthesize new DNA until it either stops or is interrupted by the start of the next cycle. In the second cycle, the extension of products also result in variable lengths. However, by the third cycle, the segment between the primers is synthesized with a defined length. From the fourth cycle onwards the target sequence is amplified exponentially. Potentially, after 20 cycles of PCR, the products should be amplified 2^{20}-fold, assuming 100% efficiency during each cycle. However, in practice, this is seldom achieved, largely because the efficiency is highly dependent on the template and primers used, as well as the cycling conditions and reactants involved.

In later cycles, the accumulation of amplified products does not expand exponentially indefinitely but levels off ("plateau effect"). There are many reasons to account for this. Some of these factors include the gradual loss of polymerase activity and hence becoming a rate-limiting step. Another factor is the reannealing of target strands as their concentration increases. The reannealing of target strands then competes with primer annealing. Practically, the process levels out after microgram quantities of products are obtained and this is sufficient for many downstream procedures, e.g. subcloning into vectors, sequencing etc.

(2) Cloning of PCR Products

PCR has allowed the isolation and amplification of any region within a complex DNA in a few hours. Cloning the PCR products into appropriate vectors allows the generation of the amplified products without having to repeat the reaction every time the products are needed. Therefore it is more convenient to amplify the gene of interest from its recombinant clone than from its original source, which could range from RNA to a DNA genome. In addition, if the PCR products are to be use in further studies, e.g. expression or the generation of single-stranded RNA probes, then it is necessary to clone the amplicons. In this section, some strategies to clone PCR products will be discussed. It is important to note that a well-planned cloning strategy is needed before starting the actual PCR reaction, as many of these strategies involve modification of the primers that are used in the PCR reaction.

(a) Blunt-end and T-A cloning

The type of DNA polymerase used in PCR will determine the type of vectors that can be used to clone the PCR products. For example, if Taq, Amplitaq or Tth DNA polymerase had been used in the reaction, the resulting amplified products will have a single A residue overhang at the 3' end. This overhang nucleotide will facilitate ligation if the vectors used to clone these PCR fragments contain a complementary T nucleotide at the 3' ends of the vector. This is called T-A cloning and is illustrated in Fig. 6.26. An example of this type of cloning vector is pGem-T.

The orientation of the DNA insert cannot be controlled, as the overhang nucleotides on the inserts and vectors are bilateral. Precautions have to be taken to avoid non-recombinants due to religation of vector molecules, which have lost the T-overhang nucleotides due to a contaminating exonuclease activity. This problem is easily overcome with cloning vectors that has the multiple cloning sites within the *lacZ'* gene, as in pGem-T. As described in Sec. 6.3.4, the presence of the DNA insert will result in the formation of a white colony, in contrast to a non-recombinant blue colony.

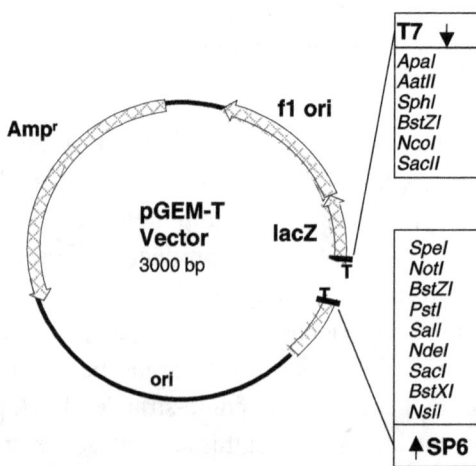

Fig. 6.26. A plasmid vector for TA cloning of PCR products.

The use of other DNA polymerases, for example *Pfu* or *Ultm*a, will however result in the formation of blunt-ended amplified products. As in the case of blunt-end ligations (Sec. 6.1.4), the ligation of blunt-ended PCR product to a vector is not as efficient as T-A cloning. However, this ligation reaction can be improved by using high amounts of starting inserts and vector. The choice of vectors that can be used for blunt-end ligation is broad, as long that the vector contains restriction sites that will result in blunt-ends, e.g. *Sma*I, making it terminally compatible with the amplified products. Alternatively, there are commercial vectors available for blunt-end cloning, for example, pCRScript and pCRScript™Direct (Stratagene).

(b) Introduction of restriction sites for cloning

To aid in efficient and unidirectional PCR cloning, restriction sites can be added to the 5' ends of one or both nucleotide primers that are to be used for PCR (Fig. 6.27). The sequences of these modified primers will be

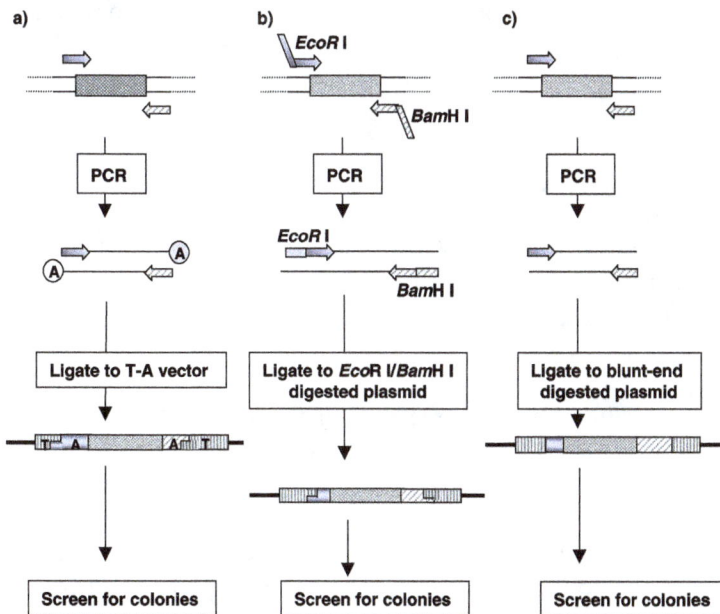

Fig. 6.27. Restriction sites can be added to the 5' ends of one or both nucleotide primers that are to be used for PCR.

incorporated into the amplified PCR products. These amplified DNA can then be digested with the appropriate restriction endonucleases to generate either blunt or cohesive ends, as required by the scientists. By the use of different restriction sites at either ends of the DNA, it is thus possible to clone a DNA insert unidirectionally into a vector.

The addition of a few extra sequences to the 5' end will not significantly affect the primer specificity, since this is determined mainly by the sequences at its 3' end. There are, however, a few points that one needs to keep in mind when using this method:

1. The restriction site(s) used in the primer sequences should not have recognition sites within the DNA of interest.
2. The restriction sites that are added to the final product via the primers often need extra bases at the 5' end to allow for efficient DNA digestion.
3. It may be essential that sequential digestions be performed on the amplified products if the enzymes chosen do not have optimal activity in a common buffer.

The techniques described above are widely used in recombinant DNA technology and provide one of the most reliable and simplest methods for PCR cloning.

6.4.3. DNA sequencing

Probably the most important technique available in recombinant technology is DNA sequencing, which enables the scientist to decode the precise order of the nucleotides within a piece of DNA. The two of the most commonly used sequencing techniques are:

- Dideoxy chain termination method (Sanger), and
- Chemical degradation method (Maxam-Gilbert).

The former is the method of choice for the high throughput sequencing. Although the two methods differ in their strategies (Sanger method is enzymatic and Maxam-Gilbert method is purely chemical-based), the fundamental principle is the same. The principle is that two single-stranded DNA molecules differing in length by just one base can be resolved into distinct bands by some form of separations. The usual way of separation is the use of denaturing polyacrylamide gel electrophoresis.

(1) Dideoxy Chain Termination Sequencing (Sanger Sequencing)
This method is conceptually rather simple. It uses polymerases to synthesize a new template by extending the primer and occasionally incorporates dideoxynucleotides (ddNTPs) instead of the usual deoxynucleotides (dNTP) during chain elongation at the 3' end (Fig. 6.28). When the dideoxynucleotide is added to the 3' end of an elongating chain, no further elongation is possible as the 3' end nucleotide lacks a 3'OH that is essential for forming the phosphodiester bond between the primer and additional dNTPs for elongation. Thus, the chain is terminated.

In manual chain-termination sequencing, each of the 4 dNTPs (dATP, dCTP, dGTP and dTTP) is added separately to one of the 4 primer extension reactions. In addition, a radiolabeled ddNTP is included in the reaction mix. The reaction with ddGTP is called the "G" reaction; the one with the ddATP is called "A" reaction, etc. Chain termination will occur in a base-specific fashion depending on which ddNTP is present during the extension reaction. For example, termination will occur only at the "G" position (complementary base of C) when ddGTP is present, or exclusively

Structures of adenosine analogues

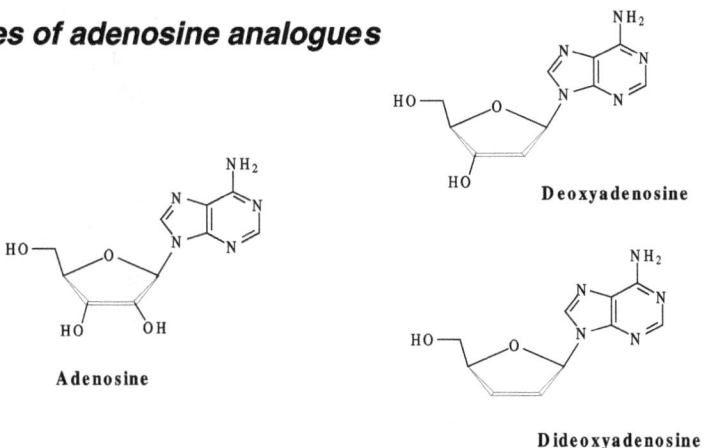

Adenosine

Deoxyadenosine

Dideoxyadenosine

Fig. 6.28. Dideoxynucleotides are used as chain terminators during DNA synthesis.

Termination with ddATP

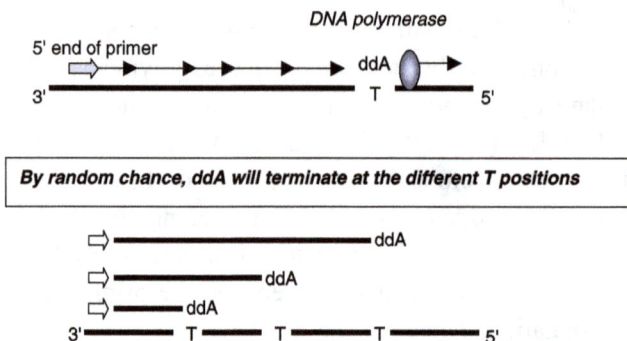

Fig. 6.29. Incorporation of a dideoxynucleotide results in chain termination.

at the A when ddATP is present (Fig. 6.29). The concentration of ddNTP is kept low relative to the corresponding dNTP. Hence, base-specific chain termination will only occur in a small percentage of the time at a given position in the sequence. As the reaction proceeds, a mixture of all possible termination products for G, A, T, or C will be generated in the 4 primer-extension reaction tubes.

The radiolabeled products from the 4 base-specific termination reactions are then resolved on a high-resolution polyacrylamide gel (Sec. 6.1.3). Up to 300–500 bases in length can be separated. Each terminated species will be visualized as a separate band on the gel with the shortest primer-extended products running fastest down the gel. The base sequence of the primer-extended products can then be read from the 5' to 3' ends by identifying which reaction mix (G, A, T, C) contains the next longest species, beginning with the shortest fastest migrating species as seen on the autoradiogram of the gel (Figs. 6.30 and 6.31).

(2) DNA Polymerases

The properties of these polymerases have been discussed in other chapters. It suffices to say that the majority of those used in manual and automated sequencing are genetically modified for longer extension (high processivity) and diminished exonuclease activities. For manual

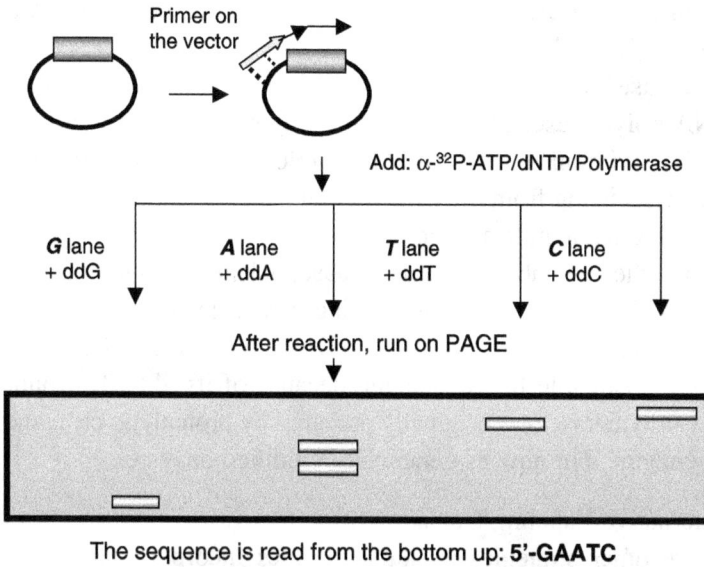

Fig. 6.30. Separation of DNA fragments during DNA sequencing.

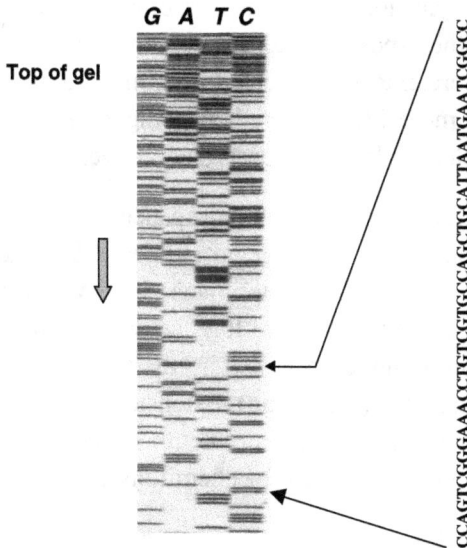

Fig. 6.31. An example of a DNA sequencing gel after Sanger dideoxy chain termination sequencing. Each lane represents a chain termination reaction with a specific dideoxynucleotide.

sequencing, some examples of the commonly used polymerases include Sequenase, *Bst* and Klenow *exo⁻*.

Sequenase is a genetically modified version of the bacteriophage T7 DNA polymerase. The unmodified polymerase (wild type) is highly processive, with an average of 2000–3000 nucleotides polymer formed (before dissociating from the template) with an elongation rate (extension) of about 300 nucleotide/second.

Before the availability of Sequenase, Klenow polymerase was the enzyme of choice for manual sequencing (Sec. 6.1). Klenow polymerase is a modified enzyme derived from *E. coli* DNA polymerase I. The unmodified version is unsuitable for sequencing because of its 5' to 3' exonuclease. Klenow polymerase was originally prepared by proteolytic cleavage of the parent enzyme, but now as genetically modified enzyme.

(3) Alternative Labeling/Detection Strategies
There are other strategies available besides incorporating radiolabeled nucleotides:

(a) End labeling of primer
Here, the primer is labeled instead of incorporating labeled nucleotides into the strands. T4 polynucleotide kinase is used, which catalyze the transfer of a ^{32}P or ^{35}S (in the γ-position) to the dephosphorylated 5' end (not the 3' end as it is where extension occurs). This method of labeling result in a chain-terminated molecule carrying just a single-labeled atom rather than the multiple labeled molecule as described above.

(b) Labeling with tags for chemiluminescence
This is an alternative to radiolabeled studies. Chemiluminescence can be as sensitive as radiolabeled approach, however, this involve more manipulations and can result in low sensitivity if mishandled (see discussion on chemiluminescence). Here the 5' end of the primer can be modified to contain a tag (e.g. biotin). Extension and termination is carried out as usual. At the end of the gel run, Southern transfer is carried out and the membrane is then processed for chemiluminescence.

(c) Fluorescent *labels*
This is extensively used in automated sequencing. Special imaging is required including a laser for monochromatic excitation and a detector with different

wavelength discrimination. Fluorescent-labeled fragments with 4 different colors can be analyzed in the same lane. As the different sizes pass the detector, one of the four colors will be recorded and deciphered using a software program.

Two main types of automated sequencers are commercially available. The first type uses single-label, 4-lane separations, similar to the above manual sequencing, whilst the second type uses 4 labels, single-lane separation (Fig. 6.32). The latter type is often the instrument of choice for large-scale genome projects.

Sequencing is carried out with dye-terminators (similar to the use of radiolabeled ddNTPs) and the samples electrophoresed on a denaturing polyacrylamide gel. Fluorescent signals are detected after excitation by a laser positioned directly in front of the detector. Automated DNA sequencing has now become a routine in many laboratories because of its ease of use and the greater amount of sequencing data that can be obtained. An example of a sequencing result using the automated fluorescent sequencing is shown in Fig. 6.33.

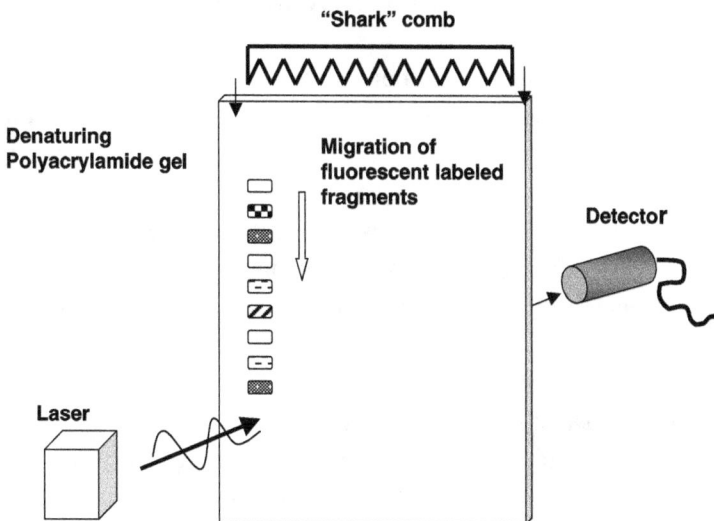

Fig. 6.32. An automated DNA sequencer which uses 4 labels, single-lane separation.

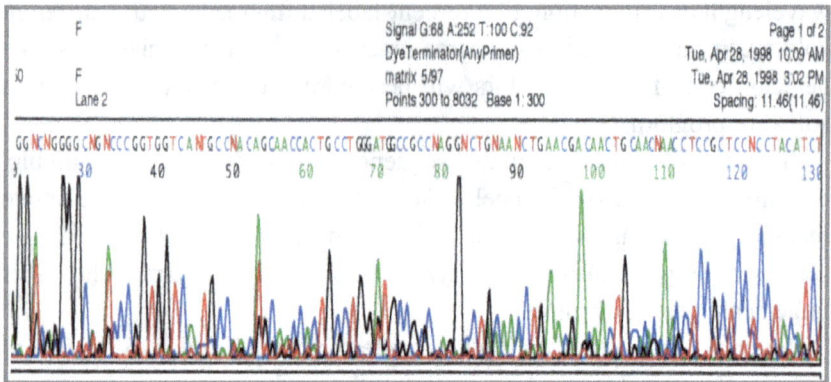

Fig. 6.33. An example of a sequencing result using the automated fluorescent sequencing is shown.

(4) Sequencing Strategies

In sequencing a large piece of DNA (10 kb), the amount of sequences that can be read from a single template is often in the 500–700 bases region. Thus, various methods have been devised to obtain the rest of the sequences. These methods can be broadly divided into two types: random and ordered sequencing.

Random sequencing is a simple method to obtain large portion of the template quickly. The template is first fragmented physically by shearing the DNA into smaller pieces, subcloned into vectors and the recombinants sequenced all at one go. This is an example of a "shot-gun" method. Another is the use of transposon-mediated sequencing method where a transposon randomly integrates into and excise parts of the DNA. Sequencing of a collection of these recombinant transposons will also yield large amounts of data at a go. An advantage is that some difficult regions (G-C rich) may be broken up during the fragmentation step, thus, facilitating the ease of sequencing. However, both methods suffer from one major drawback. They both can obtain most of the sequences quickly but the last 10% of the sequence will pose some difficulty with the presence of similar clones. Usually, to obtain the last bit of the template sequence, ordered approaches are used.

Ordered sequencing comes in many forms. The simplest but most tedious and expensive method is to "walk" the whole template. Here, the sequencing data is used to synthesize new primers that can then be used to sequence further down. Moving the template closer to the primer progressively is another approach. One of these methods is to generate "collapse clones" where internal restriction sites and those on the multiple cloning site of the vector can be used to delete off sequences. The unidirectional deletion (nested deletion) and religation of template sequences is yet another approach which is easy to carry out. This uses enzymes, e.g. *Exo* III and *Bal*31, which can delete the template in a specific direction.

Obviously, directional sequencing is more convenient and once a fragment can be localized, the rest of the fragments produced can be zoomed in and overlapping clones used for the rest of the project. Furthermore, the deleted (shorter clones) can be used for expression or mutagenesis studies.

6.5. Mutagenesis

Recombinant DNA and gene-cloning technologies allow scientists to isolate and characterize any given gene from any living organism. In addition, these genes may also be modified *in vitro* in order to produce new restriction enzyme cleavage sites, to change a particular codon, or to introduce site-specific mutations by insertion and deletion of nucleotides. This procedure has allowed scientists to dissect genes at the nucleotide level by changing one nucleotide at a time and examining the effect of each change on the function of the gene or its product. *In vivo* mutagenesis protocols, on the other hand, either makes use of bacterial transposons to either mediate mutation via random insertion of the transposon into the host genome, or the replacement of the normal gene with an altered gene sequence via homologous recombination.

6.5.1. *Random insertional mutagenesis*

A gene that has been marked with a transposon is relatively easy to map, either by genetic crosses or by physical mapping using restriction

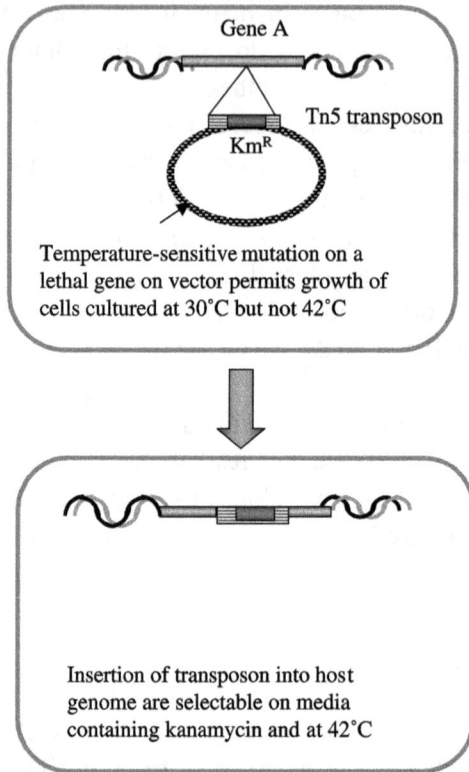

Fig. 6.34. Gene disruption using a transposon carrying a kanamycin resistance marker.

endonucleases. A typical strategy is to construct a plasmid carrying the transposon, using temperature-sensitive vector, i.e. a plasmid that is unable to replicate at an elevated temperature. When this plasmid is introduced into bacterial cells, and the cells are grown at a temperature permissive for plasmid replication (say 30°C), all the cells will carry the plasmid. When the culture is then shifted to the non-permissive temperature (say 42°C) at which the plasmid is unable to replicate, the only cells which retain the transposon (usually selectable by an antibiotic resistance marker) are those that have incorporated the transposon in its chromosome. Thus, using a transposon, such as Tn5, which carries the kanamycin resistance gene, will allow selection of transposon insertions by plating the culture on agar containing kanamycin at 42°C (Fig. 6.34).

This generates a library of transposon mutants, with the transposon inserted at a variety of chromosomal positions, and therefore causing a range of mutations. After the desired mutants have been selected using an appropriate phenotype screen, e.g. the loss of bacterial motility, the genes into which the transposon has inserted may be identified. This is achieved by cutting the total genomic DNA of the mutant with a restriction enzyme, which does not cut the transposon itself, and then cloning the collection of fragments to produce a gene library. If the library is then plated on

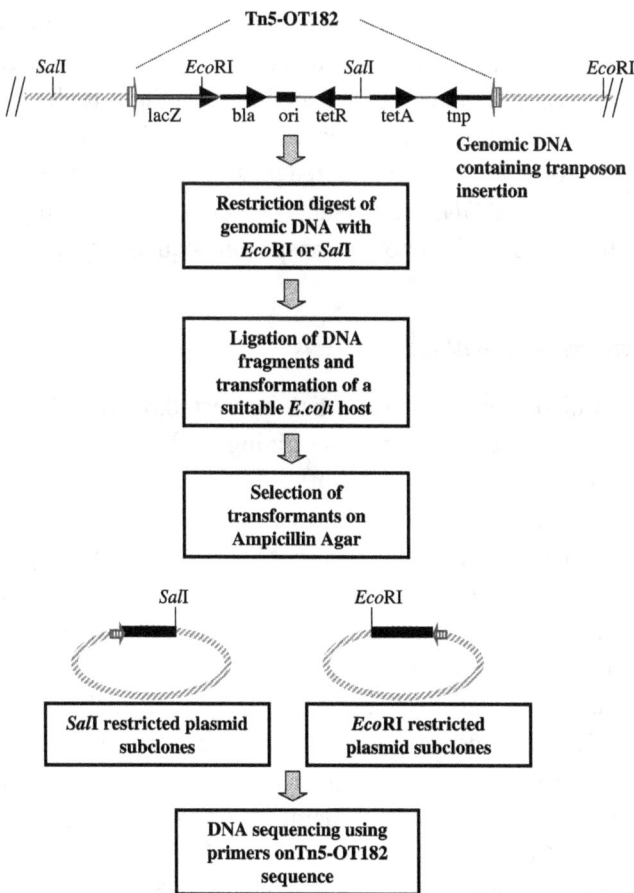

Fig. 6.35. Strategy for the identification of DNA sequences flanking the Tn5-OT182 transposon insertion site.

kanamycin agar, only those clones containing Tn5 will be able to grow. These clones will contain not only Tn5 but also some of the host sequence flanking the transposon. The cloning of DNA sequences flanking the transposon insertion is made easier when a transposon with self-cloning capabilities, e.g. Tn5-pOT182, is used. In this example, the genomic DNA is digested with a restriction enzyme that cuts once within the transposon (e.g. EcoRI or SalI), leaving its plasmid origin of replication and an antibiotic selection marker (ampicillin resistance) intact. The other EcoRI or SalI restriction site would be in the host genomic DNA sequence flanking the transposon insertion. Self-ligation of the digested DNA fragments results in recircularization of the digested genomic DNA fragments. Upon transformation of *E. coli* host cells and selection on ampicillin-containing agar, only the recircularlized DNA containing origin of replication on the transposon and the antibiotic reisistance marker would result in ampicillin-resistant transformants. The disrupted host gene is then identified by isolating the plasmid DNA from such clones and sequencing the DNA flanking either ends of the known transposon sequence (Fig. 6.35).

6.5.2. Gene replacement mutagenesis

Having identified a gene of potential interest through DNA sequence analysis, one of the best ways of determining its function and significance is to deactivate it in the host organism. This may not be easy to achieve by the classical procedures of mutagenesis and selection, especially if the phenotype associated with that gene is not known, or not readily selectable. However, if the cloned gene is inactivated *in vitro*, it can be used to replace the wild-type chromosomal gene by homologous recombination, a process also known as gene knock-out (Fig. 6.36). In this case, the central part of the cloned gene has been removed and replaced by an antibiotic resistance gene, such as kanamycin resistance. The inactivated gene is constructed on a suicide vector, which cannot replicate in the host cell. In this example, the mobilizable suicide vector, pJQ200mp18, contained the *sacB* gene. Cells carrying the intact *sacB* gene, i.e. those that had not undergone a gene replacement event, will be lethal when plated on media containing 5% sucrose. The mutated DNA is introduced into recipient cells by conjugation. After conjugation, selection on kanamycin

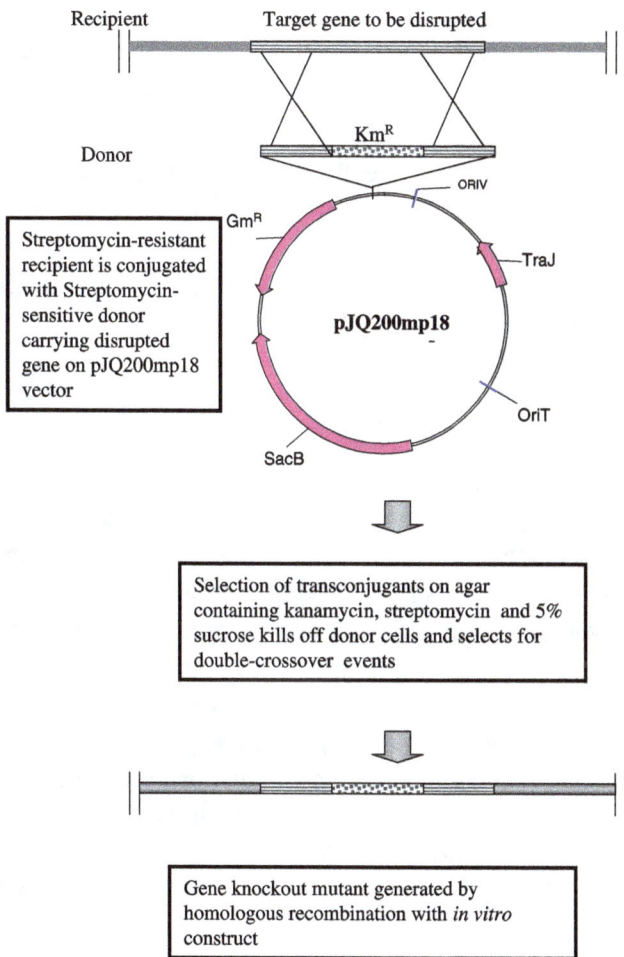

Fig. 6.36. Targeted mutagenesis using a suicide vector, pJQ200mp18, for gene replacement by homologous recombination.

and streptomycin agar containing 5% sucrose will identify cells in which the kanamycin resistance gene has been inserted into the target gene by homologous recombination, involving a double crossover. The addition of streptomycin selects against donor cells, which are streptomycin-sensitive, and the kanamycin selects against non-recombinant recipient cells. The target gene is thus deactivated. The phenotype of the knock-out mutant can then be studied using appropriate assays. A confirmation that the changed phenotype is actually due to inactivation of the gene done by introduction of the wild-type gene on a plasmid (complementation). Complementation is an important step, as the phenotypic changes could be due to polar effects on other genes, rather than directly caused by inactivation of the target gene. This not only provides a useful way of assessing the function of a cloned gene, but can also be used for producing attenuated strains of a pathogenic bacterium by inactivating a gene coding for toxin production or another virulence factor.

6.5.3. *Site-directed mutagenesis*

It is sometimes desirable to replace a gene with one with specific alterations. This requires a combination of the techniques of gene replacement and site-directed mutagenesis. The knocked-out gene can be complemented with a plasmid carrying a specifically mutated gene copy. This enables us to test, for example, the significance of a specific amino acid in the activity of an enzyme, by replacing it with a series of different residues and determining the activity of the product. Or we could look at the precise requirements for the function of a regulatory sequence on the DNA such as a promoter site, by introducing defined mutations into that region.

A number of protocols have been established for site-directed mutagenesis using hybridization of a mismatched oligonucleotide to a DNA template, followed by second-strand synthesis by a DNA polymerase. One of these is a site-directed mutagenesis procedure developed by Deng and Nickoloff that eliminated the need for subcloning and generating single-stranded DNA templates by employing double-stranded plasmid DNA (Fig. 6.37). The procedure involves simultaneously annealing two oligonucleotide primers to the same strand of heat-denatured double-stranded plasmid DNA. One of the primers (the selection primer) anneals to a region that contains a unique, nonessential restriction site. The other primer (the mutagenic primer) introduces a chosen mutation into the plasmid. Extension of these primers by T4 DNA polymerase followed by ligation of the resulting molecules with T4 DNA ligase results in a population of plasmid molecules, some of which contain the desired mutation but no longer contain the unique restriction site. The plasmid

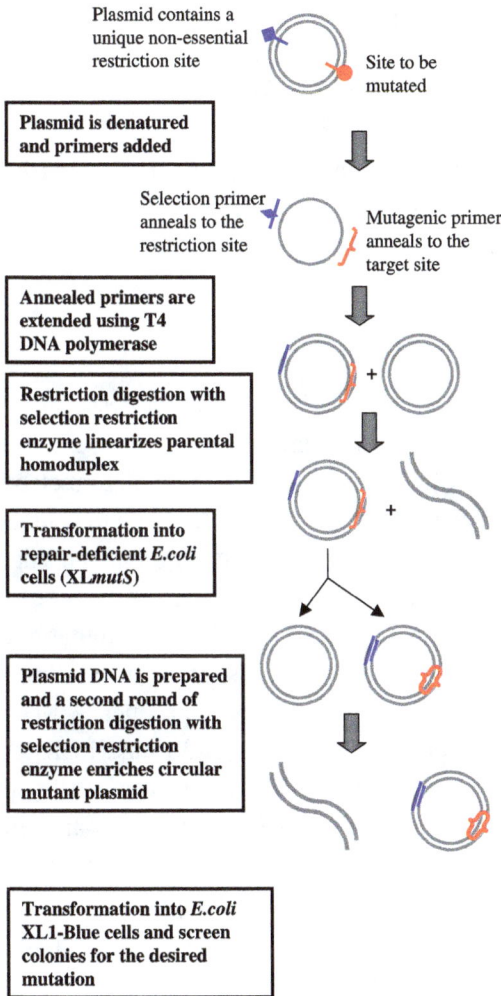

Fig. 6.37. Protocol for double-stranded, site mutagenesis.

is treated with a restriction enzyme that will linearize the parental plasmid but not the heteroduplex. Since transformation of circular DNA is many folds more efficient than that of linearized DNA, the circular heteroduplex would be preferentially transformed. Using a mismatch repair-defective strain of *E. coli* (e.g. XL*mut*S) that cannot distinguish the original unmutated strand of DNA from the newly created strand containing the desired mutation, both strands of the heteroduplex would be equally present in the newly replicated DNA. The *mut*S-deficient strain randomly selects one of these strands as "correct" and changes the other strand to be complementary to the chosen strand. DNA is isolated from the plasmids and subjected to a second round of restriction digest, which will enrich the population of mutant plasmids after transformation into an *E. coli* strain (e.g. XL1-Blue).

6.5.4. *PCR-based mutagenesis*

Various mutagenesis procedures based on polymerase chain reaction (PCR) are available. A versatile and relatively easy method is the "megaprimer" procedure (Fig. 6.38). This method uses three oligonucleotide primers and two rounds of PCR performed on a DNA template containing the cloned gene to be mutated. Primers A and B represent the "flanking" primers that can map either within or outside the cloned gene. M represents the internal "mutant" primer containing the desired mutation, which may be a point mutation, a deletion or an insertion. In the first round of PCR, primer pair A and M1 is used. The PCR product, A-M1, is purified and used as one of the primers for the second round PCR together with the flanking primer B, hence the term "megaprimer". Prior separation of the two strands in the PCR product A-M1 is unnecessary since this will be achieved during the denaturation step in the second round of PCR. The wild-type cloned gene is used as the template for both PCR reactions. The final PCR product (A-M1-B) containing the mutation can be used in a variety of standard applications, such as sequencing and cloning in expression vectors. The inclusion of restriction sites to primers A and B would facilitate the cloning of the final mutant PCR product. This is especially important when sequences upstream of A or downstream of B are essential for gene expression or regulation, and therefore, must remain unaltered.

This protocol may also be applied to the mutagenesis of protein-coding genes and the study of structure-function relationship. In this case, primers A and B can be kept constant and a variety of mutant primers (M1, M2, M3…) can be used to produce the various mutants.

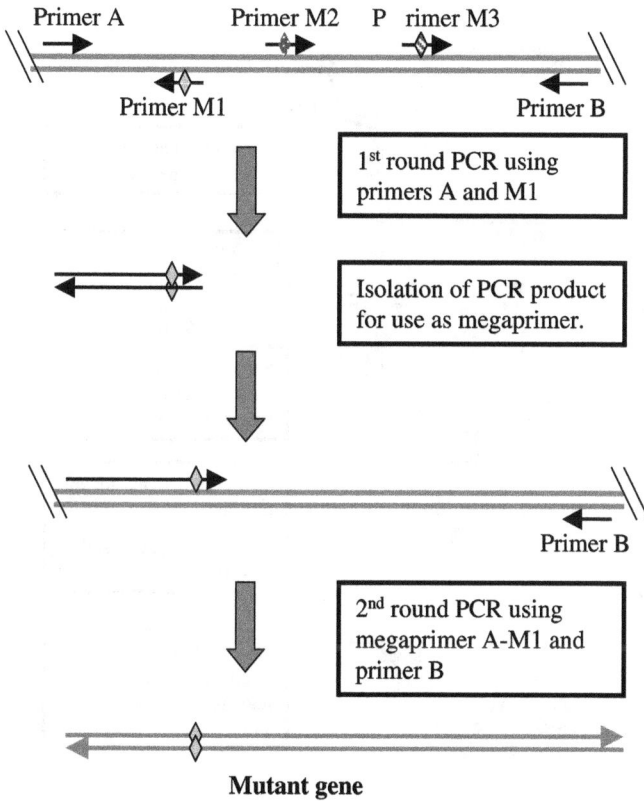

Fig. 6.38. Site-directed mutagenesis using megaprimer PCR.

6.5.5. *Ordered deletions using exonuclease III*

Ordered deletions can be used in delineating sequences that are important for the function of a gene, such as those required for transcription. Deletions are generated by limiting, unidirectional digestion of DNA with *E. coli* exonuclease III (*Exo* III). *Exo* III preferentially digests one strand of a double-stranded DNA by removing nucleotides from 3' ends if the end is blunt or has a 5' protrusion, but not if the end has a 3' protrusion of 4 or more bases (Fig. 6.39). The plasmid clone is first digested with restriction endonuclease A, which generates a 5' protrusion or blunt end next to the target sequence, and restriction endonuclease B that generates

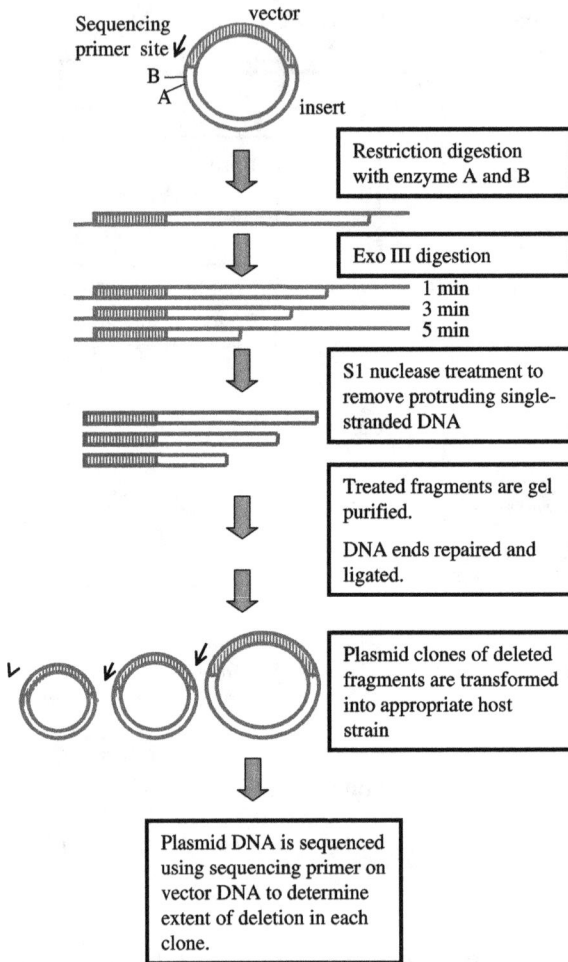

Fig. 6.39. Generation of deletion mutants using *Exo* III.

a 3' protrusion next to the sequencing primer site. The linearized plasmid DNA is then digested with *Exo* III, and aliquots of the digest are taken at a number of time points so as to yield a set of deletions of the desired lengths. These DNA fragments are then subjected to S1 nuclease digestion to remove the single-stranded DNA overhangs. The deleted DNA fragments may then be ligated to a vector with a promoterless reporter gene, such as the green fluorescent protein (GFP) or *lacZ*, to analyze the promoter activity. The cloned fragments can also be sequenced to determine the extent of the deletion.

Further Reading

1. Ausubel FM, *et al. Current Protocols in Molecular Biology.* John Wiley and Sons, Inc., 2001.
2. Balik S. Site-directed mutagenesis *in vitro* by megaprimer PCR. In: *In Vitro Mutagenesis Protocols Vol. 57*, ed. M Trower. Humana Press, New Jersey, 1996, pp. 203–215.
3. Barth PT and Grinter NJ. Map of plasmid RP4 derived by insertion of transposon C. *J. Mol. Biol.* 113: 455–474, 1977.
4. Clark D and Henikoff S. Ordered deletions using exonuclease III. In: *In Vitro Mutagenesis Protocols Vol. 57*, ed. M Trower. Humana Press, New Jersey, 1996, pp. 139–147.
5. Deng W and Nickoloff J. Site-directed mutagenesis of virtually any plasmid by eliminating a unique site. *Anal. Biochem.* 200: 81–88, 1992.
6. Ditta G, Stanfield S, Corbin D and Helinski DR. Broad host range cloning system for Gram-negative bacteria: Construction of a gene bank of *Rhizobium meliloti. Proc. Natl. Acad. Sci. USA* 76: 7347–7351, 1980.
7. Figurski DH and Helinski DR. Replication of an origin-containing derivative of plasmid RK2 dependent on a plasmid function provided *in trans. Proc. Natl. Acad. Sci. USA* 76: 1648–1652, 1979.
8. Kahn M, Kolter R, Thomas C, Figurski D, Meyer R, Remaut E and Helinski DR. Plasmid cloning vehicles derived from plasmids ColE1, F, R6K and RK2. *Methods Enzymol.* 68: 268–280, 1979.
9. Sambrook J and Russell DW. *Molecular Biology: A Laboratory Manual. Cold Spring Harbor Laboratory Press, 2001.*

Questions for Thought...

1. The Ct values of a standard curve are given in the Table below. How are Ct values determined? Determine the copy number of an unknown

sample with Ct of 30. What is the detection limit of the assay? What is the dynamic range and efficiency of the assay? NTC, non-template control; ND, not detected.

Copy number	Ct values of Replicate 1	Ct values of Replicate 2
10^7	18.1	17.51
10^6	21.32	21.93
10^5	24.98	24.37
10^4	28.54	29.3
10^3	32.13	31.87
10^2	33.56	34.13
10^1	37.87	37.55
NTC	ND	ND

2. Compare and contrast the advantage and disadvantage of sequence dependent and independent method of determining signal.
3. How do you determine the presence of primer dimer formed in a sequence independent probe experiment (e.g., Sybr Green I or Eva Green)?

Chapter 7

Applications of Bioinformatics and Biocomputing to Microbiological Research

Paxton Loke* and **Sim Tiow Suan**[†]

Ian Potter Hepatitis Research Laboratory
The Macfarlane Burnet Institute
  for Medical Research and Public Health
Commercial Road, Melbourne, Vic 3004, Australia
[†]*Department of Microbiology, Yong Loo Lin School of Medicine*
National University of Singapore
5 Science Drive 2, Singapore 117597

7.1. Introduction

This is the decade of life sciences. Indeed, the advent of recombinant DNA technology has fueled the road ahead for research and development in various disciplines. With computer-based technologies in parallel and in support of life science research, it is conceivable that much progress can be made in a shorter period of time. Bioinformatics is the application of computing to the analysis of biological data and this is in the form of manipulation of biological sequence data of DNA and proteins, as well as their structures.

Research in the field of Microbiology is no longer dependant on traditional techniques of microscopy and whole cell macro-analysis and manipulation. Since the early days of sequencing genomes of viruses and bacteriophages, microbiologists have now progressed to sequence more complex and larger genomes. With molecular biology techniques available for cloning and sequencing microbial genes, the biochemistry and physiology of a particular microbe can now be studied at the genomic level. Exploitation and improvement of genes, which may have industrial and commercial applications, are now possible in addition to the analysis of complex gene interactions.

In general, the analysis of gene sequences is easily achievable with fruitful results. On the contrary, the elucidation of structure and function is harder. As most analyzes are based on inferences, the limited information available on the biological characteristics of sequenced genes and their annotations may not be entirely helpful. In some cases, sequences that are closely related or homologous may share a common structure but not function. For example, lysozyme, which hydrolyzes bacterial cell wall polysaccharides and α-lactalbumin, which is a non-catalytic regulatory milk protein required for lactate synthase to transfer a galactose molecule from UDP-galactose to D-glucose, shares an identity of 50% (70% similarity) although both perform and are required in different functions or reactions. There are also proteins that share similar structures but have dissimilar functions, such as the serine proteases and integral membrane porins where structures like β-barrels are found. It is also interesting to note that in nature, the three-dimensional (3D) structure of proteins is more conserved than the underlying sequences. With that notion, different sequences can adopt similar folds.

Sequence analysis is a valuable tool in molecular biology as it assists in the extraction of information and biological knowledge, thus allowing us to infer biological information from sequence data. Due to the inference nature of sequence analysis, it becomes a "chicken and egg" predicament. In order to properly and accurately infer function from sequence, we must have a large starting pool of sequences that were previously obtained and closely related with known functions. This is to ensure a significant match to the unknown gene sequence, thus allowing us to make a conjecture of the function of the unknown sequence. On the other hand, we must first elucidate the function of the gene cloned. Thus, does sequence begets function or vice-versa? Nevertheless, earlier research efforts have ensured a starting pool of information, where gene function was revealed by complementation of arrays of defective mutants.

Moreover, our focus should not only be concentrated on DNA sequences and thus neglect protein sequences but both. As proteins are the main component of living cells that carry out functional tasks, their diversity in structure and function cannot be ignored. Each protein has a unique amino acid sequence that determines its 3D shape and biological activity. It is coded by a corresponding DNA sequence, which can be deciphered by the genetic code. With today's recombinant DNA technologies, it is relatively straightforward to obtain a DNA sequence and subsequently, the primary sequence of the encoded protein as the order of the amino acid residues is consequential.

However, in the early days, it took years to complete an entire protein sequence for the initial breaking of the genetic code, for example, the peptide hormone insulin in 1955 using chromatographic and labeling techniques and the ribonuclease in 1960. Nowadays, the typical approach is to obtain the coding DNA sequence at the outset followed by the usage of new conceptual translation methods to interpret or deduce the resultant protein sequence. While the primary sequence of the protein can be predicted with ease and precision, the predictions for secondary and tertiary (not to mentioned quaternary structures) structures are less accurate.

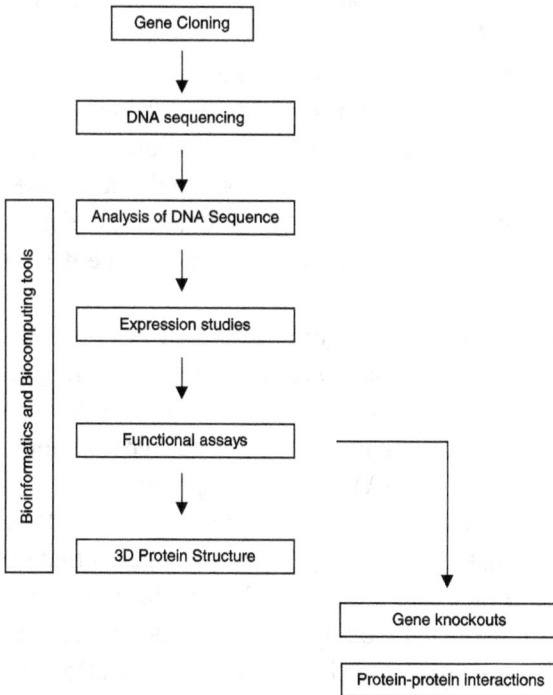

Fig. 7.1. Cloning strategies for a functional gene.

With the huge amount of microbial DNA sequence data currently available in databases, the reliability and accuracy of such data are crucial. Inaccuracies in DNA sequence information may lead to errors in translations due to frameshifts, thus generating subsequent protein sequences which may not be the actual encoded protein. In this chapter, sequence analysis using current computational programs in databases would be explored together with some of the more common tools in bioinformatics that will aid microbiological research. In essence, the standard route from cloning to elucidating the function and structure of a protein and ultimately, its natural role in the cell is shown in the flowchart (Fig. 7.1).

7.2. Gene Cloning

DNA sequences can be obtained from the partial or full cloning of the gene of interest. Traditionally, "shotgun" cloning has been performed but the process of screening to obtain positive clones of interest has been extremely tedious. Another method of cloning known as "reverse genetics" necessitates the purification of a small amount of protein (which is coded by the gene of interest). Protein sequencing of the purified protein would then reveal the order of a short fragment of about 10–20 amino acids. Thus, due to the degeneracy of the DNA code, a number of possible degenerate probes (a mixture of oligonucleotides, depending on the number of amino acids chosen initially) can then be designed and used for the hybridization of a genomic library. As prokaryotic microorganisms such as bacteria would not have introns in their genomic sequences, such methods would allow for the isolation of the complete gene of interest, and subsequently, the elucidation of the DNA sequence. For eukaryotic microorganisms, the generation of a different type of library, called the complementary DNA (cDNA) library is required.

cDNA libraries are generated from the copying of the mRNAs present in a particular cell or tissue. In this aspect, only the genes that are transcribed (exons) will be copied into the cDNA library. This will avoid the problem of introns and thus, when cloning is performed, an uninterrupted coding sequence will be obtained. One interesting usage of cDNA libraries is that the pattern of gene expression can be obtained from the various stages in the development of a cell or when a microorganism is grown under different conditions. This is because the genes contained in a cDNA library are dependant on the expression of the mRNAs, which in turn are responsive to various physiological or environmental signals. Thus, certain genes could either be turned on or off or if the genes were more frequently transcribed in response to certain signals, the corresponding amounts of mRNA would then increase.

An efficient and rapid method now being used for cloning is the polymerase chain reaction (PCR) (Box 7.1). Essentially, this is a copying method which is able to generate large amounts of DNA, depending on the number of cycles used in the replication process. Primers or short oligonucleotide sequences that are designed based on known sequences would facilitate the easy copying of the gene of interest. Even with primers designed based on related gene sequences from closely related microbial species, the amplification of the gene can also be performed. Other relevant uses of PCR today include applications in the field of molecular diagnostics, environmental monitoring and forensic medicine. The availability of sequence information from microbial genomes (completed

---

Box 7.1

How is PCR Performed?

PCR allows for the amplification of a gene in order to make numerous copies of itself. There are three major steps in a PCR reaction, namely the denaturation, the annealing and the polymerization steps. These are performed in a cyclical fashion, normally for 30–40 cycles. The first step is a denaturation step that is essential for melting the double-stranded DNA into single-stranded DNA (template) and this is performed at temperatures of 94–96°C. Following that, the annealing of a short oligonucleotide sequence (primer) to the single-stranded DNA would occur at temperatures ranging from about 50–60°C. Once the formation of the primer-template has taken place, a thermostable DNA polymerase would then begin copying the template. The third step, which is the extension step, allows for the optimal copying of the template by the polymerase at 72°C. The cycle continues with an exponential increase (2^n) in the number of gene copies generated.

Box 7.1   (*Cont'd*)

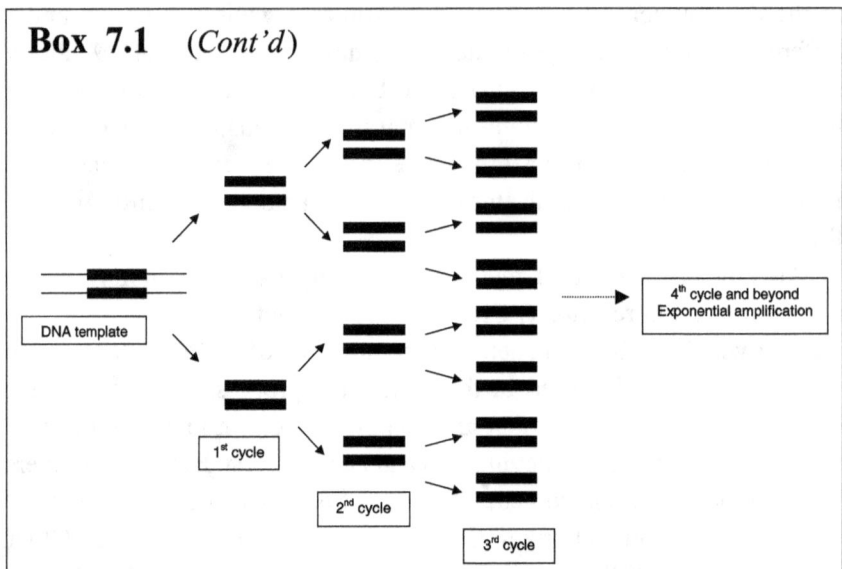

DNA template

1st cycle

2nd cycle

3rd cycle

4th cycle and beyond
Exponential amplification

Box 7.2

Microbial Genome Sequencing

The sequencing of microbial genomes has provided a large information base for the understanding of the microorganism at the genomic level in relation to its living environment, for the exploitation of genes for commercial and industrial purposes and for the understanding of microbial evolution. The sequence of the 4.6 Mb genome of *Escherichia coli*, the workhorse of microbiology, was completed in 1998. Other interesting genomes which have been completed include *Mycoplasma genitalium*, which has the smallest bacterial genome (0.6 Mb) and archaea bacteria such as *Methanococcus jannaschii*, *Methanobacterium thermoautotrophicum* and *Pyrococcus furiosus*. With the minimalist genome of *M. genitalium* available, comparisons of genes obligatory for free living existence as compared to other species of bacteria can be performed. Interestingly, when the genome of *M. jannaschii* was first examined, 65% of the putative genes sequenced had no known homology to previously discovered genes. Generally, researchers have found that approximately 35% of newly predicted genes are also not related to any other genes in the databases during the analysis and annotation of microbial genome sequences. Thus, there is a vast potential for discovering and studying new genes for basic research and utilization of these biological resources.

Box 7.2   (*Cont'd*)

Currently, 88 microbial genomes have been completed with another 20 genomes at the annotation phase and 114 more sequence projects ongoing at the various sequencing centers around the globe. With the increasing number of microbial genomes available, it is now possible to obtain similar genes across many species and this will assist not only in the phylogenetic analysis of evolutionary relatedness between species but also in the PCR cloning of closely-related gene homologs. In the analysis of microbial genomes, it is also not uncommon to encounter cases of "lateral gene transfers" between bacterial species. The most well known phenomena are the transfer of antibiotic resistance genes via mobile autonomous DNA vehicles such as plasmids or "jumping elements" known as transposons. Other areas of wide-ranging interest include the analysis of genomic organization which may shed light on gene regulation, microbial pathogenesis especially opportunistic pathogens and the interactions of different microbial species in their natural habitats. Thus, the use of bioinformatics would be relevant in light of the enormous amount of sequence data that a researcher may need to plough through before making any specific conclusions. Although genes can be identified by the means of computational analysis, ultimately "wet lab" experiments would need to be performed prior to any specific assessments. To their credit, the generation of DNA sequence data by non-profit institutions is now made available in the public domain as expeditiously as possible, thus signaling heady times for researchers in microbiology.

and unfinished) (Box 7.2) in the databases would also allow for the ease of primer design and homology searches across various microbial genomes for related genes.

7.3. DNA and Protein Sequencing

The deoxyribonucleic acid (DNA) sequencing methodology can be credited to the work of Frederick Sanger in 1974. His technique involved using a DNA polymerase to copy DNA strands with the incorporation of deoxyribonucleoside triphosphates (dNTPs) and terminating the elongation with dideoxyribonucleoside triphosphates (ddNTPs). These ddNTPs would be tagged with a radioactive label, thus allowing the identification of the various sizes of DNA fragments by X-Ray autoradiography. Four different and separate tubes, each containing the termination base of adenosine (A), cytosine (C), thymidine (T) and guanine (G) were required to generate different DNA sizes, which were then lined up and read in the exact order. This method is popularly known as the "chain termination" or "Sanger's Dideoxy" method. Another method which arose about the same time and championed by the American scientists Alan Maxam and Walter Gilbert, called the "cleavage" method, was not taken up as favorably by the scientific community.

Although this technique was initially very slow and tedious, the process was sped up using fluorescent dye-labels and thus permitting the reaction to be carried out in a single tube. This process was further improved with the inclusion of a thermostable DNA polymerase for the extension. Hence, the entire reaction can now be carried out using a PCR thermocycler. With the use of a thermostable polymerase, the formation of DNA secondary structures can be reduced as higher temperatures (just like in PCR) could be used to denature the template. The resultant cycle sequencing products (different lengths of DNA fragments terminated in a fluorescent dye-labeled dideoxynucleotide) would be resolved by size on a denaturing polyacrylamide gel. The emissions of each fluorescent dye would then be scanned and read by lasers, which are part of the automated gel readers. An electropherogram would be generated, where the different bases indicated by four different colors line up to give the ensuing sequence order (Box 7.3).

Box 7.3

Automated Fluorescent DNA Cycle Sequencing

1. A mixture of normal dNTPs (dATP, dTTP, dCTP, dGTP) together with a small amount of the relevant ddNTPs, thermostable DNA polymerase, primer and template would be subjected to thermal cycling conditions.
2. Incorporation of a particular ddNTP (for example, a ddCTP) would terminate the chain elongation reaction.
3. Different sizes of DNA fragments (which terminate in a relevant ddNTP) would be subsequently generated.

Box 7.3   *(Cont'd)*

4. The different fragments would be separated by polyacrylamide gel electrophoresis.
5. The automated gel reader would then scan through the fluorescent dye-labeled fragments and identify the final base of each fragment (base calling).
6. An electropherogram (an example only) would be generated with four colors depicting each base. The original DNA sequence would then be recreated. Depending on factors such as the quality of the template and the G+C content of the gene, reads of about 400 to 700 bases are commonly obtainable.

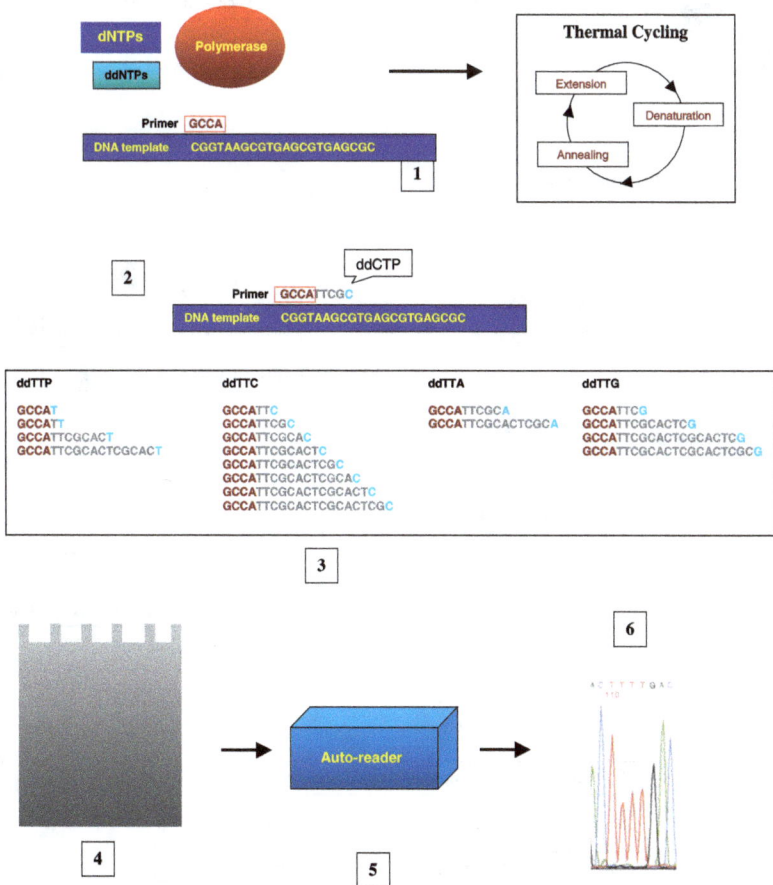

In protein sequencing, the general strategy would be to cleave the protein into smaller fragments and then sequence each fragment and determine their respective overlaps (Box 7.4). Firstly, the amino acid composition of a particular protein would be determined, followed by sequencing at the N-terminal and C-terminal portion. From then on, the

Box 7.4

Protein Sequencing Strategy

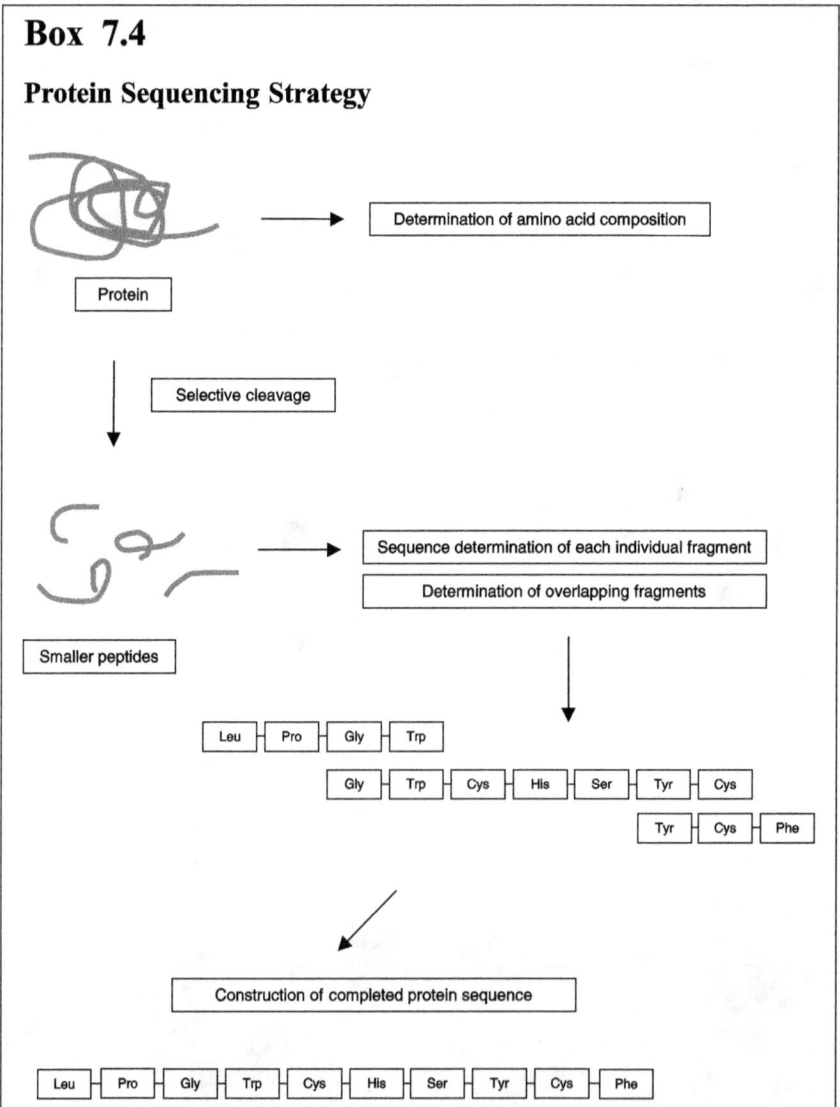

Protein → Determination of amino acid composition

Selective cleavage

Smaller peptides → Sequence determination of each individual fragment

Determination of overlapping fragments

| Leu | Pro | Gly | Trp |

| Gly | Trp | Cys | His | Ser | Tyr | Cys |

| Tyr | Cys | Phe |

Construction of completed protein sequence

| Leu | Pro | Gly | Trp | Cys | His | Ser | Tyr | Cys | Phe |

protein would be cleaved into smaller peptide fragments. The sequence of these peptide fragments would then be determined individually until an overlapping sequence is obtained. This is performed using different proteases that cleave at different locations of the peptide such as chymotrypsin, pepsin and trypsin or a chemical agent such as cyanogen bromide. With the overlapping fragments determined, the entire protein sequence is then reconstructed.

However, it has become the norm to obtain the primary sequence of a protein from the conceptual translation of the DNA sequence itself because the progress of DNA sequencing technology has exceeded that of the protein sequencing techniques. The protein product is more commonly obtained by expression of the gene sequence (a complete open reading frame) in heterologous systems and subsequently purified to obtain large amounts for analysis. On the other hand, protein sequencing is still useful and needed as direct translation methods do not tell us very much about the actual role of the protein in a microorganism. Whether the actual protein would be produced in the microorganism under certain physiological or environmental conditions or whether it is processed proteolytically or modified post-translationally, the direct cloning and sequencing of a gene and the subsequent translation to the protein product would not provide us with such information. As such, although the magnitude of DNA sequence data, due mainly to the fervor of the microbial genome sequencing projects, far exceeds that of direct protein sequencing, scientific information that could be gleaned from direct protein sequencing is still pertinent.

7.4. Outline of a Search Strategy

Before embarking on a search, it would be prudent to devise a search strategy. This will assist the researcher in determining the steps that need to be taken and the various options available to determine or to find out some clues concerning the sequence of interest (Fig. 7.2). Most of us would like to know whether a newly obtained sequence codes for any known protein or carries a complete or partial open reading frame (ORF). If it does code for some known protein homolog, we would then proceed to find out the function of the protein or whether it belongs to a certain class of tertiary structures. Thus, using biocomputing tools to help us address some of

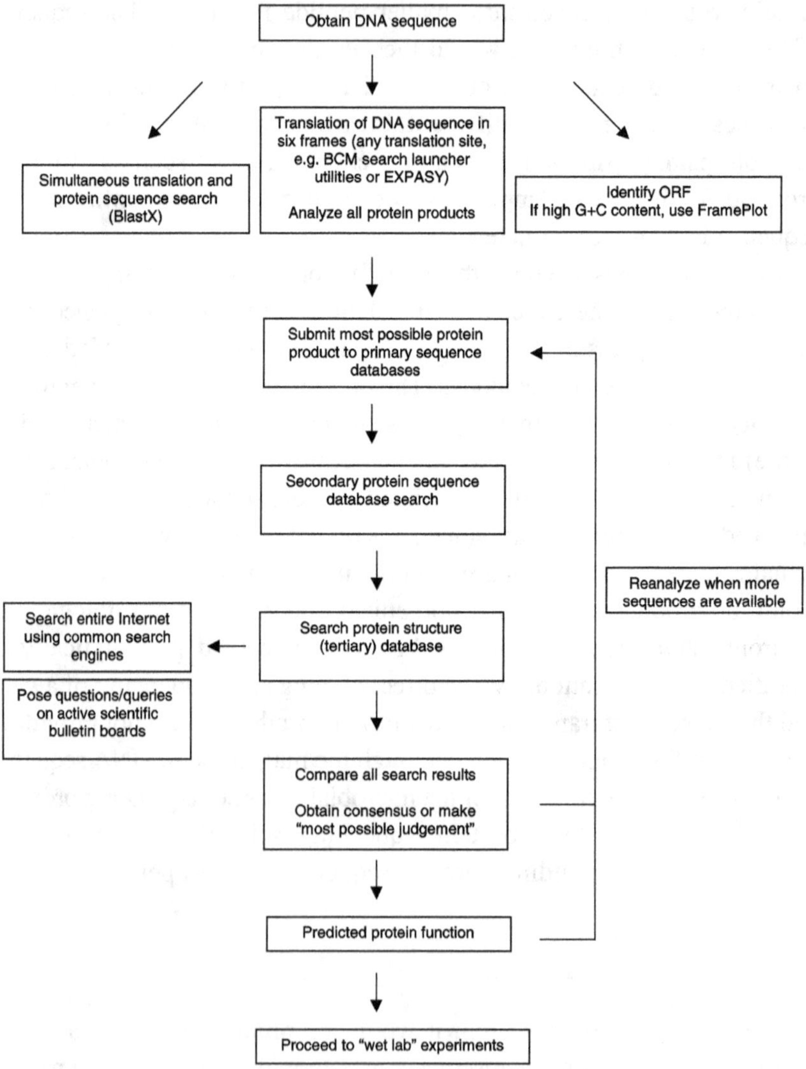

Obtain DNA sequence

Simultaneous translation and protein sequence search (BlastX)

Translation of DNA sequence in six frames (any translation site, e.g. BCM search launcher utilities or EXPASY)

Analyze all protein products

Identify ORF
If high G+C content, use FramePlot

Submit most possible protein product to primary sequence databases

Secondary protein sequence database search

Search entire Internet using common search engines

Pose questions/queries on active scientific bulletin boards

Search protein structure (tertiary) database

Reanalyze when more sequences are available

Compare all search results

Obtain consensus or make "most possible judgement"

Predicted protein function

Proceed to "wet lab" experiments

Fig. 7.2. Outline of a search strategy for a useful gene based on computational methods.

these queries would hopefully point us in the right direction before we begin to conduct "wet lab" experiments to confirm the findings predicted or suggested by computational analysis.

The initial step before conducting a search would be to confirm the accuracy of the DNA sequence obtained. This would require sequencing the DNA fragment of interest in both directions in order to remove any possible ambiguities as errors in the sequence may give rise to frameshifts or changes in the codon, which would then give us an incorrect protein sequence. Once the DNA sequence has been properly determined, the first step in a search strategy would be to obtain the translated protein product in six frames and if possible, ascertain the correct ORF. In the case of partial genes, only a partial ORF may be identified. After that, a search of the primary protein database would be performed, followed by searches through the secondary databases such as BLOCKS, Pfam, PRINTS and PROSITE. The secondary searches would most likely provide information on crucial motifs or functional features within the protein molecule itself. A search of the tertiary structure database would then be performed to see if any homologous structures are available. Sometimes, the information obtained from the database of secondary or tertiary protein folds might provide some clues in relation to the function of the protein. When all the relevant hits or matches have been obtained from the different databases, they would need to be compared to see whether all the returned query results are in agreement. In an ideal situation, all of the hits should point to the same protein or its homolog. Otherwise, one would need to make the best possible judgment from the results obtained. Thus, it would be wiser to use more methods or databases for the searches as no one method is infallible and no one database is absolute.

The return of a perfect match, especially from the primary protein database, would be excellent as it tells exactly the nature and the function of the protein. However, it also means that the queried sequence is not novel but has already been discovered. Thus, sometimes it is more valuable to obtain a significant hit which is not exactly matching as this would then be considered a novel discovery. Another consideration regarding sequence searches is the twilight zone, which is a zone of sequence identity less than 20% as this zone represents alignments or matches which may still appear credible or reasonable but are no longer statistically significant. Whenever this happens, further analyzes at the secondary or tertiary levels might be more revealing of its function.

The definitive test of function still needs to be determined experimentally although the ultimate aim of biocomputational analysis is to predict the function of a given protein from the primary sequence itself. Until that becomes a universal foolproof method, proper laboratory experiments are still very much required. We should not, however, discount the power of computing but instead make the most out of the applications available to us in our research.

7.5. Data Mining for Gene Hunting

Data mining involves the searching of databases for complementary or related sequences resembling novel DNA or protein sequences. It is important not to rely on specific methods but to explore different databases with different tools and then establish a consensus view or opinion from various approaches. This is because no database is complete. Using the Internet, a global network of computer networks, one can access different sites especially those in the United States of America, which are by far the most popular sites although other European and Asia-Pacific sites are available. The databases and computational programs that are freely available can be accessed using common web browsers such as Internet Explorer or Netscape Navigator. The leading bio-information provider in the United States of America is the National Center for Biotechnology Information (NCBI), which is part of the National Institutes of Health (NIH). GenBank, the DNA sequence database maintained by NIH is one of the three international providers of sequence information, together with the European Molecular Biology Laboratory (EMBL) database and the DNA Database of Japan (DBBJ). Across the Atlantic, a consortium of bio-information sites, which include specialist providers such as the Sanger Centre, the UK-MRC Human Genome Mapping Project Resource Centre and the European Bioinformatics Institute, form the European Molecular Biology network (EMBnet).

Since sequence information is exchanged on a daily basis among the GenBank, EMBL and DBBJ databases, a search within one of these databases would normally suffice. For example, sieving through GenBank necessitates the use of the Basic Local Alignment Search Tool (BLAST®) to assist us in going through a huge collection of sequences, whether we are starting with a DNA or protein sequence. BLAST® comes in different forms but is fundamentally a set of similarity search programs intended for the exploration of a particular database. With BLAST®, the search for a similar gene would be easier than finding a needle in a haystack although sometimes, one must still validate and confirm the accuracy of the sequence data through subsequent experiments. This is not to say that the sequence data is inaccurate as it is widely held that almost all of the sequences deposited in the databases are of excellent quality but nevertheless, the verification of the deposited sequences is the responsibility of the depositor, not the database provider itself.

If a search through the public databases results in a null return, it may be necessary to search through the raw or unannotated microbial genomic sequence data in the databases of the genome-sequencing center that is involved in the sequencing project. This kind of search is sometimes useful as full-length genes which are yet to be annotated may be discovered. In normal practice, the raw sequence data generated by these sequencing centers would first be annotated by their own computer biologists before depositing into either one of the 3 public sequence databases. The sequence data from most of the sequencing centers that are publicly funded or related to academic institutions would be easily accessible without charge whereas a fee is customarily required for accessing the sequence databases of commercial companies which are sequencing and exploiting genes discovered for profit.

At times it may be prudent to join an email subscription list where up-to-date scientific information can be received either on a daily or weekly basis. This is a good method to keep track of the latest information, which sometimes may be released unannounced except through email by database providers or sequencing centers. Scientific bulletin boards may also be useful for posing questions or asking for suggestions online, depending on how sensitive the ideas or findings may be. On the contrary, one may not get a response immediately if the chosen bulletin board is not particularly active. If scientific sites and databases fail in coming up with the relevant answers, one may decide, as a last resort, to search the entire web of information using popular search engines such as Google, Lycos or Altavista. There are plenty of such engines available and the choice of a particular engine is subjective as each of them relies on a different algorithm or search method. Nevertheless, as a vast number of sites and any databases therein are easily accessible within a "click" away, depending on the speed of the Internet line and the current Internet traffic, it is handy for the scientist to take advantage of the ever-increasing information available on the Internet.

7.6. Sequence Comparisons and Alignment

It is, *a priori*, important to obtain and ensure that the sequence data is correct and of high quality before embarking on any methods of comparison. An incorrect sequence would predict an incorrect "mutation" and it is prudent to ensure that an actual mutation has taken place rather than a fortuitous error in the sequence. Understanding mutations affecting a protein's function is significant, for example, in the hemoglobin A chain, a codon for glutamic acid (GAA), when altered to GUA, results in fatal sickle cell anemia in a homozygous individual. Likewise, the accuracy of sequence data is critical in phylogenetic analysis as "silent" mutations can only be detected at the DNA level but not at the protein level. Differences at the DNA level are likely to reflect more precisely the relationship distance between species when phylogenetic trees are drawn.

As mentioned earlier, DNA sequencing can now be performed with ease owing to the high-throughput application of automated fluorescent cycle sequencing. Although the problem or question of redundancy exists, the analysis of DNA sequences are necessary, to mention a few of the applications involved, for the inference of intron/exons, the locations of open reading frames and phylogenetic analysis. From the DNA sequence, it is imperative that the location of a possible coding region (an ORF) for a protein be determined (Box 7.5). With the predicted primary protein sequence, we can then compare with other known proteins in the databases and possibly formulate some functional and structural suggestions or even study the molecular phylogeny of this putative protein (either similarity or divergence).

There are numerous tools for sequence analysis freely available in the public domain and these are easily accessible online from different websites. Typically, text-based queries are used to identify relationships between a newly determined sequence and a known gene or protein

Box 7.5

ORF Analysis (Introns, Exons and ESTs)

ORF analysis attempts to determine the correct gene coding sequence for a putative protein product. This requires the identification of the right start codon, usually a methionine (ATG), and an uninterrupted DNA sequence in the correct frame (out of six possible choices, ending in a stop codon of either TAA, TAG or TGA). The central dogma of molecular biology states, in simple terms, that DNA begets mRNA by transcription, which is then translated into protein. In prokaryotic systems (like bacteria), the mRNA is directly translated into protein whereas in eukaryotes, the introns would be spliced out and the exons are then joined together before the mRNA gets translated. Sometimes the gene products could be in different lengths due to the alternative splicing of mRNA.

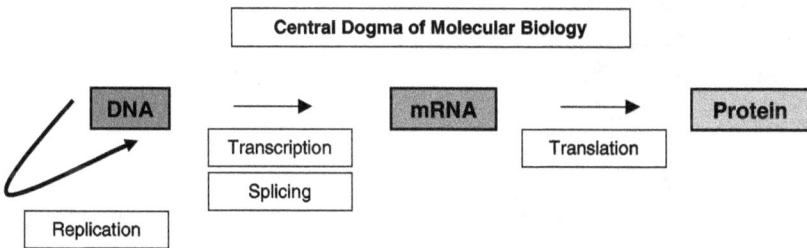

It is also possible to watch out for Kozak sequences (CCGCCAUGG) in the mRNA transcript, which may hint towards a possible starting coding sequence. Alternatively, one may compare the codon usage or detect any possible promoter or ribosome-binding sites. If a homologous protein sequence is available in the database, the likelihood of finding the correct start codon increases when a pairwise alignment or comparison is made.

Expressed sequence tags (ESTs) are short segments of DNA, which are generated from a cDNA library or mRNA from tissues or eukaryotic organisms by single-pass sequencing from either the 5' or 3' ends. They could represent coding or non-coding regions and normally consist of exon-type material. Although they are generated randomly, ESTs could be a rich source of information for finding novel genes or for identifying coding regions of a known gene that could be expressed under certain conditions. Similar to standard sequence databases, EST databases are available for searching and comparing against known ESTs generated from different organisms.

family to assess the extent of shared similarity. Some of these methods or approaches are described below.

(1) Pairwise Database Searching

In database searching, queries can either be protein, DNA or RNA. The underlying algorithm is mathematical in nature and seeks to provide the closest match between a query and the subjects available in the database. Two of the more popular programs available are FastA and BLAST®. Efficiency is necessary as databases increase rapidly and most of these searches are carried out in real-time mode. Speed is of essence although it is largely dependent on the length of the query sequence and the ever-increasing database size. Both of these programs can be run in default or "automatic" mode although more advance users may choose to modify the parameters involved. A typical output from BLAST® would show the extent of sequence identity, similarity and a statistical value (Box 7.6). Newer versions of BLAST® are able to generate gapped alignments and a hybrid of BLAST®, called Position-Specific Iterated (PSI)-BLAST®, is able to derive motifs from an initial database search.

(2) Multiple Sequence Alignment

When it is necessary to compare two or more sequences, be it protein or DNA, the use of multiple sequence alignment algorithms come into play. These algorithms are powerful tools for finding similar or even dissimilar biologically significant features. Similarities (or conventionally known as "consensus") that are extracted from the sequences are normally useful for predicting motifs and consequently, any functional characteristics and for studying the evolutionary history of these sequences. As such, a multiple sequence alignment is a basic prerequisite in order to carry out phylogenetic analysis among a group of homologous sequences.

Clustal is one of the more popular tools available and is arguably the most well known among others, and can be used with many different input formats such as FastA/Pearson, GCG/MSF and SWISS-PROT. This program can be downloaded for usage in the DOS environment but a web-based version is also available. Clustal approaches multiple sequence alignment in an evolutionary manner and the scoring of sequences are relative to each other. The output formats are normally in Clustal format (default unless specific) but the program is flexible enough to accommodate output formats such as Phylip or GCG.

Box 7.6

Example of a BLAST® Output

BLASTP 2.2.1
Reference: Altschul, Stephen F., Thomas L. Madden, Alejandro A. Schäffer, Jinghui Zhang, Zheng Zhang, Webb Miller, and David J. Lipman (1997), "Gapped BLAST and PSI-BLAST: a new generation of protein database search programs", Nucleic Acids Res. 25:3389-3402.

RID: 997544736-18339-23564

Query= (120 letters)
Database: Non-redundant SwissProt sequences 99,924 sequences; 36,332,182 total letters

Sequences producing significant alignments:

	Score (bits)	E Value
gi|121529|sp|P13006|GOX_ASPNG GLUCOSE OXIDASE PRECURSOR (GL...229	7e-61
gi|3023881|sp|P81156|GOX_PENAG GLUCOSE OXIDASE (GLUCOSE OXY... 142	6e-35
gi|3287841|sp|Q92452|GOX_TALFL GLUCOSE OXIDASE PRECURSOR (G... 130	4e-31

Alignments

```
>gi|121529|sp|P13006|GOX_ASPNG GLUCOSE OXIDASE PRECURSOR (GLUCOSE OXYHYDRASE)
(GOD)
BETA-D-GLUCOSE: OXYGEN 1-OXIDO-REDUCTASE)
Length = 605

Score = 229 bits (583), Expect = 7e-61
Identities = 114/120 (95%), Positives = 114/120 (95%)

Query: 1    TENPNISALVIESGSYESDRGPIIEDLNAYGDIFGSSVDHAYETVELATNNQTALIRSGN 60
            TENPNIS LVIESGSYESDRGPIIEDLNAYGDIFGSSVDHAYETVELATNNQTALIRSGN
Sbjct: 61   TENPNISVLVIESGSYESDRGPIIEDLNAYGDIFGSSVDHAYETVELATNNQTALIRSGN 120

Query: 61   GLGGSTLVNGGTWTRQHKAQVDSWETEFGNEGGNWDNVAAYSLQAERHRAPNAKQILAGH 120
            GLGGSTLVNGGTWTR HKAQVDSWET FGNEG NWDNVAAYSLQAER RAPNAKQI AGH
Sbjct: 121  GLGGSTLVNGGTWTRPHKAQVDSWETVFGNEGWNWDNVAAYSLQAERARAPNAKQIAAGH 180
```

Using a partial protein fragment (120 amino acids) which is highly homologous to the glucose oxidase from *Aspergillus niger* (a microbial source), the following output was obtained from BLASTP (protein database search). For the purpose of this example, only the relevant parts are shown. The following terms are explained as follows:

(i) Identity — percentage of exact sequence match (either amino acid or DNA base).

(ii) Positives — percentage of identities and similarities between the query sequence and subject sequence.

(iii) Expect (E-value or p-value) — probability or expected frequency value. Denotes statistical significance. Low values indicate greater statistical significance (less by chance), which is when the E-value approaches zero, the less likelihood that the match would have occurred by chance.

Box 7.6 (*Cont'd*)

(iv) Query — query sequence entered by the user (input).
(v) Subject — sequence in the database which is being compared to.
(vi) Length — length of the entire subject sequence in the database.

In the DOS-based Clustal program, a text file is first generated with all the sequences (typically in FastA format) needing to be analyzed. This file is the input file and it has to be entered into the program. Following that, the user can define some of the parameters involved by selecting the 'options' portion of the program. These include the type of output format, the arrangement of the sequences aligned (either according to input order or by a default output order) and whether or not the sequences are numbered. The consensus sequences (in Clustal format) obtained which are exactly identical (whether protein or DNA) will be denoted by an asterisk — "*" (Box 7.7). If shading or coloring of the sequences according to sequence identity or similarity is required, the output file (without numbering) can be entered into the BOXSHADE program, which is available online or as a DOS version that is downloadable. One drawback however, is that manual adjustment or realignment of the sequences cannot be done unless other programs are used to manipulate the outputs. Clustal is also able to generate a phylogenetic tree using the Phylip output format and provide bootstrapping values for the tree generated.

(3) Secondary Databases
Searching through secondary databases is an approach which is slanted towards recognizing motifs and patterns within a group of related sequences. Information from secondary databases is derived from primary databases and this is to ensure that we do not miss out regions which are repetitious or are part of a multi-domain sequence. Furthermore, the initial level of noise found in primary databases can be avoided. There are different secondary databases based on different methods of analysis and the diverse biological features therein.

- PROSITE is a database which consists of protein families and domains and it taps into the resources of SWISS-PROT. It primarily identifies

Box 7.7

Example of a Clustal Output

```
CLUSTAL W (1.74) multiple sequence alignment

C_ACREMONIUM       MGSVPVPVANVPRIDVSP-LFGDDKEKKLEVARAIDAASRDTGFFYAVNHGVDLPWLSRE 59
S_CLAVULIGERUS     -MPVLMPSAHVPTIDISP-LFGTDAAAKKRVAEEIHGACRGSGFFYATNHGVDVQQLQDV 58
S_LIPMANII         -MPVLMPSADVPTIDISP-LFGTDPDAKAHVARQINEACRGSGFFYASHHGIDVRRLQDV 58
S_GRISEUS          -MPIPMLPAHVPTIDISP-LSGGDADDKKRVAQEINKACRESGFFYASHHGIDVQLLKDV 58
S_JUMMONJINENSIS   -MPILMPSAEVPTIDISP-LSGDDAKAKQRVAQEINKAARGSGFFYASNHGVDVQLLQDV 58
N_LACTAMDURANS     ---MKMPSAEVPTIDVSP-LFGDDAQEKVRVGQEINKACRGSGFFYAANHGVDVQRLQDV 56
S_CATTLEYA         -MPVLMPSADVPTIDISPQLFGTDPTPRRTSRGRSTRPARGSGFFYASHHGIDVRRLQTW 59
FLAVBACTERIUM      ----MNRHADVPVIDISG-LSGNDMDVKKDIAARIDRACRGSGFFYAANHGVDLAALQKF 55
L_LACTAMGENUS      ----MNRHADVPVIDISG-LSGNDMDVKKDIAARIDRACRGSGFFYAANHGVDLAALQKF 55
A_NIDULANS         MGSVSK--ANVPKIDVSP-LFGDDQAAKMRVAQQIDAASRDTGFFYAVNHGINVQRLSQK 57
A_CHRYSOGENUM      MGSVPVPVANVPRIDVSP-LFGDDKEKKLEVARAIDAASRDTGFFYAVNHGVDLPWLSRE 59
P_CHRYSOGENUM_1    MASTPK--ANVPKIDVSP-LFGDNMEEKMKVARAIDAASRDTGFFYAVNHGVDVKRLSNK 57
P_CHRYSOGENUM_2    MASTPK--ANVPKIDVSP-LFGDNMEEKMKVARAIDAASRDTGFFYAVNHGVDVKRLSNK 57
                    *.** **:* * * :   :       ..* :***** :**::: **.

C_ACREMONIUM       TNKFHMSITDEEKWQLAIRAYNKEHESQIRAGYYLPIPGKKAVESFCYLNPSFSPDHPRI 119
S_CLAVULIGERUS     VNEFHGAMTDQEKHDLAIHAYNPDNP-HVRNGYYKAVPGRKAVESFCYLNPDFGEDHPMI 117
S_LIPMANII         VNEFHRTMTDQEKHDLAIHAYNENNS-HVRNGYYMARPGRKTVESWCYLNPSFGEDHPMI 117
S_GRISEUS          VNEFHRTMTDEEKYDLAINAYNKNNP-RTRNGYYMAVKGKKAVESFCYLNPSFSEDHPMI 117
S_JUMMONJINENSIS   VNEFHRNMSDQEKHDLAINAYNKDNP-HVRNGYYKAIKGKKAVESFCYLNPSFSDDHPMI 117
N_LACTAMDURANS     VNEFHRTMSPQEKYDLAIHAYNKNNS-HVRNGYYMAIEGKKAVESFCYLNPSFSEDHPEI 115
S_CATTLEYA         SNES-TTMTDQRSTTWRSTRYNENNS-HVRNGYYMARPGRETVESWCYLNPSFGEDHPMM 117
FLAVBACTERIUM      TTDWHMAMSAEEKWELAIRAYNPANP-RNRNGYYMAVEGKKANESFCYLNPSFDADHATI 114
L_LACTAMGENUS      TTDWHMAMSPEEKWELAIRAYNPANP-RNRNGYYMAVEGKKANESFCYLNPSFDADHATI 114
A_NIDULANS         TKEFHMSITPEEKWDLAIRAYNKEHQDQVRAGYYLSIPGKKAVESFCYLNPNFTPDHPRI 117
A_CHRYSOGENUM      TNKFHMSITDEEKWQLAIRAYNKEHESQIRAGYYLPIPGKKAVESFCYLNPSFSPDHPRI 119
P_CHRYSOGENUM_1    TREFHFSITDEEKWDLAIRAYNKEHQDQIRAGYYLSIPEKKAVESFCYLNPNFKPDHPLI 117
P_CHRYSOGENUM_2    TREFHFSITDEEKWDLAIRAYNKEHQDQIRAGYYLSIPEKKAVESFCYLNPNFKPDHPLI 117
                    .   :: :..     ** : * *** .   ::: **:*****.*  **. :
```

This is the output (partially shown only) from the alignment of 13 isopenicillin
N synthase (IPNS) isozymes from different microbial species obtained from
GenBank. IPNS is a secondary metabolism enzyme that is involved in the
penicillin and cephalosporin antibiotic biosynthetic pathway. The version
of Clustal used is ClustalW 1.74 and the meaning of the symbols are as
follows:

(i) "*" — positions are fully conserved.
(ii) ":" — one of the following "strong" groups is fully conserved: STA,
     NEQK, NHQK, NDEQ, QHRK, MILV, MILF, HY, FYW.
(iii) "." — one of the following "weaker" groups is fully conserved: CSA,
     ATV, SAG, STNK, STPA, SGND, SNDEQK, NDEQHK, NEQHRK,
     FVLIH, FYM.

biological signatures by locating a single most conserved motif or pattern. The motifs are shown as regular expression patterns.

- BLOCKS classify highly conserved regions of proteins and arrange them in a "block" fashion for comparison. These are multiple-aligned ungapped segments of sequences.
- PRINTS is a protein fingerprints' database. It consists of a group of conserved motifs and the fingerprints are determined by iterative scanning of the primary protein databases. In an analysis, more than one motif may be identified along a certain sequence, thus allowing a more thorough assessment.
- IDENTIFY is derived from BLOCKS and PRINTS. The motif obtained is displayed as a fuzzy expression pattern, which tolerates amino acid residues within a certain functional grouping.
- InterPro is another protein resource database that combines the various features of Pfam, PRINTS, PROSITE and so on. It is a useful source of information for proteomic analysis.
- ProDom is a protein domain database where the homologous domains are automatically compiled.
- Pfam runs on Hidden Markov Models and consists of multiple sequence alignments of many common domains. Families of aligned domains would then be created for analysis.
- Profile analyzes distance sequence relationships and takes into consideration sequences with very few conserved residues. The algorithm involved tolerates insertions and deletions in the sequences.

7.7. Protein Structure Prediction

One of the ultimate goals in protein structure prediction is to be able to predict the tertiary conformation from the primary sequence alone. However, due to the complexity of protein folding, one should not be over-enthusiastic about the accuracy of available programs and software packages in terms of predicting the structure of proteins. It is difficult to predict how a protein will eventually fold, if given a primary sequence, due to the numerous factors involved. Thus, structure and function cannot simply be inferred from sequence alone. Typically, there are four levels of protein structure, from the primary to the quaternary (Box 7.8) and computational

Box 7.8

The Four Levels of Protein Structure

• Primary structure — the linear polypeptide sequence of amino acids.

• Secondary structure — local regular regions such as α-helices, β-strands.

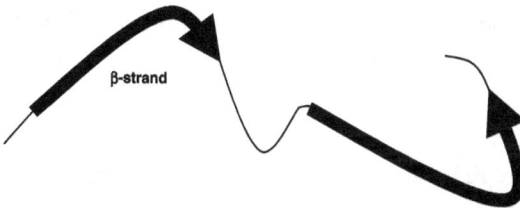

β-strand

• Tertiary structure — packing of secondary structures, giving the overall fold.

• Quaternary structure — arrangement of several polypeptide chains.

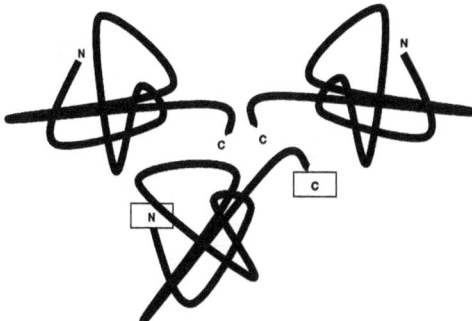

methods are now able to predict secondary and tertiary structures and hopefully, provide some insights into the function. Experimental methods such as NMR or 3D X-Ray crystallography are still very much the pillars of elucidating protein structures and they would be the final validation point of any predicted tertiary structure. Nevertheless, it has been possible to predict to a certain extent and to a fair degree of accuracy beyond merely a polypeptide backbone if a homologous structure is available in the structure database and is used as a starting template.

An approach to protein structure prediction should start with the primary protein sequence (Fig. 7.3). After which, it may be informative to obtain as much data on the primary structure as possible, such as disulfide bonds, active site residues and any available spectroscopic data. Searching the databases may also uncover known homologs or any functional domains. If a closely related structural homolog was available in the protein structure database, i.e. the PDB, then tertiary structure modeling would be relatively straightforward. The primary sequence would then be entered into the modeling program to obtain a predicted structure by homology. Modeling programs include the SWISS-MODEL and Modeller. The SWISS-MODEL

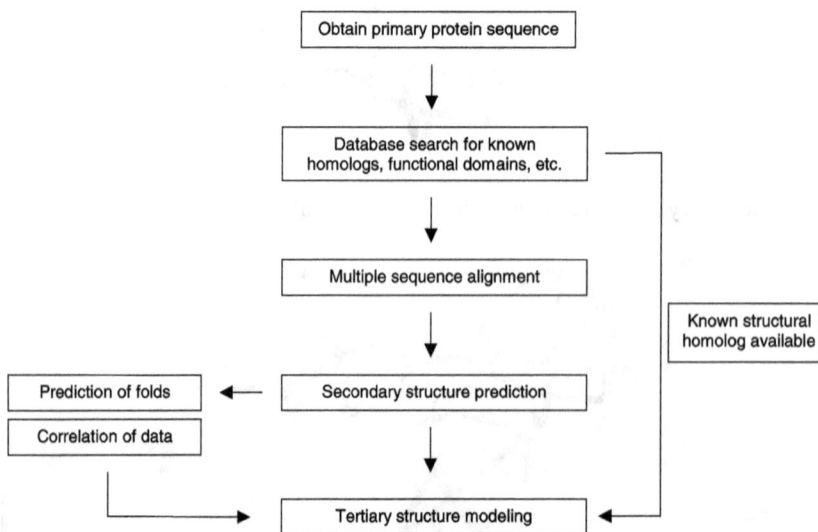

Fig. 7.3. An approach to Protein Structure Prediction.

is an automated protein-modeling program that is reasonably easy to handle and would generate a full set of atomic coordinates from a homology-modeling run.

If there are no known homologs, secondary structure predictions can then be performed although the accuracy of most programs is only approximately 50–60%. Links to most of these secondary prediction sites, which include SOPM, JPRED, PREDATOR, PHD, nnPredict and so on, are available from the ExPASy server. As an example, the PHD program, which is based on neural networks, is able to provide secondary structure predictions of a protein with locations of the α-helices and β-sheets, solvent accessibility and transmembrane helix predictions. It might also be more useful if a collection of programs is used to generate a consensus representation of all the secondary structures generated instead of relying on a single program. Analysis of folds is also another method for predicting whether or not a certain amino acid sequence can adopt a particular fold. These are carried out by fold recognition methods which will attempt to predict how known folds (available in the databases) would fit into the sequence being analyzed. Although not highly accurate and *ab initio* predictions are not possible, the analysis of folds would contribute towards predicting the overall structure. Classification of protein structures can be found in the SCOP and CATH databases.

After all the secondary structure and fold analyses have been performed, it would then be necessary to create a predicted tertiary structure. This structure would need to be refined as more information is available. Basically, homology modeling and structure prediction would be improved and made much easier if more protein structures are available for comparison. This is to ensure that the entire spectrum of structure types is covered. This would not be an easy task as comparing to the millions of sequences in GenBank, the number of protein structures in PDB is currently only approaching 20,000.

7.8. Useful Tools and Websites on the Internet for Microbiological Research

Some of the more common tools and free programs available on the Internet are listed (Table 7.1). As the information on the Internet constantly

Table 7.1. List of useful tools and websites on the Internet.

Description of sites	Name of website	Website address
Resource site for Molecular Biology information, e.g. Genbank, PubMed	National Center for Biotechnology Information	http://www.ncbi.nlm.nih.gov
International repository for the processing and distribution of 3-D macromolecular structure data	Protein Databank (PDB)	http://www.rcsb.org/pdb/ index.html
European Molecular Biology Network	EMBnet	http://www.embnet.org
European Molecular Biology Laboratories	EMBL	http://www.embl-heidelberg. de
DNA Database of Japan	DBBJ	http://www.ddbj.nig.ac.jp
European Bioinformatics Institute	EBI	http://www.ebi.ac.uk/ebi_ home.html
Genome Research Centre founded by the Wellcome Trust and MRC	Sanger Centre	http://www.sanger.ac.uk
Protein Information Resource	PIR	http://pir.georgetown.edu
Annotated protein sequence database	SWISS-PROT	http://www.expasy.ch/sprot/ sprot-top.html
Homologous Structure Alignment Database	HOMSTRAD	http://www-cryst.bioc.cam. ac.uk~homstrad/
Composite database of SWISS-PROT, PIR, GenBank and NRL-3D.	OWL	http://www.bioinf.man.ac.uk/ dbbrowser/OWL
Database of protein families and domains	PROSITE	http://www.expasy.ch/prosite
Protein fingerprints database	PRINTS	http://www.bioinf.man.ac.uk/ dbbrowser/PRINTS
Database for aligned ungapped segments of the most highly conserved regions of proteins	BLOCKS	http://www.blocks.fhcrc.org
Database of multiple sequence alignments and HMM of common protein domains	Pfam	http://www.sanger.ac.uk/ Software/Pfam
Searching both BLOCKS and PRINTS	IDENTIFY	http://dna.Stanford.edu/ identify
Profile scan server for searching against a protein database	Profiles	http://www.isrec.isb-sib.ch/ software/FSCAN_form. html

Table 7.1 (*Continued*)

Description of sites	Name of website	Website address
Structural classifications of proteins database	SCOP	http://scop.mrc-lmb.cam.ac.uk/scop
Protein structure classification	CATH	http://www.biochem.ucl.ac.uk/bsm/cath_new/
Motif search utility of GenomeNet, Kyoto University, Japan	Protein and Nucleic Acid Sequence Motifs search	http://motif.genome.ad.jp/
Restriction enzyme analysis of DNA sequences	Webcutter	http://www.firstmarket.com/cutter/cut2.html
Restriction enzyme database	Rebase	http://rebase.neb.com/ rebase/rebase.html
Analysis of DNA and protein sequences and structures (many tools and links available)	ExPASy Molecular Biology Server	http://www.expasy.ch/
Analysis of protein sequences	ExPASy Proteomics tools	http://tw.expasy.org/tools/
BLAST® server (all types available) at the National Center for Biotechnology Information	NCBI BLAST®	http://www.ncbi.nlm.nih.gov/BLAST/
Sequence search analysis	FastA server	http://www.ebi.ac.uk/fasta33/
Frame analysis of protein-coding region of high G+C content	FramePlot 2.3	http://www.nih.go.jp/~jun/cgi-bin/frameplot.pl
Promoter Prediction	Neural Network Promoter Prediction	http://www.fruitfly.org/seq_tools/promoter.html
Molecular Analysis Tools at the University of Adelaide	A Pack of Molecular Analysis Tools	http://www.microbiology.adelaide.edu.au/learn/index.htm
Pretty Printing and Shading of Multiple-Alignment files	BOXSHADE 3.21	http://www.ch.embnet.org/software/BOX_form.html
Online version of Clustal, a multiple sequence alignment program	Clustal	http://www.ebi.ac.uk/clustalw
Automated comparative protein modeling server	SWISS-MODEL	http://www.expasy.org/swissmod/SWISS-MODEL.html
Sequence utilities	BCM Search Launcher: sequence utilities	http://searchlauncher.bcm.tmc.edu/seq-util/seq-util.html
Color Interactive Editor for Multiple Alignments	CINEMA	http://www.bioinf.man.ac.uk/dbbrowser/CINEMA2.1

Table 7.1 (*Continued*)

Description of sites	Name of website	Website address
A collection of links to information and services useful to Molecular Biologists	Pedro's BioMolecular Research Tools	http://www.public.iastate.edu/~pedro/research_tools.html
ProDom with BLAST® (graphical output)	ProDom	http://prodes.toulouse.inra.fr/prodom/doc/blast_form.html
Codon usage database	Countcodon	http://www.kazusa.or.jp/codon/countcodon.html
Collection of protein secondary structure analysis and information sites	GenomeWeb Protein Secondary Structure	http://www.hgmp.mrc.ac.uk/GenomeWeb/prot-2-struct.html
Resource site for biologists	BioMedNet	http://www.bmn.com/
Resources for the microbiologist	The Microbiology Network	http://microbiol.org/
Culture Collection	American Type Culture Collection	http://www.atcc.org/
Culture Collection	Agricultural Research Service Culture Collection	http://nrrl.ncaur.usda.gov/default.htm
Comprehensive Science-Specific Search Engine	Scirus	http://www.scirus.com/
Google	Internet Search Engine	http://www.google.com
Lycos	Internet Search Engine	http://www.lycos.com
Altavista	Internet Search Engine	http://www.altavista.com

changes and web pages are frequently replaced, removed or updated, one needs to check whether these sites are still functional. One would not believe the "speed" of the Internet as a lot of changes could take place even within a few weeks of an undergraduate year, let alone the entire course. It would be useful to maintain a list of "Favorites" or "Bookmarks" (depending on your default browser) where sites which are most frequently used can be easily accessible and monitored.

7.9. Summary

From the success of sequencing the first-ever genome, the bacteriophage ϕX174, microbial genome sequencing efforts never looked back. The scientific endeavors of cloning genes, currently in bounteous amounts and the completion of many more microbial genomes have created new fields of research such as comparative genome analysis, functional genomics, proteomics and structural genomics. These are now becoming the main players in the post-genomic age. With heightening demands of increased biological information, bioinformatics will continue to play an essential role in assimilating this massive and ever-increasing amount of information. As bioinformatics is a rapidly developing field and computing processing power can only increase (according to Moore's Law), so will the sophistication and complexity of the biocomputational software programs. Both biological research and computing are intertwined and will continue to be mutually reliant as science continues to advance.

Further Reading

1. Burge CG and Karlin S. Finding the genes in genomic DNA. *Curr. Opin. Struct. Biol.* **8**: 346–354, 1998.
2. Glick BR and Pasternak JJ, eds. *Molecular Biotechnology: Principles and Applications of Recombinant DNA*. ASM Press, USA, 2003.
3. Higgins D and Taylor W. Multiple sequence alignment. *Methods Mol. Biol.* **143**: 1–18, 2000.
4. Higgins D and Taylor W, eds., *Bioinformatics: Sequence, Structure and Databanks*. Oxford University Press, UK, 2002.
5. Holm L and Sander C. Mapping the protein universe. *Science* **273**: 595–603, 1996.
6. Koel P. Protein structure similarities. *Curr. Opin. Struct. Biol.* **11**: 348–353, 2001.
7. Orengo CA, Jones DT and Thornton JM, eds. *Bioinformatics: Genes, Proteins and Computers*. BIOS Scientific Publishers Ltd, UK, 2003.
8. Westhead DR, Parish JH and Twyman RM, eds. *Instant Notes Bioinformatics*. BIOS Scientific Publishers Ltd, UK, 2002.

Chapter 8

Quantitative Real Time PCR (QPCR)

Too Heng-Phon
Department of Biochemistry, Yong Loo Lin School of Medicine
National University of Singapore
MD 7, 8 Medical Drive, Singapore 117597
Chemical and Pharmaceutical Engineering
Singapore-Massachusetts Institute of Technology
E4-04-10, 4 Engineering Drive 3, Singapore 117576
Email: bchtoohp@nus.edu.sg

Azlinda Anwar
Program in Infectious Diseases
Duke-NUS Graduate Medical School
8 College Rd, Singapore 169857

8.1. Introduction

PCR is an essential tool for quantifying nucleic acids in many clinical and biological fields of studies. A recent innovation in PCR is the development of a state-of-the art real time detection method. Since the first report of real time detection of the PCR process by Higuchi and co-workers, several thousand publications have appeared, attesting to the acceptance of this technology into the mainstream of many research and clinical disciplines (Fig. 8.1). qPCR (also known as kinetic PCR previously) has been increasingly employed in diagnostic microbiology, virology and parasitology. It is also used in the field of hematology, oncology and immunology to aid, among others, in the quantification of aberrant gene expressions, levels

241

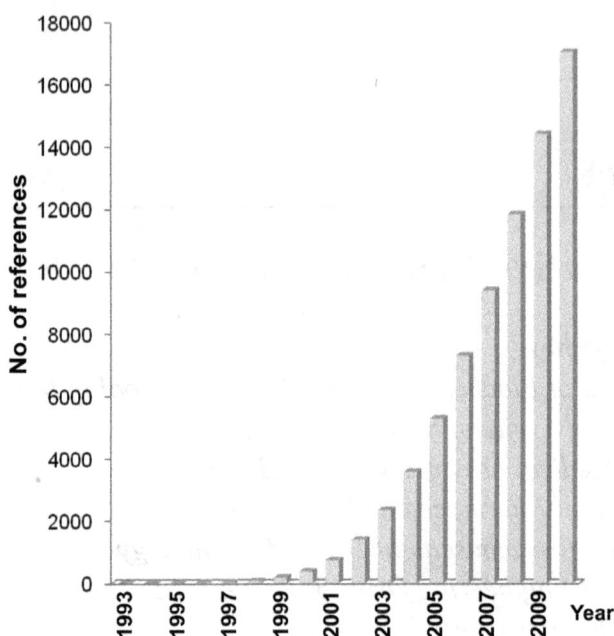

Fig. 8.1.   Number of publications with the concept "Real time PCR" or 'kinetic PCR' as reported in PubMed. The first report of real time detection of a PCR process by Higuchi and co-workers in 1993. The first commercial real time PCR instrument, the ABI Prism 7700, was introduced in 1996 by Applied Biosystems.

of circulating nucleic acid, changes in chemokines and microRNA levels, changes in transcripts levels due to RNA interference and in genotyping (copy number variations). In the field of biotechnology, qPCR has been used to genotype genetically modified organisms, monitor gene expression in bioprocesses and for evaluating the safety of biologicals. As a research tool, this technology has been employed in validating data obtained from high-throughput transcriptional profiling using DNA microarray and Next Generation Sequencing. qPCR, like DNA microarray, requires *a priori* knowledge of the query nucleic acid target.

The importance and wide use of qPCR as a quantitative tool in the life sciences is due to the inherent wide dynamic range of quantification ($10^7 - 10^9$ fold), high assay sensitivity (detection of less than 10 copies), high precision ($< 2\%$ intra-assay variation, $< 3\%$ inter-assay), with no post-PCR manipulations, thus minimizing the risk of cross-contaminations and technical variability. In addition, a distinct advantage of this technology is the ease in obtaining reliable quantitative definitions of specificity, sensitivity and efficiency of an assay as compared to the conventional end-point PCR-based assays. Thus in view of the high accuracy of measurement and reliability of qPCR, this technology will continue to be an essential tool for the quantification of gene expressions and genotyping.

8.2. What is qPCR?

The PCR process can be visualized as a cycle of three stages: denaturation, annealing and extension. During this temperatures ($>90°C$). On cooling (typically between $50–70°C$), the primers anneal onto the desired sections of the template, and are extended by a DNA polymerase at optimal temperatures (typically between $68–75°C$). Traditionally, on completion of a PCR experiment, the products are fractionated based on their sizes and detected either using DNA binding dyes (usually ethidium bromide) or by more elaborate methods, for example Southern blot-based methodologies. This end-point, conventional approach however suffers from many technical limitations, including the need for post-PCR manipulations, which increases handling time and may potentially introduce experimental errors. Several attempts have been made to develop quantitative conventional-based PCR including the use of serial dilutions of cDNA, altering the amplification cycle for each gene, using exogenous competitor templates, performing limiting dilution assays or using PCR mimics with similar primer sequences that would co-amplify with the products. All these methods however require extensive post-PCR handling and analyses. In addition, quantification of the amounts of the genes of interests is neither accurate nor reliable when determined by measuring the amounts of products made at the end point. In the initial cycles, PCR generates copies of a DNA template exponentially but this process will however not continue indefinitely. Inhibitions due to substrate depletion, loss of enzyme activity and accumulation of amplicons which compete for primer binding and other products (pyrophosphates) may be responsible for the attenuation of the exponential increase of the amplified products, resulting in the plateau phase of PCR (Fig. 8.2a). Thus, regardless of the varied amounts of starting templates, similar amounts of products may be observed at the plateau phase.

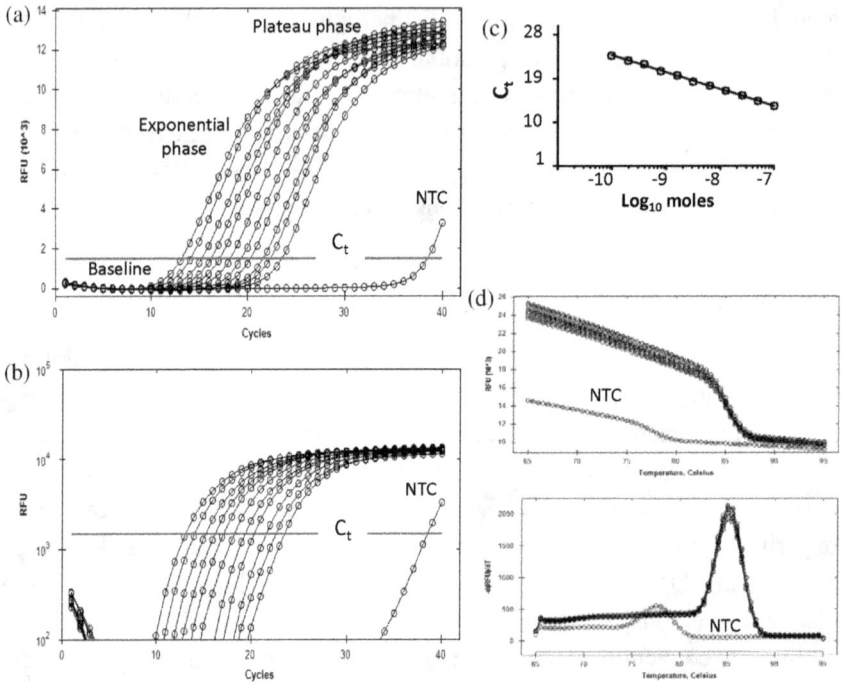

Fig. 8.2.   Plot of background subtracted relative fluorescence unit (RFU) against number of amplification cycles of two-fold diluted DNA template using SYBR Green I (a). The same data plotted as $\log_{10}$ vs. cycles (b). Plot of the C_t against the amount of initial template yields a straight line and the amplification efficiency is calculated from the slope (c). Post-PCR melt curve analyses (d). NTC, no template control. C_t, threshold cycle.

In qPCR, the accumulation of amplicons is detected and quantified periodically by fluorescence at every cycle. Since there is no requirement to isolate the probe-target hybrids from the non-hybrid probes for analysis, these assays are often referred to as "closed" or homogenous systems. A typical amplification profile of relative fluorescence unit (RFU) against the cycle number seen in qPCR is shown in Fig. 8.2a. The profile is often presented as $\log_{10}$ RFU versus cycle number (Fig. 8.2b). In the initial cycles of PCR, the fluorescent signal emitting from the sample is low and is considered to be background signal, defining the baseline for the amplification of the sample. As more amplicons accumulate in the later cycles, the signal gradually increases above the baseline. Since the intensity of the fluorescence signal is directly correlated to the concentration of amplicons, a fixed fluorescence threshold set above the baseline would therefore indicate an equal amount of amplicons in different samples. This threshold parameter is often referred to as Ct (threshold cycle) or Cp (cross-over point), and is calculated by the software of the instrument as the cycle number at which the fluorescence is significantly above a defined threshold. Most instruments calculate the average background fluorescence gathered from the first few cycles and use the readings to define statistically significant threshold signals. A linear relationship exists between the amounts of initial templates and the Ct values. Thus, a plot of the $\log_{10}$ of initial template amounts for a set of dilutions against Ct is a straight line (Fig. 8.2c). The quantification of an unknown sample is therefore carried out by comparing its Ct value against the Ct values of known amounts of serially diluted standards by interpolation.

8.3. Work Flow

Any quantitative study starts with a workflow composed of a series of interconnected and interdependent operations (Fig. 8.3). Whether measuring relative changes in gene expressions or the copy number of a particular gene, the success of the project is dictated by proper planning and management of the various steps involved. As qPCR is exquisitely sensitive, many confounding issues are known to influence the reliability of the study which include the way the study is designed (experimental design),

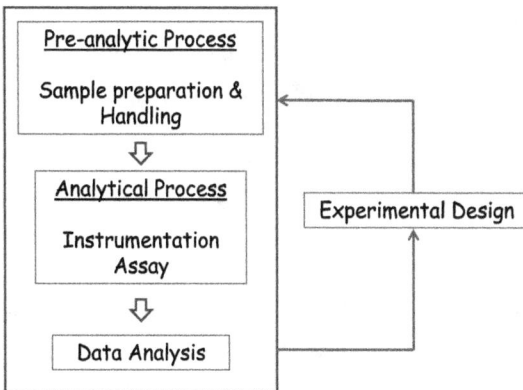

Fig. 8.3.   Work flow in a qPCR study. The various processes are interconnected and inter-related. Appropriate experimental design is necessary for the generation of reliable results.

technical and biological variability. In experimental design, proper power analysis is required to enable the assessment of the number of samples required to draw meaningful and statistical significance of the results. In addition, proper experimental setup is also an important but an often over-looked issue. Technical variability includes pre-analytics (sample prepara-tions and handling), the performance of the assay (designs of primers and probes, assay reagents) and instrumentations (variability in volume dis-pensation, qPCR thermocyclers). Biological variation is intrinsic to all organisms and may be influenced by genetic or environmental factors, as well as by whether the samples in a study are pooled or individual. Hence, the number of biological replicates required for determining the statistical significance of a study depends on the experimental design used.

8.4. Pre-Analytical Procedures

An accurate quantification of nucleic acids by qPCR is highly dependent on pre-analytical procedures. This include all processes, starting with the collection of samples (e.g. blood, tissues or cells), to the stability and quality of the nucleic acids prepared from the samples, transport, stor-age, reverse transcription (for RNA samples), right up to qPCR. There are numerous commercially available nucleic acid extraction kits. These kits are based on two common extraction procedures — liquid phase partition (e.g. TRIzol, RNAzol) and silica-based nucleic acid adsorption chromatography (e.g., High Pure and QIAamp kits). Spectra analysis by UV-Vis spectrometry (e.g. Nanodrop) is commonly used to quantify nucleic acids and identify contaminants (e.g. phenol). The quality (integrity) and quantity of the isolated nucleic acids can also be assessed with microfluidic based electrophoresis systems, e.g. 2100 Bioanalyzer (Agilent Technologies) or Experion (Biorad Laboratory). It is expected that low quality nucleic acid samples produce erroneous measurements of the true copy number and hence affect the estimation of the differ-ences in gene expressions. With RNA samples, the efficiency of the additional step to convert it into DNA (reverse transcription) may serve as yet another source of variability. Hence, the technical challenges associated with these pre-analytical issues must first be identified and optimized at the start of the study.

8.5. Instrumentation

Following the introduction of the first commercial qPCR instrument in 1996, the ABI Prism 7700 (Applied Biosystems), there is now a wide choice of instruments with varying capabilities and costs by various vendors. With the recent surge in the interest of quantifying gene expressions and the availability of economically affordable instruments, it is now quite common to find these instruments in individual laboratories. The demand for such instruments will continue and differences in costs, performances and features between instruments will gradually diminish.

Currently, qPCR detection is by fluorescence. All instruments share some common features: a module dedicated to the amplification process where the heating and cooling processes occur; a light source to excite the fluorophores; a detector to capture the emission of the fluorophores and a data analysis module where the signal is processed and displayed. The instruments either have stationary detectors with stationary sample modules (e.g. iCycler iQ, ABI 7500), rotating samples with stationary detectors (e.g. Lightcycler® 2, Rotor-gene 3000™) or mobile detectors with stationary samples (e.g. Chromo4 Opticon, MX 3005P™). Many instruments control the heating and cooling process by employing either a solid state-Peltier or a hybrid of Peltier, resistive and convective technologies, or resistive heating plates. The reaction vessels used are either polypropylene or polyethylene polymer molded into microtubes or multiwell plates, or thin glass capillaries. Most of the solid-state Peltier instruments use a 96-well plate format, while the higher throughput, automated instruments using a 384-well plate format (e.g. ABI 7900HT, Lightcycler® 480, CFX 384™). A useful feature of some, but not all qPCR instruments, is the ability to perform temperature gradient real time detection, which allows for the rapid optimization of annealing temperatures of many reactions simultaneously. In addition, many instruments have between two to six detection channels to enable multiplex studies. However, for many studies (singleplex reaction studies), the need of only one or two channels is sufficient, hence newer and more economical instruments are being produced to meet this demand. Today, qPCR themocyclers processing thousands of samples in low microliter to nanoliter volumes are commercially available (e.g. LightCycler 1536, OpenArray®, BioMark™), making qPCR a high-throughput technology.

```
                    ┌──────────────────────────────┐
                    │  QUANTIFICATION APPROACHES   │
                    └──────────────────────────────┘
                                   │
         ┌─────────────────────────┴─────────────────────────┐
         ▼                                                     ▼
┌──────────────────────────┐              ┌──────────────────────────────────────┐
│ RELATIVE METHOD(ΔΔCt)    │              │ □ ABSOLUTE □ OR STANDARD CURVE METHOD │
└──────────────────────────┘              └──────────────────────────────────────┘
         │                                                     │
         └─────────────────────────┬─────────────────────────┘
                                   ▼
                    ┌──────────────────────────────┐
                    │  EXPERIMENTAL APPROACHES     │
                    └──────────────────────────────┘
                                   │
         ┌─────────────────────────┴─────────────────────────┐
         ▼                                                     ▼
┌──────────────────────────┐              ┌──────────────────────────────────────┐
│   SEQUENCE-DEPENDENT     │              │        SEQUENCE-INDEPENDENT           │
└──────────────────────────┘              └──────────────────────────────────────┘
         │                                                     │
         ▼                                                     ▼
```

FORMATS	FORMATS
Taqman probe (hydrolysisprobe)	SYBR Green I
MGB-Taqman probes	SYBR Gold
Molecular Beacon	Ethidium bromide
Hyrbidization probes	TOTO
Scorpion	YOYO-1
Eclipse MGB probes	LC Green
Sunrise primer probes	Pico Green
Bi-probes	BEBO
Light-up probe (LUP)	BOXTO
Magiprobe	
Plextor	
LUX	
LNA	

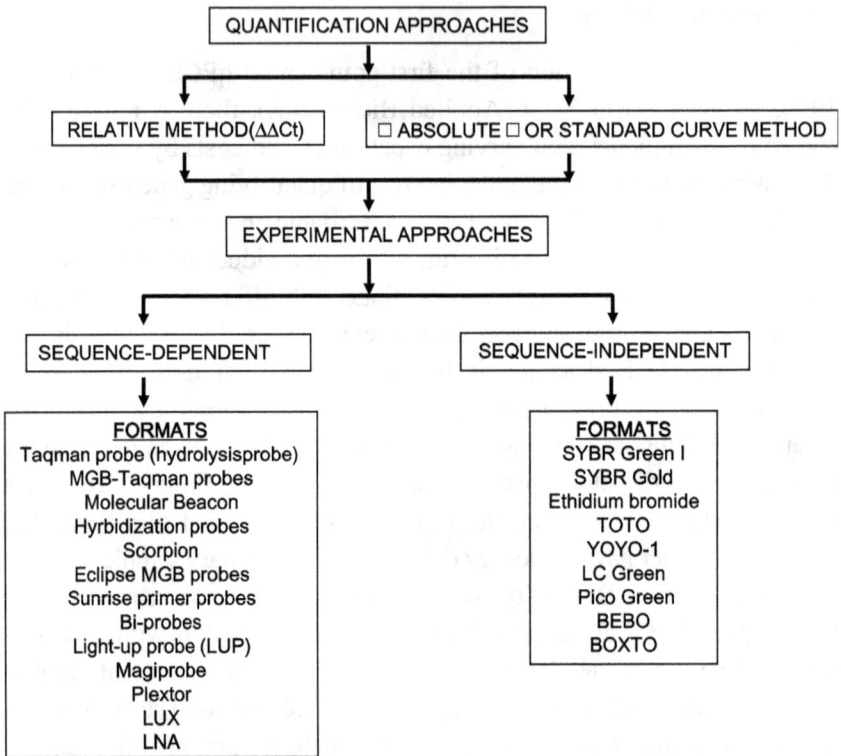

Fig. 8.4.   Overview of the quantification strategies in qPCR. A target can be quantified by either the "relative method" or "standard curve method". Different sequence-dependent and sequence-independent formats are available for quantification.

Another useful feature of most qPCR instruments is the ability to perform melt curve analysis post-PCR (Fig. 8.2d). This feature not only allows the determination of possible contributions of extraneous DNA fragments to the observed signals, but also allows for the discrimination between multiple amplicons. In well-optimized PCR conditions, a melt curve can also be used to quantify and distinguish between highly related amplicons, and to detect single nucleotide polymorphisms.

8.6. How DNA is Detected in qPCR (Assay)

Currently, a number of different experimental set-ups or formats are available for the detection of amplicons by qPCR (Fig. 8.4). One important issue is the selection of fluorescent filters with the correct excitation and emission band-pass that are suitable for the selected methods described below. There are many factors that influence the design and performance of an assay. These include the designs of specific primers/probes, the presence of potential inhibitors of PCR or reverse transcription (for RNA) in the samples, the choice of amplification reagents and the performance of the instrument used.

Online public primer and probe databases of more than 600,000 commercially pre-designed Taqman Gene expression assays (Applied Biosystems) and the availability of prevalidated genome wide real time sequence-dependent probes (ProbeLibaryTM) serve as invaluable resources for the rapid development of specific assays for a large number of genes. Similarly, less comprehensive collections can also be obtained from public databases (e.g. qPrimerDepot, RTPrimerDB, Primer Bank) or from publications. In-house assay designs are often more cost-effective but may suffer from the lack of quality controls, thus, making the assays less reproducible. To aid in the design of primers and probes, there are numerous stand-a-lone, commercial softwares (e.g. Vector NTI Software®, Beacon Designer™, Lasergene®), services offered by various oligonucleotide synthesis companies (e.g. IDT, Invitrogen, Genscript) and equipment providers (e.g. Primer Express®, LightCycler® Probe Design).

The chemical composition of the qPCR reagent resembles those used in PCR. These include deoxynucleotides, buffer, DNA polymerase, Mg2+ or Mn2+, and chemical additives to enhance the amplification reaction. There exist a bewildering number of commercially available qPCR reagent kits for the use of sequence-independent and -dependent methods. Almost all of these kits uses "hot-start", where the DNA polymerase is either chemically modified, bound by specific inhibitory antibody or by aptamers and is activated at high-temperature. Primers modified with 4-oxo-1-pentyl are also available for "hot-start" qPCR.

In selecting the mode of detection to be used for the assays, the first step is to select the type of format or "chemistry" to be used. For simplicity, these can be classified into either "sequence-independent" or "sequence-dependent" methods.

8.6.1. Sequence-independent method

In the sequence-independent method, this approach uses fluorescent molecules that bind to DNA and exhibit little or no sequence preferences

(Fig. 8.5a). Hence, detection is independent of the amplicon sequences. The binding of multiple DNA-binding fluorophores to a single amplicon will increase the sensitivity of detection. Hence, the intensity of the signal is dependent on the mass of the DNA. This widely used method is cost-effective, simple to design and easily adapted to established conventional PCR protocols. However a major disadvantage of this method is the inability of these DNA-binding fluorophores to discriminate between specific and non-specific amplicons, including primer-dimers, which may be generated during the reaction. However, in a well-designed and optimized PCR, specific amplicons will be produced in early cycles and primer-dimers, which are generated in later cycles, can be distinguished by using proper controls and by using melt-curve analysis. Although it is possible to reduce the contribution of signal from primer-dimers to the overall signal by detecting the fluorescence at temperatures higher than the melting temperature of the primer-dimer, the formation of the amplicons will be reduced. Another limitation of this sequence-independent method is its inability to be used in multiplex PCR reactions, which require simultaneous discrimination of multiple amplicons.

There is a large number of cyanine dyes used to quantify nucleic acids and only some of these are suitable for qPCR. These compounds are photosensitive chemical structures with two quaternized, nitrogen containing, heterocyclic ring linked through a polymethine bridge. The dyes have large extinction coefficients and a strong $\pi-\pi^*$ absorption which can be easily tuned from the visible to the near infra-red region by synthesizing structural modifications in the chromophore moiety. The modification of the benzazole heterocyclic moieties in cyanine dye structures influences the kind of interaction they have with nucleic acids. Substituents at position 6 of the benzothiazole moiety impart a crescent shape to the dye molecule, particularly the benzoylamino group, resulting in dyes with preference for binding to minor groove of the double-stranded DNA. Continual efforts to synthesize water soluble cyanine dyes without intrinsic fluorescence in solutions, high photostability, high binding affinities and emit only after binding to double stranded DNA will expand the repertoire of fluorescent dyes with different spectra properties.

Currently, the most commonly used fluorophore is SYBR Green I, an unsymmetric monomethine cyanine dye (Fig. 8.5d). It has several advantages over ethidium bromide (Fig. 8.5b) in that SYBR Green I is less toxic and exhibits a low intrinsic fluorescence, which is increased by 800–1000 fold on binding to ds DNA. A number of other DNA-binding fluorophores, including TOTO-1, YOYO-1 (Fig. 8.5f), SYBR Gold, Picogreen, LC Green and BEBO (Fig. 8.5e), SYTO family (e.g. SYTO-13, -82), Eva green, BOXTO (Fig. 8.5c), have been successfully used in qPCR. Although these dyes are generally thought to be non-sequences specific, recent studies however have shown that some of these dyes may exhibit some sequence specificity, which can therefore affect the accuracy of discriminating between amplicons using melt curve analysis.

8.6.2. *Sequence-dependent method*

In this approach, the identity of the desired amplicon is interrogated with a probe, which is a short specific complementary oligonucleotide labeled with one or more fluorescent dye(s). Unlike the sequence-independent fluorophores, the signal generated here is independent of the size of the amplicon. Only one molecule of the probe is hybridized to every single amplicon. With this method, it is therefore possible to perform multiplex qPCR to detect multiple targeted amplicons using distinct probes labeled with different dyes. An advantage of multiplexing is the significant reduction in the requirement for samples, labor and time. However, multiplexing is not as straightforward as singleplex assays as substrate limitations, spectra overlaps of probes, probe-probe and probe-primer interactions, and product inhibitions can pose significant challenges.

The recent progress in oligonucleotide synthesis and the availability of numerous polyaromatic quenchers and fluorescent labeled nucleobases has resulted in the development of a plethora of linear and conformationally constrained fluorescent probes for sequence-dependent methods. With some of these probes (e.g. the Molecular Beacon probes; Figs. 8.6a and 8.6b), the closed (quenched) form has the reporter (donor) and quencher (acceptor) fluorophores in close proximity to each other in space, while the open form has these fluorophores spatially separated. The fluorescence enhancement in the open form directly measures the interaction of the probe with its

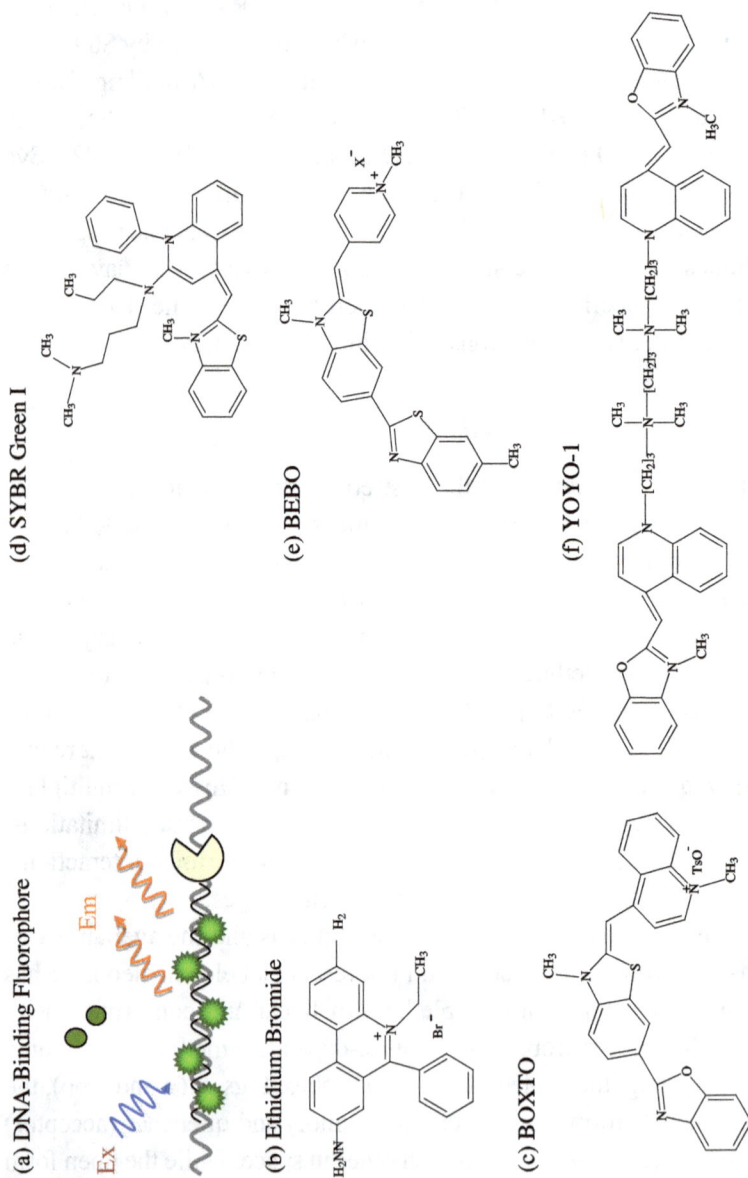

Fig. 8.5. DNA-binding dye (a). DNA-binding dyes (●) when bound to dsDNA results in the enhancement of fluorescence (✴) at the appropriate excitation (Ex) wavelength and emits at a lower energy wavelength (Em). Structures of some commonly used dyes (b–f).

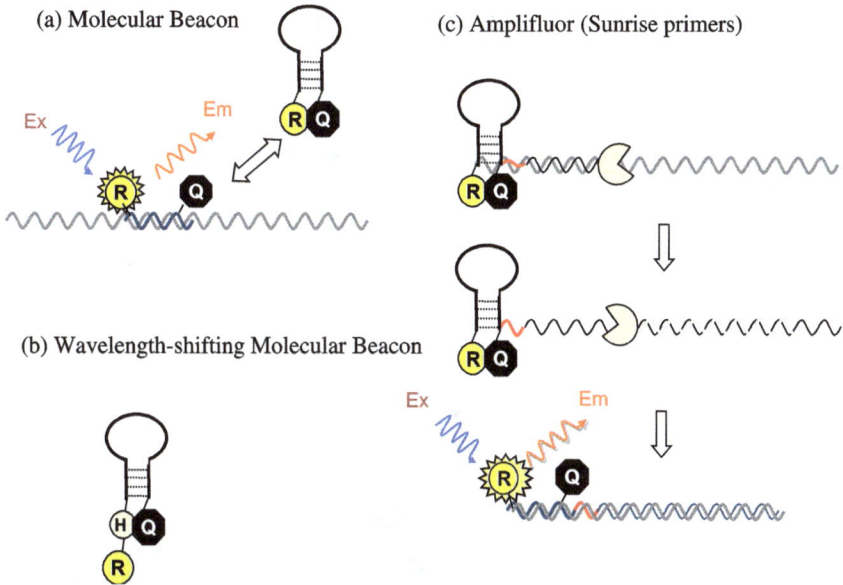

Fig. 8.6.   Designs of Molecular Beacons (a), wavelength-shifted Molecular Beacons (b) and Amplifluor primer-probes (c). In the absence of the target, both the reporter (r) and quencher (q) are kept in close contact physically by the short stem sequences in the Molecular Beacons. On binding to the target, the reporter and quencher are separated resulting in fluorescence enhancements. In wavelength-shifting Molecular Beacons (b), an additional fluorophore called the "harvester" (h), transfers the energy on excitation to the reporter (r). Fluorescent signal is only detected when the quencher (q) is physically separated from the harvester when the probe hybridizes to the complementary target sequence. Amplifluor primer-probes (c) are incorporated into the amplicons, separating the reporter (r) and the quencher (q).

target. In another format (e.g. hybridization probes; Fig. 8.7), the donor and acceptor are brought into close proximity on hybridization to the target and the emission of the acceptor measures the interaction of the probe with the target. The change of fluorescence wavelength during the close spatial proximity of these fluorophores is due to the transfer of energy between the reporter and the quencher, a process commonly described as fluorescence resonance energy transfer (FRET). In addition, depending on the mode of interaction between these dyes, part of a fluorophore labeled-DNA probe can also participate in quenching of the fluorescence.

The quenching of fluorescence can result from a number of molecular interactions including excited-state reactions, molecular rearrangements,

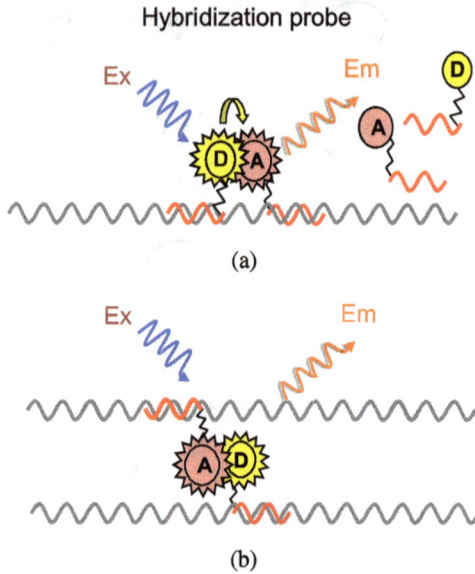

Fig. 8.7.   In the adjacent probe format, the excitation (Ex) of the hybridized probe enables the donor (d) to transfer energy to the acceptor (a) and the emission (Em) is then detected at the appropriate wavelength. The donor/acceptor can be on the same strand (a) or on opposite strands (b).

energy transfer, static (contact) and collisional (dynamic) quenching. A dynamic quenching is the result of the collisional encounters between the quencher and donor dyes in the excited state. The donor returns to the non-excited ground state without the emission of a photon. FRET (also known as Förster mechanism) is a dynamic quenching process that is used to explain the singlet-singlet transitions involved in the vast majority of donor-quencher pairs. The reporter and quencher fluorophores are physically in close proximity within a range of 20–100 Å (the Förster radius). With this mechanism of quenching, the reporter and quencher can be matched according to their spectral characteristics, i.e., they are chosen so that the emission spectra of the reporter and absorption spectra of the quencher overlap. In static (contact) quenching, a non-fluorescent dimer complex is formed between the reporter and the quencher at the ground-state. Here, the reporter and acceptor fluorophores are brought into close contact at distances less than the Förster radius. Such static quenching can be a dominant mechanism with certain reporter-quencher pairs in dual-labeled oligonucleotides, producing non-fluorescent heterodimers with their own distinct absorption spectra. In this case, pairs of donor-acceptor can also be chosen such that their absorption bands overlap in order to promote a resonance dipole-dipole interaction mechanism within a ground-state complex. Unlike FRET, fluorescence-quenched probes based on static quenching mechanisms, is less restricted to the requirements of the donor and acceptor to have an extensive emission-absorption spectral overlap. This therefore reduces the constraints on fluorescence-quenched probe design and expands on the ease of construction and the breadth of their applications. Existing assays employing FRET-based probes might significantly be improved by the use of these fluorescence-quenched probes.

Dynamic quenching is generally applicable to linear probe-based assays (e.g. adjacent hybridization probe, TaqMan), while static quenching appears to be more relevant to conformational probe-based (e.g. Molecular Beacons) assays. A number of organic molecules have successfully been utilised as quenchers, including fluorescent (e.g. TAMRA) and non-fluorescent compounds (e.g. Dabcyl, methyl red, BHQ series, QSY series, Disperse Blue, Iowa Black, Epoch Eclipse Quenchers). New organic molecules with wider absorbance spectrum will continue to be explored as the next generation of quenchers.

8.6.2.1. *Linear probes*

These probes are thought to adopt random coil conformations in solution. On hybridization to their targets, changes in fluorescence of these probes are detected as signals. Depending on the format, the probes can be reused at every cycle (e.g. Biprobe, Magiprobe, Induced FRET [iFRET] probe, Hybeacon probe, Light-up probe, Eclipse-MGB probe, Adjacent hybridization probe, LNA probes), degraded by DNA polymerase during PCR cycling (e.g. Double-Flap probe, Taqman probe, Taqman-MGB probe), or incorporated into the amplicons (e.g. G-quenching linear probe).

An example of a reusable linear probe is the adjacent hybridization probe which is the format often used in the Light Cycler instrument (Roche Applied Science) for sequence-dependent method of probing (Fig. 8.7). The principle underlying the adjacent hybridization probe format is first described by Heller & Morrison in 1985. These probes are easier to design and synthesize when compared to some conformationally restricted probes (e.g. Molecular Beacons). The assay uses two labeled probes hybridizing adjacent to each other on the target sequence (Fig. 8.7a). One probe is 5'-labeled with an acceptor fluorophore and the other at the 3'-end with a donor fluorophore. In solution, the two probes are spatially apart and the acceptor does not fluoresce. However, FRET occurs when the two probes are adjacent to one another when hybridized to the target sequence. Typically, the two probes are separated by one or more intervening nucleotides on the target sequence. FRET can still occur even when the two probes are separated by as many as 17 nucleotide bases, however with diminished fluorescence intensities. The probes can be designed to hybridize to either strands of the target (Fig. 8.7b), with the inverted position of a donor-acceptor pair (3' acceptor and a 5' donor) being equally efficient in the FRET process. In order to prevent the probes from acting as primers, the 3' end of the oligonucleotide can be modified with a phosphate, amine or biotin moiety to prevent oligonucleotide extension. An advantage of this probe format is that the concentration of the probe does not decrease, enabling the use of post-PCR melt curve for amplicon analysis.

In contrast, the commonly known Taqman probe assay uses the principle of probe degradation to generate fluorescent signal, and hence are not reusable in every PCR cycling (Fig. 8.8a). This format is an adaptation of the initial work of Holland and co-workers, where the probe is 5' terminally or internally labeled with a reporter fluorophore and the 3' terminus labeled with a quencher (Fig. 8.8b). The unbound probe adopts a random coil structure with the reporter and quencher thought to diffuse into a Förster radius by thermal motion, resulting in the quenching of the reporter fluorescence. Strand displacement of the probe occurs and the probe, which has annealed to the target, and the reporter fluorophore are subsequently cleaved off by the 5'nuclease activity of the DNA polymerase. The excitation of the cleaved reporter fluorophore thus results in fluorescense enhancement, which is directly proportional to the amount of specific amplified products. Hence, this probe format is known as either the 5' Nuclease, Hydrolysis or Taqman probe format. The 3' terminus of the probe is modified to prevent the probe from extension during PCR. As the probe is degraded with every cycle, the effective concentration of the probe decreases gradually and hence is not available for post-PCR melt curve analysis.

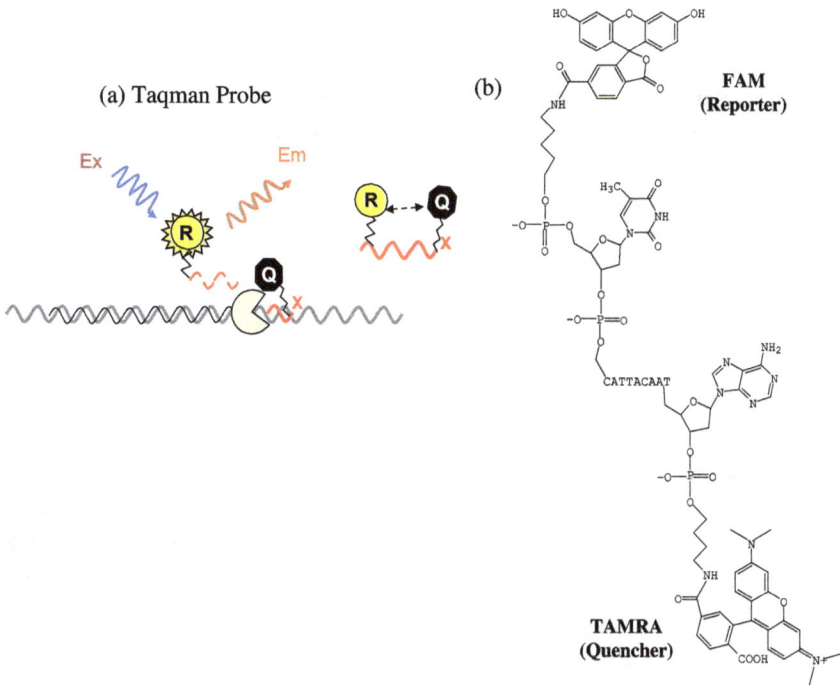

Fig. 8.8.   Taqman format (a). The reporter (r) is quenched by the quencher (q) in solution, and the eventual hydrolysis of R by the 5′ nuclease activity of the polymerase results in fluorescence enhancement. Structure of a 5′ FAM-TCATTACAATA-TAMRA-3′ Taqman probe (b).

The Taqman probe is designed to have an annealing temperature 5–10°C higher than the amplification primers to ensure probe hybridization and hydrolysis at the extension phase. The optimal distance between reporter and quencher fluorophores appears to be 6–10 bp apart and they can be placed internally when long oligonucleotide probes are used. The extent of hydrolysis and hence the accuracy of quantification, is dependent on a number of factors, including the position of the amplification primer with respect to the probe, the sequence of the 5′ nucleobase on which the reporter is attached to, and the 5′ nuclease activity of the DNA polymerase.

In instances where there is a need to detect similar targets of interest amplified using the same amplification primers, instead of producing individual probes for each target with different reporter-quencher pairs, a

Universal Taqman probe could be designed to hybridize to a universal non-target sequence which is tagged onto the 5′ end of one of the primers. Hydrolysis of the probe hybridizing onto the tag sequence thus reports the presence of these amplicons. The specificity of this approach however lies solely on the primers as the probe only detects the tag, which is incorporated into the amplicon. Hence, mispriming will result in the generation of false signals.

A recent extension of the Taqman probe assay is the development of shorter but more specific minor groove binding (MGB) probes. These probes have a 5′ reporter (fluorescein) and 3′ quencher (TAMRA) fluorophores conjugated to the minor groove binding moiety, dihydrocyclopyrroloindole tripeptide. Unlike Taqman probes, the MGB probes are shorter (10–15 bases) and have been shown to fold into the minor groove of double-stranded DNA at the terminal 5–6 base pairs. A modification of this is used by the MGB-Eclipse probe, where the quencher (Eclipse dark quencher) is conjugated to the 5′ end of an oligonucleotide, and the reporter at the 3′ end. Similarly, LNA (Locked Nucleic Acid) probes are shorter than Taqman probes and can achieve high specificity of detection. These MGB Eclipse and LNA probes are effectively insensitive to the 5′ nuclease activity of the DNA polymerase.

The underlying design of another linear probe format, the G-quenching probe, is based on previous observations that the fluorescence of a linear single fluorophore-labeled oligonucleotide can be quenched on hybridization to an unmodified complementary strand containing guanine (G) nucleobases at particular positions. These linear dequenching probes show a decrease in fluorescence on hybridization to the targets. A recent interesting modification of this approach is the incorporation of modified G (isoG) for the quenching of a fluorescent probe during PCR (Plextor™ Technology).

8.6.2.2. *Conformationally-restricted probes*

These types of probes have some degree of secondary structure in solution. On hybridizing to the target, changes in the structures result in fluorescence enhancements. Some of these probes can either be incorporated into the amplicons (e.g. Scorpion primer-probe, Amplifluor probe, dsDNA hybridization primer-probes, LUX primer-probe, Q-PNA primer-probes), or reused in subsequent PCR cycles (e.g. Molecular Beacon probe).

An example of a conformationally-restricted probe is the Molecular Beacon probe, which is a dual-labeled oligonucleotide with a fluorescent reporter at one end and a quencher at the other (Fig. 8.6a). The loop structure contains a sequence complementary to the target of interest and the ends of the probe contain a short self-complementary sequence (five to seven bases). In the absence of the target, the short self-complementary sequence form a stem-loop hairpin structure in a unimolecular reaction that serves to position the reporter in close proximity to the quencher, resulting in the quenching of the reporter. However, in the presence of the target, the sequence in the loop structure hybridizes with the complementary target sequences in a bimolecular reaction, enabling the molecule to unfold into an open form. The reporter and quencher are then physically separated and the reporter fluoresces on excitation.

The fluorescence quenching in the Molecular Beacon probes is not strictly FRET-based when dark quenchers are used (e.g. Dabcyl or BHQ), as the reporter-quencher are physically closer than the Föster radius. The increase in fluorescence intensity upon hybridization to the target can vary widely, ranging from less than 10 fold to greater than 200 fold. A modification of the Molecular Beacon probe, called the wavelength-shifting Molecular Beacon probes, has been reported where a third fluorophore (a harvester) is conjugated onto the reporter to give a substantially brighter fluorescence compared to the conventional probe (Fig. 8.6b). Yet another modification is the development of a Taqman-Beacon probe, where the 5' nuclease activity of the DNA polymerase is used to degrade the Molecular Beacon moiety. A variation of the Molecular Beacon probe format is the Cyclicon, which is a complex fluorescent-tagged pseudocyclic oligonucleotide.

The Amplifluor probe (previously known as the Sunrise probe) is another form of a conformationally-constrained probe (Fig. 8.6c). In this format, the amplification primer contains a hairpin structure with the reporter and quencher fluorophores conjugated on the stem. When this primer-probe is incorporated into the PCR product, the hairpin structure is unfolded and a fluorescent signal can be detected. Hence, this is a variation of the Molecular Beacon with an extended sequence acting as a primer. A disadvantage of this format is that mispriming of the primer-probe will result in the generation of non-specific signals. In order to increase the flexibility of this format, a universal sequence can be incorporated into the amplification primer and used for detection using Amplifuor-Uniprimer probes.

8.7. Quantification in qPCR

Depending on the objective of a study, qPCR quantification can be carried out by two approaches: a "relative quantification method" and an "absolute quantification method" (Fig. 8.4). The latter method implies that the absolute copy number of the unknown is determined. However, this is a misnomer as the quantity of the unknown sample is determined by interpolation from a set of predefined standards. Hence, the latter approach should appropriately be termed as "comparative standard curve method" or "standard curve method". The standards used in qPCR can be in the form of a cDNA, genomic DNA or plasmid DNA, and are assumed to have identical properties to the unknown samples. The accuracy of the results when using this method depends solely on the accuracy of the measurements of the standards. Thus, a highly reproducible quantification of the unknown implies a precise but not necessarily an accurate "absolute" result.

The relative quantification method is often used to measure changes of a target gene when examined in two different conditions. Quantification is carried out relative to the control or reference gene(s) by subtracting the difference in the Ct values of the reference and target genes. The resulting difference in threshold cycle numbers (dCt) of the genes in both conditions were further subtracted to generate ddCt, which is the exponent of the base 2 (due to the doubling function of PCR) and represents the fold difference between the two conditions.

There are some assumptions underlying this method, which includes the choice of a reference gene that is assumed to have a constant expression level under the experimental condition, that the amplification efficiencies of both the reference and target genes are assumed to be similar, and that the expression levels of the reference and the target genes are assumed to be similar. If all these assumptions can be satisfied, an advantage of this method is that a large number of genes can be surveyed simultaneously and, if required, further characterization can be conducted with the standard curve method.

8.8. Data Analyses

The analytical sensitivity, efficiency and specificity of a qPCR assay can be accurately determined. The analytical sensitivity or the detection limit of an assay is defined by the amplification and detection of the highest dilution of the target. It is possible to detect a single copy of a target but the quantitative accuracy decreases as the amount of template copy number approaches unity. This is partly due to technical and methodological inaccuracy during sample preparation, amplification and detection, and to sampling errors arising from stochastic distribution of low-copy-number template molecules.

The efficiency of the amplification process can be calculated from a plot of template dilution versus their respective Ct values. At the exponential phase of PCR, the accumulation of product is related to the initial amount of starting target by the following equation,

$$N_n = N_0(1 + e)^n \tag{8.1}$$

where N_n is the number of amplicons at cycle n, N_0 is the initial number of target molecules, ε is the efficiency of amplification ($\varepsilon \leq 1$) and n is the number of cycles needed to produce N_n. Thus, a 100% efficient amplification process will produce a doubling in the number of amplicons (N_n). Note that the amount of N_n present at any specific number of cycle (n) is dependent on N_0.

Taking the log of equation (8.1):

$$\log N_n = \log (N_0) + n\log (1 + e). \tag{8.2}$$

Rearranging equation (8.2):

$$n = \frac{1}{\log(1+\varepsilon)}(-\log N_0) + \frac{\log N_n}{\log(1+\varepsilon)} \tag{8.3}$$

Thus, the equation shows a linear relationship between n and $-\log N_0$. At the threshold cycle (Ct), the amount of amplicon produced is constant (as measured by the same fluorescent intensity) within the experiment. Hence, the efficiency of amplification (ε) of a target molecule can be calculated from the slope of the standard curve (a plot of Ct versus the negative log of the initial concentration of the template; Fig. 8.2c) where $\varepsilon = [10^{(-1/slope)}]-1$. Amplification of targets with high efficiencies have slopes approaching the value of 3.32 cycles for every 10-fold dilution of the target.

The accurate determination of the Ct values of an experiment is critical and relies on the identification of the exponential phases of the signal curves. A number of attempts have been made to model qPCR data by sigmoidal or logistic functions and other mathematical models so as to reduce errors in the determination of Ct values. This has allowed the development of computational algorithms to automatically calculate the amount of a target, a necessary step in high-throughput qPCR.

In order to compare the specificity of any assay, it is critical to compare the differences in the Ct values of the target to the test templates with similar but not identical sequences (ΔCt), as well as the efficiencies of the amplifications. Hence, specificity (σ) can be defined by the equation: $\sigma = (1+ \varepsilon)^{\Delta Ct}$, where $(1+ \varepsilon)$ is $10^{1/slope}$. Here, the efficiencies of amplifying the test templates are assumed to be equal to or less than that of the target. This assumption is generally true as the primers and conditions are optimized for the amplification of target and not the test templates. Hence, the larger the σ values, the more specific the assays are in discriminating the target over test templates.

8.9. Data Normalization

A number of variables can confound the accurate analyses of gene expressions and these include sample purity and integrity, reverse-transcription efficiencies, differences in transcriptional activities and technical issues. Ideally, extracted high-quality RNA can be standardized to the number of cells. However, the accurate determination of cell number is often not feasible, as in the case with solid tissues. A frequently applied normalization approach is the use of the mass of RNA. A technical drawback to the use of this approach arises when the availability of nucleic acids is limiting. A strong argument against the use of total RNA mass for normalization is that total RNA is predominantly represented by the ribosomal RNA (rRNA), and that the assumption that the ratio of rRNA to mRNA is invariant is not always true, nor is the transcription of rRNA constant under different conditions. In addition, both 18S and 28S rRNAs are not present in purified mRNA samples. For a reverse transcription reaction, an alternative to the use of input total RNA as a normalizer is to use picogreen quantified cDNA, which is more accurate.

The use of endogenous (internal reference) and/or exogenous (external reference) genes for normalization of gene expression studies has been a subject of intensive studies. For any endogenous gene (often referred to as housekeeping genes, HKG) to be considered as a reference or reference gene, it must not vary in the tissues or cells under investigation, or in response to experimental treatment. A number of studies have now shown that the expressions of many well-known HKG (e.g. actin, GAPDH, tubulin), in some but not all experimental conditions, can be altered significantly, thus, making the choice of using these HKGs for normalization uncertain without a *priori* knowledge and universal normalization genes may not exist. A variety of approaches have been employed to enable better selection of reference genes. One approach is the use of statistical algorithms, for example, geNorm, Best keeper, NormFinder, Global Pattern Recognition, and Equivalence tests, to evaluate the relative expression stabilities of genes from a pool of pre-defined lists of candidates. While this approach is certainly more robust than using the single gene methods, it too is based on potentially unfounded assumptions about which genes may be stably expressed in the conditions studied. These genes are still required to be pre-selected and incorporated into the experimental designs without any *a priori* evidence to support their use.

Hence, the use of any reference genes under any experimental conditions must be accurately reported and, if possible, standardized to allow reproducibility between laboratories. Practically, it is better to use more than one gene for normalization. If the availability of RNA is not limiting, a geometric averaging of multiple internal reference genes can increase the accuracy of normalization.

8.10. Melt Curve Analyses

At the end of qPCR, a melt-curve analysis of the products is a useful feature offered by most qPCR instruments. A melt curve measures the changes of fluorescence intensity of a sample when heated (Fig. 8.2d). When the temperature is increased, the fluorescent signal decreases gradually as a result of a temperature-dependent quenching, and more abruptly at certain temperatures. A simple way to compare and visualize multiple melt curves is to use the negative first derivative (-dF/dT) of these curves, where distinct peaks can be identified. In a melt-curve analysis using fluorophores that are not incorporated into the amplicons, the measured T_m is not that of the native ds DNA. In this case, the T_m of a product is dependent on the concentration of the fluorophore used and the rate of temperature change. Complex melting curves have also been observed with short double stranded DNAs using SYBR Green I, and dye redistribution differences between different DNA binding dyes (SYBR Green I, LC Green, BEBO) has given rise to different melt profiles of the same samples. Nevertheless, when optimized, the melt-curve analysis is an informative homogenous method for the characterization of amplicons.

High-speed and high-throughput genotyping of variants of PCR amplicons, epigenetic differences, mutations and SNPs using high resolution melt-curve analysis (HRM) at the end of qPCR is integrated into some but not all instruments. Certain formats of probes such as Taqman or Taqman-MGB probes cannot be used for melt curve analysis as the effective concentrations of the probes are reduced by degradation during PCR. Hence, by using just three colors for qPCR and the ability to discriminate four T_m, it is possible to distinguish a total of 12 different products in a single experiment, thus increasing the versatility of the study.

8.11. Standardization

Due to the lack of consensus on how best to perform and interpret qPCR studies and the lack of sufficient experimental detail in many publications, the Minimum Information for Publication of Quantitative Real-Time PCR

Experiments (MIQE) guidelines was proposed. It serves to increase the reliability of results and to promote consistency between laboratories and to increase experimental transparency. These guidelines describe the minimum information necessary for evaluating any qPCR experiments. The checklist includes complete and relevant experimental conditions and assay characteristics, such that readers can assess the validity of the protocols used. Full disclosure of all reagents, source of material, sequences, and analysis methods are included to enable other investigators to reproduce the results. Although not all investigators have adopted MIQE, it serves as a good guideline for documenting and managing qPCR studies.

8.12. Concluding Remarks

qPCR is a versatile, reliable, mature technology for the quantification of nucleic acids. Some of the limitations of the technology are being addressed currently. The integration of automated sample processing and data handling will continue to be improved, hence enabling this technology to be more accessible to many diagnostic laboratories. Newer instruments will likely to have smaller footprints, higher throughputs, lower cost and with more rapid cycling. Miniaturization of the PCR process on microfabricated devices that enable rapid reaction time, greater throughput and lower sample consumption (nanoliters to sub-nanoliters) is an interesting innovation that has been adapted for real time detection as "lab-on-a-chip" concept. Portable point-of-care (POC) systems for molecular diagnostics using qPCR are under development in various laboratories and these systems can increase the detecting infectious diseases outbreaks rapidly leading to quicker containment. The current limitation of multiplexing in qPCR, due largely to the lack of suitable organic fluorophores, may be addressed by the use of semiconductor down conversion nanocrystals (quantum dots), which have very narrow emission spectra. In view of these developments, qPCR will continue to have major impacts on many clinical and biological disciplines, and in some circumstances, replace conventional PCR.

Websites

1. Instrumentations.
   www.biocompare.com

2. qPCR resource.
   http://www.gene-quantification.de/
   qpcrlistserver@yahoogroups.com

Part II

Food Production Involving Microorganisms and Their Products

The fermented foods industry today is a big business and is the result of developments from traditional, small-scale production methods. Over the millennia, mankind has learnt to manipulate microorganisms and the environment of food to encourage the growth of desirable organisms that result in valuable foods which have good keeping quality and flavors.

Three key discoveries in the science of microbiology have advanced this area of food science. First is the discovery of microorganisms and the realization that they can be isolated and manipulated to control food fermentations, and in doing so, produce desirable fermented foods. Second is the discovery of enzymes produced by microorganisms and their role as biological catalysts that can be isolated and used to modify foods and develop non-traditional methods for modifying foods. Third is the recent development of genetic engineering, which enables a gene for a particular desirable property of an organism to be transferred to a microorganism in which the characteristic is absent.

In Chapters 9 and 10, the production methods of well-known traditional foods and alcoholic beverages are discussed. Chapter 12 describes in detail flavors and other food additives. In Chapter 13 enzymic modified food products are considered: how enzymes are extracted and applied to various foods to produce new food products.

The annex discusses protocols and procedures for assessing the safety of newly developed food products and for giving permission for their release to the market.

Chapter 9

Fermented Foods

Desmond K. O'Toole
Department of Biology and Chemistry
City University of Hong Kong
83 Tat Chee Avenue, Kowloon, Hong Kong, China
Lee Yuan Kun
Department of Microbiology, Yong Loo Lin School of Medicine
National University of Singapore
5 Science Drive 2, Singapore 117597
Email: micleeyk@nus.edu.sg

Fermented foods are those that have been transformed by microbial action or enzymes to produce desirable biochemical and physical changes. Fermentation, from the perspective of the microorganism, is an oxygen free process (anaerobic) in which an organic compound endogenous to the organism acts as the final electron acceptor. However, in food fermentations some microorganisms use oxygen as the final electron acceptor. For want of a better expression, these processes are also referred to as fermentations (Fig. 9.1). In addition, another food fermentation commonly referred to is the tea fermentation that is not a microbial process at all but an endogenous enzymic oxidation process. The microorganisms responsible for food fermentation include bacteria, yeasts and molds.

For many microbial fermentations to occur the food must contain readily fermented sugar, usually either monosaccharides (hexoses but can be pentoses) or disaccharides. A good source of sugars is starch but the polymer must be broken down into its constituent parts before the microorganisms

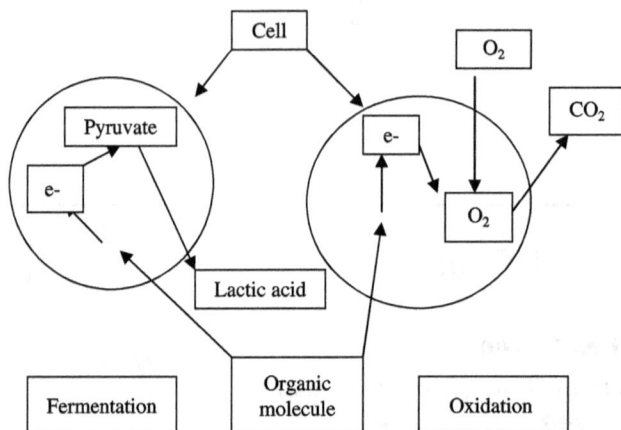

Fig. 9.1.   The difference between microbial fermentation and oxidation. In fermentation, the energy is derived from organic molecules and the electron is finally passed to another organic molecule that is then expelled from the cell. In oxidation, the energy is derived from organic molecules and the electron is finally passed to an oxygen molecule and then expelled from the cell as carbon dioxide.

can metabolize the sugars locked in the starch. Thus in some fermentations there is a preliminary step to break down starch to make it readily available for the microorganisms. This will be discussed later.

A further characteristic of many food fermentations is that they are two-stage processes that alternate between aerobic and anaerobic stages. In one form the first stage is a true microbial anaerobic fermentation but it is then followed by the second stage in which aerobic organisms grow. The aerobic organisms can be fungi or bacteria, although fungi are more common. In this second stage, the organisms produce proteolytic enzymes that breakdown the structure of the protein to soften the food and produce protein degradation products, and maybe, lipolytic enzymes that hydrolyze the fat to fatty acids resulting in esters that have strong flavors. In the other kind of two-stage process, the first stage involves an aerobic microbial growth stage, usually fungal growth, followed by an anaerobic stage.

9.1. Food Involving Acid Fermentation

The major food acids are lactic and acetic acids. However, propionic acid is also produced in small quantities as a flavoring compound. From the point of view of the microorganism, acid production is a very inefficient method for getting energy. For a small amount of growth the microorganisms produce a large amount of acid.

9.1.1. *Lactic acid bacteria*

Lactic acid bacteria are those that produce lactic acid as the sole product or the major acid from the energy yielding fermentation of sugars. Traditionally it has been easy to define the group but in recent years, it has become more difficult so that opinions differ as to what should be called a lactic acid bacterium. They can be broadly defined as Gram positive, anaerobic, micro-aerophilic or aero-tolerant bacteria; either rod or coccus-shaped, catalase negative, and fastidious in their growth requirements; many need B vitamins to grow. This definition can include an endospore forming bacterium (*Sporolactobacillus*), and a motile species but none of these are important for food production and preservation. Although considered fermentative organisms and unable to produce and use the respiratory chain found in the aerobic organisms, it is now known that *Lactococcus lactis*, when grown in the presence of hemin, produces a respiratory chain and can use O_2.

The major genera of lactic acid bacteria of importance in the food industry for food preservation (Table 9.1) and production (some cause food spoilage) are *Lactococcus* (in milk), *Lactobacillus* (in milk, meat, vegetables, and cereals), *Leuconostoc* (in vegetables and milk), *Pediococcus* (in vegetables and meat), *Oenococcus* (in grape wine), *Streptococcus* (in milk) and *Tetragenococcus* (in soy sauce). They are important for humans as well as for domestic animals in the production of many different foods that include milk and dairy products, meat products, vegetable products, grape wine, soy sauce and silage for animals.

9.1.2. *Types of lactic acid fermentation*

When these lactic acid bacteria ferment sugars, they produce varying amounts of lactic acid. This is due to two different metabolic pathways by which they may metabolize the sugars. The pathways result in different metabolic by-products. One pathway results in **homolactic fermentation,**

Table 9.1.   Characteristics of the genera of lactic acid bacteria important in the food industry.

Genus	Shape	Tetrads formed	CO$_2$ from glucose	Lactic acid[a]	Growth					
					at 10°C	at 45°C	in 6.5% NaCl	in 18% NaCl	at pH 4.4	at pH 9.6
Bifidobacterium	Pleomorphic rod	−	−	L	−	±	−	−	−	−
Enterococcus	Coccus	−	−	L	+	+	+	−	+	+
Lactobacillus	Rod	±	±	D, L, DL	±	±	±	−	±	−
Lactococcus	Coccus	−	−	L	+	−	−	−	±	−
Leuconostoc	Coccus	−	+	D	+	−	±	−	±	−
Oenococcus	Coccus	−	+	D	+	−	±	−	±	−
Pediococcus	Coccus	+	−	L, DL	±	±	±	−	+	−
Streptococcus	Coccus	−	−	L	+	±	−	−	−	−
Tetragenococcus	Coccus	+	−	L	+	−	+	+	−	+
Weissella	Short rod, coccus	−	+	DL, D	−[b]	−	+	−	+	−

[a]This refers to the optical isomer of the acid; [b]15°C.

while the other pathway leads to **heterolactic fermentation.** Consequently, the lactic acid bacteria are generally categorized on the basis of these fermentation patterns.

1. Homolactic Fermentation

The hexose is fermented exclusively to lactic acid.

$$1 \text{ hexose} \rightarrow 2 \text{ lactic acid} + 2 \text{ ATP}$$

The bacteria use the glycolytic pathway and possess aldolase, but not ketolase. Consequently, they cannot metabolize pentoses or gluconate. Organisms that use this pathway include *Lactobacillus acidophilus, Lb. delbrueckii* subsp.*bulgaricus (Lb. bulgaricus), Lb. helveticus, Enterococcus faecium* and *Streptococcus thermophilus.*

2. Heterolactic Fermentation

In this metabolic pathway, hexoses are fermented to equimolar amounts of lactate, ethanol (acetate) and CO_2.

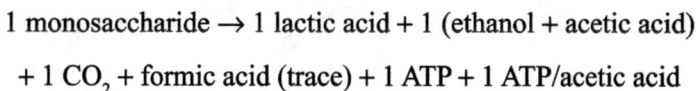

$$1 \text{ monosaccharide} \rightarrow 1 \text{ lactic acid} + 1 \text{ (ethanol} + \text{acetic acid)}$$

$$+ 1 \text{ } CO_2 + \text{formic acid (trace)} + 1 \text{ ATP} + 1 \text{ ATP/acetic acid}$$

Depending on the conditions the bacteria switch between producing either ethanol or acetate.

The phosphogluconate pathway is used and hexoses and pentoses can be metabolized through it. Organisms that use this pathway include *Bifidobacterium breve, B. infantis, B. longum, B. thermophilum, Lb. casei, Lb. fermentum, Lb. plantarum, Lb. rhamnosus* and *Lb. salivarius.*

The bacterial cultures can be categorized further on the basis of these two types of metabolic activity because in some cultures the organism may switch from using one pathway to using the other pathway. The categories are:

*   Group 1: Obligate homolactic — these organisms use the glycolytic pathway only.

- Group 2: Facultative heterolactic — these organisms use the glyco-lytic pathway and possess aldolase and ketolase that is repressed in the presence of glucose.
- Group 3: Obligate heterolactic — these organisms do not possess aldolase and use the phosphogluconate pathway.

9.1.3. *Malolactic fermentation*

A further characteristic of some lactic acid bacteria is their ability to use L-malic acid as an energy source (D-malate is not metabolized). Malic acid is present in many vegetable and plant products, including grapes and cucum-bers. Bacteria that are able to use malate include some of the species from the genera *Lactobacillus, Pediococcus, Leuconostoc* and *Oenococcus*. Utilization, however, also depends on the presence of some fermentable carbohydrate.

The enzyme responsible is L-malate:NAD$^+$ carboxylase. The reaction is as follows:

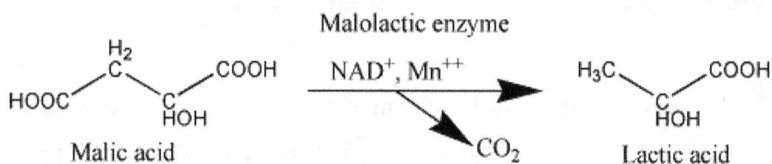

Malolactic enzyme

HOOC—C(H$_2$)—C(HOH)—COOH   NAD$^+$, Mn^{++} → H$_3$C—C(HOH)—COOH
+ CO$_2$

Malic acid   Lactic acid

This reaction usually occurs late in fermentation. It raises the pH and results in the production of some gas.

9.1.4. *Antibacterial properties of lactic acid bacteria*

The many situations in which lactic acid bacteria dominate can be explained by the production of lactic acid that lowers the pH and prevents the growth of many acid sensitive bacteria. However, in addition many lactic acid bac-teria produce substances that kill other bacteria. The most important sub-stances are the bacteriocins. The first bacteriocin described was nisin, discovered around 1947, and it is now produced commercially and incorpo-rated into many foods. Many more have been discovered and they are now placed in four broad classes. Class 1 consists of the Lantibiotics that are small peptides that contain unusual dehydroamino acids and thioether acids, lanthionin and 3-methyllanthionine. Examples are nisin A produced by *Lactococcus lactis* subsp. *lactis*, and carnosin produced by *Carnobacterium piscicola*. Class 2 bacteriocins are also peptides that are small heat stable non-lanthionine-containing membrane-active compounds. An example is lactococcin A produced by *Lc. lactis* subsp. *cremoris*. Class 3 consists of high molecular weight (>30 kDa) peptides that are heat sensitive. An example is acidophilucin A produced by *Lb. acidophilus*. On the other hand, Class 4 contains complex compounds consisting of protein, lipid and/or carbohydrate. An example is lactocin 27 produced by *Lb. helveticus*.

9.2. Acetic Acid Bacteria and Vinegar Production

This is basically a two stage microbial process with the stages being significantly separated in time. In the first stage alcohol is produced from glucose in fruit or from glucose/maltose derived from starch using plant or microbial enzymes (fungal growth). The prepared alcoholic product is then converted to vinegar.

Pure cultures of acetic acid bacteria are not used in vinegar production but studies of established commercial production systems have identified *Acetobacter aceti* (optimum temperature 29–34°C) and *A. pasteurianus* in rice wine vinegar production systems in Japan, *A. europaeus* in industrial acetators in Germany, and *A. lovaniensis* which is able to grow and produce acetic acid at 37–40°C.

Oxidation of alcohol by these bacteria results in the production of acetic acid. The biochemical event is depicted by the following equation:

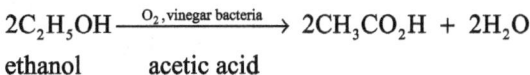

$$2C_2H_5OH \xrightarrow{\ O_2, \text{vinegar bacteria}\ } 2CH_3CO_2H + 2H_2O$$

ethanol   acetic acid

If we start with glucose then the theoretical yield of vinegar is as follows:

1 g glucose → 0.51 g ethanol → 0.67 g acetic acid

Today, vinegar is produced from a variety of sources of ethanol, but historically in the West wine was the source of vinegar, probably because of the frequency with which it is produced. Unlike other alcoholic beverages, grape wine can only be produced at one time of the year, but production of other beverages, such as beer, is not so dependent on the time of the year. Any attempt to keep alcoholic beverages exposed to air for long periods, particularly wine, results in the slow oxidative processes that produce vinegar. Commercially, the source of substrate is often indicated in the name of the product, e.g. spirit vinegar, wine vinegar, cider vinegar, malt vinegar, whey vinegar and rice vinegar. The strength of vinegar sold to the consumers is about 6% acetic acid but the strength of acetic acid that is produced by these microbial processes ranges from 10–15%.

The development of vinegar production methods through the years give us a glimpse of the advances in fermentation technology.

9.2.1. Traditional "slow" Orleans process

This is a traditional method that is still being used in Japan to produce rice wine vinegar. The bacteria used are obligate aerobes and consequently do not grow throughout the substrate but only at the air/liquid interface. If the wine is left undisturbed, a cell mat or pellicle forms on the surface of the wine (mother of vinegar or "Mycoderma aceti"). As oxygen diffuses through the mat, the bacteria use it to convert ethanol into acetic acid. Therefore, vinegar is produced at the surface and diffuses to the body of the wine (Fig. 9.2). In a similar manner, the ethanol diffuses into the pellicle where it is used by the bacteria. The pellicle is the starter culture for vinegar production. In a **batch-fed system,** wine is carefully added weekly for four weeks with minimal disturbance to the pellicle. A portion of vinegar is withdrawn in the fifth week, and replaced with the same amount of wine. The steps are repeated resulting in a **semi-continuous variable volume fedbatch culture process** (see Chapter 2, Sec. 2.4.2).

The air-liquid interface area limits the production rate because the production relies on the **diffusion** of oxygen and ethanol to the interface where acetic acid is produced.

9.2.2. The "quick" process (German process)

The principle of this process is thought to originate from a 1824 British patent but in fact, it is based on a much earlier French process dating from

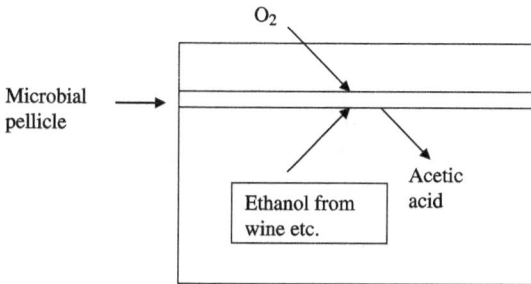

Fig. 9.2.   Fundamentally the vinegar production system depends on oxygen supply. In the first production systems, a microbial pellicle formed on the surface of the wine and the oxygen and ethanol had to be transported by diffusion to the microorganisms that occupied a relatively small space. Modern systems increase the mass of microorganisms available in a production system and increase the interaction between ethanol and oxygen.

1670. In this process (Fig. 9.3) wine is trickled down packed pumice stone, ceramic chips or wood shavings, on which acetic acid bacterial cells adhere to (**cell immobilization**), hence, this is a **fixed bed microbial film reactor.** The volume of a typical commercial vinegar generator is 50–60 m³. Air is either allowed to move freely through perforations at the bottom of the packed container, or is forced through the trickling generator to increase the oxygen transfer rate. The alcohol is converted rapidly to acetic acid while the wine is trickling down the packed vinegar generator, and 98% conversion of ethanol (12% v/v wine) into acetic acid can be achieved in five days. During the process, the temperature in the generator increases from about 28°C at the top to about 35°C at the bottom due to the metabolic activity of the bacteria. Theoretically 1 L of ethanol requires 430 L of oxygen gas at 30°C for complete oxidation to vinegar. However, in practice the volume of air required is approximately ten folds more as oxygen constitutes only 20% of the air. The vinegar generator will have to be re-packed as necessary, due to excessive cell growth and clogging of the system.

Three factors increase the rate of vinegar production when compared with the slow process. Namely, the amount of oxygen entering the wine per unit time is many times greater (higher oxygen transfer rate); the ethanol comes into contact with the bacteria more quickly due to the physical flow of the fluids; and there is a greater mass of bacteria to carry out the process. Diffusion, as in the traditional process, is a much slower process.

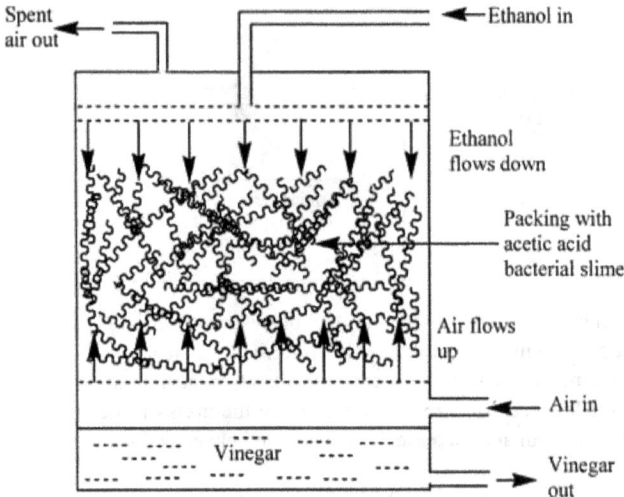

Fig. 9.3.   The "quick" process for vinegar production. The mass of bacteria is increased in this system and ethanol and oxygen are transported to the cells.

Box 9.1

How to Immobilize Acetic Acid Bacteria on Solid Supports?

A viable acetic acid bacterial culture is trickled down the packed vinegar generator. The medium is absorbed into the porous supporting material when the cells adhere onto the surface of the material. The cells then grow to form a thin film on the surface and may eventually dislodge when the film of cells becomes too thick.

The constraint remains that the bacteria are growing on a surface (the wood shavings, ceramic chips, etc.) and the oxygen and ethanol have to be brought into contact with the bacteria.

9.2.3. *Fringes acetator® or submerged culture process*

Fringes Acetator was the first submerged culture process but others developed subsequently include Yeoman's cavitator used in the US and Japan, the Bourgeois process used in Spain and Italy, and the Fardon process used in Africa.

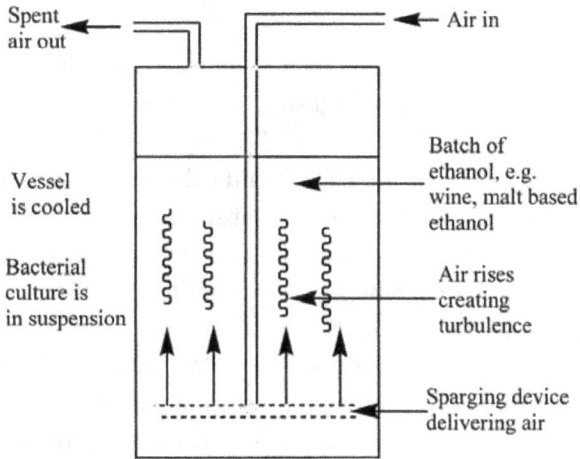

Fig. 9.4.   The submerged culture process for vinegar production. Transport of oxygen is maximized in this system.

Serious development of the **stirred tank fermentation** process for aerobic microbial production processes, including vinegar production (Fig. 9.4), arose from the success of penicillin production during World War II (1939–1945 AD). By 1993, there were more than 600 units world-wide using this principle to produce more than 135 million kL of vinegar per year. The advantages of submerged cultivation over the trickling generator are as follows:

(a)   It permits 30 times faster oxidation of alcohol, because of the high oxygen transfer rate.
(b)   It requires a smaller reactor volume (16% of that needed for the trickling generator) to produce an equivalent amount of vinegar, because the dead space occupied by the solid support in the trickling generator is removed.
(c)   It has greater efficiency with a yield 5–8% higher and >90% of theoretical yield is obtained;
(d)   It can be highly automated.
(e)   It eliminates clogging of packed wood shavings and interruption of liquid flow.
(f)   It allows an easy change in the type of vinegar being produced.

9.2.4. Quality problems

During vinegar production using the Fringes Acetator, bacteriophages can be a problem. As in the dairy industry, these bacterial viruses attack the bacteria so they no longer produce acid. Using pure culture strains results in greater problems than using mixed wild cultures. The solution is to use filtered, sterilized air and pasteurized substrate. In addition, the "quick" process vinegar producing units are a good source of bacteriophages for *Acetobacter* species, so submerged fermentation production units should be separated from them and the exchange of material or personnel between the two kept to a minimum, or preferably banned completely.

In the "quick" and submerged processes the vinegar is drawn off when the remaining ethanol reaches about 0.3%. The reason for this is twofold. Firstly, once the ethanol is depleted the *Acetobacter* strains undergo what seems to be an irreversible change in their metabolism and begin metabolizing the acetic acid to carbon dioxide and water. Consequently, the concentration of acetic acid decreases. In the presence of ethanol this metabolic pathway is repressed. If the culture undergoes this shift in metabolic activity, the only solution is to close down the generator, clean up and sterilize everything. Secondly, the presence of ethanol allows esterification to occur to improve the flavor of the final product.

Vinegar eels (*Anguillula aceti*, a small nematode) can also contaminate the vinegar and the packing in the "quick" process, but they are not so ubiquitous in modern vinegar producing plants.

9.2.5. Nata de coco

This product from the Philippines is made from coconuts and is the unique result of the growth of acetic acid bacteria. Unlike growth during vinegar production, the crucial product is not the result of electron transfer to a compound that is excreted (acetic acid); in this case the key product, cellulose, is synthesized and plays no part in the electron transport chain.

Coconut is first grated followed by steeping in water, in the ratio of 1 kg coconut to 28 L of water, strained and mixed with 2 kg of sucrose and 400 ml of glacial acetic acid before placing the mixture in a fermentation tray to give a shallow layer of about 50 mm deep. A nata starter culture, *Gluconacetobacter xylinus*, is then added and the preparation is left to ferment for up to two weeks. Under low pH conditions, in the presence of acetic acid, sugar and the coconut extract, the organism produces cellulose that forms a gel. The nata de coco is then cut into cubes, boiled several times to remove excess acetic acid, followed by boiling in sugar, at 1:1 ratio by weight, until the nata cubes become transparent.

9.3. Fermented Dairy Products

The traditional method for the production of fermented milk was by natural fermentation of surplus milk collected on the day. During the harvesting of the milk, microorganisms on the utensils, from the animals and from the environment find their way into the milk. The environmental conditions select bacteria that ferment the milk sugar, lactose. When the harvested milk leaves the animal its temperature is about 37°C and if it is not cooled quickly to about 4°C the contaminating bacteria start growing in the milk. Traditionally, the lack of proper temperature control resulted in the formation of sour milk that was tastier, possessed a thicker texture and was more stable than fresh milk. Eventually in traditional societies the primitive but simple technique was developed by keeping back a small portion of the milk, whey, etc. to add to the next day's supply of milk. It was a kind of culturing that is still used to this day for some fermentation applications.

With the development of the science of microbiology the bacteria responsible for the unique characteristics of each type of sour milk product were isolated and pure culture techniques applied to improve the quality of the sour milk products.

The chemical composition of milk used to produce fermented milk products and cheeses from a range of animals is shown in Table 9.2. Three chemical characteristics of milk aided in the development of fermented dairy products. The first is the lactose which is a disaccharide of galactose and glucose (β-D-galactopyranosyl-(1$\rightarrow$ 4)-D-glucopyranose) and is a good source of energy for lactic acid bacteria. The second is fat. Bovine milk contains from 3.5 to 5.0% fat in the form of small fat globules. The third is a protein, called casein, or κ-casein. This protein has an isoelectric point at about pH 4.6 at which it clots. The clotted protein traps the fat globules in the matrix of the clot. The protein matrix then shrinks and the

Table 9.2.   The chemical composition of milk from various animals used for production of fermented milk products and cheese.

Animal	Species	Composition, g kg^{-1}				
		Water	Fat	Casein	Whey protein	Lactose
Camel	*Camelus dromedarius*	865	40	27	9	50
Cattle	*Bos taurus*	873	39	26	6	46
Goat	*Capra hircus*	867	45	26	6	43
Horse	*Equus caballus*	888	19	13	12	62
Sheep	*Ovis aries*	820	72	39	7	48
Water buffalo	*Bubalus bubalis*	828	74	32	6	48
Yak	*Bos grunniens*	827	65	—	58[a]	46

[a] Casein plus whey protein.

Box 9.2

Lactose in Milk and its Uptake and Metabolism in Lactic Acid Bacteria

Lactose is a disaccharide comprised of a glucose and a galactose molecule. To metabolize the molecule it must be cleaved and several methods are used by these bacteria.   One method is to transport the lactose into the cell via a phosphoenol-pyruvate phospho-transferase system (PEP-PTS) that results in an internal lactose-6-phosphate that is then split into glucose and galactose-6-phosphate.   Glucose is metabolized via the EMP pathway and the galactose is shunted through the tagatose-6-phosphate pathway. Galactose-6-phosphate is isomerized to tagatose-6-phosphate, then phosphorylated to tagatose-1,6-di phosphate followed by cleaving to triose phosphate and metabolizing through the rest of the EMP pathway to pyruvic acid. The other method used by *Streptococcus thermophilus* is to transport the lactose via a permease that uses either a lactose/H$^+$ symport mechanism or a lactose/galactose exchange mechanism. The glucose is split from the galactose and metabolized via the EMP pathway and the galactose is excreted into the medium so that it accumulates in the product. Because of the rapid acid production by the yogurt bacteria, galactose concentrations up to 1% may accumulate in yogurt.

Box 9.2 (*Cont'd*)

Consumption of galactose brings about nutritional problems in humans as it is a poor sugar source. If too much is consumed, particularly in the presence of alcohol, galactose can be converted to galactitol which accumulates in the tissue, causing physiological disorders, including cataracts in the eye lens.

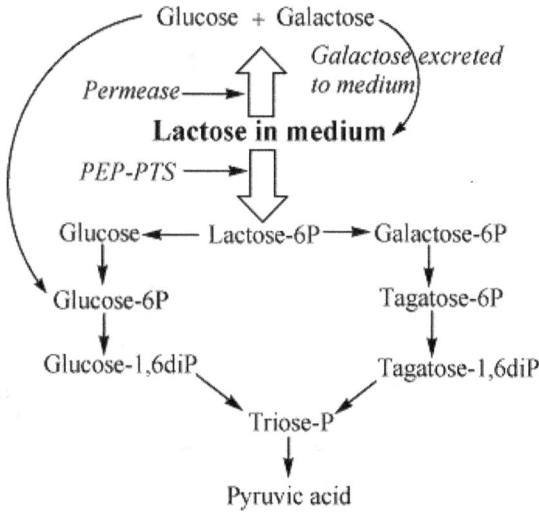

water phase is expelled from the matrix by the process of synerisis. The shrunken protein matrix can then be separated from the water phase (whey) and stored as cheese. If the matrix crumbles instead of shrinking and the whey is not removed, the clotted mixture becomes very viscous giving us the sour milk products we have today. From the perspective of flavor development, a further ingredient of interest is the citrate in milk.

9.3.1. *Fermented milk products*

The general protocol for the production of fermented milk products is shown in Fig. 9.5. Most of the naturally fermented milks contain more than one type of bacterium and occasionally yeasts (Table 9.3) because the characteristic flavors are produced through the growth of microbial consortia (a mixture of different strains of bacteria and may be yeasts) present

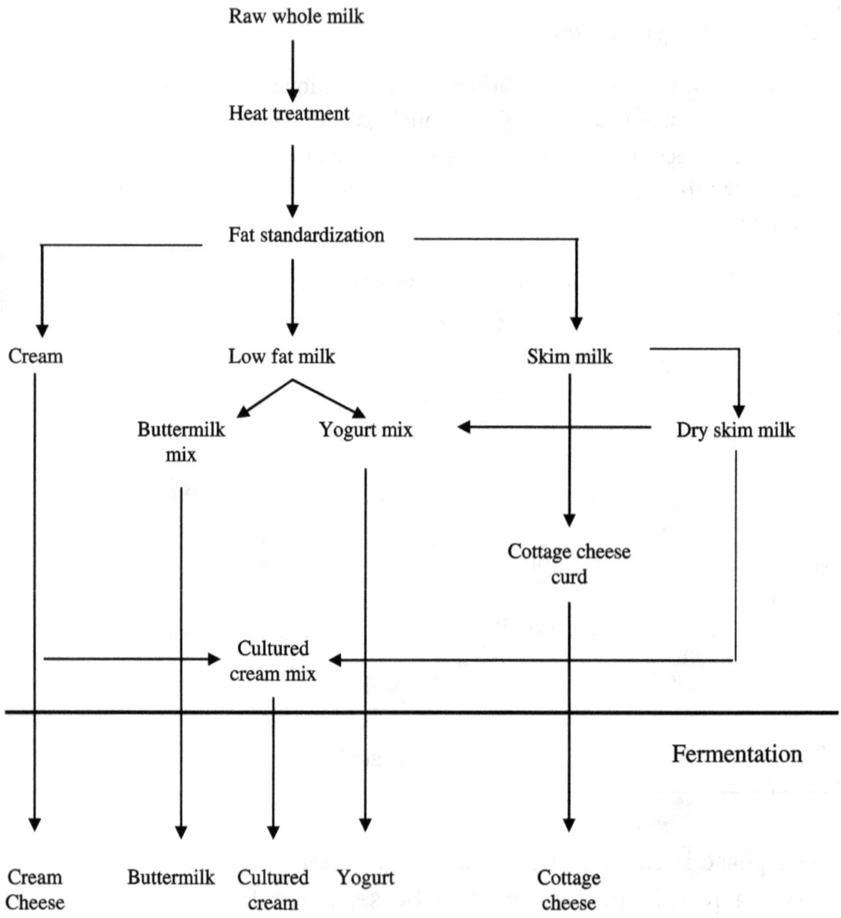

Fig. 9.5.   The general flow chart for the production of a range of fermented milk products.

naturally in the environment. A typical example is yogurt, which contains at least two lactic acid bacteria, namely *Lactobacillus delbrueckii* subsp. *bulgaricus* and *Streptococcus thermophilus*. Amino acids released by *Lb. delbrueckii* subsp. *bulgaricus*, and carbon dioxide and formic acid released by *S. thermophilus*, mutually stimulate the growth of each bacterium (Fig. 9.6). Thus, co-metabolism, or associative growth, by the consortium produces more lactic acid than the individual bacterial strains. In

Table 9.3.   Some examples of fermented milk products and the cultures responsible for their characteristics.

Fermented product	Microbial cultures	Country of origin	Source of milk
Acidophilus milk	*Lb. acidophilus*	USA	Bovine
Buttermilk[a]	*Lc. cremoris*	USA	Bovine
	Lc. lactis		
	Lc. diacetylactis		
Cottage cheese	*Lc. cremoris*	USA	Bovine
	Lc. lactis		
Cultured cream	*Lc. lactis*	USA	Bovine
	Lc. cremoris		
	Lc. diacetylactis		
	Leuc. cremoris		
Cream cheese	*Lc. cremoris*	USA	Bovine
	Lc. lactis		
	Lc. diacetylactis		
	Leuc. cremoris		
Koumiss	*Lb. bulgaricus*	Russia	Horse
	Torulopsis holmii (yeast)		
Kefir	*Lb. brevis*	Russia	Bovine, ewe, goat
	Lb. casei		
	Lb. caucasicus		
	Leuc. mesenteroides		
	Lc. durans		
	Saccharomyces delbrueckii (yeast)		
	Sacc. cerevisiae (yeast)		
	Acetabacter aceti		
Labneh	*Lb. thermophilus*	Israel	Bovine
	Leuc. lactis		
	Lb. acidophilus		
	Kluyveromyces fragilis (yeast)		
	Sacc. cerevisiae (yeast)		
Laktofil	*Leuc. cremoris*	Sweden	Bovine
	Lc. cremoris		
	Lc. lactis		
	Lc. diacetylactis		

(*Continued*)

Table 9.3.   (*Continued*)

Fermented product	Microbial cultures	Country of origin	Source of milk
Leben	*Lb. bulgaricus* *Lc. thermophilus* *Lc. lactis*	Iraq	Ewe, goat
Leben raib	*As for Leben*	Egypt	Bovine, buffalo
Long milk	*Streptococcus* sp *Leuconostoc* sp.	Scandinavia	Bovine
Quark	*Leuc. cremoris* *Lc. cremoris* *Lc. lactis* *Lc. diacetylactis*	Germany/ Europe	Bovine
Viili	*Streptococcus* sp. *Leuconostoc* sp.	Finland	Bovine
Yakult	*Lb. casei*	Japan	Bovine
Yogurt	*Lb. delbrueckii ssp bulgaricus* *Lc. thermophilus*	Turkey, Caucasus	Bovine

[a]Traditionally buttermilk was the very rich fraction left over after churning "ripened" (fermented) cream into butter and contained a high level of fat globule membrane fraction. Modern consumer products are made with milk.

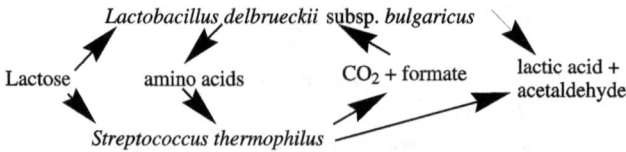

Fig. 9.6.   The yogurt fermentation. The two bacteria co-metabolize to produce more lactic acid and acetaldehyde than either of them separately. The lactobacillus produces small amounts of amino acids that stimulate the streptococcus. Meanwhile the streptococcus produces carbon dioxide and formate that stimulate the lactobacillus.

addition, the two organisms also produce acetaldehyde, the dominant flavor compound in yogurt. Another different consortium is the starter culture for kefir. Kefir is produced through the metabolism of lactobacilli, lactococci, a leuconostoc species and two yeasts. However, unique amongst milk starter cultures, the organisms grow together and cause casein to clot around them to produce soft lumps of casein mixed with the microorganisms. These lumps are called **kefir grains**. To produce kefir the grains are filtered from a previous batch and put into a new batch of milk. The consortium of organisms produces lactic acid, carbon dioxide and ethanol.

The more recently developed fermented dairy products, such as Yakult® and acidophilus milk, are produced in controlled environment fermentation vessels using selected single strains of *Lactobacillus* species.

In addition to the cultures used, modern production methods are often modified from the traditional ones to suit the purpose of economic production. An example is yogurt production (Fig. 9.7). Traditionally the milk was boiled to reduce the volume and so concentrate the milk. This produced a thicker sour milk product with no macro structure as one finds in cheese. Boiling had the effect of destroying the ability of the milk to form a smooth, firm curd. In modern production systems milk powder is added to normal whole milk to increase the concentration of milk solids in the mixture. Then the milk mixture is heated to a high temperature, as occurs in the traditional production system, to destroy the curd forming properties of the milk. For a smooth thick texture, particularly of the stirred yogurt types, a desirable thickening agent may be added. Quality control of naturally fermented milk is generally through fermentation conditions and the use of substrate (type of milk) but in modern production methods suitable pure cultures are used. Table 9.4 shows the time and temperature conditions used for various fresh fermented milk products. Generally a low inoculum percentage is used in the manufacture of fresh cheeses made using a temperature below 30°C.

9.3.2. Cheese production

Production of cheese is a huge industry. A total of about 15 million tons in 500 varieties were produced in 1994, over a third of that production being from Western Europe.

Yoghurt manufacturing

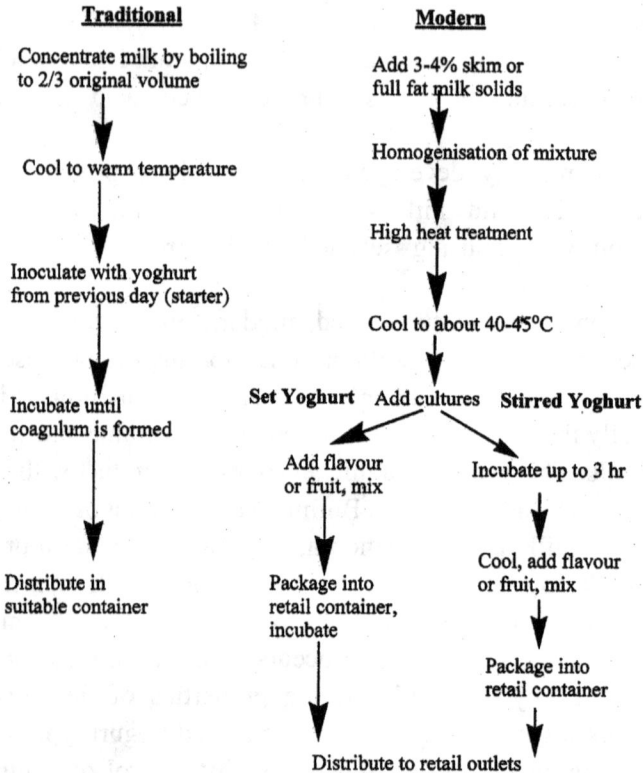

Traditional

Concentrate milk by boiling
to 2/3 original volume

↓

Cool to warm temperature

↓

Inoculate with yoghurt
from previous day (starter)

↓

Incubate until
coagulum is formed

↓

Distribute in
suitable container

Modern

Add 3-4% skim or
full fat milk solids

↓

Homogenisation of mixture

↓

High heat treatment

↓

Cool to about 40-45°C

↓

Set Yoghurt   Add cultures   **Stirred Yoghurt**

Add flavour
or fruit, mix   Incubate up to 3 hr

↓   ↓

Package into   Cool, add flavour
retail container,   or fruit, mix
incubate

                       ↓

                       Package into
                       retail container

↓   ↓

Distribute to retail outlets

Fig. 9.7.   Flow chart for the production of yogurt using modern production methods.

Table 9.4.   Incubation conditions used for various fresh cheeses or freshly fermented milk products.

Product	Process conditions
Buttermilk	22°C, 18 h
Cottage cheese	22°C, 18 h or 35°C, 5 h
Cream cheese	22°C, 18 h
Cultured cream	22°C, 18 h
Kefir	18—22°C, 12 h or 10°C, 1—3 d
Koumiss	28°C, 2 h
Quark	31°C, 5 h
Yogurt	43—45°C, 3 h

(1) Production Process

The general protocol for producing cheeses is in Fig. 9.8. In general, milk casein can be clotted in four ways.

(a)   The pH of the milk can be reduced through the growth of lactic acid bacteria that produce lactic acid throughout the milk.

(b)   One can add to the milk δ-gluconolactone that slowly hydrolyses to produce acid so producing the same effect as the growth of the lactic acid bacteria.

(c)   The pH of the milk can be reduced rapidly to about pH 5.2 with exogenous acids, like vinegar and lemon juice, that destabilizes the casein so that when heated the casein is caused to clot.

(d)   Finally, the most widely used method is to clot the casein with enzymes. The traditional enzyme source is rennet, a chymosin, obtained as an extract from calf stomach and stored in brine.

The κ -casein is responsible for the stability of the protein micelles in the milk, when treated with the enzyme part of it is hydrolyzed, and that action destabilizes the micelles and causes them to clot. The clotted curd produced by the enzyme has better handling properties than acid clotted milk and allows cheese makers to produce the various kinds of cheese that they make. Because of a shortage of animal chymosin, fungal chymosin is now available to the cheese maker. The fungal chymosin is used to make vegetarian cheeses, that is, those that contain nothing from a killed animal.

The casein fraction is important because its characteristics dictate the kind of fermented dairy product that is produced. Chymosin clots the milk from most animals but milk from the horse and the ass does not clot and so cannot be used for cheese production. Casein is also very sensitive to being denatured by heat. It is a common practice to pasteurize milk to kill spoilage microorganisms and, in particular, to kill pathogens, prior to processing into dairy products. The heating profiles (time/temperature conditions) used to kill pathogens are very close to the profiles that reduce the clotting quality of casein. A high temperature causes the clotted casein to form a soft clot that crumbles easily. Consequently, if too high a temperature is used, the milk cannot be used to make good quality cheese but it can be used to make a clotted milk product.

Raw whole milk

↓

Fat standardization

↓

Pasteurization

↓

Temperature adjustment

Add rennet and culture

Cut curd

Drain off whey

↓

Process curd

Add salt
to curd

↓

Put curd & salt
mixture in mold,
press

Put curd
in mold, press

↓

Place in brine

Put curds & whey
into mould and drain

↓

Dry salt

Add culture/rennet

↓

When milk has clotted
put into calico bags
and drain

Cheddar,
Edam, Gouda

Feta, Haloumi
Mozzarella

Roquerfort,
Camembert

Bakers' cheese

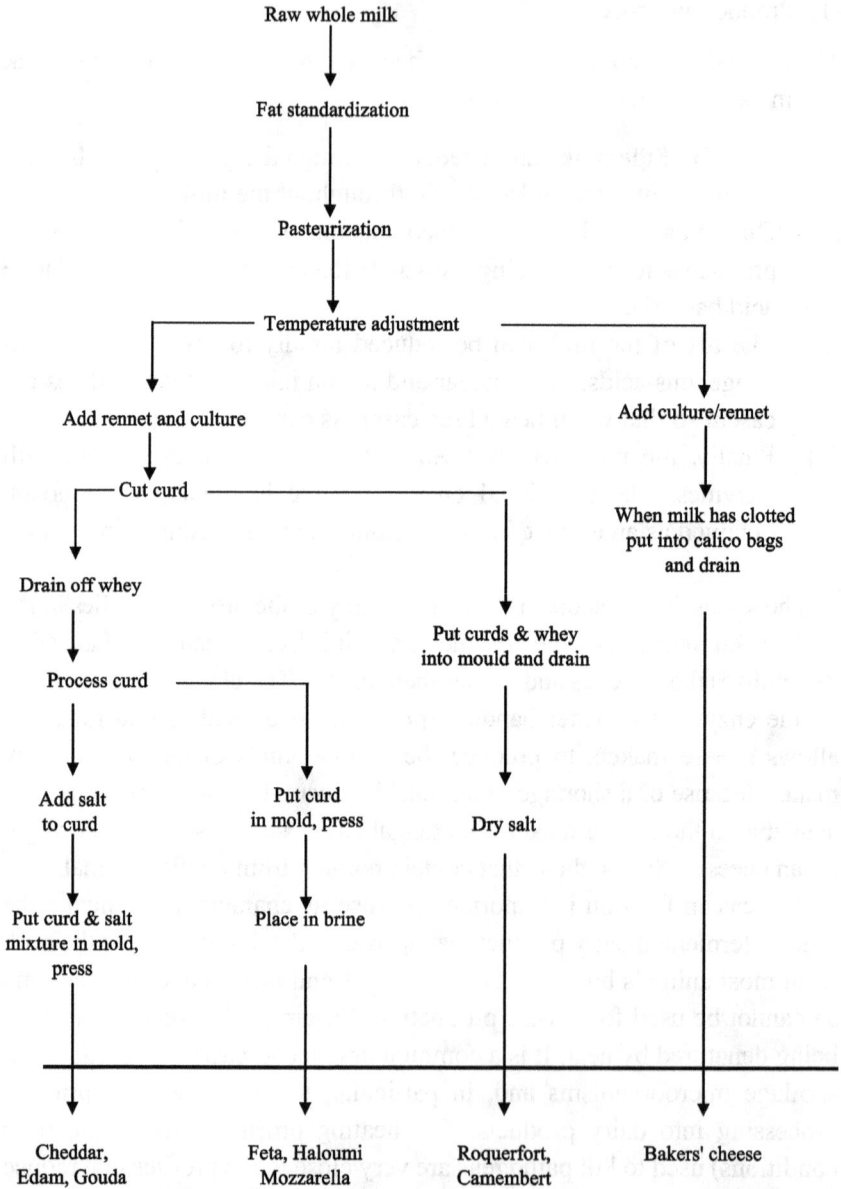

Fig. 9.8.   A general flowchart showing the production of cheeses with varying consistency and characteristics.

Cheddar cheese is the most economically significant cheese. It is a good example of the critical role that reliable lactic acid production at a predicable rate plays in production. The process begins with standardization and pasteurization of milk at 30°C, to which about 1–2% starter culture and rennet is added. The milk is allowed to stand still and it clots in about 45 minutes. At this point, it is cut with wire frames into small cubes that measure about 1.2 cm on each side. The cut curd and whey are then stirred continuously and the temperature of the curds and whey is slowly raised to about 37°C. During this "cooking" stage, the cheese maker determines the amount of lactic acid produced by titration, and the fermentation is going well if the acidity rises about 0.2% within about 90 minutes. During this cooking stage, the gradually raised temperature and lowered pH, in addition to the physical contact between the curd particles through constant stirring, increase the expulsion of the whey from the curd. The whey is then drained from the curd and the curd is piled in heaps. The whey expulsion from the curd and acid production in the curd continue at this stage. The next three to four hours is the **cheddaring stage** in which acid production continues at a steady rate so that the curd fuses and forms the right texture for the cheese, a "chicken breast meat" characteristic. The acid content of the whey running from the curd has usually risen a further 0.4% at the end of the cheddaring stage. If acid production slows or fails at any point during this cheddaring stage a poor quality cheese results, or the cheese may become a health hazard. The subsequent stages are milling of the curd (curd is cut into strips about 2.5 cm × 2.5 cm in cross section), addition of salt, and pressing of the curd. The pH of the final curd is about 5.2. The final pH alters the body of the final cheese. At about pH 5.2 the final cheese is more like typical cheddar cheese and the micelles therein are about 10 nm in diameter, but if the pH is below 5 the texture becomes "short" and crumbly and the casein micelles are about 3 nm in diameter.

(2) Microbial Maturation Processes

During the maturation stage enzymes from the starter bacteria continue to breakdown proteins and other components and in some cases produce undesirable bitter tasting peptides (Fig. 9.9). The length of this maturation stage varies with the cheese. In the case of cheddar cheese, the time varies from

Lactose	**Milk fat**	**Milk proteins**
Lactic starter bacteria	Milk lipase Added lipase enzymes	Coagulating proteases Plasmin
Lactic acid	Lactic starter lipase	**Peptides**
	Fungal lipase	Microbial proteases & peptidases

Carbonyl compounds
Acids
Alcohols
Ethyl esters

Free fatty acids
Methyl ketones
Secondary alcohols
Lactones
Derivatives of 18:2 &
18:3 fatty acids
Esters

NH_3 + amines +
amino acids

Citrate

Diacetyl
Acetoin

Volatile fatty acids
Alcohols
Sulfur compounds
Aromatic compounds
Pyrazines

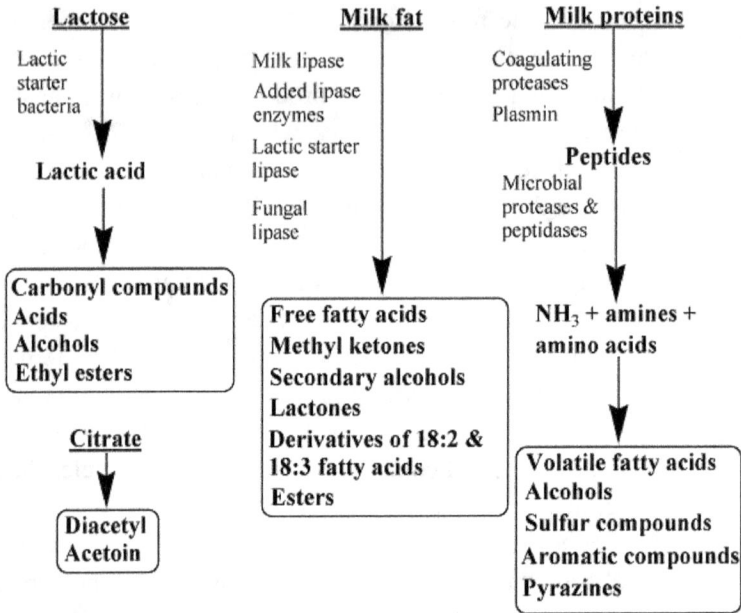

Fig. 9.9.   A generalized scheme of the breakdown of milk components to produce cheese flavors. The lactic acid bacteria produce acid from the lactose and proteolytic and other enzymes from the bacteria, or from added bacteria and fungi, further reduce the peptides to flavor compounds. Added lipase, or lipase from fungi, breakdown the fats to fatty acids that can then react with alcohols to produce fruity esters.

about three months for a mild cheese to about two years for a well-matured cheese. The whole maturation depends on the starter enzymes in the cheese and fungi play no significant role in cheddar cheese maturation. In the case of other cheeses, however, molds and bacteria grow on or in the formed cheese and produce proteolytic and lipolytic enzymes. Examples of these are blue cheeses, white mold cheeses and bacterial slime cheeses. Fungal spores are added to the milk before formation of the curd in the manufacture of blue cheeses (Roquefort, Stilton, and Gorgonzola cheese). Holes are then made through the formed curd with stiff wires to allow air to diffuse into the body of the cheese to promote the growth of the mold *Penicillium roque-forti* throughout the body of the cheese, and the production of proteolytic and lipolytic enzymes. In white mold cheese (Brie and Camembert) the mold (*P. camemberti*) grows on the surface of the formed curd and produces powerful proteolytic enzymes that diffuse into the body of the cheese.

In bacterial slime cheeses (e.g. Tilsiter cheese) a red bacterium (*Brevibacterium linens*) grows on the surface and the enzymes penetrate the cheese and produce the typical flavors.

An exception to the above rule is the propionibacteria that give Swiss cheese its distinctive flavor and physical appearance. During maturation the propionibacteria grow anaerobically on the lactic acid so that the concentrations of lactic acid decreases, propionic and acetic acids increase, and gas holes, characteristic of the cheese, are produced.

Because of the long maturation times required for many hard to semi-hard cheeses attempts to accelerate ripening have been tried. The various approaches include the following:

(i)   Raise the temperature during maturation but the potential for microbial spoilage increases.

(ii)   Add exogenous microbial enzymes. Individual enzymes added may produce a bitter flavor, but addition of lactococcal extract has produced good flavored cheese after a short maturation time, probably due to the presence of a complex mixture of proteases and peptidases. However, as enzymes are added to the milk so enzymes are lost in the whey. An encapsulation step to trap the enzymes in the curd matrix for slow release would be a better solution.

(iii) Add extra attenuated starter cells from which enzymes are progressively released. The starter cells are either sub-lethally heat treated or subjected to a freeze-thaw process to prevent growth and acid production. Encouraging results have been obtained with *Lactobacillus helveticus*.

(iv)   Genetic modification to control expression of particular enzymes so that they are increased or decreased depending on whether there are multiple copies of genes and the presence or absence of strong promoters.

(v)   Genetic modification to control release of lytic enzymes from the starter cells early in the maturation stage, or insert ripening genes into cells. For example, the gene for demethiolase from *Brevibacterium linens* can be inserted into lactococci.

(vi)   The use of cheese slurries. Before the final cheese forming stage the cut curd (milled curd in cheddar cheese making) is mixed with about

Table 9.5. Growth characteristics, acid production and diacetyl production of lactic acid bacteria used in fermented milk production.

Organism	Optimum temperature °C	Growth at		Acid production in milk, %	Diacetyl production
		10°C	45°C		
Lb. acidophilus	37	—	+	0.3–2.0	—
Lb. delbrueckii ssp. bulgaricus	42	—	+	1.5–4.0	—
Lb. casei	30–35	+	+	0.8–4.0	+
Leuc. cremoris	20–25	+	—	0.1–0.3	+
Lc. cremoris	22–30	+	—	0.8–1.0	—
Lc. diacetylactis	22–28	+	—	0.8–1.0	+
Lc. lactis	21–30	+	—	0.8–1.0	—
S. thermophilus	40–45	—	+	0.8–1.0	+/—

5% NaCl and the mixture then stored at an elevated temperature for up to nine days before pressing. The cheese slurry can also be added to fresh milled curds to accelerate ripening. This technology has been applied to a range of different cheeses including blue cheese and Swiss cheese. However, the product is very vulnerable to microbial spoilage and the final stage of the cheese making occurs after the slurry stage, unlike normal cheese making.

9.3.3. Quality considerations

The underlying control mechanism is the growth temperature of the starter organisms as shown in Table 9.5. Hence, at high incubation temperatures (> 40°C), *Lb. delbrueckii* subsp. *bulgaricus* and *Streptococcus thermophilus* are the predominant bacteria, and yogurt is produced, or mozzarella (pizza) cheese and Swiss-type cheese are produced. *Leuconostoc* and *Lactococcus* species are common bacteria found in fermented products produced by cottage industry in temperate regions. *Lb. acidophilus* and *Lb. casei* are widely distributed in fermented food in the tropics.

In many freshly made dairy products a desirable buttery taste is formed due to diacetyl that is produced by some of the lactic acid bacteria. Its production is dependent on the presence of citrate in the milk (Fig. 9.10).

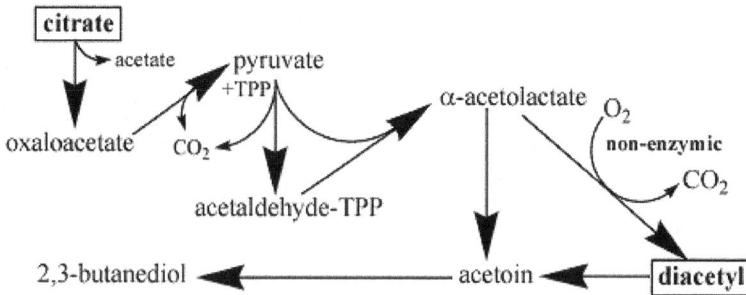

Fig. 9.10.   Pathway for the production of diacetyl, responsible for the butter flavor, from citrate.

The organisms do not use citrate as an energy source but when citrate is present with a readily fermentable carbohydrate, such as lactose, then citrate is used and diacetyl is produced if conditions are right.

The growth of the lactic acid bacteria is essential to produce the high-acidity of fermented dairy products that prevents the growth of pathogenic bacteria and other microorganisms, thus improving the shelf life and safety of fermented milk products. Raw fresh milk spoils at room temperature within 24 h, whereas fermented milk may be kept at room temperature for a week, until the product becomes too sour for consumption, due to the continuous production of lactic acid even when the bacterial cells have stopped growing (see Chapter 2, Sec. 2.2).

Besides improving the physical and sensory properties of milk, lactic acid bacteria enhance its nutritional quality by increasing the content of some vitamins (e.g. riboflavin, niacin, panthothenic acid, folacin, vitamin B12) in the fermented products as shown in Table 9.6.

In recent times the probiotic properties of lactic acid bacteria have become the subject of study. **Probiotics** are viable bacterial cell preparations or food containing viable bacterial cultures or components of bacterial cells that have beneficial effects on the health of the consumer (Table 9.7). Suitable probiotic bacteria have the properties of resistance to acid and bile, adherence to intestinal cells, transient intestinal colonization, and lack of any deleterious effects. Fermented milk has become a vehicle for the delivery of probiotic bacteria and cheddar cheese is also a suitable carrier. Studies on cheddar cheese have shown that even after 32 weeks numbers of surviving *Bifidobacterium* spp. and *Lactobaccillus casei* and *L. paracasei*

Table 9.6.   Effect of fermentation on the vitamin content of some cultured milk products.

Vitamin	Uncultured products		Cultured products		
	Low fat milk	Cream (light)	Cultured cream	Yogurt (low fat)	Cottage cheese (low fat)
Ascorbic acid (mg)	1.0	0.8	0.9	0.8	Trace
Thiamine (μg)	36	32	35	44	21
Riboflavin (μg)	140	148	149	214	165
Niacin (μg)	88	57	67	114	128
Panthothenic acid (μg)	329	276	360	591	215
Vitamin B6 (μg)	40	32	16	49	68
Folacin (μg)	5	2	11	11	12
Vitamin B12 (μg)	0.38	0.22	0.30	0.56	0.63
Vitamin A (IU)	204	720	790	66	37

remained high but numbers of *L. acidophilus* strains declined. In addition to the symptoms relieved by probiotics as shown in Table 9.7, other conditions relieved include allergic reactions like food-sensitized eczema, atopic dermatitis, allergic rhinitis and allergic asthma.

9.3.4. *Application of genetic engineering to dairy starter cultures*

9.3.4.1. *Phage control*

By far the most widely used pure bacterial cultures are the dairy starters. The application of pure bacterial cultures to the quality problems of the cheese (and butter) industries dates back to the very earliest application of the science of bacteriology. Dairy starters were being sold by Hansen's in the late 19th century and by 1929 the first pure strains of dairy starter cultures for cheese production were being used commercially in New South Wales and were subsequently adopted in New Zealand where the phage problem was discovered. With the advent of molecular biology much has been done to address dairy starter problems and their performance.

There are at least three major roles played by the lactic acid bacteria in dairy products; they are production of acid, of flavor and of texture. One

Table 9.7   Documented health benefits from the following probiotic strains.

Probiotic effect	Microorganism (strain)
Direct effect on intestinal complaints	
Rotavirus diarrhea shortened	*Lactobacillus casei* (Shirota), *L. reuteri* (SD2112), *L. rhamnosus* (GG), *Bifidobacterium lactis* (Bb12)
Relapses of inflammatory bowel disease fewer	*Escherichia coli* (Nissle, 1917), *Saccharomyces cerevisiae* (Boulardii)
Helicobacter pylori colonization reduced	*L. johnsonii* (La1)
Inflammatory bowel disease and irritable bowel syndrome relieved	*L. plantarum* (299v), *L. rhamnosus* (GG), *L. salivarius* (UCC118), *Bifid. breve*
Travelers' diarrhea incidence reduced	*Bifid. lactis* (Bb12)
Antibiotic associated diarrhea reduced	*L. acidophilus* (La5)
Effects on other aspects of health	
Oral vaccination improved	*L. johnsonii* (La1), *Bifid. lactis* (Bb12)
Modulation of immune system	*L. casei* (Shirota), *L. rhamnosus* (GG)
Treatment and prevention of allergies	*Bifid. lactis* (Bb12), *L. rhamnosus* (GG)
Recurrence of bladder cancer reduced	*L. casei* (Shirota)
LDL-cholesterol reduced	*L. plantarum* (299v)

Modified from Ouwehand *et al.* (2002).

of the characteristics of dairy starter cultures has been their inherent instability that has proved to be due to the technologically important traits being found on plasmids. These traits include lactose transport and hydrolysis, cheese maturation proteases, citrate permease and diacetyl formation, bacteriocin production and phage resistance.

Reliable acid production in the inoculated milk vat is crucial for predictable and good cheese quality and the greatest enemy of acid production is phage attack. The unpredictability of attacks, which restricts the range of strains that can be used in a given situation, is a problem so if cultures were available for which it was known that attack could not occur then cheese making would become more predictable and the range of available starter strains might be increased. Generally the ultimate source of phages is unknown but genetic studies have shown that the starter cultures themselves can be the source of the phages.

The first lactic acid bacterium for which the complete genome was elucidated is *Lactococcus lactis* subsp. *lactis* IL1403 that contains five complete prophages. In some bacterial strains the prophages may constitute 3–10% of the total genome.

The DNA sequences of over 100 phages of the lactic acid bacteria, in particular those that attack strains of *Lact. lactis* and *Strep. thermophilus*, are the most completely known. In addition, many mechanisms of resistance to phage have been characterized and as at 2002 they include: phage abortive infection systems (Abi) — 21 known; restriction modification (R/M) systems — 29 known; adsorption inhibition (Ads) mechanisms — 8 known; injection blocking systems — 2 known. These systems are predominantly linked to plasmid DNA, a characteristic that allows horizontal transfer via conjugation between starter strains so that phage resistance can be introduced to a phage sensitive strain that may have desirable flavor forming characteristics.

One plasmid encoded Abi system, namely pNP40, found in *L. lactis* subsp. *diacetylactis* DRC3, has been transferred to strains used industrially. The system seems to work by blocking phage adsorption to the cell wall by masking the receptor site. Changes in the cell wall lipoteichoic acid, or secretion of an external polymer, or altered cell wall proteins may be the reason for the blockage. A gene in a lactococcal prophage designated sie_{2009} produces a protein associated with the cell membrane. The phage attacking the bacterium adsorbs to the cell wall and transfects (in the sense of releasing its genetic material beyond the cell wall), however, plasmid transduction and phage replication are prevented by the membrane protein.

There are many phage abortive infection systems that can act in a variety of ways to prevent successful completion of a phage replicative cycle (Table 9.8) but they have only been found in *L. lactis* subsp. *cremoris*; and *L. lactis* subsp. *lactis* cultures so far. The lactococcal phages are divided into

Table 9.8.   Phage abortive systems in *Lactococcus lactis* subsp. *cremoris*; and *L. lactis* subsp. *lactis* and the action the systems have on the phage replication cycle.

Abi system	Action of the system on the phage replication cycle
A, F, K, R	Interferes with phage DNA replication
B, G	Interferes with RNA transcription
C	Reduces synthesis of the major capsid protein
D1	Interferes with a phage ORF essential for phage development
Q	Prevents maturation by accumulation of replicative form of phage DNA
S	Sequesters factors essential for phage development
U	Delays phage transcription

12 species and the phage species controlled by these systems are the species 936, P335 and c2 that are members of the *Siphoviridae*. The Abi systems involve proteins that range in size from 628 to 127 amino acids long that are mostly coded on plasmids but some are on chromosomes. All affect phage species 936, four affect species P335 (Abi's A, K, T, U) and 13 affect C2 species.

The restriction/modification systems were the earliest systems known and they are classified into three types; Types I, II and III (Table 9.9). Type I and III systems cleave the DNA at a substantial distance from the recognition site while Type II systems cleave DNA either within or very close to the recognition site (see Chapter 6, Sec. 6.1.2).

9.3.4.2. *Yogurt and neutraceuticals*

The soft texture of yogurt and a lack of whey separation are important characteristics. Extra gelling or thickening agents may be added to the mix during manufacture to ensure this outcome but a more desirable situation is for the thickening agents to come from the bacteria. Some starter bacteria can produce exopolysaccharides that could fill that function. Two strains of *L. lactis*, namely NIZO B40 and SBT 0495, produce an identical polymer (Fig. 9.11) with a regular repeating unit, →4)[α-L-Rha$p$-(1 → 2)] [α -D-Gal$p$-1-PO$_4$3]- β-D-Gal$p$-(1→4)- β-D-Glc$p$-(1→4) -β-D-Glc$p$-(1→,

Table 9.9.   Some restriction/modification systems, their types and their intracellular locations from *Lactococcus lactis* subsp. *cremoris* (LC), *L. lactis* subsp. *lactis* (LC) and *Streptococcus thermophilus* (ST).

Restriction/modification system		Bacterial host strain	Specificity[a]	Location: chromosome or named plasmid
Name	Type			
Lla1403I	I	LL IL1403	Undetermined	Chromosome
Lla90I	I	LL DPC721	Undetermined	pAH90
LlaKR2I	II	LL KR2	5′ ↓GATC 3′	pKR223
LlaFI	III	LL 42—1	Undetermined	pND801
Lla42oI*	I	LC IL420	Undetermined	pIL2614
LlaGI	I (var)	LC W10	Undetermined	pEW104
ScrFI	II	LC UC503	5′ CC‾NGG 3′	Chromosome
LlaAI	II	LC W9	5′ ‾GATC 3′	pFW094
LlaBII	—	LC W56	Undetermined	pJW565
SthSFiI	I	ST Sfi1	Undetermined	Chromosome
Sth135I	I	ST 135	Undetermined	pER35
Sth134I	II	ST 134	5′ C‾CGG 3′	Chromosome
Sth8I	II	ST 8	Undetermined	pSt08

[a] The cleavage point where known is indicated by ↓.
* Designated Lla2614I in REBASE.NEB.COM. Modified from Table 2 in Coffey and Ross (2002).

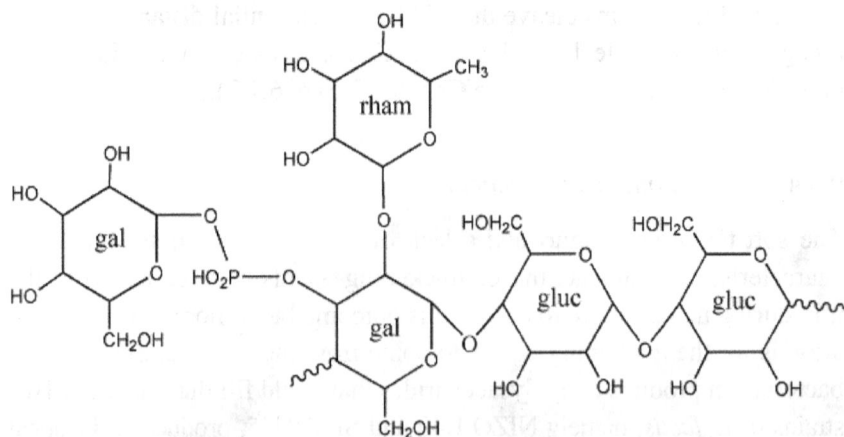

Fig. 9.11.   The repeating unit in the exopolysaccharide produced by *Lactococcus lactis* strain NIZO B40. If the gene for its production is transferred to suitable bacteria it may form the basis for a natural thickening agent in yogurt.

that is coded on a 42,180 bp plasmid, designated pNZ4000, that can be conjugally transferred between lactococcal strains. The gene cluster for the polymer is 12 kb and has 14 co-coordinately expressed genes, namely *epsRXABCDEFGHIJKL*. The polymer is synthesized on a membrane-linked lipid carrier.

Because glycosyltransferase encoding genes are available, the possibility arises of oligo- and/or poly-saccharides with different repeating units being produced to suite particular needs. For example, it has been claimed that fructose-containing oligosaccharides stimulate the growth of bifidobacteria in the human intestine so if an organism could produce such an oligosaccharide it would be useful to promote the beneficial effects of probiotic organisms.

9.3.5. *Antibacterial factors in milk*

Normally the lactic acid bacteria grow well in milk but under some circumstances they might not grow, or they might not grow properly. There are three different kinds of antibacterial factors that can interfere with lactic acid production; natural factors secreted into the milk by the mammal, antibiotics in the milk from the udders of mammals treated for mastitis or other infections, and bacteriophage.

Natural factors include the lactoperoxide-thiocyanate-hydrogen peroxide system, lactotransferrin, and agglutinins. The first two factors can prevent the production of acid but the agglutinins cause the agglutination of the bacterial cells inoculated into unrenneted milk. The agglutinated cells form larger particles that fall more quickly than smaller particles to the bottom of the vat causing more acid to be formed at the bottom of the vat when compared with the top layer of the milk. Antibiotics prevent the growth of the lactic acid bacteria so no acid is produced. The phages can cause problems at three stages. At the first stage, they may infect the milk cultures being propagated daily for eventual use in the preparation of bulk starter. At the second stage, they may infect the bulk starter unit being used to prepare inoculum for the vats of milk. At the third stage, they may infect the vat of milk and the effect observed at this stage will depend on the cheese. In those cheeses requiring a slow setting time and low inoculum level gas-producing bacteria of the *Enterobacteriaceae* family may dominate. In cheese making procedures that require rapid acid production, acid production at the beginning of the process looks fine but as time passes the acid production slows down and stops due to the viral infections of the starter bacteria. Consequently the cheese produced does not reach the pH or degree of acidity required for it. This allows other contaminants to grow, such as *Staphylococcus aureus* that produces enterotoxins. The quality of the resulting cheese is poor and can be dangerous to health.

During a production day the levels of bacteriophages built up in a factory can be very high in the whey within a relatively short period (Fig. 9.12). Bacteriophages in the cheese making industry are the major problem for the industry, and it is the classic example of an industrial problem caused by a virus.

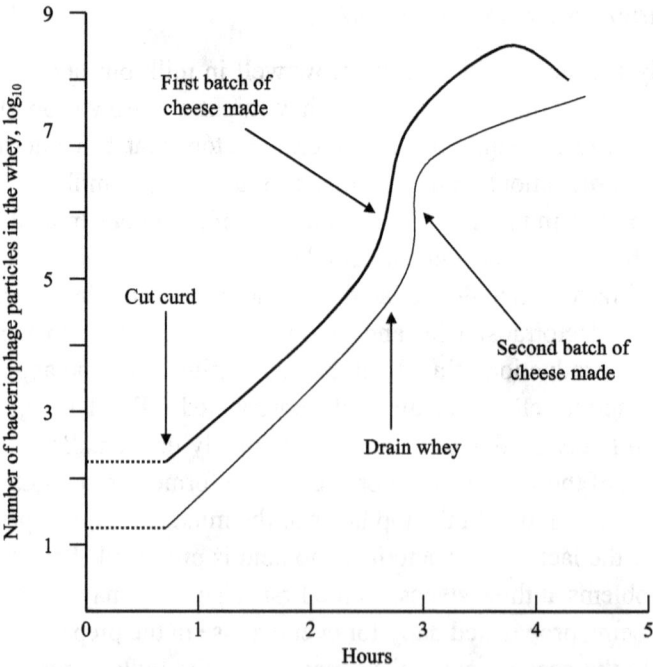

Fig. 9.12. A generalized diagram showing the growth and production of bacteriophage in whey produced during cheddar cheese manufacture. The amount of phage produced in whey from the first batch of cheese in the vat was more than that produced during production of the second batch. High concentrations of bacterial viruses build up quickly in the liquid phase and contaminate the factory environs.

Box 9.3

Traditional Techniques for Dealing with Bacteriophage Attacks in Commercial Bacterial Fermentations such as in Cheese Making

Traditionally mixed natural cultures were used but the quality of the product varied. To obtain consistent products, pure cultures were developed to make cheeses. However, when pure cultures are used, they sometimes fail due to bacteriophages that attack the bacteria. The reason for the problems is that during the growth of the fermenting bacteria, very large concentrations of bacteriophages build up in the environment of the factory. When a cheese maker tries to prepare mother cultures to make bulk starter cultures for the cheese making process, these cultures became infected with the bacteriophage and the bacteria are killed before they can grow. Where do bacteriophages come from? A primary source is the raw material (milk) that contains bacteria which are attacked by the phages that are resistant to the normal pasteurization temperatures used. Lysogenized cultures are also a source, the genomes of many starter bacteria code for prophages. Steps developed to control proliferation of the phage in a factory include: filtered air in culture preparation rooms, separation of personnel who handle cultures from those who work in the factory, daily rotations of known strains of phage unrelated cultures, the use of bulk starter media specially formulated to prevent bacteriophage attack and multiplication in the bulk starter medium, daily testing of whey for the presence of bacteriophage so as to change cultures, and purchasing of cultures from outside the factory. Frozen concentrated starter cultures are prepared elsewhere and shipped to factories for use directly in vats or for use in the preparation of bulk starter. Enclosed cheese vats and careful handling and disposal of whey also help in the control of phage.

9.4. Meat Fermentations

There are two broad applications of microbial fermentation to meat products. The first is in the production of hams and bacon; the second is in the production of fermented sausages. In the first case the portions of carcass are left in one piece while in the second case the muscle meat is comminuted.

9.4.1. *Production of ham, bacon and sausages*

Within the structure of muscle there are no bacteria. Bacteria collect on the surface of the carcass during the dressing of the carcass after slaughter. These bacteria can grow and cause spoilage, or disease, so meat is a perishable product that must be preserved for future use. The oldest technology used is salting and drying. Another method developed was to comminute the meat, add flavors and preservatives and stuff it into the cleaned intestines of animals before drying immediately, typically a process used for the production of Chinese preserved sausages. In the comminuted form, bacteria are distributed all over the surfaces of the meat particles. Even though in modern Western production systems bacteria are used in meat fermentations, partial dehydration with fermentation still plays a role in the meat preservation.

The chemical qualities of meat influence microbial growth on the meat particles and surfaces. The pH of normal meat is about 5.5 (DFD — dark firm dry meat is over pH 7.0). This arises due to the rigor mortis chain of metabolic reactions in the muscle. At death the pH of the muscle is about 7.0 and glycogen present in the meat is broken down to lactic acid and the pH drops. Some glucose remains in the meat, depending on the animal species from which the meat came (Table 9.10) as well as other nutrients such as amino acids and peptides suitable for bacterial growth. In addition, meat contains hemoglobin, an iron-containing molecule that, under the right conditions, can reduce the redox potential of the meat so that anaerobic bacteria such as *Clostridium botulinum* will grow. *Cl. botulinum*, an obligate anaerobe, produces the most toxic substance known, botulinum toxin. Nowadays, the prime purpose of using additives in meat preservation is to avoid the growth of this bacterium, while a secondary purpose is to prevent spoilage. Fortunately, heating the toxin to boiling

Table 9.10.   Typical glucose concentrations in various kinds of meat.

Meat	Glucose concentration, $\mu g \, g^{-1}$
Beef	100
Sheep meat	400
Pork	800

point will destroy it, but unfortunately, many preserved meats are not cooked before consumption. In addition to this bacterium, under aerobic conditions *Pseudomonas* species grow on the meat, and under restricted oxygen conditions lactic acid bacteria can grow and suppress *Pseudomonas* spp. Further, because meat is a good source of food, bacteria grow rapidly at high temperature so low temperatures are used during preservation,

---

Box 9.4

How does Meat Become a Good Substrate for Microorganisms to Grow on?

Energy in muscle is stored as glycogen — a long chain polymer of glucose. In a live animal, the glycogen is hydrolyzed to produce glucose. The blood flow brings O_2 to the muscle and the glucose is usually metabolized via the EMP pathway and the citric acid cycle to produce mainly energy and to convert it to CO_2 and H_2O that are taken away by the blood flow in the animal. At slaughter the blood flow stops and no further O_2 flows to the muscle. At this stage the process leading to rigor mortis begins. In this process the metabolic pathways continue operating in the muscle while the glycogen and glucose are present. Thus the glucose enters the EMP pathway and is metabolized to pyruvate but because there is no O_2 it cannot enter the citric acid cycle as in the normal respiratory pathway. Consequently, the pyruvate, to gain further energy, is converted to lactic acid that builds up in the muscle and the pH of the muscle drops. Eventually, at about pH 5.5 in the muscle, no further glycogen is converted to glucose and the metabolic process halts. If there is little or no glycogen in the muscle at death the final pH will be above 6.0 and that influences the kind of microbial growth on the meat.

During these biochemical steps to rigor mortis, other energy compounds are released and eventually a significant amount of small molecular weight compounds are available for microbial growth. One key compound is glucose. The amount of glucose in meats from various domesticated animals varies as shown in Table 9.10. Because the compounds are in the muscle, the microorganisms cannot access them unless the meat is comminuted. Once the meat has been reduced to small particles the glucose and other small molecules can diffuse to the surface to replace those being used by bacteria on the surface. Under natural conditions the lactic acid bacteria come to dominate the microflora growing in the comminuted meat.

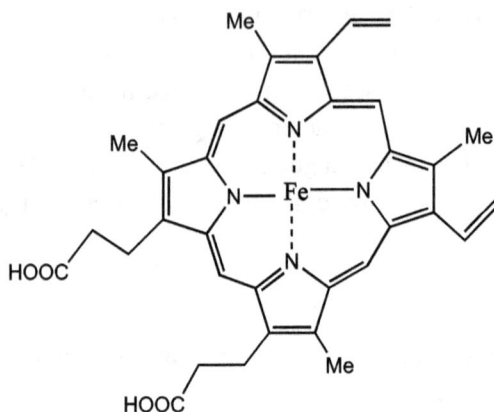

Heme; reactions with the Fe changes
the colour of cured meats

Fig. 9.13.   The heme molecule is the key in the formation of the typical cured meat color.

usually about 20°C or less. Even though glucose is commonly found in meat, glucose is commonly added to meat fermentations to encourage the growth of lactic acid bacteria and other microorganisms. It is also common to add other flavoring agents such as pepper etc.

Nitrite is the key chemical in the preservation of meat, in addition to salt. Nitrite reacts with the heme (Fig. 9.13) in meat to produce nitrosyl myoglobin that then becomes nitrosyl myochromogen when the myoglobin is denatured, e.g. by cooking. This changes the red color of the heme to the characteristic color of cured meats. Usually it is added as either potassium nitrate or nitrite and when nitrate is used, the nitrite is produced by the action of nitrate reducing organisms in the fermentation.

In the manufacture of hams and bacon the meat is placed in a saline solution containing (w/v) about 20% NaCl, and 0.5–0.7% potassium nitrate/nitrate and left for a few days to diffuse into the meat. In addition, brine solution may be injected into the muscle. The natural bacterial microflora in brine solution breakdown the nitrate to nitrite, it then diffuses into the meat reacting with the hemoglobin. Following immersion in the brine, the hams and bacon are removed and left to dry out under controlled conditions. If the ham is dried for a prolonged period, say many months, the surface may be covered with a fat layer to slow down moisture loss and yeasts and molds grow on the surface. This fungal growth contributes to the development of flavor.

To make fermented sausages potassium nitrate or nitrite is added to the meat at low concentration. When nitrate is added several things begin to happen when the sausage mixture is made. Gram-positive facultative aerobes, such as *Staphylococcus* spp. and *Micrococcus* spp. convert the nitrate to nitrite. The nitrite then reacts with the hemoglobin and fixes the red color of the meat. It also reduces the redox of the meat mixture to a level above the level at which the obligate anaerobes can grow. The nitrate/nitrite mixture renders the meat safe and free from the risk of botulinum poisoning. The presence of the nitrite does not affect the growth of the lactic acid bacteria.

9.4.2. Starter cultures in meat fermentation

The production of fermented sausages is historically a relatively recent development as it is thought that it was only about 250 years ago that the first fermented sausages were produced. Originally these fermented sausages were made using various mixtures of ingredients depending on the natural microflora for the microbial growth. However, in around 1954, lactic starter cultures were introduced for the manufacture of sausages. Today nearly all fermented sausages are made using added starter cultures (Table 9.11). In addition to bacterial cultures, fungal cultures are also used, for hams as well as sausages.

The first group of cultures used in modern fermented meats production were from the family *Micrococcaceae*, such as strains of *Kocuria varians* (formerly *Micrococcus varians*), *Staphylococcus carnosus* and *Staph. xylosus*. These organisms help in the formation and stabilization of color as well as aroma formation. Their important nitrate reductase activities are influenced adversely by NaCl concentration and low temperature. Color stabilization is related to their ability to produce catalase, the enzyme that degrades H_2O_2. The nitrosyl myoglobin, arising from the heme-nitrite reaction, is susceptible to oxidation by H_2O_2 to other colored substances such as green verdoheme and yellow biliverdin. The H_2O_2 also initiates the oxidative processes that lead to the formation of rancid flavors. The H_2O_2 arises through the growth of certain lactic acid bacteria, in the

Table 9.11.   Cultures used in meat fermentation and their useful properties for fermented sausage production.

Culture	Some characteristics and useful properties
Kocuria varians	Nitrate reduction; catalase production; color formation
Staphylococcus carnosus; *Staph. xylosus*	Nitrate reduction; catalase production; color formation; typical dry-salami aroma
Streptomyces griseus	Better aroma, darker red color; uses O_2; does not grow in meat mixture
Lactobacillus plantarum; *Lb. sakei*; *Lb. curvatus*	Homofermentative lactic acid production, but late heterofermentative metabolism occurs; hastens early moisture loss; some produce H_2O_2, destabilizes color, induces rancidity
Pediococcus pentosaceus; *P. acidilactici*	Not natural to fermented sausages; produces lactic acid homofermentatively; hastens early moisture loss; dies off rapidly
Debaromyces hansenii	Use O_2 rapidly, mainly active about meat/air interface of sausage; rapid exterior color formation; assimilates lactic acid giving slight rise in pH; produces Italian type flavor
Penicillium nalgiovense, *P. chrysogenum, P. camemberti*	Reduces late moisture loss; consumes O_2 at surface, reduces O_2 penetration; anti-oxidative effect, reduces rancidity; mold penetrates whole of sausage, influences flavor and aroma; assimilates lactate, raises pH

presence of O_2, about the air/meat interface. These cultures of *Micrococcaceae* produce catalase to break down the H_2O_2 thus preventing the formation of off colors. We have to include here another unusual bacterial culture sometimes used, namely *Streptomyces griseus sensu* Hütter. Introduced in 1977, it produces a better aroma and darker red color. The organism is proteolytic, non-lipolytic, nitrate-reducing, catalase positive and non-pathogenic. It does not grow in sausage mixtures, nor does it survive well. However, when added to sausage mixtures, a good flavor develops and nitrate levels decline, indicating metabolic activity.

The second group of cultures is lactic acid bacteria. The species are from the genera *Lactobacillus* and *Pediococcus* and include *Lb. plantarum*,

Lb. sakei, Lb. curvatus, P. pentosaceus and *P. acidilactici.* The desirable function of these bacteria is to produce lactic acid homofermentatively, that is, with no gas formation and little or no acetate/ethanol formation. The cultures of *Lb. sakei* and *Lb. curvatus*, although they grow at low temperatures, a desirable characteristic, they also tend to produce H_2O_2. On the other hand, pediococci are rarely found in naturally fermented sausages, but function well in the sausages because of their obligate homofermentative activity. The other lactic cultures used can switch metabolic pathways under certain conditions to produce formate, acetate and ethanol from lactate. The lactic acid produced lowers the pH of the meat to about 5.1–5.2 and decreases the water holding capacity of the meat so that free water is released and the sausage can dry more quickly.

The third group of cultures is the fungi — yeasts and molds. The species of yeasts commonly associated with natural fermentation are *Debaromyces, Candida* and *Torulopsis.* In commercial cultures, strains of *D. hansenii* are used, particularly for typical flavors of Italian type sausages, but they should only be used in conjunction with other cultures. The cultures when added at about $10^6 - 10^7$ cells/g, use O_2 rapidly and are mainly observed at the periphery of the sausage where O_2 is available for growth. This results in rapid exterior color formation. The yeast is able to assimilate lactic acid so a slight rise in pH occurs.

The molds most commonly found on the surfaces of fermented meats are species of *Aspergillus* and *Penicillium.* However, most of the molds found on fermented meats are toxinogenic species so care has to be taken to select non-toxinogenic strains. The first white non-toxinogenic culture made available commercially in 1972 was a strain of *P. nalgiovense.* Other species used include *P. chrysogenum* and *P. camemberti.* The effect of the mold is to reduce moisture loss due to the mold covering, and it has anti-oxidative effects due to the catalase activity of the mold, and the consumption of O_2 that reduces O_2 penetration into the sausage. Moreover, the mold penetrates the whole of the sausage and influences its flavor and aroma due to lipolytic and proteolytic activity as well as raising the pH, not because of ammonia production but because of lactate assimilation. The mold penetrates the sausage better if the casing for the sausage is natural and not made from artificial collagen.

Genetic engineering has been used to improve strains for meat fermentation. One approach has been to remove the need for the *Staph. carnosus* or *K. varians* strains by inserting the genes for catalase production into *Lb. curvatus*. The meat starter *Lb. curvatus* strain LTH683 was the base for the development. The *Lactobacillus* cloning vector pJK356 based on the cryptic plasmid pLC2 derivative of the meat starter strain LTH683 was isolated. The host was *Lb. curvatus* strain LTH1432, a plasmid cured derivative of strain LTH683. The catalase gene, *katA* from *Lb. sakei* strain LTH677, was then inserted and a recombinant strain of *Lb. curvatus*, strain LTH4002, isolated. Under aerobic conditions of growth this recombinant strain showed catalase activity four times greater than the strain of *Lb. sakei* used for the original source of the gene.

9.5. Vegetable Fermentation

A lot of vegetables are highly perishable and they are generally only available on a seasonal basis. In addition, some vegetable products are not at all highly perishable but have characteristics that make them unsuitable for human consumption. However, they can be made suitable by fermentation. In this section we will deal with these products.

9.5.1. *Fermented vegetables for preservation purposes*

These products can be broadly placed into two categories, those that are sliced before fermentation, and those that are fermented whole. Examples of the former are kimchi, the national dish of Korea, and sauerkraut, the fermented cabbage from Germany. Examples of the latter are fermented cucumbers, also from Germany and radishes from Japan.

The vegetables, as shown in Table 9.12, contain glucose, fructose and traces of sucrose. Also, the natural habitat of the lactic acid bacteria is the surfaces of vegetable matter. Under normal circumstances, the hexoses remain in the plant and the plant material repels attacks from microorganisms.

(1) Sauerkraut

Sauerkraut is fermented cabbage. The cabbage is shredded and about 2–2.5% (w/w) dry salt is incorporated into the sliced cabbage. The NaCl causes an

Table 9.12. Fermentable sugars composition (%) in vegetables that are often fermented.

Compound	Cabbage leaves	Cucumbers	Olives
TT	0.1–0.4	ND	0.1–0.2
ructose	2.0–2.1	1.0–1.4	0.1–1.1
Glucose	2.2–2.6	1.0–1.3	1.0–3.0

osmotic shock to the plant cells and the carbohydrate-laden juices flow from the cells forming brine. The shredded cabbage is then pressed down into the brined juice to exclude air. The lactic acid bacteria on the external surface of the plant materials, being halotolerant, then attack the glucose and produce lactic acid. The ideal temperature for bacterial growth is 18°C with a NaCl concentration in the brine of 1.8–2.25%. From the beginning of the fermentation, there is a distinct progression of bacterial species that carry it out. At first the heterofermentative organism *Lc. mesenteroides* dominates because of its short generation time and early initiation of growth. Due to the fermentative end products, undesirable Gram-negative bacteria die off at this stage. However, because *Lc. mesenteroides* is relatively sensitive to acid conditions, it declines in number and *Lb. brevis* and *Lb. plantarum* eventually dominate the fermentation. Initially *Lc. mesenteroides* produces lactate, acetate, ethanol and CO_2 from glucose, fructose and sucrose, but it can also use fructose as a final electron acceptor and produce mannitol. Finally *Lb. plantarum* dominates. At the end of the fermentation that is completed within 2 weeks to 2 months, the pH is ≤3.5 and total acidity about 1.6–2.3%. Fermentation temperatures above 30°C result in poor quality sauerkraut while low temperatures (≤10°C) favor *Lc. mesenteroides* exclusively and result in good quality product. Curiously, this organism produces D (–) lactic acid that is not metabolized by humans so sauerkraut is not recommended for babies and small children. Attempts to isolate and produce commercial starter cultures suitable for industrial production of sauerkraut have proved unsuccessful.

(2) Kimchi

Unlike sauerkraut Kimchi is made with Chinese cabbages and radishes as the main ingredients. Flavoring agents are also added including red pepper powder, garlic, onions, ginger, fish sauce and a small amount of sucrose (1%). There is a great variety of kimchi products which differ according to their ingredients and their regional origin. Traditionally it is produced in the cold months of the year but refrigeration is now used for year round production. About 1.5 million tons of kimchi is produced each year. The predominant lactic acid bacteria in Chinese cabbage based kimchi are species of the genera *Lactobacillus*, *Leuconostoc* and *Weissella*. The predominant organism is *W. koreensis*. Other organisms found in kimchi include *Leuc. gelidum*, *Leuc. gasicomitatum* and *Lb. sakei*.

(3) Cucumber Fermentation

Cucumbers are fermented to produce pickles. The cucumbers are immersed wholly into open brine tanks containing 5–8% NaCl and incubated at ambient temperatures, usually in the range 15–32°C. One of the main defects of concern to pickle producers is the formation of "floaters" or "bloaters" that develop internal gas causing the cucumbers to float. The solution is to ensure the removal of CO_2 or the prevention of its formation. Acetic acid is often added to the brine to reduce the pH to ≤4.5. This facilitates the expulsion of CO_2 from the brine when it is purged with N_2 or air. Lactic acid bacteria and yeasts grow in the brine. The main lactic acid bacteria that complete the fermentation are *P. pentosaceus*, *Lb. brevis* and *Lb. plantarum*, the last being the most important species (Fig. 9.14).

Sources of CO_2 vary. A prime source is heterofermentative growth by lactic acid bacteria. In addition, there is sufficient malic acid (0.2–0.3%) in cucumbers for the *Lb. plantarum* and the other lactic acid bacteria to carry out the malolactic fermentation and produce CO_2.

The major fermentation control methods used include initial low pH, purging of CO_2, and NaCl level. At present no commercial cultures have been developed for this fermentation.

9.5.2. *Fermented olives*

Olives are processed either unripe (green) or ripe (black) and contain a bitter substance, the phenolic glucoside, oleuropein, that makes olive fruit inedible. It is either removed or hydrolyzed by soaking in water or brine or through alkaline hydrolysis (1–2% NaOH). Alkaline (lye) treatment of black olives does not involve fermentation, but brine treatment of black

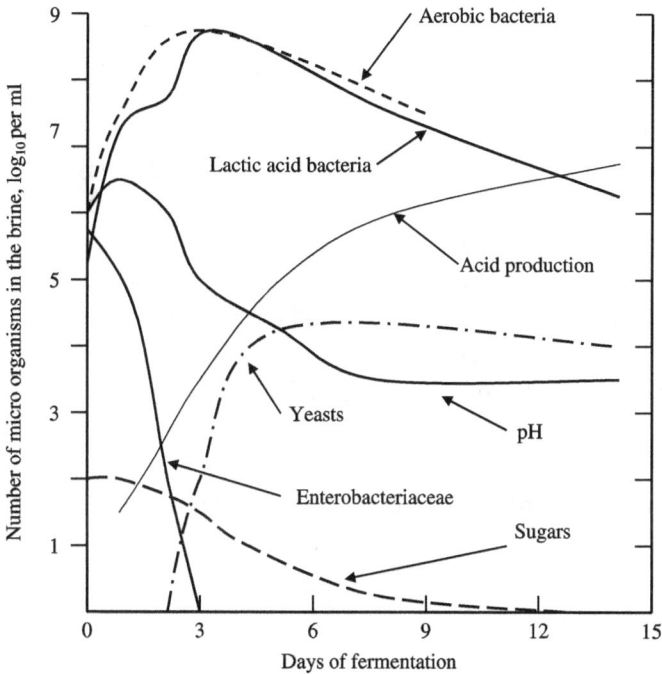

Fig. 9.14.   A generalized diagram showing microbial growth and acid production in pickled cucumber. Lactic acid bacteria eventually dominate but the acid conditions also allow the yeasts to grow.

olives does. On the other hand lye treatment of green olives (for 5–12 hours) is followed by fermentation.

As a consequence of the lye treatment the microbial load is reduced and the pH is in the range 7.5–8.5. The fermentation then goes through three stages. In the first stage, because of the pH, a wide variety of Gram-negative aerobes (e.g. species of *Pseudomonas*, *Flavobacterium* and *Aeromonas*) and facultative aerobes (e.g. species of *Enterobacter*, *Citrobacter* and *Klebsiella*) predominate for up to two days. The time is reducible by lowering pH to 6.0 by bubbling CO_2 through the liquid or by adding lactic or acetic acids. At this pH, the initial microflora die off and lactic acid bacteria start to grow in the second stage. The pH is lowered to about pH 4.5 over 10–15 days during which time *Lb. plantarum*, along with some yeasts and occasionally *Lb. delbrueckii*, dominate and reach a peak in numbers after about a week. Once pH 4.5 is reached, the third stage of the fermentation begins and lasts until the fermentable carbohydrates are depleted. *Lb. plantarum* dominates but also present are high populations of fermentative and oxidative yeasts that produce ethanol, ethyl acetate and acetaldehyde, that may contribute to the final flavor. Some of the yeast species involved include *Hansenula anomala*, *H. subpelliculosa*, *Candida krusei*, *C. parasilopsis* and *Saccharomyces chevalieri*. A fourth fermentation stage can occur but it usually results in spoilage.

Starter cultures for olive fermentation are generally obtained by the old method of re-inoculation from a previous fermentation. *Lb. plantarum* dominates in these fermented brines and the pH is usually about 4. Efforts to isolate suitable *Lb. plantarum* cultures have resulted in one culture, LPCO10, being isolated. When used in fermentation, its success depended on its ability to produce two bacteriocins: plantaricins S and T that prevented the growth of clostridia, a significant concern in olive fermentations, and propionibacteria as well as other lactic acid bacteria.

9.6. Soybean Fermentation

Soybeans are a good source of protein and are able to meet the essential amino acid needs of humans. However, like all legumes, the proteins are not readily digestible and contain anti-nutritional factors that can be destroyed by cooking and processing before consumption. When the cooked beans or processed beans are subjected to microbial fermentations a variety of diet-enhancing soybean products are produced. Fermented soybean products are most common in Asia.

9.6.1. *Sauces and pastes with meat-like flavor (soy sauce)*

Soy sauce is almost as old as recorded Chinese history. It was introduced to Japan in 552AD. During the 17th and 18th centuries, the Dutch exported large quantities of soy sauce from Japan to other parts of Asia and to Europe. The Japanese established a soy sauce manufacturing plant in the United States of America in 1909.

The essential ingredient for soy sauce production is soybean. After mold growth on the grain, it is steeped in a brine solution. Soy sauce is the liquid extract from the mixture. However, the varieties of soy sauces on the market owe their differences to the addition of wheat, a source of carbohydrate. Soybean is high in oil and protein and contains about 30% carbohydrate that is comprised of a soluble and an insoluble fraction (Table 9.13). The soluble fraction, about 10% of the bean, is a mixture of sucrose (5%), raffinose (1%) and stachyose (4%), and the insoluble fraction consists of structural carbohydrates such as pectin, hemicellulose and cellulose, and less than 1% starch.

Starch added as wheat in the soybean mix promotes the growth of the fungus and production of enzymes. This is desirable as a greater amount of enzymes enable better breakdown of the protein. The soybean:wheat ratios used vary from 100:0 to 25:75. In Japan, these ratios are formalized and the resulting soy sauces have various names (Table 9.14). Although the main Chinese type of soy sauce is similar to tamari, ratios of 66:33 and 50:50 are also common in China.

Table 9.13   Composition of soybeans, defatted soy flour, wheat grain and wheat flour, g kg^{-1}, used in the manufacturing of soy sauce.

Component	Soybean		Wheat	
	Whole bean	Defatted flour	Whole grain	Flour
Water	86	72	115—140	—
Protein	343	470	153	140–180
Oil	187	12	19	9–24
Ash	51	62	18.5	4–14
Carbohydrate	316	384	717	650–740

Table 9.14.   Some of the major types of soy sauce, and the soybean: wheat ratios used in their production.

Soy sauce	Soybean:wheat ratio	Comment
Tamari	90:10	Most common type in South East Asia
Saishikomi	50:50	Koji is steeped in previously made raw soy sauce
Koikuchi	50:50	Most popular soy sauce in Japan, 83% of market
Shiro	25:75	High sugar content and very light color
Usukuchi	About 33:66	Color lighter than koikuchi
Kecap asin	100:0	Indonesian soy sauce

Box 9.5

Traditional Soy Sauce Production in Hong Kong

The photos show how soy sauce is prepared traditionally in Hong Kong. The soybeans are cooked, cooled and flour is added. The floured soybeans are then placed in a thin layer on trays to allow the mold to grow. Then the koji is added to the clay jars with brine and covered. They are left in the open for the moromi fermentation to take place. When the fermentation is judged to be complete the liquid is drawn off and more brine added. This is repeated several times so that up to five collections are made.

The carbohydrate in soybean and wheat is broken down into fermentable sugars with the aid of fungal enzymes from either *Aspergillus oryzae* or *A. sojae*. The soy sauce making process for koikuchi soy sauce is summarized in Fig. 9.15.

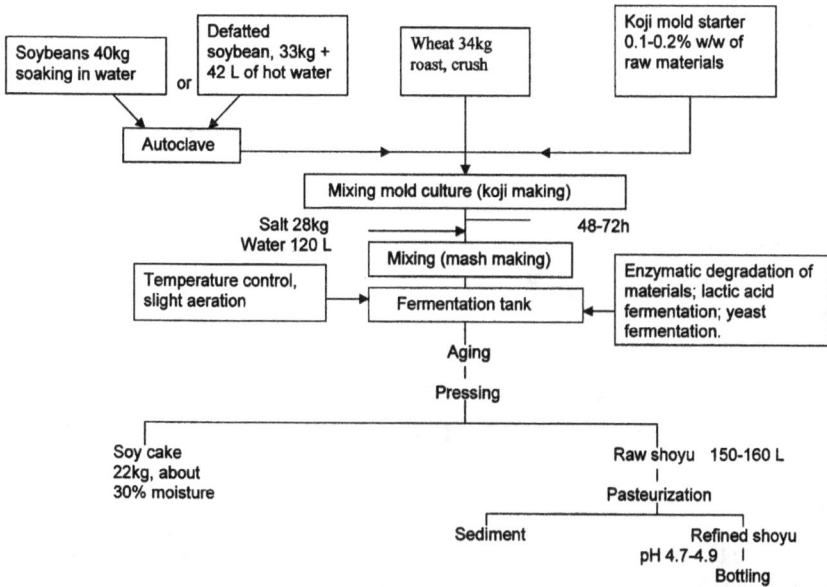

Fig. 9.15. The flow diagram setting out the process used to produce soy sauce. The process depicted here is for the production of koikuchi soy sauce. The other types depend on variations in the soy: wheat ratio.

An important step is the cooking of the soybean. The yield of soluble nitrogen from the soybean increases with increasing cooking temperature and decreasing cooking time. Cooking temperatures of 118°C for 45 min and 170°C for 15 s result in soluble nitrogen yields of about 82% and 88%, respectively (Table 9.15). Defatted soybean is often used instead of whole soybean because the oil plays no significant part in the soy sauce. Wheat bran is often preferred over wheat kernels as it serves as a source of ferulic acid that is converted to 4-ethylguaiacol. In addition, the pentose in wheat bran serves to enhance darker coloration and color stability, but results in reduced alcohol content in the soy sauce.

The vocabulary in soy sauce production is dominated by Japanese terminology. The molded soybean mixture is called "koji" which in Japanese means "bloom of mold". When the koji is mixed into the brine the resulting mixture is called "moromi".

As a preliminary step in koji making, steamed polished rice or wheat bran is often used to produce tane koji, the equivalent of a starter culture, to

Table 9.15.   Effect of cooking pressure, time and temperature of soybeans on the yield of total nitrogen, amino nitrogen, percent nitrogen yield, and protein digestibility in soy sauce.

Pressure, kg cm^{-1}	Cooking temperature, °C	Time, min	Effect of cooking on nitrogen fractions in soy sauce and on protein digestibility			
			Total N, mg mL^{-1}	Amino N, mg mL^{-1}	Yield, %	Protein digestibility, %
0.9	118	45	16.53	8.17	82.05	86
2.0	132	5	16.97	8.37	84.24	92
6.7	151	2	17.52	8.39	86.87	—
7.0	170	0.25	17.78	8.78	88.26	95

inoculate the prepared soybean and wheat for sauce production. The seed koji (0.1–1.0% w/w) is mixed with cooked soybean and wheat. The mixture of tane koji, soybean and wheat is spread out to a depth of about 3–5 cm and left to incubate for 2–3 days to allow mold growth and production of enzymes.

The temperature of the koji increases rapidly and reaches about 38°C. The optimal temperature for the growth of the mycelium of *A. oryzae* is between 30 and 35°C. The fungi produce amylase in the temperature range 35–40°C. In order to achieve faster mycelium growth and to prevent over production of amylases that cause sticky koji, the substrate needs to be cooled by mixing twice at around 18 hours and 27 hours of incubation (Fig. 9.16). Good ventilation is important, as it helps to prevent overheating. However, over-cooling of koji can be detrimental, as it encourages the growth of undesirable yeasts (e.g. *Zygosaccharomyces rouxii* is undesirable at this stage but desirable in the moromi stage) and other fungi (e.g. *Rhizopus nigricans*). Common bacterial contaminants are *Bacillus subtilis,* and species of *Micrococcus, Leuconostoc* and *Lactobacillus.*

The growth of the mycelium eventually slows at around 36 hours, and the temperature in the koji gradually cools down to the room temperature. During the last stage of koji making, proteases begin to accumulate. The amount of proteases accumulated in koji is important as it determines the flavor of soy sauce. The process is stopped before fungal spore formation. Enzymes produced by koji molds are shown in Table 9.16.

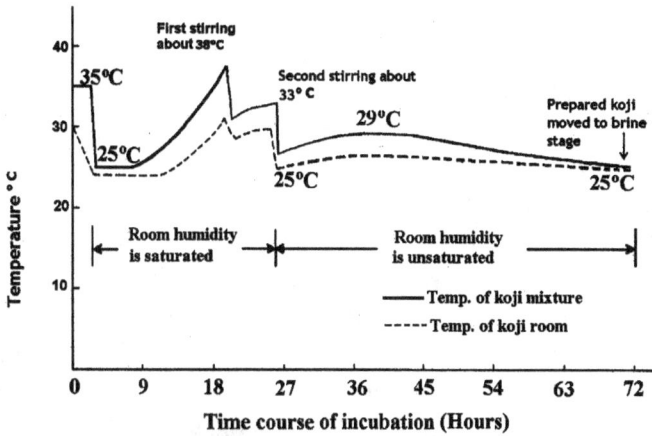

Fig. 9.16. The temperature and relative humidity in a traditional small tray method for preparing koji. The temperature increases and must be reduced by stirring the mixture. The optimum levels of enzymes are produced after about 48 hours. The harvested koji is then steeped in brine (after Fukushima).

Table 9.16. Enzymes produced by koji mold and the levels of enzymic activities recorded.

Enzyme	Activity, units g^{-1} koji
Total protease	1129
Acid protease (pH 3)	56
Neutral protease (pH 7)	89
Alkali protease (pH 7)	984
α-amylase	3920
Acid carboxypeptidase	0.46
Leucine aminopeptidase	0.36
CM-cellulase	21.6
Pectin transeliminase	12.3

The koji is subsequently steeped in brine water to give a salt concentration of 17–19% w/v. The high salt content inhibits the growth of *Micrococcus* and *Bacillus* species. Use of water containing low iron and copper to make the brine is important, as high mineral contents reduce color stability. The brine water is first adjusted to pH 6.5–7.0 and cooled to 0°C to maintain the temperature of the freshly prepared moromi mash at about 15°C for several days. The initial low temperature prevents a rapid drop in pH due to lactic acid fermentation. Alkaline proteases remain active for a longer period to further digest the 1–3% w/w of protein in the mash. Thereafter, the temperature is allowed to increase gradually to 28–30°C after 20–30 days. The temperature level is important because it controls the production of ethanol in the soy sauce. Japanese consumers prefer soy sauce containing about 3% ethanol. To obtain this ethanol, a starch source, e.g. wheat, must be included in the koji, and the temperature must rise to 30°C. If the temperature is kept below this level, ethanol is not produced. At the end of 2–3 months, most of the fermentable carbohydrate has been converted to lactic (1% w/v), ethanol (2–3% v/v) and simple sugars (e.g. glucose, xylose).

9.6.1.1. *Enzymic processes*

The protein in the soybean and wheat is released by macerating enzymes such as cellulase, hemicellulase, and pectinase (Fig. 9.17). Then the alkaline and neutral proteinases release polypeptides that are further digested by acid carboxypeptidase and leucine aminopeptidase to release amino acids. Glutaminase breaks down the glutamine to glutamic acid (naturally produced MSG), preferably at temperatures below 25°C and in a low salt condition. Pectin transeliminase serves to reduce viscosity due to acidic polysaccharides derived from soybeans, so that the soy sauce is more easily pressed from the mash and clogging of the filter is avoided. The amylase enzymes release glucose from the starch.

9.6.1.2. *Microbial processes in the brine stage*

During the moromi stage, a complex microbial process occurs. The brine kills off aerobic bacteria and with the release of the glucose, lactic acid bacteria start to grow and lower the pH. When the pH drops to about 5.5, yeasts start to grow, in particular *Z. rouxii,* and its growth is followed by the growth of further important flavor producing yeasts (Fig 9.18).

The soy sauce may be allowed to incubate a long time for aging, which involves darkening of color due to browning reactions.

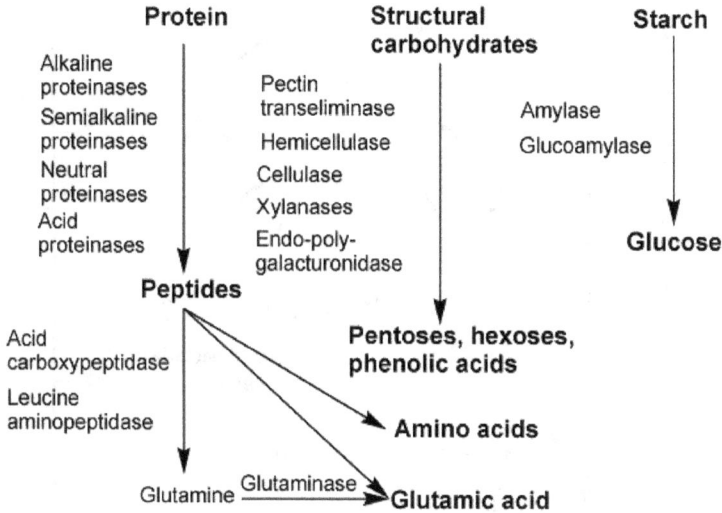

Fig. 9.17. The suite of enzymes produced by the fungi in the koji act on the three main components, the complex polysaccharides holding soy and wheat cells together, the proteins in the soybeans and wheat, and the starch in the wheat. The pentoses, hexoses and phenolic acids are further broken down to flavor compounds.

Flavor compounds (Table 9.17) come via a variety of mechanisms and sources. *Z. rouxii* and *Candida* species such as *C. versatilis* produce 2-phenylethanol from *p*-hydroxy cinnamic acid and phenylalanine; 4-ethylguaiacol from ferulic acid; and isoamyl alcohol, iso- and n-butanol, 3-methylthio-1-propanol from amino acids; 2- and 3-methyl-1-butanol, 2- and 3-methylbutanoic acids from isoleucine and leucine; and methionol and methional from methionine. Roasting of the wheat produces furfural from which the yeasts produce furfuryl alcohol, and heating of the raw soy sauce promotes Strecker degradation reactions from which arise isovaleraldehyde, acetaldehyde and methional from dicarbonyl compounds and amino acids, as well as HMMF via the Maillard reaction, and HDMF from pentoses and amino acids (Fig. 9.19). The yeasts produce glycerol and *Tetragenococcus halophilus* produce acetic acid.

9.6.1.3. *Biotechnological approaches to soy sauce production*

One of the disadvantages of soy sauce production is the amount of time required to produce the final product, a problem shared with all

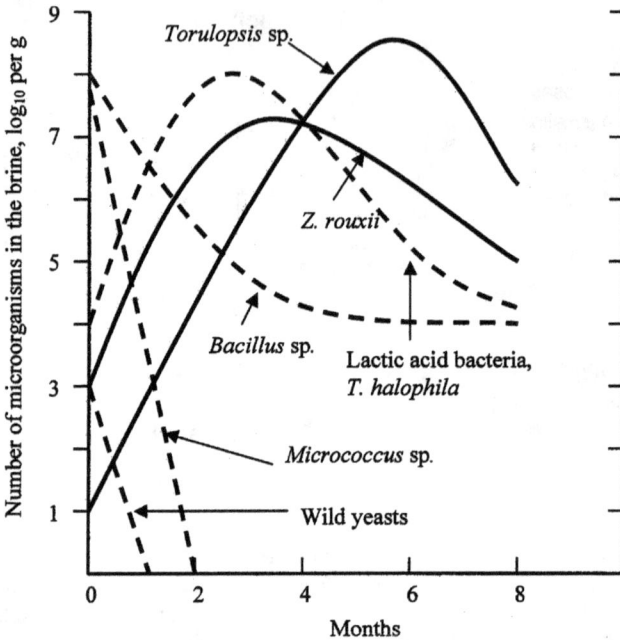

Fig. 9.18.   The time course of the brine fermentation in Japanese style shoyu. The brine kills wild yeasts and species of *Micrococcus*, and any vegetative bacilli are killed off leaving the endospores, and the lactic acid bacterium, *Tetragenococcus halophilus*, begins to grow lowering the pH. This then allows the yeast *Zygosaccharomyces rouxii* to begin to grow followed by the *Torulopsis* species, such as *Candida etchellsii*.

fermented foods with a long maturation time. Continuous processes and bioreactors to speed up the manufacturing time have been examined (Fig. 9.20).

One continuous method that ran for 80 days through a 280 L reactor involved immobilization of cells of *T. halophilus*, *Z. rouxii* and *C. versatilis* in calcium alginate gel to ferment the moromi. The soybean/wheat mixture was hydrolyzed enzymatically and passed through a column bioreactor containing immobilized *T. halophilus* cells for the lactic fermentation. Then the production stream was split into two, one stream went through a *Z. rouxii* reactor and the other was put through a *C. versatilis* reactor. Each stream was then heated, filtered and combined to give a production time of two weeks, but the aroma pattern did not match traditional sauce. When the gel, unstable to heat and a high salt concentration,

Table 9.17. Some major flavor components in Japanese soy sauce.

Compound	Quantity, mg kg^{-1}	Compound	Quantity, mg kg^{-1}
Ethanol	31,501	Isobutanol	11
Lactic acid	14,346	Furfuryl alcohol	11
Glycerol	10,208	Isopentyl alcohol	10
Acetic acid	2,107	Acetoin	10
HMMF	256	n-Butanol	9
2,3-Butanediol	238	HDMF	5
Isovaleraldehyde	233	Acetaldehyde	5
HEMF	239	2-Phenylethanol	4
Methanol	62	n-Propanol	4
Acetol	24	Acetone	4
Ethyl lactate	24	Methionol	4
2,6-Dimethoxyphenol	16	2-Acetylpyrrole	3
Ethyl acetate	15	4-Ethylguaiacol	3
Isobutyraldehyde	14	Ethyl formate	3
Methyl acetate	13	γ-Butyrolactone	2
		p-Ethylphenol	Trace

HMMF = hydroxy-2(or 5)-methyl-5 (or 2) methylfuran-3(2H)-one.
HEMF = hydroxy-2(or 5)-ethyl-5 (or 2) methylfuran-3(2H)-one.
HDMF = 4-hydroxy-2,5-dimethylfuran-3(2H)-one.

The alcohols contribute to the volatile flavor of soy sauce. The HEMF is considered the *character impact substance* because of its intense sweet aroma. HDMF provides a caramel-like flavor. The characteristic soy sauce taste is also attributed to 4-ethylguaiacol produced by torulopsis yeasts. Methionol has a synergistic effect with 4-ethylguaiacol and it appears that 3.9 mg L^{-1} of methionol and 0.8 mg L^{-1} of 4-ethylguaiacol are the optimum concentrations.

was replaced with a ceramic bead carrier for the whole cells, a satisfactory soy sauce was produced in eight days. A further advance was to use chitopearl to immobilize glutaminase, and alginate gel to immobilize the microorganisms. The koji mixture was then hydrolyzed in 8.5% NaCl at 45°C for three days, filtered and then put through the glutaminase reactor followed by the microbial reactors. Production took about two weeks and the system ran continuously for 100 days, but the product still did not match completely the natural product.

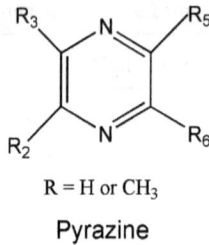

R = H or CH$_3$

Pyrazine

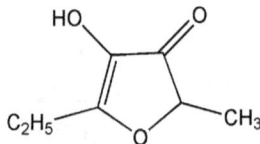

4-hydroxy-2-methyl-5-ethyl-
3(2H)-furanone (HEMF)

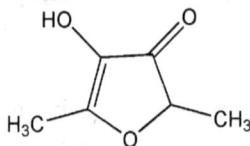

4-Hydroxy-2,5-dimethyl-
3(2H)-furanone (HDMF)

Fig. 9.19. Important flavoring compounds in soy sauce (HEMF and HDMF) and natto (pyrazine).

9.6.1.4. Variations in traditional soy sauce making

The above process is basically the Japanese method of making shoyu, however, in other countries there is some variation in those methods used. In Indonesia only soybeans are used, the incubation period in the moromi (baceman) stage may be as little as one day but usually longer. Sugar may then be added to give a sweet soy sauce, kecap manis, as opposed to the normal sauce, kecap asin. In traditional soy sauce production in Hong Kong, the moromi stage is conducted in ceramic jars of about 40 L capacity and when the fermentation is completed the sauce is drawn off and fresh brine added, left for a few days and then the liquid drawn. Liquid is drawn about five times. This results in various grades of soy sauce. Further, the sauce is exposed to the sun for up to six weeks, a

Conventional Method

| Soybeans | Wheat |

Autoclave   Roast
water

| Fungal starter |

| Koji |

| Salt solution |

| Temperature control, slight aeration | → | Moromi mash |

Hydrolysis

| Lactic acid and alcohol fermentation |

Aging

Filtration

Pasteurization

| Soy sauce |

Bioreactor Method

| Defatted soybean | Wheat |

Steam   Roast

| Continuous broth culture of Koji mold |

Hydrolysis
(40°-50°C for 3 days)

| Moromi mash |

Filtration

| Raw liquid |

| Salt |

| Immobilized glutaminase |

| Immobilized lactic acid bacteria T. halophila |

| Immobilized yeast Z. rouxii, C. versatilis |

| Soy sauce |

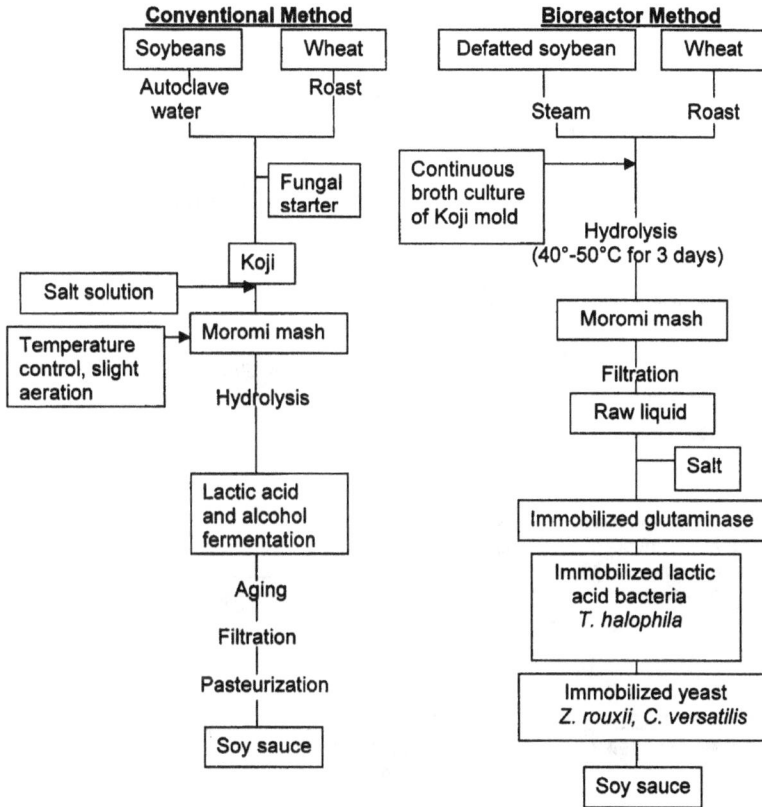

Fig. 9.20. A comparison of the traditional and a biotechnological approach for accelerating shoyu production. The difference between the traditional method and this biotechnological method is that the former requires up to eight months while the latter takes about two weeks.

practice also in parts of China. In Korea, the koji is produced as a kind of cake (meju or meiju). The soybeans are cooked and crushed to form a cake shape and inoculated with mold. The cakes are then incubated in the koji room at 15–20°C and 40–50% relative humidity up to 30 days. Better kanjang (Korean soy sauce) results if a temperature of 25–30°C and 80–90% relative humidity up to 30 days is used.

9.6.2. *Fermented soybean curd (furu, sufu)*

This is a cheese-like fermented bean curd, popular amongst Chinese people. It is also known as chao in Vietnam, tahuri in Philippines, taokaoan in

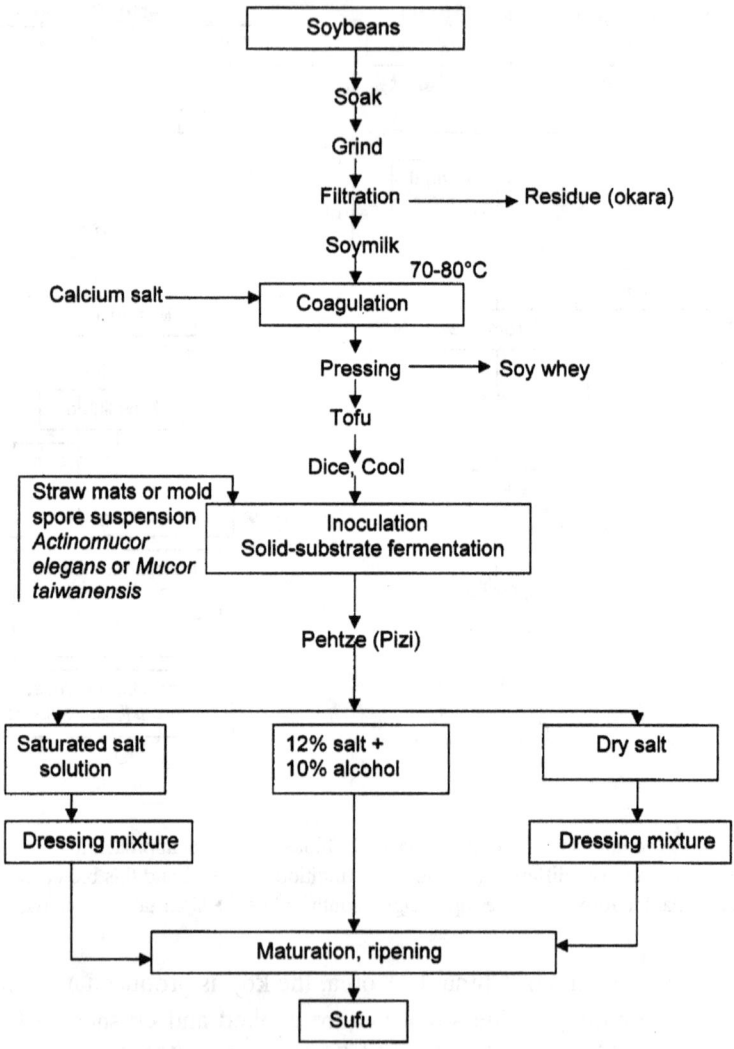

Fig. 9.21. The production of furu. The raw material is soy extract from soybean (soy milk) that is precipitated and then fermented with microorganisms. The product has a very strong aroma.

Indonesia and tao-hu-yi in Thailand. Its production, outlined in Fig. 9.21, depends on the characteristics of the protein extracted as soymilk from soybeans. Beans are soaked overnight and then ground up into fine particles to form a slurry. The slurry is then filtered to produce soymilk and a solid waste product, okara. The soymilk contains protein and fat globules. Like casein, soymilk protein can be clotted to form a curd (tofu), but the clotting agent is usually Mg or Ca salts.

The dried surface of tofu encourages growth of fungi on the surface and hydrolysis of protein in the surface layer. The fungi used in commercial production of furu may include *Actinomucor elegans*, or *Mucor dispersus*, or *M. silvaticus*, or *Rhizopus chinensis*. However, in natural fermentations the furu fungi may, in addition, include *M. corticolus, M. praini, M. racemosus*, and *M. subtilissimus*, all of which are common microbial isolates found on the surface of rice or wheat straw.

During the ageing period in brine, Na ions cause release of fungal proteases and lipases. The proteases penetrate the curds and hydrolyze the soybean protein component that results in a soft cheesy furu, and the lipase activity produces fatty acids. When ethanol is included in the brine, proteases activity is reduced and the fatty acids react with the ethanol to produce fruity flavored esters (Table 9.18).

Furu produced in summer is generally inferior to that produced at times of cooler temperature. In a comparison of production at 26°C and 32°C, a fermentation temperature of 26°C favored β-glucosidase production by the fungi, which resulted in more isoflavone aglycons being produced from the tofu that may be associated with better quality. In addition,

Table 9.18. Biochemical changes occurring during the maturation in brine or in brine with added ethanol, of furu made with *Actinomucor taiwanensis*.

Maturation time, days at 25°C	Brine, 12% w/v NaCl		Brine, 12% w/v NaCl, plus ethanol, 10% w/v	
	N ratio, amino N:total N	Free fatty acids, mg g^{-1} lipid	N ratio, amino N:total N	Free fatty acids, mg g^{-1} lipid
0	0.05	45	0.05	45
15	0.09	70	0.06	80
30	0.15	120	0.08	110
45	0.18	180	0.09	80
60	0.20	160	0.11	70
75	0.27	170	0.13	75

research has shown that Chinese sufu samples exhibit anti-mutagenic activity, e.g. against benzo(α)pyrene in the Ames test.

9.6.3. *Whole bean fermentation*

For a substantial proportion of the human population, plants and seeds are the major sources of protein as animal products are not readily available. However, plant proteins are lacking in taste and texture. Consequently, these populations often lack a plentiful supply of good quality protein. In addition, the cost of protein is high so to provide an adequate supply for the poor is a priority. One traditional source is fermented whole soybeans.

9.6.3.1. *Tempeh*

Tempeh kedelle is a popular fermented soybean product and meat substitute in Indonesia. After preparation it requires just 3–4 minutes to deep fry or 10 minutes to cook.

The production of tempeh is outlined in Fig. 9.22. In the first step the soybeans are soaked in water or acidified water at room temperature.

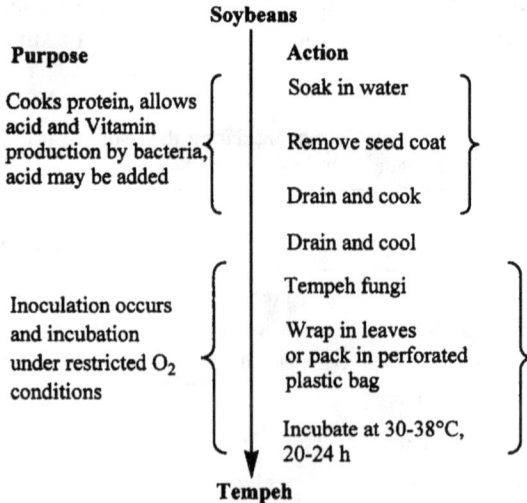

Soybeans

Purpose	Action
Cooks protein, allows acid and Vitamin production by bacteria, acid may be added	Soak in water
	Remove seed coat
	Drain and cook
	Drain and cool
Inoculation occurs and incubation under restricted O$_2$ conditions	Tempeh fungi
	Wrap in leaves or pack in perforated plastic bag
	Incubate at 30-38°C, 20-24 h

Tempeh

Fig. 9.22.   An outline of the production of tempeh, an Indonesian soy based food produced using fungi. Careful control of oxygen supply during the fungal growth phase is critical for a good tempeh product.

During this stage two things happen, firstly a partial germination of the soybeans may occur depending on the amount of O_2 available to the seed, and secondly, acid is produced by bacteria in the soak water and suppresses the growth of undesirable microorganisms. If a partial germination occurs, it can impact on the protein properties of the soybean and the subsequent fungal fermentation. Further, the bacteria that grow in the steeping water produce vitamin B_{12}, a significant nutrient in the final tempeh product. The most desirable bacterial species for this stage is *Klebsiella pneumoniae*, one of the *Enterobacteriaceae*, but by using pure cultures, other bacteria can perform the same function.

The soaked beans are then dehulled and cooked, a critical process in which a balance must be struck between over-cooking and under-cooking the beans. The cooked beans are then drained and cooled below 35°C followed by dusting with wheat flour to provide a good source of carbohydrate for the fungal inoculates. The desirable fungal species for successful tempeh production, whether from environmental inoculation or pure starter inoculation, are *Rhizopus oligosporus* (e.g. NRRL-2710), *R. stolonifer, R. arrhizus, R. oryzae, R. formosaensis,* and *R. achlamydosporus.* The spores of *R. oligosporus* are produced commercially in Indonesia for industrial scale tempeh production. During the fungal growth stage, the O_2 level must be controlled at a reduced level otherwise the fungus will grow too quickly and form black spore masses that are an indicator of poor quality. The traditional way to control O_2 is to wrap the inoculated beans in banana leaves but a modern innovation is the use of micro-perforated polyethylene plastic. The mycelia then grow and knit the beans into a firm cake, thus giving tempeh its meaty texture.

Proteases are most active at 50–55°C and pH 3.0–6.0. The enzymes from the fungi transform the soybeans making them more nutritious as shown in Table 9.19. Tempeh must be consumed quickly after purchase.

Defects include black patches due to sporulation, slime due to excessive bacterial growth because of too little O_2 or a temperature $\geq 42°C$, or a yellow color due to growth of mycotoxic fungi. The yellow color indicates the tempeh is highly toxic and should not be eaten.

An Indonesian product similar to tempeh is ontjom in which groundnuts are used and the preferred fungus is *Neurospora sitophila* that has orange-red spores. The process differs in that sporulation to produce the color is encouraged.

Table 9.19   Chemical changes in cooked soybean due to its conversion to tempeh.

Component	Cooked bean, %	Tempeh, %
Hemicellulose	2.0	1.1
Fibre	2.0	1.5–4.3
Total nitrogen	7.5	7.5
Soluble nitrogen	0.5	2.0
Free fatty acids	Trace	10
Vitamin B_{12}	Trace	30 μg g^{-1a}

[a] Vitamin B_{12} is produced by *Klebsiella pneumoniae* in the mixed inoculum.

9.6.3.2. *Natto*

This is produced in Japan where soybeans are used. In Korea it is known as chung kook jang, and in Thailand as thua nao but the Thais sun-dry the product and a similar product is found in India. Traditionally the beans are soaked, boiled, cooled and wrapped in rice straw and left alone for one to two days at a warm temperature to allow bacteria to grow on the beans. The straw is said to impart a straw aroma to the fermented beans as well as being the source of the bacteria for the fermentation, and it absorbs some of the unpleasant aroma of ammonia released during the bacterial growth. During growth, the bacteria produce a viscous, sticky slime that pulls out into long strings when two soybeans from the final product are pulled apart. The viscous substance increases to about 2% of the natto after 48 hours and its crude protein content is about 80% of which about 20% is mainly the D isomer of γ-poly-glutamic acid. Free amino acids make up a large proportion of the remaining crude protein in the viscous material. The bacterium responsible for the characteristics of natto is an obligate aerobe *Bacillus subtilis* var. *natto*. The bacterial fermentation results in a high concentration of vitamin K_2 (menaquinone–7) that is important for carboxylation of osteocalcin, a bone protein. Some evidence suggests that natto consumption by Japanese women reduces the risk of hip fracture in postmenopausal women.

A further result of the fermentation is a reduction of the "beany flavor" and aroma of soybeans and an increase in alkyl pyrazines that are responsible for the characteristic odor of natto (Fig. 9.19). The amount of volatile compounds develops from around 35 μgkg^{-1} wet weight in cooked soybeans to 2–3.5 mg/kg wet weight in 72 h. Predominant volatiles in Japanese natto after 72 hours of fermentation are 3-hydroxybutanone (acetoin), 2,5-dimethylpyrazine, trimethylpyrazine, and tetramethylpyrazine. Some compounds (acetoin) appear early in fermentation and then decrease, whereas others appear only towards the end of fermentation.

Further Reading

1. Beech GA, Melvin MA and Taggart J. Food, drink and biotechnology. In: *Biotechnology: Principles and Application*, eds. IJ Higgins, DJ Best and J Jones. Blackwell Scientific, Oxford, 1985, pp. 73–110.

2. Beuchat LR. Indigenous fermented foods. In: *Biotechnology Vol. 5*, Chapter 11, ed. G Reed. Verlag Chemie, Weinheim, 1983, pp. 477–528.

3. Bottazzi V. Other fermented dairy products. In: *Biotechnology: Food and Feed Production with Microorganisms Vol. 5*, ed. G Reed. Verlag Chemie, Weinheim, 1983, pp. 315–364.

4. Cho J, Lee D, Yang C, Jeon J, Kim J and Han H. Microbial population dynamics of kimchi, a fermented cabbage product. *FEMS Microbiol. Lett.* **257**: 262–267, 2006.

5. Coffey A and Ross RP. Bacteriophage-resistance systems in dairy starter strains: Molecular analysis of applications. *Antonie van Leeuwenhoek* **82**: 303–321, 2002. (Note: Tables 1 and 2 were not included in the published paper).

6. Ebner H and Follmann H. Vinegar. In: *Biotechnology Vol. 5*, Chapter 9, ed. G Reed. Verlag Chemie, Weinheim, 1983, pp. 425–446.

7. Eck A and Gillis J-C, eds. *Cheesemaking: From Science To Quality Assurance,* 2nd ed. Lavoisier Publishing Inc. Paris, Intercept Ltd, London, 2000.

8. Fleming HP, McDonald LC, McFeeters RF, Thompson RL and Humphries EG. Fermentation of cucumbers without sodium chloride. *J. Food Sci.* **60**: 312–315, 1995.

9. Fox PF, ed. *Cheese: Chemistry, Physics and Microbiology, Vol.1, General Aspects, and Two Major Cheese Groups*. Chapman & Hall, London, 1993.

10. Frazer WC and Westhoff DC. *Food Microbiology*. McGraw-Hill, New York, 1988.

11. Friend BA and Shahani KM. Fermented dairy products. In: *Comprehensive Biotechnology: The Principles, Applications and Regulations of Biotechnology in Industry, Agriculture and Medicine Vol. 3*, eds. HW Blanch, S Drew and DIC Wang. Pergamon, Oxford, 1985, pp. 567–592.

12. Ghose TK and Bhadra A. Acetic acid. In: *Comprehensive Biotechnology Vol. 3*, Chapter 36, eds. HW Blanch, S Drew and DIC Wang. Pergamon, Oxford, 1985, pp. 701–730.

13. Han B-Z, Rombouts FM and Nout MJR. A Chinese fermented soybean food. *Int. J. Food Microbiol.* **65**: 1–10, 2001.

14. Jay JM. *Modern Food Microbiology*. Van Nostrand Reinhold Co., New York, 1992.

15. Lee YK, Salminen S, Nomoto S and Gorbach S. *Handbook of Probiotics*. John Wiley & Sons, New York, 1999.

16. Nout MJR and Aidoo KE. Asian fungal fermented food. In: *The Mycota: Industrial Applications Vol. X*, ed. HD Osiewacz. Springer-Verlag, Berlin, 2002, pp. 23–47.

17. O'Toole DK. Factors inhibiting and stimulating bacterial growth in milk: An historical perspective. *Adv. Appl. Microbiol.* **40**: 45–94, 1995.

18. O'Toole DK. Soy-based fermented foods. In: *Encyclopedia of Grain Science*, eds. C Wrigley, H Corke and C Walker. Elsevier, 2004, pp.174–185.

19. Ouwehand AC, Salminen S and Isolauri E. Probiotics: An overview of beneficial effects. *Antonie van Leeuwenhoek* **82**(1–2): 279–289, 2002.

20. Rogers P, Chen J-S and Zidwick MJ. Organic acid and solvent production. Part I: Acetic, lactic, gluconic, succinic and polyhydroxyalkanoic acids. In: *The Prokaryotes: A Handbook on the Biology of Bacteria Vol. 5*, 3rd ed., Chapter 3.1, ed. M Dworkin, S Falkow, K-H Schleifer, E Rosenberg and E Stakebrandt. Springer Science, 2006, **1**: 511–755.

21. Salminen S, Deighton M and Gorbach S. Lactic acid bacteria in health and disease. In: *Lactic Acid Bacteria*, eds. S Salminen and A von Wright. Marcel Dekker, New York, 1998, pp. 211–254.

22. Steinkraus K, ed. *Handbook of Indigenous Fermented Foods*. Marcel Dekker, New York, 1995.

23. Wood BJB, ed. *Microbiology of Fermented Foods, Vols. 1 and 2*. Blackie Academic and Professional, 2–6 Boundary Row, London, 1998.

24. Yokotsuka T. Traditional fermented soybean foods. In: *Comprehensive Biotechnology Vol. 3*, Chapter 19, eds. HW Blanch, S Drew and DIC Wang. Pergamon, 1985, Oxford, pp. 396–427.

Websites

http://www.soyfoods.com/index.html
http://en.wikipedia.org/wiki/Main_Page
http://www.japanweb.co.uk/listing/index.htm
http://www.foodreference.com/index.html
http://www.marukome.co.jp/mehome.html
http://www.saitoku.com/
http://www.ebs.hw.ac.uk/SDA/cheese1.html
http://biology.clc.uc.edu/Fankhauser/Cheese/Cheese.html
http://www.dairyfoods.com/CDA/ArticleInformation/features/BNP__
Features__Item/0,6775,144097,00.html

Questions for Thought...

1. Use vinegar production as an example to illustrate the technological development in microbial fermentations. Highlight the technologies developed at each stage to overcome the rate-limiting steps in the vinegar production processes.

2. Design a continuous flow fermentation process for the production of cultured milk. Incorporate a mechanism to use a portion of the final product as the inoculum.

3. Bacteriophages are viruses that attack bacteria. If you were setting up a new factory to produce a new product using a bacterial culture,

what would you do to avoid the production problems that a bacterio-phage attack may cause to the production system?

4. Discuss the measures taken in the various steps of soy sauce produc-tion to ensure the involvement of the right kind of microorganisms and enzymes.

5. "The production of fermented foods relies entirely on microbial anaerobic processes". Discuss this statement.

6. "Ultimately we will produce all of the fermented foods using enzymes alone". Discuss this statement.

Chapter 10

Food Involving Yeast and Ethanol Fermentation

Desmond K. O'Toole
Department of Biology and Chemistry
City University of Hong Kong
83 Tat Chee Avenue, Kowloon, Hong Kong, China
Lee Yuan Kun
Department of Microbiology, Yong Loo Lin School of Medicine
National University of Singapore
5 Science Drive 2, Singapore 117597
Email:micleeyk@nus.edu.sg

Yeasts are a natural part of the microbial population on fruits and on ground cereal grains. When fruit juice is released and ground cereal grain is wetted, the yeasts (as well as other organisms) grow over a few days and produce ethanol. From these natural processes humans have learnt to harness the activity of the yeasts.

Alcoholic beverages in ancient times were an important part of the diet. They provided a way of storing the food value in otherwise perishable agricultural products and provided important nutrients, including vitamins, in the diet. Today when these products are consumed, they are consumed as much for their flavors as for their alcoholic content and its effects, and food value. The quality of the final beverage rests as much on the ethanol production as it does on the development of flavoring compounds in the final product. Production of ethanol may be the prime purpose of production methods but the final flavor does not depend on the microbial ethanol producers alone. Bacteria, in addition to yeasts other than *Saccharomyces cerevisiae*, play a role in the development of the final desirable flavor of an alcoholic beverage.

Ethanol is also distilled from yeast fermented products and is used for a variety of purposes. These include:

- The production of alcoholic beverages to which flavors may be added or flavors may be allowed to develop through maturation processes;
- To preserve fruits and wines;
- To use as an industrial product;
- To use as a biofuel to replace up to 10% of the petrol (gasoline) used in internal combustion engines.

Ethanol is the end product of the anaerobic (without oxygen) glycolytic pathway (Embden-Meyerhof-Parnas cycle) for energy metabolism of yeast (*S. cerevisiae*) and a bacterium (*Zymomonas mobilis*):

$$1 \text{ glu cose} \xrightarrow{\text{Yeasts}} 2CO_2 + 2ATP + 2 \text{ ethanol}$$

Common alcoholic beverages include:

- Beer, containing about 5% alcohol v/v, the important substrate is barley;
- Wine, containing about 12% alcohol v/v, the important substrates are grape, rice and palm sap;
- Spirits (liquor), containing about 40% alcohol v/v, the important substrates are grape, barley, rice and/or sorghum;
- Fortified wine, distilled spirit is added to wine raising the ethanol content to about 18%, e.g. port and sherry are fortified wines.

The production of ethanol depends on the availability of glucose or other hexoses, and disaccharides for fermentation. Glucose is readily available in many fruits, for example grapes. However, by far the vast majority of glucose is in the form of polymers, namely the starches and cellulose. Starch is the energy storage compound of plant life but yeasts and *Z. mobilis* generally cannot use the glucose units unless they are removed from the starch molecules. This requires another step in the production system to make the glucose available to the yeasts and bacteria.

Two basic systems have been developed to release fermentable sugars from starch for the production of alcoholic beverages. The first is to begin sprouting the living cereal seed by steeping it in water and hydrating it under aerobic conditions. During sprouting the seed produces amylases to break down the starch so as to make the energy in the starch available to the growing seedling. A high level of amylase builds up. The sprouted grain is then dried to preserve the amylase activity before the seedling has grown significantly. The second system involves wetting or cooking the grain and then letting mould grow on the grain. The mold produces amylase (in addition to other enzymes) that is used to break down the starch in the grain.

Box 10.1

Starch is an Important Source of Glucose for Various Fermentations

By far the greatest source of glucose for alcohol production is starch. However, starch is also the base for a range of other food products including MSG. Natural starch occurs in starch granules in plant material such as tapioca (cassava), maize, potato, barley, wheat, rice and sago palm, to name some examples of industrial sources of starch. The starch is not soluble in water and the granules are surrounded by a protein membrane. In most cases the granule must be destroyed and the starch gelatinized (hydrated) before the starch can be metabolized. The temperature of **gelatinization** varies from 65–80°C, depending on the plant source. In the presence of smaller quantities of water the gelatinization temperatures may be much higher.

There are two forms of starch. Starch mainly consists of **amylose** that is comprised of long chains of unbranched glucose units with α-1,4 links. **Waxy starch** contains a high proportion of **amylopectin**, in addition to amylose, that consists of chains of glucose units from which glucose units frequently branch through an α-1,6 link to form additional chains with α-1,4 links. Natural starch sources may also contain some β-glucans, such as β-1,3 and β-1,4-linked glucan polymers that are viscous and gum-like.

A range of enzymes from sprouting seeds and from a variety of fungi and some bacteria attack the links in the glucose chains. The enzymes from plant sources include α-amylases, β-amylases and α-1,6-glucosidases. The α-amylases break-down the internal α $(1\rightarrow4)$ linkages in the chains. The β-amylases hydrolyze the

Box 10.1   (*Cont'd*)

penultimate α (1→4) linkages at the non-reducing ends of the chains thus producing maltose. Finally, the α-1,6-glucosidases breakdown the branch points resulting in a mixture of sugars including glucose, maltose, and maltotriose. On the other hand, fungi and bacteria produce amyloglucosidases that remove glucose units from non-reducing ends and can hydrolyze the α-1,6-glucoside links.

These enzymes are more formally known as glucanases. The α (1→4) glucanases are further split into endo α (1→4) glucanase that hydrolyze the α (1→4) linkages anywhere in a chain and exo α (1→4) glucanase that hydrolyzes the non-reducing end. There are six enzymes that are classed as an exo α (1→4) glucanase.

Enzymes hydrolyzing the α (1→6) linkages are also known as the pullulanases and are α (1→6) glucanases of which there are two endo-α (1→6) glucanase and one exo-α (1→6) glucanase.

The expression of the **amylase complex**, as it is called in fungi, depends on catabolic induction from the substrate. Low basal levels of the amylase are produced that generate low concentrations of oligomers that feedback into the cell to induce amylase complex formation. Induction occurs due to the presence of maltose, isomaltose, and panose but glucose and glycerol repress complex induction.

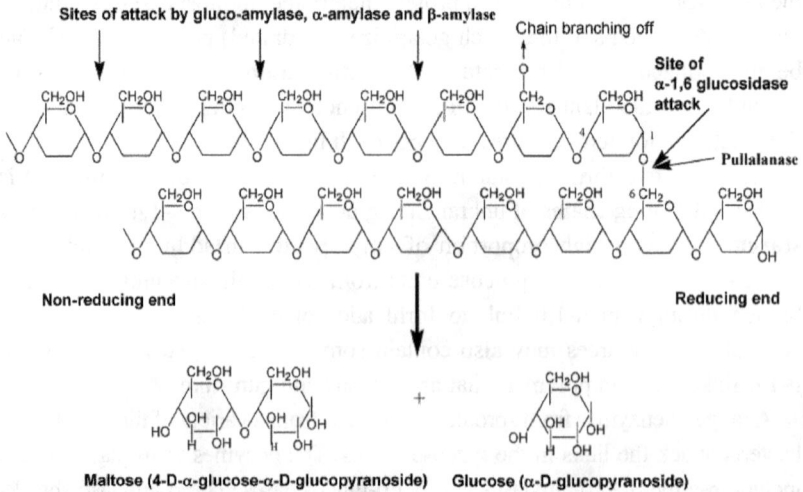

Sites of attack by gluco-amylase, α-amylase and β-amylase   Chain branching off   Site of α-1,6 glucosidase attack   Pullalanase

Non-reducing end   Reducing end

Maltose (4-D-α-glucose-α-D-glucopyranoside)   +   Glucose (α-D-glucopyranoside)

10.1. Beer Production

Beer production originated in about 6,000 BC in ancient Babylonia, but today, German terms dominate in the vocabulary of the beer brewing industry. Barley is the essential base material for the amylase and starch sources. After processing the barley, the sprouted and dried barley is called **malt**. The malting process used is expensive because of the heat energy needed to dry the sprouted barley. Brewers in the 19th century discovered that they did not need to use only malted barley to make beer because there are two kinds of barley, two-row and six-row, the latter produces more "diastase", has higher protein content and higher β-glucan level than the other. This allowed brewers to reduce the amount of malted barley they used and to increase the amount of other sources of cheap starch that are called **adjuncts**. There is sufficient amylase in the malted barley to break down starch in other grains, but the grain has to be cooked and the starch gelatinized.

Beer making is a complex process and is summarized in Fig. 10.1.

10.1.1. *Malting*

In the malting process, barley is soaked in water (steeped) to give 42–46% moisture content to release germination hormones, gibberellins, and to remove dirt, husk phenols, soluble organic matter, and soil microorganisms. Oxygen is required by respiratory enzymes in the barley seed and is provided by forcing cool humid air through the grain bed. Germination occurs in the aerated germination chamber during 4–5 days at 13–16°C. Hot dry air, in a process called **kilning**, is then pulled through the grain bed to stop and kill the sprouting plant but retain enzyme activity. At the same time, malt looses its gherkin-like taste and obtains the typical malt aroma. Kilned malt can be kept for some time.

The quality of barley used for malting is dependent on a number of factors. Apart from the barley being of good quality (often defined in terms of enzyme production), there should be low fungal contamination of the grain by *Fusarium* species, such as *F. graminearum*, *F. culmorum* and *F. poe*. Heavy *Fusarium* contamination of malting barley may lead to the presence of deoxynivalenol and other mycotoxins, all water-soluble compounds that are removed during steeping. However, fusaria remain trapped within the mass of germinating grain, continue

Barley   Adjunct (barley, rice, corn, wheat)

  ↓ H₂O (Steeping)   ↓

Malting   cereal cooker

  ↓ 4–5 days

Kilning

  └──→ Amylase, proteinase ──────→ mash mixer (70°C)
                                      ↓
                                    Lauter tun ──────────→ Spent grain
                                      ↓
                                    Wort
                                      ↓
Hops ──────────────────────────→ kettle
                                      │
                                    O₂, then anaerobic
                                      ←──────────── Yeast (Pitching)

                                    10°C then 5°C for US lager
                                    6°C then 5°C for European lager
                                    15–20°C for ale
                                      ↓
                                    Green beer
                                      │
                                      ↓ Aging, up to 4 months, 2–6°C
                                    filtration
                                      ↓
                                    carbonation (0.45–0.52% CO₂)
                                      ↓
                                    pasteurization (60–61°C)*
                                      ↓
                                    Beer

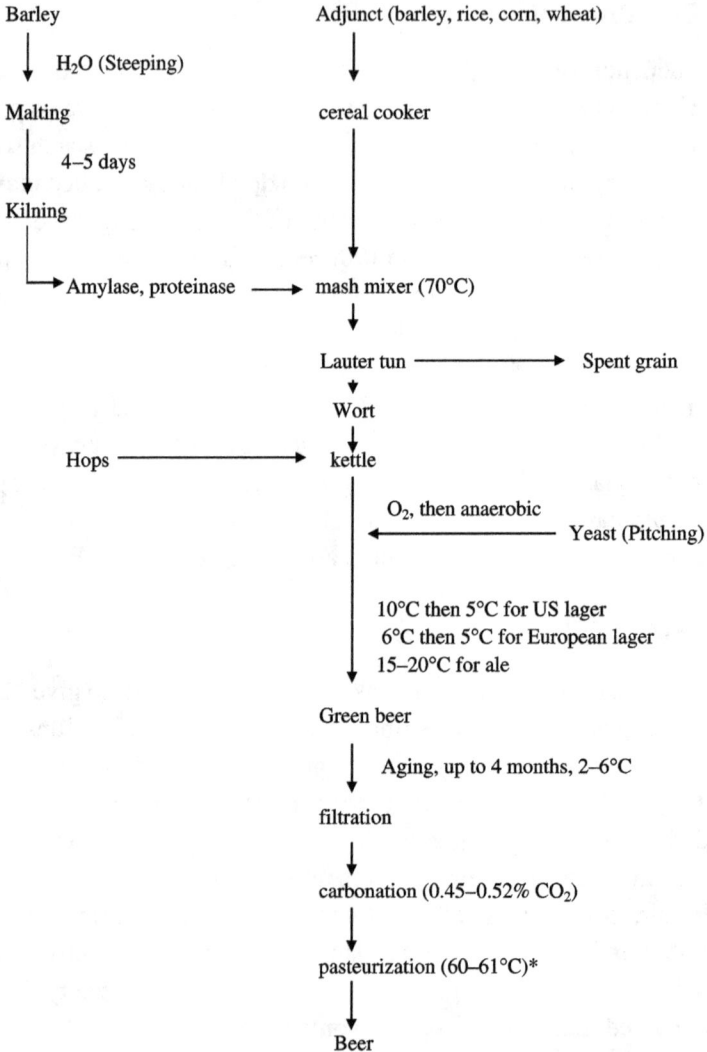

Fig. 10.1.   Flow diagram for the production of beer.

*In some more advanced production systems beer is sterilized by filtration and packed aseptically.

production of these heat resistant mycotoxins during germination of the barley, and can be found in the final beer. This leads to the fault in beer known as "gushing of beer", a quick uncontrolled spontaneous over-foaming immediately after opening the bottle or can. *Fusarium* con-taminated barley is banned in some countries to avoid the problems associated with it. However, a new method for reducing the effects of fungal contamination and to improve malt quality is to add a **starter culture**, either lactic acid bacteria or *Geotrichum candidum*, a yeast, to the steeping barley to control microbial growth. Certain strains of *Lb. plantarum* and *Pediococcus pentosaceus* are especially efficient when added to the steeping water of barley at the level of about 10^7 cells g^{-1}. Addition of starter cultures ensures high quality malt, regardless of the natural variation of the microflora of barley, due to the microbiocidal compounds produced by the lactic acid bacteria and other enzyme activities. Addition at the steeping stage is important because fusaria grow rapidly during the first hours of steeping. The starter effect on this defect is proportional to the level of original fungal contamination in the barley, due to the negative effect on the fusaria. If there are no fusaria in the barley then there is no gushing defect. However, the added starters have other effects on other stages of the brewing process, and the lactic acid starter cultures also restrict the growth on the grain of harmful aerobic bacteria. In addition, the use of lactic acid bacteria in malting has led to significant improvements in the physical and chemical quality of malt including improved filtering (lautering) of wort (pronounced "wert") and beer, and reduced formation of beer haze during finished beer storage. The effect of lactic starter cultures on lautering performance can be explained partly by their own enzyme activity, namely the production of thermostable (60°C) β-glucanase. Further, the xylanase activities of malts produced with starter are higher. The net result is a higher extract yield.

The quality of malting barley is judged by the activities of various endogenous barley enzymes. These enzymes include α-and β-amylase, proteases and β-glucanase. While amylases are important in starch degradation, β-glucanase is also important because β-glucans in the grain are soluble and are extracted into the wort. The β-glucans interfere with the filtration of the wort and the beer and contribute to the formation of haze during the storage of finished beer.

10.1.2. *Adjuncts*

An adjunct, as the source of fermentable carbohydrate, can be corn free of germ and skin, rice, sorghum, barley, wheat grit, enzymically converted

Table 10.1.   The main components in wort used for beer production.

Compound	Concentration, g L^{-1}
Maltose	40–55
Maltotriose	10–15
Glucose	5–10
Sucrose	1–3
Fructose	1–2
Non-fermentable carbohydrate	22–25
Free amino acids	1–2
Peptides and protein	1–3
Organic acids	0.175
Inorganic ions	1.5–2
Hops concentrate	0.05–0.1

corn syrup or sucrose. Adjunct is mixed with kilned malt to produce mash and is incubated at 70°C. This allows hydrolysis of carbohydrate and protein in adjunct by enzymes present in the malt. The **wort** (liquid part of the mash) is then separated from the **spent grain**. The wort contains the simple fermentable sugars as shown in Table 10.1 that are ready for beer fermentation.

10.1.3. *Hops*

Hops are then added into the wort. Hops, whose quality is important for the final beer flavor, were first introduced to beer brewing by St Hildegard (1098–1197AD) in Germany who reported that hops promoted the stability of beer, and was good for health. Hops are unfertilized dried female cones of the hop plant *Humulus lupulus*. The small lupulin glands contain bittering resins and essential oils that contribute to the unique bitterness and flavor of beer. Polyphenolic compounds in bracteoles enhance the astringency and tea-like character of beer. The traditional baled hops have now been replaced by various concentrates, pellets etc. The preferred hops extract is that extracted using super-critical carbon dioxide or liquid carbon dioxide.

10.1.4. *The fermentation*

An important technical advance in brewing has been the use of high gravity or very high gravity brewing. The wort is produced and a sugar rich syrup is added to the wort kettle to give a wort with a substantially higher gravity, over 20° Plato (°Plato = weight of extract in a 100 g solution at 17.5°C), than needed for the final beer. The wort is then fermented after which the beer is adjusted to the desired concentration with oxygen-free water. This process increases the brewing capacity of a given brew-house.

The actual fermentation and maturation is the next stage in the process. Initial presence of oxygen is desirable for the yeasts because it is essential for membrane lipid synthesis and rapid cell growth before anaerobic growth begins. During ethanol formation and growth of the yeasts, the amino acids valine and leucine are produced and one of the intermediates in the metabolic pathway is α-acetolactate. This compound is excreted from the yeast cells into the brewing wort where it is then converted by a slow chemical process to diacetyl, a compound with a strong buttery flavor that has a very low taste threshold in beer of 0.05 mg L^{-1} or less. The diacetyl is undesirable in beer and is subsequently degraded enzymically to acetoin and then to 2,3-butanediol. The long maturation process of lager beer is designed to allow the depletion of diacetyl in the beer.

The yeasts in beer do not breakdown α-acetolactate but some bacteria possess α-acetolactate decarboxylase so the exogenous enzyme can be added during fermentation. Alternatively, a gene encoding α-acetolactate decarboxylase can be inserted into the yeast genome. The genes of *Klebsiella terrigena*, *Enterobacter aerogenes* and *Acetobacter xylinum* have been inserted using either PGK or ADH promoters. Pilot scale studies have confirmed that such genetically modified yeasts can reduce the diacetyl concentration below the taste threshold at the end of the main fermentation thus making it possible to reduce the maturation time which is usually about fourteen days.

Box 10.2

Killer Yeasts

These are strains of yeasts that produce anti-yeast factors, also known as zymocins, killer factors or killer toxins that kill other yeasts including strains of the same species. They are produced by yeast strains from many genera but industrially those produced by *Saccharomyces* strains are most important. The presence of the zymocin producing ability in a yeast strain confers resistance to zymocins produced by other yeast strains. In yeast cultures contaminated with a killer yeast strain, the latter becomes the dominant strain and will cause the production of off flavors in addition to other defects. It is difficult to detect the presence of killer strains in a culture. In the case of the best known *Saccharomyces* zymocin, KT_1, to be effective the zymocin requires a very narrow pH range (pH 4.3–4.7) as well as the right concentrations of some metal ions (K^+, Ca^{2+} and Mg^{2+}).

The KT_1 zymocin consists of a double stranded heterodimer protein in which the α chain (105 amino acids long) and the β chain (83 amino acids long) are linked by three disulfide bonds. The zymocin exerts its toxicity by absorbing to the cell wall of sensitive cells at a specific receptor-binding site containing a $(1 \rightarrow 6)$ β-D-glucan. This process requires no energy. Once bound to the cell wall the next step requires energy to bring the toxin into contact with the plasma membrane. If the organism has no energy available then the cell membrane cannot be disrupted. Once in contact with the plasma membrane ion-permeable channels open up that are large enough to allow leakage of ions. Consequently, protons are extruded which dissipate the membrane potential so that the proton-amino acid co-transport system is inhibited, potassium ions leak from the cell, and leakage of large molecules such as ATP occurs.

A yeast strain's ability to produce the toxin, designated K^+ phenotype, and the ability to resist it, designated R^+ phenotype, is due to a dsRNA called M (M = medium size). As there is more than one killer toxin, the others are designated with a subscript 1, 2 etc. To understand the yeast killer system, a very complex system, only the M_1 system will be briefly described here. The M_1 dsRNA is a 1.8 kb unit that codes for the polypeptide and a resistance function. The strains carrying it are designated genotype KIL-k_1. For the killer system to work the cell needs to contain the cytoplasmic gene HOK (*helper of k*iller) that is located on certain forms of L-A dsRNA (L = large size). Further, maintenance of the killer system in a cell requires possession of some chromosomal genes designated MAK (*ma*intenance of *K*IL-k_1), comprised of at least 29 genes. The L-A and M dsRNA strands are found in an intracellular virus-like particle (ScV particle)

Box 10.2   *(Cont'd)*

whose major coat protein (capsid) is coded for by the L-A strand. For the killer toxin to be expressed in a cell, the cell must also contain a further two chromosomal genes, designated *KEX*, needed to process the toxin precursor.

   The ability of a killer strain to resist killer toxin offers the advantage of controlling wild infections of cultured strains. Consequently, attempts have been made to transfer the killer system to a desirable strain for wine or beer making. In one experiment, the killer system was inserted into the protoplasts of a non-killer strain of *S. cerevisiae* by electro-injection. The transformed culture was stable in culture for many generations. In a further successful experiment, a DNA copy of one of the killer systems was cloned and put into an ADH1 expression cassette that was subsequently integrated into the yeast chromosome.

   In summary, killer yeasts have been used:

(i)   to control wild-type yeasts that contaminate fermentations during the production of wine, beer and bread;
(ii) as bio-control agents to preserve foods;
(iii) to bio-type medically important pathogenic yeasts and related fungi;
(iv) in the development of novel antimycotic substances to treat human and animal fungal infections.

10.1.5. Types of beer

There are basically three categories of beer; the bottom fermented lager beer brewed by *Saccharomyces carlsbergenesis* and *S. uvarum,* top fermented ale beer brewed by *S. cerevisiae*, and stout. At the end of beer fermentation, lager beer is siphoned off and the yeast cells settled at the bottom are discarded, whereas yeast cells floating on the surface of ale beer are skimmed off and re-used.

   Ale beer tends to be sweet because *S. cerevisiae* ferments only the fructose moiety of the trisaccharide raffinose. The ale yeast forms a relatively high level of esters due to higher levels of coenzyme A activity. Esters constitute an important group amongst the aroma compounds in beer due to their strong, penetrating fruity flavors.

   Lager yeast is able to utilize most of the sugars in the wort, thus lager beer is generally dry (not sweet). Moreover, rapid reduction of acetaldehyde to ethanol by alcohol dehydrogenase and reduced nicotinamide adenine dinucleotide coenzyme, results in a low level of acetaldehyde in the beer.

During aging, the flavor of beer matures. There is a reduction in the concentration of hydrogen sulfide, acetaldehyde, diacetyl and other undesirable flavors. Aging at high temperature, in the presence of yeast cells, results in autolysis of lager yeast and a change in the flavor profile of beer.

Recent developments in brewing science include brewing with unmalted cereals digested by microbial enzymes and production of low carbohydrate beer. As the malting process is difficult to control, use of microbial enzymes allows selection by the brewer of appropriate enzymes in respect to nature and quantity. Moreover, addition of these enzymes can be programmed, and it is not necessary for them to be present at the beginning of mashing. One possible consequence of the use of adjuncts is the presence of hemicelluloses that inhibit complete starch extraction or cause the formation of hazes in stored beer. Pentosanases that breakdown arabans, galactans, mannans and xylans, are added during either fermentation or maturation to eliminate them. Microbial enzymes used include bacterial or fungal α-amylases, in particular a thermostable one from *Bacillus licheniformis*, amylo-glucosidases, proteases, pullulanase and β-glucanases.

A modern trend in beer brewing is the production of beer with a caloric value that is a third, or more, lower than conventional beers. The beer may have reduced alcohol content (20–50% of normal beer), or a reduced content of non-fermentable sugars, e.g. dextrins that emanate from the malt and adjunct. There are a number of ways to reduce the amount of non-fermentable sugars in beer.

(1)   Dilution of the regular strength beer with water, but this would unavoidably dilute the flavor of the beer;

(2)   Addition of microbial or cereal α- or β-amylase or glucoamylase, β-glucanase and pullulanase to the wort during mashing or fermentation;

(3)   Use of a totally fermentable sugar such as flucose, fructose or sucrose as an adjunct;

(4)   Use of a brewing yeast strain with amylolytic activity for production of low carbohydrate beers and for maximizing fermentation efficiency.

The last approach has been quite extensively studied. The yeast *S. diastaticus* produces the extracellular enzyme glucoamylase (regulated by genes *DEX1, DEX2* and *STA3*). Using classical hybridization techniques, a diploid strain containing the *DEX* and *STA* genes in the homozygous condition was constructed. The *DEX*-containing strain of *S. diastaticus* ferments wort to a greater extent than the brewing strain due to the partial hydrolysis of dextrins by glucoamylase, and is desirable in the production of light beer. However, the beer produced by this strain has a characteristic phenolic-off-flavor, due to the presence of 4-vinylguaiacol. A single dominant gene *POF* (phenolic-off-flavor) codes for a ferulic acid decarboxylation enzyme that is responsible for the conversion of ferulic acid, a wort constituent derived from the grain, to 4-vinylguaiacol. Thus, a haploid strain that was *DEX* and *POF* positive was mated with a *DEX* and *POF* negative haploid strain. When tetrad dissection of the resultant diploid strain was performed, the *DEX* and *POF* genes segregated independently of each other. Thus, a haploid strain that was *DEX* positive and *POF* negative was obtained, and subsequently a diploid strain with the genotype *DEX2/DEX2 pofo/pofo* was constructed.

Box 10.3

The Role of Ferulic and *p*-Coumaric Acids in Food and Beverage Flavors

Ferulic and *p*-coumaric acids are widely distributed in plants, grains and fruits. Consequently they are commonly found in a variety of fermented products including wine, beer and soy sauce and they are the source of various flavoring constituents in these fermented food products.

The flavors or odors, often described as strong spicy, smoke-like, medicinal, clove-like, woody or phenolic, arise through the metabolic activity of yeasts and bacteria. The flavors are associated with the production of 4-ethylguaiacol, 4-ethylphenol, 4-vinylguaiacol and 4-vinylphenol as well as vanillin.

Ferulic and *p*-coumaric acids, originating from lignin, are transformed during the manufacturing process, to flavored compounds that

Box 10.3   *(Cont'd)*

may or may not be desirable in the product. Most wine yeast strains and some beer brewing strains of *Saccharomyces cerevisiae* are able to decarboxylate ferulic and *p*-coumaric acids to produce phenolic off-flavors (POF) during fermentation. In one study, 81 to 95% of 116 wine yeast strains produced these compounds while growing in spiked grape juice. However, in practice during wine making, *Brettanomyces* yeasts (sporulating form is *Dekkera*) are mainly responsible for the development of a phenolic character in red wines. The *Brettanomyces* species multiply by fermenting the very small quantities (275 mg of sugars L^{-1}) of residual sugars (glucose, fructose, galactose, and trehalose), much less than normal levels at the end of fermentation, but sufficient for the production of an excessive quantity of ethyl phenols. Other yeast genera that also decarboxylate ferulic and *p*-coumaric acids include *Rhodotorula*, *Candida*, *Cryptococcus*, *Pichia* and *Hansenula*. Some strains of *Lactobacillus brevis* and *Pediococcus pentosaceus* are able to decarboxylate *p*-coumaric acid to form 4-vinylphenol as actively as *S. cerevisiae*, but are unable to breakdown ferulic acid. In practice the lactic acid bacteria have no impact on the development of phenolic off-flavors in wine. The critical concentration of these free forms of these hydroxyphenolic acids in grape required for the production of the phenolic off-flavors is estimated to be more than 10 mg L^{-1}. The lag phase of growth of *S. cerevisiae* is increased by as little as 50 mg L^{-1} ferulic acid, and higher concentrations can inhibit the organism.

In high salt conditions (15% NaCl) *Debaryomyces hansenii*, *Pichia subpelliculosa*, *P. anomala*, and *Candida famata* convert ferulic and *p*-coumaric acids to 4-vinylguaiacol and 4-vinylphenol in static culture, and *C. versatilis*, *C. halophila*, and *C. mannitofaciens* convert them to 4-ethylguaiacol and 4-ethylphenol. *C. etchellsii*, *C. nodaensis*, *C. halonitratophila*, *P. farinosa* and *Zygosaccharomyces rouxii* do not produce volatile phenols. These cinnamic acids are converted from ferulic acid and *p*-coumaric acid to vinylphenols by cinnamate decarboxylase, and successively from vinylphenols to ethylphenols by vinylphenol reductase. The soy sauce spoilage bacteria, such as

Box 10.3   (Cont'd)

Bacillus subtilis, B. licheniformis, and *B. megaterium*, as well as some *Staphylococcus* sp. (from koji), produce vinyl phenols at cell concentrations greater than 10^7 cells mL^{-1}.

Although some microorganisms can metabolize them, they are also inhibitory for some yeasts. Ferulic acid has an inhibitory effect at 100 mg L^{-1} on strains of *P. anomala, D. hansenii* and *S. cerevisiae*. At 500 mg L^{-1} it appreciably inhibits *P. anomala, D. hansenii* and *S. cerevisiae* and prevents detectable growth of odd strains of *P. anomala* and *D. hansenii*. Coumaric acid is less inhibitory than ferulic acid.

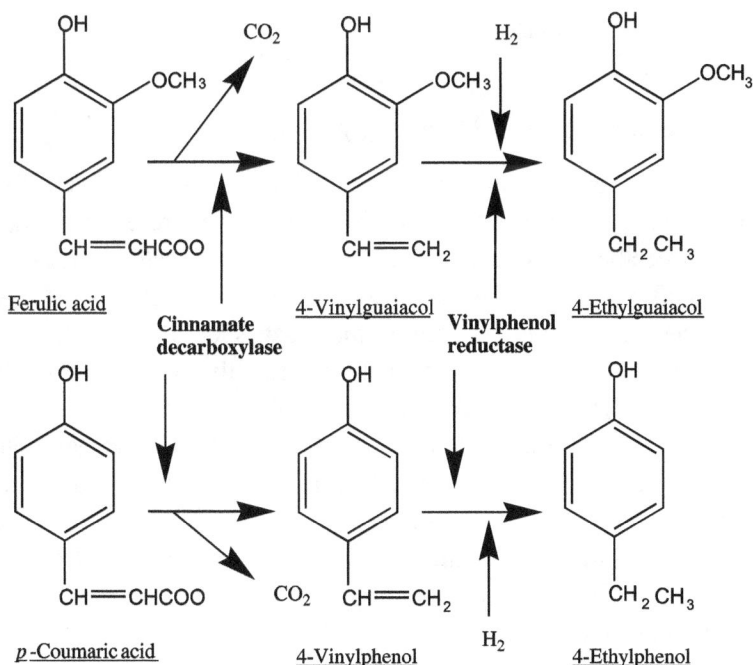

The hydroxycinnamic acid decarboxylase enzyme, with an apparent molecular mass of 39.8 kDa by gel filtration, catalysing the decarboxylation of ferulic and *p*-coumaric acids was isolated from the

Box 10.3   (*Cont'd*)

yeast, *Brettanomyces anomalus* (NCYC 615). Its activity was optimal at 40°C and pH 6.0 and was enhanced by EDTA, Mg^{2+}, and Cr^{3+} while Fe^{3+}, Ag^+, and sodium lauryl sulfate completely inhibited the activity. The para-hydroxy group on the phenol ring is essential for enzyme activity. On the other hand, a strain of *Lb. plantarum* has shown substrate-inducible decarboxylase activity on *p*-coumaric and ferulic acids indicating that at least two phenolic acid decarboxylases are produced in this bacterium. On SDS-PAGE the enzyme band is 23.5 kDa but the native molecular mass is 93 kDa indicating that the enzyme is a homotetramer. This *p*-coumaric acid specific enzyme exhibited maximum activity at 30°C and at pH 5.5–6, and did not need cofactors or metal ions.

10.1.6. *The flocculating properties of yeast*

When brewing lager beers the precipitation of the yeast cells is an important characteristic. Cells that flocculate are more easily removed from the green beer. However, it is not a simple relationship as ideally the yeast should remain in suspension while the sugar is fermented and at the end (the stationary growth phase) the yeasts should flocculate, that is, the yeast cells should adhere and sediment rapidly to the bottom of the brewing vessel. This helps reduce the cost of filtration or centrifugation to clarify the beer. The process of flocculation depends on an initial input of mechanical energy (stirring) to promote physical contact, and the presence of Ca^{2+}. Sensitivity to Ca^{2+} varies between bottom and top fermenting yeasts, possibly due to cell surface hydrophobicity. Bottom fermenting strains flocculate in the presence of high Ca^{2+} concentration and low sugar concentrations; mediated by a lectin-like mechanism. Top fermenting strains flocculate without added Ca^{2+}, but need a high ethanol concentration. However, the flocculating activity of some top fermenting strains is inhibited by mannose, but not by sucrose or galactose.

During yeast flocculation a lectin-like protein binds to a mannose chain receptor on neighboring cell walls. The lectins are synthesized continuously in the exponential phase of growth and are inserted into the cell wall, where they remain inactive for up to 14 hours, before being activated at flocculation onset by an unknown mechanism. The binding is inhibited by mannose but not by glucose. In addition, flocculation can be affected by the final pH as shown in experiments where flocculation was induced with a pH-upshift and de-flocculation occurred with a pH-downshift.

The flocculating properties are genetically controlled and are expressed as changes in the cell wall components. Experiments have shown a protein component is responsible and cell wall proteins have been identified. The gene *FLO1* that is located on chromosome I, is the major gene as shown when a nonflocculent industrial polyploid yeast strain, *S. cerevisiae* 396–9–6V, was converted to a flocculent one by introducing the *FLO1* gene at the *URA3* locus. A homologue of *FLO1*, named *Lg-FLO1*, was isolated from a flocculent bottom fermenting yeast strain in which flocculation is inhibited by both mannose and glucose (mannose/glucose-specific flocculation).

The DNA sequence of *FLO1* showed that the open reading frame (*ORF*) of the intact gene, composed of 4611 bp, codes for a protein of 1537 amino acids. This FLO1 protein contains four families of repeated sequences and it has a large number of serines and threonines. The N- and C-terminal regions are hydrophobic. Another gene isolated from flocculating strains was *FLO5*. This gene has been shown by restriction mapping, Southern analysis, and chromosome mapping to be related to the *FLO1* gene.

10.2. Distilled Spirits

Once wines and beer are made they are subject to deterioration due to the activity of vinegar bacteria or chemical oxidation unless stored properly. Distillation of the ethanol is one way of preserving the ethanol content, that can also be applied to preserving wines and fruits. Historically distillation of alcoholic spirits from fermented products developed in both the East and the West. Western spirits will be dealt with here as sorghum liquor, an Eastern product, is dealt with in Sec. 10.4.3.

Western liquors can be broadly categorized as grape or fruit based (e.g. brandy, cognac, calvados), molasses based (e.g. rum) or grain based (e.g. whisky, bourbon). The grain based liquors are split into malt whisky and grain whisky, the former being made exclusively with malted barley kilned with smoke from peat fires, and the latter being based on other grains like maize and rye grain. The starch in malted barley and grains are gelatinized and hydrolyzed in a manner similar to the beer making process except that no hops are added, the wort is not boiled, and the spent grains may not be separated from the wort before fermentation and distillation. For the actual fermentation a heavy inoculum ($5-20 \times 10^6$ cells mL^{-1}) of a cheap source of yeast is often used.

The secret of success in the production of whisky, rum and some brandies is thought to be in the actual distillation process followed by maturation in wooden casks, particularly for whisky. Pure ethanol boils at 78.3°C but the ethanol/water mixture results in a non-ideal mixture so that during distillation an azeotropic mixture comes off at a lower temperature of 78.15°C containing 96.5% ethanol. Generally higher alcohols in the mixture distill off more easily and methanol distills off less readily than ethanol. These characteristics are taken into account during the final distillation stage.

For whisky production the brew is distilled twice using two batch based **pot stills** in sequence. These pot stills are always made of copper, with a spherical base, a long conical neck and a long horizontal arm that takes and condenses the spirit mixture through a coiled tube cooled with water. The shape and size of these pot stills, and the high polish on the copper that reduces heat losses during distillation, are considered critical to maintain the flavor characteristics of a whisky from a single distillation unit. The first still is the 'wash still' from which the output contains about 21% ethanol. (Spent grains from the first still are dried for cattle feed.) This output is then introduced to the "spirit still" and the output here is split into three portions, the first and the last portions are fed back into the spirit still but the middle fraction that contains about 60% ethanol, is collected as the product for maturation.

Whisky by law must be matured for at least three years and it is placed in wooden casks that may be new or have been used for aging other wines or liquors. During aging the water penetrates more rapidly than ethanol through the wood so that the ethanol becomes more concentrated. The flavor changes during maturation due to chemical reactions and to flavors seeping from the wood. Color, acids, esters, furfural and tannins all increase, while aliphatic sulfide compounds, e.g. dimethyl sulfide, decrease rapidly. Vanillin, syringaldehyde and β-methyl-γ-octolactone are extracted from the wood.

The final product for sale usually contains caramel to produce a desirable color and it is diluted to about 40% ethanol.

Distilled liquor is also produced for which no maturation period is used. Flavors, usually from plant sources, e.g. juniper berries used for gin, are added and blended to obtain a desired flavor.

10.3. Grape Wine

Fruit juices contain readily available glucose and fructose for the growth of yeasts. In addition, the pH of fruit juice is generally below pH 5. In the case of grape juice, the pH is in the range of pH 3.0–4.0, due mainly to the natural tartaric, the dominant acid, and malic acids in the juice. Further, the natural habitat of sugar fermenting yeasts is the outer layer of the fruits. The most commonly fermented fruit is grape juice, which is used to produce grape wine, but other fruits include apples that are used to make cider. Records of wine production go back over four thousand years to ancient Egypt, and the ancient wine making steps are depicted on the walls of the tombs of the Pharaohs.

The traditional way of making wine is as follows. The grapes are picked at the optimum ripeness and crushed. The juice is collected and placed in a suitable vessel. The natural yeasts on the grape skins begin to grow and ferment the glucose and fructose to ethanol and carbon dioxide. The fermentation period is usually 2–4 weeks. The fermented juice is then drawn off (yeasts cells are separated from the finished wine) to avoid yeast cell autolysis that can cause off flavors, and the finished wine stored in a wooden cask or stainless steel tank until bottling.

A critical factor in the process is the amount of sugar in the grape juice and that depends on the amount of sunlight that the grape vine received during the growing season. In more northern climes, this becomes the critical factor restricting the places where wine can be produced. Other critical factors for wine quality are soil type and rainfall.

Wines are categorized according to a range of factors, one factor being color. Red wines are produced using red or black grapes. The color is in the skin, not in the juice so the color (anthocyanins) of the skin has to be extracted to color the red wine. The skins of the crushed grapes are not removed from the **must** (the grape juice for fermentation), and the fermentation begins while the skin is in the juice. When sufficient color is extracted, the skins are removed. White wines are produced from grape juice separated from the skins, but it can be produced on the skins if the grapes are white.

After the initial fermentation has been carried out, various other procedures are used to produce the wide variety of wines we see.

10.3.1. *The fermentation*

One key to the modern wine industry is the careful control of the fermentation and this begins at harvest. Two factors are important, chemical oxidation and the microbial activities of wild yeasts and bacteria. In the modern industry, grapes are harvested mechanically at night to ensure that the fruit temperature is kept low. In the field, at this stage the harvested grapes are swathed in sulfur dioxide to kill wild yeasts, a step only possible with mechanical harvesting. Following transport to the winery, the grapes are de-stemmed. After this, the steps depend on the wine being produced. Grapes for white wine are crushed and the must separated and pumped to the fermentation vessel where it is inoculated. Crushed grapes and juice for red wine are pumped into the fermentation vessel and inoculated. Volumes in the red wine fermentation vessel are kept to about 6,000 L so that the skins can be kept in contact, usually by pumping, with the fermenting must during the fermentation. During the crushing stage sulfur dioxide may be added to kill wild yeasts and bacteria, and to act as an antioxidant. Modern wine making yeasts, usually *S. cerevisiae* var. *ellipsoideus* are resistant to this chemical and may even produce significant quantities of sulfur dioxide.

Another key factor is the temperature of fermentation. White wines are fermented at 7–16°C rising to 20–23°C to retain the fresh bouquet of the grapes. Red wines are fermented for a longer period at 21–27°C.

In both cases the vats should be cooled to counteract the heat produced by the metabolising yeasts. At the end of the fermentation the red wine must is crushed to remove skins in the same way that grapes are crushed for white wine.

The fermentation and subsequent flavor of the wine is not due entirely to the fermentative activity of *S. cerevisiae*. When wild fermentations of grape juice are studied, there is a consistent pattern of microbial growth that occurs. Within the first two to three days the fermentation is dominated by *Candida* and *Kloeckeral/Hanseniaspora* species (Fig. 10.2). These organisms are obligate aerobes and poor fermenters and so are eventually overcome and die off due to the lack of oxygen and *S. cerevisiae*, which is a good fermenter, eventually dominates. In due course, the glucose and fructose are used up and the fermentation stops. At this stage a lactic acid bacterium, *Oenococcus oeni*, may or may not increase in number and begin fermenting the malic acid in the wine in the malolactic fermentation. This metabolic process raises the pH of the wine a little, produces carbon dioxide, and changes its flavor. If the malolactic fermentation occurs in a bottle of still wine ("still" or "table wine" means there is no gas in the wine) gas will be produced and give the impression that the wine has spoiled. One solution to this problem is to add *O. oeni* to the wine during the bulk storage stage to promote the malolactic fermentation so as to stabilize the wine before bottle storage. While the malic acid remains in a bottled wine, the possibility exists for a bacterium to ferment the malate and result in the formation of gas.

After the completion of the fermentation the wine is stored in suitable vats or wooden barrels. Flavoring compounds from the wooden barrels add further complexity to the flavor of the finished wine. Residual tartaric acid may remain in the wine. The tartaric acid precipitates out in the storage vessels and can be sold as food acid for use in baking. The wine maker then bottles the wine and the bottled wine can be stored for further maturation, or consumed immediately, depending on the wine.

10.4. East Asian Alcoholic Beverages

Thousands of years ago, the Chinese developed a second method of processing the starch to make alcoholic beverages, namely a

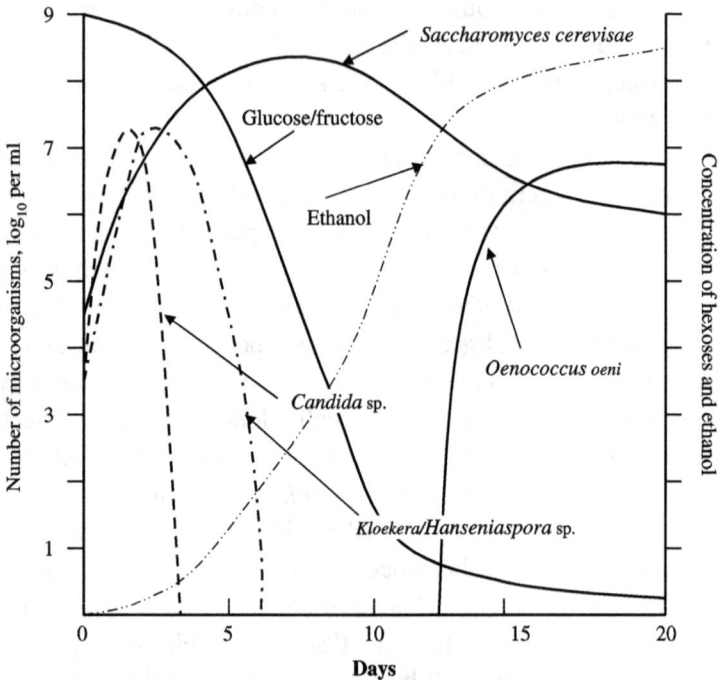

Fig. 10.2.   A generalized diagram showing the relationship between the various microorganisms involved in wine fermentation, and the depletion of the fermentable hexoses, glucose and fructose, and production of ethanol. The *Candida* sp. and *Kloekera/Hanseniaspora* sp. are poor fermenters and grow in the early stage of the fermentation when oxygen is available. *Oenococcus oeni* begins growing, if the batch of wine supports its growth, after the ethanol has been produced and it ferments the malate using the malolactic fermentation.

Box 10.4

The Fermentation of Glucose by *Saccharomyces cerevisiae*

Yeasts are grown for three fundamental purposes, to produce gas, to produce ethanol and to produce yeast cell mass. When glucose is fermented for these purposes some unusual metabolic activity is seen. Glucose is metabolized by *S. cerevisiae* via the EMP pathway and tricarboxylic acid cycle so under aerobic conditions a large amount of cell mass is produced per gram of glucose because the metabolic pathway is very energy efficient. The major by-products from the metabolism are carbon dioxide and water. However, when

Box 10.4   *(Cont'd)*

grown under anaerobic conditions these yeast cultures do not produce a lot of cell mass because this ethanol producing pathway is not very energy efficient so mass production per gram of glucose is low. Although these yeasts can grow anaerobically they can only do so for a while, for a maximum of four to five generations, before they require a further brief period of aerobic growth. The reactions describing the major energy relationships are:

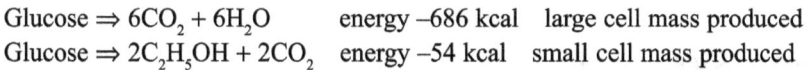

Glucose $\Rightarrow 6CO_2 + 6H_2O$ energy -686 kcal   large cell mass produced

Glucose $\Rightarrow 2C_2H_5OH + 2CO_2$ energy -54 kcal   small cell mass produced

If sufficient oxygen is pumped into a broth medium culture so that it is saturated with oxygen, then the organism will only respire and produce no ethanol. This effect was observed by Pasteur and is known as the **Pasteur effect.** However, *S. cerevisiae*, along with other *Saccharomyces* species, is sensitive to a glucose concentration greater than 1.0 g L^{-1} and consequently under those conditions respiration is repressed for a while and the organism reverts metabolically to anaerobic growth and ethanol production, in the presence of oxygen. This is known as the **Crabtree effect**, or more generally as a case of glucose repression or catabolite repression. This effect does not occur with maltose, the main sugar in wort.

molded-cereal based system using environmental fungi. The starting point of the process is wet grain, cooked or uncooked, on which fungi have been allowed to grow to produce carbohydrate degrading enzymes, the key constituent. In Japan, this fungal product is called **koji** and in China it is called **jiuqu**. Apart from alcoholic beverages, koji is used to produce a variety of foods throughout East Asia. Early in Chinese history, the equivalent of beer brewing was carried out; however it seems to have been dropped in favor of the microbial based system. (Modern beer making methods were introduced to Harbin in China around 1886 by the Russians.) This microbial-based starch degrading system has a long history, dating back with certainty to the Xia Dynasty 4,000 years ago but archeological artifacts suggest an origin over 5,000 years ago. Specific mention of jiuqu in Chinese literature dates to the Zhou Dynasty (770–221 BC). As in the case of beer in the West, in the early days in China, and even now, alcoholic beverages were perceived to have some beneficial effects for healthy living.

In China this fungal based system is used to make grain based distilled liquors and wines, e.g. from rice, corn, millet, and it is also used in Japan to make saké, the Japanese rice wine. In the Philippines a wet form of koji, called binubudan, is mixed with sugar cane juice to ferment, as opposed to using rice. Study of the science of rice wine production is most advanced for the production of Japanese saké.

10.4.1. *Saké brewing*

Japanese rice wine production began around 300 BC and the system of production now depends on the use of pure fungal and yeast cultures as opposed to the traditional rice wine making systems mostly used in China. Saké production has three stages, the koji making stage, the moto making stage and the final brewing stage which is a **solid-liquid state fermentation.**

To make saké the rice must be polished resulting in reduction in the rice grain weight by 25–50%, and the degree of polishing affects the quality of the finished saké so saké is classified according to the degree of polishing (Table 10.2). Generally, ginjo saké is made with highly polished rice while honjozo saké is made with rice not polished below 60%. Less highly polished rice results in a higher level of inosinic acid that is responsible for the characteristic flavor of saké.

Table 10.2.   The types of saké are based on the Seimaibuai classification that is related to the percentage of the original rice left after the rice has been polished.

Type of saké	Rice grain left after polishing, %	Comment
Junmai Daiginjo	50	Highest grade of saké, light dry type
Junmai Ginjo	60	Sweeter, heavier than above
Tokubetsu junmai	60	"Special Junmai", a marketing product
Honjozo	70	Some distilled alcohol added, smooth, light more fragrant flavor
Junmai	70	"Pure saké", no added alcohol, has a heavier, fuller flavor

10.4.1.1. *Koji production for saké*

Koji is rich in fungal enzymes and is produced using a **solid-state cultivation** method. The enzymes breakdown starch external to the koji as well as in the koji so we can think of the added steamed rice as equivalent to adjuncts in beer production. Choice of rice variety is important as large rice grains with soft kernels are chosen so that water uptake is rapid and fungal mycelia can penetrate the kernels more easily. Water quality influences the final saké quality, low iron and potassium levels and the presence of phosphorus and magnesium promote fungal growth, as does a neutral to alkaline pH. The presence of calcium and high chlorination levels in the water improve fungal amylase extraction during the production.

Rice grains are steeped to take up water equal to 25–30% of the milled rice weight, then drained and steamed in shallow open vessels for half to one hour, in batch systems, or for a shorter period in continuous systems, so that the weight of the grains increases a further 8–12%. The cooled rice (34–36°C) is transferred to a koji making room with temperature and relative humidity control, and piled on an insulated bed and covered with cloth to equilibrate to the room conditions. The inoculum (10–100 g 100 kg^{-1} of polished rice), a seed culture of *Aspergillus oryzae* that has been grown for up to six days to maximize sporulation, is then spread evenly over the pile that is reheaped and covered with cloth. The temperature rises as the fungus grows so the inoculated rice must be stirred and reheaped to control temperature.

The way the fungal mycelia grow through the resulting molded grain influences the quality of the saké produced. If fungal growth is confined to the outer layers of the grains excessive breakdown of grain protein reduces saké quality. The fungus produces about 50 kinds of enzymes and of particular importance are α-amylase (amylo-1,4-dextrinase) and S-amylase (amyloglucosidase). The former breaks down the starch molecules to smaller fragments to decrease viscosity during the wine making stage, the latter removes glucose units (Fig. 10.3).

10.4.1.2. *Moto production*

This is the preliminary stage of the brewing process in which conditions for the growth of the yeasts are optimized. Although the process

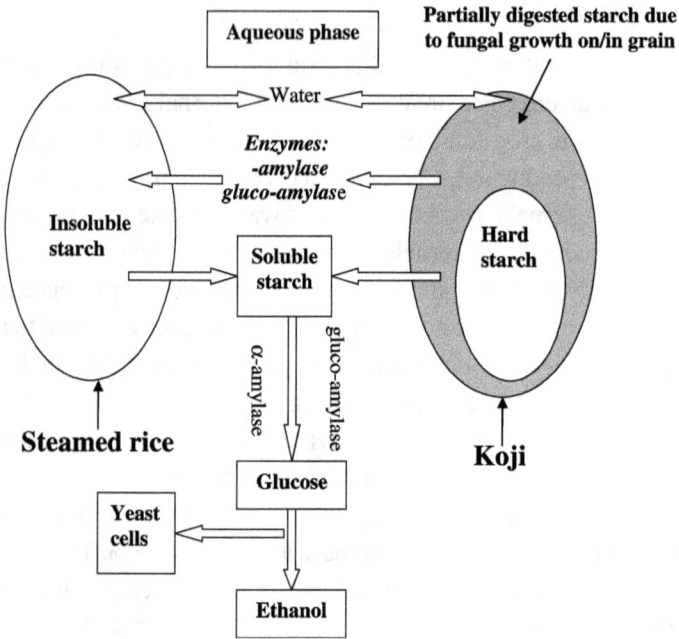

Fig. 10.3. During koji production the fungal enzymes in the koji penetrate the outer layers of the rice grain and begin hydrolyzing the starch. On mixing the koji and steamed rice in water, the water begins penetrating the rice and koji grains and absorbing into the starch and the mixture becomes very thick in consistency. The fungal enzymes and soluble starch are released from the koji and the enzymes begin hydrolyzing the steamed rice grains. As the soluble starch is released the water absorbed into the grains is also released and the mixture becomes more fluid.

seems complex it is simpler than traditional methods. Water (200 L), rice-koji (60 kg) and steamed rice (120 kg) are mixed and held at 13–14°C for three to four days with occasional stirring, during which the temperature drops to about 7–8°C. The koji enzymes begin the breakdown of the starch in the rice, and some wild bacteria and yeasts begin growing in the mix. However, conditions favor growth of the lactic acid bacteria (*Leuconostoc mesenteroides* var. *saké* and *Lactobacillus saké*) which lower the pH. These acid conditions, along with the high levels of sugars, eventually kill off the wild bacteria and yeasts. The temperature rises to about 15°C and by about 15 days a pure culture of saké yeast, a strain of *Saccharomyces cerevisiae* with

Table 10.3.   Some cultural characteristics that distinguish saké yeast and *Saccharomyces cerevisiae*.

Characteristic	Response	
	Saké yeast	*S. cerevisiae*
Growth in vitamin free medium	+	—
Growth in biotin free medium	+	—
Tolerance to ethanol % concentration	20–23	16–19
Habitat	Saké brewing only	Widely distributed in fruits, grains, etc.

some unique characteristics (Table 10.3), is added to give a population of about 10^5 to 10^6 g^{-1} that rises to about 108 g^{-1} as incubation continues, and temperature rises to 20–23°C. Lactic acid may be added to the moto mix instead of relying on microbial growth to bring the pH down to 3.6–3.8, followed by the saké yeast inoculum. At the inoculation stage the water phase has a specific gravity of 1.124–1.128 due to the presence of 26–28% reducing sugar, 0.5–0.8% of amino acids as glycine, and a total acidity of 0.3–0.4% as lactic acid. The traditional production time for moto is 25–30 days.

10.4.1.3. *Saké production process*

The wine making process is basically a **batch-fed** system conducted as a parallel fermentation, preferably at low temperatures, beginning with the prepared moto. In a **parallel fermentation** while the enzymes are releasing the glucose from the starch, the yeasts are fermenting it to produce ethanol. This is in contrast to beer where the soluble sugars are first released and then the fermentation is conducted.

Moto, at 12–13°C, is placed in a vessel 6–18 kL in size along with steamed cool rice, rice-koji and water (Table 10.4) giving a mixture called **moromi**. The saccharolytic enzymes from the koji release the maltose/glucose from the rice starch thus making it available for the yeast to ferment. After the first addition of water, rice-koji and steamed rice to the moto there are two further additions to the vessel, and as each addition is made the temperature is lowered by adjusting the temperature of the

Table 10.4.   The raw materials added progressively to the prepared moto to ferment the rice for saké.

Preparation stage	Day	Total rice (kg)	Cumulative totals		
			Steamed rice (kg)	Rice-koji (kg)	Water (L)
Moto[a]	0	180	120	60	270
1st addition	1	510	360	150	500
2nd addition	3	1,020	830	210	1,230
3rd addition	4	1,710	1,350	360	2,329
Total		3,420	2,640	780	4,329

[a]Moto production takes two to four weeks to complete before the actual wine production begins.

steamed rice and water so that by day 4 the temperature is 7–8°C. The lower temperature controls wild microbial growth. The moromi is agitated twice daily and by three to four days the specific gravity of the water phase is 1.051–1.058, the ethanol level is 3–4%, and total acidity, as succinic acid, is 0.06–0.07%. At each addition of steamed rice, koji and water, the ethanol and reducing sugar levels drop (Fig. 10.4).

The yeast concentration reaches about 2.5×10^8 cells g^{-1} and the fermentation becomes more vigorous and temperature rises to about 13–18°C in six to nine days where the temperature is held for a further five to seven days until the ethanol content is 17.5–19.5% and the temperature begins to decline.

The release of the sugars initially from the steamed rice occurs rapidly but the increased glucose concentration suppresses α-amylase activity so release of glucose declines until significant yeast activity reduces glucose and produces ethanol so that enzyme activity increases again (Fig. 10.5). The breakdown of the rice-koji in the moromi is more complex. The mold growth in the outer layers of the rice grain solubilizes the outer layer of starch rather quickly leaving variable amounts of a core of the rice grain that breaks down much more slowly under the action of the α-amylase. The breakdown of this inner core is dependent on the variety of rice used in the manufacturing.

In the final stage pure distilled ethanol may be added to the moromi to bring ethanol to 20–22%, and 7–8% steamed rice may be added to encourage production of excess glucose to sweeten the wine by a natural process. However, pure glucose may also be added as well as lactic,

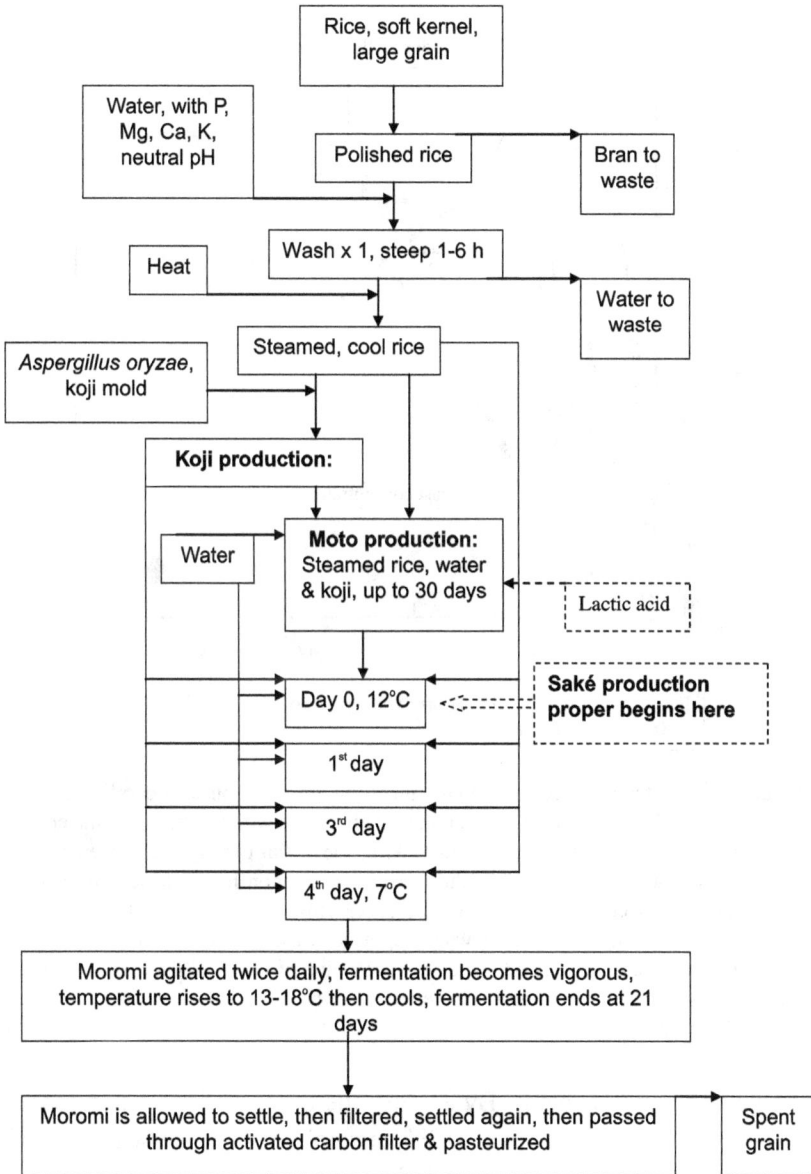

Fig. 10.4.   An outline of the production stages and parameters used in the production of saké.

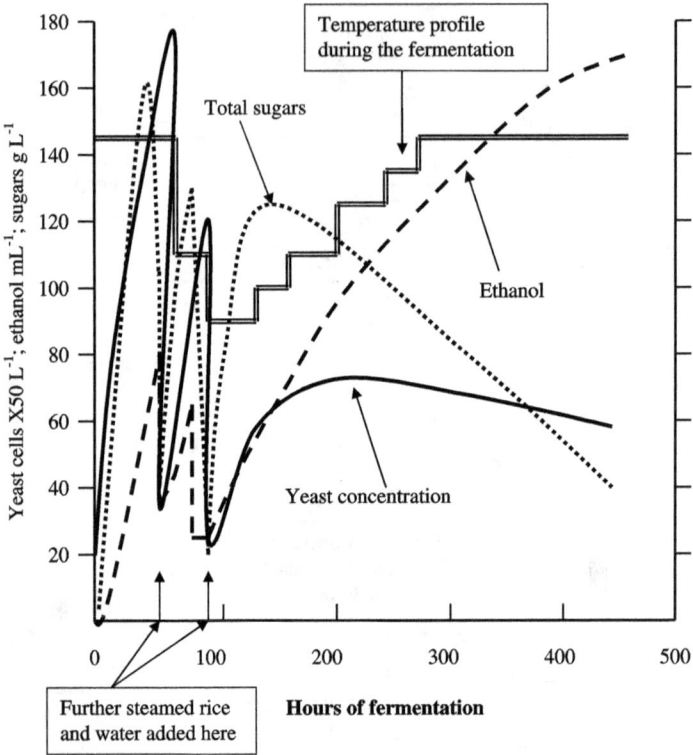

Fig. 10.5.   After steamed rice and koji are mixed in the water phase a parallel fermentation takes place. The koji, steamed rice and water are mixed with the moto and the enzymic
and microbial processes begin accelerating. After the initial mixing the concentration of
total sugars increases rapidly, before the yeast concentration has significantly changed.
The high sugar concentration temporarily reduces amylase activity until the yeast can
significantly reduce the sugar level where upon amylase activity resumes. As further
steamed rice, water and koji are added the concentration of yeast cells, sugars and ethanol
are temporarily reduced accordingly. The temperature is controlled to minimize wild yeast
and bacterial growth by adding cooled water and streamed rice.

succinic and glutamic acids. By 21 days the density of the water phase is
1.004–1.000, reducing sugar is 3–2%, ethanol is 18.5–19.5% and the acid
content as succinic acid is 0.14–0.16%.

The moromi is allowed to stand for a few days to settle and it is then
filtered and again allowed to settle for five to ten days. If the period is too
long the saké quality begins to deteriorate. It is then passed through an
activated carbon filter, pasteurized, and stored at 13–18°C to mature
before bottling.

Table 10.5.   The Chinese classification of rice wines.

Wine type	Sugar g 100 mL^{-1}, as glucose	Ethanol %, v/v	Total acid g 100 mL^{-1}, as succinic acid
Dry type	<0.5	>14.5	0.45
Semi-dry type	0.50–3.0	>16.0	0.45
Semi-sweet type	3.0–10.0	>15.0	0.55
Sweet type	>10.0	>13.0	0.55

10.4.2. *Chinese rice wine manufacture*

Although Chinese rice wines are usually made with rice they are also made with maize and millet. When made with rice, about 230 kg of rice wine is produced from 100 kg of rice. Generally they are classified according to the residual sugar level, namely dry type, semi-dry type, semi-sweet type, sweet type and extra-sweet type (Table 10.5). They are also classified according to the production method. The four basic production methods and their names (Table 10.6) are:

(1)   Ling Fan Jiu, e.g. Yuan Hong wine, a dry type: the steamed rice is cooled by rinsing the rice with cold water then mixing it with a jiuqu starter where upon saccharification begins. It is commonly used as starter for inoculating more mash. It can be consumed after fermentation for several days.

(2)   Tang Fan Jiu, e.g. Jia Fan wine, a semi-dry type: the steamed rice is spread onto bamboo rafts before mixing with jiuqu, and acidified rice-steeping liquid.

(3)   Wei Fan Jiu, e.g. San Niang wine, a semi-sweet type: this rice wine is produced with a **fed-batch** fermentation method. The fermentation is begun as in Ling Fan Jiu but steamed rice is added intermittently during the fermentation period, usually in three batches.

(4)   Fortified rice wine, e.g. Xiang Xue wine, a sweet type: aged rice wine or spirits distilled from rice wine are added before the main fermentation ceases. The final alcohol concentration is above 20%. Consequently, the microorganisms in the mash are inhibited, but the saccharification to sugar of the remaining dextrin in the main mash continues slowly, resulting in a high residual sugar content.

Table 10.6. The ingredients and chemical composition of four kinds of Chinese rice wine.

	Type of rice wine			
Component	Yuan Hong	Jia Fan	San Niang	Xiang Xue
Proportions of components used to make rice wine				
Glutinous rice, kg	144	144	144	100
Wheat koji, kg	22.5	25	25	10
Seed mash, kg	8–10	8–9	15	—
Acidified rice-steeping water, kg	84	50	50	—
Fresh water, kg	112	68.6	100[a]	—
Distilled water, kg	—	5	—	100[b]
Composition of rice wines				
Specific gravity	0.992	0.995	1.0349	1.073
Ethanol, %, v/v	15.6	17.8	16.7	19.4
Extracts, g 100 mL^{-1}	3.325	4.45	15.65	24.44
Sugar, g 100 mL^{-1}	0.38	0.78	6.50	20.0
Total acids, g 100 mL^{-1}	0.48	0.46	0.46	0.28
Volatile acids, g 100 mL^{-1}	0.06	0.027	0.054	0.056
Non-volatile acids, g 100 mL^{-1}	0.42	0.43	0.406	0.22

[a]This is Yuan Hong wine. [b]This is 50% distilled spirit.

10.4.2.1. *Jiuqu*

Unlike koji, the jiuqu usually has two functions, the saccharifying function, due to the enzymes, and the fermenting function, due to the presence of yeasts in the final jiuqu. So during the many different procedures used to make jiuqu, the fungi grow through the jiuqu producing saccharolytic enzymes that hydrolyze starch and produce soluble sugars that encourage yeasts to also grow in the jiuqu (similar to a parallel fermentation).

The actual number of types of jiuqu used for the production of alcoholic beverages in China is unknown. However, there are three broad types of jiuqu, based on the raw material used, and within each broad type there are subtypes based on the color, presence of additives and size of the molded masses or molded pancakes produced. The raw materials are rice, wheat and wheat bran. The manufacturing processes also vary, and this also influences the resulting jiuqu. These processes include the use of: pure or mixed cultures; high or medium temperature; incubation of jiuqu in a thin or a thick layer; hanging in the air in a suitable small container; and wrapping in straw. Consequently, for a given mass of jiuqu the biological activity may vary widely.

The most active wheat based jiuqu, shenqu, is made by milling wheat then splitting it into three portions, one of which is roasted, another steamed, and the other left uncooked. The three portions are then mixed thoroughly with added water to a consistency that is less than slurry. The mixture is transferred to an incubation room, and spread on the floor. Moisture content and temperature are critical and these are traditionally managed by opening windows to lower temperature and heaping the mixture to raise temperature. The first three days require high moisture content to promote growth of mycelia, and lower water content is required for the subsequent incubation period up to 49 days. After this, the shenqu has a yellowish to yellowish-green color. The primary fungi for this preparation are *Aspergillus* species and pure cultures have been isolated and are used. In the preparation method with pure cultures a compartment similar to those used to germinate malting barley is used. The cooked wheat grits (coarsely ground grain) are inoculated with fungal spores and piled in a heap for three hours. The wheat is then spread in a 30 cm layer over the perforated plate in the compartment. The temperature and moisture content are controlled, and oxygen is supplied by blowing suitably conditioned air (temperature and humidity) through the layer. Generally the temperature is kept to a maximum of about 30°C.

Daqu, meaning "large qu", is used for the manufacture of some Chinese spirits and its manufacture is different from shenqu. There are two basic kinds, depending on the temperature reached during manufacture. One kind is high temperature that is allowed to rise above 55°C, perhaps as high as 65°C, and the other is kept below 55°C. The raw materials are wheat flour, barley flour and pea flour mixed together in proportions, depending on the manufacturer. One manufacturer, Yanghe Distillery, uses the following proportions and procedures. The wheat:barley:pea proportion is 5:4:1; the moisture content is brought to about 40% and the resulting dough kneaded and eventually packed into

A picture showing a whole process of DA-QU preparation

Fig. 10.6.   An old Chinese depiction of the process of making daqu. A line of ten workers treading the dough can be seen and the box like mould in which it is treaded is seen at the top.

a wooden mold ($195 \times 138 \times 75$ cm). The packing is done by human treading by ten people with a gradation in body weight to get the dough at just the right firmness in the mold (Fig. 10.6). The block is then removed from the wooden mold and placed in the incubation room, with other blocks, in two layers on bamboo rods giving a space of 2 cm between the layers covered with straw. Incubation is about four weeks and has three phases. In the first phase of about nine days, the temperature rises rapidly due to the growth of fungi and yeasts. The dominant fungi are *Aspergillus* species and *Rhizopus* species, namely *R. oryzae* and *R. chinensis*. The humidity is initially high and is reduced by opening room windows. The temperature of the whole batch is controlled within the range of 30–45°C by turning and re-arranging blocks about six times in the nine-day period. In the second phase, which lasts about eight days, the temperature of the batch is allowed to rise smoothly a little further but within the range of 30–45°C. In this phase, more moisture evaporates and its loss is controlled so that it is not lost too quickly. In the third phase, the room temperature is allowed to decline and moisture loss from the blocks continues. If at the end the moisture content has not been reduced to about 12%, the blocks are piled and re-covered with straw to raise the temperature and accelerate moisture loss.

Xiaoqu is another important jiuqu. The traditional method is similar to that used for daqu production except only steamed and powdered rice is used, the water used is flavored from steeped herbs, and the freshly made blocks are cut into smaller cubes and formed into balls. Then selected old jiuqu is used as a seed culture to inoculate the outer surfaces of the balls. As the dominant fungi involved have been isolated, a pure culture method is now used in its preparation as follows. The *Rhizopus* species are grown on rice meal or wheat bran. Yeast, *Saccharomyces* species, is cultured on an extract of a special jiuqu and harvested by centrifugation. Then the *Rhizopus* culture and yeast concentrate are mixed in a 50:1 ratio. This mixture is then used to make a seed mash for rice wine. This approach significantly reduces the time to make the rice wine. Dried yeast of pure cultures is also produced and used in many rice wine producing factories.

10.4.2.2. *Rice wine making*

Ideally the water should have a hardness in the range of 2–6, pH around 7, and have a low iron content. The acidified rice-steeping liquid is traditionally prepared as follows. The polished rice is steeped in the water for a period of 10–20 days. During the steeping, lactic acid bacteria grow and lower the pH. This liquid is separated out and reduced in volume by boiling; when added to the final brew, it helps to inhibit germination of bacterial spores and the growth of vegetative cells; provides amino acids and vitamins, lowers the pH for yeast growth and helps improve the flavor of rice wines. However, alternatives to this include using diluted vinegar solution or a broth made with wheat flour, but lactic acid is also commonly added.

The actual production process begins with the preparation of **seed mash.** Traditionally the ingredients used to make this seed mash are 125 kg of glutinous rice that is soaked for several days and then steamed. The steamed rice is then acidified to pH 4.0–4.5 with lactic acid or acidified steeping water, and after it cools to about 27–30°C it is spread in a ceramic vat and half of the xiaoqu jiuqu is mixed through the rice. The rice is then stacked around the sides of the vat to form a hollow in the middle and the rest of the xiaoqu powder is sprinkled onto the surface (0.187–0.25 kg total). This maximizes the surface exposure to encourage growth of *Rhizopus* species that produce lactic and fumaric acids to lower the pH, discourage the growth of undesirable organisms and encourage yeast growth. In addition, the amylase and glucoamylase in the jiuqu saccharify the rice. After about two days, a pool of liquid containing sugars etc., gathers in the hollow. Then 19.5 kg of jiuqu is added, mixed through the fermenting rice mix and fresh water is added to bring the total water, including that taken up by the rice during steeping and steaming, to 230 kg. A pure culture of yeast (1–2%) may be added to the seed mash instead to promote the development of yeast activity.

By **fed-batch** and **parallel fermentation** processes the fermentation proceeds for about seven days. The timing of the first stirring is critical and it is followed fairly quickly by three more stirrings. After seven days the fermented liquid is transferred to narrow-necked jars and sealed tightly to keep out O_2 where a **secondary fermentation** takes place over a 70–80 day period. The wine is then filtered, clarified, pasteurized and packaged.

10.4.3. *Sorghum liquor*

Maotai, Chu Yeh Ching (Bamboo Leaf Green), Ta Che, Fen Gue and many other Chinese liquors are sorghum spirit (53–55% alcohol) with characteristic aromatic fragrance due to the presence of 4-ethylguaiacol, vanillic acid and vanillin. Starch is hydrolyzed using exogenously produced fungal amylases in a **solid-state cultivation.** The production process as outlined in Fig. 10.7, varies. In Taiwan, three cycles of cultivation of the substrate and distillation are conducted. The fermentation stage is usually about ten days long.

A typical wine (yeast) cake (or jiuqu), produced with cracked wheat, contains five yeasts (including *S. cerevisiae, S. fibuligera*), four molds (including *Mucor* and *Rhizopus* species) and six bacteria, including a *Clostridium* species. The wine cake contains some hydrolytic enzymes, but activities are low. To inoculate the steamed sorghum, it is spread out on the floor to a thickness of about 5 cm and powdered wine cake is mixed through it. The moisture content of the sorghum is about 55%. The fungi

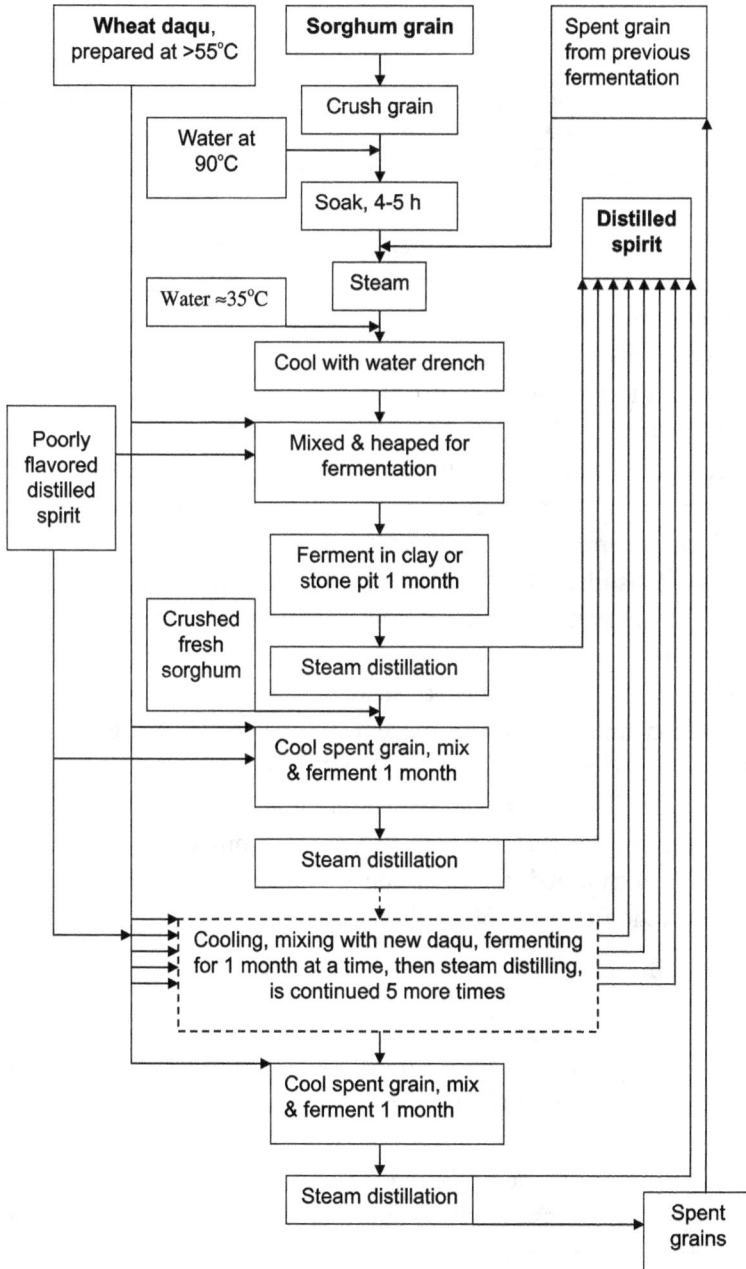

Fig. 10.7.   An outline of the production stages and parameters used in the production of sorghum liquor. Production depends on a repeated solid-state fermentation.

grow on the cool and dry surface, whereas bacteria and yeasts grow within the wet mass of sorghum. After 24 hours the sorghum is piled into heaps that encourage growth of yeast in an anaerobic and warm condition (pre-fermentation). The size of the pile is restricted because the internal temperature of the pile should not be allowed to rise above 40°C, and the heat produced metabolically is not easily dispersed. In Taiwan, the inoculated sorghum is placed into stainless steel tanks or boxes and covered with polythene sheet to prevent moisture loss and reduce aeration. Microbial growth and metabolism is allowed to progress for up to ten days after which the mixture is steamed to distil the ethanol. The pile of sorghum is re-inoculated to encourage further growth and when it is ready, the sorghum is distilled again. Heat treatment (steaming) causes partial hydrolysis of carbohydrate and protein. Steaming also serves to remove ethanol, which prevents the inhibitory effect of ethanol on cell growth in the thin film of water surrounding sorghum grains. Alcohol is collected by condensation of distillate. The yield of ethanol is about 22.5 kg per 100 kg of sorghum. In some procedures the processes of fermentation and steaming need to be repeated up to seven times to achieve complete conversion of carbohydrate to alcohol. However, in Taiwan, repeating a ten-day fermentation period and distillation twice is sufficient to maximize ethanol production.

Microbial cells grow slowly on sorghum due to the low availability of fermentable substrates, and this leads to the production of secondary metabolites mentioned above, which contribute to the characteristic strong aromatic fragrance of sorghum liquor.

10.5. Dough Fermentations

The desirable qualities of baked flour dough products like bread depend on added yeast. Traditionally prepared dough was kept from day to day to provide active yeast but this encourages growth of lactic acid bacteria. Consequently two products of interest have arisen. Firstly, technology for producing baker's yeast to add to the flour was developed, and secondly, sour dough baked products are still being produced based on our understanding of the role of lactic acid bacteria.

10.5.1. *Bread and cracker biscuit making*

Most baker's yeast is used in the making of normal bread and one relatively rapid commercial process called the Chorleywood Bread Process which allows bakers to use wheat with lower protein level. In general bread making involves adding yeast to flour, salt and water and mixing vigorously (kneading) to form a dough. The dough is then incubated to encourage the yeast to grow and produce gas as well as flavoring substances. After the dough has doubled in size it is "punched down" to work the gas out of the dough and is then formed into suitable loaf sizes, placed in suitable baking pans, then allowed to rise further before being baked. The final product has the soft sponge-like texture typical of bread. On the other hand, cracker biscuits are also made from baker's yeast but they are thin with a flat bubbly appearance, and brittle. Cracker manufacture begins with flour, yeast, salt but with less water, that is formed into a dough and fermented to form a sponge-like texture. Then fat, salt, sodium bicarbonate and water are added to the spongy dough and mixed into the dough which is allowed to incubate further. Following further mechanical kneading of the dough it is then "sheeted" to form the cracker biscuit before baking.

10.5.2. *Baker's yeast*

Each year about two million tonnes are produced and sold as either cream yeast (solids 16–20%), compressed yeast (solids 30–35%), both of which may be frozen, or dried yeast. Consequently the fermentation to produce yeast for bakers (who prefer cream and dried yeast) must be controlled to produce cells that are very active and able to withstand the stresses involved. A broth medium of molasses and simple N sources as well as other nutrients are heat treated and added to fermenters with agitation and aeration. Optimal growth conditions rely on (i) nutrient feed rates (a batch fed system is used) that ensures carbohydrate levels are kept low, (ii) very high aeration levels that ensure respiration only occurs so maximizing yeast mass production, and (iii) pH, temperature and ethanol level. The respiratory quotient, $CO_2:O_2$ ratio, is also important.

At the end when nutrient addition ceases aeration is continued for up to an hour during which trehalose accumulates in the cells as a storage carbohydrate. Depressed carbohydrate level, high aeration rate and the removal of N sources favor the gluconeogenic pathway (glucose is synthesized from nonsugar sources, in particular amino acids and TCA cycle molecules) and synthesis of trehalose. The trehalose protects the yeast cells from stresses involved in freezing and drying.

10.5.3. *Sough dough*

Sour dough is produced when flour and water are mixed and left for up to 24 hours during which yeasts and lactic acid bacteria grow. The soured dough is then used to make bread and other products. The production of rye bread relies on it. In Europe and other parts of the world, wheat based sour dough products are made. One example of a wheat dough product is a Cantonese deep fried product called Yau Chau Kwai. It is usually prepared in the mornings and people buy it for breakfast. In the USA, the San Francisco sour dough French bread process is used to make bread. The Cantonese product illustrates the basic process, used for thousands of years around the world. The maker of Yau Chau Kwai keeps the very wet dough aside and each day he takes out half of the mixture and adds an equal mixture of wheat flour and water back into the remaining wet dough. This stimulates the lactic acid bacteria and yeasts in the mixture to grow to produce sufficient microbial growth for the next day's production.The process depends on the chemical and microbiological characteristics of the flour. The flour from wheat and rye grains contains monosaccharides and disaccharides, particularly maltose, amino acids, minerals, and fatty acids (Table 10.7). In addition, they contain lactic acid bacteria and yeasts from the growing environment of the grains. When the flour and water are mixed, the lactic acid bacteria begin to grow rapidly. They lower the pH of the mixture and this encourages the growth of yeasts that grow more slowly than the bacteria. In normal dough the yeasts constitute about 1% of the total number of microbial cells in the mixture. Factors that affect the process include temperature, water activity, redox potential (oxygen) and fermentation time.

Table 10.7. Mono- and di-saccharides
in wheat flour.

Sugar	Percent, w/w
Glucose	0.02
Fructose	0.04
Sucrose	0.26
Maltose	0.12

The dominant organisms in sour dough are *Lactobacillus sanfranciscensis*, and *Lb. pontis*, and others found include *Lb. fructivorans*, *Lb. fermentum* and *Lb. brevis*. Most of them grow better on maltose. Homofermentative and heterofermentative rods and cocci are involved.

The yeasts associated with sour dough include *Candida milleri*, *C. holmii*, *Saccharomyces exiguous* and *S. cerevisiae*.

In the case of rye bread, the lowering of the pH of the dough is necessary for it to be made into bread. It inhibits amylase activity and improves the swelling of pentosans that take over the function of gluten in wheat bread. The pentosans in the flour at low pH (optimum pH 4.9) have improved water-binding and gas-retaining properties that produce better quality bread.

In the case of rye and wheat based sour dough, the resulting breads show better keeping quality, and improved nutritional value. The sour dough prevents the production of ropey bread and the development of fungi on the stored bread.

The fermentation improves flavor of the bread through non-enzymic browning during baking, fatty acid oxidation and the products of microbial fermentation.

Further Reading

1. Ayres JC, Mundt JO and Sandine WE. *Microbiology of Foods*, Chapter 7: Alcoholic yeast fermentations. WH Freeman, San Francisco, 1980.
2. Beech GA, Melvin MA and Taggart J. Food, drink and biotechnology. In: *Biotechnology: Principles and Application*, eds. IJ Higgins, DJ Best and J Jones. Blackwell Scientific, Oxford, 1985, pp. 73–110.

3. Fugelsang KC. *Wine Microbiology*. Chapman and Hall, New York, 1997.

4. Hardwick WA. Beer. In: *Biotechnology Vol. 5*, Chapter 3, ed. G Reed. Verlag Chemie, Weinheim, 1983, pp. 165–230.

5. Jakobsen M, Larsen MD and Jespersen L. Production of bread, cheese and meat. In: *The Mycota: A Comprehensive Treatise on Fungi as Experimental Systems for Basic and Applied Research,* eds. K Esser and PA Lemke. X, *Industrial Applications*, ed. HD Osiewacz. Springer-Verlag, Berlin, 2002, pp. 3–22.

6. O'Toole DK. Beverages: Asian alcoholic beverages. In: *Encyclopedia of Grain Science*, eds. C Wrigley, H Corke and C Walker. Elsevier, 2004, pp. 86–96.

7. Priest FG and Campbell I, eds. *Brewing Microbiology*, 2nd ed. Chapman and Hall, London, 1997.

8. Springham DG. The established industries. In: *Biotechnology: The Science and the Business*, 2nd ed., Chapter 16, eds. V Moses and RE Cape. Harwood Academic Publishers, The Netherlands, 1999, pp. 261–305.

9. Stewart GG and Russell I. Modern brewing biotechnology. In: *Comprehensive Biotechnology Vol. 3*, Chapter 17, eds. HW Blanch, S Drew and DIC Wang. Pergamon, Oxford, 1985, pp. 335–382.

10. Tuite MF and Oliver SG, eds. *Saccharomyces*: *Biotechnology Handbooks Vol. 4*. Plenum Press, New York and London, 1991.

11. Vondrejs V, Janderova B and Valasek L. Yeast killer toxin K1 and its exploitation in genetic manipulations. *Folia Microbiol.* **41**: 379–394, 1996.

12. Xu GR and Bao TF. Grandiose Survey of Chinese Alcoholic Drinks and Beverages, at http://www.jiangnan.edu.cn/zhgjiu/umain.htm.

13. Yoshida T. Technology development of saké fermentation in Japan. In: *The First International Symposium on Insight into the World of Indigenous Fermented Foods for Technology Development* and *Food Safety*, organized by the Department of Microbiology Faculty of Science Kasetsart University, and the National Research Council of Thailand International Cooperation, Ministry of University Affairs, 13–14 August 2003 at the Microbiology-Genetics (MG) Building, Department of Microbiology, Kasetsart University, Bangkok, Thailand.

Websites

http://www.sytu.edu.cn/zhgjiu/umain.htm
http://www.brewingtechniques.com/
http://www.foodreference.com/
http://www.internationalrecipesonline.com/recipes/
http://www.beerinfo.com/vlib/commercial.html
http://dmoz.org/Business/Food_and_Related_Products/Beverages/Beer/
Tools_and_Equipment/
http://www.winepros.org/wine101/enology.htm
http://www.sake.nu/
http://www.esake.com/
http://www.beerinfo.com/rfdb/
http://www.esake.com/index.html

Questions for Thought...

1. Production of sorghum liquor by solid state fermentation is a slow process. How could the process be speeded up without compromising the flavor and fragrant quality of the final product?
2. Compare and contrast the various ways that starch is used to produce alcoholic beverages.
3. Summarize the enzymatic processes occurring in rice grains during the manufacture of Japanese saké. How do these processes differ from traditional Chinese rice wine making processes?
4. What two environmental factors are used to control the growth of wild yeasts and bacteria during rice wine production? Why are they not used for that reason for beer and grape wine production?
5. Consider the advantages and disadvantages of producing rice wine using purely enzyme based methods or using microbes, either natural or genetically engineered, to carry out the processes.
6. Rice wine production in general is a batch fed system. What are the reasons for this?
7. Why is sulfur dioxide used in grape wine production but not in the production of other alcoholic beverages?

Chapter 11

Fungal Solid State Cultivation

Desmond K. O'Toole
Department of Biology and Chemistry
City University of Hong Kong
83 Tat Chee Avenue, Kowloon, Hong Kong, China

A large number of industrial microbial fermentations are conducted in liquid media. Liquid media have the advantage that the growth medium is homogeneous and can be kept homogeneous, and is easily manipulated and handled in an industrial setting. However, other forms of microbial cultivation, solid state cultivations, are possible and have been used for centuries. The term "fermentation" is often applied to this form of microbial cultivation but the word has a specific meaning in microbial metabolism related to anaerobic growth, and as most of the processes dealt with here are aerobic in nature, the word "cultivation" is used here. However, even some of the liquid broth fermentations are aerobic in nature.

11.1. Fungal Solid State Cultivation

Fungal solid state cultivation can be defined as the growing of "microorganisms on moist, water-insoluble solid substrates in the absence or near absence of free liquid." The solid state is characterized by its "porous substrate with a gaseous continuous phase" that is usually aerobic in nature. It is used to produce microbial by-products, particularly enzymes (Table 11.1), and it is the basis of a number of Asian food production

Table 11.1. Enzymes and the fungi used to produce them on a commercial or pilot scale and the substrates used in their production by solid state cultivation.

Enzyme	Organisms	Solid substrate
Pectinases	*Aspergillus carbonarius, A. sojae, A. saito, A. niger*	Wheat bran
Pectinases	*A. flavus, A. oryzae, A. fumigatus, Rhizopus* sp.	Wheat bran and other solids
α-amylase	*A. oryzae*	Wheat bran
Glucoamylase	*Rhizopus* spp.	Bran
Glucoamylase	*A. niger*	Wheat bran and corn flour
Rennet	*Mucor pusillus*	
Rennet	*M. meihei*	Wheat bran
Rennet	*R. oligosporus*	Wheat and rice bran
Cellulase	*Trichoderma reesi*	Wheat bran
Cellulase	*T. harzianum*	Wheat straw and bran
Xylanase	*A. terreus*	Wheat bran
β-xylosidase	*A. terreus*	Wheat bran

systems, e.g. soy sauce, rice wine and vinegar in China. The system is ideal for aerobic, mycelium-producing organisms, like fungi and streptomycetes, and it is not commonly used for culturing single-celled organisms such as yeasts and bacteria. Two exceptions are: an aerobic *Bacillus* species cultured on soybeans to produce natto, and in silage fermentation with wild lactic acid bacteria that grow anaerobically on chopped maize plants and other plant crops so O_2 plays no significant part. Composting and commercial mushroom growing are further examples of solid state cultivations. However, the fundamental objectives of each of these processes differ (Table 11.2). During composting the objective is to **transform**, using natural microflora, a substrate that may be objectionable (e.g. animal manure) or of no other value so the end product can be used for another purpose. In mushroom growing the objective is to **grow** a particular fungus on the substrate so as to generate fungal fruiting bodies for human consumption. The substrate is of no direct value. (The mushroom industry is substantial because each year about 1,223,000 tons of *Agaricus bisporus* are produced and when the commercial production of all other fungal fruiting bodies is added in about 2,182,000 tons are produced.) On the other hand, in the solid

Table 11.2.   A comparison between applications of the principles of solid state cultivation.

Characteristic	Solid state cultivations	Composting	Mushroom growing
Substrate	Processed vegetative substrate, often of one kind, that favors fungal growth	Mixed material of vegetable origin; may include animal waste	Mixed or single type vegetative material that favors mold growth and fruiting body formation
Fungus/ microorganism	A pure strain, or wild mixture, usually not grown aseptically	Wild mixture depending on the environment	A pure strain, but other organisms may be needed to promote fruiting
Substrate for human consumption	Yes, or an extract of the fermented substrate	No	No
Fungus consumed	No	No	Fruiting body only
By-product isolated and used	Yes, in some cases	No	No

substrate cultivations considered here the objective is either to transform the substrate so it can be **consumed completely** or **extracts** for human consumption or for human use can be produced, or to prepare and isolate **by-products** for further use. Much of what follows can be applied to all three broad applications because aeration, water activity and temperature level are common parameters that impact the outcome of the processes.

11.1.1. *Substrate characteristics*

Ideally the solid substrate should consist of solid particles (Table 11.3) that are of small dimensions, say 3 to 10 mm, either oblong or ovoid in shape, e.g. soybean, rice, or long in shape, that when packed together result in gaps between the particles to allow air to pass easily. Sliced material or flat material tends to pack down and close off gaps bringing about a reduction in the porous nature of the solid mass. However, geometry of the substrate particles affects the volume of the porous continuum and the movement

Table 11.3. Substrates commonly used in fungal solid substrate cultivations.

Agricultural waste	Processed food waste	Grains	Other materials
Wheat straw	Wheat bran	Rice	Wood
Corn stover*	Sugar beet pulp	Soybean	Cassava
Rice stover	Banana meal	Corn	
Feedlot waste	Rye meal	Buckwheat seeds	
		Sorghum	

*Dry lignocellulose material that is not straw.

of air in a substrate mass and eventually microbial growth as well. For example, in one comparative experiment the growth of *Aspergillus koppan* was improved on a rice medium as the mesh size of the particles was reduced from 8 mesh (2 mm) to 30 mesh (0.5 mm). However, when the mesh size was further reduced allowing closer packing of the particles and reduced air volume, growth was suboptimal.

The advantages of this system for producing microbial products, when compared with liquid fermentations, are claimed to include better product quality, and higher product yield and concentration that results in less expensive downstream processing, low capital and energy costs, and low wastewater output. Disadvantages include the heterogeneity of the substrate, restricted transport of the mass and heat energy which causes local overheating and dehydration, and, within the substrate mass, local depletion of oxygen and increase in carbon dioxide. In addition, when a solid state cultivation is leached to obtain the product, the extract is often viscous which cannot be easily vacuum concentrated resulting in a product that cannot be converted to a powder and is not easily blended or re-constituted.

11.1.2. *The characteristics of fungal growth*

The hyphal mode of growth by fungi gives the fungi a large advantage over the growth of single-celled organisms like yeasts and bacteria. Fungi grow by apical extension of the hyphal tips and by the generation of new hyphal tips by branching. The growth at these tips, however, is due to the transport of nutrients in the hyphae to the tip, and the growth is so quick that the hyphal tips extend at a greater rate than the rate at which nutrients in the substrate can diffuse to the hyphae. The rate of branching is related to the density of nutrient available in the substrate, but the rate of extension of mold hyphae, that is their colonization rate, is unaffected by nutrient availability. The highest hyphal extension rates are observed in **non-septate fungi** because the lack of septa allows rapid transport of nutrients from the rest of the mycelia to the growing apical tip. In **septate fungi** the nutrients must pass through septal pores that separate fungal cells and slow down nutrient transport. Streptomycetes, which have a similar growth habit to fungi and form mycelia, cannot grow as fast due to the formation of individual separate cells in the hyphal chain through which nutrient cannot be delivered to the growing tip consequently their expansion is limited.

11.1.3. *The fungal inoculant*

The usual form of the inoculum is a preparation of mold conidiospores or sporangiospores (the bacterial streptomycetes produce exospores) and not vegetative molds. The spores are produced aseptically by inoculating sterile rice or wheat bran in flasks with fungal spores and incubating them until growth terminates at maximum sporulation. The spores are of uniform size and are easily harvested and stored for a long time while being resistant to mishandling; vegetative mycelial fragments, on the other hand, are heterogeneous and lumpy and would require considerable force (that could damage the hyphae) to obtain a reasonable mixture that could be easily mixed with the substrate.

Following spore inoculation into the prepared substrate, the fungal growth is delayed because of the lag time required for the spores to germinate and start producing the hyphae which produce enzymes to act on the substrate from which nutrients for the growing fungus arise. Further, the optimum germination temperature may differ from the optimum growth temperature for the hyphae that must spread through the substrate to produce the mycelium.

The density of the inoculum should be optimized for each substrate and organism. Spore inoculation levels used range from 10^2 to 10^7 spores g^{-1} substrate. If inoculation levels are too high or too low, lag time, mold growth rates and product yield can be influenced. Early fungal sporulation during the production stage must also be avoided in some cases, e.g. for tempeh, but in other cases it does not matter.

11.1.4. *Growth conditions in solid state cultivations*

Three factors must be controlled in the midst of the substrate during the growth phase for a successful outcome, namely: temperature, water activity and air or oxygen. Another factor that might affect growth is pH, but filamentous fungi can grow over a wide pH range (2–9), and optimally from pH 3.8–6.0. As a consequence of the fungal tolerance to environmental pH, the pH of the substrate can be manipulated to a low level to restrict bacterial growth, but the pH level of the growth medium can influence enzyme production by the fungi, e.g. *Aspergillus niger* produces protease optimally in the pH range 3.55–4.3.

11.1.4.1. *Temperature*

In general, the fungi used grow in the temperature range 20–55°C but the optimum temperature for growth, and for product formation, can differ. The temperatures used for some production systems are shown in Table 11.4.

Temperature control is a key issue. Firstly, because of the aerobic metabolism high levels of heat are generated by the microbial growth. The maximum and average rates of heat generation by *Aspergillus oryzae* in a soybean/wheat mixture for soy sauce production was measured at $130\,kJ\,kg^{-1}\,h^{-1}$ and $53\,kJ\,kg^{-1}\,h^{-1}$, respectively, and the maximum and average rates of heat generation for *A. niger* grown on a starch/rice bran

Table 11.4. Examples of optimum temperatures used in some fungal solid state cultivations.

Microorganism	Substrate	Product	Temperature, °C
Rhizopus oligosporus	Wheat bran	Acid protease	25
Aspergillus niger	Wheat bran	Glucoamylase	30
Humicola sp.	Agribusiness waste	α-galactosidase	45
Penicillium candidum	Wheat bran	Lipase	29
Aspergillus carbonarius	Canola meal	Phytase	53

mixture for citric acid production was measured at $71\,\mathrm{kJ\,kg^{-1}\,h^{-1}}$ and 33 $\mathrm{kJ\,kg^{-1}\,h^{-1}}$, respectively. The fungi penetrate the mass of the substrate completely and grow over and into the particles making up the substrate. However, the substrate does not have good heat conducting properties, and convection currents to take away heat may be weak, nor can refrigeration be used as it can with broth-based cultivations. Consequently, temperature gradients may build up in the solid substrate and some parts of the substrate mass may reach temperatures sufficient to kill any fungus and stop product formation. The excessive heat may also dry out the substrate as evaporative moisture loss, through the latent heat of evaporation, is one of the key ways in which the temperature can be reduced.

Control in traditional systems is by using shallow trays so that the heat generated has only a short passage to travel to be dissipated to the environment. In modern systems, mechanically driven moisturized air is pushed through the solid state material but the air flow rate is another important parameter affecting temperature, water activity and oxygen levels. In reality, it is not easy to fully control the whole process.

11.1.4.2. Water

If water content is too high the substrate particles may agglomerate and diffusion of air throughout the substrate may be restricted thus allowing undesirable bacteria to grow. However, water content alone is not a measure of the best moisture level for microbial growth. **Water activity** is the best measure because the level of water activity affects mold growth (Table 11.5) and in these systems the water activity is finely balanced

Table 11.5. Some optimum water activity levels used in fungal solid state cultivations.

Microorganism	Substrate	Product	Water activity
Trichoderma viride	Sugar beet pulp	β-xylanase	0.995
Candida rugosa	Coconut oil cake	Lipase	0.92
Penicillium roquefortii	Buckwheat seeds	Growth and sporulation	>0.96
Aspergillus oryzae	Various	Protease	0.982–0.986
Rhizopus oligosporus	Chickpea	Fermented food	0.92

between the optimum and actual water activity level that may change during growth.

The water activity in a substrate is not directly related to the moisture content. It reflects the free water available for microbial activity and it ranges from 0 to 1 with 1 being the optimum for most organisms, and about 0.6 is the minimum at which any microorganism will grow. It is equivalent to **relative humidity** but it is expressed as the **ratio** of the relative humidity over a substrate in an enclosed container, to the equivalent relative humidity over pure water, both at equilibrium with the air phase. The amount of water that evaporates into the air depends on the number of soluble particles that can bind water in the water phase, or the concentration of bound water molecules at the water surface; a low concentration of soluble particles that bind water molecules results in a high water activity, a high concentration of solutes results in a much lower water activity. In many of the solid substrates used in these cultivations the levels of available solute are relatively low so relatively low water content can result in a high water activity which favors the growth of most microorganisms. For example, when the water activity is just at 1 the moisture contents of grain sorghum, wheat straw and sugar beet pulp are about 40%, 63% and 77%, respectively.

A further factor is the flow of air through the substrate along with a rising temperature. The air flow and rising temperature increase the evaporation of water from the substrate causing the water content to go down which may further lower the water activity.

To control water activity levels in primitive systems the substrate is stirred and doors and windows of the incubation room may be opened and closed but in modern systems using mechanical delivery of air through the substrate, the relative humidity of the air can be controlled to balance all the factors to maintain optimum water activity levels.

11.1.4.3. Oxygen

Oxygen control methods are really about air control and depend on the design of the bioreactor system. Shallow substrate layers, stirring, tumbling and mechanically driven air are four possible ways to control the access of air to the substrate.

11.2. Bioreactor Systems for Fungal Solid State Cultivations

Five broad classes of bioreactors are used. The first is the traditional tray system, the second is the traditional small package system, the third is the packed column system, the fourth is the rotating drum system, and the fifth is the stirred substrate system.

11.2.1. *Trays*

In the **tray system** the bottom of the tray is perforated to allow natural air circulation through the inoculated substrate that is placed in a layer 2 to 4 cm thick in the tray that is incubated in a suitable room. The conditions, including temperature, aeration and relative humidity, in the cultivation room must be controlled during the incubation period. Temperature and humidity gradients develop in the substrate so, if needed, the substrate can be stirred to reduce temperature.

The main advantage of the tray system is its simplicity. During the cultivation period it is easily observed, and experimentation is easily conducted.

11.2.2. *Small packages*

The traditional **small package system** is similar to the tray system except the solid substrate is inoculated and then formed into small, flat packets, e.g. as used in tempeh production, or flat round cakes, e.g. as used for Korean soy sauce making and Chinese rice wine making, or small round balls (about 15–40 mm diameter) on and through which the fungi are allowed to grow.

11.2.3. *Packed columns*

The **packed column system** can be an enclosed system and it depends on forced air flow through the substrate in the column. Some advantages of the packed column system are:

(i)   variations in temperature and gas concentration throughout the substrate can be eliminated;

(ii) water activity levels can be maintained with moisturized air;

(iii) aseptic conditions can be maintained, unlike other systems; and
(iv) the column can be converted to a **trickle bed extractor** for recovery of soluble product.

11.2.4. *Rotating drums*

In the **rotating drum system** the drum is laid horizontally. After the substrate is introduced the drum is rotated either continuously or intermittently so that the substrate is put in motion which promotes heat and mass transfer (see Fig. 20.5). Three disadvantages of the system include the development of particle agglomeration over time, mechanical damage to the vegetative hyphae of some fungi and the difficulty of controlling temperature. The addition of baffles in the drum facilitates aeration because they lift the substrate and drop it as opposed to allowing it to roll or slump.

11.2.5. *Stirred substrates*

The **stirred substrate systems** are more modern. They include a **fluidized bed system** which costs, it is claimed, only one-sixth of that for other types of solid state cultivation systems for culturing *Aspergillus oryzae*. **Continuous systems** with mechanical stirring have been developed for koji production for making Japanese shoyu products.

11.3. Modeling in Solid State Cultivation Systems

Various aspects of solid state substrate systems can be modeled; e.g. the effect of fungal mycelium on mixing in these systems, fungal growth and metabolism inside solid particles, and temperature and water relations. However, a further complication is the fact that fungal growth is also three-dimensional.

11.3.1. *Fundamentals of fungal growth*

The growth cycle of the filamentous fungi is far more complex than bacteria and yeasts. Generally they have a vegetative growth stage where the hyphae extend and branch into a complex three-dimensional mycelial structure. During this stage the fungi may produce enzymes and metabolic by-products that are linked directly to the growth, or the by-products of interest may be produced as a consequence of **secondary metabolism** that is not linked directly to vegetative growth. Most of the industrially important fungi then enter a sporulation stage where aerial spores are formed, often on specialized fruiting structures. The cytoplasm in some hyphae may migrate leaving portions of the mycelium empty — cell walls surrounding empty space — resulting in some metabolically inactive hyphal biomass along with the spores formed that also show no metabolic activity. These non-metabolizing components of fungal origin are known as **necromass** and can, in time, become a significant component of fungal-derived material in the substrate. Consequently, measuring the mass of fungal material and relating it to fungal growth is not as simple as for organisms that form single cells and grow in liquid culture. Furthermore, estimations of fungal mass may be confined to total biomass and necromass (includes all mycelia and sexual and asexual bodies), or growing biomass, or biologically active biomass, or reproductive biomass. When fungal growth occurs in a solid substrate the estimation of fungal growth becomes even more difficult.

During fungal growth in solid substrate cultivations other parameters are also changing. In an incubating mass of substrate the total dry mass declines as the total biomass increases. This is due to the consumption of O_2 and production of CO_2, the production of other volatile products, the evaporation of metabolically generated water and the consumption of water in the hydrolysis of polysaccharides, assumed to be 1 mol of water chemically bound per mol of glucose produced. Also, as the fungi grow and produce enzymes, the enzymes hydrolyze the proteins, starch, cellulose and hemicelluloses to water soluble solutes that then lower the water activity of the solution in the substrate.

11.3.2. Measuring fungal growth

Before any models can be developed to describe fungal growth in solid substrates, a measure of fungal mass is essential. Some of the methods devised include:

(a) Total biomass measurement; suitable for some simulation models based on substrates that can be solubilized.

(b) Total hyphal length; as km g^{-1} dry substrate, can be converted to bio-volume with estimates of hyphal diameter.

(c) Propagule concentration; by a viable plate counting method, suitable for examining the inoculants used as starters.

(d) Cell components, chitin/glucosamine, this may also include necromass; glucosamine concentrations of about 114 mg g^{-1} fungal dry mass have been reported and this number is used in some studies, but glucosamine concentrations vary according to the age of the culture and growth medium; a conversion factor of 12g dry mass g^{-1} glucosamine has been suggested for *Rhizopus oligosporus*.

(e) Cell components, ergosterol from cell membrane; ergosterol contents are usually of the order of 5 µg mg^{-1} fungal dry mass, but this value is affected by the lipid content of the substrate, O_2 level during growth, the analytical method used, growth conditions and growth phase.

(f) Cell component, protein (N × 6.25); N in the substrate complicates the estimate.

(g) Enzyme activity; laccase, reductase and esterase.

(h) Direct measure of cell activity; ATP, DNA.

(i) Compression/penetration pressures; the response of the matted mycelium to pressure is related to the mass of mycelium.

(j) Metabolic activity; O_2 consumption and CO_2 release can be used to follow the progress of the cultivation for process control purposes.

11.3.3. *Mathematical models*

The mathematical models suggested for fungal growth in solid substrate fungal cultivations are classified into four kinds.

(1) Branching models based on mycelial growth using a combination of elongation of hyphal tips and branch formation. This model has limited application.

(2) Black-box models in which the kinetics of experimentally observed growth are not related to any cause and effect relationships.

These models rely on a modified form of the logistic law used to describe skewed sigmoidal curves;

$$r_x = \mu_{max}[1 - (X/X_{max})^n]X$$

where r_x is the biomass production rate (kg h^{-1}), μ_{max} is the maximum specific growth rate (h^{-1}), i.e. the limit of the specific growth rate as the amount of biomass approaches zero, X is the amount of biomass present (kg), X_{max} is the maximum amount of biomass (kg) that can develop in the system, and n is an exponent. The results obtained depend on the measure of biomass chosen, and this approach does not take into account the gradual decline in the biomass production rate.

(3) Particle-level models, mechanistic in nature, in which intrinsic growth and maintenance kinetics, and space and transport limitations on or in the substrate particle are used to predict experimental results. The advantage of these models is that they improve understanding of solid substrate cultivation processes, and allow evaluation of substrate particle geometry and any added liquid nutrients.

(4) Reactor-level models, that predict the macroscopic behavior of the bioreactor based on reaction kinetics and conservation laws.

Satisfactory models for the mathematical description of the growth dynamics and product formation still need further work before they can rival the growth models developed for cultivation of single celled and other types of organisms growing in liquid media. Mitchell has summarized models published since 1980. Amongst the models, the following have proved useful for various aspects of microbial growth in solid state cultivations.

A. An empirical logistic equation was used to describe the growth of *A. oryzae* in koji.

$$dm/dt = \mu m[1 - m/N]$$

where m = mycelial dry weight g^{-1} dry matter, μ = specific growth rate during logarithmic growth, and N = maximum mycelial dry weight g^{-1} of dry matter.

Because of the difficulties of measuring dry weight, the model then assumed that CO_2 production, an easily measured parameter, was due to both growth and maintenance and wrote the following equation:

$$dA/dt = K_1 dm/dt - K_2 m$$

where A = CO_2 evolved mg^{-1} of dry matter, K_1 = CO_2 evolved mg^{-1} of dry mycelium formed, and K_2 = CO_2 evolved for endogenous respiration mg^{-1} of dry mycelium h^{-1}.

By combining both equations and using a respiratory quotient of 1.0 the O_2 consumption rate could be predicted.

B. Another model was based on a consideration of energy balance, to control water content during koji production with *A. oryzae* by taking into account the heat capacity of the air used to aerate the bioreactor. To maintain constant water content, temperature is taken into account:

$$T_{surr} = T_{out} + 1.03(T_{out} - T_{in}) - 131 PCO_2$$

where T_{surr} = temperature of the surroundings of the bioreactor, T_{out} = temperature of outlet air, T_{in} = temperature of inlet air, and ΔPCO_2 = difference in partial pressure of CO_2 between the inlet and outlet air streams, i.e. correlated with metabolic heat production.

By controlling the temperature of the surrounding air, according to the equation, the water content of the koji remains constant.

C. However, Mitchell claimed that only the logistic model and that developed by him and his co-workers actually describe a model based on growth kinetics. The model was based on the growth of *Rhizopus oligosporus* on starch-containing substrate and took into account five key components, namely biomass, glucoamylase, starch, glucose and distance within the substrate. From an understanding of what happens during cultivation, the events surrounding each of these components can be described mathematically, namely:

(i)   Release of glucoamylase by the mycelium at the surface of the substrate.
(ii) Diffusion of the glucoamylase to the starch, using Fick's Law.

(iii) Hydrolysis of the starch to release glucose, using Michalis-Menton kinetics.

(iv) Diffusion of glucose to the surface mycelium.

(v) Absorption of the glucose and its conversion into mycelial mass, using Monod kinetics.

When all combined 13 parameters must be measured or estimated, however, when applied to real growth experiments the model satisfactorily described glucose and starch concentrations, and glucoamylase activities during growth.

Further Reading

1. Doelle HW, Mitchell DA and Rolz CE, eds. *Solid State Cultivation.* Elsevier Applied Science, London and New York, 1992.
2. Gowthamam MK, Krishna C and Moo-Young M. Fungal solid state fermentation — An overview. In: *Applied Mycology and Biotechnology*, eds. GG Khachatourians and DK Arora. Elsevier, Amsterdam, New York, 2001, pp. 305–352.
3. Mitchel DA. Growth patterns, growth kinetics and the modeling of growth in solid-state cultivation. In: *Solid State Cultivation*, eds. HW Doelle, DA Mitchell and CE Rolz. Elsevier Applied Science, London and New York, 1992, pp. 87–114.
4. Nout MJR, Rinzema A and Smits JP. Biomass and productivity estimates in solid state fermentations. In: *The Mycota: A Comprehensive Treatise on Fungi as Experimental Systems for Basic and Applied Research*, eds. K Esser and PA Lemke; *Environmental And Microbial Relationships, Vol. 4*, eds. DT Wicklow and BE Söderström. Springer-Verlag, Berlin, 1997, pp. 323–346.
5. Rahardjo YSP, Tramper J and Rinzema A. Modeling conversion and transport phenomena in solid-state fermentation: A review and perspectives. *Biotechnol. Adv.* **24**: 161–179, 2006.
6. Raimbault M. General and microbiological aspects of solid substrate fermentation. *Elect. J. Biotechnol.* **1**(3): 174–188, 1998.
7. Terebiznik MR and Pilosof AMR. Biomass estimation in solid state fermentation by modeling dry matter weight loss. *Biotechnol. Tech.* **13**: 215–219, 1999.

Websites

http://www.wcfs.nl/webdb/OverviewProjectPages?readform&Page=
C005&Prog=C
http://www.ejbiotechnology.info/content/vol1/issue3/full/9/bip/

Questions for Thought...

1. Compare and contrast the growth of a mycelial fungus, like a *Penicillium* species, when cultured in a broth medium and a solid state cultivation.
2. Describe how a fungus might grow in a solid substrate cultivation and think about what would happen if the fungus produced an enzyme that broke down the substrate.
3. Why do manufacturers of soy sauce not use the packed column system of solid state cultivation to make soy sauce?

Chapter 12

Food Ingredients

Desmond K. O'Toole
Department of Biology and Chemistry
City University of Hong Kong
83 Tat Chee Avenue, Kowloon, Hong Kong, China

Lee Yuan Kun
Department of Microbiology, Yong Loo Lin School of Medicine
National University of Singapore
5 Science Drive 2, Singapore 117597
Email: micleeyk@nus.edu.sg

With a better understanding of food flavor, texture, rheological properties, nutritional value and human food preferences, the food ingredient industry has developed to enhance the qualities of many foods. The ingredients industry is a huge industry, e.g. the food flavors market in 2009 was estimated to be worth US$20 billion, and food manufacturers rely on the ingredients to enhance food quality.

12.1. Flavors

Microorganisms play an important role in the flavor and aroma formation of many foodstuffs. For example, the flavor of mold-ripened cheeses is determined by the break-down products of fungi (e.g. *Penicillium roqueforti* in blue cheese), such as methyl ketones, short chain fatty acids, acetic and lactic acids, alcohols, aldehydes, lactones, ammonia, amines, sulfides and mercaptans. On the other hand, the earthy notes of pond cultured fresh water fish, e.g. tilapia, is due to geosmin (*trans*-1,10-dimethyl-*trans*-9-decalol) produced by actinomycetes (e.g. *Streptomyces odorifer*), myxomycetes and cyanobacteria. Lately, there is increasing interest in using microorganisms as the source of "natural" flavor and fragrance, and some examples are cited in Table 12.1.

12.1.1. γ-Decalactone

An important flavor component in the flavors of strawberry, raspberry and peach/apricot is γ-decalactone. A combination of biological and chemical methods was developed in the early 1960s to produce this component. Baker's yeast (*S. cerevisiae*) was found able to reduce β-ketocapric acid to hydroxydecanoic acid, a precursor of *γ-decalactone*. Subsequently, species of *Candida* (e.g. *C. tropicalis*, *C. parakrusei*), and a strain of *Escherichia coli* were found that could produce 4-hydroxydecanoic acid from ricinoleic acid that is found in relatively high concentrations in castor oil. The process depends on releasing ricinoleic acid from the triacylglycerol through the action of a lipase. Then the yeasts, through an aerobic metabolic process of β-oxidation, reduce the C18 ricinoleic acid to the C10 acid, 4-hydroxydecanoic acid, that is then lactonized by heating at a low pH to γ-decalactone. Yields are low, at 1 g L^{-1}, even after seven days fermentation.

More recent studies have targeted a yeast species of the genus *Sporidiobolus*, namely *S. ruinenii*. The metabolic pathway is the same as before but cultural conditions where a small amount of hydrogen is incorporated into an aeration stream (2.5% H$_2$ and air [19.8% O$_2$]) resulted in a jump in γ-decalactone production unconnected with actual microbial growth. The addition of H$_2$ lowered the Eh$_7$ of the growth medium to about +2.5 mV. It seems that the addition of a reducing agent (H$_2$) results in more rapid production of γ-decalactone to levels significantly higher than those seen in the control conditions.

Table 12.1. Flavors and their character impact substances produced by various microorganisms.

Flavor	Character impact substance	Microorganisms
Banana, peach, pear	γ-, δ-decalactone	*Ceratocystis moniliformis*
Apricot, banana, pineapple	Butyric acid pentyl ester	*Clostridium acetobutylicum*
Pear, banana, pineapple	Butyric acid isobutyl ester	*Clostridium acetobutylicum*
Apple	Isovaleric acid ethyl ester	*Peptostreptococcus anaerobius* *Pseudomonas aeruginosa*
Apple, raspberry	Isovaleric acid isopentyl ester	*Peptostreptococcus anaerobius* *Pseudomonas aeruginosa*
Coconut	γ-Octalactone, γ-nonalactone, 6-pentyl-2-pyrone	*Trichoderma viride*
Musk precursor	Ustilagic acids A & B	*Ustilago zeae*
Butter	Diacetyl	*Lactococcus lactis, Lc. diacetilactis, Lc. cremoris,* *Enterobacter cloacae*
Rose	β-phenyl ethanol, furan-2-carboxylate	*Ascoidea hylacoeti*
Rose	6-methyl-5-hepten-2-ol acetate, citronellol, linalool, geraniol geranyl acetate	*Ceratocystis virescenes*
Pine, rose, mushroom	Thujopsene, 3-octanone, 1-octen-3-ol, nerolidol, β-phenyl ethanol	*Penicillium decumbens*

12.1.2. *Monosodium glutamate (MSG)*

The most common flavor compound produced by a microbial process is L-glutamic acid. The establishment of fermentative production of the flavor enhancer, L-glutamic acid, by bacteria belonging to the genera

Corynebacterium and *Brevibacterium,* is a milestone in the development of industrial production of **microbial primary metabolites**. Primary metabolites are biomolecules directly involved in cell growth and reproduction, such as amino acids and tricarboxylic acid (TCA) cycle intermediates, whose production is tightly regulated in cells. Industrial production of microbial primary metabolites was considered impractical due to their low concentration in microbial cultures but today, almost all of the amino acids are industrially produced by fermentation.

Development of the production of L-glutamic acid by fermentation was based on the understanding that accumulation of L-glutamic acid in a culture is determined by its excretion from cells rather than by its biosynthesis. Thus, by limiting the supply of biotin, or including C_{16}–C_{18} saturated fatty acids and their esters, penicillin or Tween-60 in the broth medium results in an increase in the permeability of the cell membranes in the growing cells, and excretion of L-glutamic acid and other amino acids (Fig. 12.1). Overproducing mutants were subsequently isolated

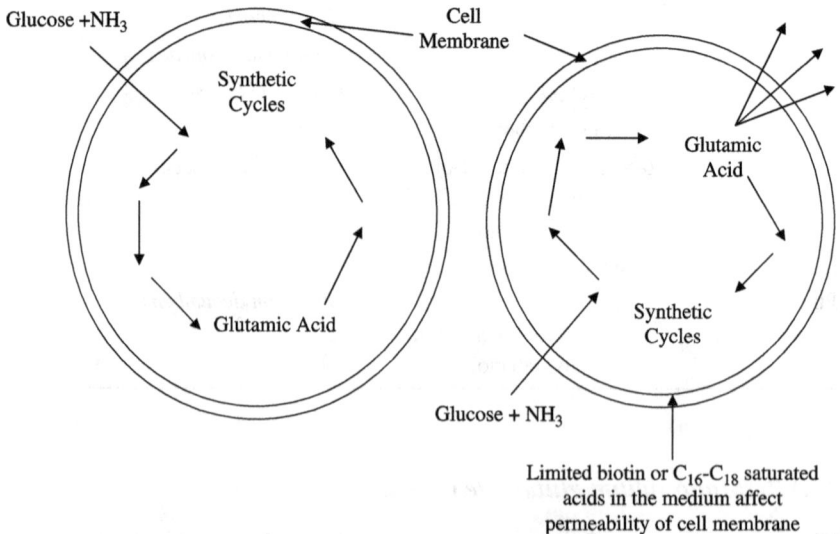

Fig. 12.1.   The production of glutamic acid. By manipulating growth conditions, the cell membrane of the bacterium is weakened so that the permeability of the membrane increases. As the glutamic acid is produced, much of it is excreted from the cell. This helps to promote glutamic acid production in the cell.

which further increased the production of glutamic acid. Examples include monofluoroacetate resistant mutants derived from *Corynebacterium glutamicum* (formerly known as *Brevibacterium lactofermentum*) that have increased phosphoenol pyruvate carboxylase activity. Other mutants with decreased isocitrate lyase activity lead to increased carbon dioxide fixation and increased L-glutamic acid production.

To produce MSG a cheap source of starch is hydrolyzed enzymically to glucose. With ammonia as the N source and other nutrients the fermentation is conducted under aerobic conditions. Glutamine is produced and converted to glutamate under alkaline conditions. The MSG is then crystallized from the medium.

12.2. Antioxidants

One of the major causes of food deterioration during long term storage is the chemical oxidation of **unsaturated fatty acids** in the lipids present in all foods. These fats are present in fat deposits and in cell membranes. End products of lipid oxidation, mainly aldehydes and ketones, contribute to the rancid off-flavors and odors in the foods. Other consequences of oxidation include discoloration; changes in the texture, flavor and taste; and the loss of nutritional value due to the destruction of essential fatty acids as well as vitamins.

The oxidation of a lipid involves the initiation of catalytic oxidation of lipid by the presence of oxygen and heavy metals, thermal or photodecomposition of peroxides, or by ultraviolet irradiation, resulting in the formation of a lipid radical ($L^{\bullet}$). Propagation of a chain reaction sees the formation of a lipid peroxyl radical (LOO^{-}) and lipid hydroperoxide ($LOOH$). Lipid peroxyl radicals react with each other to form the complex $LOOL$. The addition of an antioxidant minimizes the extent of lipid oxidation by inhibiting the initiation of propagation steps, thereby prolonging the shelf life of the food products.

There is now a growing interest in the development of "natural antioxidants" for foods, because of consumer preference and some queries regarding the toxicity of chemical antioxidants such as butylated hydroxytoluene (BHT). The presence of highly unsaturated fatty acids in microbial cells and their relative stability have prompted many researchers to turn to microorganisms as a possible source of natural antioxidants.

The phospholipid fraction of algae showed antioxidant activity similar to that of BHT. Some of the phospholipids identified were phosphatidyl ethanolamine and phosphatidyl inositol. The bromophenols in the acetone-soluble fraction have been implicated.

Enzyme systems that could be used as antioxidants in food include superoxide dismutase (SOD) and glucose oxidase. SOD has been shown to prevent the oxidation of ascorbic acid, anchovy fat, and the enzymic darkening of sliced mushroom, apple, and potato. The enzyme scavenges oxygen free radicals and dismutates them into oxygen and hydrogen peroxide:

$$2O_2^- + 2H^+ \xrightarrow{\text{SOD}} H_2O_2 + O_2$$

The SODs isolated from the yeast *S. cerevisiae* var. *ellipsoideus* and *Lac. lactis* are relatively heat stable, and are particularly suitable for foods that are mildly acidic to alkaline.

Glucose oxidase catalyses the oxidation of D-glucose in the presence of oxygen to D-gluconolactone and hydrogen peroxide. In the presence of catalase, the hydrogen peroxide, which may not be desirable in the food products, is broken down.

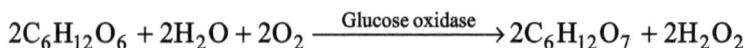

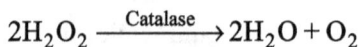

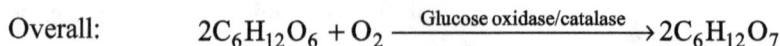

$$2C_6H_{12}O_6 + 2H_2O + 2O_2 \xrightarrow{\text{Glucose oxidase}} 2C_6H_{12}O_7 + 2H_2O_2$$

$$2H_2O_2 \xrightarrow{\text{Catalase}} 2H_2O + O_2$$

Overall: $$2C_6H_{12}O_6 + O_2 \xrightarrow{\text{Glucose oxidase/catalase}} 2C_6H_{12}O_7$$

Due to its mild action and a Generally Regarded as Safe (GRAS) status, glucose oxidase is used widely in the food industry. It is used in the production of dried foods like instant coffee, cake mixes, milk powder and active dried yeast; alcoholic beverages, such as beer and wine; citrus drink; soft drink; mayonnaise and salad dressing.

12.3. Essential Amino Acids

Plant proteins are relatively low in essential amino acids, such as L-lysine and L-tryptophan. These amino acids are extensively used in the food and the animal feed industry as additives to achieve nutritionally balanced diets and animal feeds. Amino acids (e.g. L-arginine, L-isoleucine, L-leucine, L-phenylalanine, L-proline, L-tryptophan, L-valine) are used in medicine as ingredients of infusion solutions, and in the chemical industry as starting materials for the manufacturing of biopolymers. The latter includes polyalanine fibres and lysine isocyanate resins; poly-γ-methylglutamate is used as a surface layer for synthetic leather; N-acyl-serine is used as a surface-active agent in cosmetics.

Demand for amino acids is estimated by Ajinomoto to be about two million tons per annum. Production by fermentation of major amino acids for food applications is summarized in Table 12.2.

Table 12.2.   The organisms and their genetic characteristics, and substrates used for the production of various amino acids for various applications.

Amino acid	Microorganisms	Genetic characteristics	Major substrates	Applications
L-lysine	*Corynebacterium glutamicum*	Hom$^-$Leu$^-$AEC^r	Sugar cane molasses	Feed additive
	Brevibacterium flavum	Hom^l Thre$^-$	Acetate, NH$_4^+$, biotin, thiamine	
L-tryptophan	*C. glutamicum*	Phe$^-$Thyr$^-$5MT^{r}TrpHx^r 6FT^{r}4MT^{r}PFP^{r}PAP^r	Glucose	Antioxidant
L-aspartic acid	*Pseudomonas fluorescens*	Wild type	Fumaric acid, NH$_4^+$	Flavour, aspartame
	Escherichia coli	Aspartase mutant	Fumaric acid, NH$_4^+$	
L-phenylalanine	*C. glutamicum*	Tyr$^-$PFP^{r}PAP^r	Glucose	Aspartame

Resistance: AEC: S-(β-Aminoethyl)-L-cysteine, 6FT: 6-Fluorotryptophan, 4MT: 4-Methyltryptophan 5MT: 5-Methyltryptophan, PAP: *p*-Amino-phenylalanine, PFP: *p*-Fluorophenylalanine PheHx: Phenylalanine hydroxamate, TyrHx: Tyrosine hydroxamate.

Box 12.1

What is Aspartame?

Aspartame is an ester of L-aspartic acid and L-phenylalanine methyl ester. It is used as a low-calorie sweetener in drinks and beverages. On a per weight basis, aspartame is as sweet as cane sugar (sucrose) but has lower energy content. It is readily digested in the human digestive tract and absorbed as amino acids. Biotechnologically it is an interesting industrial compound because patents are lapsing and a number of methods are being explored for its commercial production. L-phenylalanine and L-aspartic acid can be produced fermentatively, the former using *trans* cinnamic acid, ammonia, and phenylalanine ammonia lyase. In addition, L-phenylalanine can also be produced from phenylpyruvate, formic acid and ammonia via ACA acylase. The gene for this enzyme has been cloned and sequenced and a promoter found for use in a mutant of *Corynebacterium*. L-aspartic acid is produced from fumaric acid with vermiculite immobilized cells of *Escherichia coli* with aspartase activity. In more recent research, genetically engineered organisms, based on *E. coli* and yeasts have been used. Complementary nucleotides have been designed to code for Asp-Phe-Asp-Phe. Following successful insertion into *E. coli,* one clone directed the synthesis of the fused poly(Asp-Phe) as inclusion bodies calculated to be 11.2% of the total cell protein. A chemically synthesized gene, which encodes a 64 or 128 times-repeated tripeptide aspartyl-phenylalanyl-lysine, has been cloned into a yeast that was able to synthesize the polypeptide so that it made up to about 30% of the total cell protein.

   More recently, a search of soil isolates found a strain of a fungal endophyte which is able to produce a specific dipeptidase able to produce aspartame, without the protection of amino acid side chains, from L-aspartic anhydride with L-phenylalanine methyl ester.

12.4. Essential Fatty Acids

Polyunsaturated fatty acids (PUFAs) are an important component of the human diet. In particular, omega-3 fatty acids (ω3FAs) are important dietary components required for the proper development of brain in humans, and vision in infants. Some of the ω3FAs, e.g. γ-linolenic acid (GLA, 18:3), eicosapentaenoic acid (EPA, C20:5) and docosapentaenoic acid (DHA, C22:6) have been recognized as dietary components which prevent cardiovascular diseases. These PUFAs are also precursors for the industrial production of eicosanoids (e.g. prostaglandin, prostacyclins, thromboxane and leucotrienes), which are used in the treatment of high blood pressure, asthma, menstrual pain and induction of labor. The main traditional source of GLA is plant seeds (e.g. Perilla, linseed and red currant), whereas EPA and DHA are extracted from aquatic animals (e.g. oils of Atlantic mackerel, Atlantic salmon, Pacific herring and cod).

GLA in plants is synthesized from hexadecatrienoic acid ($16:3\omega3$) in the chloroplast membrane. Photosynthetic microorganisms, e.g. *Spirulina major*, a cyanobacterium, that accumulate approximately 1% GLA in dry biomass, are being used as an alternative source, especially in the health food sector. EPA and DHA are synthesized by phytoplankton, which is consumed by fish, molluscs and crustaceans, and thereby are concentrated in the aquatic food chain. Thus, microorganisms are being considered sources of EPA and DHA. Some of the omega-3 fatty acid producing microorganisms are listed in Table 12.3. Another group of organisms, not listed in the table, are the *Thraustochytrium* sp., marine protists of Kingdom Chromista.

Table 12.3. Fermentation times and production rates of omega-3 fatty acids by bacteria, microalgae and fungi.

Organism	$\omega3$ fatty acid content, % dry biomass	Fermentation time, day	$\omega3$ fatty acid productivity, $gL^{-1}d^{-1}$
Bacteria			
Vibrio sp. (EPA only)	0.7	0.3	0.13
Microalgae			
Schizochytrium sp.	10.0	2.0	1.0
Crypthecodinium cohnii (DHA only)	9.0	2.5–4.0	0.8–1.2
Nitzschia alba (EPA only)	2.3	2.7	0.25
Fungi			
Mortierella alpina	5.8	7.0	0.2
Saprolegnia sp.	0.4	3.0	0.1
Pythium ultimum	1.9	6.0	0.03

12.5. Vitamins

Vitamins are essential in both human and animal nutrition and their industrial production is a huge business. The global market for vitamins in 2003 was estimated to be valued at about US$2.3 billion. The economic production method of choice is chemical synthesis but recently microbiological methods for the production of some of these compounds have become economically competitive (Table 12.4).

Ergosterol, converted to vitamin D_2 under UV irradiation, is produced using *Saccharomyces cerevisiae*. The C source can be a carbohydrate or ethanol but adequate aeration in the fermentation is essential so that ergosterol may comprise up to 10% of the cell dry mass. To extract the ergosterol, cells are treated with hot alkali to saponify the lipid followed by ether extraction.

The vitamin riboflavin is used in pharmaceutical products, in animal feeds and as a colorant (orange-yellow) in human foods. Four organisms are used for production and two of them are fungi, namely *Eremothecium ashbyii* and *Ashbya gossypii*. A biotechnology based production system

Table 12.4.   Vitamins and vitamin-like compounds and the organisms used to produce them and their estimated yearly production levels.

Compound	Organism	Yearly world production, tonnes
Ergosterol[a]	*Saccharomyces cerevisiae*	?
γ-lenolenic acid[a]	*Mortierella isabellina*	1,000
Riboflavin	*Ashbya gossypii, Candida famata, Bacillus subtilis*	1,000–10,000
Pantothenate	*Escherichia coli*	1,000–10,000
Cobalamin[a]	*Pseudomonas denitrificans, Propionibacterium shermanii*	10,000
Orotic acid[a]	*Brevibacterium* sp.	100
Ascorbic acid	*Gluconobacter oxydans* ssp. *suboxydans*	100,000

[a] Only produced microbiologically, others in Table are produced chemically and, microbiologically. Table modified from Stahmann (2002).

for riboflavin failed in the late 1960s but in 1974 Merck began using the fungus *Ashbya gossypii*. Following improvements in cultures, production levels up to 15 g L⁻¹ riboflavin have been reported. When *A. gossypii* is grown using glucose as a C source it produces ethanol until the glucose is exhausted. During this stage lipid droplets accumulate in the cells. On the exhaustion of the glucose riboflavin production begins and both the lipid and ethanol are metabolized. Lipid levels in the cells (up to 20% of the dry mass) can be enhanced, and riboflavin production increased by adding vegetable oil to the growth medium; and the composition of the lipid can be changed by adding free fatty acid to the fermentation. High lipid levels can be added to a fermentation vessel without altering the water activity level of the medium. Production can be further enhanced by adding precursors, such as glycine to the fermentation mixture. A neutral pH and high O_2 levels are also beneficial to maximize production.

During the riboflavin production phase the lipid is catabolized via β-oxidation in the peroxisomes to acetyl units that are transported to the mitochondria. The actual synthesis begins with guanosine triphosphate (GTP) (precursors of GTP in the growth medium, such as glycine and hypoxanthine enhance production) and ribulose-5-phosphate (R5P).

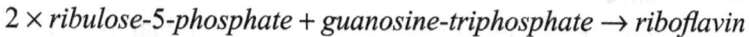

$$2 \times ribulose\text{-}5\text{-}phosphate + guanosine\text{-}triphosphate \rightarrow riboflavin$$

Following biosynthesis the riboflavin is either transported to vacuoles in the cells or it is exported to the cellular environment.

Gene manipulation has further enhanced production in three ways.

(a) Injection of a second copy of the *ICL*1-gene, that codes for isocitrate lyase, improves production when soybean oil is used in the fermentation.

(b) Delivery of glycine was improved. A copy of the *GLY*1-gene, that codes for threonine aldolase, under the control of the *TEF* promoter and terminator was inserted. Then when the medium was supplemented with threonine instead of glycine, riboflavin production increased ten-fold. This was due to the more rapid uptake of threonine by comparison with glycine uptake.

(c) Transport to internal cellular vacuoles instead of direct secretion into the medium slows production. To prevent vacuolar accumulation the vacuolar ATPase subunit gene, the VMA1-gene, was inactivated which ensured all the riboflavin was excreted into the growth medium.

For six years from 1990 BASF ran chemical and biological production methods side by side using *A. gossypii* and eventually dropped the former method. The other two cultures used for production are under patent protection. Strains of *Candida famata* have been genetically improved by Coors Biotech, Inc to produce >20 gL^{-1} riboflavin. Roche (Switzerland) are using a strain of *Bacillus subtilis* genetically engineered by Omnigene (USA).

Genetically engineered ascorbic acid (vitamin C) production has also been achieved. A gene for the enzyme 2,5-diketo-D-gluconate reductase from a *Corynebacterium* species was inserted into a culture of *Ewinia herbicola* resulting in a one step fermentation process. The culture could then ferment glucose to produce 2-keto-gulonate at a concentration of 20 gL^{-1}, after 72 hours, to give a 50% yield based on the glucose input. The gulonate could then be chemically converted to ascorbate.

12.6. Polysaccharides

Polysaccharides are used as thickening and stabilizing agents in the food industry, and plants and seaweed are the traditional sources for these polysaccharides (e.g. starch, agar and alginate). However, microbial polysaccharides have become more widely used in recent years. Table 12.5. shows some of the important commercial microbial polysaccharides. Alginate is not included as it is currently extracted mainly from seaweed, but industrial production of alginate by the bacteria *Pseudomonas aeruginosa* and *Azotobacter vinelandii* is being investigated because the chemical structure of seaweed alginate is very variable and microbial production may allow genetic engineering of the bacteria to provide tailor-made alginates for particular applications.

The extracellular polysaccharides are recovered by precipitation with alcohols, such as isopropanol or methanol. The precipitated polysaccharides are then dried and ground.

12.7. Microbial Meat Substitute — RHM Mycoprotein

A more recent modern source of a meatlike protein is fungi grown using a completely modern scientific approach to produce the commercial product Mycoprotein, also known as Quorn™. In the 1960s, a lack of protein

---

Table 12.5. Production of polysaccharides from carbon substrates by microorganisms and the monomer moieties and structures of the polymers.

Polysaccharide	Sugar moiety/ Structure	C-substrate	Organism
Bacteria			
Cellulose	β-1→4 glucan	Glucose	*Gluconacetobacter xylinus*
Curdlan	Glucose/ β-1,3-glucan	8% glucose	*Agrobacterium* spp.*
Dextran	Glucose/α-1,6-,1,2-, 1,3-, 1,4-glucans	Sucrose	*Leuconostoc mesenteroides, Leuc. dextranicum, Streptococcus mutans*
Xanthan	Glucose, mannose, glucuronic acid, acetate, pyruvate/ pentasaccharide	4% glucose/ sucrose/ starch hydrolysate	*Xanthomonas campestris*
Fungi			
Scleroglucan	Glucose/β-1,3-, 6,1- glucans	Glucose	*Sclerotium glucanicum, S. delphinii, S. rolfsii*
Pullulan	Glucose/α-1,4-, 1,6- glucans	5% glucose	*Pullularia pullulans*

*Includes *Agrobacterium radiobacter*, formerly attributed to *Alcaligenes faecalis* var. *myxogenes*.

for human consumption seemed a real future possibility. Microorganisms can use simple nutrients, reproduce rapidly, and produce a far greater amount of protein per unit of energy input when compared with traditional land based protein sources (Table 12.6). Therefore, microorganisms were developed to supply microbial protein as a solution for future food supply for humans. However, unicellular microbial cells lack meaty texture. In addition, microbial cells contain high levels of RNA, which could cause nutritional problems when consumed in large quantities. The organism chosen for the development of this product was the fungus *Fusarium graminearum* ATCC 20334 (isolated from soil). The meat analogue developed from the fungal mycelium is now called Mycoprotein.

Table 12.6.   Feed conversion efficiency rates in protein formation.

| Feed stock | Biomass | |
	Protein	Total
Cattle	1 kg feed	14 g	68 g beef
Pig	1 kg feed	41 g	200 g pork
Chicken	1 kg feed	49 g	240 g meat
F. graminearum	1 kg $C_6H_{12}O_6$ + N	136 g	1080 g wet cell

Table 12.7.   Composition of Mycoprotein compared to beef.

Component	Mycoprotein, % dry wt.	Raw lean beefsteak, % dry wt.
Protein	47	68
Fat	14	30
Dietary fiber	25	Trace
Carbohydrate	10	0
Ash	3	2
RNA	1	Trace

Mycoprotein is produced by a continuous fermentation process at a dilution rate of up to 0.2 h^{-1}, temperature of 30°C and pH 6, and controlled by the addition of gaseous ammonia into the inlet air stream, and using glucose, ammonia, ammonium salt and biotin. Cell concentrations of 15–20 $g L^{-1}$ can be maintained.

After fermentation, the culture is heated (64°C for 20–30 min) to reduce the RNA content from 9% to <2%. The mycelial mass gives a meat-like texture to the product, which is flavored and colored accordingly.

Microbial cells have lower fat and protein, but higher fiber and carbohydrate contents (Table 12.7). Mycoprotein is available in supermarkets as packaged vegetarian hamburger steaks.

12.8. Organic Acids

It is possible to produce all organic acids using microbial fermentation processes. Examples of practical microbiological processes for organic acid production are listed in Table 12.8. Two acids of interest are citric acid and lactic acid and their production is illustrated.

Table 12.8.   Practical production and use of organic acids, the organisms, and substrates used for production, and some aspects of the process.

Organic acid	Organism	C-substrate	Process	Uses
Acetic acid	*Acetobacter aceti* *Gluconobacter oxydans*	Ethanol	35 h, 30°C	Food
Citric acid	*Aspergillus niger*,	Molasses, starch,	8 d, 30°C	Food, and many others
	Yarrowia lipolytica	n-alkane, glucose		
	Asp. wentii	Sucrose		
Fumaric acid	*Rhizopus oryzae*	Glucose	3 d, 33°C	Resin
	Candida sp.	n-alkane	7 d, 30°C	
Gluconic acid	*Aspergillus niger,* *Gluconobacter suboxydans*	Glucose	24 h, high O$_2$ pressure, 12–25% glucose	Food
Lactic acid	*Lactobacillus delbrueckii*	Glucose	72 h, 40°C	Food
	Lb. delbrueckii ssp. *bulgaricus*	Whey		
	Lb. pentosus	Sulfite liquor		
Propionic acid	*Propionibacterium* sp.	Lactose, glucose, starch	8–12 d 30°C	Perfume, fungicide
Malic acid	*Leuconostoc brevis*	Fumaric acid, n-Alkane	24 h	Food
α-keto-glutaric acid	*Candida hydrocarbofumarica*	n-Alkane	40 h	
	Aerobacter sp.	Glutamic acid		
5-keto-gluconic acid	*Gluconobacter suboxydans*, formerly *Acetobacter suboxydans*	Glucose	5–6 d	L-tartaric acid
2-keto-gluconic acid	*Serratia marcescens*	Glucose	16 h	Isoascorbic acid

12.8.1. *Citric acid*

It is estimated that in 2004 about 1.4 million tons of citric acid-1-hydrate, or anhydrous citric acid, were produced globally, with growth in demand of about 4% per annum. China accounted for 35–40% of that production. Citric acid is useful because of its acidity and buffering capacity, its flavor, and its ability to chelate metal ions. It is mainly used in the food and beverage industry (60%), for enhancing flavor of fruit juices, and for addition to marmalade, soft drinks, candy and ice cream, and as a food preservative. Over 90% of citric acid is manufactured by microbial fermentation and its fermentation is an example, along with vinegar production, of a highly refined process based on scientific know-how. Fungi are used in the fermentation and while many fungi, and some bacteria, can also produce citric acid, the organisms primarily used in commercial production are selected strains of *Aspergillus niger*, although *A. wentii* and the yeasts *Yarrowia lipolytica*, *Candida guilliermondii* and *C. oleophila* have also been used. *A. niger* produces citric acid more rapidly with minimal production of undesirable side products, resulting in economic yields. It can produce about 70% of the theoretical yield from carbohydrate sources. The actual yield of citric acid from a production process depends on the type and concentration of carbon source as well as nitrogen and phosphate limitations in the medium, the presence of trace elements, initial pH, aeration and incubation temperature.

Substrates used include sucrose, glucose and fructose as well as a wide range of food industry by-products such as molasses from cane and beet sugar production, and starch from various sources. A high substrate concentration of sugars (140– 20 gL^{-1}) is required to obtain satisfactory rates of production and yields of citric acid. In solid state cultivation a wide range of agricultural by-products are used including fruit pomaces (fruit residue from juice extraction), rice and wheat bran and sugar cane bagasse.

12.8.1.1. *Metabolic processes*

Metabolically, citric acid production involves glycolysis and the TCA cycle. Hexoses are broken down to pyruvate via the hexose monophosphate pathway. The pyruvate is then decarboxylated to acetyl-CoA and CO_2 that is recycled to pyruvate via pyruvate carboxylase to produce oxaloacetate. Oxaloacetate and acetyl-CoA are then combined to produce citrate. To excrete citric acid, however, the feedback mechanism from excess intracellular citric acid, which acts on the phosphofructokinase enzyme, has to be inactivated. This occurs via an increase in the intracellular ammonium levels. Hence control of the nitrogen metabolism is also crucial for production. In **submerged culture** production systems, the nitrogen concentration is held between 0.4 to 0.6 gNL^{-1}. Other nutrients needed for growth include Mg^{2+} (20–100 mgL^{-1}), K^+, Zn^{2+} (0.3–0.5 mg L^{-1}) and phosphate (0.4–1 gL^{-1}).

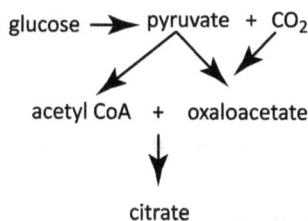

$$glucose \longrightarrow pyruvate + CO_2$$

$$acetyl\ CoA + oxaloacetate$$

$$citrate$$

Control of trace metals in the fermentation medium is also necessary for good citric acid production. A limiting concentration of manganese, <1 mgL^{-1}, particularly in submerged culture conditions, is required. This requirement is linked to increased protein turnover, impaired DNA biosynthesis, and alterations to the composition of cell walls and plasma membranes in the cells. Iron, as a cofactor for aconitase, at a concentration of 1.3–1.5 mgL^{-1}, is necessary to achieve optimal growth for *Aspergillus* culture. But for optimal production of citric acid lower levels of iron are needed. Thus, when invert sugar or starch hydrolysate is used as the carbon source, the high iron content of the raw material results in growth without citric acid production. In this case, a preliminary treatment with either a precipitant, e.g. calcium hexaxcyanoferrate, sodium ferrocyanide, or cation exchangers, is carried out to remove iron and manganese. Likewise, molasses, a common cheap raw material, is treated with precipitants to remove metals. Otherwise, a large amount (100 mgL^{-1}) of copper is added to the molasses mixture to reverse the inhibitory effect of manganese on citric acid formation during the production phase.

The optimal level of acidity for growth of *Aspergillus* culture (trophophase) is about pH 5, but the pH is kept below 2 during the citric acid production phase (idiophase), in order to suppress oxalic and gluconic acid formation. If too much oxalic acid is produced, a yellowish pigment is formed, which later hinders the purification process of citric acid. Thus, ammonium is often used as the nitrogen substrate and the initial pH is set at 5–7 to allow germination of spores and optimal initial growth. In the first 30–48 hours, during the trophophase, and as a result of the metabolism of ammonium ions, the pH of the culture falls to below 3. However, at that low pH heavy metals leach from normal steel fermenter walls and inhibit citric acid formation. Consequently fermenters must have plastic liners or be made of food grade stainless steel.

12.8.1.2. *Cultivation methods*

Three different cultivation methods are used in production, **surface liquid cultivation, solid state cultivation** and **submerged culture**. During surface liquid cultivation, the traditional production method, the fungus is grown on the surface of fermentation medium in shallow trays. Although low volumes are produced this way, its advantages are low capital investment (the fermenter) and low energy costs because cooling, agitation and aeration are not needed. In the process, spores are inoculated onto the surface of solid substrates, such as wheat bran or sweet potato pulp, or onto a liquid medium containing molasses, with depths of between 5–20 cm. Ventilation is essential to carry away CO_2 (>10% v/v inhibits citric acid production) and fermentation heat (to maintain fermentation at 30°C). Yields from surface processes amount to 1.2 to 1.5 kg citric acid monohydrate per m^2 of fermentation surface per day.

The solid state cultivation method is the simplest method of production and it was developed in Japan so they could use fruit wastes and rice bran in a koji like process. Various solid state cultivation formats are used, e.g. packed bed column, single layer packed bed etc. Yields of citric acid up to 263 $g kg^{-1}$ of cassava bagasse in 4 cm deep beds have been obtained.

The submerged culture method, by which about 80% of citric acid is now produced, can be managed as a **batch**, a **fed batch** or a **continuous culture** process. A yield value of about 900 g citric acid m^{-3} culture medium can be obtained in the submerged process (usually in 120–220 m^3 fermenters). The form of mycelium appears to be an important factor in the production of citric acid by the submerged process. If the mycelium is loose, filamentous and has no chlamydospores, little citric acid is produced. Abnormally short, forked and bilbous hyphae are brought about by a deficiency of manganese in the culture medium. The deformed hyphae aggregate to form small (0.2–0.5 mm), hard, smooth surfaced pellets, and produce a high level of citric acid. In practice the culture is grown from spores in submerged culture in the laboratory to ensure that the pellet form of the culture is initiated. Once in the pellet form the culture is added to the production culture.

In advanced modern systems of production, agitation and aeration in deep fermenters is necessary to ensure a minimum dissolved oxygen concentration of 75% saturation of the growth medium with air (equal to 150 mbar partial pressure of O_2) below which citric acid production ceases. The culture is low in viscosity due to the formation of mycelial pellets, so an **airlift bioreactor** without stirring can also be used. In an airlift system, fine air bubbles are injected into the bottom of the culture. As air bubbles rise to the top, they are replaced by growth medium free of air bubbles. Larger air bubbles tend to rise faster than the water column falls, thus creating turbulence around them.

12.8.1.3. Genetic modification

Apart from the classical genetic approach to strain improvement, some improvement attempts have been made using genetic engineering. Generally molasses is used as a cheap hexose source but reducing Mn^{2+} concentration to a minimum to produce at a pH of <2 is essential. However, if the enzymes glucose oxidase and oxaloacetate hydrolase are inactivated citric acid accumulates at pH 5 with no need to control Mn^{2+}. The goxC gene in A. niger prevents the production of gluconic acid from glucose, and another gene, prtF, results in a lack of oxaloacetate hydrolase activity. When both genes are combined in one strain citric acid is produced at pH 5.

12.8.2. Lactic acid

The annual rate of world lactic acid production in 1982 was 24,000–28,000 tons and by 2006 it was about 68,000 tons. About 90% is produced fermentatively with yields of about 90 g lactic acid per 100 g glucose, and stereospecific lactate production, D(+) and L(−), is possible. Lactic acid is also produced chemically but it results in the two isomers. It is used in food as an acidulant, a preservative (50%), as stearoyl-2-lactylate (20%) (an emulsifier for the baking industry), and in the pharmaceutical industry. It is also used for the production of biodegradable and biocompatible polylactate polymers as a plastic substitute.

The desirable characteristics of the organisms used include an ability to completely ferment a cheap feedstock with minimal nitrogenous materials, and to give high yields of stereospecific lactic acid under conditions of low pH and high temperature with the production of low cell mass and negligible other by-products. The species used to produce from glucose include *Lb. delbrueckii,* and *Lb. leichmannii.*

Energy sources include glucose and starch. The starch is frequently degraded enzymatically to glucose before fermentation. This breakdown step complicates the process so recent research efforts have been directed to developing bacterial cultures that can breakdown starch to glucose and then to lactic acid. Further, starch has to be gelatinized by cooking; so further research has been directed towards the use of raw starch sources as the feedstock for the fermentation.

An organism able to break down pure starch is *Lb. amylovorus* B-4542. After gelatinization and enzyme thinning of the starch, the organism metabolizes the starch directly in a saccharification step without the need for enzymes. This reduces the energy input, fermentation tank requirements and time needed. The rate of reaction and final level of lactic acid produced is the same as for *Lb. delbrueckii* B-445 using glucose as the substrate. More recently, raw cassava and potato have been converted to lactic acid, either directly by the lactic acid bacterium used, or via an intermediate enzyme step using fungal enzymes.

Another modern approach, using starch as the starting material, is to use a continuous combined enzyme and culture process involving a reversibly soluble auto-precipitating amylase (D-AS) whose solubility is pH dependent. The D-AS is precipitated by the low pH of the lactic acid produced, separated and then fed back into the starch mixture. The culture used is *Lb. casei* entrapped in κ-carrageenan. The yield obtained is 3.1 g $L^{-1}h^{-1}$ at a dilution rate of 0.1 h^{-1}, which is better than batch processes.

12.9. Colorants

Food colorants from chemical synthetic sources have been controversial for some years so many have sought biological sources, particularly from fermentation processes, because many microorganisms are known to produce a variety of pigments. Riboflavin is one such pigment whose fermentative production has been covered in Sec. 12.5. Vitamins. In Europe fermentative production of β-carotene, and in Asia pigments from the fungus *Monascus*, are well developed. Other colorants are already in commercial production or under development for industrial production (Table 12.9).

Table 12.9.   Pigments produced by microorganisms that are in industrial production or in development for production for use in the food industry.

Pigment molecule produced	Color	Microorganism, fungi (f), bacteria (b)
Pigments that are industrially produced		
Ankaflavin	Yellow	*Monascus* sp. (f)
Anthraquinone	Red	*Penicillium oxalicum* (f)
Monascorubramin	Red	*Monascus* sp. (f)
Riboflavin	Yellow	*Ashbya gossypi* (f)
Rubropunctatin	Orange	*Ashbya gossypi* (f)
β-carotene	Yellow-orange	*Blakeslea trispora* (f)
Pigments under development		
Astaxanthin	Pink-red	*Xanthophyllomyces dendrorhous*, formerly *Phaffia rhodozyma* (f)
Lycopene	Red	*Blakeslea trispora* (f)
Rubrolone	Red	*Streptomyces echinoruber* (b)
Torularhodin	Orange-red	*Rhodotorula* sp. (f)
Zeaxanthan	Yellow	*Flavobacterium* sp. (b)
β-carotene	Yellow-orange	*Mucor circinelloides* (f)
Unknown	Red	*Penicillium purpurogenum* (f)

Data extracted from Table 1 in Dufossé L. (2006)

12.9.1. *Fermentative production of* β-*carotene*

Selected strains of *Blakeslea trispora* are used to produce β-carotene in Europe. This fungus occurs in two mating types designated (+) and (−) and the (+) type synthesizes trisporic acid, which is a precursor of β-carotene. However if the two types are mated in a specific ratio, the (−) type which results produces large amounts of intracellular β-carotene.

The production system relies on a two-stage process. Glucose and corn steep liquor are used to provide sufficient carbon and nitrogen. Seed cultures produced from the original strain cultures are added to the reaction vessel and growth promoted in an **aerobic submerged batch fermentation** resulting in biomass rich in intracellular β-carotene. In the second recovery stage the biomass is separated, transformed into a suitable state and the β-carotene extracted with ethyl acetate. After purification and concentration the β-carotene is crystallized or micronized in a vegetable oil.

12.9.2. *Monascus pigments*

Monascus pigments have been used for centuries in the Orient, particularly China (known as ang-kak), Japan (known as beni-koji, aga-koji), Indonesia and Philippines, to color food. Chinese red rice wines, red soybean cheese (fu-ru), pickled vegetable, fish and salted meats are some examples of Oriental food colored red by monascus pigments; in the case of fish and meat the pigments preserve the flesh. In addition, consumption of monascus colored foods are said to have health benefits. However, pigments from these organisms have not been embraced in the West where they are regarded with some suspicion. The main microbial producers used for commercial production of monascus pigments are the fungi, *Monascus purpureus* and *M. ruber.* Strain selection to screen out producers of citrinin, a nephrotoxic mycotoxin, is important, as some strains of *Monascus* have been reported to produce it. In one study, food grade ang-kak did not show death or toxic phenomena in mice at the level of 18 g kg^{-1} body weight. In the same study, the LD$_{50}$ for peritoneal injection was 7 g kg^{-1}, and subacute toxicity tests revealed no abnormal phenomena. Some strains also produce antibiotics, an undesirable component for a food additive, but antibiotic formation can be reduced and pigment formation enhanced by using media containing acetate. The monascus pigments are essentially a mixture of the red pigments monascorubin ($C_{23}H_{26}O_5$) and rubropunctatin ($C_{21}H_{26}O_5$), the purple pigments monascorubramine ($C_{23}H_{27}NO_4$) and rubropunctamine ($C_{21}H_{23}NO_4$), and the yellow pigments monascin ($C_{21}H_{26}O_5$) and ankaflavin ($C_{23}H_{30}O_5$) (Fig. 12.2). The red and yellow pigments are normal secondary metabolites of the growth of the fungus and purple pigments are thought to arise from chemical or enzymic modification of those pigments.

The monascus pigments are relatively stable to heat treatment, and stable in the pH range 3–10. The high affinity of the pigments for protein, particularly the red pigments that react easily with amino groups, makes them particularly attractive as colorants for protein rich food, such as meat, sausages, processed seafood (surimi) and dairy products.

The traditional method of pigment production involves the inoculation of *Monascus* seed culture (dry spores on rice grains) onto solid substrates, such as steamed rice. The resulting mass is then dried and ground

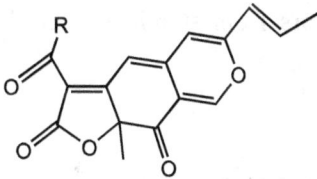

Yellow color compounds
monascin: R = C_5H_{11}
ankaflavin: R = C_7H_{15}

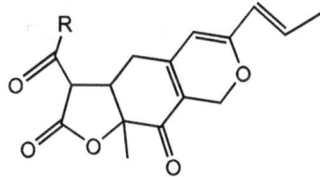

Red color compounds
rubropunctatin: R = C_5H_{11}
monascorubin: R = C_7H_{15}

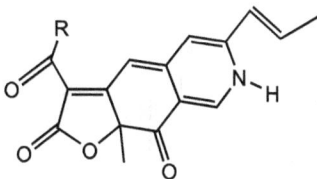

Purple color compounds
rubropunctamine: R = C_5H_{11}
monascorubramine: R = C_7H_{15}

Fig. 12.2.   Structure of the pigments produced by *Monascus*.

to powder to be used as the colorant. In recent years, the pigments have been produced in a submerged culture of *Monascus*. Rice powder and tapioca starch are suitable carbon substrates for maximum pigment production, as simple sugars are found to suppress the formation of pigments. Optimum conditions for pigment production are polished long grain rice at 50% moisture and 97–100% relative humidity, pH 5–6 and incubation at 30°C. Supplementation with Zn leads to increased red and yellow pigment by factors of 2.5 and 2.0, respectively. To overcome the initial high viscosity of the starchy medium and high oxygen demand for growth in the fungal culture as well as pigment production, starchy substrate can be partially digested by enzymes (controlled digestion is important to avoid formation of excessive simple sugars). Otherwise, suitable processes such as **solid-liquid state fermentation** can be used. In the solid-liquid state fermentation system, the solid starchy substrate is first coated onto the fermenter wall surface, by spraying onto it hot gelatinized non-water-soluble starch such as tapioca, and allowing it to solidify at a lower temperature. Then the liquid culture medium is added to the fermenter (Fig. 12.3). Amylolytic enzymes produced by the culture result in the release of fermentable sugars from starch into the liquid

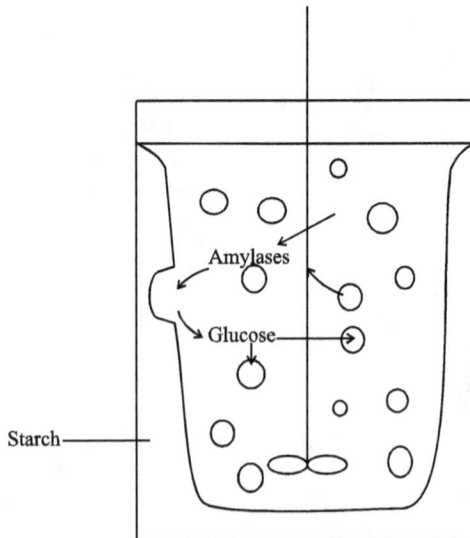

Fig. 12.3.   A solid-liquid state fermentation system for the cultivation of *Monascus*.

culture medium. The strain *M. purpureus* (ATCC 16365) is very suitable for these solid-state fermentations.

A recent novel approach to the production of these pigments involved a modified solid/liquid system in which a 400 g L^{-1} gelatinized tapioca starch cake in a liquid was used to increase yield. The organism grew in the liquid and slowly digested the starch block. A maximum cell concentration of about 37.5 g L^{-1} was achieved and a yield of about 145 OD units of both red and yellow pigments was obtained. This growth yield is significantly higher than on tapioca starch at 50 g L^{-1} on which the biomass usually reaches about 8 g L^{-1}.

12.9.3. *Algal pigments*

The need for natural, as opposed to synthetic, food colorants has led to the development of production systems based on microalgae. The pioneering work has been done with the halophilic green alga *Duniella salina* as a source of β-carotene. Other pigments include chlorophyll, of which *Chlorella* species are the major commercial source, as well as red and blue phycobiliproteins. These phycobiliprotein pigments are produced by red algae (Rhodophyta), cyanobacteria (Cyanophyta) and crytomonads (Cryptophyta). Phycobiliproteins are more stable in the pH range 5 to 9, and tend to precipitate at lower pH. However, stability of the pigments at low pH can be achieved by subjecting the pigments to a hydrolytic reaction using proteolytic enzymes, such as pronase. While pigments extracted from most algae are relatively sensitive to heat, pigments obtained from thermophilic algae are quite stable to heat.

A blue colorant extracted from the cyanobacterium *Spirulina platensis*, is commercially produced in Japan, Australia, and America. Uses of the colorants in chewing gum, soft drink, alcoholic drinks and fermented milk products, such as yogurt, have been patented in Japan. In the wild *S. platensis* grows at 27°C, in the pH range 7.2–9.0 and at a salt concentration in excess of 30 g L^{-1}. The red alga *Porphyridium cruentum* is grown in artificial seawater with added potassium nitrate at a temperature of about 21°C. Both organisms can be grown in either open ponds or closed polyethylene tubular reactors. Salinity and/or alkalinity restrict the growth of other microorganisms that might otherwise grow, particularly in open ponds.

Box 12.2

Open Pond vs Enclosed Photobioreactor for Microalgae Cultivation

Open culture systems are usually race-way-like ponds with a divider at the centre, typically 100 m x 20 m x10 cm. The cultures are circulated and agitated by a paddle wheel located at one end of the ponds. The cell concentration is typically $0.5\,g\,L^{-1}$.

The construction of the systems is simple and low cost, but there is no control over the contamination of biological (microorganisms, insect fragments, bird matter, and plant parts) and non-biological (sand, dust) contaminants. The Association of Official Analytical Chemists, USA, published a method to detect and enumerate this "light filth" in algal powder in 1989 (AOAC Official Method 970.66).

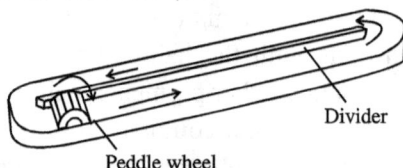

Divider

Peddle wheel

An enclosed photobioreactor can be a tubular or flat plate system made of glass or transparent plastic materials connected to a degasser. The culture is circulated by means of a pump or an airlift system. The system is closed, thus free from foreign materials. The area biomass output rate (g cells m^{-2} illuminated area) is comparable with that achievable in an open pond system, but the cell concentration is 10 to 40 times higher (i.e. $5–20\ g\,L^{-1}$), thus saving on down-stream processing costs.

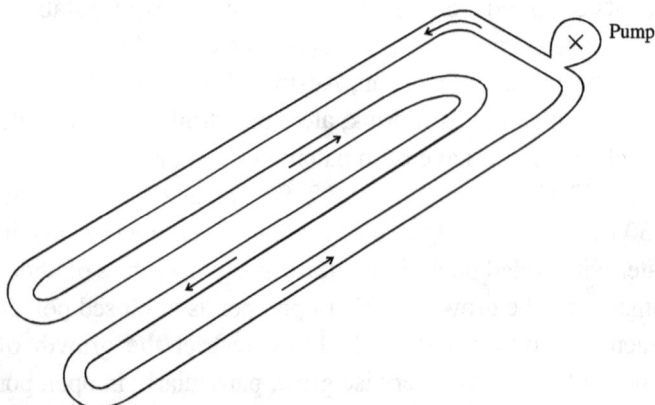

Pump

References

1. Dufossé L. Microbial production of food grade pigments. *Food Technol. Biotechnol.* **44**: 313–321, 2006.
2. Stahmann K–P. Vitamins. In: *The Mycota. X. Industrial Applications.* Springer-Verlag, Berlin-Hiedelberg, 2002, pp. 231–246.

Further Reading

1. Francis FJ. Less common natural colorants. In: *Natural Food Colorants*, eds. GAF Hendry and JD Houghton. Blackie, London, New York, 1996, pp. 10–342.
2. Jay JM. *Modern Food Microbiology.* Van Nostrand Reinhold Co., New York, 1992.
3. Lin Y-L, Wang T-H, Lee MH and Su N-W. Biologically active components and nutraceuticals in the Monascus-fermented rice: A review. *Appl. Microbiol. Biotechnol.* **77**: 965–973, 2008.
4. Solomons GL. Production of biomass by filamentous fungi. In: *Comprehensive Biotechnology Vol. 3*, Chapter 22, eds. HW Blanch, S Drew and DIC Wang. Pergamon, Oxford, 1985, pp. 483–506.

Websites

http://ift.confex.com/ift/2001/techprogram/paper_7150.htm
http://aem.asm.org/cgi/content/abstract/60/7/2627
http://en.wikipedia.org/wiki/Lactic_acid
http://www.fgsc.net/asilomar/citric.html
http://www.cyanotech.com/pdfs/axbul60.pdf
http://www.biomatnet.org/secure/Fair/S1015.htm

Questions for Thought...

1. What was the reason that industrial production of microbial primary metabolites had been considered impractical?
2. Today, almost all the amino acids, which are primary metabolites, are industrially produced by microbial fermentation. Outline the biotechnological developments that have made this possible.

3. Discuss the scientific know-how, which lead to the industrial production of citric acid by microbial fermentation.
4. Many ingredients are produced by modifying cultural conditions to maximize production, discuss and compare cultural methods used in the production of riboflavin and monosodium glutamate.
5. Discuss the metabolic processes that are the foundation for the production of riboflavin and consider the genetic engineering required to improve production.
6. Compare and contrast the biology of the microorganisms used in the production of citric acid and lactic acid.

Chapter 13

Enzyme Modified Food Products

Desmond K. O'Toole
Department of Biology and Chemistry
City University of Hong Kong
83 Tat Chee Avenue, Kowloon, Hong Kong, China

Lee Yuan Kun
Department of Microbiology, Yong Loo Lin School of Medicine
National University of Singapore
5 Science Drive 2, Singapore 117597
Email: micleeyk@nus.edu.sg

The discovery of enzymes and their isolation opened up the possibility of using the enzymes to modify foods. Microbial enzymes can be produced cheaply in large quantities and because microorganisms grow under a wide range of environmental conditions, enzymes that can be used in relatively extreme conditions are available. Enzymes have been isolated to modify fats, carbohydrates and proteins as well as in other applications.

13.1. Oils and Fats

Oils and fats are triacylglycerols (three fatty acyl groups bound to a glycerol molecule by ester bonds). In the presence of water, microbial lipases (glycerol ester hydrolases EC 3.1.1.3) catalyze the hydrolysis of oils and fats to yield free fatty acids, partial glycerides and glycerol. The reaction is reversible under the condition of a limited amount of water, where lipases catalyze the formation of acylglycerols from glycerol and

Action of group I lipases

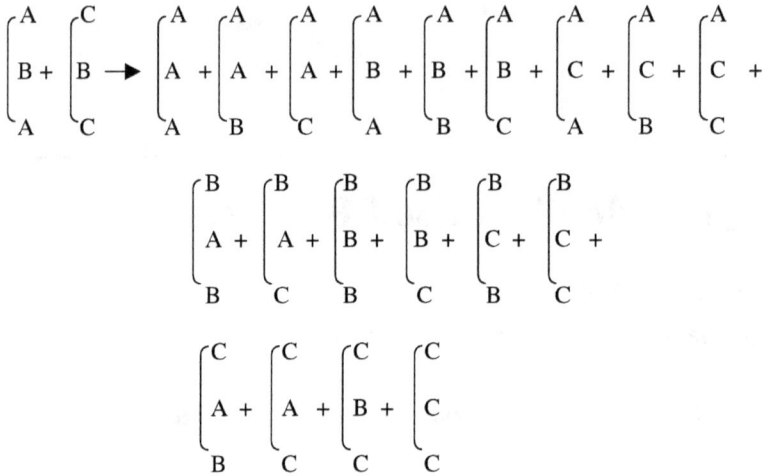

$$\begin{bmatrix} A \\ B \\ A \end{bmatrix} + \begin{bmatrix} C \\ B \\ C \end{bmatrix} \rightarrow \begin{bmatrix} A \\ A \\ A \end{bmatrix} + \begin{bmatrix} A \\ A \\ B \end{bmatrix} + \begin{bmatrix} A \\ A \\ C \end{bmatrix} + \begin{bmatrix} A \\ B \\ A \end{bmatrix} + \begin{bmatrix} A \\ B \\ B \end{bmatrix} + \begin{bmatrix} A \\ B \\ C \end{bmatrix} + \begin{bmatrix} A \\ C \\ A \end{bmatrix} + \begin{bmatrix} A \\ C \\ B \end{bmatrix} + \begin{bmatrix} A \\ C \\ C \end{bmatrix} +$$

$$\begin{bmatrix} B \\ A \\ B \end{bmatrix} + \begin{bmatrix} B \\ A \\ C \end{bmatrix} + \begin{bmatrix} B \\ B \\ B \end{bmatrix} + \begin{bmatrix} B \\ B \\ C \end{bmatrix} + \begin{bmatrix} B \\ C \\ B \end{bmatrix} + \begin{bmatrix} B \\ C \\ C \end{bmatrix} +$$

$$\begin{bmatrix} C \\ A \\ B \end{bmatrix} + \begin{bmatrix} C \\ A \\ C \end{bmatrix} + \begin{bmatrix} C \\ B \\ C \end{bmatrix} + \begin{bmatrix} C \\ C \\ C \end{bmatrix}$$

Action of group II lipases

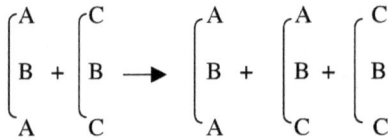

$$\begin{bmatrix} A \\ B \\ A \end{bmatrix} + \begin{bmatrix} C \\ B \\ C \end{bmatrix} \rightarrow \begin{bmatrix} A \\ B \\ A \end{bmatrix} + \begin{bmatrix} A \\ B \\ C \end{bmatrix} + \begin{bmatrix} C \\ B \\ C \end{bmatrix}$$

Action of group III lipases

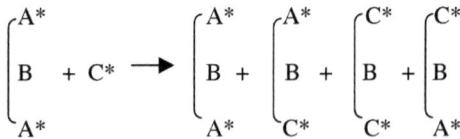

$$\begin{bmatrix} A^* \\ B \\ A^* \end{bmatrix} + C^* \rightarrow \begin{bmatrix} A^* \\ B \\ A^* \end{bmatrix} + \begin{bmatrix} A^* \\ B \\ C^* \end{bmatrix} + \begin{bmatrix} C^* \\ B \\ C^* \end{bmatrix} + \begin{bmatrix} C^* \\ B \\ A^* \end{bmatrix}$$

A, B, C, represent different fatty acyl groups, A* and C* represent long-chain fatty

acids containing a *cis* double bond in the 9 position.

Fig. 13.1. The various reactions carried by the three groups of lipases that attack lipids resulting in inter-esterification of the fats. Through the action of these enzymes, fats can be tailored to meet specific characteristics.

free fatty acids. At a water content of about 10% of the weight of glycerol and free fatty acids, lipases catalyze the hydrolysis and re-synthesis of ester bonds but for some lipases the reactions can also occur under virtually anhydrous conditions. The process achieves an exchange of fatty acyl groups between glycerol molecules in a mixture of different fats or a mixture of fat and free fatty acids, and the reaction is called **inter-esterification**. Inter-esterification is used in the food industry to alter the composition and therefore the physical and nutritional properties of oils and fats.

The carbons on the glycerol in natural triacylglycerols are numbered due to the chiral nature of the molecules and the stereospecific reaction products. In natural triacylglycerols at *sn*-1 (stereochemical number 1) saturated fatty acids predominate, at *sn*-2 unsaturated fatty acids are found and at *sn*-3 a variable range of fatty acids occur.

Based on their substrate specificity, microbial lipases can be classified into three groups, as shown in Fig. 13.1.

Group I

These are lipases that show no specificity both as regards to the position on the glycerol molecule and the nature of the fatty acid they attack. These lipases catalyze random exchange of fatty acyl groups between glycerol molecules in a mixture of two fats. Organisms known to produce them include *Candida cylindracae*, *Propionibacterium acnes* or *P. parvum* (formerly *Corynebacterium acnes*) and *Staphylococcus aureus*.

Group II

These are lipases that catalyze reactions only at the *sn*-1 and *sn*-3 positions of the acylglycerols. That is they catalyze selective exchange of fatty acyl groups at the 1- and 3-positions. Organisms known to produce them include *Aspergillus niger*, *Mucor javanicus* and *Rhizopus arrhizus*.

Group III

These are lipases that selectively attack esters of long-chain fatty acids containing a *cis* double bond in the 9 position, such as oleic, palmitoleic, linoleic and linolenic acids. An organism known to produce them is *Geotrichum candidum*.

Other organisms whose lipases have proved useful include *Can. antarctica, "Chromobacter viscosum"* (invalid name but published as strain NRRL B-3673; maybe *"Chromobacterium viscosum"* ATCC 6918), *Rhizopus delemar* and some *Pseudomonas* species.

Lipase catalyzed inter-esterification reactions can be performed either batch wise or continuously in **stirred tank reactors** or continuously in **packed-bed reactors** and **membrane bioreactors**. The catalyst is often immobilized on solid supports, such as kieselguhr, and the fat is extracted by acetone precipitation or by solvent evaporation. The catalyst particles can be reused for up to ten successive batch reactions or up to 600 hours in a continuous process, as long as the oils are saturated with water. Solvents such as petroleum ether are sometimes added to facilitate mixing in the oil phase.

13.1.1. *Cocoa butter equivalent*

An example in which lipase-catalyzed inter-esterification is used to modify edible fats and oils is in the production of the high value cocoa butter equivalent from a cheap starting material such as palm oil. Cocoa butter from cocoa bean has a high content of 1(3)-palmitoyl-3(1)stearoyl-2-monoolein (POS) and 1,3-distearoyl-2-monooleine (SOS) which gives the fat a sharp melting point of around 32–35°C. Consequently a bar of chocolate candy made with this fat will only melt in the mouth but not at room temperature or in one's hand. Supply of cocoa butter is limited so it is expensive. By mixing the mid-fraction of palm oil, which is rich in POP with the solid fraction of palm oil, which is rich in SSS (tristearine) in a suspension with Group II lipase, a product that is enriched in POS and SOS is produced as shown below. POS and SOS can be isolated by conventional fat fractionation techniques such as countercurrent liquid-liquid extraction and crystallization from solvent.

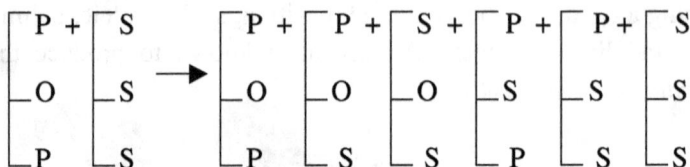

$$
\begin{bmatrix} P \\ O \\ P \end{bmatrix} + \begin{bmatrix} S \\ S \\ S \end{bmatrix} \rightarrow \begin{bmatrix} P \\ O \\ P \end{bmatrix} + \begin{bmatrix} P \\ O \\ S \end{bmatrix} + \begin{bmatrix} S \\ O \\ S \end{bmatrix} + \begin{bmatrix} P \\ S \\ P \end{bmatrix} + \begin{bmatrix} P \\ S \\ S \end{bmatrix} + \begin{bmatrix} S \\ S \\ S \end{bmatrix}
$$

A mixture of POP and stearic acid can also produce the same results (Fig. 13.2).

13.1.2. *Betapol*

This is the name of a human milk fat substitute developed by Loders Croklaan. The fat in cow's milk incorporated into infant milk formula is not as nutritious as human milk and results in reduced mineral and fat absorption as well as hard stools that can cause the development of constipation and blocked bowels in bottle-fed babies. The fat in human milk does not cause this because it contains a high quantity of palmitic acid at the *sn*-2 position and at the *sn*-1 and *sn*-3 positions the unsaturated oleic acid predominates, represented as OPO. Betapol is rich in palmitic acid at the *sn*-2 position and rich in oleic acid.

To produce OPO, tripalmitin rich vegetable oil and oleic acid are mixed and passed through a packed-bed reactor loaded with an immobilized *Rhizomucor miehei* lipase at 70°C under practically anhydrous conditions. At equilibrium, three triacylglycerols result, namely PPP, OPP and OPO, the desired product. The time to equilibration depends on the substrate

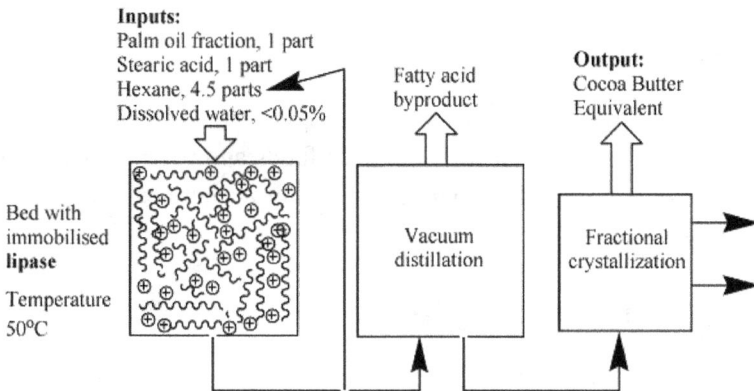

Inputs:
Palm oil fraction, 1 part
Stearic acid, 1 part
Hexane, 4.5 parts
Dissolved water, <0.05%

Bed with immobilised **lipase**

Temperature 50°C

Fatty acid byproduct

Vacuum distillation

Output:
Cocoa Butter Equivalent

Fractional crystallization

Fig. 13.2. The manufacture of cocoa butter equivalent using a Group II Lipase immobilized on a substrate. The palm oil fraction and stearic acid dissolved in hexane with a minimum amount of water are passed through the bed at 50°C with a residence time of about 10 min. The hexane is recovered and free fatty acids from the palm oil separated out by vacuum distillation. Then using fractional crystallization the product is separated from other fats produced during the reaction.

molar ratio (oleic to tripalmitin); the higher the ratio the longer the time and the higher the maximum incorporation. In the industrial setting two consecutive packed-bed reactors are used and after the first pass the free fatty acids released are removed by distillation and the product is passed through the second reactor with the addition of further oleic acid. Distillation, fractionation and refining follow to produce the final product.

13.1.3. *Low joule (calorie) fats*

Not all fatty acids are easily absorbed during digestion. A reduced calorie fat product has been produced by replacing some of the long chain fatty acid, behenic acid (C22:0) with short chain fatty acids. Procter and Gamble Company inter-esterify coconut, palm kernel and rapeseed oils to produce Caprenin. The resulting oil contains C8:0, C10:0 and C22:0 fatty acids. Because the behenic acid interferes with fat absorption, the actual calories delivered nutritionally are only 5 kcal per g versus 9 kcal per g for normal fat.

Diacylglycerols cannot be deposited in the adipose tissues. Designer cooking and salad oil based on diacylglycerols may prevent accumulation of fat in the body. Diacylglycerols can be produced by time-controlled hydrolysis of oil using Group I or Group II lipases.

13.2. Polysaccharides

Enzymes are used to breakdown the major digestible polysaccharide, namely starch, for use in a wide range of foods and applications, as well as to modify and produce non-digestible polysaccharides with desirable physical properties.

13.2.1. *High fructose corn syrup*

Sweeteners derived from starch, e.g. cornstarch, contribute to more than a third of the worldwide sweetener market. The process involves hydrolysis of starch by α-amylase and glucoamylase to dextrose monomer (glucose) and subsequent isomerization of glucose to fructose using immobilized glucose isomerase (Fig. 13.3). Fructose is twice as sweet as sucrose from cane and beet sugar.

In the first step, the heat gelatinized starch slurry (30–35% w/w) is liquefied and dextrinized by a bacterial thermostable α-amylase in a

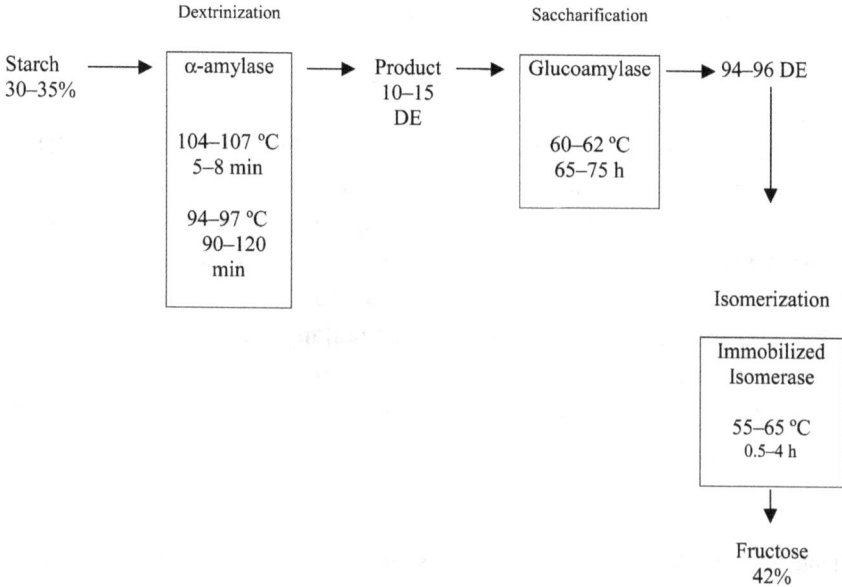

Fig. 13.3. Production of high fructose syrup from corn starch.

continuous two-stage reaction. Starch slurry (0.05–0.1%w/w starch) together with the enzyme is incubated at 104–107°C for 5–8 minutes, resulting in the formation of 0.5–1.5 dextrose equivalent (DE) hydrolysate. Pure dextrose solution has a unit of 100. Lowering the temperature to 94–97°C and holding for 90–120 minutes, results in a hydrolysate of 10–15 DE.

During saccharification, the hydrolysate is further converted to dextrose by the action of a fungal glucoamylase (at 1 L ton^{-1} starch). Commonly, a temperature of 60–62°C and a residence time of 65–75 hours are used to obtain hydrolysate containing 94–96% dextrose.

The 94–96% dextrose hydrolysate is then passed through a column of immobilized isomerase (55–65°C), with a residence time of 0.5–4 hours. The activity of the enzyme decays in a nearly exponential manner over a period of several months (140–240 days). Thus, the residence time for a new enzyme column is short, and then the flow rate through the column has to reduce progressively to give a longer residence time compensating for the lower enzyme activity. As the conversion of glucose to fructose is a reversible reaction with an equilibrium constant of about 1.0 at 60°C, a practical conversion level of 42% fructose can be achieved.

13.2.2. Non-digestible oligosaccharides (NDOs)

Oligosaccharides are saccharides containing between 3 and 19 sugar moieties. Some of these large saccharides cannot be hydrolyzed or absorbed in the upper part of our gastrointestinal tracts. These NDOs eventually reach the colon and become substrates for particular bacteria, such as species of *Bifidobacterium* and *Lactobacillus*. Thus, NDOs are **prebiotics** that selectively enhance the growth of indigenous probiotic bacteria in the colon. They are classified, for food labeling, as dietary fiber, and are used by the food industry as bulking agents, sugar substitutes and fat replacers besides being functional food ingredients. Examples of commercially available NDOs and the enzymes used in their production are shown in Table 13.1.

Table 13.1. Production of oligosaccharides and the substrates and enzymes used to produce them.

Oligosaccharide	Substrate	Enzyme used
Produced by hydrolysis of polysaccharides:		
Fructo-oligosaccharides, Fructofuranosyl (Fn) type and non-reducing glucosyl (GFn) type	Inulin	β-fructofuranosidase
Produced by transglycosylation reactions:		
Fructo-oligosaccharides (GFn) type	Sucrose	Fructosyltransferase
β-galacto-oligosaccharides	Lactose	β-galactosidase
Genito-oligosaccharides	Starch/maltodextrins	β-glucosidase
Glucosylsucrose	Maltose/sucrose	Cyclomaltodextrin-glucanotransferase
Isomalto-oligosaccharides	Starch/maltose	α-glucosidase
Lactosucrose	Lactose/sucrose	Levansucrase
Palatinose (isomaltulose)	Sucrose	Sucrose-6-glucosylmutase

13.2.3. *Plant pectins*

Pectin consists of linear polymers of α-1,4-linked galacturonic acid residues of which up to 70% are methoxylated and which have neutral sugar side branches. Pectins act as adhesives in plant cell walls and lamellae. Three groups of enzymes degrade native pectins; (i) polygalacturonases are hydrolytic enzymes that cleave linkages randomly and release galacturonate from non-reducing ends; (ii) pectin lyases that cleave, by a trans-elimination mechanism, 1,4-α-glycosidic linkages; and (iii) pectin esterases that cleave off the methoxy side groups.

Fruit juices and fruit by-products from the fruit processing industry contain pectins. Natural juice, e.g. from apple, cherry and raspberry, contains pectins that render the juices viscous and turbid so that the juices cannot be easily filtered. To change these characteristics fungal enzymes have been added to juices since the 1930s. The enzymes of choice, particularly from *Aspergillus* sp., are the pectin lyases because methanol (a toxin) is produced by pectin esterases and polygalacturonases tend to produce pectin precipitates.

Pectin extracted from apple pomace and citrus peel is used as a gelling agent and can be extracted by boiling in weak acid and eventually precipitating the pectin with salts. However, chemical extraction of pectin from mandarin orange peels is too harsh so the peels are inoculated with a fungus, *Trichosporin penicillatum*, that releases the pectin within about 24 hours.

13.3. Proteins

Proteases that catalyze the hydrolysis of protein change the solubility and functional properties of protein and food. Three groups of microbial proteases are widely used in the food industry; they are the alkaline, acid and neutral proteases.

13.3.1. *Alkaline proteases*

Alkaline proteases are serine proteases that contain both serine and histidine at their active site. Alkaline proteases for food use are produced by bacilli, e.g. *Bacillus licheniformis*, and by aspergilli. The enzymes have an optimum pH of about 8.0, are used to produce highly functional soy protein hydrolysate with good whipping characteristics, and a highly soluble hydrolysate, used in protein fortified soft drinks as a dietetic beverage. The enzymic reaction is stopped by reducing the pH to 4.0.

13.3.2. *Acid proteases*

The acid or carboxyl proteases generally have pH optima in the range of 3–4. An example is the microbial rennet produced by *Mucor miehei* and *M. pusillus*. Rennet causes coagulation of milk for cheese making, by hydrolysis of a specific bond in κ-casein. The traditional source of rennet is the inner lining of the fourth stomach of bovine calves.

Other important acid proteases used in food processing are pepsin and pepsin-like enzymes from *Aspergillus* and *Rhizopus* species. Pepsin in combination with alkaline proteases is being used to produce soy hydrolysates with good whipping characteristics.

13.3.3. *Neutral proteases*

Neutral proteases of *Bacillus subtilis* and *Aspergillus oryzae* are metallo-proteases with a metal ion involved in their catalytic reaction. The enzymes have a pH optimum near 7.0. *B. subtilis* neutral protease enhanced flavor intensity of cheese without producing bitterness. Thus, this protease in combination with lipases is used to produce strongly flavored cheese by adding the enzymes to scalded curds and then curing at 10–25°C for 1–2 months. This flavor enhancing approach that uses proteases has led to the development of enzyme modified cheese (EMC). This is a cheese flavor enhancing process in which proteases and lipases are added to a slurry of cheese and incubated for several weeks under controlled conditions. EMC pastes of Blue, Cheddar, Cream, Edam, Mozzarella, Parmesan, Provolone and Romano cheeses are commercially available. The EMC pastes are incorporated into processed cheeses (cheese produced by mixing and heating natural cheeses with flavors and emulsifiers) to improve flavor.

Proteases of *A. oryzae* are also used in bread making. The fungal proteases include both endo- and exo-peptidase. Endopeptidase modifies the viscoelastic properties of the dough by hydrolyzing interior gluten peptide bonds, thus improving the dough handling properties and increasing loaf volume. Exopeptidase releases amino acids from gluten, that react with sugars through the Maillard reaction during baking, thereby contributing to flavor and crust color.

13.4. Amino Acid Production

Apart from the straight fermentative production of amino acids, they can also be produced from non-food sources using microbial enzymes. Amino acids are racemic and contain a chiral center and the biologically important amino acids are the L- isomers. When chemical synthesis is used to produce amino acids, a racemic mixture inevitably results. Using microbial enzymes the racemic mixture can be converted to the L-isomer (Fig. 13.4).

13.4.1. Methionine

The first application of this approach was developed in the 1950s, to produce L-methionine. N-acetyl-DL-methionine is synthesized using conventional chemical methods. Cells of *A. oryzae* that produce an N-acylase specific for the L-isomer of methionine were used to hydrolyze the racemic mixture to produce L-methionine and N-acetyl-D-methionine. L-methionine was then precipitated from the supernatant by adjusting the

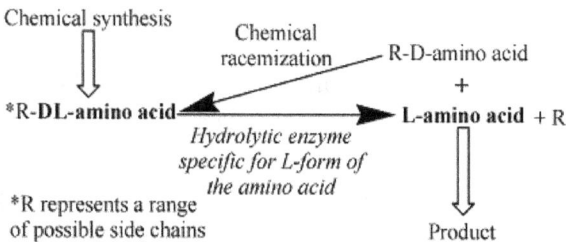

Fig. 13.4. The production of L-amino acids from a racemic mixture using an enzyme. Following chemical synthesis during which the DL mixture of the amino acid is produced linked to an acyl group, the product is then hydrolyzed by an N-acylase enzyme that releases the acyl group from the L-amino acid. The R-D-amino acid is unchanged but it can be isolated and chemically racemized resulting in the formation of another racemic mixture on which the N-acylase can act to release more L-amino acid. Spontaneous racemization is also possible thus eliminating a separate chemical racemization step.

pH to the isoelectric point of the amino acid. The remaining N-acetyl-D-methionine was then chemically racemized and the whole process repeated. The whole cells produced many more enzymes than the specific enzyme required so whole cells were abandoned and the requisite enzyme isolated and immobilized on a basic ion-exchange resin in a column. When the reaction mixture is passed through the column the pH of the mixture drops as the free amino acid is released, consequently the pH of the column must be managed to prevent precipitation of the free L-methionine in the column. A similar approach is used for transforming other racemic amino acid mixtures.

13.4.2. *Cysteine*

L-cysteine has traditionally been produced by acid hydrolysis of keratin, mostly from hair. This method is being replaced by a chemical synthesis method involving stereospecific hydrolysis followed by racemization. In the chemical synthesis stage methyl-2-chloroacrylate is converted to DL-amino-Δ_2-thiazoline-4-carboxylate. An enzyme from either *Pseudomonas thiazolinophilum* (invalid name), *Delftia acidovorans* (formerly known as *P. desmolytica*) or *Micrococcus luteus* (formerly known as *Sarcina lutea*), is then used to hydrolyze the compound and release L-cysteine. Racemization of the D-form occurs spontaneously in the reaction mixture. The gene for the enzyme involved in the hydrolysis has been isolated, characterized and cloned into an *Escherichia coli* strain.

Further Reading

1. Akoh CC and Xu X. Enzymatic production of betapol and other specialty fats. In: *Lipid Biotechnology*, eds. TM Kuo and HW Gardner. Marcel Dekker, New York, 2002, pp. 461–478.
2. Crittenden RG and Playne MJ. Production, properties and applications of food-grade oligosaccharides. *Trends Food Sci. Technol.* 7: 353–361, 1996.
3. Marangoni AG and Rousseau D. Engineering triacylglycerols: The role of interesterification. *Trends Food Sci. Technol.* 6: 329–335, 1995.
4. Panyam D and Kilara A. Enhancing the functionality of food proteins by enzymatic modification. *Trends Food Sci. Technol.* 7: 120–125, 1996.

5. Poutanen K. Enzyme: An important tool in the improvement of the quality of cereal foods. *Trends Food Sci. Technol.* **8**: 300–306, 1997.
6. Roberts SM, Turner NJ, Willetts AJ and Turner MK. *Introduction to Biocatalysis Using Enzymes and Microorganisms.* Cambridge University Press, Cambridge, New York, 1995.
7. Voragen AGJ. Technological aspects of functional food-related carbohydrates. *Trends Food Sci. Technol.* **9**: 328–335, 1998.

Questions for Thought...

1. Public health authorities are becoming more concerned about the presence of trans fatty acids in hydrogenated oils. Consider how you might overcome this problem using enzymes.
2. In this chapter we show how enzymes are used to produce sweeteners from corn starch. However, where else are the same enzymes used for other processes in this part of this book?
3. What are the roles of the enzymes that breakdown protein in food production systems?
4. If you invented a new food based on the introduction of a foreign gene into a different microorganism that coded for enzymes that broke down proteins, what would you have to do to have it accepted in the USA and Europe?

Annex

Regulation of Foods Involving Genetically Engineered Microorganisms

Desmond K. O'Toole and Lee Yuan Kun

The European Union **Novel Food Regulation** (1997) provides a framework for the safety evaluation of foods developed and manufactured using novel technologies and also for new food sources. The regulation covers all foods and food ingredients that "have not hitherto been used for human consumption to a significant degree within the Community," even if they have a long history of safe use in other parts of the world. Thus, novel foods involving microorganisms can be categorized into three groups.

(1) Genetically modified viable microorganisms defined by the Directive 90/220/EEC. An example is genetically modified lactic acid bacteria used in fermented milk.
(2) Foods produced from, but not containing, genetically modified microorganisms (Art. 1.2. a). An example is extracellular products of genetically modified microorganisms.
(3) Foods isolated from microorganisms (Art. 1.2. d). These include new products based on single-cell protein, such as the fungus *Fusarium graminearum*, and omega-3 fatty acids derived from algae, such as *Crypthecodinium cohnii* and *Nitzschia alba*.

Novel foods can be put into the market without formal approval if they can be shown to be substantially equivalent to their existing counterparts as regards to their composition, nutritional value, metabolism, intended use and content of undesirable substances. Substantial equivalence can be demonstrated either on the basis of generally recognized scientific evidence or on the basis of an opinion delivered by one of the EU member states' competent food assessment bodies.

Non-equivalent novel foods would need the authorization of the member state in which the product is intended to be marketed, and a copy of the request must be submitted to the EC Commission (Art. 4). The request must contain all relevant information on studies performed to demonstrate that the food does not present a danger to, or mislead, the consumer, and also that it does not differ from food that it is intended to replace to such an extent that its normal consumption would be nutritionally disadvantageous. Thus, a case-by-case evaluation of their safety, to include toxicological, microbiological and nutritional considerations, is necessary.

The Novel Food Regulation cross links the Deliberate Release Directive for Genetically Modified Microorganisms (90/220/EEC Art. 9) and the Plant Variety Directives (70/457/EEC and 70/458/EEC Art. 3) to avoid duplication of procedures.

The Novel Food Regulation requires specific labeling of "any characteristic or food property such as composition, nutritional value or nutritional effects, or intended use which renders the food no longer equivalent" to its conventional counterpart (Art. 8). Thus, if a genetic modification leads to a novel protein, all foods and ingredients containing this protein should be labeled as such. However, labels are not required for hydrolyzed protein preparations, in which the modified protein is degraded, or for refined oils, which would no longer contain significant amounts of the protein.

The Directive 90/220/EEC also implies that a new *Lactobacillus* strain developed using traditional genetic tools (e.g. mutation and selection) falls completely outside the scope of the regulation and would require neither a formal safety evaluation nor a label. A strain with the same modification, but achieved using targeted genetic modification (e.g. recombinant DNA technology), would need both an authorization and a label.

The US Food and Drug Administration (FDA) requires that genetically modified foods meet the same safety standards for all other foods, based on existing food law. US law places responsibility on producers and sellers to offer only safe food products to consumers. Thus, pre-market approval for any new substance that is added to a food, irrespective of whether or not it is a product of genetic engineering is not required.

Further Reading

1. Ager B, revised Moses V. The regulation of biotechnology in Europe. In: *Biotechnology: The Science and the Business*, 2nd ed., Chapter 9, eds. V Moses and RE Cape. Harwood Academic Publishers, The Netherlands, 1999, pp. 123–132.

2. Huggett AC and Conzelmann C. EU regulation on novel foods, consequences for the food industry. *Trends Food Sci. Technol.* **8**: 133–139, 1997.

3. Nicholas RB, revised Springham DG. The regulation of biotechnology in the United States. In: *Biotechnology: The Science and the Business*, 2nd ed., Chapter 8, eds. V Moses and RE Cape. Harwood Academic Publishers, The Netherlands, 1999, pp. 113–122.

4. Robinson C. Genetically modified foods and consumer choice. *Trends Food Sci. Technol.* **8**: 84–88, 1997.

5. Robinson C. Understanding the commercial and regulatory issues for genetically modified and novel foods and food ingredients. *Trends Food Sci. Technol.* **9**: 83–86, 1998.

6. Turner SM, Turner NJ, Willetts AJ and Turner MK. *Introduction to Biocatalysis Using Enzymes and Micro-organisms*. Cambridge University Press, New York, 1995.

Websites

- Agriculture and Agri-Food Canada
  http://aceis.agr.ca
- Australia CSIRO Library
  http://www.dfst.csiro.au/fdnet20a.htm
- Codex Alimentarius Home page
  http://foodnet.fic.ca/regulat/codex.html
- Colorants: An opinion on riboflavin as a coloring matter authorized for use in foodstuffs produced by fermentation using genetically modified *Bacillus subtilis*
  http://europa.eu.int/comm/food/fs/sc/scf/out18_en.html
- EU COST Action 99 (Food Consumption and Composition Data Research)
  http://food.ethz.ch:2000/cost99.html
- FAO 50th Anniversary Internet Forum on Food Security
  http://FAO50.FSAA.ULAVAL.CA/english/start.html
- Food and Agriculture Organization of the United Nations (FAO)
  http://www.fao.org
- FoodCom News
  http://www.foodcom.com/foodcom

- FoodNet, Canada
  http://foodnet.fic.ca
- INFOODS, the UN University Food and Nutrition Program
  http://www.crop.cri.nz/crop/infoods/infoods.html
- International Food Information Council
  http://ificinfo.health.org
- International Organization for Standardization (ISO)
  http://www.hike.te.chiba-u.ac.jp/ikeda/ISO/home.html
- Food composition data from around the world
  http://food.ethz.ch/
- Leatherhead Food Research Association, UK
  http://www.worldserver.pipex.com/lfra
- MATFORSK (Norwegian Food Research Institute)
  http://128.39.177.3/Matforsk/engelsk/summary.htm
- MEATNET(links to other websites of interest to the meat sector)
  http://www.meatnet.nl/links.htm
- New Scientist
  http://www.newscientist.com
- Paul Singh's On-line Food Engineering Teaching Course
  http://nachos.engr.ucdavis.edu/-rpsingh/index.html
- Royal Society of Chemistry's Nutrition page
  http://chemistry.rsc.org/rsc/nuts.htm
- UK Government Information Service (with links to all Government departments)
  http://www.open.gov.uk/index.htm
- UK Institute of Food Research
  http://www.ifrn.bbsrc.ac.uk/information/IFR.html
- UK Institute of Food Research (food safety research information)
  http://www.ifrn.bbsrc.ac.uk/Information/safety.html
- UK Institute of Food Science and Technology's Home Page
  http://www.easynet.co.uk/ifst
- US Department of Agriculture (USDA) Food and Nutrition Information center
  gopher://gopher.nalusda.gov:70/11/infocntr/fnic
- USDA Food and Nutrition Software List
  gopher://gopher.nalusda.gov:70/11/infocntr/fnic/software

- US Institute of Food Technologists (IFT)
  http://www.ift.org
- US Food and Drug Administration (FDA) Center for Food Safety and Nutrition
  http://vm.cfsan.fda.gov/list.html

Part III

Microbes in Agrobiotechnology

Molecular biological methods have becoming increasingly applicable to the diagnosis of infectious diseases and vaccines development in veterinary medicine. The commonly employed microbiology and virology assays and the conventional serological test to identify microorganisms are often not specific and sensitive enough to make the diagnosis in a short period of time. However, with advances in diagnostic methods based on immunology and molecular biology technology, these methods are rapidly invading into laboratories as the preferred choice for diagnosis of animal diseases. The reasons are that these methods are user friendly, safe, cost effective, reproducible and some are automated to facilitate the evaluation of large numbers of specimens.

Molecular biology procedures also have a major impact on the preparation of veterinary vaccines. These "modern" vaccines did not only overcome the limitations of the "classical" vaccines but are also much safer and efficacious. It also stimulates an immune response distinguished from that of natural infection, so that immunization and eradication may proceed simultaneously. Such a vaccine is also commonly known as marker vaccines since it is able to "mark" the vaccinated animals from the naturally infected animals.

Plant biotechnology has opened a new era of agricultural revolution by introducing genes from different organisms to generate new varieties of plants — genetically modified (GM) plants. Among them, the new food crops (such as grain crops, vegetables and fruits) will have higher yield, better nutritional qualities, resistance against pests (insects, diseases, nematodes, weeds and abiotic stresses), others will help us to produce new industrial raw materials (oils, starches and biodegradable plastics) and pharmaceutical products. Despite the great economical and humanitarian potential, plant biotechnology has caused great concerns about the safety of the genetically modified (GM) foods, the environmental impact of GM plants, and moral issues. In this chapter, both technical and regulatory aspects of plant biotechnology are discussed in detail. The future success of plant biotechnology depends on science-based facts, proper governmental regulations and public education efforts.

Chapter 14

Microbes and Livestock

Jimmy Kwang and He Fang
Institute of Molecular Agrobiology
National University of Singapore
1 Research Link, Singapore 117604

We have charted an amazing growth rate in the human population in the last two centuries. From 1804 to within a hundred years or so, the world population had increased from one billion people to more than two billion. However, by 1974, within a span of only 50 years, the human population had doubled to four billion people. To date, there are more than seven billion people in the world, with the figure still growing daily (Fig. 14.1). This rapid increase signals some major problems, like the shortage of food to meet human demand.

Backyard farming used to be a common practice, but with the high cost of maintenance and spatial constraints on an individual, it is no longer practical for one to raise his own livestock for food. Due to these reasons, modernization has made way for large-scale industries to produce the necessary sustenance to cater for the growing population. Animals bred for consumption are kept in large-scale space-conserved housing, with automated animal farming to provide a more cost efficient means of rearing. However, this is where potential problems set in. When animals are kept densely in close proximity to one another, the entry of an infectious pathogen into these facilities can result in rapid transmission of the pathogen from one animal to the next within a short

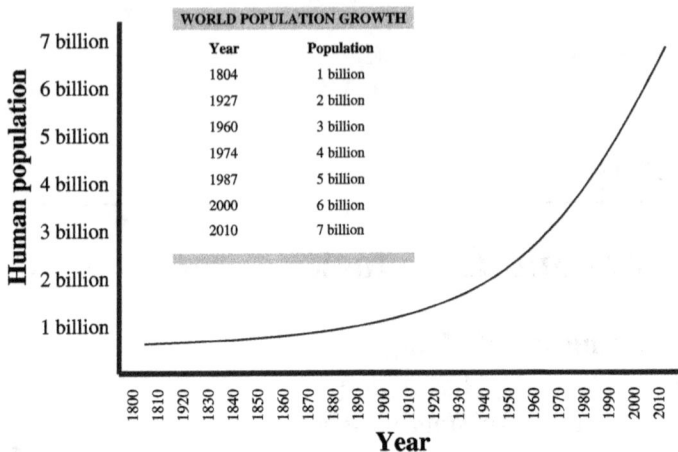

WORLD POPULATION GROWTH	
Year	Population
1804	1 billion
1927	2 billion
1960	3 billion
1974	4 billion
1987	5 billion
2000	6 billion
2010	7 billion

Fig. 14.1.   World population growth from year 1804 to 2010.

time, even causing mass mortality. This can result in big losses for the farms involved and also a major setback to the livestock industry. In 1996, the infamous mad cow disease or Bovine Spongiform Encephalopathy (BSE) led to the slaughter of six million cattle and inflicted worldwide paranoia over the consumption of beef from the UK. In 1997 alone, several epidemics had occurred, for instance Foot and Mouth Disease (FMD), which had resulted in the slaughter of four million pigs in Taiwan. Classical Swine Fever (CSF) caused millions of pigs in the Netherlands to be culled, and *Escherichia coli*: O157:H7 and *Salmonella* led to the loss of billions of dollars to the meat industry. In 1998, the emerging Nipah virus outbreak in Malaysia led to more than one million pigs being culled, and the banning of pork export to Singapore and other countries. The outbreak of FMD in the United Kingdom during early 2001 struck a devastating blow to the farm sector leading to the culling of millions of animals with a slump in exports and also the threat of spread to other European countries. This is in addition to the mad-cow disease that has already plagued some European countries.

These catastrophes could have been prevented with good management practices including effective vaccination procedures associated with the early detection of diseases before their transmission. The importance of vaccination and disease detection will be discussed.

14.1. Vaccination for Animals

Ancient Chinese observed that individuals affected by smallpox were resistant to repeated attacks of the disease. Therefore, attempts to protect against smallpox in those times had been practiced by inoculation using vesicle fluid from persons with mild forms of smallpox or by deliberate contact with diseased individuals. Subsequently, this technique called variolation spread westwards. "Variola" is the Latin word for "smallpox". During the second half of the 18th century, when smallpox claimed many victims during the epidemics in Europe, some of the milkmaids were found to be protected from smallpox. It was discovered that their immunity against smallpox was a result of their exposure to cowpox virus while milking the cows. Many years later, due to the efforts of the English physician Edward Jenner, who attempted the first immunization with cowpox in 1796 on a brave young man named James Phipps, smallpox was completely prevented. Jenner's experiments were followed up by Louis Pasteur who developed vaccines against fowl cholera and also a successful vaccine against rabies. The term "vaccination" derived from the Latin word "vacca" which means cow was introduced to replace the term variolation. Vaccination, which is fundamentally based on the concept that prevention is better than cure, was employed long before the mechanism of immune protection was known. To date, vaccination is the most successful veterinary measure, with more animal production safeguarded than through all other veterinary measures combined.

A range of microorganisms, including bacteria, ricketesia, viruses, fungi and protozoa, cause infectious diseases to animals. Within any one group of these pathogens, there is variation in the mechanism of disease production. Therefore, knowledge of the infectious mechanism of any pathogen is critical to understanding of the mechanisms of acquired resistance for formulating of an effective vaccine. As the pathogenesis and epidemiology of each disease varies, so does the role and efficacy of vaccination. While some vaccines may be highly efficacious, inducing an immunity that prevents clinical symptoms of the disease as well as infection and replication of the pathogen, others may only prevent clinical disease, but not infection and/or the development of the carrier state. In other cases, immunization may be completely ineffective or only able to reduce the severity of the disease. Therefore the application of vaccination as part of animal disease control strategy requires a thorough knowledge of the characteristics of the pathogen and its epidemiology, along with the characteristics and capabilities of various vaccines.

14.1.1. *The immune response*

The concept of immunity dates back many centuries, but the broad application of immunization is a more recent development and still a subject for continued experimentation. Vaccination is intended to induce an immunological defense response that will provide protection to the respective infectious agent. Therefore, as a general rule, the immune response should be of the same nature as that which results from a natural infection.

There are various cells in the blood, which include erythrocytes, granulocytes and lymphocytes. These cells not only have different morphology, but they also have totally different functions. Immune responses are carried out by lymphocytes, and the number of these cells only constitutes a small percentage (<0.2%) of the total number of blood cells. Immunological protection is provided primarily by humoral-mediated (B lymphocytes) and cell-mediated (T lymphocytes) arms of the immune system (Fig. 14.2). Two other major components of the immune system are complement and phagocytes.

All lymphocytes appear similar under the light microscope, ranging from 7 to 15 μm in diameter. The morphology, structure and staining characteristic of the cells do not provide clues to their complex roles within the body. Lymphocytes can only be differentiated and defined on the basis of their ontogeny, surface receptors, function and characteristic cell surface antigens. T cells are the predominant circulating lymphocytes, constituting 65–80% of the lymphocytes seen in peripheral blood. T cells are relatively long-lived cells, while B cells, in contrast, have a short life span.

B cells perform two important functions:

(1) They differentiate into plasma cells and produce antibodies. Antibody molecules are proteins called immunoglobulin (Ig) (Table 14.1). Ig may be classified on the basis of their electrophoretic mobility, their molecular weight, and their antigenic structure. Soluble antibodies (Fig. 14.3) are found in many body fluids but are present in highest concentrations and are most easily obtained from blood serum.

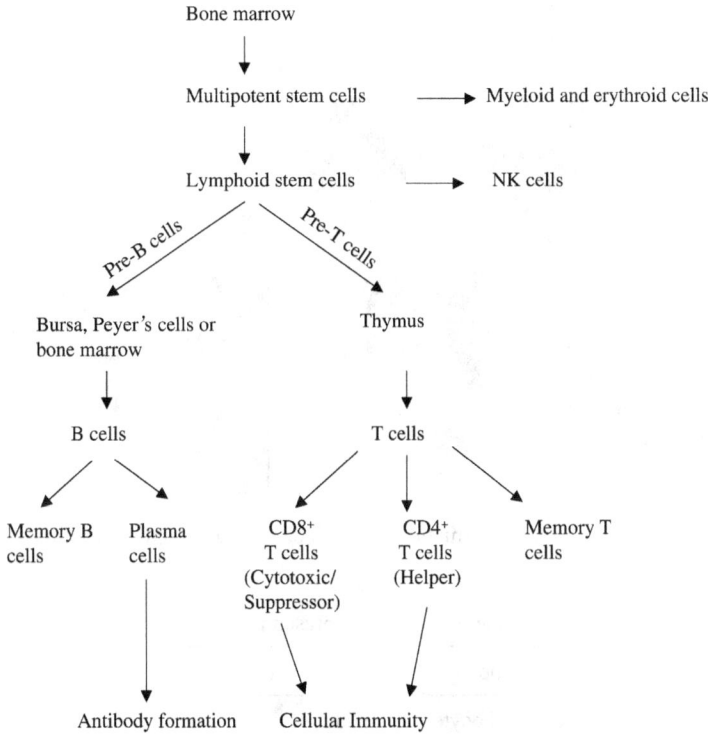

Fig. 14.2. Development of T and B lymphocytes in the body, which make up the arms of the immune system.

Table 14.1. Characteristics of the immunoglobulin isotypes in domestic mammals.

| Property | Immunoglobulin isotype | | | | |
	IgM	IgG	IgA	IgE	IgD
Sedimentation coefficient	19S	7S	11S	8S	7S
MW	900,000	160,000	360,000	200,000	180,000
Heavy chain	μ	γ	α	ε	δ
Mainly produced by	Spleen, lymph nodes	Spleen, lymph nodes	Intestinal, respiratory tracts	Intestinal, respiratory tracts	Spleen, lymph nodes

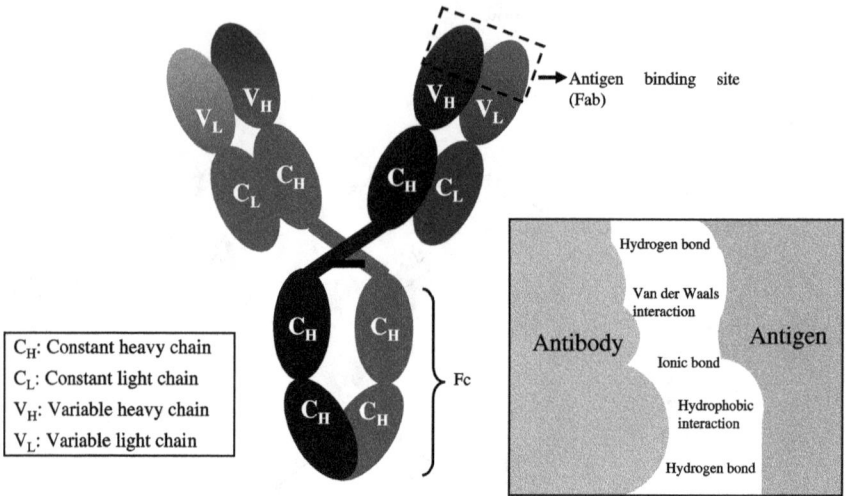

Fig. 14.3.   Antibody structure and antibody-antigen binding reaction.

Table 14.2. Antigen presenting cells.

Cell type	Distribution
B lymphocyte	Wide
Macrophage	Wide
Dentritic cells	Lymphoid tissues
Langerhans' cells	Skin
Kupffer's cells	Liver
Microglia cells	Central nervous system (CNS)

(2) They are antigen presenting cells (APC) (Table 14.2). B cells carry
    cell membrane-bound immunoglobulin molecules that function as
    antigen receptors to bind antigens (Fig. 14.4). In order for an antibody
    response to occur, the recognition of cell-membrane-bound antigens
    by T lymphocytes is required.

There are two predominant types of phagocytic cells: neutrophils and
macrophages. Macrophages exist both as free cells, e.g. monocytes and
fixed in tissues, e.g. Kupffer cells of liver. Neutrophils and macrophages

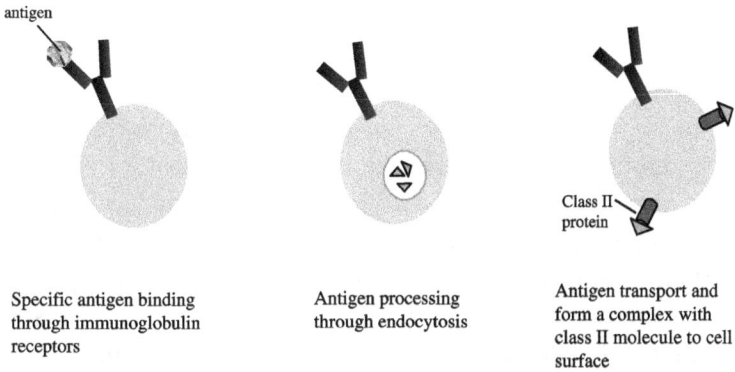

Specific antigen binding through immunoglobulin receptors

Antigen processing through endocytosis

Antigen transport and form a complex with class II molecule to cell surface

Class II protein

Fig. 14.4.   Antigen processing by antigen presenting cells.

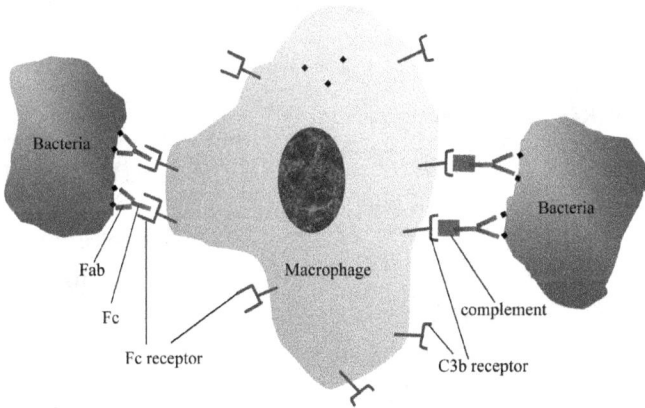

Bacteria

Bacteria

Fab

Macrophage

Fc

complement

Fc receptor

C3b receptor

Fig. 14.5.   Two mechanisms by which antibodies enhance phagocytosis of bacteria: (1) Antibodies link the bacteria to macrophage by means of their Fc region. (2) Binding of the antibodies to bacteria initiate activation of the complement (C3b), which can then bind directly to the surface of the bacteria and enhance its phagocytosis.

have been termed phagocytes because their membranes are equipped with two specialized receptors:

(1) receptor for the Fc region of IgG molecules;
(2) receptor for the third component of complement (C3).

These receptors facilitate the process of phagocytosis more efficiently by assisting in the ingestion of microorganisms with IgG or activated C3 on their surface, a phenomenon known as opsonization (Fig. 14.5).

T cells perform several important functions, which can be divided into two main categories, namely regulatory and effector. The regulatory functions are mediated primarily by CD4$^+$ helper T cells, which have the ability to augment B cell response and to amplify the cell-mediated response through a group of mediators called interleukins (Table 14.3). The term interleukin (IL-1 to IL-20) is applied to the molecules that function as communicators between lymphocytes. Another regulatory function is mediated by suppressor CD8$^+$ T cells, which down-regulate the immune response. Failure of such regulation may cause autoimmune diseases.

The effector functions are carried out primarily by cytotoxic (CD8$^+$) T cells, which kill virus-infected cells, tumor cells and allografts, as well as CD4$^+$ cells, which mediate delayed-type hypersensitivity (DTH) against intracellular organisms. T cells, B cells and macrophages must interact

Table 14.3. The interleukins.

Name	Mw (KD)	Mainly produced by	Major targets
IL-1	17	Macrophages	T cells, B cells
IL-2	15	T cells, NK cells	T cells, B cells
IL-3	28	T cells	Hematopoietic cells, stem cells
IL-4	20	T cells	T cells, B cells, mast cells
IL-5	18	T cells	B cells, T cells
IL-6	26	Fibroblasts, T cells	B cells
IL-7	25	Stromal cells	Immature lymphocytes
IL-8	8	Macrophages	T cells, neutrophils
IL-9	39	T cells	T cells, mast cells
IL-10	19	TH2 cells, B cells	T_H1 cells, macrophages
IL-11	24	Stromal cells	B cells
IL-12	75	Macrophages, dendritic cells	T cells, NK cells
IL-13	12	T cells	B cells
IL-15	17	Wide variety of cells and tissues	NK cells, T cells
IL-16	14–17	CD8+ T cells	CD4$^+$ T cells
IL-17	17.5	T cells	T cells
IL-18	22.3	Kupffer's cells	T_H cells
IL-19	?	Monocytes	T_H2 cells and/or B cells
IL-20	?	?	Keratinocytes

very closely to ensure that immune response occurs. In order for the cooperation between the three kinds of cells to occur, they must have common major histocompatibility complex (MHC) antigens. MHC antigens, encoded by three classes of genes, are highly polymorphic membrane glycoproteins expressed on body cells. Class I and II genes code for antigens associated with self or non-self discrimination and antigen recognition, and class III genes code for complement proteins. The complement is a collective term used to designate a group of chemically and immunologically distinct serum proteins that play a key role in the host defense process. Complement proteins are capable of interacting with each other, with antibodies and with cell membranes. These interactions lead to the generation of biological activity. The biological consequences of complement activation include:

(1) lytic destruction of cells, bacteria and viruses, allografts and tumor cells,
(2) regulation of the inflammatory and immune response, and
(3) opsonization.

Almost all nucleated cells express class I MHC molecules. On the other hand, class II molecules are only expressed in a few cell types, predominantly in antigen-presenting cells (APC) such as macrophages and B cells (Table 14.4). Although MHC class I and II molecules have subtle differences in their structural features, both molecules have highly polymorphic outer extracellular domains that form a long cleft in which peptide fragments are bound to. MHC class I molecules present peptides derived from endogenous proteins (e.g. fragments of viral proteins in infected cells) while class II molecules present peptides derived from exogenous proteins (e.g. pathogens that are internalized by phagocytic cells or pathogens living in vesicles of macrophages or B cells).

Class I and II MHC molecules are recognized by different subsets of functional T cells. Foreign peptides bound to MHC class I molecules on the cell surface are recognized by the CD8+ T cells, which in turn differentiate into cytotoxic T cells (CTL) that kill the infected cells. Cytotoxic T cell response or cell mediated immune response is very important in host defense against pathogens, predominantly viruses that survive in the cytosol. Cytotoxic T cells kill infected targets with great precision to limit the spread of these pathogens.

Table 14.4. Characteristics of T cells, B cells, and macrophages.

Features	T cell		B cell	Macrophage
	Helper	Cytotoxic		
Surface Markers				
CD4	+	−	−	−
CD8	−	+	−	−
Immunoglobulin (Ig)	−	−	+	+
MHC class I	+	+	+	+
MHC class II	−	−	+	+
Antibody synthesis	−	−	+	−
Cellular immunity	+	+	−	+
Receptor for IgG Fc	−	−	+	+
Receptor for C3b	−	−	−	+

Peptides presented on the MHC class II molecules are recognized by the CD4$^+$ T cells that can differentiate into T$_H$1 and T$_H$2 cells. Peptides derived from pathogens that are multiplying inside the intracellular vesicles tend to stimulate the differentiation of T$_H$1 cells, whereas peptides derived from extracellular pathogens tend to stimulate the differentiation of T$_H$2 cells (Figs.14.6 and 14.7). T$_H$1 cells activate the microbicidal properties of macrophages by inducing the fusion of their lysosomes with the intracellular vesicles housing the pathogens and at the same time induce B cells to produce IgG antibodies that are effective at opsonizing extracellular pathogens for uptake by other phagocytic cells. T$_H$1 cells also release cytokines that attract macrophages to the site of infection and thus are also known as inflammatory CD4 T cells, or delayed-type hypersensitivity T cells. T$_H$2 cells play an essential role in humoral response by activating naive B cells to produce IgM antibodies and subsequently stimulating B cells to proliferate and differentiate into cells capable of secreting different antibody isotypes. Hence, T$_H$2 cells are also termed helper T cells for their ability to help B cells make antibodies in response to antigenic challenge.

Although helper T cells are required to induce B cell response to protein antigens, certain microbial antigens, such as bacterial polysaccharides, are able to activate B cells in the absence of helper T cells. These antigens

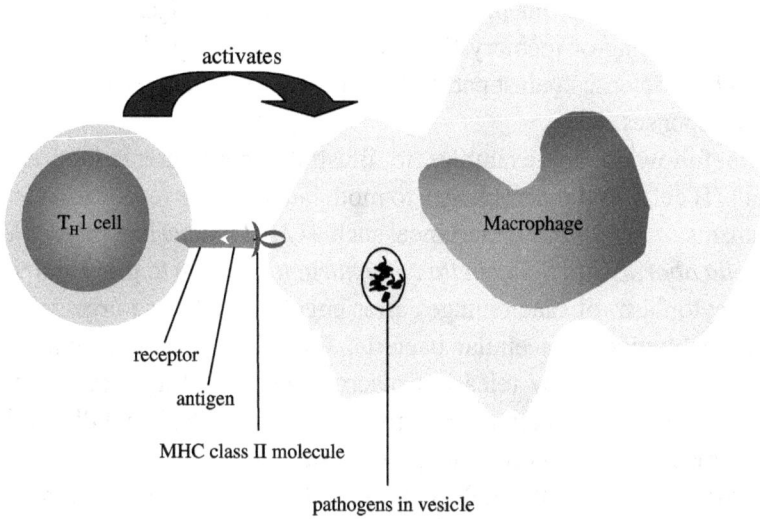

Fig. 14.6. $T_H 1$ cell recognizes fragment presented by MHC class II molecules on the surface of a macrophage containing pathogens, and activates the macrophage.

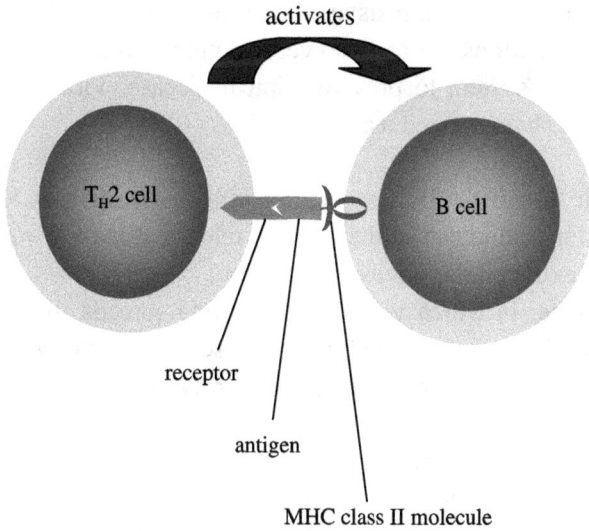

Fig. 14.7. $T_H 2$ cell recognizes fragment presented by MHC class II molecules on the surface of B cells, and activates the B cell.

are known as thymus-independent or TI antigens. TI antigens do not induce isotype switching or memory B cells but they certainly play an important role in host defense against pathogens whose surface antigens cannot elicit T cell responses.

The following are examples to illustrate these interactions among T cells, B cells, and macrophages to modulate immune responses. Certain organisms of veterinary importance, such as *Mycobacterium tuberculosis*, *Brucella abortus* and *Salmonella typhimurium*, are able to grow and thrive in the cytoplasm of macrophages after engulfment. These organisms are called facultative intracellular bacteria. When T_H1 cells encounter these bacterial antigens, they release a macrophage-activating factor that can enhance macrophage activity. Macrophages then destroy and digest these organisms into small pieces, and present the processed antigen in association with class I MHC molecules to T lympocyte. In addition, macrophages release biological mediators (such as interleukin), which further amplify the response of the T lymphocyte and thus greatly stimulate the immune response. Cell-mediated immunity associated with the activation of macrophages limits the process of infection. Since activated macrophages show a non-specific enhancement of their microbicidal activities, they confer enhanced resistance to bacteria in general.

Virus-host relationships are also very complex, which leads to pressure on the immune system to prevent clinical disease. Viral infections can sometimes be characterized as:

(1) cytolytic, in which viral association with a host cell may lead to virus replication with death of the cell and release of mature virus particles into the extracellular environment; or

(2) steady-state, in which virus replicates and spreads from cell to cell without killing the host cell. In this case, virus may cause persistent infection, latency, or neoplastic transformation.

In general, humoral immunity mediates immunity against cytolytic viruses whereas cell-mediated immunity is of primary importance with steady-state viruses. However, a role for antibodies, especially in preventing an initial infection, in resistance to steady-state viruses is recognized. For example, antibodies can effectively prevent influenza virus from

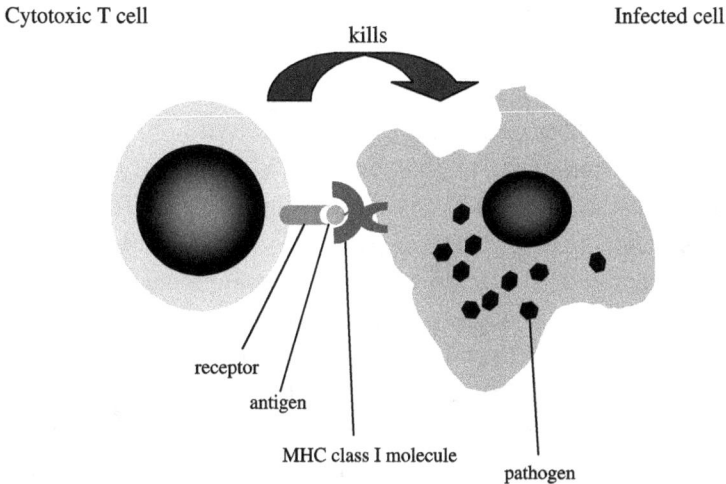

Fig. 14.8.   CD8⁺ cytotoxic T cell recognizes antigen presented by MHC class I molecules on the surface of an infected cell, and kills the infected cell.

infecting cells of the respiratory tract, but cell-mediated responses usually develop following infection and play a role in killing influenza virus infected cell. After cells are infected with the influenza virus, viral envelope glycoproteins, namely hemagglutinin and neuraminidase, appear on the surface of the infected cell in association with class I MHC protein. A cytotoxic T cell binds via its antigen-specific receptor to the viral antigen-class I protein complex and is stimulated to grow into a clone of cells by interleukin-2 produced by helper T cells. These cytotoxic T cells can then specifically kill influenza virus-infected cells by recognizing viral antigen-class I protein complexes on the cell surface (Fig. 14.8). Thus, in order to develop a vaccine against the flu virus, the viral envelope glycoproteins associated with the viral pathogenicity and would be a crucial ingredient.

The synthesis of antibodies directed against determinants on the pathogenic organism can protect the host in several ways. Sometimes the simple interaction of antibodies with the antigen accomplishes the response such as toxin neutralization, virus neutralization and opsonization of bacteria. In other cases, the interaction of antibodies and antigens in turn activate complements. In this case, the function of the antibodies is to confer specificity on the immune response, and the actual effectors of the response are the complement proteins, which induce lethal injury to the membrane of the invading organism. A few examples of these mechanisms are discussed below.

Humoral immunity by antibodies alone is of primary importance in preventing diseases caused by toxin-producing bacteria. Toxins are major virulence factors in many bacterial pathogens and are either secreted (exotoxins) or derived by bacterial autolysis (endotoxins). Exotoxins are among the most toxic substances known. A disease-causing exotoxigenic bacteria, for example, *Clostridium perfringens* (*C. perfringens*), is widely distributed in the soil and gastrointestinal (GI) tract of animals and is characterized by its ability to produce potent toxins that are responsible for enterotoxemia. The major toxins implicated in this infection have been identified and these are highly immunogenic. When treated with formaldehyde, these toxins lose their toxicity but retain their antigenicity. The inactivated toxins called toxoids stimulate protective antitoxin antibody production after injecting into animals. Vaccines based on toxoids are highly effective in controlling *Clostridal* infection in animals.

Another important virulence factor that determines the establishment of a disease is the ability of bacteria to adhere to surface of cells. The adherence mechanisms are essential for organisms such as enterotoxigenic *Escherichia coli* (*E. coli*) that attach to the specific receptors on mucosal epithelial cells. Humoral immunity expressed locally by antibodies can block adherence of *E. coli* to epithelial cell receptors preventing colibacillosis in nursing and weaning pigs. Bacterial adherence to epithelial cells is by means of pili or fimbriae which is the first step for successful establishment by *E. coli* on intestinal vili. Non-pathogenic strains do not possess pili and are washed by GI fluids and peristalsis and thus do not cause disease. There are multiple antigenic types of fimbriae of which K88, K99, 987P and F41 are most commonly recognized. Subunit vaccines containing the fimbrial proteins, responsible for attachment of the bacteria to the intestinal mucosal cells, when expressed and immunized to animals will promote antibody production to block bacterial adhesion. This is a novel approach to *E. coli* vaccination and indicates a shift away from the conventional whole bacterial vaccine to a more specific and defined product.

Mucosal immunity mediated by IgA antibodies is of particular importance as most animal pathogens initiate their infectious processes at mucosal surfaces of conjunctiva, nasopharynx, respiratory, gut, and genito-urinary tracts. The antibody response to the pathogens (e.g. viral particles) when directed against surface proteins can be an effective way to neutralize infectivity. This is brought about by binding of the neutralizing antibody to the surface protein responsible for adsorption of the virus to the target cells. A specific example to highlight the importance of mucosal immunity in the protection of piglets is the passive immunity derived from the sow against Transmissible Gastroenteritis Virus (TGEV). The virus destroys villous epithelial cells of the small intestine and causes vomiting, profuse diarrhea, with mortality reaching 100% especially in piglets less than two weeks of age.

Transmissible gastroenteritis is caused by TGEV, a member of the *Coronaviridae* family. The virus is enveloped and pleomorphic and has three major structural proteins: a nucleocapsid protein (N), a trans-membrane glycoprotein (M), and a peplomer glycoprotein (S). The S protein is of importance. It binds to cellular receptors, causes membrane fusion, and induces the production of complement-independent virus neutralization. Total immunity against TGEV is dependent on the infection of the small intestine, which stimulates antibody production. Active immunity against TGEV in the gut relates to stimulation of the secretory IgA (SIgA) immune system, with the production of intestinal SIgA antibodies by lymphoid cells within the lamina propia of the gut. In order to protect piglets, passive immunity from sow thus becomes critical against TGEV infection. However, conventional available vaccines against TGEV are usually given parenterally (e.g. intramuscular injection, IM) which are relatively difficult in achieving effective mucosal immunization. The humoral antibodies produced by IM injection mainly are of the IgG class, which protect the newborn piglet against systemic infection but not intestinal infection. The concentration of IgG antibodies is high in the colostrum but decreases rapidly during the first week of lactation, while SIgA concentration declines marginally and becomes the predominant Ig in milk. Oral administration of attenuated TGEV vaccine thus offers better protection because infection of the gut elicits SIgA antibodies. These antibodies in the lumen of the intestine of suckling pigs will neutralize ingested TGEV and thus interfere with the virus adherence to the villous epithelial cells of the small intestine in an immunogenic mechanism called the lactogenic immunity.

Lactogenic immunity mediated by mucosal immunization provides greater protection against TGEV than parenteral inoculation because there is a higher level of IgA class in milk. IgA antibodies are more resistant to proteolysis and they selectively bind to gut enterocytes and confer protection. After exposure to the viral antigen in the gut, IgA lymphocytes from the lamina propia migrate to the mammary gland where they localize and secrete IgA antibodies into the colostrum and milk. This concept called the gut-mammary immunologic axis is important in vaccine design.

14.1.2. *Immunization procedures*

A non-immune animal can be artificially made immune to an infectious disease either by passive immunization or active immunization. Briefly, artificial passive immunization is produced by transfer of preformed antibodies from a resistant animal to a susceptible animal, resulting in a temporary protection against the pathogen (Fig.14.9 and Table 14.5). For instance, antibodies against toxins of *C. perfringens* raised in horses are used to passively protect non-immune piglets, which have been exposed and are consequently at risk. These passively transferred antibodies give

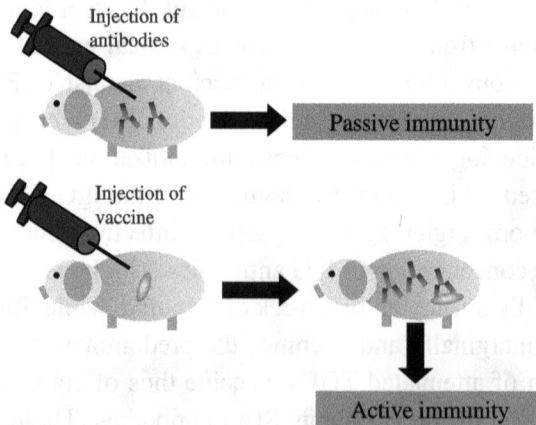

Fig. 14.9.   Passive versus active immunity.

Table 14.5. Passive and natural immunity.

| Type | Acquired | |
	Natural	Artificial
Passive immunity	Maternal antibody	Serum antibody and cells
Active immunity	Infection	Vaccination

temporary immediate protection, but since they are gradually catabolized, the recipient will eventually become susceptible to reinfection.

Artificial active immunization, on the other hand, involves administering antigen in the form of vaccine to the animal so that it induces a protective immune response without causing the disease itself (Fig. 14.9 and Table 14.5). In any case, vaccination is based on the principle of artificial exposure of the body to an immunogenic agent that is harmless but shares antigenic determinants with the intact, fully virulent disease agent. Immunization with vaccines then provokes a primary immunological defense response of the body without causing the disease itself. Hence, exposure to this same antigen later in life will provoke a "ready to fight" secondary immune response. This response may be either antibody-mediated or cell-mediated, or both. Primary vaccination doses are usually followed by regular booster doses to maintain a continued immunity to the relevant pathogens (Fig.14.10). Although protection conferred by active immunization is not immediate, the main advantages of active immunization over passive immunization are the prolonged period of protection achieved and the ability of recalling and boosting of this protective response by repeated injections of antigen or by exposure to infection.

14.1.3. *Vaccines*

An ideal vaccine for current needs will not only require to overcome the limitations posed by the classical vaccine, but should also be safer and more efficacious as compared to the classical vaccines. Besides, it should also stimulate an immune response distinguishable from that due to a natural infection, so that immunization and eradication may proceed simultaneously. Such a vaccine is also commonly known as marker vaccines, since it is able to "mark" the vaccinated animal from the naturally infected animal.

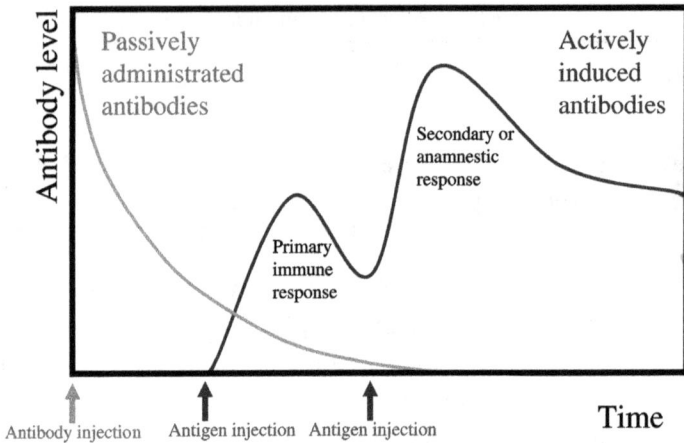

Fig. 14.10.   Passsive immunity, primary and secondary immune response.

Vaccination procedures usually require two injections; i.e. the first injection to sensitize the immune system and the second injection to boost the immune response. For the benefit of easy management, the vaccine should be capable of inducing a strong immunity after a single injection to save time, labor and cost for vaccination as well as induce less stress on the animal. Last but not least, it will be ideal if the vaccine can be administrated orally. This is especially useful for the vaccination of fishes as conventional vaccination procedures by injection can be quite a challenge.

14.1.3.1. Conventional vaccines

The market for veterinary biologicals in the US and Europe was approximately US$2.4 billion in 1998, an estimate that does not include large markets in Asia, such as China and India. Markets in the US and Europe are quite mature and future growth will most likely to be outside these regions. Animal vaccines are responsible for preventing diseases and reducing losses and thus have a major impact on animal industry; both attenuated and inactivated conventional vaccines have contributed enormously to animal health worldwide. Unfortunately, two prerequisites of an ideal vaccine, namely high antigenicity and absence of adverse side effects, tend to be mutually incompatible for two types of vaccines, namely attenuated and inactivated. Although attenuated vaccines stimulate a better immune response, it may present shedding hazards of residual and reverted virulence. On the other hand, inactivated vaccines may be safer but they are usually poorer immunogens. Table 14.6 gives a comparison of the primary characteristics of these two types of vaccines.

(a) Live or attenuated vaccines

Attenuated vaccines are usually formulated from alternative or mutant strains of the pathogenic organism that display reduced or no virulence

Table 14.6.   Comparison of different vaccines.

Vaccine	Advantages	Disadvantages
Live attenuated vaccines	Stimulate good immune responses	Risk of residual virulence
	Can be administrated in low doses	Shedding of live organisms
	Require no adjuvants	Risk of reversion to virulence
		Unstable and require constant refrigeration
Inactivated vaccines	No risk of residual virulence	Low immunogenic
	Stable and easy to store	Need for adjuvants
		Requires administration of high multiple doses
Recombinant protein vaccines	No risk of residual virulence	Low immunogenic
	Stable and easy to store	Need for adjuvants
	Lower production cost and no safety concern	Requires administration of high multiple doses
Recombinant surface-displayed vaccines	No risk of residual virulence	Mass production method has not been fully established
	No safety concern during production	
	Dual effectiveness as protein and vectored vaccines	
VLPs	No risk of residual virulence	Mass production method has not been fully established
	No safety concern during production	
DNA and vectored vaccines	Broad range with multiple genes	Safety issue need to be further studied during the use
	Lower developing cost	

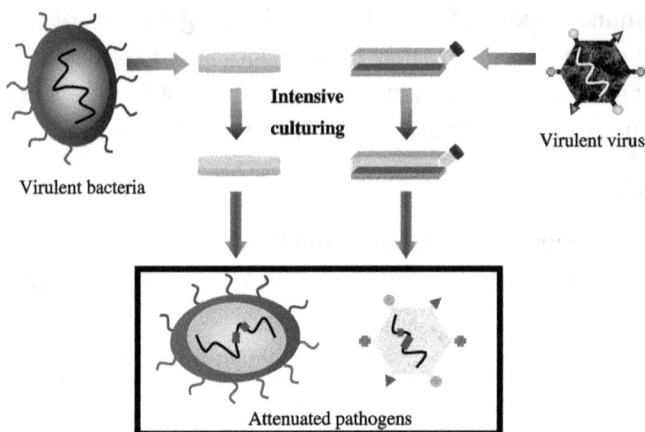

Fig. 14.11.   Attenuation of virulent bacteria and virus.

(attenuated) in the host, while maintaining their ability of transient growth and antigenic integrity to induce an immune response in the vaccinated host.

Attenuated mutations can be introduced by adapting pathogens for prolonged periods under unusual conditions, which results in the pathogen losing its specificity for, and thus reducing its virulence in its target animal species or tissue (Fig.14.11). Bacteria can be attenuated by culture under abnormal conditions. For instance, the attenuated strain of anthrax currently used in vaccines was rendered avirulent by growing in 50% serum agar in a CO_2 rich atmosphere, resulting in its inability to form a capsule essential for infection.

Microbial pathogens are highly adaptable microorganisms that express complex and multifactorial virulence phenotypes. The factors required by a pathogen to cause disease are termed determinants of virulence. These determinants can be involved in toxicity, colonization, invasion and avoidance of host defenses. In order to design prevention strategies, it is important to identify virulence determinant genes and be able to differentiate them from normal housekeeping genes. In order to identify the proteins that are involved in interaction with the host cells, it becomes necessary to mutate the gene coding for that protein or to change the conformation of that protein such that interaction with the host is affected.

This is quite possible with the use of transposons. A number of bacterial pathogens like *Salmonella typhimurium*, *Vibrio cholerae* and *Edwardsiella tarda* have been effectively mutated using transposons.

In the case of *V. cholerae*, transposon mutagenesis was used to create a mutant with an altered viable but non-culturable response. Mutants defective in swarming motility as a result of transposon mutagenesis were also found to be defective in the synthesis of lipopolysaccharides that are crucial genetic components involved in virulence. Other mutants deficient in an outer membrane protein (OmpC) were found to be defective in adhesion to macrophages. Macrophages recognize, adhere to, and phagocytose *Salmonella typhimurium*. The major outer membrane protein OmpC is a candidate ligand for macrophage recognition. Transposon mutagenesis was used to develop an OmpC-deficient mutant, which was compared along with the wild type for macrophage adherence and association assays. Radiolabeled wild type *S. typhimurium* bound to macrophages at levels five-fold higher than did the OmpC mutant. The deduced OmpC amino acid sequence of *S. typhimurium* shares 77% and 98% identity with OmpC amino acid sequence of *E. coli* and *S. typhimurium*, respectively. Evidence from this study supports a role for the OmpC protein in initial recognition by macrophages and distinguishes regions of this protein that potentially participate in host-cell recognition of bacteria by phagocytic cells. With extensive experiments carried out to validate the efficacy of using transposon mutagenesis, the significance of the work lies in the extreme handiness of this tool. Transposons are highly mobile genetic elements that have unique inverted repeats at their ends. Apart from carrying an antibiotic resistance marker, transposons also bear a transposase gene that helps in the insertion of this transposon into the chromosome in the right frame. Mutations can be induced when transposons or insertion sequences are integrated into the bacterial chromosomes. These newly inserted pieces of DNA can cause profound changes in the gene into which they insert and perhaps even in the adjacent gene. These mutations may or may not have obvious defective phenotypes, which can be effectively screened using various assays.

Tn*phoA* is a transposon which, upon correct insertion into an open reading frame of a protein containing the N-terminal export leader sequence, will produce a fusion protein with bacterial alkaline phosphatase (PhoA).

This can be detected by using a PhoA-specific chromogenic detection reagent (5-bromo-4-chloro-3-indolyl-phosphate) leading to the formation of a blue bacterial colony. The sequences flanking the Tn*phoA* can be amplified by PCR, cloned and sequenced to enable further genetic studies. Tn*phoA* is designed as a probe to detect genes whose products are exported across the cell membrane, as PhoA is only active when in the periplasm and hence very useful for identifying exported gene products and protein regions that are periplasmic. A common factor of many virulence determinants is that they are exported across the bacterial cytoplasmic membrane to interact with the host. Thus, Tn*phoA* is an excellent tool for the identification of genes whose products are exported and therefore most likely to be putative virulence determinants (Fig. 14.12).

Edwardsiella tarda is a gram-negative bacterium that causes a systemic infection, edwardsiellosis, in fish. This pathogen also affects reptiles and is also responsible for stray incidences of gastroenteritis in man. Though some of the virulence factors of *E. tarda* have been elucidated like motility, adhesion, invasion and cytotoxicity, systematic studies regarding the pathogenicity and the genetic factors responsible for the virulence are lacking. Certain virulence factors of *E. tarda* were identified using Tn*phoA* mutagenesis. Mutants deficient in siderophore production, catalase secretion, motility and serum resistance were identified using transposon mutagenesis. Moreover, some of these mutants were also found to be effective as vaccine strains. There is sufficient evidence to show that transposon mutagenesis is a convenient tool to obtain mutants, which are deficient in major virulence factors, as well as to identify virulence genes and to study gene and protein functions.

Viruses, on the other hand, can be attenuated by growing in a non-target cell culture line or species to which they are not naturally adapted (Fig. 14.11). For instance, canine distemper virus rendered avirulent by growing in birds was shown to be safe and immunogenic in gray foxes, bush dogs, maned wolves and fennec foxes. The flurry strain of rabies that underwent 178 passages in eggs lost its virulence for normal dogs and cats. In the case of the fowl influenza virus that normally infects chickens, attenuation was achieved by culturing in pigeon eggs. Nonetheless, the most commonly used method of virus attenuation is by prolonged culturing in cell types to which it is not adapted. For instance, canine distemper

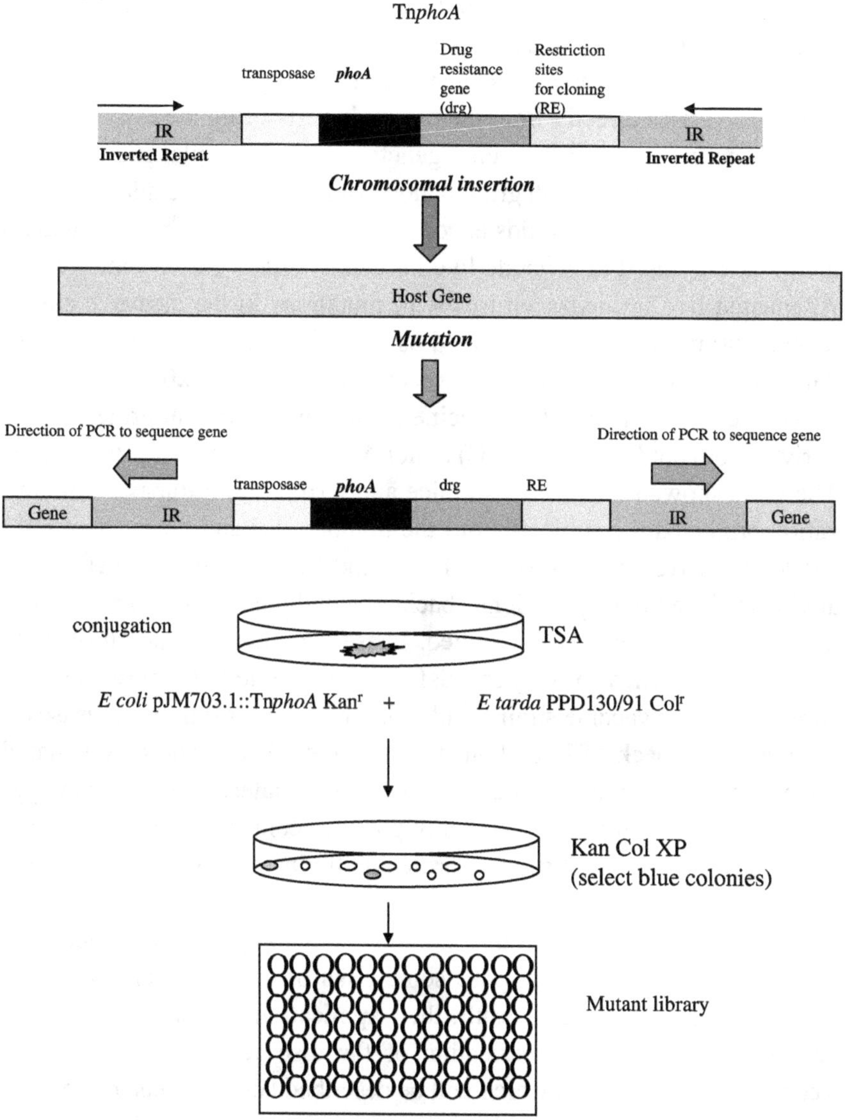

Fig. 14.12.   TnphoA is a Tn5 derived transposon that can be used to identify mutants that lack major secreted virulence factors. Upon insertion into the host genome, PCR can be used to amplify the gene that has been interrupted due to transposon insertion.

virus that normally infects lymphoid cells is attenuated by passage extensively in canine kidney cells.

Advancements in the reverse genetics technology allows to derive live influenza viruses directly from the plasmids harbouring the essential viral genes. The eight-plasmid reverse genetic system (six plasmids encodes the internal proteins of high growth, non-pathogenic or cold adapted influenza strains and two plasmids encodes the HA and NA genes of interest) is now widely used to generate live influenza viruses for vaccine studies. Attenuated live influenza viruses with mutations in the cleavage site of hemagglutinin have been described as one of the most promising candidates of the live vaccine in future. Recently Joseph *et al.* (2008) developed a live attenuated H7N3 virus vaccine using the six internal protein genes derived from cold adapted (25°C) donor A/ Ann Arbor/6/60 (H2N2) virus. The virus showed attenuation in mice and ferrets and intranasal immunization protected the subjects from the homologous and heterologous H7 challenges. Live vaccines against H9N2 and H5N1 have also been generated with the cold adapted donor backbone and the safety and efficacy in mice and ferrets have been verified. Phase I clinical evaluation of these vaccines is currently in progress. Safety concern due to the possible reassortment of live vaccine strains with other circulating influenza viruses is the main drawback of live attenuated vaccines. Furthermore, attenuated live viruses require to be cultured in the embryonated pathogen-free eggs under sterile conditions to avoid any possible contaminations with other pathogens. This is an additional hurdle that limits the advancement of live attenuated vaccines.

Although attenuated vaccines are still commonly used for control of animal diseases today, they pose some risks of continued circulation at subclinical levels and as a result, their applications are not permitted in certain countries where eradication of the disease is being attempted. Recombinant technology, which will be described later, permits irreversible attenuation of pathogen by selectively removing genes that are necessary for virulence.

(b)   Gene-deleted vaccines

Gene-deleted vaccines are engineered by deleting specific gene sequence, usually coding for its pathogenicity and virulence, from the genetic material of a pathogen, resulting in the gene-deleted organism to be irreversibly attenuated. Furthermore, given that such organism is permanently "marked" from the pathogenic strain, when it is used as a vaccine, antibodies can be distinguished from that induced by the pathogenic strain when tested serologically.

A classical example is the use of gene-deleted vaccine for the eradication of Aujeszky's disease. Aujeszky's disease or pseudorabies as it is otherwise called is found in most pig producing regions, is a herpes viral infection that affects the central nervous system leading to high mortality in piglets. There is an urgent need for its control in many countries. Vaccination with gene-deleted vaccine against pseudorabies is commonly practiced, and along with companion diagnostic tests, has successfully eradicated the disease in some countries. The pseudorabies vaccine is based on the deletion of the thymidine kinase gene from the virus, leading to its inability to replicate in the nerve cells to cause disease. The major glycoproteins gX, gI, gp50 and gp63 of pathogenic pseudorabies are not essential for either growth or virulence, and the virus appear to function normally when these genes are deleted (Fig. 14.13). Thus, this gene-deleted vaccine not only confers effective protection, but it also blocks

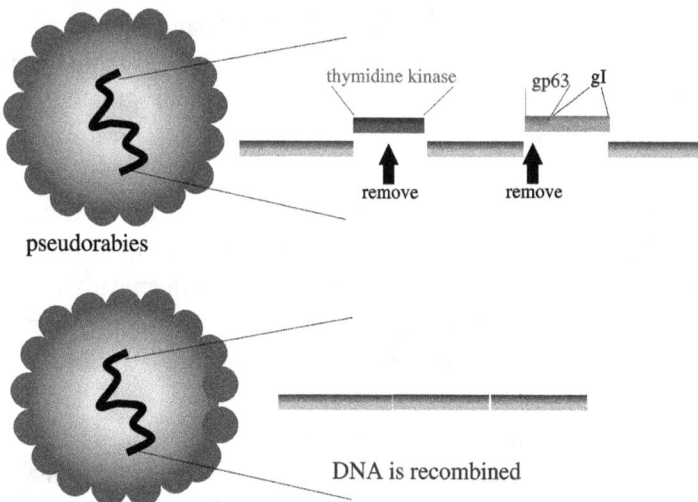

Fig. 14.13.   Gene-deleted vaccine.

cell invasion by virulent pseudorabies virus and prevents development of a persistent carrier state in the vaccinated animal. Besides, this vaccine induces an antibody response distinguishable from that caused by the pathogenic strains, enabling diagnostic differentiation for disease eradication.

(c)   Killed or inactivated vaccines

Another common vaccine technology for control of animal diseases is to inactivate the pathogen so that it is no longer capable of replication in the host. It is important that the inactivated pathogen be antigenically similar to the living organisms to induce immune response similar to natural infection. For this reason, inactivation methods should maintain the structure of epitopes and surface antigen, hence crude methods of killing like heating are usually unsatisfactory because it causes extensive protein denaturation and some functional epitopes of may possibly be altered.

Knowledge of the molecular structure of antigens and the level of infectivity of pathogen is essential to determine the method for inactivation of organism. Chemical inactivation with various alkylating agents like ethylene oxide, ethylenimine, acetyl-ethylenimine and beta-propiolactone are among the most common choices in veterinary vaccines. These alkylating agents act by cross-linking nucleic acids chains while leaving the surface protein antigens intact, and hence do not interfere with antigenicity of the organism. Formaldehyde on the other hand acts on amino and amide groups of proteins and on non-hydrogen-bonded amino groups in bases of nucleic acids to form cross-links and confer structural rigidity. Although reduced antigenicity has always been a concern for inactivated vaccines, some of these vaccines confer immunity comparable to that induced by attenuated vaccines at their best. This is further enhanced by the use of special adjuvant formulations due to their synergistic activity. However, human carelessness in the production of chemically inactivated vaccines has been subjected to safety concerns. The vaccine against Foot and Mouth Disease virus, though efficiently protects livestock from the disease, is linked to outbreak due to the release of virus from vaccine production plants, or the presence of residual live virus.

The production source of inactivated viral vaccines is one of the most important factors to determine the cost and the quality of vaccines. For example, inactivated influenza vaccines are traditionally produced in embryonated eggs based on the circulating virus strains followed by inactivation and purification. Kim *et al.* (2008) recently demonstrated the efficacy, cross-protectiveness and immunogenicity of recombinant inactivated whole-virus influenza vaccines comprising the clade 1 and clade 2 in ducks. The vaccines were able to completely protect the subjects against the cross-clade and cross-subclade influenza virus challenge though the vaccines were not able to induce proper hemagglutination inhibition (HI) titers and neutralization antibodies. Another group from Tokyo university addressed the cross-clade immunity conferred by the inactivated vaccines individually comprising the influenza viruses from clade 1, clade 2.1, 2.2, 2.3.4. Clade-dependent protection was obtained when challenged with heterologous viruses. The inactivated vaccines have already been evaluated clinically. However, it has been documented that inactivated viral vaccines are not well immunogenic and requires high contents of immunogen to attain the desired level of antibody response. A recent study revealed that only 54% of participants have a neutralizing serological response to a high intramuscular dosage (90 µg HA) of unadjuvanted H5N1 inactivated vaccines. Hence, the use of adjuvants such as aluminium, oil–in–water emulsion etc. is often necessary to enhance the immunogenicity of inactivated vaccines. Further, the requirement of high-level bio-containment facilities and stable supply of SPF (specific pathogen free) eggs for vaccine production are other important considerations that restrict the progress of inactivated viral vaccines to the next level.

The current problems associated with the egg-based technology can be overcome when cell culture, in place of embryonated eggs, is employed for the production of live influenza vaccines with required attenuation (e.g. cold adapted backbone). Ghendon *et al.* (2005) developed MDCK cell culture based cold adapted live influenza vaccines that were produced in a serum free medium in a bioreactor showing the feasibility of large-scale production of influenza vaccines in cell culture. Another report from Romanova *et al.* (2004) demonstrated the use of vero cells to produce live cold adapted influenza vaccines and evaluated the safety and immunogenicity of the vaccines in young adult volunteers. The vaccines were found to be safe, protective and well tolerated. Cell culture based method will promote the timely production of safe and effective influenza vaccines.

14.1.3.2. *Recombinant subunit vaccines*

In simple terms, recombination is a process in which genetic material is broken and fragments of the genetic material are rejoined to give a new combination. In recent years, the application of recombination technologies resulted in the development of novel vaccines that overcome the limitations of classical vaccines. Besides addressing the safety issue and other limitations in relation to conventional vaccines, considerable improvements have already been achieved in certain fields, and some new approaches widely commercialized. It has long been known that in controlling the spread and eradication of any disease, it is important to develop procedures that distinguish vaccinated from convalescent animals. This is because convalescent animals can become carriers of the pathogen, which consequently can become the source of new outbreaks, and possibly introduce antigenic variants into the population. A recombinant vaccine is able to stimulate an immune response distinguishable from that due to a natural infection, and hence is also known as a marker vaccine. This is something, which a conventional vaccine is incapable of. Therefore, the use of a marker vaccine allows vaccination and eradication to proceed simultaneously. Indeed, the ability to differentiate between animals vaccinated with recombinant vaccines and those infected with natural strains of pathogen has already played a key role in eradication of certain diseases, like pseudorabies due to the development of companion diagnostic tests.

The following are types of recombinant vaccines that are either available for use or are under development against different animal diseases and have shown promising results. These are subunit, vector, gene-deleted and DNA vaccines. Each has different characteristics that allow it to be used for different diseases and situations (Fig. 14.14).

(a) Subunit protein vaccines

Production of large quantities of purified antigen economically has often been difficult. Therefore genetic material encoding an immunogenic protein is recombinantly propagated in expression systems for effective extraction and purification of large and pure quantities of protein for use as subunit vaccines. To date, the most commonly used expression systems include the bacterial *E. coli* expression system, yeast expression system

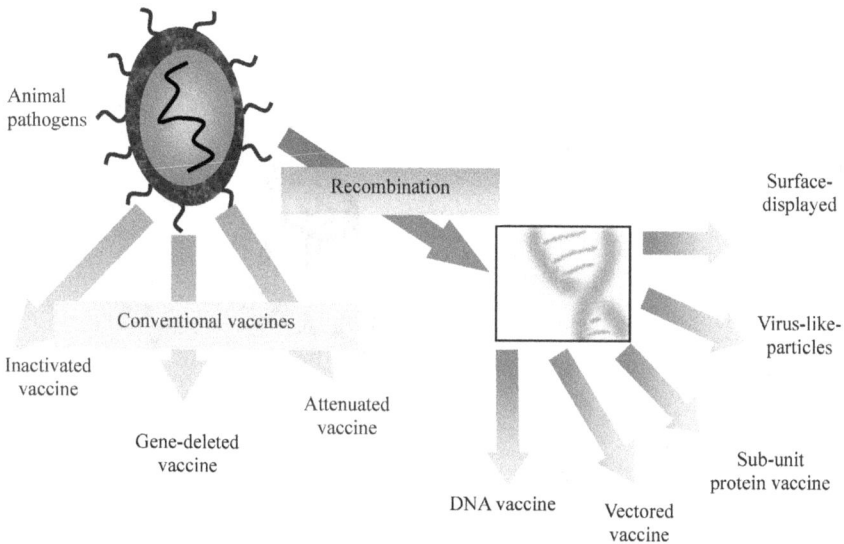

Fig. 14.14.   Different types of vaccines.

and the baculovirus expression vector system (BEVS). *E. coli* is a more preferred system, because it is easier to work with and the expression level is high. Subunit vaccines are engineered with genes that code for part of the pathogenic organism, which are then introduced into the expression host system. The host cell manufactures the protein and if all goes well with the clinical trials, this recombinant protein will be available as a subunit vaccine. For instance, in order to prepare subunit vaccine for FMD, the first stage is to identify which portion of the virus can confer immunogenic protection. This genetic material can then be cloned into an expression system to produce the specific antigen for vaccination.

The first commercially available veterinary subunit vaccine was developed against Feline Leukemia Virus (FeLV). The major envelope protein of FeLV, gp70, is the main epitope that induces protective immunity in cats against feline leukemia virus. The gene encoding gp70 is expressed and the protein is produced in huge quantities from recombinant *E. coli*. Once harvested and purified, the recombinant protein is mixed with a saponin adjuvant and used as vaccine (Fig. 14.15). Another example is the enterotoxin of enteropathogenic *E. coli*, which consists of two subunits, the alpha subunit that is toxic, and the beta subunit that

Ligate into a expression plasmid

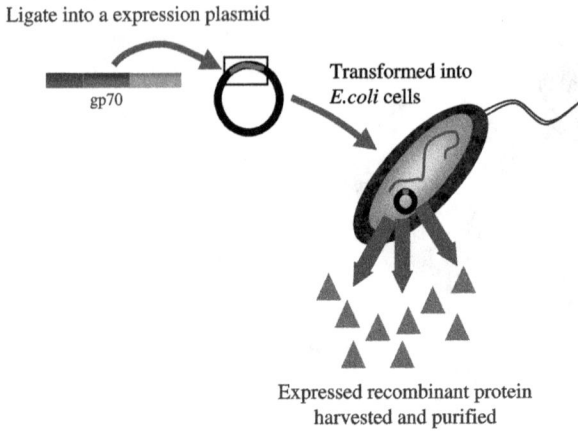

gp70

Transformed into
E.coli cells

Expressed recombinant protein
harvested and purified

Fig. 14.15.   Subunit-FeLV vaccine.

binds the alpha subunit to enteric cells. The purified beta subunit is immunogenic and functions as an effective toxoid. The attachment pili protein of enteropathogenic *E. coli* is currently used as a vaccine to protect animals from infection by preventing bacterial attachment to the intestinal wall.

The recombinant expression system gives the capacity to produce large quantities of various bacterial or viral proteins that may be used as subunit vaccines. Vaccination with subunit vaccine produces antibodies only to the protein delivered. However, natural infection by a disease agent elicits responses to a variety of proteins. Therefore, the subunit vaccine itself should carry a marker, which distinguishes vaccinated animals from naturally infected animals (Fig. 14.16).

(b) Surface displayed subunit vaccines

Surface display strategy is based on the mechanism by which baculovirus acquires its envelop protein (GP64) during the virus budding. In the baculovirus replication cycle, GP64, a class I transmembrane protein found on baculovirus surface, is expressed at the earlier stage of infection, processed and translocated to the plasma membrane of the insect cells. Newly synthesized virions harbor GP64 on its surface during the budding process. This property of the baculovirus has been explored to display the foreign proteins or peptides on the viral surface. The system enables the

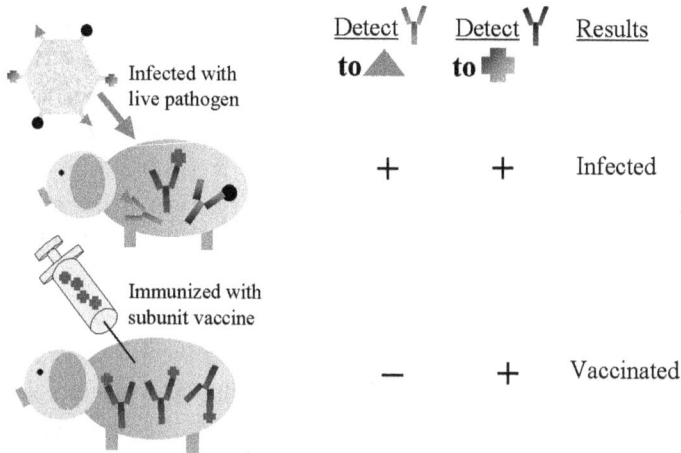

Fig. 14.16.   Marker vaccines allow distinguishing vaccinated animals from naturally infected animals by serological methods.

presentation of large complex proteins on the surface of baculovirus particles when substituted with suitable signal peptides and transmembrane domains, the key components of surface display technology. Antigen presentation on the baculovirus surface makes it readily accessible to cellular components of the immune defense cells system. Antigens could be internalized by macrophages, dendritic cells and coupled to the Class I MHC molecules for the recognition by T-cells for cell mediated immune response. Display of foreign proteins can also be performed by fusing a heterologous protein (or peptide) in-frame to the N-terminus of the GP64 gene or at specific sites within native GP64 without altering its function. Baculoviruses with single-chain antibody fragment displayed on the surface are specifically targeted by mammalian cells. Baculoviruses harboring synthetic IgG binding domains on their surface were used to construct an eukaryotic mab library.

Similar to GP64, HA is a class I transmembrane protein with signal peptide (SP) located at the amino-terminus and the transmembrane (TM) domain at the carboxy-terminal end. The nature of HA, being a surface glycoprotein, supports its display when expressed in baculovirus — insect cell system. Besides, influenza NA, a Class II transmembrane protein where both the signal peptide and the TM domain are located at the amino terminal, has been reported to enable EGFP expression on baculovirus surface.

Immunogold labeling and EM demonstrated that HA and NA could be properly expressed, translocated and incorporated onto the baculovirus particles. Surface displayed HA and NA sustain their properties as indicated by the functional assays . It has been proven that the biological activities of surface displayed proteins are similar to those that are in soluble form confirming the structural integrity of the surface displayed proteins. A few studies have also investigated the use of SP, TM and cytoplasmic domain (CTD) derived from gp64 to enhance the surface display efficiency. Influenza HA, with SP and CTD derived from either HA or GP64, is translocated to the plasma membrane of insect cells, and incorporated into the baculoviral surface. However, as compared with HA CTD, the baculovirus construct with gp64 CTD significantly improved the gene delivery and transgene expression in mammalian cells, which might result from the promoted incorporation of HA into the baculoviral envelope with the gp64 CTD.

The efficacy of surface displayed H5 vaccines in baculovirus has been evaluated in different animal models, such as mouse, chicken and duck (unpublished data). The dosage of the surface displayed vaccine at 500ng, based on HA contents (HA titer 2^8), could elicit significant anti-HA immune response and protects the animals from lethal H5N1 challenge with both homologous and heterologous strains via intramuscular or intranasal routes. Surface displayed vaccines have presented their advantages in immunogenicity and transduction efficiency over any of the conventional vaccines, though the large-scale production system has not been tested yet. Further vector engineering and proper adjuvant formulation will certainly promote the development of this antigen-displaying baculovirus to be an efficient vaccine for influenza.

(c) Virus-like- particle (VLP)

Virus-like particles resemble viruses, but are non-infectious because they do not contain any viral genetic material. The expression of viral structural proteins, such as Envelope or Capsid, can result in the self-assembly of virus like particles (VLPs), which presents immunogenicity as vaccines. VLPs have been produced from components of a wide variety of viruses including influenza virus, HIV and Hepatitis C virus.

The flexibility of baculovirus to accommodate large DNA molecules makes baculovirus an ideal tool for synthesizing virus-like particles which require simultaneous expression of multiple viral structural proteins for self-assembly. So far, various VLPs of different viruses, including HIV, herpes simplex virus, human papillomavirus, polyomavirus, parvovirus, infectious bursal disease virus, hepatitis C virus (HCV), and enterovirus have been expressed using the baculovirus/insect cell system. Together with the role of matrix protein M1 in protein assembly, influenza virus-like-particles have been generated in baculovirus system by coexpressing HA and NA with M1. These VLPs can fully mimic the influenza virus structure and induce protective immunity against the viral infections as indicated in preclinical studies. Several reports indicated that influenza virus-like particles could elicit better immune responses than either inactivated whole virus vaccine or recombinant hemagglutinin. With the VLPs expressing HA, NA, and M1 from H3N2 strain, immunized BALB/c mice (3 µg–24 ng based on HA contents) had high anti-HA antibody titers. Antibodies against NA were also detected in animals vaccinated with the high doses of same VLPs (3 µg based on HA contents). Antibodies elicited by VLPs recognized a broader panel of antigenically distinct H3N2 viral isolates compared to rHA or whole inactivated virus. Another report from Bright et al. assessed the cross-clade protective cell mediated and humoral immune responses elicited by influenza virus like particles with HA and NA from clade 1 and clade 2 isolates. Though no significant level of mucosal immunity was observed, mice vaccinated with virus-like particles (3 µg–600 ng based on HA contents) were protected from the H5N1 homologous and heterologous challenge, while mice vaccinated with rHA only showed considerable weight loss and death. Baculovirus based VLP vaccines present significant advantages in terms of immunogenicity over traditional protein vaccines, such as rHA vaccines. However, the purification of VLPs meets some challenges during the large scale production as it currently relies on ultracentrifugation, which has not been well established for the industrial manufacture.

14.1.3.3. *Nucleotide based-recombinant vaccines*

(a) Vectored vaccines

Genes encoding the protective antigens can be recombinantly inserted into an attenuated viral or bacterial "vector" that acts as a vaccine delivery system for the antigen into the animal body. The vectored vaccine is host-restricted so that it will only replicate itself within the vaccinated animal. Since this vectored vaccine can replicate in the host cells they overcome the limited immunogenicity associated with subunit vaccines. Vectored vaccines possess many of the characteristics required for an "ideal" vaccine. These include free of adverse side effects, high levels of stability, and suitable for mass vaccination program. In addition, such vaccines exclude the use of adjuvants, and allow differentiation between vaccinated and naturally infected animals. Early data collected on these vaccines indicate strong immunity, no side effects, and no shedding into the environment when used as a "vector" to distribute antigen into the habitat of wildlife species.

Vaccinia virus, derived from the naturally occurring cowpox strain, is a large complex virus containing a genome of about 200 genes. It can carry several dozen foreign genes without impairing its capacity to infect host cells and replicate. However since vaccinia virus is not species-specific, concerns regarding the risk of subsequent expression in non-target animal were questioned. This problem can be overcome with the species-specific pox virus, in which vectored vaccine with a stable recombinant fowlpox virus had been made available recently, offering protection against Newcastle Disease (ND) and Infectious Bronchitis (IB). The canary pox vector protects dogs from canine distemper virus, and with canary pox vector containing a gene encoding rabies glycoprotein that offers protection for dogs and cats against rabies. Other vectors under investigation include the adenovirus vector systems, which have lower cloning capacity than pox viruses, but are generally stable, enabling ease of handling and induction of mucosal immunity of both respiratory and intestinal surfaces when administered orally. This makes them ideal candidates for the development of mucosal vectored vaccines.

To engineer a vectored vaccine against ND virus, the fusion gene and the hemagglutinin gene from ND virus are incorporated into a vaccine strain of fowlpox. The immune system of the host reacts against the recombinant poxvirus, and the proteins encoded by the inserted genes of ND virus, inducing protection against both diseases (Fig. 14.17). In the recombinant vectored rabies vaccine, the gene encoding for the rabies glycoprotein was inserted into an attenuated vaccinia virus. Since this virus is not species-specific, cells of many different wildlife species can take up this vaccine non-selectively for protection against rabies. However,

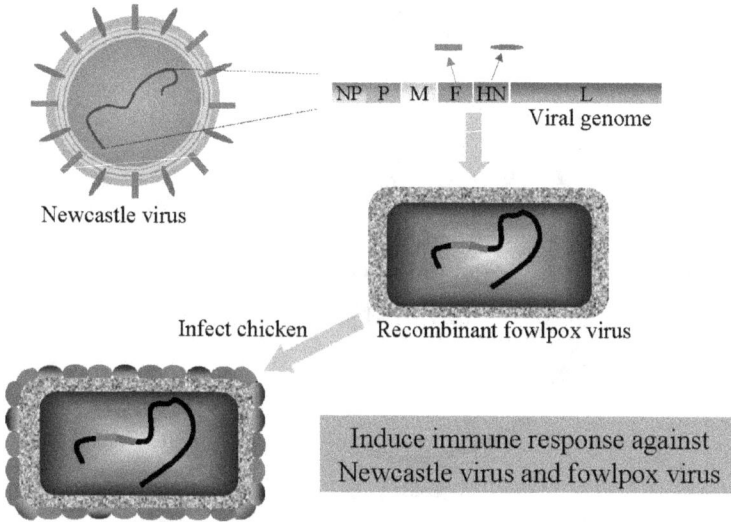

Fig. 14.17. Engineering a vectored vaccine against ND virus.

there are concerns about the use of vectored vaccines in immune deficiency animals, which even an attenuated vaccine can be potentially fatal.

(b) DNA vaccines

Reports in 1990 that naked DNA encoding viral genes could induce protective immunity when delivered into the host came as a surprise to most scientists. In just a few years, the expanding number of examples where recombinant plasmid DNA induces immune response against the respective pathogens has developed from an interesting observation to a highly promising approach for the development of vaccine technology. DNA vaccination refer to the direct inoculation of eukaryotic expression vector encoding the antigenic protein(s) into the animal resulting in an in situ production of the encoded antigen within the tissue cells of the host and hence inducing immunity in the animal without disease. Commercially available mammalian expression vectors such as pcDNA 3.1 (Invitrogen, San Diego, CA) has been extensively used for DNA vaccine studies (Fig. 14.18). Over the last few years, DNA vaccines have been demonstrated to induce both humoral and cellular immunity against many different disease agents, and to be effective in a variety of bacterial, viral, and parasitic animal models. The

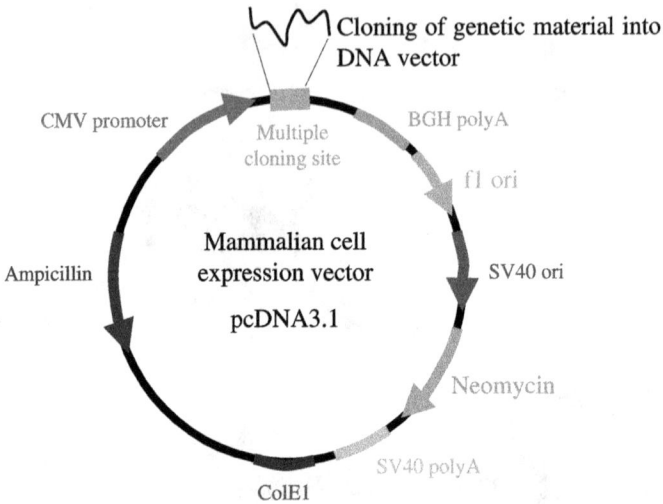

Fig. 14.18.   A commonly used mammalian cell expression vector.

use of pure plasmid DNA offers a number of potential advantages over the other commonly used methods of producing vaccines (such as genetic modification of pathogens, production of recombinant subunit proteins or attenuation of live organisms). These include simpler production, quality control, avoidance of contaminating pathogens and viability without cold storage during transport, as DNA material is stable over a wide temperature range. Other advantages included the ease of construction of recombinant DNA molecules expressing foreign genes, elimination of risk associated with use of live agents and problems associated with preexisting immunity to the agent or vector, as well as low cost (Fig. 14.19). However, the antibody titer induced by this DNA vaccine is comparatively low to that induced by conventional vaccines and repeated injections at higher dosages are required to obtain the necessary levels of immunity.

An alternative strategy to boost the titer of antibodies induced by DNA vaccines is to deliver pathogen specific antibodies termed "intracellular antibodies" into the host instead of delivering pathogen-specific antibodies. The concept of intracellular antibodies is based on expression of an antibody fragment inside the cell that can bind to and inactivate a pathogen (Fig. 14.20). The intracellular antibodies delivered by DNA plasmids have shown significant potential against Human Immunodeficiency Virus (HIV)

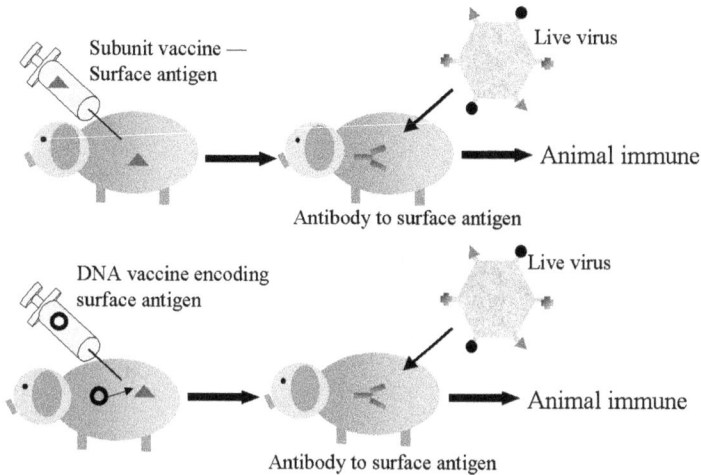

Fig. 14.19.   Subunit vaccine and DNA vaccine.

and a fish pathogenic rhabdovirus causing Viral Hemorrhagic Septicemia (VHS). To accomplish this task, a hybridoma cell line that produces a neutralizing monoclonal antibody was first raised against a neutralizing epitope on a pathogenic organism. The variable domain genes of the monoclonal antibody were then cloned into a mammalian expression vector, which was then injected into the host. In an experiment with VHS virus, administration of DNA vaccine encoding the pathogen specific antibody demonstrated that intramuscularly injected fish produced high levels of recombinant antibodies with neutralizing activity that protected immunized fish from viral challenge. This approach circumvents the potentially hazardous use of live recombinant viral vectors that are often used for transfer of antibody genes.

14.1.4. *Immunological adjuvants*

Recombinant vaccines offer many advantages over traditionally used vaccines. However, most are poorly immunogenic when administered alone, especially for subunit and DNA vaccines. Therefore, a great need exists for immunological adjuvants that are potent, safe and compatible with the new-generation vaccines. Moreover, they must have additional features like biodegradability, stability, ease of manufacturing, cost, and

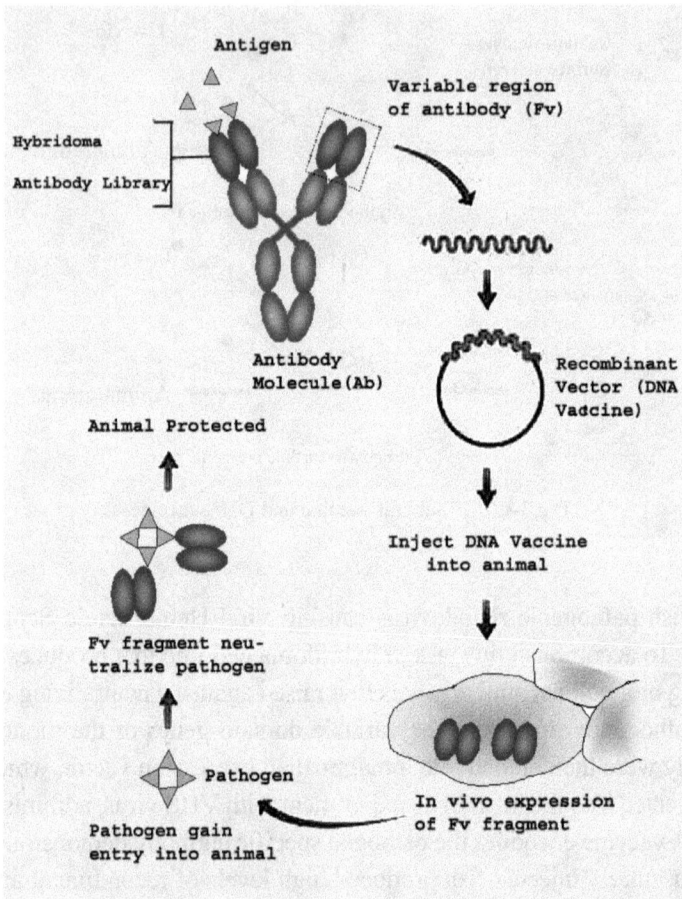

Fig. 14.20.   Illustration to show the making of intracellular antibodies. PCR is used to amplify the variable region of the immunoglobulin chains (Fv fragment) from the hybridomas or antibody libraries. The sequences cloned into the mammalia expression vector allow *in vivo* expression of the Fv fragment. A pathogen gaining entry into the immunized animal will be neutralized by the expressed Fv fragment, thereby protecting the animal from infection.

applicability to a wide range of vaccines. However, there is no universal adjuvant as the actions of different adjuvants are not yet clear and they probably rely on different mechanisms. They must be adapted according to several criteria like the target species, the antigen, the type of immune response, the route of inoculation, and the duration of immunity. Thus, different types of adjuvants have been developed for different purposes.

Immunological adjuvants are materials that enhance the immune response when administered along with an antigen. These encompass a wide range of materials. Although the mechanisms of action of adjuvants are often poorly understood, based on their mode of action, they can be classified into immunostimulatory adjuvant such as BCG, muramyl peptide (MPP), saponin etc. and depot adjuvants such as aluminium salts, mineral oil etc. Although many adjuvants have been administered experimentally, few have been accepted for use in domestic animals. In following section, emphasis will be made on depot adjuvants (Table 14.7).

Table 14.7. Adjuvants currently used in veterinary vaccines (from SEPPIC, Paris, France, with permission).

	Advantages	Disadvantages
Emulsions		
Water in oil	Long term immunity Stable Cost effective	Local reactions with reactive antigens (crude extracts) Safety is improved with metabolizable oils
Oil in water	Short term immune response (outbreak) Well tolerated cost effective	Possible pyrogenic effect
Water in oil in water	Short and long term immune responses Fluid, well tolerated	Less stable at 37°C
Mineral salts		
Al(OH)$_3$	Well tolerated Cost effective	Poor efficacy with purified Ag No Th1 cell response No CTL response
Plant extracts		
Saponin extracts	CMI	Variability from batch to batch
Others		
Bacterial extracts	Induces cellular immunity	High cost Local reactions
Liposomes		Large scale production High cost

Depot adjuvants function as a slow-release deposit for antigen at the injection site into the body and thereby prolong the immune response. The aluminum based salts such as aluminium hydroxide, aluminium phosphate and potassium aluminium sulphate (alum) have been widely used in animal vaccination. In this technique, the antigen is precipitated with the salt before inoculation. Upon injection, the salt/antigen mixture forms a small deposit in the tissue that slowly releases the antigen into the body and provides a prolonged antigenic stimulus. Antigens that normally persist for only a few days may be retained in the body for several weeks by this technique. In addition, the adjuvants may influence their ability to cause inflammation, thereby intensifying the general reaction to antigen exposure. A combination of the above factors probably account for their activity. Aluminum salts emulsion has a good safety record, but is a weak adjuvant for antibody induction by protein subunits and cell-mediated immunity. This problem can be overcome by the use of oil-based adjuvants. Adjuvants based on highly refined pure mineral oil will render water in oil (w/o), oil in water (o/w) and water in oil in water (w/o/w) emulsions (Fig. 14.21). Varying percentages of emulsifying agents like Tween (polyoxyethylene sorbitan monoleate) or eumulgin and non-ionic emulsifying agents are added depending on the viscosity of final product

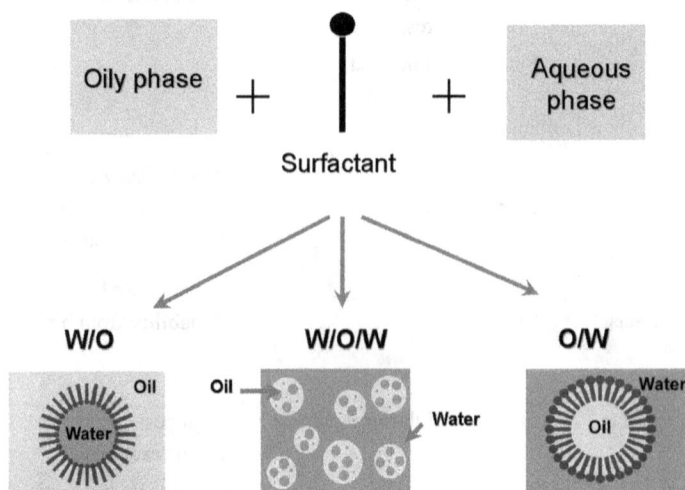

Fig. 14.21.   Oil adjuvant (from SEPPIC, Paris, France, with permission).

required. Field trials conducted with FMD vaccines in swine using oil-adjuvants indicated their superiority when compared to vaccines using aluminium hydroxide adjuvants where the antigen was increased three to ten times. In the control of major poultry diseases since 1965, oil-adjuvant based vaccines have proven to be effective and hence are widely used. Adverse reactions to oil-adjuvants include pro-inflammatory reactions at the site of injection, sterile abscesses, cysts and granulomas but most of these reactions can be overcome by the use of highly refined standard ingredients.

In recent years, natural mediators of the immune systems such as cytokines were studied for use in new adjuvant technology. Most cytokines have the ability to modify and redirect the immune response. IL-12 performs a critical role in modulating the T_H1/T_H2 balance and in providing protective immunity in the host defense system against bacteria, viruses and intracellular protozoa. Co-administration of vaccine with exogenous IL-12 as an adjuvant has resulted in the induction of protective immunity against *Leishmania*, *Listeriosis* and *Schistosoma*. In the case of immunization of PRRSV killed vaccine, the vaccine alone was insufficient to induce a specific immune response against PRRSV. However, co-administration of the killed vaccine with porcine recombinant IL-12 induced a strong humoral and cell-mediated immune response. Other examples of adjuvants, such as recombinant carp interleukin-1β, when used as an immunoadjuvant for vaccination of fish, elicited a significantly higher antibody titer at three weeks of post-vaccination than the formalin-killed pathogen alone. These results reveal the potential application of cytokine in the role of immunoadjuvant in fish vaccination. Molecules that have been evaluated most extensively as adjuvants include IL-1, IL-2, IFN-gamma, IL-12 and granulocyte-macrophage colony stimulating factor (GM-CSF). It may be expected that these natural mediators will become successors of the empirically developed adjuvant.

14.1.5. *Vaccine delivery systems*

The simplest and most widely practiced method of delivering vaccines to livestock is primarily by intramuscular or subcutaneous injection. Though this conventional approach is excellent for relatively small number of animals and for diseases in which systemic immunity is important, it usually requires multiple doses, is time consuming and labor intensive and is stressful to the animal. Improved delivery systems require only a single injection, subsequently reducing the cost of vaccination and labor. Moreover, vaccination during early stages of life when the animal is young is easier to handle and thus reduces the stress level in the animal.

Mucosal immune response plays an important and effective role in protecting the host from various infectious diseases. The mucosal system links the lungs, upper respiratory tract, gastro-intestinal tract, the urogenital tract, and the eyes. As more than 90% of animal pathogens gain access to their host cells through these surfaces, the ability to provide mucosal immunity offers enhanced disease protection. Furthermore, these responses may be circulated into the bloodstream to other tissues. For example, intranasal vaccines are effective in protecting cattle against systemic Infectious Bovine Rhinotracheitis (IBR) and poultry against Infectious Bronchitis (IB) and Newcastle Disease (ND). It has been shown that immunization of the gut or bronchial mucosa generates disseminated as well as local responses. For example, intranasal immunization of chicken with IBV vaccine induced a secretory antibody response in the GI tract and at other mucosal sites. When dealing with viral respiratory diseases such as influenza, antibody response in mucosal tissues is essential in preventing the infection. Mucosal immunization, which is shown to stimulate both mucosal and systemic immune response, would be an effective way to control the infection by influenza viruses. Oral and intranasal vaccinations are the two main options for induction of mucosal immune response. Intranasal administration of recombinant baculovirus displaying HA (Bac-HA) elicited high level of HA specific secretory IgA and serum IgG antibody response in mice. It was also reported that the combination of H5 HA expressed on baculovirus surface and recombinant CTB mucosal adjuvant form an effective mucosal vaccine against H5N1 infection. Log 2^8 HA titer of Bac-HA combined with 10 μg? rCTB provided 100% protection against 10MLD50 of homologous and heterologous H5N1 strains. This study also indicated that with the same dosage of rCTB and HA, baculovirus surface vaccine provided better protective immune response against lethal H5N1 challenge than the inactivated whole viral vaccine.

Unfortunately, immunization techniques by the oral or intranasal route require handling of each individual animal. This induces the stress on the animal and also compounds labor and cost problems. Aerosolization of vaccines enables them to be inhaled by all the animals in a herd, group, or flock — an obvious advantage when the unit is large. Alternatively, vaccine may be administered in feed or drinking water, which reduces some of the problems associated with the other methods. Edible or oral vaccines are particularly useful especially in the case of fish farms where herd vaccination can be very difficult and it would be ideal if the antigen may be added to the water in which they live.

Edible vaccines provide a viable strategy for the immunization of animals, as it facilitates compliance in the animals, and the process is less laborious and time consuming in comparison to the conventional method of injection immunization. This is especially evident for the case of mass vaccination of fishes where conventional method is neither practical nor cost effective.

Oral vaccination is considered as another option to stimulate mucosal immunity with increased patient compliance. Oral immunization is non-invasive and affordable with improved logistics and mass coverage during pre-pandemic and pandemic situation. Moreover, there is evidence to prove the ability of oral vaccination to prevent infection of the lungs and cause transcytosis of the molecule across the cells into the circulation. Oral vaccination has been previously reported to induce mucosal immune response (IgA) in the respiratory tract to confer protection against influenza viruses. More recently, oral administration of WSSV ie1 promoter based baculovirus displaying HA was evaluated in a mouse model. This recombinant baculovirus was able to induce significant level of HA specific mucosal IgA without the inclusion of any mucosal adjuvants. In addition, this vaccine was also able to induce HA-specific serum IgG antibody response, suggesting that the vaccine vector was transported into the circulation upon gastrointestinal delivery. Importantly, live baculovirus vaccine induced strong cross-clade neutralization against 100 TCID50 of heterologous H5N1 strains (clade 1.0, 2.1, 2.3 and clade 8.0) compared to its inactivated counterpart, indicating the possible role of functional HA in host immune response. Indeed, unlike live baculovirus, inactivated baculovirus failed to deliver HA genes both *in vitro* (human adenocarcinoma cell lines: HT29 and HCT116) and *in vivo* (mouse intestine). Based on the results obtained, it is reasonable to speculate that functional oligomeric HA displayed on the live baculovirus could have played a role in binding to the receptors expressed in the intestinal cell membrane resulting in gene delivery and stimulation of non-classical MHC-I molecules to present antigenic peptides to cytotoxic T- lymphocytes (CTL). Intestinal epithelial cells can act as non-professional antigen-presenting cells with the act as non-professional antigen-presenting cells with the help of these non-classical MHC class 1 molecule and mediates the constant crosstalk between the intestinal lumen and the mucosal immune system. In fact, it is evident from immunohistochemistry results that live baculovirus was able to transduce and express HA in the intestinal epithelial cells in mice, which in turn would have activated the mucosal immunity.

The use of transgenic plants as vaccine production systems has been researched extensively, and there have been some initial successful studies on edible vaccines delivered by transgenic plants. Through the delivery of antigens via genetically modified plant material, cheap and simple vaccination of livestock species may be possible in the future. There is also increased safety in terms of production, as plants do not generally serve as hosts for animal pathogens. Edible vaccines can be produced in plants in many ways. Antigens from infectious bacterial or viral diseases have been introduced into plants through plant virus-mediated infection or *Agrobacterium tumefaciens*-mediated stable transformation methods. Oral immunization with transgenic plant tissues that contain vaccine antigen proteins stimulates both systemic and mucosal immune responses in animals. Recent documentation in edible vaccine includes the expression of the *E. coli* heat labile enterotoxin (LT) in transgenic potato leaf and TGEV large spike peplomer glycoprotein in transgenic potato tuber.

14.1.6. *Vaccine microencapsulation systems — microsphere*

Development of single dose vaccines by microencapsulation of vaccines in biocompatible and biodegradable polymers has been investigated recently. It has been evident that recombinant vaccines encapsulated in biocompatible and biodegradable polymers, known as microsphere are released in a continuous or pulsate manner with reliable and reproducible kinetics, concomitantly stimulating the immune system and obviating the need for booster immunizations in animal models. Results have shown that a single dose of antigen loaded microspheres induce antibody responses equivalent to those obtained with multiple doses of conventional vaccines. Microsphere vaccines have been shown to be effective in stimulating both humoral and cell-mediated immune response. To date, among the several polymers tested, co-polymers of glycolic and lactic acids, poly(DL-lactide-co-glycolide) (PLG), have yielded the most promising results with extensive record of safe use in human and veterinary medicine for drug-release. This polymer degrades by non-enzymic hydrolysis to yield the normal body metabolites lactic and glycolic acids that are easily resorbed or eliminated, and the rate of degradation can vary from a few days to a year. PLG can be processed as micropheres in a range of sizes that are optimal for uptake by macrophages or other antigen presenting cells (Fig.14.22).

Orally ingested proteins are inefficiently taken up by the immune inductive tissues in the gastrointestinal tract, largely because they are degraded by both acid hydrolysis and proteolysis, and are rapidly excreted. Attempts to overcome these problems by increasing the amount of ingested protein, or the frequency of dosage, rapidly leads to the state of oral tolerance, i.e. unresponsiveness to vaccine. Particles size of less than 10 μm are known to be taken up by specialized epithelial cells in the GI

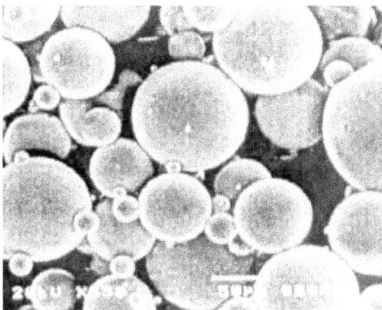

microspheres for intramuscular use (80 μm)

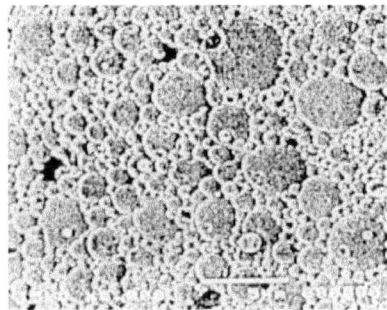

microspheres for oral use (5 μm)

Fig. 14.22.   Microsphere delivery systems.

tract, M cells, which transport the particles directly to the specialized immune inductive tissues, the Peyer's patches. Microspheres have also shown potential as carriers for oral vaccine delivery due to their protective effects on encapsulated antigens and their ability to be taken up by the M cells of Peyer's patches in the intestine. Following their uptake into Peyer's patches, a proportion of microparticles disseminate within macrophages to mesenteric lymph nodes and the spleen, where the vaccines released in a controlled and sustained manner continues, stimulating the systemic immune response. Thus the oral delivery of microencapsulated vaccine can simultaneously induce mucosal and systemic immunity.

Although microsphere encapsulated vaccines have not been used in animal vaccination commercially, many studies have indicated promising results in both human and animal medicine. Studies on PLG encapsulated formalin-inactivated *Salmonella enteritidis* demonstrated over six months of high serum anti IgG in chickens, in contrast to the short-lived response in those immunized with oil-emulsified *S. enteritidis* (Fig. 14.23). Both mucosal and systemic immune responses were detected very early after vaccination when this vaccine formula was immunized orally. Besides, both humoral and cell-mediated immune responses were demonstrated from microencapsulated vaccines such as HIV-1 gp120 to mucosal

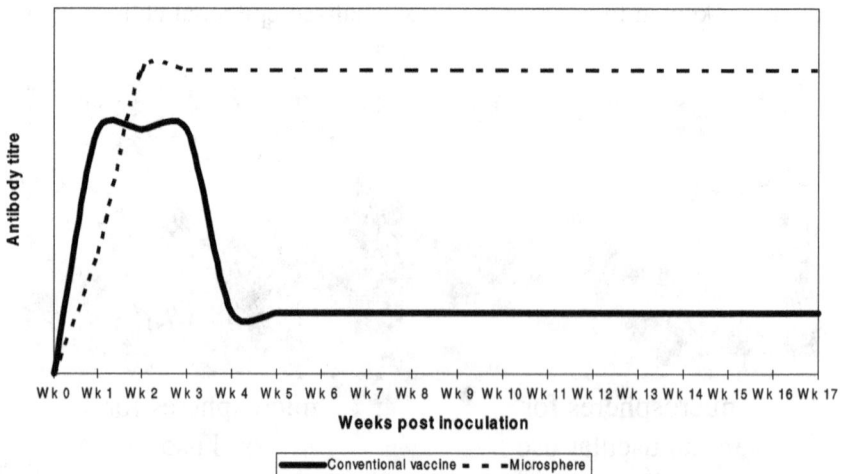

Fig. 14.23.   Comparison of antibody response between conventional vaccination versus microsphere vaccination.

surfaces, and a single oral dose of PLG-encapsulated pertussis fimbriae protects against *Bordetella pertussis*-induced disease following intra nasal challenge. Thus, development of new microencapsulated delivery systems has made killed and subunit vaccines more practical options for parenteral and mucosal immunization.

14.2. Diagnosis of Animal Pathogens

Diagnosis of animal pathogens has been one of the important methods recommended for the control of animal diseases. For example, as the demand for poultry products increases, the issue of public health and food safety assumes greater priority. *S. enteritidis* recorded outbreaks have occurred with frequency worldwide, including in the UK, Europe, and the United States. Whilst such occurrences and studies have been reported in Thailand, which has a large poultry export industry, information on outbreaks in Asia remain limited in terms of acquired information. *S. enteritidis* is accountable for a large percentage of food poisoning cases diagnosed, with some infections resulting in the death of young and old. Food poisoning cases such as these are attributed to the consumption of undercooked and infected poultry meat as well as contaminated eggs (Fig. 14.24).

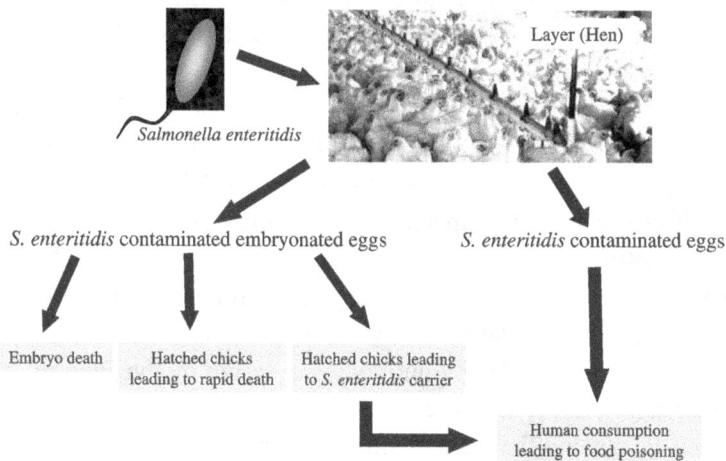

Fig. 14.24.   Food poisoning caused by consumption of poultry meat and eggs contaminated with *S. enteritidis*.

Diagnosis and surveillance of infected flock have thus taken important roles in the management to reduce the likelihood that contaminated meat and eggs will reach consumers. Current techniques employed in the identification of *S. enteritidis* include bacteriological isolation and agglutination assays. In *S. enteritidis* infection, the pathogen colonizes the intestine of the chicken and causes brief mild diarrhea. The fecal material shed from the chicken is thus a means of detecting the pathogen. However, some *S. enteritidis* strains are able to persist in the intestinal tract for a long period. This results in intermittent shedding of fecal material. This makes the conventional bacteriological isolation technique of *S enteritidis* from environmental and clinical samples less reliable in identifying all infected flock. Moreover, the method requires lengthy procedures. On the other hand, the widely used *S. pullorum* slide-agglutination test has been adapted for use with *S. enteritidis*. However, the agglutination-based system has been proven less sensitive. It is advantageous to develop a more specific and purified antigen preparation for the detection of antibodies against *S. enteritidis* infection. Recombinant proteins of high purity are now accepted as reagents for serodiagnosis of *S. enteritidis*. Assays based on recombinant proteins provide a simple, highly sensitive, non-infectious and inexpensive tool for antibody detection. Compared to the conventional bacteriological examination and agglutination assay, this newly developed recombinant protein assay based on ELISA (which will be described in later sections) makes it a valuable tool for the reliable detection of *S. enteritidis* infection.

In another example, Ovine Progressive Pneumonia (OPP), also known as maedi virus or Ovine Lentivirus (OLV) pneumonia, is caused by a non-oncogenic, exogenous retrovirus of the *Lentiviridae* subfamily. The disease is characterized by increasing respiratory distress, chronic weight loss and wasting, and the inevitable death of infected sheep. The disease has been recognized in most of the sheep-producing countries and is endemic in the United States. OPP virus (OPPV) carriers are usually identified by detecting anti-OPPV antibodies in serum by the agar gel immunodiffusion (AGID) test. However this test employs a viral antigen that is produced by an inefficient and time-consuming tissue culture system. Therefore, the AGID test is expensive and insensitive for routine clinical laboratory use.

Second-generation of recombinant protein assays and synthetic oligopeptide assays have been developed to circumvent these problems. Properly selected and prepared recombinant and synthetic peptides used in solid-phase ELISA formats will increase the density of diagnostically relevant epitopes compared with viral lysates, resulting in increased test sensitivity. High compositional purity decreases the likelihood of non-specific (false-positive) reactions. Furthermore, well-defined, uniform, and reproducible recombinant or synthetic peptides have distinct quality control advantages over virus culture-derived proteins when used as anti-gens in serodiagnostic assay. The requirement to maintain the virus itself is also eliminated by recombinant or synthetic antigen production.

While the initial development of recombinant proteins is a biotechno-logically intensive and time-consuming process, it represents a one-time investment of resources and effort. After that, it is possible to produce and purify sufficient recombinant antigen in a week to test several thousand serum specimens. Moreover, the growth and technical requirements for bacterial culture of *Escherichia coli* to produce recombinant proteins are simpler and less expensive than those for tissue culture of viruses to pro-duce conventional antigens. Thus, given the inherently faster and simpler *in vitro* culture of bacteria than that of viruses, recombinant proteins offer the potential for reduced cost per animal tested. Similarly, identification of diagnostically relevant synthetic peptide epitopes is a one-time resource investment. Thus overall conclusion is that second-generation recombi-nant proteins offer the best combination of economy and test performance compared with conventional tests.

There are generally four types of detection for animal diseases, namely direct isolation of pathogens, detection of pathogens, detection of antibod-ies and antigens from pathogen-infected animals, and detection of DNA or RNA sequences from pathogens. Many immunological methods have been developed to detect for antibodies or antigens. Among these meth-ods, the Enzyme-linked immunosorbent assay (ELISA), which detects for antibody or antigen is most commonly used for animal disease diagnos-tics. Foot and mouth disease of cattle is most effectively diagnosed by the presence of antibodies against the virus. Molecular biological methods have used to detect for pathogen nucleic acid. For example, the polymer-ase chain reaction (PCR), which detects for nucleic acid is gradually gaining popularity among clinical laboratories. The technique has been able to diagnose shrimp population for the presence of the White Spot Syndrome Virus (WSSV) by the amplification of a portion of a WSSV genome.

Progress in understanding the molecular biology of animal pathogens and the role of individual proteins in infection and immunity has been slow. In terms of viral pathogens, the low level of viral protein production in tissue culture systems and the difficulty in purification of the virus are probable reasons. As for bacterial pathogens, the probable reason lies in that batch production of antigens from crude extraction of bacteria requires specific growth conditions, and the purified antigens from the cultures are difficult to standardize. These factors make it difficult to obtain adequate quantities of pure antigens for experimentation.

Over the past several years, advances in diagnostic methods based on immunology and molecular biology technology are rapidly invading into laboratories to replace microbiology and virology assay methods, as the choice for diagnosis of animal diseases. Immunological and molecular biological assays greatly improved the diagnosis of many animal infections, especially diseases that cannot be practically detected in most clinical settings due to the difficulties in culturing and isolating of the pathogens. Besides, the main appeal of these methods over microbiology and virology assay systems is the speed that it can offer, in which identification of pathogens can be completed in a very short time, usually less than a few hours in certain cases. Aside from that, these methods are easy to perform, safe, reproducible and some are automated to facilitate the evaluation of large numbers of specimens.

14.2.1. *Direct isolation of animal pathogens*

Early diagnostic applications of animal diseases involved isolation of pathogens followed by various methods for identification and differentiation, in which clinical laboratories are generally able to isolate and determine the most routinely encountered pathogens. To date, direct isolation of pathogens remains the most accurate method of diagnosis for animal infection, with a few exceptions. The system used for pathogen isolation, which can either be *in vitro* or *in vivo*, depends on the nature of the organism being sought. Generally, *in vitro* methods are preferred for the obvious reason of convenience. Briefly, microbiological identification of bacterial diseases included isolation of the pathogens from clinical specimens by microbial culture, followed by identification of organisms, for instance by growing on selective and differential media. On the other hand, diagnosis of viral diseases included isolation of pathogens from clinical specimens by tissue culture systems and examination by electron microscopy. When the identity of the pathogen is uncertain, multiple systems are used for identification. Several animal diseases are still preferably being detected by this method.

Direct isolation of pathogens, however is extremely laborious, time-consuming and not practical for large-scale diagnosis. Besides, pathogens like certain viruses, bacteria and protozoa present diagnostic problems since these organisms are difficult to culture *in vitro*, and efforts to isolate the pathogens are enormous. For instance, the *in vitro* isolation of shrimp pathogens like the White Spot Syndrome Virus (WSSV) is difficult due to the lack of cell lines compatible with the shrimp system. This has been a major impediment to the molecular biology of WSSV and the role of individual proteins in infection. Although many viruses have distinctive morphologies, and diagnostic virology can be done by electron microscope in major specialized laboratories having the appropriate equipment, it is not a confirmatory method and additional experiments need to be performed to confirm the identity of the virus. Also, for organisms like the mycobacterium species, *in vitro* culture systems exhibit extreme slow growth rate (positive cultures are obtained after two to four months). These concerns have been an impetus for the development of diagnostic methods that can accurately identify the pathogen without involving culturing work at all. Immunological diagnostic methods (such as ELISA) are developed for direct detection of pathogen related antigens in the animal and antibody responses induced against the pathogen by the animal. In addition, recent advances in the development of molecular biological technologies (such as PCR and specific nucleic acid probes) have been applied to the detection of pathogen genomes with sensitivity that is comparable to that achieved with detection by direct isolation of pathogen. Needless to say, immunological and molecular assays can function as a surrogate for detection of the infectious agent itself.

14.2.2. *Histopathological detection*

For the detection of the pathogen, a classical case study would be that of the Bovine Spongiform Encephalopathy (BSE). The disease is believed to be caused by infectious agents called prions that have not been isolated as yet. A prion refers to a proteinaceous infectious particle, which is composed only of protein with no detectable nucleic acid. This is unlike all previously known pathogens like bacteria and viruses, which contain nucleic acids to enable them to reproduce.

Prion diseases, caused by a 33–35 kDa glycoprotein prion protein (PrP), are sometimes referred as transmissible spongiform encephalopathies (TSE), as the brains of infected animals were riddled with holes. Other forms of prion diseases include scrapie and Creutzfeldt-Jakob Disease (CJD). In mammals, PrP exists as a normal cellular protein in a stable shape. The cellular function of the PrP protein is not known, but it has been speculated that PrP involves in Cu^{2+} transport and Cu^{2+} homeostasis. The protein is easily degraded by cellular proteases and hence does not cause disease (cellular PrP). However, a variant form of PrP (scrapie PrP) exists, which is resistant to protease degradation. This form of PrP is believed to cause TSE. Further studies revealed that the difference between a cellular PrP and a scrapie PrP lies in the conformation. A scrapie protein contacts normal PrP proteins and somehow causes them to unfold and flip from their usual conformation to the abnormal shape of the scrapie protein. This initiates a cascade, in which the newly converted molecules change the shape of other normal PrP molecules. In this way, the scrapie protein propagates itself. Thus prion diseases arise from aberrations of protein conformation.

It was also discovered that prions prefer to interact with PrP of homologous or similar compositions. Therefore, if the amino acid sequence of a variant PrP molecule resembles the PrP molecules in that of a host, it is likely that the host will acquire prion disease. Sheep PrP and bovine PrP share much homology in their amino acid sequence. This probably explained why cattle in UK contracted the prion disease when they were fed with sheep tissue-tainted feed.

The development of a diagnostic test for BSE test is particularly difficult because the specific infectious agent has not been isolated and there is no detectable immune response, which are the two prerequisites for developing diagnostic tests. Another problem, which hinders the diagnosis of BSE, is that it takes at least two and half years for abnormal PrP to accumulate and eventually cause enough brain damage for the symptoms to be visible. Under such circumstances, there would be little point in testing the brain tissue of slaughtered cattle at abattoirs at between 18 and 24 months of age, because there would be no detectable infectivity even if they are infected. The test would give a false negative result and provide false reassurance. The current methods of diagnosis are histopathology and immunocytochemistry. The latter is adopted if histopathology does not give a definitive diagnosis. Hence, immunocytochemical tools using polyclonal or monoclonal antibodies against the prions are employed as an effective diagnosis for the presence of the pathogen.

14.2.3. *Detection of antigens and antibodies from pathogen-infected animals*

Detection of antibody response induced against pathogen by the animal is based on the fundament of antibody-antigen interactions. The interaction between antibody and antigen is highly specific, where each antibody molecule is able to detect for one molecule of the corresponding antigen out of more than 10^8 similar molecules. This profound specificity of antigen-antibody interactions has lead to the development of a variety of assays for animal disease detection. These diagnostic methods differ in speed, sensitivity, and application, with some solely for detection of diseases, while others being able to detect as well as quantitatively determine the antigens or antibodies detected in the animals. Some of these diagnostic methods can only be used to detect for either antigens or antibodies, while others can detect antigens as well as antibodies at any one time, based on the presented format of the assay. The more commonly applied assays for animal disease diagnostics including immunodiffusion, immunofluorescence, and ELISA will be discussed.

14.2.3.1. *The nature of antigen-antibody interaction*

The interaction between an antibody and an antigen involves non-covalent interactions between the epitope of the antigen molecule, and the variable (V_H/V_L) domain of the antibody molecule, particularly the hypervariable complementarily-determining regions (CDRs). These non-covalent interactions that form the basis of antigen-antibody binding include hydrogen bonds, ionic bonds, hydrophobic interactions, and van der Waals interactions. Each of these non-covalent interactions generally occurs in close proximity, less than 1×10^{-7} mm in most cases, resulting in a very close fit between the antibody and antigen, which reflect the high degree of specificity characteristic of antibody-antigen interactions. Furthermore, in order to form a strong bond, large number of non-convalent interactions is involved in any one antibody-antigen binding, since the strength of each non-covalent interaction is weak (Fig. 14.3).

Although antigen-antibody reactions are highly specific, an antibody elicited by one antigen can cross-react with an unrelated antigen in some cases. Such cross-reactions occur if two different antigens have shared epitope(s) or if antibodies specific for one epitope also bind to an unrelated epitope possessing similar chemical properties. For such reasons, detection for specific antigens is sometimes performed with monoclonal antibodies instead for polyclonal antibodies, while the detection of specific antibodies is done with an oligopeptide coding for the epitope of the pathogen, or a recombinant expressed polypeptide, which is derived from the pathogen. Generally, the level of specificity required for disease detection depends largely on the pathogen of interest. Several viruses and bacteria possess antigenic determinants identical to or similar to the normal host-cell components. Indeed in some cases, these microbial antigens have been shown to elicit antibody that cross-reacts with host-cell components, resulting in a tissue-damaging autoimmune reaction. For instance, the bacterium *Streptococcus pyogenes*, expresses cell-wall proteins, known as M antigens. Antibodies produced against these antigens have been shown to cross-react with the glycoproteins of heart valves, antigens of joints and soluble components of glomerular basement membrane. This immunologic reaction resulting from the cross reaction has been implicated in rheumatic fever and acute glomerulonephritis following streptococcal infections.

14.2.3.2. *Assays based on immunodiffusion*

The purpose of an immunodiffusion technique is to detect the reaction of an antigen with an antibody by a precipitin reaction. Precipitin reaction is based on the formation of visible precipitate in solution from the antibody-antigen lattice, which involves the interaction of antibody-antigen complexes. Formation of an antibody-antigen lattice depends on the number of specific binding site on both antibody and antigen, in which the antibody must be bivalent, and the antigen must have multiple epitopes. When the ratio of antigens to antibodies is in optimal proportion (equivalent zone), extensive cross-linking formation occurs. Thus, formation of lines of precipitin in any immunodiffusion system is highly dependent on relative concentrations of the antigen and antibody. In the region of excess antibodies (prozone phenomenon), each antigen molecule is covered with antibodies, preventing cross-linkage and thus there is no precipitin reaction. In the region of antigen excess, each antibody molecule is bound to a pair of antigen molecules. Further cross-linking is impossible and thus no precipitin reaction can occur (Fig. 14.25).

Amount of antibody precipitation

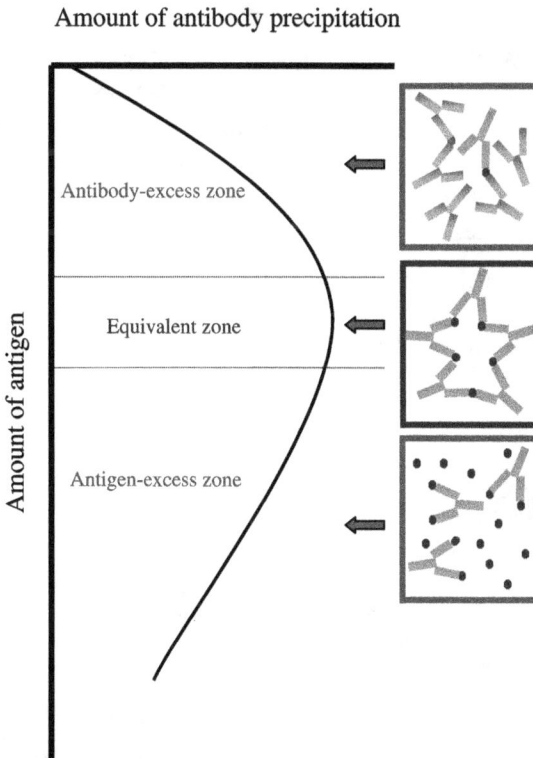

Fig. 14.25.   A quantitative precipitation curve showing the effect of adding increased amounts of antigen to a constant amount of antibody.

Many diagnostic methods have been developed based on the precipitin reaction for qualitative and quantitative detection of either antibodies or antigens. The most basic format of precipitin reaction based assay is the agar gel immunodiffusion (AGID) and radial immunodiffusion. AGID is a qualitative assay, while radial immunodiffusion is a quantitative assay. The application of radial immunodiffusion is based on the principle that a quantitative relationship exists between the amount of antigen placed in a well that is cut in an agar antibody plate and the resulting ring of precipitin. In this method, antibodies are incorporated into an agar gel and precisely measured amounts of antigens are added into the wells. Following diffusion of the antigen into the agar containing antibodies, a ring of precipitin will form at the equivalent zone if the incorporated antibodies are specific for the antigens. The diameter of the ring formed represents the antigen concentration determined by comparing with antigen standards (Fig. 14.26). One major drawback of immunodiffusion assays is that at least 18–24 hours are required before precipitin lines appear. This limitation is overcome by using countercurrent electrophoresis, as seen in assays like electroimmunodiffusion, which exhibit greater speed and sensitivity than in immunodiffusion assays. In the electroimmunodiffusion assay,

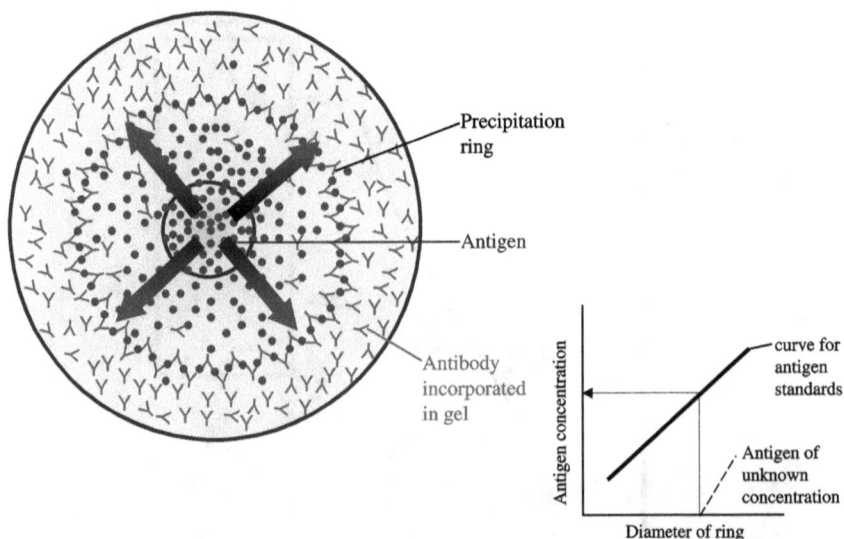

Fig. 14.26.   Precipitin reaction — radial immunodiffusion, a quantitative assay.

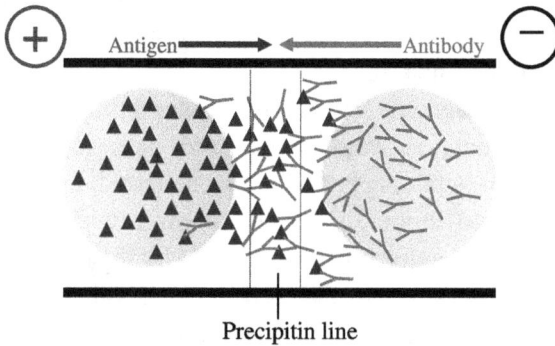

Fig. 14.27. Precipitin reaction — electroimmunodiffusion. Migration of antigen and antibody towards each other in an electrical field results in the formation of a precipitin line.

positively charged antibody and negatively charged antigen are first added into separate wells in the gels. When directed by an electrical field, the antigen and antibody migrate towards one another, resulting in a sharp precipitin line within minutes (Fig. 14.27). Although these assays are seldom used for the detection of animal diseases today, a few pathogens like the causative agent of Rinderpest, Equine Infectious Anemia, Ovine Lentivirus, Bluetongue, Lumpy Skin Disease and Sheep Pox still rely on this assay as one of their diagnostic methods.

14.2.3.3. *Enzyme-linked immunosorbent assay*

Enzyme-linked immunosorbent assay, commonly known as ELISA, is the most popular diagnostic method among the existing immunological assays for animal disease detection, favored by its simplicity and sensitivity. In principle, ELISA employs an enzyme label for detection of antibody-antigen complexes formed on a solid phase. Alkaline phosphatase and horseradish peroxidase are the most commonly employed enzymes, which can be chemically coupled to antibodies under conditions that retain the biological properties of both components of the conjugate. Detection of antibody-antigen complexes is based on the enzyme catalytic activity of an appropriate colorless substrate to give a colored product, whose intensity is measured by the optical density. A variety of ELISA formats have been developed for detection and quantification of either antigen or antibody. The following are the most commonly used formats for the detection of animal diseases.

Indirect ELISA diagnoses for animal diseases through the detection and quantification of serum antibodies specific for pathogen related antigens. Serum or samples containing antibodies to be tested, referred as primary antibody, is added to an antigen-coated microtiter well to react with the bound antigens. If the antigen is being sought by the test sample, specific antibodies will be attached to the well, with unbound antibodies being washed away. Following this antibody-antigen interaction, secondary enzyme-conjugated antibody (e.g. horseradish peroxidase) specific for the primary antibody is added for reaction. After unbound reporter antibodies are washed away, the respective substrate for the enzyme is added, and the converted colored product is measured by specialized spectrophotometric plate readers, which can measure the absorbance of a plate of 96 samples in less than a minute (Fig. 14.28). The indirect ELISA has been the method of choice to detect for serum antibodies against many animal pathogens, including Porcine reproductive and respiratory syndrome virus (PRRSV), the causative agent of the porcine reproductive and respiratory syndrome which causes major losses in the pig industry. In this assay, either the whole virus or recombinant expressed capsid protein (ORF 7)

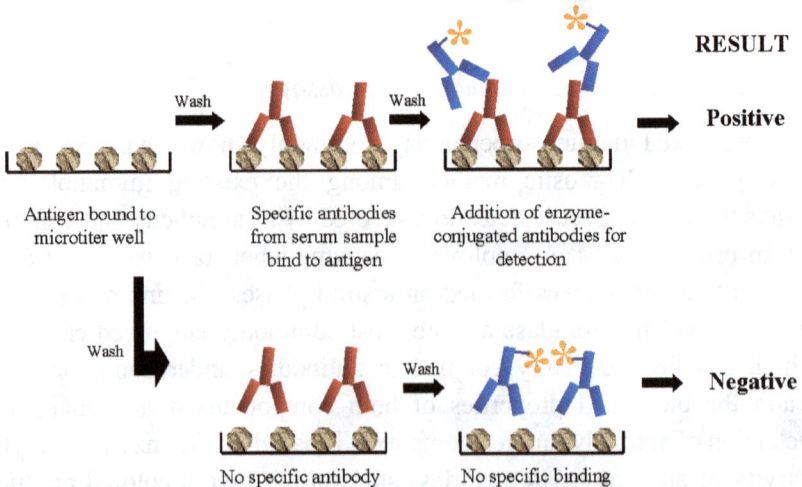

Fig. 14.28.   Procedures for indirect ELISA.

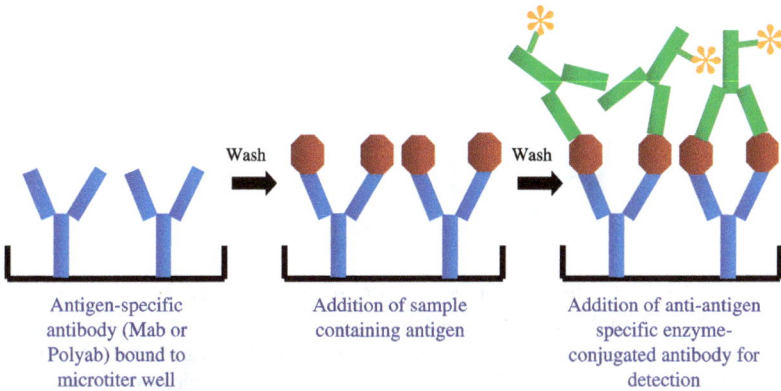

| Antigen-specific antibody (Mab or Polyab) bound to microtiter well | Addition of sample containing antigen | Addition of anti-antigen specific enzyme-conjugated antibody for detection |

Fig. 14.29.   Sandwich ELISA procedures for antigen detection.

of PRRSV is adsorbed as solid-phase antigen to microtiter wells. When an animal is infected with PRRSV, it may provoke the production of specific antibodies to epitopes on the whole virus, resulting a positive signal upon measurement.

A sandwich ELISA allows detection and quantification of pathogen related antigen. In this case, the antibody specific for the antigen is coated to the surface of microtiter wells to which the antigen containing test sample is added. After unbound antigens are washed away, the antibody-antigen complexes are detected by an enzyme-linked antibody, which is specific for a different epitope on the antigen. Next, unbound reporter antibodies are being washed away, the substrate is added and the colored reaction product is measured (Fig. 14.29). The sandwich ELISA has been the method of choice for the detection of many animal pathogens such as Avian Leukosis Virus (ALV), Classical Swine Fever (CSF) and Feline Leukemia Virus (FeLV) (Fig. 14.30).

An epitope-blocking ELISA allows detecting specific antibodies to certain pathogen in human or animal sera, such as H5N1 influenza virus. The assay relies on an H5 specific monoclonal antibody recognizing a conserved epitope in H5 viruses. The monoclonal antibody will compete with H5 specific antibodies from sera samples to bind to H5 antigen coated. The blocking efficiency will be evaluated, which corresponds to the level of specific antibodies in the sera samples. Though the H5N1

strong positive   negative

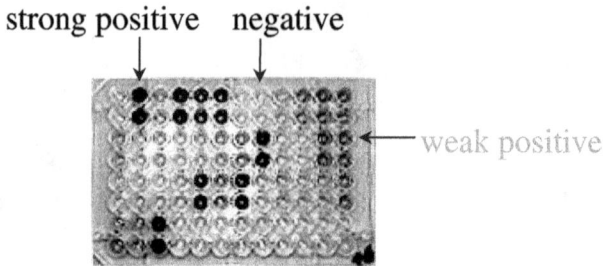

weak positive

Fig. 14.30.   Color development in an ELISA reaction correlates to the amount of antigen or antibody in the sample. A reaction that gives a colored product of strong intensity indicates a positive sample, while a reaction that gives no colored product indicates a negative sample. A colored product of intermediate intensity indicates a weak positive sample.

ELISA (indirect ELISA) has been widely used in serologic surveillance of chicken and turkey flocks, cross-reacting antibodies elicited by infection or vaccination with seasonal influenza virus can yield false positive H5N1 test results that reduce the value of indirect H5N1 ELISA in animals and humans. The sensitivity and specificity of the epitope-blocking ELISA for H5N1 were evaluated using chicken antisera to multiple virus clades and other influenza subtypes as well as serum samples from individuals naturally infected with H5N1 or seasonal influenza viruses. The epitope-blocking ELISA results were compared to those of hemagglutinin inhibition (HI) and microneutralization assays. Antibodies to H5N1 were readily detected in immunized animals or convalescent human sera by the epitope blocking ELISA whereas specimens with antibodies to other influenza subtypes yielded negative results. The assay showed higher sensitivity and specificity as compared to HI and microneutralization (Fig. 14.31).

14.2.3.4. Immunoblotting assay

Immunoblotting assay, or Western blotting (WB), is a diagnostic method to identify either a pathogen-related antigen in a complex mixture of proteins, or an antibody specific for a pathogen related antigen of precise molecular weight. As its name implies, the principle of WB is similar to Southern blotting, which detects for DNA fragments, and Northern blotting which for detects mRNAs. In WB, proteins, following separation by

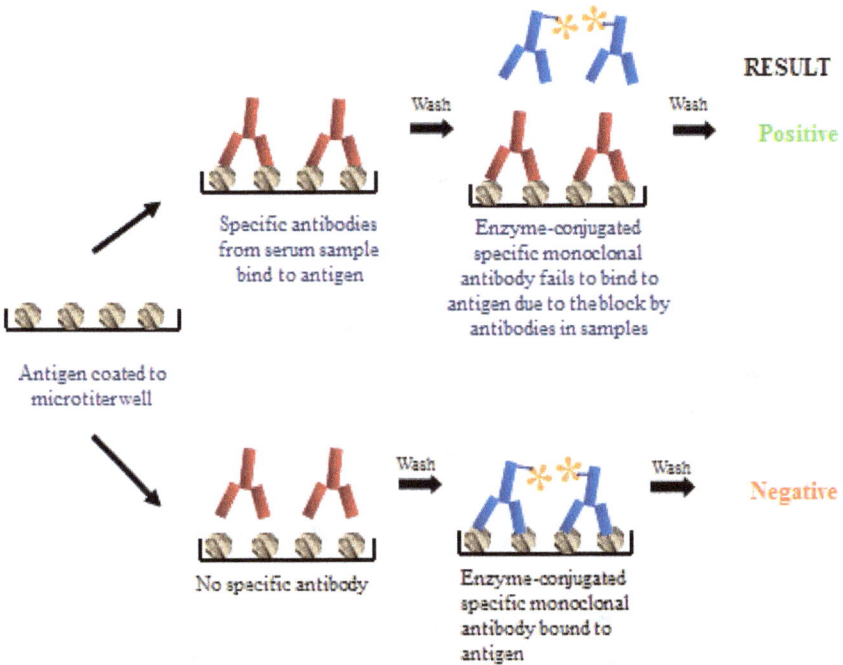

Fig. 14.31.   Blocking ELISA procedures for antibody.

electrophoresis under denaturing conditions, are transferred or blotted onto a solid support, usually a nitrocellulose or nylon membrane prior to being detected by either enzyme-labeled or radioisotope-labeled antibodies. The antigen-antibody complexes are then visualized either by a substrate conversion signal or autoradiography, depending on the reporter antibody used (Fig.14.32). Similar to ELISA, indirect immunoblotting can be performed to detect antigen-antibody complexes by a secondary labeled anti-isotype antibody with improved sensitivity. Many different reporter systems are available for detecting antigen-antibody complexes in immunoblotting, with enzymatic-catalyzed methods being widely preferred over autoradiography. Enzymatic detection systems included chemiluminescence detection that provides signal that is captured by film darkening, and chromogen detection where enzyme-labeled reagents is used to catalyze a chromogenic substrate to give a color product, generating a visual signal. One drawback of the WB assay is that it maybe inconvenient

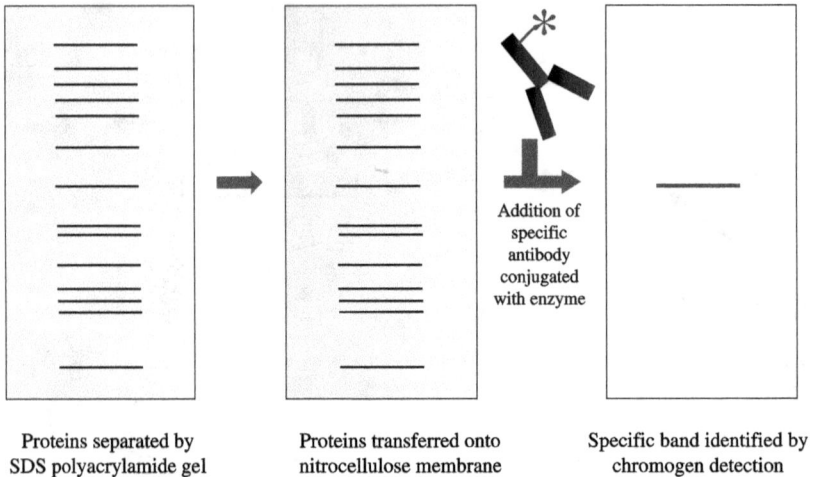

| Proteins separated by SDS polyacrylamide gel | Proteins transferred onto nitrocellulose membrane | Specific band identified by chromogen detection |

Fig. 14.32.   Western blotting procedure.

to screen large numbers of samples for epidemiologic studies. However, WB is a sensitive and specific alternative to the ELISA. If an ELISA result disagrees with the known immunologic status of an animal, then WB or other serologic tests of equal or higher sensitivity should be used to validate the ELISA result. Pathogens like Human Immunodeficiency Virus (HIV), the presumptive diagnosis of the infection is made by the detection of antibodies in patients by ELISA. However, as there are some false-positives with this test, the definitive diagnosis is made by WB.

Mycoplasma hyopneumoniae is the etiological agent of enzootic pneumonia of swine, a chronic non-fatal disease resulting in poor food conversion, retarded growth and predisposition to bacterial pneumonia. The disease is spread worldwide and causes large economic losses in swine production. Transmission occurs through direct contact with respiratory secretions from carrier animals and via sow-to-piglet transmission.

Methods developed for the detection of *M. hyopneumoniae* have been based mainly on whole cell or membrane preparations and gene fragments. These include culture of *M. hyopneumoniae* from pig lungs, immunofluorescence assay, DNA probes, ELISA and PCR. Most of these methods lack specificity due to cross-reactions with the closely related swine mycoplasmas, *M. flocculare* and *M. hyorhini*. In addition, these

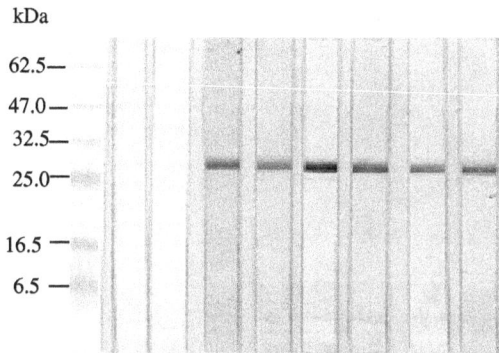

Fig. 14.33.   Western blotting detection of *Mycoplasma hyopneumoniae* infection in pigs using ABC transporter ATP binding protein. The brown band indicates that the animals have been infected. No band indicates that the animals are not infected. The blue bands on the left belong to the protein marker which are used to estimate the size of the protein.

methods have limited sensitivity and involve sophisticated laboratory procedures. Proteins that are predominantly antigenic in pigs infected with *M. hyopneumoniae* are important targets for the development of specific diagnostic tools. Research has shown that detection for *M. hyopneumoniae* specific antibodies by western blotting technique using either the L-lactate dehydrogenase or the ABC transporter ATP-binding protein analogue/ multidrug resistance protein homologue, PR2 of *M. hyopneumoniae* were highly specific for *M. hyopneumoniae* (Fig.14.33). These two proteins were chosen as potential diagnostic antigens based on their strong antigenicity and their roles in virulence of *M. hyopneumoniae*.

14.2.3.5. *Immunofluorescence assay*

Immunofluorescence assay is a technique where pathogen related antigens in tissue sections or cells are identified and localized by antibodies labeled with a fluorochrome (Fig. 14.34). Fluorescein isothiocyanate and rhodamine isothiocyanate are the most commonly employed fluorochrome, which can be conjugated to an antibody molecule under conditions that retain the biological properties of both components of the conjugate. Each of these dyes can be excited by a unique wavelength to emit a longer wavelength that can be viewed with the fluorescence

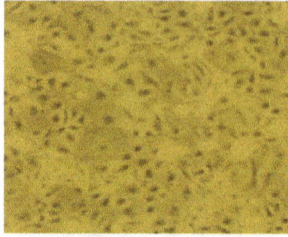

Cells viewed under regular light.

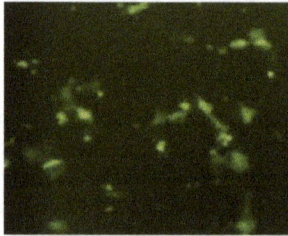

Cells viewed under fluorescent light. The cells show fluorescent staining, indicating a positive sample.

Cells viewed under fluorescent light. The cells do not show any fluorescent staining, indicating a negative sample.

Fig. 14.34.   Immunofluorescence assay.

microscope. Fluorescein absorbs blue light (490 nm) and emits an intense yellow-green fluorescence (517 nm), while rhodamine absorbs in the yellow-green range (515 nm) and emits a deep red fluorescence (546 nm).

Like ELISA, antigens in tissue section or cells reacted with primary unlabeled antibody can be detected with secondary fluorescein-labeled anti-isotype antibody by the indirect immunofluorescence method for increased sensitivity. Another reagent used for indirect immunofluorescence is the fluorochrome-labeled protein A of *Staphylococcus aureus,* which has high affinity for the Fc region of IgG, and can be used if the binding complex contains IgG antibody (Fig.14.35).

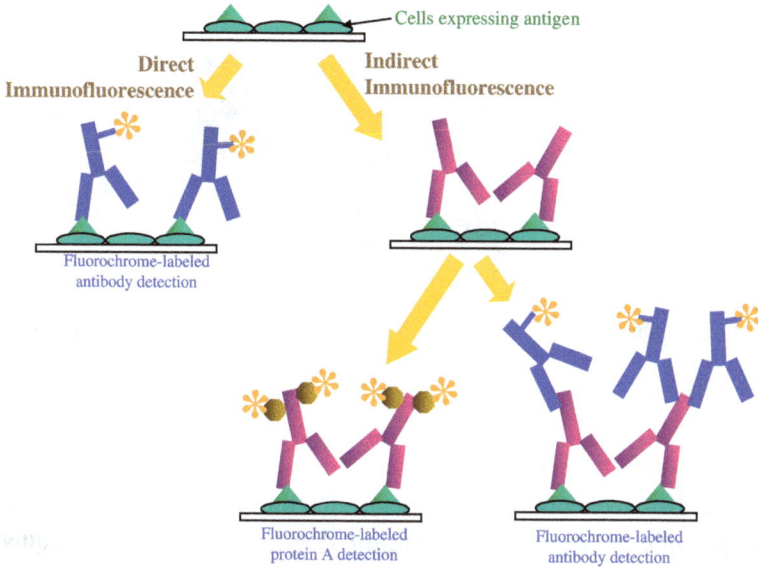

Fig. 14.35.   Immunofluorescence assay using fluorescent antibodies for antigen detection.

Although not often used as the main assay, immunofluorescence has been applied as a confirmatory detection for several animal pathogens, including African Swine Fever virus, Pox Virus Disease of cattle and Equine Piroplasmosis. African Swine Fever (ASF) is an infectious disease of domestic pigs caused by a virus that produce a range of syndromes varying from acute to chronic. ASF cannot be differentiated from classical swine fever by either clinical or post-mortem examination and differential diagnosis between the two is essential. Direct immunofluorescence has been used to detect ASF antigen in tissues of suspected pigs in the field or those inoculated at the laboratory, and indirect immunofluorescence has been used to detect for ASF-specific antibodies.

14.2.3.6. *Quartz crystal microbalance*

Microgravimetric quartz crystal microbalance (QCM) is a measuring device that is based on the most frequently used acoustic wave sensor, the Piezoelectric quartz crystal. This combination of piezoelectric quartz

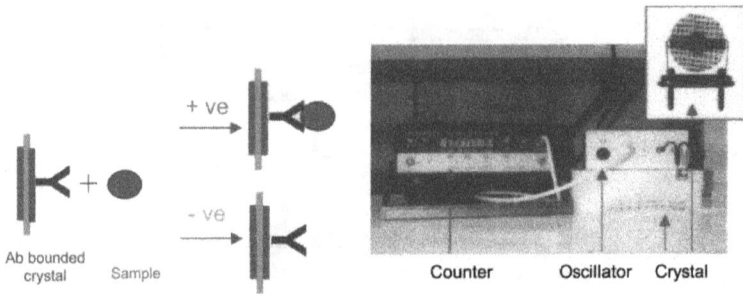

Fig. 14.36.   Quartz crystal microbalance — the principle and instruments involved.

crystal devices with biological active elements has brought about a new generation of immunoassay. This technique consists of a quartz crystal wafer sandwiched between two metal electrodes (Fig.14.36). The electrodes provide means of connecting the device to an external frequency oscillator circuit that drives the quartz crystal at its resonant frequency. This frequency is dependent on the mass of the crystal and hence any changes in the mass on the surface of the electrode can be detected as a digital reading. For the detection of animal infection, antigen-antibody interaction results an alteration to the crystal mass, thereby causes a shift in frequency. Based on the frequency information generated from the immuno-interaction, QCM devices enable detection and quantification of antigen or antibody.

Most diagnostic methods for animal diseases are conducted in a laboratory environment and hence require the sample to be delivered prior testing. Some animal diseases like *Salmonella enteritidis (S. enteritidis)* and Porcine Reproductive and Respiratory Syndrome Virus (PRRSV) require regular screening at the various stages of the growing cycles. Therefore, the QCM technique provides an on-site analysis that can be performed routinely with relatively low overheads. Like many immunoassay techniques, alternative assay designs and formats, for instance direct assay, indirect assay and sandwiched assay, have been developed with the aim to increase detection sensitivity, specificity and the range of QCM applications. The veterinary applications of QCM developed included the Africa swine fever virus detection in pigs, adult worm antigen of *S. japonicum* detection in rabbits, and PRRSV infection in pigs.

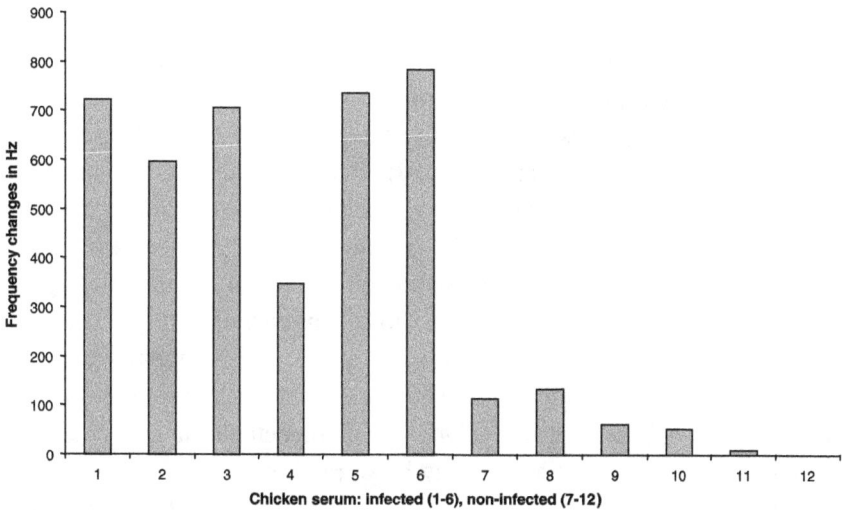

Fig. 14.37.   Chart depicting the frequency changes in Hertz for infected and non-infected chicken serum. Lanes 1–6: serums from infected chickens; Lanes 7–12: serums obtained from non-infected chickens.

S. enteritidis is a major problem for the poultry industry and has been one of the dominant pathogens that causes food poisoning in humans. The QCM technique developed for the diagnosis of *S. enteritidis* infection in chicken were immobilized with recombinantly expressed *S. enteritidis* flagellin antigens that served to capture the pathogen-specific antibodies in infected sera. Binding of the specific antibodies to the immobilized antigens causes changes to the crystal mass, and thus resulting a shift in frequency, which is detectable by the oscillator. This procedure was used in a screening of chicken sera, non-infected and infected with *S. enteritidis*, and was found that the serological readings obtained from infected chicken and those from non-infected chickens could be distinguished (Fig. 14.37). Thus, this technology could provide another method of the screening of *S. enteritidis* infected flocks.

14.2.4. *Detection by pathogen nucleic acid*

The progress made in nucleic acid technology for diagnostics is enormous in relation to the period of time during which it has been applied to the development of assay systems as compared to that during which phenotypic *in vitro* analysis had developed. Indeed to date, nucleic acid technology has assumed an essential role in various areas of *in vitro* diagnosis, including animal diseases. Basically, this approach employs genotypic materials (DNA and RNA) as diagnostic targets for identifying suspected pathogens. The rapid advancement of nucleic acid diagnostic is a consequent of two essential advantages: (1) nucleic acids can be rapidly and sensitively measured, and (2) the sequence of nucleotides in a given gene is highly specific and it can be used to distinguish closely related serotypes. Besides, nucleic acid detection offers many advantages over immunological assays. Nucleic acids are more stable than proteins to high temperatures, high pH, organic solvents, and other chemicals, hence clinical samples can be treated in a relatively harsh manner without destroying the nucleic acid for detection. Tools use for nucleic acid detection application like DNA primers and probes are more defined entities than antibodies and antigens, and the composition of these tools can be accurately checked by sequence analysis, and produced in DNA synthesizers whenever necessary. The more commonly applied nucleic acid assay technology in animal disease diagnostics include polymerase chain reaction and nucleic acid probes, and will be discussed in the following sections.

14.2.4.1. *Polymerase chain reaction*

The polymerase chain reaction (PCR) diagnostic method detects for specific nucleic acid sequence found in the genome of pathogens. It is based on DNA replication *in vitro* to produce large quantities of a target sequence of pathogen isolated from a complex mixture of heterogeneous DNA molecules (e.g. host genome, nucleic acids from other pathogens present). The reaction is set up using the following reagents: reaction buffer, DNA template, deoxynucleotides (dNTPs), primers, and thermostable DNA polymerase. A pair of primers is used for the reaction. The promers are single-stranded DNA molecules of about 20–30 nucleotides long (oligonucleotides), are designed from the sequences flanking the two ends of the target genome. These primers are specific to the pathogen to maximize the specificity of the assay. The amplification of DNA by the PCR is accomplished through cycling a succession of incubation steps of different temperatures optimized for DNA replication. Briefly, a sample containing mixture of DNA molecules is heat-denatured to separate complementary strands of DNA into single-stranded molecules for reaction. Next, the reaction is brought down to a lower temperature to allow pathogen-specific primers search and anneal to the complementary sequence on opposite strands of the target DNA. This is followed by DNA extension at a higher temperature, with DNA polymerase adding on dNTPs using the annealed DNA sequence as a template to produce copies of the targeted sequence. The three steps make up one cycling reaction, with each cycle producing duplicates of the DNA targets. The cycle of steps is then repeated, in which the newly synthesized DNA strands can also serve as templates for primer extension. These steps are repeated 20–40 times, yielding exponential amplification of the target DNA sequences (Fig. 14.38). The amplified products can then be visualized on a DNA gel, or further assayed by nucleic acid probing for increased sensitivity and specificity. The technique offers a powerful approach to the diagnosis of infectious diseases that are defined at the nucleotide sequence level.

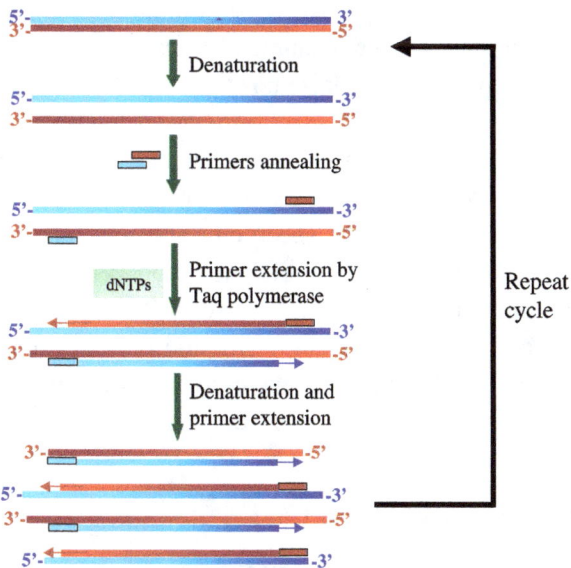

Fig. 14.38.   Steps in each cycle of polymerase chain reaction.

For RNA viruses, a complementary DNA (cDNA) copy of the RNA genome must be made with reverse transcriptase (RT) prior to the performing of PCR. One such example is the detection of nodavirus in marine fish. The virus is gaining more importance due to the high mortality it causes in larvae and juveniles of marine fish. In addition, it is widespread in water bodies including the Pacific, Mediterranean and Atlantic Oceans. In general, the earlier the commencement of the disease, the greater is the rate of mortality, which may even reach 100%. Therefore, it is essential to develop a specific and sensitive method for the detection of very early stage of nodavirus infection in diagnostic and epidemiological work. A reverse transcription polymerase chain reaction (RT-PCR) method to detect nodavirus based on specific primer sets has been developed and found to be more sensitive than histopathological and immunological methods.

PCR offers a major advantage of detecting pathogens in a single cell or viral particle. This saves the hassle of isolating the pathogens in culture. PCR is a highly sensitive procedure and may prove to be very useful in the diagnosis of chronic-persistent infections such as those caused by retroviruses. This disease presents serious problems in terms of diagnosis and prevention, as the infected animal often does not demonstrate clinical signs until it is at an advanced stage of disease. Furthermore, infected animals appear to be a constant potential source of transmission. However, due to the exquisite sensitivity of PCR, false-positives misidentified from contamination of samples can be possible. To overcome such problem, a second set of primers that are complementary to the internal sequence of the target DNA can be used to amplify a subfragment from the PCR product of the first reaction. This increases the specificity and sensitivity of the assay, since spurious and non-specific amplification products are eliminated. This procedure is commonly referred to as "nested PCR". Alternatively, diagnostic methods like nucleic acid probes, restriction fragment length polymorphism, or sequence analysis can also be used to confirm the identity of the PCR product and to further characterize the genome.

14.2.4.2. *Real time quantitative PCR*

Nowadays, a new method of PCR quantification has been invented. It is called "real-time PCR" because it allows the researchers to actually view the increase in the amount of DNA as it is amplified. Several different types of real-time PCR are being marketed to the scientific community at this time, each with their advantages. TaqMan® real-time PCR and SYBR® Green are two major types used in recently years. The TaqMan® probe is required during TaqMan® real-time PCR. The probe consists of two types of fluorophores, which are the fluorescent parts of reporter proteins. While the probe is attached or unattached to the template DNA and before the polymerase acts, the quencher (Q) fluorophore reduces the fluorescence from the reporter (R) fluorophore. The reporter dye is found on the 5' end of the probe and the quencher at the 3' end. Once the TaqMan® probe has bound to its specific piece of the template DNA after denaturation (high temperature) and the reaction cools, the primers anneal to the DNA. Taq polymerase then adds nucleotides and removes the Taqman® probe from the template DNA. This separates the quencher from the reporter, and allows the reporter to give off its emit its energy. This is then quantified using a computer. The more times the denaturing and annealing takes place, the more opportunities there are for the Taqman® probe to bind and, in turn, the more emitted light is detected. The SYBR® Green dye was the first to be used in real-time PCR. It binds to double-stranded DNA and emits light when excited. Therefore, the more PCR products are made, the more fluorescence signals are emited. Specific probe is not required for SYBR® Green real time PCR. Unfortunately, it binds to any double-stranded DNA which could result in inaccurate data, especially compared with the specificity found in the Taqman® probe.

Quantification of targets mainly relies on special softwares to measure the fluorescence signals. The specifics in quantification of the light emitted during real-time PCR are fairly involved and complex. The values obtained do not have absolute units associated with them, but a fraction or ratio of the sample relative to the standard. The significantly increased sensitivity of real time PCR, as compared to other diagnosis methods, makes it be applied to rapidly detect nucleic acids that are diagnostic of infectious diseases, though at a high cost. The introduction of real-time PCR assays to the clinical microbiology laboratory has significantly improved the diagnosis of infectious diseases, and is deployed as a tool to detect newly emerging diseases, such as flu, in diagnostic tests.

14.2.4.3. *Nucleic acid probes*

The fundament of the specific association of nucleic acid sequences by complementary base pairing through hydrogen bonding, which centralizes on the basis of adenine complementary to thymine and uridine, while guanine complementary to cytosine, has allowed the development of several nucleic acid diagnostic methods for the detection of animal diseases. Nucleic acid probes are short fragments of nucleic acid sequences that detect specific sequences associated with pathogen by nucleic acid hybridization through the use of highly conserved sequences. If pathogen contains sequences complementary to the probe, the two sequences can hybridize to form a double-stranded molecule. The probe can be chemically synthesized according to the target gene sequence, which can be derived if a short amino acid sequence of the protein encoded by the target gene is known. During synthesis of the probe, labeled nucleotides are incorporated, so as to enable detection of the probe after hybridization. The types of labeled probes include radioactive (which gives a radioactive signal), biotin (which generates a color) and chemiluminescent (which is enzyme-linked).

Nucleic acid hybridization is often performed in the form of Southern blotting which detects for DNA and Northern blotting which detects for RNA. As its name implies, the principle is similar to that of Western blotting, which detects for protein. In both Southern blotting and Northern blotting, denatured DNA and RNA samples are separated by electrophoresis and immobilized onto a solid support system, usually nitrocellulose or nylon membrane respectively, prior to detection with radioisotope-labeled or enzyme-labeled sequence-specific probe(s). Hybridization is allowed to occur at an optimized temperature in which considerable sequence homology between the target sequence and the probe is necessary to form a stable duplex. Following washing to remove unhybridized probe, the hybridization is detected according to how the probe was labeled.

Nucleic acid probes can be designed for identifying all serotypes or some important serotypes of a pathogen. However, probes used alone may not be very sensitive. Although not practiced in the diagnosis of animal diseases, nucleic acid hybridization procedures can be combined with PCR to provide a powerful tool, which exhibit remarkable specificity and sensitivity. It is especially useful when the target pathogen cannot be cultivated, or is difficult to cultivate. On the other hand, nucleic acid probes in diagnostics is designed for use in the food industry, such as for the detection of *Salmonella* and *Staphylococcus* employ probe dipsticks, which remove hybridized DNA from liquid solution. For this method, a two-component probe is used, in which one serves to bind to the specific sequence of the pathogen, and the other serves to bind to the dipstick. Following hybridization of the two-component probe to target sequence, the dipstick is inserted into the hybridization solution to remove the hybridized DNA for measurement.

14.3. Conclusion

Animal diseases continue to affect global animal industries. In 2001, Foot and Mouth Disease (FMD) outbreaks in UK resulted in an estimated loss of ten billion pounds (about 1.1% of the country's GDP). A better known example is the mad cow disease or Bovine Spongiform Encephalopathy (BSE), which caused great alarm among the consumers. The disease caught attention with worldwide paranoia over the consumption of British beef. Though the spread of the disease hit headlines across the globe, it was the grim tales of lost lives due to human variants of the disease that caught the attention of people. The UK witnessed about 100 cases of the medically relevant form of BSE, called the vCDJ or variant Creudzfeldt-Jakob Disease that affects humans.

In Asia, the deadly H5N1 Fowlpox influenza virus killed six people in 1998 and wreaked havoc in Hong Kong's poultry industry. The Nipah virus cost the Malaysian government about one billion US dollars. All pig farms were cleared and a number of countries banned imports of pork from Malaysia. All these examples illustrated that disease attacks on farm animals do not just simply affect the agricultural sector alone. In some cases, the massive outbreaks had rendered the farm products unfit for human consumption, leading to bans on the imports of these products in many other countries. Considering the vast market for cattle, swine and poultry products, the economic losses were definitely severe.

The reason for such losses in most instances was due to the absence of fast and early diagnosis of disease. If the infected animals were detected before clinical symptoms of disease were visible, action could be taken before the rapid spread of disease had occurred. Early diagnosis and detection is a real need to maintain a disease-free and healthy environment for animals reared for human consumption. Reliable diagnostic tests that are rapid, sensitive and affordable should be developed for use by farmers and veterinarians. The veterinary diagnostic test kit market has experienced tremendous growth during the last ten years. However, diagnostic testing is still not keeping pace with the magnitude of the infectious disease problem in either the efficiency required to economically test large numbers or in the flexibility to accurately test for different diseases in the veterinary market. Therefore, improved testing must include simplicity and economy, which integrates into a reduced reliance in laboratories and skilled technicians. In this respect, emphasis has to be placed on the development of user friendly approaches like those based on recombinant technology. By monitoring which pathogens persist in a herd and when transmission occurs, producers and their veterinarians can make informed decisions about health control of the herd.

Vaccines play a major role for disease control programs because highly intensive livestock farming has created problems associated with disease spreading, which contributes to the increase of new infections in a herd. In the United States, losses of cattle, swine and chicken cost the farmers billions of dollars annually. When talking about such a tremendous loss, one will then realize the importance and necessity of such vaccines that can salvage the lives of livestock. Vaccines that could prevent losses from infectious diseases would have a tremendous impact on animal industries. The new technology opens doors to biologicals that were never available before. Current advancement of recombinant technology has greatly improved vaccinology to meet today's need for an ideal vaccine. These recombinant vaccines include gene-deleted vaccine, DNA vaccine, subunit vaccine, vector vaccine, and so forth. In addition, adjuvant technology, vaccine delivery system and associated testing technology have since progressed drastically to help fulfill the criteria of an ideal vaccine. With these new technologies merging together, the search for the "new-age" vaccinology and companion diagnosis technology, which are essential for today's agricultural needs, should be near. There is a general acceptance that major advances in veterinary vaccine over the next few years will result from such approaches.

When a proper program of vaccination and early disease diagnosis is in place, its impact is tremendous. First, vaccination allows the animals to build up strong lasting immunity against prevalent diseases. With the present advances in vaccine technology, alternative methods of vaccination such as microencapsulated vaccines can be adopted to reduce stress in the animal. Secondly, a good farm management system should incorporate reliable diagnostic tests as a routine to screen the farm population for early signs of diseases. In actual fact, the preventive measure of vaccination, as well as early diagnosis of the disease through rapid testing are crucial in limiting the risk of diseases. Farmers will therefore benefit greatly as their profits are increased. For the consumers, they can enjoy a constant supply of safer animal products and need not fear contracting the disease directly from the products they eat.

References

1. Bright RA, Carter DM, Daniluk S, Toapanta FR, Ahmad A, Gavrilov V, Massare M, Pushko P, Mytle N, Rowe T, Smith G and Ross TM. Influenza virus-like particles elicit broader immune responses than whole virion inactivated influenza virus or recombinant hemagglutinin. *Vaccine* **25**(19): 3871–3878. Epub 2007 Feb 15.

2. Ghendon YZ, Markushin SG, Akopova, II, Koptiaeva IB, Nechaeva EA, *et al.* Development of cell culture (MDCK) live cold-adapted (CA) attenuated influenza vaccine. *Vaccine* **23**: 4678–4684, 2005.

3. Joseph T, McAuliffe J, Lu B, Vogel L, Swayne D, *et al.* A live attenuated cold-adapted influenza A H7N3 virus vaccine provides protection against homologous and heterologous H7 viruses in mice and ferrets. *Virology* **378**: 123–132, 2008.

4. Kim JK, Seiler P, Forrest HL, Khalenkov AM, Franks J, *et al.* Pathogenicity and vaccine efficacy of different clades of Asian H5N1 avian influenza A viruses in domestic ducks. *J. Virol.* **82**: 11374–11382, 2008.

5. Romanova J, Katinger D, Ferko B, Vcelar B, Sereinig S, *et al.* Live cold-adapted influenza A vaccine produced in Vero cell line. *Virus Res.* **103**: 187–193, 2004.

Further Reading

1. Aiello SE, ed. *The Merck Veterinary Manual,* 8th ed. Merck and Co, New Jersey, 1998.
2. Calnek BW, Barnes JH, Beard CW, McDougald RL and Saif YM, eds. *Diseases of Poultry.* Iowa State University Press, Ames, 1997.
3. Leman AD, Straw BE, Mengeling WL, D'Allaire S and Taylor DJ, eds. *Diseases of Swine,* 7th ed. Iowa State University Press, Ames, 1992.
4. Lists of A and B: Diseases of Mammals, Birds, and Bees. *Manual of Standards for Diagnostic Tests and Vaccines, 3rd ed.* World Organization for Animal Health, Paris, 1996.
5. Schultz RD, ed. *Veterinary Vaccines and Diagnostics*, Advances in Veterinary Medicine, Vol. 41. Academic Press, San Diego, 1999.
6. Tizard I. *Veterinary Immunology: An Introduction*, 4th ed. W.B. Saunders Company, Philadelphia, 1992.
7. Wesley T. *Veterinary Vaccines: Products and Markets*, Animal Pharm Reports. PJB Publications, UK, 1999.

Websites

*   Surveillance
    http://www.fao.org/WAICENT/FAOINFO/AGRICULT/AGA/AGAH/EMPRES/Info/other/surveill.him#Early%20Detection
*   Animal Diseases
    BSE
    http://w3.aces.uiuc.edu/AnSci/BSE/
    http://sparc.airtime.co.uk/bse//welcome.htm
    FMD
    http://www.thepigsite.com/FeaturedArticle/Default.asp?AREA=FeaturedArticle&Display=305
    Poultry
    http://www.ianr.unl.edu/pubs/AnimalDisease/g1039.htm

*   Directory to Veterinary Science
    http://www.mic.ki.se/Diseases/c22.html

Immunology
  http://www-micro.msb.le.ac.uk/MBChB/ImmGloss.html
  http://golgi.harvard.edu/BioLinks/Immunology.html
  http://www.whfreeman.com/kuby/
  http://esg-www.mit.edu:8001/esgbio/imm/monoclonal.html*Vaccine*
  http://www.sciam.com/1999/0799issue/0799weiner.html
  http://www.sciam.com/2000/0900issue/0900langridge.html

Recombinant DNA Technology
  http://esg-www.mit.edu:8001/esgbio/rdna/rdnadir.html

Immunological Tests
  http://www.1rc.arizona.edu/courses/mmiweb/lake/immunology_modules/intro/immunology.html

---

Questions for Thought...

1. How do the humoral-mediated and cell-mediated arms of immune response limit the spread of viral infection in the body?
2. When a chicken is infected by *Salmonella enteriditis*, how does the animal's defense mechanism function? Describe the interactions between B cells, T cells and macrophages that will bring about the immune response, as well as the other molecules are involved.
3. What are the important considerations in developing an effective vaccine for use in animal farms? Compare and contrast the differences between conventional and recombinant vaccines.
4. *Edwardsiella tarda* is a Gram-negative bacterium that causes a systemic infection, edwardsiellosis, in fish. In order to develop suitable vaccines, how can the virulence factors of this organism be identified and studied?
5. Describe how a recombinant vaccine was developed for the eradication of Aujeszky's disease or pseudorabies in pig farms.
6. Explain how large amounts of purified antigen can be produced economically for vaccination.
7. What are the principles behind the use of DNA vaccines?

8. Recombinant vaccines offer many advantages over conventional vaccines. How can they be administered into the animal to ensure their effectiveness?

9. With the use of examples, explain how cytokines may serve as immunoadjuvants.

10. What are the problems faced by fish farms in carrying out mass vaccination? How can they possibly be overcome?

11. What were the problems in diagnosing the White Spot Syndrome Virus (WSSV) in shrimps in the past? What are the alternative methods adopted nowadays?

12. Explain with examples why some methods of detecting pathogen in animals are time-consuming and may not be effective.

13. How can the principle of antibody and antigen interaction be applied to the immunological diagnosis of animal diseases?

14. How do the relative concentrations of antibody and antigen molecules determine the formation of precipitin lines in an immunodiffusion reaction? How can this be applied to the quantitative detection of antigen or antibody?

15. How is the enzyme-linked immunosorbent assay applied in other formats for the detection and quantification of either antigen or antibody?

16. Describe how the principle of immunoblotting can serve as a sensitive and specific diagnostic test for *Mycoplasma hyopneumoniae* infection in pigs.

17. Compare and contrast between enzyme-linked immunosorbent assay and immunofluoresence assay.

18. How can the quartz crystal microbalance method be used in screening of chicken sera for *Salmonella enteriditis* infection?

19. What advantages does polymerase chain reaction offer in the diagnosis of nodavirus in marine fish?

20. What is the significance of performing early diagnosis of disease and mass vaccination in managing an animal farm?

Chapter 15

Transgenic Plants

Hong Yan and Bu Yun Ping
Temasek Life Sciences Laboratory
National University of Singapore
1 Research Link, Singapore 117604

15.1. Increasing Adoption of Agrobiotechnology

World population is projected to grow from 6.1 billion in 2000 to 8.9 billion in 2050, an increase of 47% (United Nations, 2004). On the other hand, there is less farming population and less farm land available with the rapid urbanization of developing countries like India, China and Brazil. Facing this challenge of utilizing less farmland to feed more population on earth, agrobiotechnology is without doubt the way forward if not the only way. Agrobiotechnology will not only increase productivity to achieve food supply sustainability, but also it will generate health care products and industrial raw materials and contribute to sustainable usage of natural resources. Since the first commercially grown genetically modified (GM) crop FLavrSavr tomato (modified to ripen without softening) in 1991, there have been numerous GM crops approved for commercialization. GM crops quickly gained larger market every year and spread from the US, Canada and Europe to many developing countries. If success is measured by increases in global acreage, GM crop would most certainly been successful. In 2011, 160 million hectares of GM crops were planted in 29 countries, with an 80-fold increase from 1996 to 2009 and year-to-year growth of 12 million hectares or 8% in acreage (Clive, 2011). Share of global biotech crop by developing countries was around 50% in 2011 and is expected to increase further. The global market value of GM crops was US$10.5 billion (US$8.2 billion was in the industrial countries and US$2.3 billion was in the developing countries). To date, the predominant GM crops are maize, soybean, cotton, canola and potatoes. In addition, there have been field trials of GM plants from at least 52 species including all the major field crops, vegetables, and several herbaceous and woody species. An increasing area of transgenic crops with higher yields, improved quality and enhanced resistance against herbicide, diseases and pests is being grown in the US and other parts of the world. If agrobiotechnology is able to gain more consumer acceptance, it will undoubtedly lead to the second green revolution.

15.2. Plant Transformation

15.2.1. *Transgenic technology*

There are several techniques to introduce foreign genes into plant cells to generate transgenic plants (Table 15.1) but only two — *Agrobacterium* and biolistics — are widely used.

Agrobacterium-mediated transformation is the most widely used method for plant transformation. *Agrobacterium tumefaciens* is a soil phytopathogenic bacterium causing crown gall diseases in some plants. Wild type agrobacterium contains a circular and double-stranded plasmid (the tumor-inducing or Ti plasmid) with the size of 200 kb. Ti plasmid contains virulence genes necessary for transformation (vir genes), an origin of replication and the part to be transferred to host genome (T-DNA, ~25 kb), which consists of sequences for auxin and cytokinin production and nopaline or octopine production. These elements in T-DNA are bordered by two border sequences (Right and Left borders). Since the border sequences define the region to be translocated into host genome, the border sequences were cloned into a plasmid vector and the auxin, cytokinin and nopaline/octopine producing elements and border sequences were deleted from Ti-plasmid. A simple vector (binary vector) with border sequences, selection markers in *E. coli* / Agrobacteria as well as origins or replication in *E. coli* and Agrobacteria is commonly used to clone expression cassettes of target genes and a selection marker within the border sequences. Agrobacteria containing the binary vector is then used to infect plant cells (explant), which were allowed to regenerate whole plants in the presence of a

Table 15.1.   Transformation techniques.

Technique	Features/Advantages	Disadvantages
Indirect DNA delivery		
Agrobacterium-mediated (Herrera-Estrella *et al.*, 1983)(Herrera-Estrella, 1983 #222)(Fromm, 1985 #225)	Higher chance of single copy insertion, higher transformation efficiency for dicotyledonous plants	Many plant species are recalcitrant to agrobacterium mediated transformation
Direct DNA delivery		
1) Biolistics or particle bombardment (Herrera-Estrella, 1983 #222) (Klein *et al.*, 1987)	Easier to perform, can be used for transformation of monocotyledons and difficult plant species; can be also used for transformation of mitochondria and chloroplasts	Unpredictable integration pattern, often with multiple copies that may result in instability of transgene and gene silencing
2) Protoplast systems: including PEG fusion (Krens *et al.*, 1982)		
3) Electroporation (Fromm *et al.*, 1985) (Fromm, 1985 #225)	Higher transformation efficiency	Regenerating protoplasts is difficult
4) Microinjection (Crossway *et al.*, 1986)		
5) Pollen tube-mediated approach (Wemmer *et al.*, 1994)	Free of tissue culture and germplasm-independent	Multiple insertion, difficult selection and inefficiency
Chloroplast transformation (Boynton and Gillham, 1993)	Multiple copy insertions, stable transgene and high level expression; maternally inherited, can prevent gene spread through pollination. Foreign genes are inserted via targeted homologous recombination into the plastid genome	The absence of the tissue-specific control mechanism, successful only for few plant species

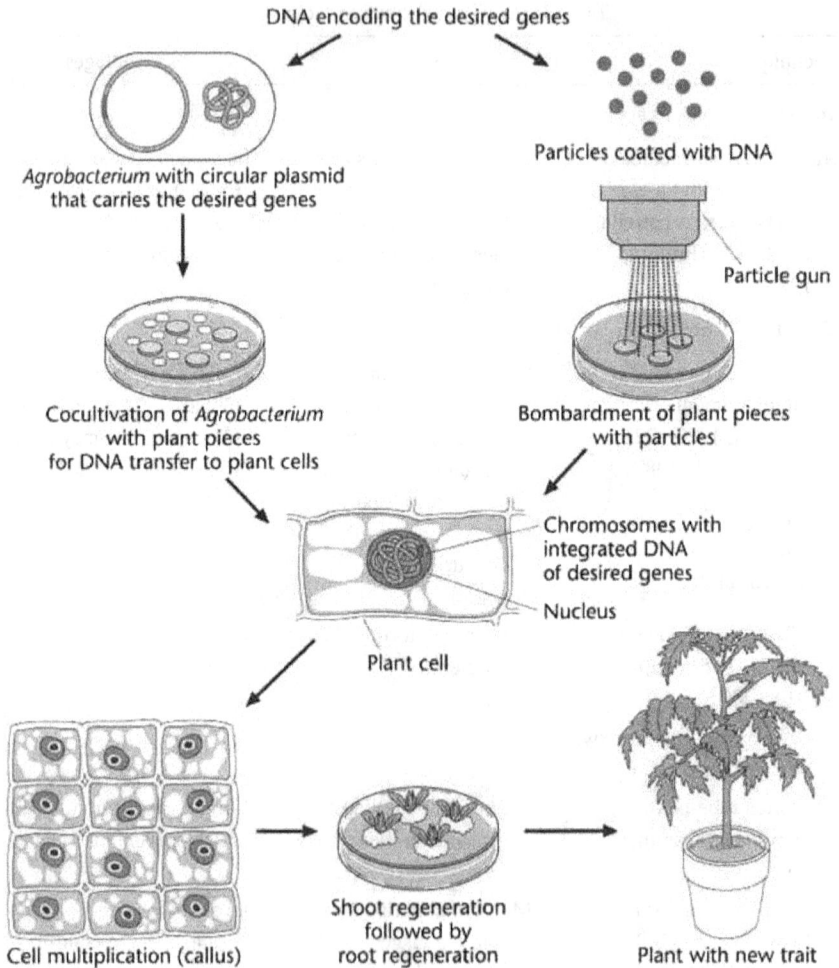

Fig. 15.1.   Schematic representation of two different methods to create transgenic plants. In the *Agrobacterium* method (left), DNA carrying desired genes is inserted into agrobacterium and, when the bacterium infects wounded tissue, this DNA is transferred to a cell nucleus and integrated into the chromosome. In the particle gun method, metal particles coated with DNA are fired into plant cells and the DNA becomes integrated into the plant chromosome. When a new plant is regenerated from a single plant cell, all the cells in the plant carry the new genes (reproduced from Chrispeels and Sadava, 1994 with permission).

selection agent. Regeneration of whole plant can be achieved either through formation of new shoots and roots (organogenesis) or formation of new embryos (somatic embryogenesis). Regeneration procedures are generally species or even variety specific and it takes time and effort to develop.

Biolistics or particle bombardment, also known as the "gene gun", is another way to introduce desired genes into plants. A target gene cassette together with a selection marker cassette is firstly cloned into a plasmid. Biolistic transformation procedure involves firing plasmid DNA-coated gold or tungsten particles (about 0.6 to 1.0 μm in diameter) at high speed by the rapid release of high pressure (100–2,200 psi or 7–152 bar) helium gas into the target cells. Some of the metal particles enter a plant cell without causing lethal damage and deliver the construct so that the foreign DNA may stay inside a plant cell and incorporate into plant genome. The transformed cells are then induced to form plants under selection which only permits regenerates expressing the selectable marker to grow.

A transgene consists of expression cassettes, each cassette has three components: a promoter that drives the transcription of a downstream target gene, a target gene that will confer a desirable trait followed by a terminator sequence. Two cassettes are usually necessary in one transgene: one allows selection of transgenic tissue or plants (marker gene) while the other will give a desired phenotype in transgenic plant. Transgenes with multiple cassettes can also be used. For agrobacterium mediated transformation, two border sequences are essential to define the region to be transferred. Other additional elements like replication initiation site and selection marker in agrobacterium are also required, which are typically outside of the two borders.

15.2.2. *Source of genes*

As genetic engineering has broken down the species boundary, genes and promoters for transformation can be obtained from a wide range of sources, such as another plant species from a different genus or different family, an animal species, a fungus or bacterial and even synthesized *de novo*. Up to now, the complete genome sequences of many organisms are available (Tzotzos *et al.*, 2009). Whole genome sequence provides scientists vast information about gene structure and possible functions.

Furthermore, functional genomics based on random T-DNA insertion, gene trap and enhancer trap technologies with transposons, microarray and proteomics will provide detailed and definitive information on functions of individual genes in the development of a particular organism and its interactions with different environmental factors. With spatial and temporal patterns of promoters and gene functions characterized, scientists will have endless source of genes and promoters to create a lot of tailor made plants with traits like higher productivity, reduced use of toxic chemicals, production of more nutritious and more tasty foods, production of raw materials for different industrial and pharmaceutical production as green factory, clean up organic and heavy metal pollution in the environment, and producing food-medicine nutraceuticals to reduce the medical bills and taking care the "grey population".

15.2.3. *Promoters for gene expression*

The gene of interest is assembled into a construct (usually a plasmid) which comprises all the elements required for successful transformation and expression of the target genes. To express transgenes in plant cells, appropriate promoter sequences have to be introduced upstream of a gene to ensure efficient transcription of mRNA at locations and time periods when the gene expression is required. A promoter may be constitutive (expressing the gene in most, if not all, cells), tissue dependent (expressing only in certain tissues such as roots, leaves, stems, fiber, or seed) or inducible (expressing only when a certain inducer, usually a chemical, is applied).

The first generation of transgenic plants almost all used constitutive promoters like 35S promoter, ubiquitin promoter and actin promoter. CaMV 35S (or derivatives of it) originated from the cauliflower mosaic virus and, although not completely constitutive, it produces continuous gene expression in most tissues of the plant. However, levels of gene expression have been reported to vary between different species of plant and different parts of the plant. Continuous gene expression in all plant tissues is likely to increase the risk of the pests developing resistance, resulting in yield penalties as the plant directs more resources than necessary to its defense, causing gene silencing, and raising concerns about the safety of the foreign genes inside the edible parts of GMO crops, such as allergic reactions. Using constitutive promoters like CaMV35S to control the synthesis of antimicrobial or insecticidal products, plant growth and development are occasionally affected.

For the second generation of transgenic plants, tissue-specific promoters will be widely used for leaves, roots, stems, fiber, grain and so on to improve efficiency for disease and pest control, to reduce the waste of resources in host plants, to enhance agronomic traits and/or to minimize or avoid foreign gene products in parts consumed as food. An example is the deployment of a phloem-specific promoter for genes providing resistance to phloem-sucking insect pests such as aphids. With the use of such tissue specific promoter, foreign gene expression is restricted to the area that is required. There is no foreign gene expression in seed that is consumed as food. Another example is the fiber-specific promoters isolated from cotton, which will be widely used to increase cotton yield, to improve fiber quality, and to produce naturally colored cotton fibers. Other specific promoters include seed-specific and pollen-specific promoters, and promoters especially suited for use in monocotyledonous plants.

There is also the movement towards using inducible promoters, which are very useful to stress tolerance. One example is the use of wound induced promoters, which lead to gene expression only when the plant is actually attacked.

The use of chemical inducible system allows the control of the activity of the target gene in a reversible and temporally defined manner. This is especially useful when studying gene function that would interfere with the regeneration process. With the ability to control the expression of transgene with chemical, scientist can restrict the target transgene expression to specific organs, tissue and cell type. In addition, with careful and judicious application of inducers, it is possible to regulate gene expression in transformed plants at temporal specific manner, such as a specific developmental stage or a controlled duration of time. A chemically inducible promoter with its inducer should only affect the expression of the transgene. This strategy favors the use of well characterized regulatory elements from distant and well studied organism like Drosophila, *E. coli* or mammalian cells. An ideal chemically inducible promoter should have

Table 15.2.   A list of chemical-inducible systems in plants (Zuo and Chua, 2000).

System	Transcription	Inducer
Repression systems	TetR	Tetracylcine
	tTA	Tetracylcine
Activation systems	GVG	Dexamethasone(dex)
	AlcR	Ethanol
	GVGEc	RH5992
	ER-C1	β-estadiol
	XVE	β-estadiol
Dual-control	TGV	Dex and tetracycline

a low background activity in the absence of the inducer and demonstrate high expression in the presence of the inducer to allow a significant change in the transgene expression. In general, chemical inducible promoter can be activation system or repression system.

15.2.4. *Marker genes*

As only a proportion of bacterial cells (in which the construct is amplified) and plant cells are transformed, in order to select the successful transformants, selectable marker genes are introduced alongside the target gene to allow the separation of cells that have incorporated the new genes from untransformed cells. There are different selection marker genes used (Table 15.3). The most commonly used marker is the antibiotic-resistance genes, such as neomycinphosphotransferase-II gene (nptII) and hygromycin-phosphotransferase gene (hpt, hph or APHIV). The second commonly used marker is herbicide-resistant genes, such as BAR, EPSPS, PAP, ALS etc., which use different herbicides to select the true transgenic plants. To alleviate safety concern for using antibiotic resistance or herbicide resistance marker genes, genes that allow metabolic selection for transgenic plants have been developed and utilized. One such example is PMI gene coding for phosphomannose isomerase, which will allow a transgenic plant utilize mannose as the carbon source. This marker gene was used to generate new generation of Golden Rice (Hoa *et al.*, 2003).

Table 15.3.   Examples of marker genes.

Antibiotic resistance

  aad gene (resistance to streptomycin and spectinomycin) (B, P)

  bla gene (resistance to ampicillin) (B, P)

  hpt gene (resistance to hygromycin) (B, P)

  *npt*II gene (resistance to kanamycin and neomycin) (B, P)

  tet gene (resistance to tetracycline) (B, P)

  cat gene (resistance to chloramphenicol) (B, P)

Herbicide tolerance

  ALS gene (tolerance to chlorosulpharan) (P)

  BAR gene (tolerance to phosphotricine, glyphosinate, bialophos) (P)

  EPSPS (tolerance to glyphosate) (P)

  PAP (tolerance to glufosinate-ammonium) (P)

Metabolic genes

  Phosphomannose isomerise (PMI) gene metabolizes mannose to give
  fructose (P)

B indicates that selection operates in bacteria, **P** selection operates in plants.

15.2.5. *Marker free transgenic plants*

Selection markers are necessary in transgenic plants for selection purposes but they are no longer required in mature plants. The continuous presence of marker genes in plant and seeds often raises concern and it is regarded as unnecessary, undesirable and unacceptable. There are questions about biosafety of the selectable marker which includes medical implications of consuming GM food and environmental implication of growing GM crops. For example, if the marker genes code for toxic product or allergen, if they will be transferred to its related species or gut microorganisms in human and animals by horizontal transfer? This has prompted development of strategies of deleting marker genes from transgenic plants. One strategy is site-specific excision, which takes advantage of the ability of site-specific recombinases from microbial systems to cleave a marker gene within two specific sites (Hare and Chua, 2002). Recombinase can be under control of a chemical inducible promoter, which could activate recombinase genes after the selection phase of plant production to create marker free transgenic plants. One variation of this strategy is to introduce recombinase in hybrid plants between one transgenic plant with markers genes between excision sites with another plant expressing a recombinase. Another strategy is to conduct co-transformation of two gene constructs, one containing a utility gene and the other containing a marker gene. A transgenic event with both constructs can be selected, allowed to grow and selfed. Marker free transgenic plants can be screened from F1 population.

15.3 Food with Better Nutrition

15.3.1. *Vitamins and minerals*

15.3.1.1. *Golden rice*

Cereals are generally high in energy and food yield but low in other important nutrients, such as essential amino acids and vitamins needed for normal body functions. Rice centered diet caused serious public health problems among people in many countries in Africa, Asia, and Latin America, whose populations (around three billion people, approximately half of the global population) rely exclusively on rice as their staple food. As many as 127 million people and 25% of pre-school children in the developing countries suffer from Vitamin A deficiency (VAD), around 250,000 to 500,000 of them eventually become irreversibly blind annually (Clive, 2009). This is because human and vertebrate animals do not synthesize carotenoids, and they depend on dietary carotenoids (mainly β-carotene) for making their retinoids (such as the main human visual pigment retinal, the vitamin A retinol and retinoic acid, the substance controlling morphogenesis). The Recommended Dietary Allowance (RDA) for vitamin A is 1000 retinol equivalents, equal to 6 mg β-carotene per day. However, rice endosperm does not normally contain any carotenoids. Vitamin A supplementation in developing countries is conducted by the FAO, but it is expensive (costing around US$500 million a year), not sustainable, and it cannot reach remote areas.

GM rice with β-carotene (called Golden Rice for its yellow color) offers a practical biotech crop remedy that provides cost-effective and efficient protection against VAD. In 1984, Dr. Peter Jennings, a rice breeder at IRRI, conceived the Golden Rice initiative because he wanted to alleviate Vitamin A deficiency in rice consuming populations. Initial success of Golden rice was accomplished in 2000 by introducing three transgenes (daffodil phytoene synthase, a lycopene β-cyclase and a carotene desaturase from bacteria) into rice endosperm under the control of the endosperm-specific glutelin and CaMV 35S promoters (Ye *et al.*, 2000). However, its beta carotene content was low at 1.6 to 1.8 μg/g, only a tenth of the RDA assuming daily intake of 300 g rice. Further technology improvement resulted in development of Golden Rice 1 with higher β-carotene content (6 to 8 μg/g). Success in both elite Indica and Japonica rice varieties was achieved using a single construct as well as the use of metabolite selection marker instead of antibiotics selection marker (Hoa *et al.*, 2003). In 2005, Syngenta researchers developed Golden Rice 2, a GM Japonica rice line contained a better phytoene synthase from maize that produced up to 36.7 μg/g beta carotene — more than a four-fold increase compared with Golden Rice 1 (Paine *et al.*, 2005). The Golden Rice 2 lines were donated by the developer to the Humanitarian Board. A single copy Golden Rice 2 event GR2G was selected for further breeding into the most promising and popular rice varieties in VAD prone areas. Philippines, India, Bangladesh, Vietnam, and Indonesia were identified as the countries where the GR2G would be the only event to move forward through regulatory approvals and eventually released. It is expected that Golden Rice will be released in the Philippines and Bangladesh as early as 2012, followed by India, Indonesia and Vietnam. The genetically engineered rice lines make it possible to provide sufficient vitamin A in a typical Asian rice diet. Using similar technical approach, β-carotene has also been introduced into maize, canola seeds and tomato fruits (Zhu *et al.*, 2008).

15.3.1.2. *Food with more minerals*

Minerals, in the context of the human diet, are inorganic chemical elements (or more properly their dissociated ions) that are required for biological or biochemical processes including the accumulation of electrolytes. Carbon, hydrogen, nitrogen and oxygen are excluded from the list as these are found in common organic molecules. There are 16 essential minerals, but 11 of them are required in such small amounts and/ or are so abundant in food and drinking water that deficiency arises only in very unusual circumstances. The remaining five are present in limiting amounts in many foods, so a monotonous diet can easily result in deficiency. These minerals are iodine, iron, zinc (Zn), calcium (Ca) and selenium (Se) (Table 15.4). Staple food crops, in particular cereal grains, are poor sources of the five key mineral nutrients. As a result, the world's poorest people, generally those subsisting on a monotonous cereal diet, are also those most vulnerable to mineral deficiency diseases, although Ca deficiency is also widespread in the industrialized world. Various strategies have been proposed to deal with micronutrient deficiencies including the provision of mineral supplements, the fortification of processed food, the biofortification of crop plants at source with mineral-rich fertilizers and the implementation of breeding programs and genetic engineering approaches to generate mineral-rich varieties of staple crops.

Genetic engineering is the latest weapon in the armory against mineral deficiency. It uses advanced biotechnology techniques to introduce genes directly into breeding varieties. The genes can come from any source (including animals and microbes) and are designed to achieve one or more of the following goals (Ghandilyan et al., 2006; Bauer and Bereczky, 2003; Zhu et al., 2007):

- Improve the efficiency with which minerals are mobilized in the soil
- Improve the efficiency with which minerals are taken up from the soil into the roots of the plant
- Improve the transport of minerals from the roots to storage tissues, such as grain
- Increase the capacity of storage tissues to accumulate minerals in a form that does not impair plant vegetative growth and development, but remains bioavailable for humans
- Reduce the level of antinutritional compounds such as phytic acid, which inhibit the absorption of minerals in the gut
- Increase the level of nutritional enhancer compounds such as inulin, which can enhance the absorption of minerals in the gut.

Table 15.4.   Average reference daily intakes (RDIs) and tolerable upper limits (ULs) for the five key minerals — iodine, iron, zinc, calcium and selenium — in US adults.

Element	RDI	UL
Iodine (μg)	150	1100
Iron (mg)	8–18	45
Zinc (mg)	8–11	40
Calcium (mg)	1,000–1,200	2500
Selenium (μg)	55	400

Reproduced from USDA data (http://fnic.nal.usda.gov/).

The major advantages of genetic engineering over conventional breeding are the diversity of the source of genetic information, the speed with which modified elite varieties can be generated and, perhaps most important for the future, the fact that nutritional traits for different vitamins and minerals can be stacked in the same plant without highly complex breeding programs.

Most of the transgenic approaches for mineral biofortification reported in the literature thus far have focused on improving Fe accumulation and bioavailability. Since mineral Fe in the soil exists predominantly as Fe, which cannot be absorbed by plants, two major strategies have been developed to enhance Fe uptake from the rhizosphere. Strategy I involves the expression of Fe reductases, allowing the subsequent absorption of Fe, whereas strategy II involves the absorption of a complex made up of secreted phytosiderophores and Fe (Ishimaru et al., 2006). Plants have therefore been transformed with genes encoding transporters, reductases and enzymes involved in phytosiderophore biosynthesis (Bauer and Bereczky, 2003; Ghandilyan et al., 2006). The greatest success has been achieved by overexpression of the iron storage protein ferritin in transgenic rice seeds under the control of an endosperm-specific promoter (Goto et al., 1999). Fe levels increased in edible tissues, with the highest levels exceeding 35 mg kg^{-1}. The combined expression of ferritin and phytase (to increase absorption) had been achieved in rice and maize thus increasing Fe levels and availability in simulated digestion/absorption trials (Drakakaki et al., 2005).

Since Fe and Zn share certain channels and transporters, efforts to improve Fe levels in plants often have concurrent positive effects on Zn levels (Suzuki *et al.*, 2006; Ghandilyan *et al.*, 2006). Zn accumulation improved in cereal grains overexpressing ferritin, in the best cases reaching nearly 50 mg kg⁻1 (Vasconcelos *et al.*, 2003).

Genetic engineering strategies for increasing the Ca content of crops have focused on the overexpression of Ca^{2+}/H^+ antiporters located in the vacuolar membrane. The Arabidopsis cation exchanger 1 antiporter (CAX1), which is important for Ca^{2+} homeostasis in plant cells, enhances the level of bioavailable Ca in transgenic potato tubers (Park *et al.*, 2005), carrots (Jeong and Guerinot, 2008) and lettuce leaves (Park *et al.*, 2009).

15.3.2. *Proteins and amino acids*

Plant storage proteins, a major food reservoir for developing seeds, roots and tubers, determines the nutritional value of plants and grains when they are used as foods and feed for human and domestic animals. Human and many farm animals are incapable in synthesizing certain amino acids *de novo* and it has long triggered tremendous interest in increasing the level of these essential amino acids in crop plants. Most seeds do not provide a balanced source of protein because of deficiencies in one or more of the essential amino acids in the storage proteins. It has been challenging to increase the level of protein and essential amino acids significantly by utilizing classical plant breeding approaches. Therefore, plant biotechnology can make important contributions to food nutritional improvement. Among the 20 essential amino acids, Lysine, Tryptophan and Methionine has gained most attention mainly because they are limiting in major food sources such as cereals (Lysine, Threonine and Tryptophan) and legume crops (Methionine and Cysteine).

The seed storage globulins of legumes are low in sulfur-containing amino acids cysteine, and methionine. Molecular techniques have been employed to improve the concentration of sulphur-containing amino acids in both legume and non-legume plant seed proteins. One widely used approach has been to express the sulphur-rich 2S albumin gene from Brazil nut (*Bertholletia excelsa*) or sunflower (*Helianthus annuus*) in target plants. The Brazil nut 2S protein (BN2S) contains about 18% methionine and 8% cysteine, respectively. The BN2S gene has been successfully expressed in potato, rapeseed, narbon bean, lupin and soybean (*Glycine max*) (Townsend and Thomas, 1996). In each case, the BN2S gene was expressed and its protein product correctly processed, with a concomitant increase in the methionine and cysteine contents in transgenic plants. However, the Brazil nut protein proved to be a major allergen, which prevented further commercial development (Streit *et al.*, 2001). By contrast, the seed albumin gene AmA1 from golden Amaranth (*Amaranthus hypochondriacus*) is non-allergenic in nature and is rich in all essential amino acids, and the composition corresponds well with the World Health Organization standards for optimal human nutrition. In an attempt to improve the nutritional value of potato, the AmA1 coding sequence was successfully introduced and expressed in potato tuber-specific and constitutive manner. There was a striking increase in the growth and production of tubers in transgenic populations and also of the total protein (up to 60% increase) content with an increase in most essential amino acids (lysine, tyrosine, and sulfur amino acid) (Chakraborty *et al.*, 2010). The same *ama1* gene (coding for AmA1 protein) was also successfully transformed into wheat and resulted in improvement of functional quality and increase of essential amino acids (Tamas *et al.*, 2009).

In cereal grains, the major deficiency is lysine, which contributes to the poor nutritional value of foods to human beings and feeds to domestic animals. Transgenic plants expressing the lysine feedback-insensitive enzyme(s), AK and/or DHPS (aspartate kinase and/or dihydrodipicolinate synthase), are capable of accumulating more lysine in seeds. When a gene was designed to code for a protein containing 31% lysine, its expression in tobacco seeds resulted in a significant increase in lysine (by up to 30%) (Shaul and Galili, 1993). With the same strategy, drastic boost of free lysine was achieved in canola and soybean (Falco *et al.*, 1995) and maize (Falco, 2001). But when similar method was used in barley (Brinch-Pedersen *et al.*, 1996) and rice (Lee *et al.*, 2001), only slight increases in free lysine were detected in their mature seeds. More fruitful results were obtained when the degradation pathway was engineered together with the biosynthetic one. In the investigation of an Arabidopsis LKR/SDH (lysine-ketoglutaric acid reductase/saccharopine dehydrogenase) reductase knockout mutant, expressing a bacterial DHPS in seed-specific manner caused a synergis-tic ~80-fold elevation of free lysine (Zhu and Galili, 2003). The result was much better than those Arabidopsis plants having the bacterial DHPS or LKR/SDH knockout gene alone (only ~12-fold or ~5-fold higher free lysine). Comparable results were also found in maize that the combined effect of feedback-insensitive DHPS and LKR/SDH knockout resulted in two- to three-fold higher level of free lysine than the effect of DHPS alone (Falco, 2001).

Tryptophan (Trp) synthesis is strongly feedback regulated by inhibiting its own biosynthetic enzyme, anthranilate synthase. An Arabidopsis mutant with a point mutation in alpha-subunit of anthranilate synthase is widely used in a large number of studies to improve Trp content in crop plants. This mutant plant is insensitive to feedback inhibition by Trp and thus enhances Trp accumulation in the seed. Transgene OASA1D which encodes an analogous feedback-insensitive a-subunit of the rice anthra-nilate synthase driven by constitutive ubiquitin promoter was transformed into rice. This has let to a significant increase of free Trp in rice seeds with minimal side effects on germination and yield (Wakasa et al., 2006). This transgene was also proven to be able to raise the free Trp level in trans-genic potato (Yamada et al., 2005), adzuki bean (Hanafy et al., 2006) and Arabidopsis (Ishihara et al., 2006). Similar transgene encoding feedback-insensitive tobacco anthranilate synthase driven by CaMV 35S promoter was shown to increase free Trp in transgenic soybean plants (Inaba et al., 2007). This entails the fact that there is an evolutionarily conserved mechanism for the regulation of Trp synthesis in plants.

The storage proteins of maize are a group of alcohol-soluble polypeptides called zeins. These proteins are synthesized in the developing endosperm, where they form protein bodies within the rough endoplasmic reticulum. Because they account for more than half of the total seed protein, zeins are the primary determinants of the amino acid composition of the seed. All of the zeins are devoid of lysine, an essential amino acid for monogastric animals. One approach is to modify zein genes so that they encode proteins that contain lysine and tryptophan. Analysis of the synthesis and process-ing of these modified zein proteins in transgenic maize indicates that the content of lysine and tryptophan was increased without interferring with their association into protein bodies (Davies, 2010). Alternatively, zean genes can be down regulated to increase more nutritionally balanced non-zein proteins, and therefore enhance the overall quality of corn protein (Frizzi *et al.*, 2010).

Expression of the human milk protein β-casein in transgenic potato plants opened the way for reconstitution of human milk in edible plants for replacement of bovine milk in baby foods for general improvement of infant nutrition, and for prevention of gastric and intestinal diseases in children (Chong *et al.*, 1997). Bicar *et al.*, has developed transgenic maize with the milk protein α-lactalbumin in its endosperm (Bicar *et al.*, 2008).

Proteins also confer functional properties, which allow seeds to be processed in food systems. These are particularly important in soybean (for texturing and gel (tofu) formation) and in wheat. Such functional properties are biophysical rather than biochemical and may involve complex interactions with other components as well as change with temperature. They are more difficult to define in molecular terms than other targets for manipulation.

There is considerable progress been made in altering the quality of wheat for making bread and other processed foods. The processing properties of wheat are largely determined by the gluten storage proteins. These form a continuous network in dough, conferring the viscoelastic properties necessary to entrap carbon dioxide released during the proving of leavened bread. The protein network also provides the cohesiveness required for other foods. The quality of wheat is determined by genetic and environmental factors, with poor quality generally resulting from low gluten elasticity. Although gluten comprises over 50 proteins, one group of proteins — the high molecular weight subunits of glutenin — are particularly important (Alvarez *et al.*, 2000; Altpeter *et al.*, 1996; Barro *et al.*, 1997; Blechl and Anderson, 1996; Zhang *et al.*, 2003). There is convincing evidence from biochemical and genetic studies that variation in gluten elasticity is associated with differences in the number, amount (about 6%–12% of the total protein), and properties of these high molecular weight subunits. Attention has therefore focused on manipulating the amount and properties of this group by inserting additional genes encoding mutant and wild-type proteins. Increasing the number of expressed subunits from two to three and four results in stepwise increases in dough elasticity, mirroring the effects of manipulating gene dosage by classical breeding. However, it has been possible to go beyond the gene dosage obtainable by conventional means. This results in gluten that contains over 20% of high molecular weight subunits and a massive increase in dough elasticity. The resulting flour is actually too "strong" to be used for bread making but may be valuable for blending to fortify poor quality wheat. The genetic modification of potato has also been used to improve the functional properties of tuber-derived flour through the overexpression of the wheat low-molecular-weight glutenin gene (Benmoussa *et al.*, 2004).

In addition to improvement of human food, transgenic plants can also improve quality and quantity of animal feeds, which can translate to better products from animal, such as sheep wool. The post-ruminal supply of the sulfur-containing amino acids, methionine and cysteine, has been reported to be a major limitation to wool growth in sheep. Wool growth rates were significantly increased when feed was supplemented with approximately 1 to 2 g of sulfur amino acids daily. A modified albumin rich in sulfur amino acids was engineered into alfalfa by plant transformation. The highest level of sunflower seed albumin found in transgenic alfalfa leaves was estimated to constitute 1% of soluble leaf protein. This level of accumulation of the foreign protein would be predicted to supply an extra 40 mg of sulfur amino acids daily to sheep fed the modified forage. Another approach to increase plant nutritive value of pasture is to reduce the content of indigestible lignin using similar approaches in engineering forest trees.

15.3.3. *Fats and oils*

It is now possible to engineer "designer oilseeds" tailored to specific end-uses. Plants produce a wide diversity of fatty acids, the majority of which accumulate in the seed as triacylglycerols (three fatty acids linked to a glycerol backbone). Plant lipids have different end-uses, from industrial applications as lubricants and detergents to food and nutrition. Plant lipids also have important roles as nutraceuticals and pharmaceuticals. The possibility of "designer oilseeds" stems from groundbreaking work that allowed the definition and isolation of key genes involved in the biosynthesis and modification of plant fatty acids.

Here, we focus on production with GM plants of abnormal fatty acids, which are deviations from the normal ones due to differences in chain length (i.e., greater than 18 carbons or shorter than 14), position or number of double bonds (i.e., polyunsaturated), or having modifications other than simple double bonds (i.e., hydroxylation, epoxy groups, etc.). There are three types:

(a) Fatty acids with functional groups such as hydroxyl, epoxy, or acety-lenic bonds. One example is as the C-12 hydroxylated fatty acid ricinoleic acid
(b) Short fatty acids with chain length less than 14 carbons, such as lauric acid (a 12-carbon saturated fatty acid) and
(c) Polyunsaturated fatty acids (PUFAs) normally found in fish oils and microorganisms.

For industrial fatty acids, such as the hydroxylated fatty acid ricinoleic acid (12-hydroxyoctadec-cis-9-enoic acid; 12-OH 18:1Δ9), they can serve as a precursor for chemical conversion to many useful products, ranging from lubricants, emulsifiers, and inks to biodiesel formulation and nylon precursors. There have been many efforts toward obtaining high-level accumulation of these products in transgenic plants (Scarth and Tang, 2006; Jaworski and Cahoon, 2003).

Laurate, which is used in confectionery, is normally obtained from either coconut or palm oil. Although both plants yield relatively high levels of the fatty acid, they are limited in their agricultural utility. In transgenic canola, lauric acid (a 12-carbon saturated fatty acid) can be increased to 60% in triglycerols (Eccleston and Ohlrogge, 1998). The gene encoding an acyl-ACP thioesterase from the California bay plant was expressed in canola seeds. This enzyme prevents the production of long-chain fatty acids by cleaving the fatty acids from the enzyme complex that synthesizes it after it reaches the 12-carbon stage. This results in the accumulation of lauric acid. Calgene has produced a number of canola lines that accumulate over 40% laurate. These oils are now marketed for use in confectionery in North America under the Laurical trademark. This work demonstrated the feasibility of producing large amounts of transgene-modified plant oils to supplement or replace fluctuating "natural" sources.

For PUFAs, although they have a much more complex biosynthesis, accumulation of some target fatty acids by transgenic plants has been remarkably successful. The primary biosynthesizing sources of PUFAs are marine microbes such as algae, which form the base of an aquatic food web that culminates in the accumulation of these fatty acids in fish oils. Much interest has recently been focused on the importance of omega-3 PUFAs, such as eicosapentaenoic acid (EPA) (20:5Δ5,8,11,14,17), docosapentae-noic acid (DPA) (22:5Δ,7,10,13,16,19) and docosahexaenoic acid (DHA) (22:6Δ4,7,10,13,16,19), and omega-6 PUFAs, such as gamma linolenic acid (LA) (20:3Δ6,9,12) and arachidonic acid (ARA) (20:4Δ5,8,11,14), as cardi-ovascular-protective components of the human diet. It is also known that the ARA and DHA are important PUFAs for optimal neonatal health and development. So there has been considerable interest in the production of PUFAs in transgenic plants to provide a sustainable, clean, and cheap source of these fatty acids (Venegas-Caleron et al., 2010).

The biosynthesis of PUFAs is catalyzed by an alternating sequence of fatty acid desaturation and elongation and therefore requires two distinct types of primary biosynthetic activities (desaturases and elongases). The biosynthetic pathway for PUFAs has been extensively reviewed over the past few years (Sayanova and Napier, 2004) and genes encoding all the primary biosynthetic activities have been isolated from many different organisms. The primary activities required for converting LA and ALA to either ARA, EPA, or DHA are shown in Fig. 15.2. Δ6-desaturation repre-sents the first committed step on the PUFAs biosynthetic pathway, and expression of a Δ6-desaturase isolated from a cyanobacteria resulted in the accumulation of low levels of γ-linolenic acid (Drakakaki et al., 2005) (18:3Δ6,9,12) and stearidonic acid (SDA) (18:4Δ6,9,12,15) in transgenic tobacco plants. Much higher accumulation of GLA and SDA was obtained by expression of a Borago officinalis (common borage) Δ6-desaturase (Sayanova et al., 1997). Co-expression of Δ6-desaturases from plant or fungal sources with FAD2 Δ12-desaturases resulted in elevated levels of GLA (Huang et al., 1999), whereas co-expression with FAD3 Δ15-desaturase resulted in high levels of SDA (Eckert et al., 2006). There is now evidence that GLA are essential for healthy human metabolism. GLA is also a registered pharmaceutical for the treatment of eczema and mastalgia, and may also have wider applications for antiviral and cancer therapies. There is, therefore, a good deal of current research devoted to trying to produce GLA in conventional oil crops like soybean, canola and rice that are easier to grow and more widely consumed.

As shown in Fig. 15.2, biosynthesis of the two important PUFAs ARA and EPA requires three primary enzyme activities $\Delta6$-desaturase, $\Delta6$-elongase, and $\Delta5$-desaturase. Several studies have now evaluated the cumulative performance of these activities from different species in transgenic plants, resulting in the accumulation of ARA and EPA to varying levels (Abbadi et al., 2004). Obtaining transgenic plants with relatively high levels of EPA allowed the further engineering of these lines for the synthesis of DPA and DHA by adding another elongase ($\Delta5$-elongase and $\Delta7$-elongase) and desaturase ($\Delta4$-desaturase) (Fig. 15.2) (Napier, 2007).

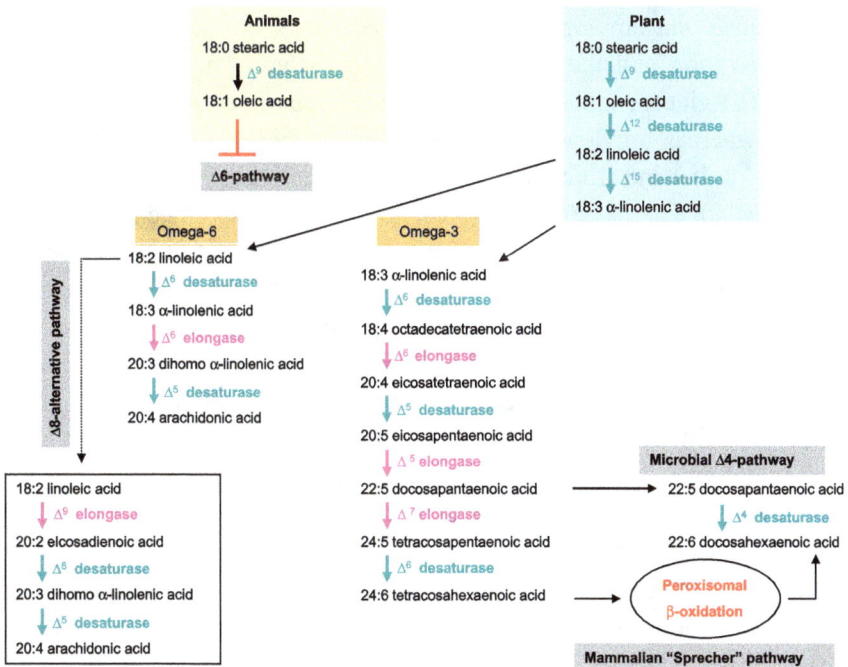

Fig. 15.2. Schematic representation of PUFA biosynthesis (Napier, 2007). The various routes for synthesis of arachidonic acid (ARA), eicosapentaenoic acid (EPA), and docosahexaenoic acid (DHA) are shown, as mediated by the key enzyme activities (desaturases, elongases). The predominant $\Delta6$-pathway (via the $\Delta6$-desaturase is shown, as is the alternative $\Delta8$-pathway. Two routes for DHA synthesis are shown, although only the simpler $\Delta4$-desaturase pathway is described in relation to transgenic plant expression.

15.3.4. *Improvement of starch*

Cereal starches are not only widely used in the food industry to produce glucose and fructose syrups for beverages, conferring functional properties for food processing, but also used as a raw material for the production of non-food products in the paper, plastic, adhesive, textile, binder, gelling agent, film former, medical and pharmaceutical industries. At present, the demand for variation in functional properties is satisfied by using starches from different plant sources (potato and cassava tubers, wheat and maize grain); by chemical modification; and, in the case of maize, by exploiting naturally occurring genetic variants.

Although knowledge of the biophysical and molecular basis for the variation in functional properties of starch is still incomplete, it is generally accepted that several factors are important. The first is the proportion of amylose and amylopectin, the two polymers differing in degree of polymerization and the number of side branches. Amylose is a linear chain of D-glucose molecules with a low degree of polymerization ($< 10^4$ units), whereas amylopectin shows a higher degree of polymerization (10^5–10^6 units) and which has important implications for function (Sestili *et al.*, 2010). The basic pathways of amylose and amylopectin synthesis are well understood and the genes were cloned. Amylose and amylopectin are synthesized by two different pathways using a common substrate (ADP-glucose). A granule bound starch synthase (GBSSI) is involved in amylose synthesis, whereas amylopectin is produced by the concerted action of starch synthases (SS I, SS II, SS III), starch branching enzymes (SBE I, SBE IIa and SBE IIb) and starch debranching enzymes of isoamylase- and limit dextrinase-type (ISA and LD).

High amylose starch has attracted particular interest because of its correlation with the amount of resistant starch (RS) in food. RS plays a role similar to fibre with beneficial effects for human health, providing protection from several diseases such as colon cancer, diabetes, obesity, osteoporosis and cardiovascular diseases (Nugent, 2005; Sestili *et al.*, 2010). Amylose content can be modified by a targeted manipulation of the starch biosynthetic pathway. In particular, the inactivation of the enzymes involved in amylopectin synthesis, silenced using the RNA interference (RNAi) technique, can lead to the increase of amylose content. In rice, mutations eliminating the expression of SBE IIb lead to increased amylose content (up to 35% in an Indica background, and up to 25–30% in a Japonica background) (Mizuno *et al.*, 1993). In wheat, RNAi suppression of the expression of SBE IIa and SBE IIb yielded a very high amylose content (>70%), which improved the large bowel health in rats (Regina *et al.*, 2006). Sestili *et al.* (2010) increased the amylose content of durum wheat through silencing of the SBEIIa genes, using two different methods of transformation (biolistic and Agrobacterium). Expression of RNAi transcripts was targeted to the seed endosperm using a tissue-specific promoter. In barley, a significant reduction in amylopectin to a level >65% was achieved when both SBE IIa and SBE IIb were reduced. It was established that suppression of both isoforms of SBE II was required to elevate the amylose content in barley (Regina *et al.*, 2010).

Besides the amylose rich starch, there is the potential to produce cereal starches with a wider range of structures and properties. The targets are varied but include "mutant" wheat starches to mimic those from maize, phosphorylated starch, resistant starches for healthy diets, thermoplastic and biodegradable starches for packaging, eliminating the need for chemical modification. In addition, those in the food processing industry may want to fine-tune the amylose and amylopectin structures to produce new functional properties for food. This may be aided by the use of enzymes from other sources.

15.4. Herbicide Resistance

Weeds compete with crops and forest trees, especially at the early stage of plantation, for nutrients, water and sunlight. Therefore, extensive weed control throughout the whole growth season is vital for the success of crop production and re-forestation. For herbicide resistance, direct herbicide detoxification and reduction of target enzyme sensitivity to the applied herbicide are the two major approaches for herbicide resistance in transgenic trees and crops. Three different strategies for protecting plants from the action of specific herbicides through genetic modification include: (1) introducing an additional insensitive form of the gene for the target site of the herbicide; (2) over-producing a sensitive target site enzyme to swamp out or titrate out the effects of the herbicide; and

(3) introducing a novel detoxification system into the plant to break down or metabolize the herbicide. Glyphosate inhibits aromatic amino acid biosynthesis, and is the main active component of many branded herbicides that are used for pre-emergence weed control. It is a specific competitive inhibitor of the enzyme 5-enol-pyruvylshikimate phosphate synthase (EPSPS), the sixth enzyme in the shikimic acid pathway leading to the aromatic amino acids tryptophan, tyrosine and phenylalanine. Glyphosate herbicide tolerant is commonly produced by introducing a gene encoding the enzymes 5-enolypyruvylshikimate-3-phosphate synthase (EPSPS) from the CP4 strain of *Agrobacterium tumefaciens*, of which EPSPS variant is insensitive to glyphosate. Glyphosate insensitive mutants also served the same purpose. Glyphosate oxidase from *Ochrobactrum anthropi* that can detoxify glyphosate was also introduced into several crops. A gene from *Achromobacter* sp that degrades glyphosate by conversion it to aminomethylphosphonic acid (AMPA) and glyoxylate was also utilized to confer glyphosate resistance. Glyphosate tolerance became a major commercial success for transgenic crops. Another class of very potent and broad-spectrum herbicides is the sulfonylurea herbicides inhibiting the enzyme acetolactase synthase (ALS), the first enzyme in the pathway for biosynthesis of the essential amino acids isoleucine, leucine and valine in plants. Sulphonylurea tolerant tomato, tobacco and cotton were developed by using a mutated resistant form of ALS from tobacco by changing two amino acids in the encoded protein. The bialaphos resistance gene (*bar* and *pat*) from the *Streptomyces* species, on the other hand, can detoxify any activated glufosinate herbicide. When expressing the *bar* or *pat* gene under the 35S promoter, transgenic tobacco, potato, canola exhibited strong resistance to glufosinate ammonium. A specific nitrilase enzyme gene isolated from the soil bacterium *Klebsiella ozaenae* has also been used to develop a resistant cotton variety against the broadleaf herbicide bromoxynil (BXN™).

15.5. Plant Disease Resistance

Microbes cause a lot of different diseases in plants. Plant pathogens include viruses, fungi, bacteria, nematodes, mycoplasma and protozoa. Pathogenic microbes use different strategies to get into plants and propagate by using nutrients from plants. To ensure agricultural harvest, agro-chemicals are heavily used in the field to protect crops from microbial pathogens in modern agriculture.

Plants contain a vast arsenal of defenses whose strong and rapid activation after microbial infection is frequently governed directly or indirectly by a gene-for-gene interaction between the products of a plant resistance (R) gene and a pathogen avirulence (Avr) gene. If either the plant R gene or the pathogen Avr gene is lacking, these defenses are not induced in a timely and effective manner and the pathogen colonizes the plant. This gene-for-gene specificity appears to be mediated, at least in some of the better-characterized systems, such as the tomato *Pseudomonas syringae* pv. *tomato* system, by a direct interaction between the products of the R gene *Pto* and the cognate *Avr* gene. Over 20 R genes with recognition specificity for defined *Avr* genes have been isolated from several plant species, including both monocots and dicots. These genes are effective against bacterial, viral, and fungal pathogens and in one interesting case, the Mi gene from tomato, against both nematodes and an aphid species. There is great diversity in the lifestyles and pathogenic mechanisms of disease-causing organisms. However, R genes were found to encode proteins with certain common motifs. Four classes of R genes are now recognized: the NB-LRR (nucleotide binding leucine rich repeat) genes, Ser/Thr kinases such as Pto, receptor-like kinases (RLKs) and receptor-like proteins (RLPs). It should be noted that most of these naturally R genes occurr in wild varieties instead of cultivated varieties. The pathogen resistance is presumably lost during the yield focusing breeding process.

R gene related disease resistance can be transferred to other varieties of the same species (intraspecies transfer) or other plant species (interspecies transfer). Susceptible crops are transformed with cloned R genes, which will generate novel disease resistant lines rapidly. Such specific introduction of one single gene by GM avoids the genetic "drag" of negative characteristics, a frequent problem with classical breeding approaches. An inducible defense system would also confer higher-level resistance and utilize fewer plant resources than many of the constitutive defenses discussed previously. This approach is particularly attractive in crops in which no effective resistance genes have been identified and allows interspecies transfer of R genes. One such example is the introducing one *R* gene *Bs2* from tomato to pepper. The transgenic pepper acquired the resistance to bacterial spot disease (Tai *et al.*, 1999). However, interspecies transfer of *R* genes, particularly between more distantly related species, often failed, suggesting that *R* genes isolated from one species may not necessarily be compatible with the resistance signaling components in another species.

As more becomes known about the function of the different *R* gene motifs, it is possible to engineer recombinant genes with novel resistance specificities. Of particular interest would be the creation of durable *R* genes that recognize conserved features of a broad spectrum of pathogens. A good example is the production of rice resistant to the bacterium *Xanthomonas oryzae* pv. *oryzae* by the introduction of four different *R* genes (Li *et al.*, 2001).

Efforts to engineer broad-spectrum resistance are not limited to *R* genes but also include approaches around the plant defense responses elicited by these R-genes: the rapid and localized hypersensitive response (HR) and the subsequent establishment of the systemic acquired resistance (SAR) response. "Hypersensitive response" was a term first applied by Stakman in 1915 to describe the rapid and localized plant cell death induced by rust fungi in rust-resistant cereals. HR, like the apoptosis in mammals, is characterized by a rapid and localized cell death at the point of pathogen attack/recognition. One elegant strategy aims to alter regulation of the HR in such a way that this response is induced by both virulent and avirulent pathogens.

A promising approach to engineering disease resistance is to express a pathogen component in the plant. Recognition of these elicitor molecules leads to the activation of a full defense response that is sufficient to inhibit the pathogens. The long list of pathogen component includes but not limited to *Avr* genes, structural components like flagellin, toxin, Hrp proteins, cell wall components such as chitin and melanin as well as virus coat proteins. The difficult part of this approach is to put a highly active protein elicitor under the control of a promoter that is specifically inducible by a virulent pathogen because the production of these elicitors will result in an activation of the plants defenses and even possibly cell death. One successful example was the introducing a pathogen induced elicitor (cryptogein) under the control of a pathogen inducible *hsr203J* promoter into transgenic tobacco (Keller *et al.*, 1999). Under non-induced condition, the transgene was silent. After infection by the virulent oomycete *Phytophthora parasitica* var. *nicotianae*, expression of the elicitor resulted in a hypersensitive response.

RNAi strategy can also be used for inhibiting the expression of pathogen genes at both transcriptional and post transcriptional level in plants. Most successful stories came from virus resistant plants (melon, squash, tomato, eggplant, tobacco and papaya) through introducing virus coat proteins into the plants. The 1994 papaya ringspot virus (PRSV) crisis in Hawaii led to the production and subsequent commercial plantation of transgenic papaya that expresses the PRSV coat protein and thereby eliminate expression of this essential protein upon infection (Fermin, 2005).

Another idea is to manipulate the expression of "master switch" genes like those for kinases and transcription factors, which regulate multiple genes in pathogen induced signaling network for the purpose of increase in disease resistance. A potential issue of this approach is that manipulation of some master switch genes can be detrimental to plant development. Several transcription factor families have roles in plant defense and hence are targets of manipulation. They include WRKY transcription factors, ERF1, Pti4 and MYB30, CGCG box binding proteins and DNA binding protein DBP1. Overexpression of AtWRKY18 in Arabidopsis potentiated defense responses like the expression of pathogenesis related genes and resistance to the bacterial pathogen *Pseudomonas syringae* (Chen and Chen, 2002). On the other hand, MAP kinase (MAPK) signaling cascades are integral parts of many defense-signaling pathways. Overexpression of the MKK4a and MKK5a resulted in enhanced resistance to virulent *P. syringae* and *Botrytis cinerea* (Asai *et al.*, 2002). Besides transcription factors and kinases, PR genes that increase the level of pre-formed barriers against pathogen invasion and genes that are involved in the biosynthesis of hormones SA and JA can also be targeted.

Upon challenge by fungi, plants often produces antifungal proteins, hytoalexins (low molecular weight antimicrobial compounds), enzymes involved in plant cell reinforcement or in the breakdown of pathogen infection structures. A longstanding strategy for engineering durable resistance has therefore been to express proteins with antimicrobial activities in plants (van der Biezen, 2001). Plant antifungal proteins (AFPs) include defensins and other small cysteine-rich peptides, 2S albumins, chitin-binding proteins, lipid-transfer proteins, and hydrogen-peroxide-generating enzymes. Thionins are antimicrobial proteins that accumulate in a variety of plants resisting pathogen infection. Overexpression of the Arabidopsis Thi2.1 gene has been shown to enhance resistance to both bacteria wilt and Fusarium wilt in tomato (Chan *et al.*, 2005). One interesting further development of this strategy was the use of pathogen specific antibody fused with antimicrobial peptides. Peschen *et al.* (2004) used a Fusarium surface antigen specific antibody linked to antifungal peptides. Expression of the fusion proteins in transgenic Arabidopsis plants conferred high levels of protection against *Fusarium oxysporum* f.sp. *matthiolae*, whereas plants expressing either the fungus-specific antibody or AFPs alone exhibited only moderate resistance.

It should be noted that, despite the many strategies and efforts, development of commercial crops that are resistant to fungal and bacterial diseases by introduction of transgenes has generally been unsuccessful. The main problem was the finding that over-expressed transgenes were generally adversely affected plant size and/or seed production. The answer to this problem is to express transgenes only when and where they are needed. The challenge is the identification of promoters that respond tightly, rapidly and in a cell autonomous manner to infection.

15.6. Insect Resistance

Insect-resistance transgenes, whether of plant, bacterial or other origin, can be introduced into plants to increase the level of insect resistance, a technology that has dramatically extended the scope of resistance genes available to plant breeders. The adoption of insect resistant transgenic crops has been increasing annually at double rates since the commercial release of first generation maize and cotton expressing a single modified *Bacillus thuringiensis* toxin (Bt) in 1997. More than 40 different genes conferring insect resistance have been incorporated into crops so far. Insect tolerance through GM technology is seen as an additional tool for the control of crop pests and could offer certain advantages over conventional insecticides, such as more effective targeting of insects protected within plants, greater resilience to weather conditions, fast biodegradability, reduced operator exposure to toxins and financial savings. The use of transgenic plants may lead to a reduction in the use of broad-spectrum insecticides, thereby extending the useful life of these compounds and reducing the ecological damage they cause. For *Bt* cotton, documented benefits include a 70% reduction in insecticide applications in *Bt* cotton field in India, resulting a saving of up to US$30 per hectare in insecticide costs, and an 80–87% increase in harvested cotton yield (Qaim and Zilberman, 2003).

Bacillus thuringiensis (Bt) is a ubiquitous spore-forming soil bacterium that produces insecticidal protein crystals, also called Bt toxins, endotoxins or crystal proteins, within its cells during the sporulation process. Spores and protein crystals of several strains of this bacterium have been used as microbial insecticides since the 1950s and, owing to their selectivity, now have an established role in some integrated pest-management systems. For the past 20 years, endotoxins from different *Bacillus thuringiensis* (Bt) subspecies have been used to protect crops against insects. After ingestion by susceptible insects, the protein is solubilized and activated by proteinases in the insect midgut. The precise mode of action is not fully understood, but it appears that the activated toxins bind to receptors in the insect midgut epithelium and insert into the midgut membrane, leading to the disruption of the electrical, K^+ and pH gradients by creating pores, resulting in irreversible damage to the midgut wall. Separate strains of Bt produce a variety of crystal toxins with distinct host ranges. At least ten genes encoding different Bt toxins have been engineered into plants: cry-1Aa, cry 1Ab, cry 1Ac, crylBa, crylCa, crylH, cry2Aa, cry3A, cry6A and cry9C. Most Cry proteins, even within the CrylA subfamily, have a distinctive insecticidal spectrum. The crylA and crylC genes code for the CrylA and CrylC proteins, respectively, which are specific to larvae of lepidopteran pests such as the codling moth (*Cydia pomonella*), the European corn borer (*Ostrinia nubilalis*) or heliothine bollworms. In contrast, the Cry3A protein is toxic to coleopteran pests, including the Colorado potato beetle (*Leptinotarsa decemlineata*).

Bt toxins have been continuously improved to confer higher level of insect resistance or resistance to a wider spectrum of insects. The commercialized maize event MON863 was developed using a synthesis (codon optimized) variant of the wild type Cry3Bb1 gene from *Bacillus thuringiensis kumamotoensis* that encodes a protein that has eight times more insecticidal activity than the wild type (Hibbard *et al.*, 2005). To date, codon-optimized genes have been transferred into some of these crops, including cotton, maize, potato, broccoli, cabbage and alfalfa. Generally, these plants will express Bt toxins at levels sufficient to cause high mortality of target pests in the field. The second strategy is the use of multiple resistance genes (pyramiding transgenes) to confer wider spectrum of insect resistance. Three insecticidal proteins, Cry1Ac, Cry2A and GNA were introduced into Indica rice to control three major pests: rice leaf folder, yellow stemborer and the brown planthopper (Maqbool *et al.*, 2001). The Bt proteins target the leaf folder and the stem borer, and the GNA protein targets the planthopper. Another strategy is domain swapping in Cry toxins. Various investigators demonstrated that hybrid Cry toxins exhibited substantially enhanced toxicity or host range.

Lectins are plant derived carbohydrate-binding proteins, some of which are toxic to insects. Various lectins have shown some toxic activity against species of the insect orders *Homoptera, Coleoptera, Lepidoptera* and *Diptera.* The mode of action of lectins against insects remains unclear, but it has been shown that at least some bind to midgut-epithelial cells. However, some insecticidal lectins also show significant mammalian toxicity, including lectins from *P. vulgaris,* winged bean, soybean and wheat germ, hence not suitable for creating transgenic crops. Other lectins, for example those from pea and snowdrop *(Galanthus nivalis),* have demonstrated insecticidal activity but are less toxic to mammals. Recent interest has mainly concentrated on the mannose-specific snowdrop lectin *(Galanthus nivalis* agglutinin: GNA). When fed to insects, GNA binds to the gut epithelium and passes into the haemolymph. GNA has been expressed in nine different crops, including potato, oilseed rape and tomato. Laboratory tests with modified potatoes showed that GNA did not increase the mortality or development time of the glasshouse potato aphid *(Aulacorthum solani)* but reduced fecundity considerably, an effect that led to a much slower population buildup of this aphid in a glasshouse trial. The effect of GNA is antifeedant rather than insecticidal. A recent strategy is the use of GNA to serve as a carrier protein to deliver other insecticidal peptides and proteins (Fitches *et al.*, 2004). GNA-SF11 (an insecticidal spider venom neurotoxin) fusion protein expressed in *Pichia pastoris* was found insecticidal to lepidopteran lavae while none of the individual components showed oral toxicity.

Another major group of plant-derived genes useful to confer insect resistance is inhibitor of digestive enzymes or proteinase inhibitors, which are part of the plant's natural defense system against herbivores. Proteinases in insects include serine, cysteine, aspartic and metallopro-teinases that catalyze the release of amino acids from dietary protein and so provide the nutrients crucial for normal growth and development. Different proteinases predominate in different insects; for example, serine-like proteinases are dominant in lepidopteran larvae, whereas coleopteran species have a wider range of dominant gut proteinases. Serine- and cysteine-proteinase inhibitors have been reported to inhibit the growth and development of a range of insects, mainly lepidopteran and coleopteran species, respectively. The anti-metabolic mode of action of these inhibitors is not fully understood: direct inhibition of digestive enzymes is not considered to be the main effect, and a more important factor might be the hyper secretion of digestive enzymes caused by the presence of the inhibitors, resulting in depletion of essential amino acids. So far, at least 14 different plant proteinase-inhibitor genes have been introduced into crop plants. However, proteinase inhibitors are generally found to be only marginal effective against pests. To date, no crop expressing proteinase-inhibitor transgenes has been commercialized. Reasons for this lack of effectiveness include the adaptive capacity of gut proteolysis in phytophagous insects, based on genetic diversity of protein-ases, and low potency of specific proteinase inhibitors. Even combined use of two such inhibitors, the potato PI-II and the carboxypeptidase (PCI) inhibitors, was not adequate to prevent this compensatory response.

Because Bt-endotoxins are not effective against all insects, there have been efforts to identify other insecticidal proteins. Vip (vegetative insecti-cidal proteins) is produced by *Bacillus thuringiensis* during its vegetative growth. They have insecticidal activity towards a wider spectrum of insect pests (Yu *et al.*, 1997). Transgenic cotton expressing such a Vip was approved for commercialization in USA and Australia in 2005. Several such proteins have recently been identified, including Vip3A, cholesterol oxidases and the Al amylase inhibitor. Another new group of insect toxins are from *Photohabdus* and *Xenorhabdus* bacteria, which kill insects through septicaemia (inflammatory response). Arabidopsis plants express- ing toxin A gene from *Photorhabdus luminescens* showed strong insecti- cidal activity against one lepidopteran and moderate activity against a coleopteran pest (Liu *et al.*, 2003). Insect transcription factors can also be targeted. Transcription factor EcR serves as a molting hormone (ecdysone) receptor, initiates the expression of genes involved in the molting process. Transgenic plants expressing these transcription factors caused insect larvae feeding on them to undergo faulty and/or lethal molting (Retnakaran *et al.*, 2003).

Despite the wide adoption and commercial success of insect resistant GM crops, the sustainability and durability of pest resistance continues to be discussed. It has been predicted that insect populations would develop resistance towards Bt proteins soon, due to the presence of mutations present at low frequency in "wild" pest populations. However, no such tolerance to the toxin has been observed. An eight-year monitoring study (1997–2004) of pink bollworm resistance to Bt toxin with laboratory bio- assays of strains derived annually from ten to 17 cotton fields in Arizona (US) showed no net increase of bollworm resistance to Bt toxin (Tabashnik *et al.*, 2005). The reasons for this delayed resistance include the refuge deployment strategy (part of the total plantation is given to plants that are susceptible to the pest), possible extremely unfitness of resistant individu- als or that Bt might have multiple targets. Looking forward, plants expressing multiple toxins are likely to have the potential to delay resistance in target insect populations more effectively than single toxin containing plants.

15.7. Nematode Resistance

Parasitic nematodes predate on many crops worldwide, and the economic losses caused are estimated to be in the order of US$100 billion each year. Plant-parasitic nematodes are either ectoparasites (living outside the root) or endoparasites (living inside the root). In addition, nematodes create opportunities for fungal infection. Although several different groups of nematodes are plant parasites, the most damaging are the root knot nema- todes (genus *Meloidogyne*) and cyst nematodes (including the genera

Heterodera and *Globodera*). These tiny parasites obtain their nutrients from the cytoplasm of living plant root cells while evading or suppressing host defenses. Nematicides have been successfully developed to control nematodes but the expense and environmental toxicity are deterring factors for its wide application. Host resistance is a preferred option.

Several plants use R-gene based resistance mechanisms to protect themselves from nematodes. Nematode resistant genes exist in several crop species and are an important component of many breeding programs for tomato, potato, soybeans and cereals. The HslPro⁻1 was the first nematode resistance gene (Nem-R) cloned from sugar beet. Its encoded protein confers resistance against the beet nematode *Heterodera schachtii* and does not have obvious similarities to known plant genes. So far around 15 Nem-R genes have been cloned from tomato, potato, wheat, rice, pepper and plum, most of them resemble plant *R* genes in their domain structure (Williamson and Kumar, 2006). Resistance to nematodes in plant is generally characterized by failure of the nematodes to produce functional feeding sites in the host after invasion and to develop subsequently as reproducing females. However, the timing and localization of the response varies with the particular Nem-R gene-nematode interaction. For example, Mi-1.2 from tomato confers rapid localized cell death near the anterior end of the nematode in the region of the root where feeding sites initiate, hence no development of feeding site or nematode develop. Some other interactions lead to atrophy or abnormal development of the feeding site or high ratio of male nematodes.

GM technology makes it possible to transfer isolated Nem-R genes to economically important crop plants where resistance is not available. GM nematode resistance will have the clear benefit of reducing use of highly toxic nematicides. So far intraspecies transfer of Nem-R genes by transgenic techniques has been successful. For example, Mi-1 could confer effective resistance when transferred into susceptible tomato varieties, but didn't function when introduced into tobacco or Arabidopsis. The reason for the poor transferability of nematode resistance to another species could be the lack in the host organism other gene products that are required for recognition of the pathogen or signaling the defense response.

An alternative molecular strategy for establishing nematode resistance in plant species has been pursued with the development of artificial resistance. The aim of this approach is to generate a durable and efficient resistance that controls a broad range of plant-parasitic nematodes. Each different step of the attack of the nematode on the root tissue that is dependent on the individual life cycle of that species could serve as a putative target for the construction of an artificial defense system. Thousands of nematode ESTs and many genes encoding secretion products of nematode have been identified, providing a rich source of candidate genes to intercept. Transgenes can encode enzymatic inhibitors that block physiological processes within the nematode, or toxic compounds that are then ingested, or enzymes that interact with the nematode and substances that cause the breakdown of specific feeding structures. One example is the introduction of a proteinase inhibitor cystatin, of which expression in transgenic plants prevented female nematodes from developing properly, causing reduced size and fecundity (Lilley *et al.*, 2004). However, it still has to be proven whether the reduced fecundity is sufficient to confer complete resistance against plant-parasitic nematodes in the field. The many other approaches include introducing Bt-toxins from *Bacillus thuringiensis* strains, expression within the plant of monoclonal antibodies that are directed against nematode-specific proteins and expression of chitinase and collagenases into plants.

15.8. Tolerance to Abiotic Stresses like Drought and Salinity

Under the natural environment, plants have to endure different environmental stresses such as heat, drought, cold and salination. These abiotic stresses especially salinity and drought are the primary causes of crop loss worldwide. Human activities have made the situations even worse. For example, fume gas produced by cars and other industrial activities from burning fossil oil and coal has significantly contributed to the "greenhouse" effect on the climate, i.e. the average temperature in the atmosphere and on earth has increased dramatically. Modern agriculture has also contributed to the worsened situation of drought and salination by pumping out ground water with higher mineral content for irrigation that caused the decrease of ground water level and increased mineral content in soil after evaporation. Drought and salinity are already widespread in many regions, and are expected to cause serious salinization of more than 50% of all arable land by the year 2050.

Plant adaptation to environmental stresses is controlled by cascades of molecular networks, which activate responsive mechanisms to re-establish homeostatis and to protect and repair damaged proteins and membranes. Responses to abiotic stresses are generally multigenic and thus more difficult to control and engineer. There are three general strategies to confer abiotic stress tolerance by genetic modfication: (1) expression of genes that are involved in signaling and regulatory pathways; (2) expression genes that encode proteins conferring stress tolerance; and (3) expression enzymes for synthesis of functional and structural metabolites.

Genes involved in signaling cascades and in transcriptional control include mitogen activated protein (MAP), salt overly sensitive (SOS) kinase, phospholipases and transcription factors like C-repeat/dehydration-responsive element binding protein (CBF/DREB) and ABA responsive element binding factor. It was reported that constitutive expression in maize of a tobacco mitogen activated protein kinase kinase kinase (MAPKKK/NPK1) activated an oxidative signal cascade and led to cold, heat and salinity tolerance (Shou *et al.*, 2004).

Heat shock proteins (Hsps) and molecular cheparones, as well as late embryogenesis abundant (LEA) proteins are also involved in plant abiotic stress tolerance. They control proper folding and conformation of proteins, hence protect them from denaturation and dysfunction caused by stresses. Overexpression of HSP101 from Arabidopsis in rice resulted in a significant improvement of growth performance during recovery from heat stress (Katiyar-Agarwal *et al.*, 2003). Similarly, overexpression of one LEA protein was correlated with drought tolerance in transgenic rice under field conditions (Xiao *et al.*, 2007). Stress induced production of reactive oxygen species (ROS) is another aspect of environmental stress in plants. Introduction of antioxidants and ROS scanvengers can enhance plant resistance to salt and drought. Aldehyde dehydrogenase catalyzes the oxidation of toxic aldehydes, which accumulate as a result of side reactions of ROS with lipids and proteins. Overexpression of an aldehyde dehydrogenase *AtALDH3* gene in Arabidopsis conferred tolerance to both drought and salt stress (Sunkar *et al.*, 2003).

Abiotic stresses lead to osmotic stress, which can cause detrimental changes in cellular components. A wide range of metabolites can prevent these detrimental changes. They include amino acid proline, other amines like glycine-betaine and polyamines, a variety of sugars and sugar alcohols like mannitol and trehalose. These metabolites can be up regulated for abiotic stress tolerance. Glycine betaine is a widely studied osmoprotectant. Upregulation of this metabolite was achieved in several plants through introduction of a bacterial gene for choline dehydrogenase (codA), which catalyzes the oxidation of choline to glycine betaine. Transgenic plants were found to have enhanced abiotic stress tolerance. Trehalose, a rare, non-reducing sugar, is present in many bacteria and fungi and in some desiccation-tolerant higher plants. Overexpression of trehalose conferred tolerance to multiple abiotic stresses in rice (Jang *et al.*, 2003).

15.9. Phytoremediation

Phytoremediation is the use of plants to clean up polluted soil and water resources. Although some plants have the inherent ability to detoxify environmental pollutants, they generally lack the catabolic pathway for the complete degradation compared to microorganisms or mammals. Genetic modification provides the possibility of engineering plants with superior degradation abilities by overexpression of plant genes and/or introducing genes from other organisms. Phytoremediation with transgenic plants may offer an effective, environmental non-destructive and cheap remediation method.

Phytoremediation involves several processes: pollutants in water and soil can be taken up inside plant tissue (phytoextraction) or absorbed to the roots (rhizofiltration); pollutants inside plant tissues can be transformed by plant enzymes (phytotransformation) or can volatile into the atmosphere (phytovolatization); pollutants in soil can be degraded by microbes in the root zone (root zone bioremediation) or incorporated in soil material (phytostabilization).

Historically, transgenic plants for phytoremediation were firstly developed to improve heavy metal tolerance. For example, tobacco plants (*Nicotiana tabacum*) expressing a yeast metallothionein gene for higher tolerance to cadmium, or Arabidopsis overexpressing a mercuric ion reductase gene for higher tolerance to mercury. Recent years, genes involved in metal uptake, translocation and sequestration and transfer were utilized to create transgenic plants that detoxify or accumulate cadmium, lead, mercury, arsenic and selenium. One example is the introducing a bacterial enzyme glutamylcysteine synthetase into mustard (*B. juncea*). The transgenic plant could extract more Cd, Cr, Cu, Pb and Zn than wild plants (Zhu *et al.*, 1999). Bacterial MerA and MerB genes were introduced into plants and resulted in plants more tolerant to mercury and volatilized elemental mercury that was converted from more toxic organic mercury (Bizily *et al.*, 2000). Transgenic Arabidopsis was developed by expression bacterial arsenate reductase (ArsC) and glutamyl-cysteine synthetase, which transported oxyanion arsenate to aboveground, reduced to arsenite and sequestered it to thiol peptide complexes (Dhankher *et al.*, 2002).

Environmental pollution by organic compounds like pesticides, pharmaceuticals, petroleum compounds, Polycyclic aromatic hydrocarbon (PAH) and polychlorinated biphenyl (PCB) is a global problem. Transgenic plants have also been created for enhanced biodegradation and phytoremediation of organic pollutants. The first attempt of using transgenic plants for phytremediation of organic compounds targeted explosives. Transgenic tobacco expressing a bacterial pentaerythritol tetranitrate reductase was found be able to degrdate explosive glycerol trinitrate (GTN) (French *et al.*, 1999). A degrading protein can also be introduced into rizhosphere for degradation of pollutants. This strategy has the obvious advantage of not taking up pollutants in order to detoxify them. A fungal enzyme, the laccase of *Coriolus versicolor* was introduced into tobacco plants, which produced laccase that was secreted into the rhizosphere. The transgenic plants was able to remove bisphenol A or pentachlorophenol from soil (Sonoki *et al.*, 2005). A recent publication described the development of transgenic poplars overexpressing a mammalian cytochrome P450, a family of enzymes commonly involved in the metabolism of toxic compounds. The transgenic plants showed enhanced metabolism of trichloroethylene and the ability of removal of a range of other toxic volatile organic pollutants, including vinyl chloride, carbon tetrachloride, chloroform and benzene (Doty *et al.*, 2007).

Despite the many efforts, phytoremediation with transgenic plants still lacks wide application. One reason is the low efficiency. The second concern is the accumulation of toxic compounds in plants that could be later released into environment or contaminate the food chain. For pollutants that accumulate in plants, the question of disposal is also a serious concern. Going forward, there is the need to create more efficient transgenic plants for phytoremediation, preferably completely degrade pollutants into non-toxic forms. On the other hand, control of dispersal of transgene in environment is also necessary for regulatory approval and better public acceptance.

15.10. Forest Biotechnology

Forest trees have two major contributions to human: commercial use as wood products (such as timber for construction and furniture, raw material for paper and pulp production and as an energy source) and maintaining and preserving the Earth ecosystem. The demand for wood is expected to increase at 20% in the next decade while the world's forest cover declines at about 13 million hectares due to converted to other uses or lost through natural causes in the last decade (FAO, 2010). There is increasing interest in the development of fast growing, short rotation forest trees to meet the increasing demand for wood products, also as a second generation bioenergy crop. Conventional silviculture and breeding techniques alone are no longer sufficient to meet these requirements because of long juvenile periods and the lack of native plant genes that are needed to impart commercially important traits. Recent advances in forest tree molecular biology, including gene discovery, transcript profiling, genome sequencing and genetic mapping have all paved the way to genetic engineering of trees for high yielding clonal plantations. Genetic modification becomes an important avenue to accelerate the domestication of forest trees with the main advantage of adding commercially important traits to elite clones. The second advantage is the reduction of long generation times that are typical for most forest trees.

The largest effort in genetic engineering of trees has been devoted to the modifying the amounts and composition of lignin in tree to improve lignin extractability during pulping. Lignin contributes up to 25% wood biomass and is therefore the second most abundant organic compound on earth after cellulose. For the pulp and paper industry, lignin must be removed in an energy consuming process that uses environmental polluting chemicals. GM trees have been developed to use less chemicals in pulping process (to separate lignin from cellulose) and hence reduce energy consumption and ecological impact on the environment. Overexpression of ferulate-5-hydroxylase (F5H) in popular resulted in a less condensed lignin and improvements in lignin extractability and bleaching while fiber quality remained the same (Huntley *et al.*, 2003). Cinnamyl alcohol dehydrogenase (CAD) is another key enzyme for lignin biosynthesis. Downregulation of this enzyme improved wood quality for chemical pulping, as demonstrated with wood harvested from four-year-old field trials. Less chemical was used for a higher pulp yield (Pilate *et al.*, 2002). Importantly, no adverse phenotypes on plant growth and health were noticed during the four-year trial period in the field. Co-transformation of poplar with two constructs was also successful. Increasing of F5H expression and reducing of CAD together resulted in reduced lignin, more cellulose and less condensed lignin (Li *et al.*, 2003). This research demonstrated the feasibility of stacking transgenes in trees, a strategy that could reduce many generations of conventional breeding.

Besides quality, wood productivity is also an important trait. Many genes involved in different processes have been found to impact on growth in transgenic poplar. Overexpression of cytosolic pine glutamine synthase (GS), a key enzyme involved in nitrogen assimilation, increases plant tree height by 41% and stem diameter by 36% in three-year-old field grown trees. Enhanced growth and cellulose production and reduced lignin have been obtained through overexpression of xyloglucanase in poplar (Park *et al.*, 2004). This enzyme breaks the linkage between hemi-cellulose and cellulose and promotes cell expansion. Many other genes like endoglucanase, horseradish peroxidase are also used with various level of success in increasing tree growth.

There is also the interest to enhance tolerance to various abiotic stresses like drought, salinity and low temperature, which all contribute to decreases in forest productivity. Overexpression of a pepper ERF/AP2 transcription factor in eastern white pine resulted in increase in tolerance to drought, freezing and salt stress (Tang *et al.*, 2007). The increased tolerance was associated with polyamine biosynthesis. Increase salt tolerance was achieved by overexpression of the choline oxidase (codA) gene from *Arthrobacter globiformis* in several lines of *Eucalyptus globulus*

(Yu, 2009). The *ThCAP* gene from *T. hispida* led to greater resistance to low temperature when expressed in transgenic poplar (Guo *et al.*, 2009).

On the other hand, genetic engineered insect resistance can not only increase forest productivity but more environmentally beneficial because of the reduced need for synthetic pesticides. Bt toxin and proteinase inhibitors have been engineered into forest trees. Bt engineered black polar (*P. nigra*) was approved for commercialization in China in 2002 (Ye *et al.*, 2000).

The many other GM traits in development for trees include disease resistance, herbicide tolerance, enhanced root development, flowering control and phytoremediation of environmental pollutants.

Tree genetic engineering provides an opportunity for sustainable production of forest products. However, achievements so far are mostly on genus *Populus*, a group of fast growing tree species with small genome and amenable to genetic modification. It will take some time to reach the same level of success on other forest trees. On the other hand, commercialization of GM tree has been hindered by regulatory and social hurdles. Two major concerns are transgene stability in trees and the possible gene flow. More research and development are needed to address these issues. It was recently reported that a glyphosate tolerance transgene, *CP4*, was stable for over eight years in poplar under field conditions (Li *et al.*, 2008). There have also various efforts on flowering control for the purpose of transgene confinement. Besides these technical issues, it is also important to have greater transparency and outreach in GM tree development to gain public acceptance.

15.11. Green Vaccines and Biopharmaceuticals

15.11.1. *Current status*

Plant made pharmaceuticals (PMP) is a term used to describe the use of transgenic plants to produce vaccine antigens and pharmaceutical proteins. Globally, approximately 15 million of 57 million deaths per year (>25%) are related to infectious diseases (Morens *et al.*, 2004). Vaccination is considered the most efficient and cost effective means to combat infectious diseases. However, the high costs of vaccine production and delivery make it unaffordable for most people living in developing countries. The high cost of vaccine and the low affordability by many calls for developing alternative approaches. On the other hand, there has been the trend to move towards subunits vaccines using one or two proteins instead of avirulent or killed pathogens. Production of vaccine antigens by transgenic plants followed by oral delivery becomes an attractive option with the many possible advantages such as no need for expensive fermentation, purification, cold storage, transportation and sterile delivery. Oral vaccines act by stimulating the immune system at effector sites (lymphoid tissue) located in the gut. Hepatitis B surface antigen (rHBsAg) was the first vaccine candidate, which was expressed initially in tobacco leaf. Potato tuber with rHBsAg expression was orally administered to volunteers previously vaccinated and was found to increase serum anti-HBsAg titre in more than half of the volunteers who ate two doses of transgenic potato (Thanavala et al., 2005). Recently, HBsAg was also expressed in rice grain (Qian et al., 2008). Other viral vaccine antigens that have been expressed include Rabies virus glycoprotein in tomato, Norwalk virus capsid protein, SARS CovS protein (S1) and one domain of Dengue virus. The first vaccine antigen to reach clinical trial stage was the enterotoxigenic E. coli labile enterotoxin B (LT-B) expressed in potato (Tacket et al., 1998). In healthy volunteers, the vaccine antigen resulted in both mucosal and systemic immune responses. Since LT-B remains stable in gastric fluid and highly immunogenic, it has also been fused to other proteins to improve stability and immunogenicity for oral delivery. Other bacterial vaccine antigens expressed in plants include anthrax protective antigen and Tuberculosis antigen. Even protozoan antigens have been tested in transgenic plants. Malaria AMA1 and MSP1 antigens were expressed in tobacco and lettuce chloroplasts. Oral or subcutaneous administration to mice generated high titers of antibodies that completely inhibited proliferation of the malaria parasite (Davoodi-Semiromi et al., 2010).

Similarly, transgenic plants can be used for production of pharmaceutical proteins. The possible lower production cost has been the main driver. The first pharmaceutical protein made in plants was human growth hormone, which was expressed in transgenic tobacco in 1986. Afterwards in 1989, there was the first success in tobacco leaves of synthesis of a secretive antibody against dental cavity causing bacteria *Streptococcus mutans*. Subsequently, high-affinity monoclonal antibody synthesized in plant was found to be able to neutralize anthrax toxin activity *in vitro* and *in vivo* (Hull *et al.*, 2005). Recently, the red algal protein griffithsin was produced in tobacco and it was found of capable of blocking cell to cell HIV transmission. Other examples include human basic fibroblast growth factors, human epidermal growth factor and human growth hormone (Bai *et al.*, 2007; Ding *et al.*, 2006; Kim *et al.*, 2008). In all the cases, plant produced human proteins were found to be comparable in biological activities to those commercial counterparts. One prominent example is the production of human glucocerebrosidase enzyme (GCD) in carrot cells. Recombinant GCD was found to be biologically active and showed no significant innate or humoral immune reactions (Aviezer *et al.*, 2009). A Phase III clinical trial is ongoing for treatment of patients of Gaucher's disease because of a hereditary deficiency of the enzyme glucocerebrosidase. The list of medically important recombinant molecules that can be expressed in transgenic plants keeps growing rapidly.

15.11.2. *Production platform*

Potato tuber was firstly used to produce and administer oral vaccine antigens to humans. It was also used for the production of human serum albumin, novel vaccine candidates, tumor necrosis factor α (TNF-α) and antibodies. Other production hosts that have been used to express vaccine antigens include tomatoes, bananas, carrots, lettuce, maize, alfalfa, white clover and Arabidopsis. For delivery of recombinant vaccines in edible plant organs, it would be advantageous to use locally grown plants for vaccination campaigns. For the bulk production of biopharmaceuticals, many of the early, plant-derived recombinant proteins were produced in transgenic tobacco plants and were extracted directly from harvested leaves. Tobacco is a well established host for which robust transformation procedures are available. The high biomass yields and rapid scalability make tobacco very suitable for commercial molecular farming. It is also a non-food, non-feed crop, and so carries a reduced risk of transgenic material or recombinant proteins contaminating feed and human food chains. However, tobacco has high content of nicotine and other toxic alkaloids, which must be removed. Other alternative production platforms include lettuce, alfalfa and *Chlamydomonas reinhardtii*, a unicellular alga. Proteins that are expressed in leaves tend to be unstable, which means the harvested material has a limited shelf life and must be processed immediately after harvest. By contrast, proteins that are expressed in cereal seeds are protected from proteolytic degradation and remain stable for longer periods. Several different cereals, including rice, wheat, barley and maize, have been investigated as potential hosts for recombinant protein production. Main criteria for a suitable production platform include ease of transformation, high yield and scalability.

15.11.3. *Three systems of production*

The earliest research using plants for the recombinant expression of vaccine antigens was performed using stable transformation of the nuclear genome of tobacco, which was the dominant strategy for several years. Advantages of stable transformation of the nuclear genome include the ability to scale up, production of large amounts of proteins in transgenic seed and the possibility of expression in fruits or other edible plant organs, enabling oral delivery of minimally processed materials. However, stably integrated nuclear transgenes typically yield relatively low levels of target protein expression (<1% total soluble protein, TSP), which can be variable from plant to plant and generation to generation, probably owing to gene silencing or position effect.

In one alternative chloroplast system, foreign genes are integrated into the chloroplast genome by homologous recombination (Verma and Daniell, 2007). Advantages include little variation of expression among independent transgenic lines, no gene silencing and high levels of expression of proteins are facilitated by >10,000 copies of transgenes in each transformed plant cell. Chloroplast expression also minimizes the risk of foreign gene transfer via pollen from genetically modified crops to other related crops or weeds owing to maternal inheritance. Chloroplast system is gaining popularity. Examples of chloroplast expressed proteins are anthrax protective antigen in tobacco chloroplast (Koya *et al.*, 2005) and Malaria antigens in tobacco and lettuce chloroplasts (Davoodi-Semiromi *et al.*, 2010). One disadvantage of chloroplast system is the lack of glycosylation, hence not suitable to those proteins that require glycosylation for activity.

A third production system is the plant viral vector system for transient expression, it allows for rapid expression of recombinant proteins at levels higher than with stably integrated nuclear transgenes. Tobacco mosaic virus, although an RNA virus that replicates in the cytosol can be delivered as a DNA construct that is transcribed in the nucleus to produce viral RNA. In this system, vectors are generally introduced by vacuum infiltration of Agrobacteria, the products must be purified for delivery as vaccines. One recent example was the production of griffithsin (GRFT), one of the most potent HIV entry inhibitors. This red algal protein was produced in multigram quantities after extraction from *Nicotiana benthamiana* plants transduced with a tobacco mosaic virus vector expressing GRFT (O'Keefe *et al.*, 2009).

15.11.4. *Challenges ahead*

It has been 20 years since the first production of an antibody in a plat expression system. So far, there have been more than 800 peer-reviewed publications on plant made pharmaceuticals (PMP) and about 100 PMP related patent families (Faye and Gomord, 2010). However, there is still no definitive success in molecular farming. None of the plant made vaccines have advanced beyond Phase I human clinical trials, largely because they either represented proof-of-principle targets or impractical oral delivery vehicles. Only two plant produced biopharmaceuticals have been approved for human use. One was the secretory IgA used for prevention of tooth decay. There are several challenges ahead. Technically, expression levels in stable nuclear transformation system are often inadequate for commercial development. Chloroplast system lacks glycosylation that is often required for activity of human proteins. Viral expression system is generally not suitable for oral delivery. If proteins need to be purified from transgenic plants, costs associated with purification, cold storage, transportation and sterile delivery will become major challenges. The ultimate success of plant derived vaccine antigens and pharmaceutical proteins will depend on successful clinical trials and commercial viability of products.

15.12. Regulatory Approval for GM Plant and Products

GM plants need to go through comprehensive pre-market safety evaluation processes before obtaining approval for commercial release. First of all, a GM plant variety must grow in the field and perform better than its non-GM counterpart. Furthermore, the new GM variety will subject to a wide array of compositional and functional tests. The new protein introduced is subjected to extensive biochemical characterization to confirm the protein produced is the intended one. Other tests include acute toxicity and allergenicity evaluations, extensive agronomic, performance and yield analyses, molecular and biochemical characterization of the expressed protein to assure specificity, compositional analyses of key metabolites and nutrients, and animal nutrition and feed-performance studies. Potential impact on the environment will also need to be addressed. Regulation of transgenic crops is currently based on specific "events" (specific transgenic insertions into the host genome). The rationale for event-specific regulation is that transgenic insertion can happen randomly in the genome, which might inactivate or alter the expression of endogenous genes or interact with different genetic backgrounds, thereby resulting in unexpected consequences. In addition, different insertion event often vary in transgene expression levels, patterns or stability. As a general practice, site specific data for the entire inserted DNA together with adjacent genomic sequences near the insertion site are submitted to regulatory agencies. Such information is required for event specific tracking purposes as part of the European Union's traceability and labeling requirements for post marketing surveillance. Regulatory systems for GMO approval vary from country to country, but they are generally aligning either with US regulatory framework or EU regulatory framework.

15.12.1. *US Regulatory framework for GMO approval*

In 1986, the White House Office of Science and Technology Policy finalized the *Coordinated Framework for the Regulation of Biotechnology*, and established a biotechnology working group — the Biotechnology Science Coordinating Committee — which specified EPA (Environmental Protection Agency), USDA (US Department of Agriculture), and FDA

(Food and Drug Administration) as the three primary regulatory agencies for regulating biotechnology. The FDA is responsible for biotechnologically-derived medical products; the USDA for transgenic plants and the EPA for pesticidal plants and genetically-engineered microbial pesticides. According to the Coordinated Framework, new regulations were not necessary since current laws provided adequate statutory authority for biotechnology regulation. A subsequent report from the National Research Council concluded that "the product of genetic modification and selection constitutes the primary basis for decisions and not the *process* by which the product was obtained," and this became the basis for regulatory American policy. In May, 2000 a panel of the National Academy of Sciences issued a report endorsing the safety of those biotech foods currently on the market and opining that the process of inserting genes from one species into another was not inherently dangerous. It thus endorsed the central principle underlying the American government's existing biotech regulations, namely that genetically engineered foods pose no special risk to consumers simply because they are produced by a new process. Once a GM-crop is found "substantial equivalence" with its non-GM counterpart, it can be rapidly deregulated. Afterwards, it can be grown and marketed under normal food safety regulations. The FDA determined that labeling was not required on the basis of the method of food production (i.e. genetic engineering), but only if the new food itself posed safety problems for consumers. To date, the FDA has imposed no labeling requirements for any genetically modified foods.

15.12.2. *EU system*

The EU has the probably strictest and most complicated regulations in the world for GM crops and foods. In the 1990s these was an attempt to develop a uniform EU-wide policy for approvals and trade in GM crops and foods, which was growing increasingly controversial in a number of member states in the mid and late 1990s. Despite EU approvals for commercialization of several GM crops under Directive 90/220/EEC, a number of member states invoked a "safeguard clause" to ban the use of the approved GM crops in their respective countries. In 1998, a number of EU member states, led by France, vowed to block approval of GM crops unless existing labeling and safety regulations were further tightened. As a result, no new GM foods or crops were approved beginning in 1998 through 2004, constituting a *de facto* moratorium on GMO approvals while the EU was working to develop new EU-wide legislation more acceptable to the member states. In 2003, the EU approved new legislation governing approval of GM food and feed for commercialization. It went into effect in April 2004. The independent European Food Safety Authority (ESFA) was established under Regulation (EC) 178/2002. EFSA implements Directive 2001/18/EC on the deliberate release into the environment of genetically modified organisms and Regulation (EC) 1829/2003 on genetically modified food and feed. Under Regulation (EC) 1830/2003, the EU legislation expanded the existing labeling requirements significantly and also required "traceability" — the ability to track a GM product from the farm through all of the distribution, processing, and manufacturing steps to the final consumer product.

15.12.2.1. *Labeling*

All food and feed consisting of GMOs or produced from GMOs are required to be labeled if the level of approved GMO exceeds 0.9% of unintentional adventitious presence. Products required to be labeled, must state that "This product contains genetically modified organisms" or that it has been "produced from genetically modified (name of organism)." For the first time, refined products, like soy oil or high fructose corn syrup, are required to be labeled, even in the absence of any detectable amounts of GM DNA or proteins because they are "produced from" GMOs. The regulations also require animal feed to be labeled along the same principles as for GM food, but do not require labeling of products such as meat, milk or eggs obtained from animals fed GM feed or treated with GM medicinal products.

15.12.2.2. *Traceability*

The EU legislation also requires businesses that grow, store, transport, or process GM products to track them throughout the commercial food chain, from "farm to fork." Under these rules, industry must ensure that systems are in place to identify to whom and from whom GM products are made available and to retain records for five years. All foods require documentation demonstrating whether they contained ingredients derived from GM crops, even if the presence of GM-derived material cannot be detected in the final product.

Another complication of EU approval for GM crops and foods is the requirement for both the European commission and the Member States. Individual countries maintain the right to domestically ban particular transgenic crops approved for cultivation by the European Commission (Meldolesi, 2009).

15.12.3. *Trend in development*

The disparity between the very strict EU regulations and less stringent regulatory systems adopted by US and several other countries has created some peculiar situation when EU imports foods and animal feeds from North and South American countries where more GM crops have been approved without requirements for labeling and segregation. One such example was the rejection of a cargo with 180,000 tons of non-GM soy flour in July 2009 (Davison, 2010). Trace amount of MON88017 maize found in the cargo was probably due to inadvertent admixing somewhere along the supply chain. It should be noted that application for approval was filed in EU for MON88017, a GM event previously approved in many countries. It was not approved despite the positive recommendation by EFSA-GMO panel. Since current methods of cultivation, storage and transport do not permit complete segregation of GMO and non-GMO crops, some mixing is expected. Based on EU regulations, unlabeled cargos containing more than 0.9% approved GMOs or any non-approved GMOs will be returned to the port of origin. The very strict implementation of EU regulations might result in food or animal feed shortages and price increases of meat and poultry. The solution is to either to modify EU regulations or to synchronize approvals on an international level.

It is clear that worldwide harmonization of GMO approval procedures is urgently required. The Codex Alimentarius Commission (CODEX) is facilitating such movement. CODEX was created in 1963 by FAO and WHO to develop food standards, guidelines and related texts under the

Joint FAO/WHO Food Standards Program, of which the main purposes are protecting health of the consumers and ensuring fair trade practices in the food trade, and promoting coordination of all food standards work undertaken by international governmental and non-governmental organizations. The Ad Hoc Intergovernmental Task Force on Food Derived from Biotechnology of the Codex Alimentarius Commission was formed. In response to the increased delivery of genetically modified (GM) foods to international markets, CODEX agreed in March 2003 on principles for the human health risk analysis of GM foods. These principles dictate a case-by-case premarket assessment that includes an evaluation of both direct and unintended effects. They state that safety assessment of GM foods needs to investigate direct health effects (toxicity), tendency to provoke allergic reactions (allergenicity), specific components thought to have nutritional or toxic properties, the stability of the inserted gene, nutritional effects associated with genetic modification and any unintended effects that could result from the gene insertion. Of particular note, the task force broadens risk assessment to encompass not only health-related effects of the food itself, but also the indirect effects of food on human health (e.g., potential health risks derived from out crossing).

It should be noted that the more flexible US system is approaching the rigid EU system due to the recent changed situations in US. First of all, more flaws in the functioning of the US systems have been found, whereby non-authorized (sometimes experimental) GM crops have been mixed with the food and feed chain. It is clear that low level of GMO admixing cannot be avoided using current procedures. Secondly, countries other than US are producing new GM-crops about which the US (and EU) authorities have little or no information. Thirdly, new food safety concerns arise from the proposals to cultivate GM crops for pharmaceutical or industrial productions. It is possible that US GMO regulations may converge with the EU regulations in areas of authorizations, co-existence of supply chains, detection and traceability in the near future.

15.13. Conclusion

From the research and development in the past 20 years, benefits of GM plant to both agricultural producers and consumers have been clearly demonstrated. So far, we have seen no clear evidence that GMO will cause any adverse effects to both human beings and the environment. Regulatory systems generally functioned and addressed all safety concerns. With better technology development and proper scientific education to gain public confidence, genetic engineering will become a powerful tool for our societies to further increase agricultural production, improve nutritional values, reduce environmental pollution or even offer alternatives for environmental clean-up, and setting up green factories to produce pharmaceutical proteins and molecules.

References

1. Abbadi A, Domergue F, Bauer J, Napier J and Welti R. Biosynthesis of very-long-chain polyunsaturated fatty acids in transgenic oilseeds: Constraints on their accumulation. *Plant Cell* **16**: 2734–2748, 2004.
2. Altpeter F, Vasil V, Srivastava V and Vasil IK. Integration and expression of the high-molecular-weight glutenin subunit 1Ax1 gene into wheat. *Nat. Biotechnol.* **14**(9): 1155–1159, 1996.
3. Alvarez ML, Guelman S, Halford NG, Lustig S, Reggiardo MI, Ryabushkina N, Shewry P, Stein J and Vallejos RH. Silencing of HMW glutenins in transgenic wheat expressing extra HMW subunits. *Theor. Appl. Genet.* **100**(2): 319–327, 2000.
4. Asai T, Tena G, Plotnikova J, Willmann MR, Chiu WL, Gomez-Gomez L, Boller T, Ausubel FM and Sheen J. MAP kinase signalling cascade in Arabidopsis innate immunity. *Nature* **415**(6875): 977–983, 2002.
5. Aviezer D, Brill-Almon E, Shaaltiel Y, Hashmueli S, Bartfeld D, Mizrachi S, Liberman Y, Freeman A, Zimran A and Galun E. A plant-derived recombinant human glucocerebrosidase enzyme — a preclinical and phase I investigation. *PLoS One* **4**(3): e4792, 2009.
6. Bai JY, Zeng L, Hu YL, Li YF, Lin ZP, Shang SC and Shi YS. Expression and characteristic of synthetic human epidermal growth factor (hEGF) in transgenic tobacco plants. *Biotechnol. Lett.* **29**(12): 2007–2012, 2007.
7. Barro F, Rooke L, Békés F, Gras P, Tatham AS, Fido R, Lazzeri PA, Shewry PR and Barceló P. Transformation of wheat with high molecular weight subunit genes results in improved functional properties. *Nat. Biotechnol.* **15**(12): 1295–1299, 1997.

8. Bauer P and Bereczky Z. Gene networks involved in iron acquisition strategies in plants. *Agron* **23**: 447–454, 2003.

9. Benmoussa M, Vézina LP, Pagé M, Gélinas P, Yelle S and Laberge S. Potato flour viscosity improvement is associated with the expression of a wheat LMW-glutenin gene. *Biotechnol. Bioeng.* **87**(4): 495–500, 2004.

10. Bicar EH, Woodman-Clikeman W, Sangtong V, Peterson JM, Yang SS, Lee M and Scott MP. Transgenic maize endosperm containing a milk protein has improved amino acid balance. *Transgenic Res.* **17**(1): 59–71, 2008.

11. Bizily SP, Rugh CL and Meagher RB. Phytodetoxification of hazardous organomercurials by genetically engineered plants. *Nat. Biotechnol.* **18**(2): 213–217, 2000.

12. Blechl AE and Anderson OD. Expression of a novel high-molecular-weight glutenin subunit gene in transgenic wheat. *Nat. Biotechnol.* **14**(7): 875–879, 1996.

13. Boynton JE and Gillham NW. Chloroplast transformation in Chlamydomonas. *Methods Enzymol.* **217**: 510–536, 1993.

14. Brinch-Pedersen H, Galili G, Knudsen S and Holm PB. Engineering of the aspartate family biosynthetic pathway in barley (*Hordeum vulgare* L.) by transformation with heterologous genes encoding feedback-insensitive aspartate kinase and dihydrodipicolinate synthase. *Plant Mol. Biol.* **32**: 611–620, 1996.

15. Chakraborty S, Chakraborty N, Agrawal L, Ghosh S, Narula K, Shekhar S, Naik PS, Pande PC, Chakrborti SK and Datta A. Next-generation protein-rich potato expressing the seed protein gene AmA1 is a result of proteome rebalancing in transgenic tuber. *Proc. Natl. Acad. Sci. USA* **107**(41): 17533–17538, 2010.

16. Chan YL, Prasad V, Sanjaya, Chen KH, Liu PC, Chan MT and Cheng CP. Transgenic tomato plants expressing an Arabidopsis thionin (Thi2.1) driven by fruit-inactive promoter battle against phytopathogenic attack. *Planta* **221**(3): 386–393, 2005.

17. Chen C and Chen Z. Potentiation of developmentally regulated plant defense response by AtWRKY18, a pathogen-induced Arabidopsis transcription factor. *Plant Physiol.* **129**(2): 706–716, 2002.

18. Chong DKX, Roberts W, Arakawa T, Illes K, Bagi G, Slattery CW and Langridge WHR. Expression of the human milk protein β-casein in transgenic potato plants. *Transgenic Res.* **6**(4): 289–296, 1997.

19. Clive J. *Global Status of Commercialized Biotech/GM Crops 2011.* ISAAA Brief 43, ISAAA, Ithaca, NY, 2011.
20. Crossway A, Oakes JV, Irvine JM, Ward B, Knauf VC and Shewmaker CK. Integration of foreign DNA following microinjection of tobacco mesophyll protoplasts. *Mol. Gen. Genet.* **202**: 179–185, 1986.
21. Davies HM. Commercialization of whole-plant systems for biomanufacturing of protein products: Evolution and prospects. *Plant Biotechnol. J.* **8**(8): 845–861, 2010.
22. Davison J. GM plants: Science, politics and EC regulations. *Plant Sci.* **178**(2): 94–98, 2010.
23. Davoodi-Semiromi A, Schreiber M, Nalapalli S, Verma D, Singh ND, Banks RK, Chakrabarti D and Daniell H. Chloroplast-derived vaccine antigens confer dual immunity against cholera and malaria by oral or injectable delivery. *Plant Biotechnol. J.* **8**(2): 223–242, 2010.
24. Dhankher OP, Li Y, Rosen BP, Shi J, Salt D, Senecoff JF, Sashti NA and Meagher RB. Engineering tolerance and hyperaccumulation of arsenic in plants by combining arsenate reductase and gamma-glutamylcysteine synthetase expression. *Nat. Biotechnol.* **20**(11): 1140–1145, 2002.
25. Ding SH, Huang LY, Wang YD, Sun HC and Xiang ZH. High-level expression of basic fibroblast growth factor in transgenic soybean seeds and characterization of its biological activity. *Biotechnol. Lett.* **28**(12): 869–875, 2006.
26. Doty SL, James CA, Moore AL, Vajzovic A, Singleton GL, Ma C, Khan Z, Xin G, Kang JW, Park JY, Meilan R, Strauss SH, Wilkerson J, Farin F and Strand SE. Enhanced phytoremediation of volatile environmental pollutants with transgenic trees. *Proc. Natl. Acad. Sci. USA* **104**(43): 16816–16821, 2007.
27. Drakakaki G, Marcel S, Glahn RP, Lund EK, Pariagh S, Fischer R, Christou P and Stoger E. Endosperm-specific co-expression of recombinant soybean ferritin and Aspergillus phytase in maize results in significant increases in the levels of bioavailable iron. *Plant Mol. Biol.* **59**(6): 869–880, 2005.
28. Eccleston VS and Ohlrogge JB. Expression of lauroyl-acyl carrier protein thioesterase in *Brassica napus* seeds induces pathways for

both fatty acid oxidation and biosynthesis and implies a set point for triacylglycerol accumulation. *Plant Cell* **10**(4): 613–622, 1998.

29. Eckert H, LaVallee B, Schweiger BJ, Kinney AJ, Cahoon EB and Clemente T. Co-expression of the borage Δ6 desaturase and the Arabidopsis Δ15 desaturase results in high accumulation of stearidonic acid in the seeds of transgenic soybean. *Planta* **224**: 1050–1057, 2006.

30. Falco SC. Increasing lysine in corn. *Amino Acids* **21**: 57–58, 2001.

31. Falco SC, Guida T, Locke M, Mauvais J, Sanders C, Ward RT and Webber P. Transgenic canola and soybean seeds with increased lysine. *Biotechnology* **13**(6): 577–582, 1995.

32. FAO. *Global Forest Resources Assessment 2010*. Rome, 2010.

33. Faye L and Gomord V. Success stories in molecular farming-a brief overview. *Plant Biotechnol. J.* **8**: 525–528, 2010.

34. Fermin G. Comparative development and impact of transgenic papayas in Hawaii, Jamaica, and Venezuela. *Methods Mol. Biol.* **286**: 399–430, 2005.

35. Fitches E, Edwards MG, Mee C, Grishin E, Gatehouse AM, Edwards JP and Gatehouse JA. Fusion proteins containing insect-specific toxins as pest control agents: Snowdrop lectin delivers fused insecticidal spider venom toxin to insect haemolymph following oral ingestion. *J. Insect Physiol.* **50**(1): 61–71, 2004.

36. French CE, Rosser SJ, Davies GJ, Nicklin S and Bruce NC. Biodegradation of explosives by transgenic plants expressing pentaerythritol tetranitrate reductase. *Nat. Biotechnol.* **17**(5): 491–494, 1999.

37. Frizzi A, Caldo RA, Morrell JA, Wang M, Lutfiyya LL, Brown WE, Malvar TM and Huang S. Compositional and transcriptional analyses of reduced zein kernels derived from the opaque2 mutation and RNAi suppression. *Plant Mol. Biol.* **73**(4–5): 569–585, 2010.

38. Fromm M, Taylor LP and Walbot V. Expression of genes transferred into monocot and dicot plant cells by electroporation. *Proc. Natl. Acad. Sci. USA* **82**: 5824–5828, 1985.

39. Ghandilyan A and Vreugdenhil D, MGM A. Progress in the genetic understanding of plant iron and zinc nutrition. *Physiol. Plant* **126**: 407–417, 2006.

40. Goto F, Yoshihara T, Shigemoto N, Toki S and Takaiwa F. Iron fortification of rice seed by the soybean ferritin gene. *Nat. Biotechnol.* **17**(3): 282–286, 1999.

41. Guo XH, Jiang J, Lin SJ, Wang BC, Wang YC, Liu GF and Yang CP. A ThCAP gene from *Tamarix hispida* confers cold tolerance in transgenic *Populus* (*P. davidiana* x *P. bolleana*). *Biotechnol. Lett.* **31**(7): 1079–1087, 2009.

42. Hanafy MS, Rahman SM, Khalafalla MM, El-Shemy HA, Nakamoto Y, Ishimoto M and Wakasa K. Accumulation of free tryptophan in azuki bean (*Vigna angularis*) induced by expression of a gene (OASA1D) for a modified α-subunit of rice anthranilate synthase. *Plant Sci.* **171**(6): 670–676, 2006.

43. Hare PD and Chua NH. Excision of selectable marker genes from transgenic plants. *Nat. Biotechnol.* **20**(6): 575–580, 2002.

44. Herrera-Estrella L, Depicker A, Van Montagu M and Schell J. Expression of chimaeric genes transferred into plant cells using a Ti-plasmid-derived vector. *Nature* **303**: 209–213, 1983.

45. Hibbard BE, Vaughn TT, Oyediran IO, Clark TL and Ellersieck MR. Effect of Cry3Bb1-expressing transgenic corn on plant-to-plant movement by western corn rootworm larvae (Coleoptera: Chrysomelidae). *J. Econ. Entomol.* **98**(4): 1126–1138, 2005.

46. Hoa TT, Al-Babili S, Schaub P, Potrykus I and Beyer P. Golden Indica and Japonica rice lines amenable to deregulation. *Plant Physiol.* **133**(1): 161–169, 2003.

47. Huang YS, Chaudhary S, Thurmond JM, Bobik EGJ and Yuan L. Cloning of Δ12-and Δ6-desaturases from *Mortierella alpina* and recombinant production of γ-linolenic acid in *Saccharomyces cerevisiae*. *Lipids* **34**: 649–659, 1999.

48. Hull AK, Criscuolo CJ, Mett V, Groen H, Steeman W, Westra H, Chapman G, Legutki B, Baillie L and Yusibov V. Human-derived, plant-produced monoclonal antibody for the treatment of anthrax. *Vaccine* **23**(17–18): 2082–2086, 2005.

49. Huntley SK, Ellis D, Gilbert M, Chapple C and Mansfield SD. Significant increases in pulping efficiency in C4H-F5H-transformed poplars: Improved chemical savings and reduced environmental toxins. *J. Agric. Food Chem.* **51**(21): 6178–6183, 2003.

50. Inaba Y, Brotherton JE, Ulanov A and Widholm JM. Expression of a feedback insensitive anthranilate synthase gene from tobacco increases free tryptophan in soybean plants. *Plant Cell Rep.* **26**(10): 1763–1771, 2007.

51. Ishihara A, Asada Y, Takahashi Y, Yabe N, Komeda Y, Nishioka T, Miyagawa H and Wakasa K. Metabolic changes in *Arabidopsis thaliana* expressing the feedback-resistant anthranilate synthase α subunit gene OASA1D. *Phytochemistry* **67**(21): 2349–2362, 2006.

52. Ishimaru Y, Suzuki M, Tsukamoto T, Suzuki K, Nakazono M, Kobayashi T, Wada Y, Watanabe S, Matsuhashi S, Takahashi M, Nakanishi H, Mori S and Nishizawa NK. Rice plants take up iron as an Fe^{3+}-phytosiderophore and as Fe^{2+}. *Plant J.* **45**(3): 335–346, 2006.

53. Jang IC, Oh SJ, Seo JS, Choi WB, Song SI, Kim CH, Kim YS, Seo HS, Choi YD, Nahm BH and Kim JK. Expression of a bifunctional fusion of the *Escherichia coli* genes for trehalose-6-phosphate synthase and trehalose-6-phosphate phosphatase in transgenic rice plants increases trehalose accumulation and abiotic stress tolerance without stunting growth. *Plant Physiol.* **131**(2): 516–524, 2003.

54. Jaworski J and Cahoon EB. Industrial oils from transgenic plants. *Curr. Opin. Plant Biol.* **6**(2): 178–184, 2003.

55. Jeong J and Guerinot ML. Biofortified and bioavailable: The gold standard for plant-based diets. *Proc. Natl. Acad. Sci. USA* **105**(6): 1777–1778, 2008.

56. Katiyar-Agarwal S, Agarwal M and Grover A. Heat-tolerant basmati rice engineered by over-expression of hsp101. *Plant Mol. Biol.* **51**(5): 677–686, 2003.

57. Keller H, Pamboukdjian N, Ponchet M, Poupet A, Delon R, Verrier JL, Roby D and Ricci P. Pathogen-induced elicitin production in transgenic tobacco generates a hypersensitive response and nonspecific disease resistance. *Plant Cell* **11**(2): 223–235, 1999.

58. Kim TG, Baek MY, Lee EK, Kwon TH and Yang MS. Expression of human growth hormone in transgenic rice cell suspension culture. *Plant Cell Rep.* **27**(5): 885–891, 2008.

59. Klein TM, Wolf ED, Wu R and Sanford JC. High-velocity microprojectiles for delivering nucleic acids into living cells. *Nature* **327**: 70–73, 1987.

60. Koya V, Moayeri M, Leppla SH and Daniell H. Plant-based vaccine: Mice immunized with chloroplast-derived anthrax protective antigen survive anthrax lethal toxin challenge. *Infect. Immun.* **73**(12): 8266–8274, 2005.

61. Krens FA, Molendijk L, Wullems GJ and Schilperoort RA. *In vitro* transformation of plant protoplasts with Ti-plasmid DNA. *Nature* **296**: 72–74, 1982.

62. Lee SI, Kim HU, Lee YH, Suh SC, Lim YP, Lee HY and Kim HI. Constitutive and seed-specific expression of a maize lysine-feedback-insensitive dihydrodipicolinate synthase gene leads to increased free lysine levels in rice seeds. *Mol. Breed* **8**: 75–84, 2001.

63. Li J, Meilan R, Ma C, Barish M and Strauss SH. Stability of herbicide resistance over 8 years of coppice in field-grown, genetically engineered poplars. *West. J. Appl. For.* **23**: 89–93, 2008.

64. Li L, Zhou Y, Cheng X, Sun J, Marita JM, Ralph J and Chiang VL. Combinatorial modification of multiple lignin traits in trees through multigene cotransformation. *Proc. Natl. Acad. Sci. USA* **100**(8): 4939–4944, 2003.

65. Li ZK, Sanchez A, Angeles E, Singh S, Domingo J, Huang N and Khush GS. Are the dominant and recessive plant disease resistance genes similar? A case study of rice R genes and *Xanthomonas oryzae* pv. *oryzae* races. *Genetics* **159**(2): 757–765, 2001.

66. Lilley CJ, Urwin PE, Johnston KA and Atkinson HJ. Preferential expression of a plant cystatin at nematode feeding sites confers resistance to *Meloidogyne incognita* and *Globodera pallida*. *Plant Biotechnol. J.* **2**(1): 3–12, 2004.

67. Liu D, Burton S, Glancy T, Li ZS, Hampton R, Meade T and Merlo DJ. Insect resistance conferred by 283-kDa *Photorhabdus luminescens* protein TcdA in *Arabidopsis thaliana*. *Nat. Biotechnol.* **21**(10): 1222–1228, 2003.

68. Maqbool SB, Riazuddin S, Loc NT, Gatehouse AMR, Gatehouse JA and Christou P. Expression of multiple insecticidal genes confers broad resistance against a range of different rice pests. *Mol. Breed.* **7**(1): 85–93, 2001.

69. Meldolesi A. EU impasse over GM deepens. *Nat. Biotechnol.* **27**(4): 304, 2009.

70. Mizuno K, Kawasaki T, Shimada H, Satoh H, Kobayashi E, Okumura S, Arai Y and Baba T. Alteration of the structural properties of starch components by the lack of an isoform of starch branching enzyme in rice seeds. *J. Biol. Chem.* **268**(25): 19084–19091, 1993.

71. Morens DM, Folkers GK and Fauci AS. The challenge of emerging and re-emerging infectious diseases. *Nature* **430**(6996): 242–249, 2004.

72. Napier JA. The production of unusual fatty acids in transgenic plants. *Annu. Rev. Plant Biol.* **58**: 295–319, 2007.

73. Nugent AP. Health properties of resistant starch. *Nutr. Bull.* **30**: 27–54, 2005.

74. O'Keefe BR, Vojdani F, Buffa V, Shattock RJ, Montefiori DC, Bakke J, Mirsalis J, d'Andrea AL, Hume SD, Bratcher B, Saucedo CJ, McMahon JB, Pogue GP and Palmer KE. Scaleable manufacture of HIV-1 entry inhibitor griffithsin and validation of its safety and efficacy as a topical microbicide component. *Proc. Natl. Acad. Sci. USA* **106**(1): 6099–6104, 2009.

75. Paine JA, Shipton CA, Chaggar S, Howells RM, Kennedy MJ, Vernon G, Wright SY, Hinchliffe E, Adams JL, Silverstone AL and Drake R. Improving the nutritional value of Golden Rice through increased pro-vitamin A content. *Nat. Biotechnol.* **23**(4): 482–487, 2005.

76. Park S, Elless MP, Park J, Jenkins A, Lim W, Chambers ET and Hirschi KD. Sensory analysis of calcium-biofortified lettuce. *Plant Biotechnol. J.* **7**(1): 106–117, 2009.

77. Park S, Kang TS, Kim CK, Han JS, Kim S, Smith RH, Pike LM and Hirschi KD. Genetic manipulation for enhancing calcium content in potato tuber. *J. Agric. Food Chem.* **53**(14): 5598–5603, 2005.

78. Park YW, Baba K, Furuta Y, Iida I, Sameshima K, Arai M and Hayashi T. Enhancement of growth and cellulose accumulation by overexpression of xyloglucanase in poplar. *FEBS Lett.* **564**(1–2): 183–187, 2004.

79. Peschen D, Li HP, Fischer R, Kreuzaler F and Liao YC. Fusion proteins comprising a Fusarium-specific antibody linked to antifungal peptides protect plants against a fungal pathogen. *Nat. Biotechnol.* **22**(6): 732–738, 2004.

80. Pilate G, Guiney E, Holt K, Petit-Conil M, Lapierre C, Leple JC, Pollet B, Mila I, Webster EA, Marstorp HG, Hopkins DW, Jouanin L, Boerjan W, Schuch W, Cornu D and Halpin C. Field and pulping performances of transgenic trees with altered lignification. *Nat. Biotechnol.* **20**(6): 607–612, 2002.

81. Qaim M and Zilberman D. Yield effects of genetically modified crops in developing countries. *Science* **299**(5608): 900–902, 2003.

82. Qian B, Shen H, Liang W, Guo X, Zhang C, Wang Y, Li G, Wu A, Cao K and Zhang D. Immunogenicity of recombinant hepatitis B virus surface antigen fused with preS1 epitopes expressed in rice seeds. *Transgenic Res.* **17**(4): 621–631, 2008.

83. Regina A, Bird A, Topping D, Bowden S, Freeman J, Barsby T, Kosar-Hashemi B, Li Z, Rahman S and Morell M. High-amylose wheat generated by RNA interference improves indices of large-bowel health in rats. *Proc. Natl. Acad. Sci. USA* **103**(10): 3546–3551, 2006.

84. Regina A, Kosar-Hashemi B, Ling S, Li Z, Rahman S and Morell M. Control of starch branching in barley defined through differential RNAi suppression of starch branching enzyme IIa and IIb. *J. Exp. Bot.* **61**(5): 1469–1482, 2010.

85. Retnakaran A, Krell P, Feng Q and Arif B. Ecdysone agonists: Mechanism and importance in controlling insect pests of agriculture and forestry. *Arch. Insect Biochem. Physiol.* **54**(4): 187–199, 2003.

86. Sayanova O and Napier JA. Eicosapentaenoic acid: Biosynthetic routes and the potential for synthesis in transgenic plants. *Phytochemistry* **65**: 147–158, 2004.

87. Sayanova O, Smith MA, Lapinskas P, Stobart AK, Dobson G, Christie WW, Shewry PR and Napier JA. Expression of a borage desaturase cDNA containing an N-terminal cytochrome b5 domain results in the accumulation of high levels of delta6-desaturated fatty acids in transgenic tobacco. *Proc. Natl. Acad. Sci. USA* **94**(8): 4211–4216, 1997.

88. Scarth R and Tang J. Modification of Brassica oil using conventional and transgenic approaches. *Crop Sci.* **46**(3): 1225–1236, 2006.

89. Sestili F, Janni M, Doherty A, Botticella E, D'Ovidio R, Masci S, Jones HD and Lafiandra D. Increasing the amylose content of durum wheat through silencing of the SBEIIa genes. *BMC Plant Biol.* **10**: 144, 2010.

90. Shaul O and Galili G. Concerted regulation of lysine and threonine synthesis in tobacco plants expressing bacterial feedback-insensitive aspartate kinase and dihydrodipicolinate synthase. *Plant Mol. Biol.* **23**(4): 759–768, 1993.

91. Shou H, Bordallo P, Fan JB, Yeakley JM, Bibikova M, Sheen J and Wang K. Expression of an active tobacco mitogen-activated protein kinase kinase kinase enhances freezing tolerance in transgenic maize. *Proc. Natl. Acad. Sci. USA* **101**(9): 3298–3303, 2004.

92. Sonoki T, Kajita S, Ikeda S, Uesugi M, Tatsumi K, Katayama Y and Iimura Y. Transgenic tobacco expressing fungal laccase promotes the detoxification of environmental pollutants. *Appl. Microbiol. Biotechnol.* **67**(1): 138–142, 2005.

93. Streit LG, Beach L, Register JC III, Jung R, Fehr WR. Association of the Brazil nut protein gene and Kunitz trypsin inhibitor alleles with soybean protease inhibitor activity and agronomic traits. *Crop Sci.* **41**: 1757–1760, 2001.

94. Sunkar R, Bartels D and Kirch HH. Overexpression of a stress-inducible aldehyde dehydrogenase gene from *Arabidopsis thaliana* in transgenic plants improves stress tolerance. *Plant J.* **35**(4): 452–464, 2003.

95. Suzuki M, Takahashi M, Tsukamoto T, Watanabe S, Matsuhashi S, Yazaki J, Kishimoto N, Kikuchi S, Nakanishi H, Mori S and Nishizawa NK. Biosynthesis and secretion of mugineic acid family phytosiderophores in zinc-deficient barley. *Plant J.* **48**: 85–97, 2006.

96. Tabashnik BE, Dennehy TJ and Carriere Y. Delayed resistance to transgenic cotton in pink bollworm. *Proc. Natl. Acad. Sci. USA* **102**(43): 15389–15393, 2005.

97. Tacket CO, Mason HS, Losonsky G, Clements JD, Levine MM and Arntzen CJ. Immunogenicity in humans of a recombinant bacterial antigen delivered in a transgenic potato. *Nat. Med.* **4**(5): 607–609, 1998.

98. Tai TH, Dahlbeck D, Clark ET, Gajiwala P, Pasion R, Whalen MC, Stall RE and Staskawicz BJ. Expression of the Bs2 pepper gene confers resistance to bacterial spot disease in tomato. *Proc. Natl. Acad. Sci. USA* **96**(24): 14153–14158, 1999.

99. Tamas C, Kisgyorgy BN, Rakszegi M, Wilkinson MD, Yang MS, Lang L, Tamas L and Bedo Z. Transgenic approach to improve wheat (*Triticum aestivum* L.) nutritional quality. *Plant Cell Rep.* **28**(7): 1085–1094, 2009.

100. Tang W, Newton RJ, Li C and Charles TM. Enhanced stress tolerance in transgenic pine expressing the pepper CaPF1 gene is associated with the polyamine biosynthesis. *Plant Cell Rep.* **26**(1): 115–124, 2007.
101. Thanavala Y, Mahoney M, Pal S, Scott A, Richter L, Natarajan N, Goodwin P, Arntzen CJ and Mason HS. Immunogenicity in humans of an edible vaccine for hepatitis B. *Proc. Natl. Acad. Sci. USA* **102**(9): 3378–3382, 2005.
102. Townsend JA and Thomas LA. Method of Agrobacterium-mediated transformation of cultured soybean cells. U.S. Patent, 1996.
103. Tzotzos GT, Head GP and Hull R. *Genetically Modified Plants: Assessing Safety and Managing Risk.* Elsevier Inc., 2009.
104. United Nations. *World Population to 2300.* New York, 2004.
105. van der Biezen EA. Quest for antimicrobial genes to engineer disease-resistant crops. *Trends Plant Sci.* **6**(3): 89–91, 2001.
106. Vasconcelos M, Datta K, Oliva N, Khalekuzzaman M, Torrizo L, Krishnan S, Oliveira M, Goto F and Datta SK. Enhanced iron and zinc accumulation in transgenic rice with the ferritin gene. *Plant Sci.* **164**: 371–378, 2003.
107. Venegas-Caleron M, Sayanova O and Napier JA. An alternative to fish oils: Metabolic engineering of oil-seed crops to produce omega-3 long chain polyunsaturated fatty acids. *Prog. Lipid Res.* **49**(2): 108–119, 2010.
108. Verma D and Daniell H. Chloroplast vector systems for biotechnology applications. *Plant Physiol.* **145**(4): 1129–1143, 2007.
109. Wakasa K, Hasegawa H, Nemoto H, Matsuda F, Miyazawa H, Tozawa Y, Morino K, Komatsu A, Yamada T, Terakawa T and Miyagawa H. High-level tryptophan accumulation in seeds of transgenic rice and its limited effects on agronomic traits and seed metabolite profile. *J. Exp. Bot.* **57**(12): 3069–3078, 2006.
110. Wemmer T, Kaufmann H, Kirch HH, Schneider K, Lottspeich F and Thompson RD. The most abundant soluble basic protein of the stylar transmitting tract in potato (*Solanum tuberosum* L.) is an endochitinase. *Planta* **194**(2): 264–273, 1994.
111. Williamson VM and Kumar A. Nematode resistance in plants: The battle underground. *Trends Genet.* **22**(7): 396–403, 2006.
112. Xiao B, Huang Y, Tang N and Xiong L. Over-expression of a LEA gene in rice improves drought resistance under the field conditions. *Theor. Appl. Genet.* **115**(1): 35–46, 2007.

113. Yamada T, Tozawa Y, Hasegawa H, Terakawa T, Ohkawa Y and Wakasa K. Use of a feedback-insensitive α subunit of anthranilate synthase as a selectable marker for transformation of rice and potato. *Mol. Breeding* **14**(4): 363–373, 2005.

114. Ye X, Al-Babili S, Kloti A, Zhang J, Lucca P, Beyer P and Potrykus I. Engineering the provitamin A (beta-carotene) biosynthetic pathway into (carotenoid-free) rice endosperm. *Science* **287**(5451): 303–305, 2000.

115. Yu CG, Mullins MA, Warren GW, Koziel MG and Estruch JJ. The *Bacillus thuringiensis* vegetative insecticidal protein Vip3A lyses midgut epithelium cells of susceptible insects. *Appl. Environ. Microbiol.* **63**(2): 532–536, 1997.

116. Yu X. Establishment of the evaluation system of salt tolerance on transgenic woody plants in the special netted house. *Plant Biotechnol.* **26**: 135–141, 2009.

117. Zhang X, Liang R, Chen X, Yang F and Zhang L. Transgene inheritance and quality improvement by expressing novel HMW glutenin subunit (HMW-GS) genes in winter wheat. *Chin. Sci. Bull.* **48**(8): 771–776, 2003.

118. Zhu C, Naqvi S, Breitenbach J, Sandmann G, Christoua P and Capell T. Combinatorial genetic transformation generates a library of metabolic phenotypes for the carotenoid pathway in maize. *Proc. Natl. Acad. Sci. USA* **105**: 18232–18237, 2008.

119. Zhu C, Naqvi S, Gomez-Galera S, Pelacho AM, Capell T and Christou P. Transgenic strategies for the nutritional enhancement of plants. *Trends Plant Sci.* **12**: 548–555, 2007.

120. Zhu XH and Galili G. Increased lysine synthesis coupled with a knockout of its catabolism synergistically boosts lysine content and also transregulates the metabolism of other amino acids in Arabidopsis seeds. *Plant Cell* **15**: 845–853, 2003.

121. Zhu YL, Pilon-Smits EA, Tarun AS, Weber SU, Jouanin L and Terry N. Cadmium tolerance and accumulation in Indian mustard is enhanced by overexpressing gamma-glutamylcysteine synthetase. *Plant Physiol.* **121**(4): 1169–1178, 1999.

122. Zuo J and Chua NH. Chemical-inducible systems for regulated expression of plant genes. *Curr. Opin. Biotechnol.* **11**(2): 146–151, 2000.

Further Reading

1. Bhojwani SS and Soh WY. *Agrobiotechnology and Plant Tissue Culture*. Science Publishers, Inc., 2003.
2. Montpetit É, Rothmayr C and Varone F. *The Politics of Biotechnology in North American and Europe*. Lexingeon Books, 2007.
3. Nagata T, Lörz H and Widholm JM. *Biotechnology in Agriculture and Forestry*. Springer, 2007.
4. Xu Z, Li J, Xue Y and Yang W. Biotechnology and sustainable agriculture 2006 and beyond. In: *Proceedings of the 11ᵗʰ IAPTC&B Congress, August 13–18, 2006 Beijing, China*. Springer, 2007.

Websites

- www.isaaa.org
- www.gmac.gov.sg
- www.fao.org/ag/agn/agns/biotechnology_en.asp
- www.geneticallymodifiedfoods.co.uk/
- www.cera-gmc.org/?action=gm_crop_database
- www.bio.org/foodag/faq.asp
- http://agribiotech.info/
- www.usda.gov/agriculture
- www.cgiar.org/biotech/rep0100/contents.htm
- www.gmo-compass.org/eng/home/
- http://ec.europa.eu/food/food/biotechnology/index_en.htm

Questions for Thought...

1. In what aspects is agrobiotechnology important?
2. In ten years' time, what traits and products by agrobiotechnology can we expect?
3. Is agrobacterium mediated genetic transformation intrinsically harmful?
4. How to strike a balance between having regulations to address safety concerns and facilitating technology development?

5. In what areas can agrobiotechnology help human healthcare?
6. How to transfer agrobiotechnology from where it is developed (mostly developed countries) to where it is mostly needed (mostly developing countries)?

Part IV

Microbes in Medical Biotechnology

The staggering breakthroughs in medical and microbial biotechnology, notably in molecular biology, have led to great strides in the understanding and treatment of human diseases. Monoclonal antibodies, gene probes and polymerase chain reaction offer improved diagnosis of infectious diseases and other ailments. Microbial and animal host cells, transgenic animals and plants generate genetically-engineered, high value pharmaceuticals and vaccines in large quantity. Sophisticated structural studies and computer modeling of complex molecules permit the design of novel drugs. Previously intractable infections are now preventable by novel immunization strategies using recombinant DNA technology, and amenable to new anti-microbial agents discovered from nature. Complemented by the enormous Human Genome Project, gene therapy is at the threshold of a new dimension in medical science. Recognizing the impact of these advances on human health and economic development, scientists are harnessing these enabling technologies to meet the new challenges in medicine, including the disciplines of medical microbiology and infectious diseases.

Chapter 16

Diagnostic Clinical Microbiology

Vincent T. K. Chow

Infectious Diseases Program
Department of Microbiology, Yong Loo Lin School of Medicine
National University Health System
National University of Singapore, Kent Ridge, Singapore

Tim J. J. Inglis
School of Pathology and Laboratory Medicine
University of Western Australia
Crawley WA 6009 Perth, Australia

Song Keang Peng
School of Science, Monash University
Sunway Campus, Selangor, Malaysia
Email: song.keang.peng@sci.monash.edu.my

The 21st Century is hailed as a potentially significant milestone in human history that will witness the next new wave of technology, namely the Biotechnological Revolution whose seeds were sown with the advent of many critical advances in molecular biology. Biotechnology is the integrated and practical application of bioscientific and engineering disciplines to the industrial processing of materials by biological agents to provide useful products and serve other desirable purposes. Recent advances in biotechnology have tremendous impact on numerous fields of human endeavor, including the food and agricultural industry, environmental management, and notably medicine. Fundamental scientific research in

587

medicine, genetics, immunology and communicable diseases accelerated the discovery of new biotechnological innovations, especially in molecular biology over the last 30 years culminating in this "golden age of biology". This leads to a better understanding of the principles of human health, disease etiopathogenesis, prevention and management, with important implications in clinical medicine and health care. For example, significant advances in unraveling the molecular mechanisms of cancer, a major global health problem, have been accomplished through the study of oncogenic viruses.

16.1. The Challenges of Emerging and Re-Emerging Infectious Diseases

About two fifths of all deaths worldwide are caused by diseases of the circulatory system and cancers which are more prevalent in developed countries with increasingly aging populations, and are linked to genetic and lifestyle changes such as dietary habits, obesity and stress. These factors also contribute to the rising incidence of other noncommunicable ailments such as Alzheimer's disease, allergic and autoimmune disorders in developed societies.

The myth that infectious diseases have all been eradicated still persists in this era of high technology medicine. In fact, these diseases remain the leading cause of mortality, accounting for one third of deaths worldwide, especially in underdeveloped countries.

Multiple predisposing factors, often in combination, provoke new epidemics of communicable diseases, including:

(a) Evolution and adaptation of human pathogens (e.g. mutability to virulent strains), evasion of host defences, and decreasing herd immunity to certain microbes.

(b) Socio-political disruptions, demographic, lifestyle and behavioral changes such as explosive world population growth, rampant urbanization, poverty, malnutrition, ignorance, poor healthcare infrastructure, uncontrolled migration (e.g. refugees), mass transportation, international air travel, tourism and sexual promiscuity.

(c) Environmental deterioration and ecological degradation such as water pollution, adverse climatic conditions (e.g. global warming, El Nino and La Nina weather phenomena), man's encroachment into virgin jungles where unknown pathogens lurk, and deforestation facilitating contact between humans and animals, especially rodents.

Of great concern is the global resurgence of "old" diseases such as dengue, measles, malaria, cholera, meningitis, plague, pneumonia, tuberculosis, typhoid, gonorrhea, and sexually transmitted diseases. Equally alarming is the emergence of new diseases such as those that are thought to jump across species (e.g. human immunodeficiency virus or HIV; Ebola virus from primates serving as reservoir of the virus in the wild; variant Creutzfeldt-Jakob disease from "mad cow" disease; "bird flu" attributed to the avian influenza H5N1 strain which can cause multi-organ failure; severe acute respiratory syndrome (SARS)), and of diseases spread from animals (e.g. hantavirus pulmonary syndrome resulting from increased rodent population). There is also the potential risk of new retroviral infections arising from xenotransplantation, the practice of transplanting organs and tissues from animals such as pigs, to humans. More than 5 million new HIV infections occur annually, indirectly giving rise to HIV-related opportunistic infections, e.g. herpes simplex, cytomegalovirus infection, multi-drug resistant tuberculosis (MDRTB), *Pneumocystis carinii* pneumonia, candidiasis, penicilliosis (caused by *Penicillium marneffei*), toxoplasmosis, cryptosporidiosis, and isosporiasis (parasitic infection of intestines).

Anti-microbial resistance is spreading inexorably at a frightening pace, arising chiefly from the abuse by doctors, improper compliance by patients, high usage in immunocompromised patients and widespread antibiotic use in animal feed to promote livestock growth, culminating in the selection of drug-resistant strains (e.g. *Salmonella*). Methicillin-resistant *Staphylococcus aureus* (MRSA), vancomycin insensitive *S. aureus* (VISA), vancomycin-resistant enterococci (VRE), penicillin-resistant pneumococci, penicillinase-producing *Neisseria gonorrhoeae* (PPNG), MDRTB, resistant *Acinetobacter baumannii* strains, enterobacteria producing extended spectrum beta-lactamases (ESBLs), cotrimoxazole-resistant *Shigella*, and chloroquine-resistant *Plasmodium* parasites are causing enormous problems both at hospital and community levels. Also appearing with alarming rapidity are drug-resistant HIV genotypes that harbor mutations in protease and reverse transcriptase genes. To deal with the rising incidence of MDRTB, the World Health Organization is implementing DOTS (directly observed treatment, short course) as part of its tuberculosis control strategy.

Other examples of recently discovered or re-emerging microbes and related diseases include flesh-eating streptococcal bacteria; *E. coli* O157:H7 causing food poisoning outbreaks; *Helicobacter pylori* associated with inflammation, ulcer and cancer of the stomach; *Chlamydia pneumoniae* and its association with coronary heart disease; *Clostridium difficile* and antibiotic-associated colitis; *Burkholderia pseudomallei* and melioidosis; *Borrelia burgdorferi* and Lyme arthritis; *Bartonella henselae* and cat scratch disease, bacillary angiomatosis; oncogenic human papillomaviruses associated with genital cancers; hepatitis B and C viruses with primary liver cancer; human herpesvirus 8 with Kaposi's sarcoma; Rift Valley fever; encephalitides associated with enterovirus 71, West Nile, Hendra and Nipah viruses; and *Cyclospora cayetanensis* protozoa with diarrheal disease (Table 16.1). The potential spectre of biological terrorism and germ warfare (e.g. with deadly smallpox virus and anthrax spores) cannot be disregarded. New outbreaks of old enemies and of undiscovered diseases are likely to occur in the foreseeable future.

Humankind ignores the threat of these killer microbes at its own peril. It would be prudent to prepare ourselves for the worst case scenario — the appearance of a disease of high infectivity and lethality. There is thus an urgent need for politicians, doctors, and scientists to concentrate resources on fighting infectious diseases. In recognition of the importance of public awareness, surveillance of and contingency plans for emerging infections throughout the world, the World Health Organization's focus for World Health Day on 7 April 1997 was "Emerging Infectious Diseases: Global Alert, Global Response". In addition, pharmaceutical companies should continue to invest in anti-microbial drug and vaccine development, harness molecular biotechnological advances, and devise new strategies to combat microbial diseases.

16.2. Medical and Diagnostic Bacteriology

Most larger laboratories are able to provide a comprehensive diagnostic bacteriology service, though many centers offer a restricted range of virology, parasitology and mycology tests. The bulk of diagnostic clinical microbiology comprises bacteriological tests. Most bacterial pathogens are relatively easy to cultivate and manipulate under laboratory conditions. Susceptibility testing of antibacterial agents can quickly identify acquired

Table 16.1. The emergence of significant microbial pathogens and clinical infectious diseases (1965–2005).

Viruses	Diseases
Ebola virus	Ebola hemorrhagic fever
Enterovirus 71	Hand, foot and mouth disease Encephalitis
Guanarito virus	Venezuelan hemorrhagic fever
Hantavirus	Hemorrhagic fever with renal syndrome Hantavirus pulmonary syndrome
Hendra virus (equine morbillivirus)	Encephalitis
Hepatitis A	Infectious hepatitis
Hepatitis B	Serum hepatitis Chronic hepatitis, cirrhosis Primary liver cancer
Hepatitis C	Non-A, non-B hepatitis Liver cirrhosis Primary liver cancer
Hepatitis E	Enteric-transmitted hepatitis
Human herpesvirus 6	Exanthem subitum (Roseola infantum)
Human herpesvirus 8	Kaposi's sarcoma
Human immunodeficiency viruses 1 and 2	Acquired immunodeficiency syndrome
Human papillomaviruses 16, 18 and other oncogenic types	Genital cancers
Human T-cell lymphotropic virus I	Adult T-cell leukemia and lymphoma
Human T-cell lymphotropic virus II	Hairy cell leukemia
Lassa virus	Lassa hemorrhagic fever
Marburg virus	Marburg disease
Nipah virus	Encephalitis
Parvovirus	Fifth disese Fetal death Aplastic crisis in chronic hemolytic anemia
Prion	Variant Creutzfeldt-Jakob disease
Rotavirus	Infantile diarrhea
Sabia virus	Brazilian hemorrhagic fever
SARS coronavirus	Severe acute respiratory syndrome

Table 16.1 (*Continued*)

Bacteria	Diseases
Bartonella henselae	Cat scratch disease Bacillary angiomatosis
Borrelia burgdorferi	Lyme disease
Campylobacter sp.	Diarrheal disease
Clostridium difficile	Pseudomembranous colitis Antibiotic-associated colitis
Ehrlichia chaffeensis	Ehrlichiosis
Escherichia coli O157:H7	Hemorrhagic colitis Hemolytic uremic syndrome
Helicobacter pylori	Gastritis Peptic ulcer
Legionella pneumophila	Legionnaires' disease/pneumonia
Staphylococcus aureus *(toxigenic strains)*	Toxic shock syndrome
Vibrio cholerae O139	Epidemic cholera

Parasites	Diseases
Cryptosporidium parvum	Enterocolitis
Cyclospora cayetanensis	Diarrheal disease
Microsporidia	Immunodeficiency-related microsporidiosis

forms of antibiotic resistance and therefore predict an increased risk of treatment failure. Moreover, some bacteria have become prominent causes of hospital-acquired infection and dictate the use of supplementary analytical methods to support hospital infection control. Paradoxically, molecular biology has been relatively late in influencing diagnostic bacteriology, and is only now beginning to alter the practices that have been laboratory routine for many years. These and other technologies present new challenges for laboratory quality control. Finally, information technology opens up a range of opportunities for making better use of diagnostic laboratory-generated data.

16.2.1. *The purpose of diagnostic bacteriology*

You may think that the phrase "diagnostic bacteriology" is self-explanatory. But it fails to recognize the complex nature of diagnostic laboratory work, which is much more than just the performance of laboratory tests. The best diagnostic laboratory can be described as a problem-solving toolbox whose components co-ordinate a process that starts with a clinical problem (i.e. a bacterial infection) and ends with a solution (i.e. a causal agent or prospective pathogen, recommendations for antibiotic therapy or guidance for infection control). Understanding what bacterial agent caused the infection in question is crucial to this process, but establishing a causal relationship can be difficult (Fig. 16.1), and the subsequent parts of the problem-solving process are the ones that make the more important contribution to the patient's eventual outcome. A problem-oriented focus is doubly important now that physicians are graduating in increasing numbers from new, problem-based medical school curricula. How effectively they integrate diagnostic bacteriology into their clinical practice will depend all the more on how effectively microbiologists can develop a "can-do" laboratory environment.

16.2.2. *General objectives and schema*

The principal product of diagnostic bacteriology is information. The process is usually initiated by a written test request and completed by a laboratory

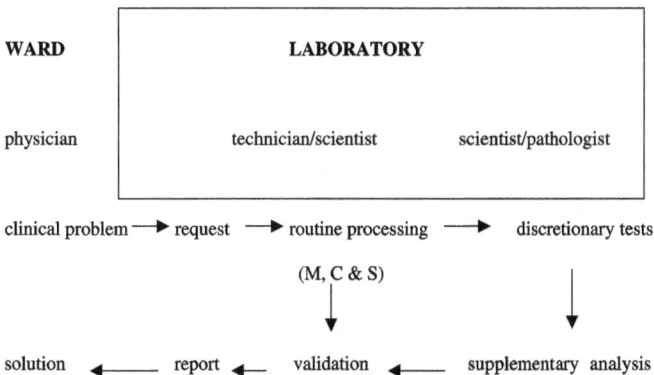

Fig. 16.1.   The general schema and process of diagnostic bacteriology.

report. Although the sequence can theoretically be entirely electronic, the physical actions are bracketed by those critical information events. The information product or report can be judged against time, relevance, accuracy and consistency. The latter, accuracy and consistency, are key issues to be incorporated in quality control and assurance systems. Time, measured as turnaround, is a measurable often tracked by laboratory management. Relevance is more difficult to quantify and usually requires feedback from laboratory users. This is the province of the medical microbiologist and is perhaps one of the toughest tests of the laboratory's relationship with the corresponding clinical service.

16.2.3. *General methods*

Laboratory request forms often contain the phrase "microscopy, culture and sensitivity" (often shortened to "M, C and S"). These words summarize the core of traditional bacteriological testing of clinical specimens. This is shorthand for a generic range of tests that is applied to the majority of clinical microbiology specimens.

(1) **Microscopy** normally equates with Gram stain. This stain is used to assess the range of bacteria and inflammatory cells present in an inflammatory exudate. Though crude, the speed at which a Gram stain can be performed means that it is one of the truly rapid diagnostic tests available. Care must be taken not to over-interpret the data that can be obtained from a Gram stain, e.g. calling clusters of Gram-positive cocci in exudate full of neutrophil polymorphonuclear cells "*Staphylococcus aureus*" until definitive identification procedures have been performed. Other important microscopic techniques include examination of an unstained urine specimen for inflammatory cells, or of a Ziehl-Neelsen stained sputum specimen for the acid-fast bacilli characteristic of tuberculosis (Table 16.2).

(2) **Culture** is the term used to describe the biological amplification of cultivatable bacteria from clinical specimens using artificial growth media. These media may be liquid (broth), solid (due to the addition of agar, and used for their growth-supporting surface), enriched (with growth-promoting substrates), selective (with inhibitors to reduce

Table 16.2. Staining for direct microscopic detection of bacterial cells.

Method	Dye	Application	Processing time
Direct film	None	Urine deposit	circa 5 min
Gram stain	Methyl violet carbol fuchsin	Bacterial wall Shape and structure	circa 10 min
Ziehl-Neelsen Stain	Carbol fuchsin	Mycobacterial detection	circa 30 min

growth of unwanted species) or contain specific indicators to reveal growth of species of particular interest. Bacterial growth also requires the right temperature and atmospheric conditions. Many bacterial pathogens grow best at or near human body temperature (37°C). The more common atmospheric conditions used to grow bacteria are in room air, anaerobically (without oxygen), microaerophilically (low oxygen) and in the presence of carbon dioxide. The temperature and atmospheric conditions required for bacterial growth are usually provided in a purpose built incubator, but occasionally alternative systems such as gas jars or glove boxes are used. The bacteria recovered from clinical specimens in or on artificial media are said to have been "isolated" because, to be useful to the laboratory worker and requesting physician, they have first to be recovered in pure culture. This requires growth of each, individual species in single colonies (hence "isolated").

In order to identify bacteria isolated from clinical specimens a process of **identification** is followed, beginning with the simplest of test and moving onto increasingly complex, costly methods. The first identification tests performed include observation of bacterial colony morphology, alteration of the surrounding agar medium, the Gram stain (on single colonies, so no human cellular material or exudate is present), so-called benchtop tests such as coagulase, catalase and oxidase reactions and specific antibody agglutination reactions. In many cases, these tests will provide sufficient information without having to go further. As a rule of thumb, it is good practice to identify only those colonies which suggest the presence of bacteria that are capable of causing infection at the site from which the specimen was collected. Non-sterile surfaces will have a commensal flora that does not require laboratory work-up. Further identification procedures often involve a panel of substrate utilization tests, each of which will require some period of incubation to promote bacterial growth. A popular method of performing these tests efficiently is to use a multiple test panel. Results of any phenotypic identification system are subject to variation and possibly even interpretation. Additional identification procedures for difficult bacterial groups such as fastidious anaerobes, the actinomycetes, non-fermentative Gram-negative bacilli and mycobacteria may require more advanced methods including gas chromatography, nucleic acid amplification, molecular subtyping and even gene sequencing. At present these methods are beyond the means of all but the largest laboratories. But as the expense of providing these services is falling steadily, they can be expected to be in more widespread use soon.

(3) **Susceptibility testing** of bacteria isolated from a clinical specimen is one of the most important contributions that clinical microbiology has made to patient care, since the results can have considerable impact on the choice of antibiotics used to treat the patient's infection. However, it is important not to overestimate the significance of antibiotic susceptibility results. The time taken to generate a susceptibility report by conventional testing methods is too long for the attending physician to wait for the results before starting initial antibiotic therapy. The physician will therefore make a presumptive choice based on the antibiotics most suitable for a particular infection and modify this initial choice later according to laboratory results. Thus, if the problem is a wound infection thought from the Gram stain to be caused by staphylococci, the presumptive choice of antibiotic might be flucloxacillin. This would be satisfactory unless subsequent susceptibility tests show the isolate to be a methicillin- resistant *Staphylococcus aureus*, in which case the physician would probably want to substitute flucloxacillin with vancomycin.

Given the clinical importance of antibiotic susceptibility testing, it is essential that all stages of the process are strictly quality controlled. Standard operating procedures must be based on widely recognized methods. Result interpretation must be consistent and a method of dealing with indeterminate results should be agreed. An intermediate result (as opposed to sensitive or resistant) in most cases is likely to confuse requesting physicians. The most widely used methods are (a) the Kirby-Bauer disk diffusion method in which measurement is made of the diameter of an inhibition zone produced by an antibiotic-containing disk placed on agar seeded with a lawn of test bacteria; and (b) the growth of suspensions of test bacteria on agar plates containing dilutions of antibiotic (sometimes known as the "break-point" method). These and other methods must be conducted with control strains whose results are already known. Both methods have their weaknesses such as variation due to differences in the density of inoculation of the bacterial lawn in the disk diffusion method, or results close to the specific dilution chosen for the break-point method. Careful attention to results with control strains and regular review of out-of-range results (e.g. new antibiotic resistance patterns) will help minimize these sources of potential error. It should be noted that antibiotic susceptibility testing provides only a guide antibiotic therapy with the range of agents tested. Many other factors affect the outcome of antimicrobial chemotherapy and in some cases (e.g. mucoid strains of *Pseudomonas aeruginosa* and facultative intracellular bacterial pathogens such as *Listeria monocytogenes* and *Legionella pneumophila*) conventional susceptibility testing methods can be a very poor guide to clinical outcome.

In some circumstances (such as bacterial endocarditis) a more definitive susceptibility result may be required. This is usually obtained by determination of the minimum inhibitory concentration (MIC) to a given antibiotic compound, which represents the lowest concentration at which growth of the bacterial species is inhibited under standardized conditions. Several methods are in common use including broth macrodilution (the test tube version), broth microdilution (using microtiter trays), and the application of an antibiotic-impregnated plastic strip onto a bacterial lawn (E-strip). The extra work required for MIC determinations means they are conducted more selectively than conventional disk diffusion or break-point methods.

(4) **Molecular** and other bacterial **subtyping** methods are progressively displacing more traditional typing systems such as phage typing, whose historical role will prevent their complete abandonment. The typing systems now coming into wider use include gas chromatography, nucleic acid amplification-based techniques, molecular subtyping and even gene sequencing. At first glance, the choice of methods seems bewildering; an alphabet soup of acronyms. However, there are some essential criteria that can help sort out an order of priority. The ideal typing system will reproducibly subdivide or discriminate between different strains of a given species. The typing system should be easily performed at little expense and work reliably with any strain. These features are known respectively as reproducibility, discrimination, ease, cost and typability. No system fulfils all these criteria and inferences derived from one typing system may need confirming by use of a second dissimilar system. Genetic typing relies on the assumption that all strains of a given species derive from a common ancestor. All genetic variation accumulated by successive variation from the ancestral clone can be represented in terms of genetic distance. The degree of dissimilarity detected between two isolates is therefore thought to represent a measure of distinctness. The question of how much difference is necessary to call two isolates different often arises and is usually resolved by considering the number of separate genetic events (mutations) necessary to cause the differences observed. So in a DNA macrorestriction analysis (pulsed field gel electrophoresis or PFGE), a single band difference between two lanes would be regarded as insignificant, a two to four band difference borderline, and a six band difference significant.

A decade ago, molecular epidemiology was performed by relatively cumbersome methods such as plasmid gel electrophoresis (PGE), and such methods were subjected to a range of potential errors. The arrival of nucleic acid amplification techniques led to the development of systems based on polymerase chain reaction (PCR) including random amplified polymorphism detection (RAPD) and repetitive extragenic palindromic sequence analysis (REPS). These and similar methods were limited by a lack of reproducibility, and in some instances, poor typability. At present, DNA macrorestriction is widely used as a definitive molecular typing technique for medically important bacteria. This system uses infrequently cutting restriction enzymes to divide the bacterial chromosome into large fragments that are then focused in an agarose gel by a current running from hexagonal electrodes. Switching of current direction allows the migration of relatively large molecules such as DNA fragments. The bands formed by these fragments can be seen by staining the gel with ethidium bromide. The resulting ladder pattern resembles a bar code and can be scanned for computer-based analysis with interpretative software. DNA macro-restriction is time-consuming and requires a high level of laboratory skill. In larger diagnostic laboratories these methods may be used to support infection control or public health investigations. PFGE is slow and cannot be used to transfer data between laboratories efficiently. A recent development has been the introduction of an automated ribotyping device (RiboPrinter, Dupont-Qualicon, DE), which uses a restriction enzyme on an *E. coli* cDNA probe. The final result is a digital ladder or "ribotype" which can be used, stored for future reference or compared against an external digital library. In theory, the Internet should allow networking of these devices with further gains from collating multi-site information. Some bacteria are just too complex to permit conventional molecular epidemiology. Members of the CMN group can be difficult, but analysis of fatty-acid methyl ester (FAME) gas chromatograph retention times can be useful for definitive identification and typing purposes. In the event of conflicting results from other typing systems or a lack of typing data, sequencing of part of the bacteria chromosome at the 16SrRNA locus can be helpful. Gene sequencing is being increasingly used to arrive at a definitive identification of mycobacteria.

Molecular and other subtyping methods have yet to become a part of the regular repertoire of most service laboratories. Establishing advanced bacterial typing systems as a part of standard bacteriology operating procedures remains a challenge. Most molecular subtyping is performed by research laboratories and lacks the consistency expected of clinical service laboratories. Access is limited by availability. As this improves, strategies will need to be developed that triage the requests for typing. It is already apparent that not all bacterial strains that can be subtyped should be. A minimum requirement should be a prior knowledge of the genetic population structure of the species in question (how much variation can be expected in a random collection of isolates), baseline epidemiological data (person, time and place) with which to formulate a working hypothesis, and a clear line of communication/accountability (who will receive results and initiate further action). Where these basic requirements cannot be satisfied, subtyping of bacterial collections must be regarded as exploratory and will be less likely to produce epidemiologically useful results.

16.3. Bacteriological Emergencies

Many specimens are submitted to diagnostic microbiology laboratories marked "urgent", but only a small proportion will have any direct bearing on true bacteriological emergencies. Even when the patient's condition is unequivocally life-threatening, the results of any test that requires bacterial growth cannot be expected to contribute to the immediate clinical management of the patient. There is a surprisingly small collection of rapid tests which when applied to a small number of infections can help alter the course of the disease. The infections concerned are all potentially fatal and progress rapidly, necessitating prompt decisions on antibiotic choice and other therapeutic interventions.

16.3.1. *Acute bacterial meningitis*

Despite being less common than its viral counterpart, bacterial meningitis has a much higher rate of major neurological complication and of fatal outcome. The sporadic occurrence of this condition in previously healthy young adults and children underlines the potential virulence of the bacterial species that cause meningitis. Unfortunately, in centers where the condition is seen very rarely, this can lead to inappropriate and sometimes irrational requests for laboratory investigations that have little clinical justification.

The technological gains of the last decade have added little to the initial examination of cerebrospinal fluid (CSF) from patients with suspected bacterial meningitis. The most useful analyses are total and differential cell count, Gram stain on a spun deposit and biochemistry (glucose and

Table 16.3. CSF cell content and biochemistry in central nervous system infections.

CNS infection	Neutrophils (no./μL)	Lymphocytes (no./μL)	Protein (g/L)	Glucose (CSF/blood)
Normal	0	<5	<0.5	2/3
Viral meningitis	<100	10–10,000	0.5–1	2/3
Bacterial meningitis				
Not TB	>200–10,000	<100	>1	<1/2
Tuberculous	<100	50–1,000	1–5	<1/3
Brain abscess	5–100		>1	2/3

Table 16.4. Issues and exceptions in CSF analysis.

Cell count
  No detectable cells   Some cases of early meningitis
  Low neutrophils, turbid CSF   Some *S. pneumoniae* meningitis
  Artificially low cell count   Glass CSF container
  Raised but not meningitis   Seizure, systemic viral
                                 infection,   infective endocarditis

Biochemistry
  Low glucose but no bacteria   Consider mumps or
                                 lymphocytic   choriomeningitis virus
  Raised protein but no cells, bacteria   Consider brain abscess

Gram stain
  Neutrophilia but no bacteria   Bacteria below limit of detection in
                                 >10% cases
                                 H. influenzae easily missed
                                 against   background of cell debris
L. monocytogenes can be mistaken for *S. pneumoniae* in CSF Gram stain

protein). When carefully performed on CSF from a previously untreated patient, a presumptive etiological diagnosis should be possible in the majority of cases. The expected results are given in Table 16.3. Important but less common exceptions from these ranges are given in Table 16.4.

As in all other areas of diagnostic bacteriology, the quality of clinical information available to the laboratory has a substantial bearing on how the results are reported. Key points to ascertain include whether the patient has a community-acquired infection or has been in hospital for some time (e.g. following a neurosurgical procedure), and whether they have recently received antibiotics that could inhibit the growth of meningitis-causing bacteria.

CSF should be cultured by inoculation of 5% horse blood agar and chocolated blood agar in 5% CO_2. It is common practice to incubate any remaining drops of CSF in glucose broth to enrich a low inoculum or resuscitate sublethally damaged bacteria. Blood cultures should be collected from any patient suspected of having bacterial meningitis to improve the diagnostic yield in view of the frequency of concurrent meningitis, and the incomplete overlap of bacterial meningitis and septicemia.

Most of the less commonly isolated bacterial causes of meningitis (*Escherichia coli, Listeria monocytogenes* and the staphylococci) can be recovered using standard bacterial isolation procedures. *Mycobacterium tuberculosis* is the exception. When tuberculous meningitis is suspected on the grounds of clinical signs or CSF biochemistry, a centrifuged deposit of CSF should be examined microscopically following acid-fast stain and inoculated into mycobacterial culture medium (e.g. Middlebrook's broth). Nucleic acid amplification can be used to obtain a quicker positive result than is possible by culture, but in the absence of a positive acid-fast stain, several days of culture-based enrichment may be necessary to yield a positive result. Few centers provide out-of-hours nucleic acid amplification for tuberculous meningitis.

Direct antigen detection of the commoner bacterial causes of meningitis immediately after receipt of CSF has been used to obtain a rapid etiological diagnosis. Latex particle and staphylococcal co-agglutination is still used in some laboratories, but is now generally out of vogue since large surveys showed that antigen detection methods add little to the clinical management of patients with suspected meningitis. Surprising though this may seem at first, it reflects the relative efficiencies of CSF Gram stain and antigen detection systems. Antigen detection loses much of its specificity when bacteria cannot be seen in the CSF by Gram stain. Moreover, the relatively poor immunogenicity of group B *Neisseria meningitidis* means that effective antigen detection reagents are more difficult to prepare against this common cause of meningococcal meningitis. Whether antigen detection reagents should be reserved for use on CSF specimens already shown to contain bacteria by Gram stain needs to be addressed by senior laboratory staff. Their decision will need to take into account the number of specimens likely to be tested during the working shelf-life of the reagent batch.

Nucleic acid amplification by PCR is coming into widespread use for the diagnosis of meningococcal and tuberculous meningitis. A general lack of 24 h access molecular diagnostic laboratories currently prevents the use of PCR-based protocols for truly rapid etiological diagnosis of bacterial meningitis. However, steadily falling costs of molecular reagents, faster thermal cycling and electrophoresis-free systems can be expected to change this in the near future. At present, the principal contribution that nucleic acid amplification makes to the diagnosis of bacterial meningitis is the detection of non-culturable or slow-growing bacterial pathogens. In the setting of meningitis, this translates into *N. meningitidis* from antibiotic treated patients and *M. tuberculosis* before conventional cultures turn positive. Both of these advances seem small gains in view of the fact that the patients will in most cases have received appropriate antibiotic therapy before an etiological diagnosis is made. However, confirmation of the diagnosis can help the attending physician narrow down on more specific antibiotic therapy, reducing the risk of toxic and other adverse effects. The falling rate of culture-positive meningococcal meningitis means that we are likely to become more reliant on nucleic acid amplification tests for *N. meningitidis*, but this should not be seen as an opportunity to ease off on attempts to culture the causal pathogen from CSF, blood or skin lesions. Not only would this be expected to reduce the overall diagnostic yield but it might also lead to missed opportunities to confirm pneumococcal and other bacterial infections. Of equal importance is the role of positive CSF culture isolates in antibiotic resistance and molecular epidemiological studies. Antibiotic susceptibility and subtyping studies can only be performed with live cultures. Clinical isolates in properly catalogued culture collections form an important resource for laboratory-based studies on antibiotic resistance trends, vaccine design and outbreak investigation. Given the importance of bacterial meningitis, these studies are performed by major reference laboratories in many countries. In smaller countries that lack the resources required to maintain a specialist reference laboratory for meningococci, pneumococci and *Haemophilus* species, collaboration with centers in neighboring countries can produce useful data on regional trends.

Given the variety of susceptibility methods used for this group of bacteria, it is essential that testing be performed under controlled conditions using a method recognized by a clinical laboratory standard or accrediting organization. Clear guidelines must be provided for difficult-to-interpret situations such as mucoid pneumococcal strains. A pathologist or infectious diseases physician should also review the susceptibilities reported against each pathogen for clinical appropriateness.

Larger clinical laboratories with an interest in molecular epidemiology may choose to perform studies on collections of bacterial CSF isolates as part of wider epidemiological investigations. These results (e.g. DNA macrorestriction by pulsed-field gel electrophoresis) must be set in the context of descriptive epidemiological information and the population structure of the bacterial species under study. For service purposes, molecular epidemiological studies will normally be reserved for investigations of case clusters.

16.3.2. *Bacterial myonecrosis*

Often known by its colloquial name, bacterial myonecrosis is another uncommon but potentially fatal infection that occurs sporadically usually in the setting of penetrating trauma, but also in patients with diabetic and other arteriopathic conditions. The infection is characterized by a rapidly progressive destruction of striated muscle in the affected limb. The speed of progression is so fast in some cases that the only prospect of halting tissue destruction is surgical debridement or even amputation. Under these circumstances, culture-based diagnostic methods have little to offer. The diagnosis is primarily clinical, but Gram stain of the thin exudate that sometimes leaks from the surface of the affected body part can be instructive providing the procedure is performed quickly, competently and communicated directly back to the requesting physician. The causal species, usually *Clostridium perfringens*, may be evident as a Gram-positive bacillus. However, it does not typically produce spores under these conditions. Although the clinical specimen is an inflammatory exudate, polymorphonuclear cells are often notable by their absence.

Confirmation of clostridial toxin production and other procedures that require culture isolates of this obligate anaerobic species have little more than a confirmatory role in the overall management of clostridial myonecrosis.

16.3.3. Necrotizing fasciitis

Necrotizing fasciitis is another infection that can cause rapid destruction of soft tissues. The name indicates that the principal pathological process is located in the layer overlying striated muscle, rather than the muscle itself as in clostridial myonecrosis. Necrotizing fasciitis also occurs in patients with diabetes and other arteriopathic conditions, but may also occur following minor trauma such as insect bites or in children with chickenpox. The potential for continued tissue destruction and even death after commencement of appropriate antibiotic therapy has earned the bacteria that cause this condition a reputation for "flesh eating". The specific factors responsible for setting in train the series of events that leads to unstoppable tissue destruction are still poorly understood, but clearly involve a substantial patient contribution.

Recent studies indicate that the best outcome for patients with necrotizing fasciitis are obtained when a surgical opinion is sought early in the course of the infection, when laboratory specimens are collected during a surgical exploration and when a second surgical procedure is performed to ensure physical removal of infected tissue. The most significant contribution the clinical microbiology laboratory can make is a competently performed Gram stain on a macerated tissue specimen. The commonest single bacterial cause of necrotizing fasciitis is *Streptococcus pyogenes* (group A beta-hemolytic streptococcus). But other Gram-positive cocci, enteric Gram-negative bacilli and obligate anaerobic bacteria can be involved. The principal purpose of the Gram stain is therefore to determine whether the infection is likely to be caused by a single species or is polymicrobial. Nevertheless, the requesting clinician will usually opt for presumptive antibiotic therapy based on recommendations in published series.

When necrotizing fasciitis is caused by *S. pyogenes*, the condition may be a part of the more generalized invasive streptococcal infection in which case other potentially relevant body sites should be sampled, including peripheral blood.

16.3.4. *Severe community-acquired pneumonia*

It has been recognized in recent years that the overlap in the clinical features of pneumonias caused by the major pulmonary pathogens is too great to allow a presumptive etiological diagnosis from X-ray and baseline clinical information. Community-acquired pneumonia is now managed during its initial stages on a syndromic basis, i.e. decisions are made according to how severely ill the patient happens to be, rather than from the presumed bacterial pathogen. The potential role of the clinical microbiology laboratory is reduced further by the poor quality of many respiratory specimens due to difficulty obtaining respiratory secretions uncontaminated by oropharyngeal contents, particularly in breathless or confused patients. There have been many attempts to improve the bacteriological diagnosis of pneumonia, but with little impact on early therapeutic decisions. The contribution the clinical microbiology laboratory makes to management of the patient with community-acquired pneumonia is therefore modest.

A Gram stain on a carefully collected specimen of respiratory secretions can sometimes demonstrate many Gram-positive diplococci suggestive of *Streptococcus pneumoniae*, or clusters of Gram-positive cocci consistent with *Staphylococcus aureus*. When either of these is accompanied by many polymorphonuclear neutrophils and no buccal epithelial cells, the stain results may help towards a faster etiological diagnosis. Unfortunately, this occurs too infrequently to justify the amount of work performed in Gram staining unselected respiratory secretions in many clinical laboratories.

The laboratory can make a useful contribution in the setting of severe community-acquired pneumonia by raising the possibility of Legionnaires' disease and tuberculosis. Both conditions are probably underdiagnosed. Both often require a specific request to the laboratory since investigations for the respective pathogen are rarely performed as a routine procedure.

Both can be detected by culture, though there are problems when relying entirely on this approach. In the case of Legionnaires' disease, cough is only rarely productive of sputum so there may be no specimen for culture in the first place. Even when one does reach the laboratory, culture may be negative due to the general difficulty of recovering *Legionella* species from clinical material or to prior antibiotic treatment. A rapid alternative to culture is the urinary antigen test (UAT) currently gaining wide acceptance as the most sensitive means of diagnosing Legionnaires' disease. To provide a truly rapid diagnosis, the batch test version of the UAT needs to be used. At present only *L. pneumophila* serogroup I can be detected by UAT.

As indicated above, *M. tuberculosis* is a slow growing species. Diagnosis of pulmonary tuberculosis has benefited considerably from the introduction of molecular methods. Nucleic acid amplification is widely available to rapidly confirm presumptive clinical diagnosis. To date, PCR-based and similar protocols have not replaced acid-fast stains as the principal rapid diagnostic test. However, acid-fast stains cannot establish which species of mycobacteria are present. PCR-based methods have come into their own as an effective means of demonstrating that the acid-fast bacillus present in a sputum specimen is *M. tuberculosis* or one of the other commoner mycobacteria. Nucleic acid amplification has been disappointing as means of demonstrating mycobacteria in smear-negative sputa. Molecular detection methods have not replaced culture-based techniques for detection of mycobacteria since culture isolates are still required for subsequent manipulation to determine the definitive identification, and antibiotic susceptibilities and for subtyping. These additional tests remain the province of specialist reference laboratories. Methods are currently in a state of flux. The changing epidemiology of tuberculosis has put pressure on clinical microbiologists to speed up susceptibility testing (to detect multidrug resistant strains), definitive identification (to confirm atypicals which require very different antibiotic choices) and subtyping (for epidemiological investigations into case clusters). Automated culture systems are being put into use for rapid susceptibility testing and some resistance-conferring gene loci have been identified for future PCR-based routine detection in clinical laboratories. Definitive identification can be performed rapidly by automated bacterial DNA sequencing followed by comparison with databases of known sequences. These results can be used to compare clinical isolates for epidemiological studies. However, the established molecular typing method in current use is a standardized Southern blotting method using insertion sequence 6110. Other methods available include DNA macrorestriction analysis and bacterial cell wall fatty-acid methyl ester analysis by gas chromatography.

16.3.5. *Septicemia*

Septicemia is a clinical infection caused by multiplication of micro-organisms in the bloodstream, and is usually accompanied by fever and other features of severe sepsis such as chills, sweating and confusion. In severe cases, shock, renal failure and loss of consciousness may follow. The clinical features of septicemia are common to the wide range of species that may cause the condition, though specific features localized to the skin or internal organ may suggest a bacterial etiology at the time of initial diagnosis. The generalized infective processes that constitute septicemia often follow an initial localized infection in the soft tissues or a specific organ system.

It follows that laboratory procedures used to investigate a case of septicemia form part of a wider investigative process that includes the identification of localized foci of initial infection such as the urinary tract, surgical incisions or invasive medical devices. The laboratory should expect additional clinical specimens from these sites and will gain from collating the results of all such related specimens. It is the prompt examination of these related specimens that will prove more rewarding when diagnostic results are needed urgently to assist in the immediate management of septicemic patients.

Blood culture, by comparison, is relatively slow due to its reliance on bacterial growth for a positive result. In exceptional circumstances (24 h staffing of the laboratory reception area, immediate entry of new blood cultures, automated detection of new positives, immediate Gram stain and availability of requesting physician) it is possible to return Gram stain results of positive cultures to the clinical service within 3–4 h of culture collection. From experience, patients generating very fast positive blood cultures usually have severe sepsis and cannot be left untreated for 3–4 h for the sake of a Gram stain result. However, the value of a promptly delivered preliminary blood culture result should not be underestimated. In many clinical centers, a pathologist follows up these preliminary results in a blood culture–specific ward round, using the opportunity to advise the attending physician on further bacteriology tests and antibiotic selection. The initial presumptive choice of antibiotics can often be refined at this point, even before definitive antibiotic susceptibility results are known. This contact with the clinical service may also provide the foundation for team management of patients with serious complications of infection.

Several automated blood culture systems are widely available. Bacterial growth detection is performed automatically on cultures which can remain undisturbed in the incubator cabinet until a positive has been registered. The more recent models allow entry of patient identifier data by bar code readers and interfacing with laboratory computer information systems. Many large hospital laboratories have automated blood culture detection in the last decade, mainly for logistic reasons, in particular the ability to free senior laboratory staff from the highly repetitive processes required to subculture large numbers of mainly negative blood cultures. Automation has also reduced the potential for laboratory contamination during subculture and has possibly standardized blood culture processing. However, automation has brought a number of challenges. Reliance on automated systems places a considerable strain on the laboratory in the event of computer or blood culture analyzer failure. When the equipment is running optimally, its full potential can only be realized if staff are on hand around the clock to confirm a machine positive. Very few centers have been able to justify this level of priority for blood culture processing in view of the time it takes to obtain even a fast positive.

In view of these considerations, there has been increased interest in improving the pre-analytical stage of automated blood culture processing. Studies indicate that a maximum of three blood culture sets is required per septic episode and in many cases, two will be satisfactory in view of the improved quality of modern blood culture media. Only in unusual circumstances such as truly subacute bacterial endocarditis is a further set justified on epidemiological grounds. A minimum of two types of media is required, necessitating two bottles per set. Most manufacturers include anion exchange resins to absorb out antibiotics from the patient's blood in spite of a lack of evidence that this approach significantly increases the isolation rate. Large blood volumes are preferable for inoculation of each bottle i.e. 8–10 mL, except in pediatric cases. It is common practice to review the rate of isolation of bacteria judged to be probable contaminants from the venesector's or patient's skin, in order to focus attempts to improve blood culture collection technique. This has become increasingly important in view of the rising incidence of medical device-associated infections caused by coagulase-negative staphylococci and other common skin organisms. Unless great care is taken to avoid contamination of the blood culture with skin organisms during venesection and inoculation of the culture bottle, patients may receive unnecessary and possibly toxic antibiotics.

16.3.6. *Other bacterial infections*

(1) Urinary Tract Infection
Infections of the urinary tract probably generate more clinical specimens than any other type of bacterial infection. The methods in common use are well established and have changed little in several decades of clinical microbiology. In brief, the inflammatory cell and bacterial content of a midstream specimen of urine are determined by a combination of microscopy and culture on selective agar (usually cysteine lactose electrolyte deficient or CLED agar). Pure growth of a single bacterial species at greater than or equal to 10^5 colony-forming units per mL in the presence of >500 polymorphonuclear cells per mL and in the absence of squamous epithelial cells indicate a clinically significant result. These numerical criteria were derived from studies on otherwise healthy adult women with uncomplicated bacterial cystitis and should not be regarded as absolute. There are many circumstances in which these criteria can be regarded as little more than a guide, including recurrent cystitis, pyelonephritis, urinary tract infection (UTI) in patients with neutropenia, infants or staphylococcal UTI. Nevertheless, growth of mixed bacterial species, an absence of inflammatory cells or the presence of squamous epithelial cells usually indicates a poorly collected specimen, especially if the result has been obtained a second time from a repeat specimen. Conversely, very low counts of bacteria in urine specimens have increasing

clinical significance in settings such as recurrent cystitis each time the same result is obtained. A poorly collected specimen rate of 10% or greater should lead to attempts to improve patient education and other aspects of specimen collection. Laboratory staff need to remain alert to unusual but important results that may suggest contamination at first but which have greater significance to the patient. Examples include mixed bacterial growth due to a pathological connection between the bladder and the colon (vesico-colic fistula is occasionally first suspected because of an unexpected laboratory result) or bacterial growth in a catheter specimen of urine from a patient who has catheter-related septicemia.

Probably the most significant development in this overlooked area of clinical microbiology is the introduction of nitrate and leucocyte esterase rapid tests for near patient testing. These tests have gained wide acceptance as a way of supporting clinical decisions. In clinical settings where a very low rate of positive cultures are expected, e.g. antenatal screening, these tests can be used to dispense with culture altogether providing the urine specimen is visually inspected. Surveys have found that culture of clear urine specimens negative for both nitrate and leucocyte esterase adds very little to patient care.

(2) Soft Tissue Infections
A comprehensive review of bacterial infections of the skin and underlying soft tissues is beyond the scope of this chapter. Key points in the laboratory investigation of soft tissue infections that constitute a medical emergency are covered above. The majority of soft tissue infections that generate a specimen request are in hospital patients who have suffered traumatic injury or who have undergone a surgical procedure. The work of the laboratory is thus biased towards hospital practice, though well-collected specimens from infections encountered in general practice or the outpatient clinic can provide an explanation for infections that fail to respond to first choice antibiotics or that show unusual clinical features.

Specimens collected for the diagnosis of soft tissue infections are obtained from a non-sterile surface, and may therefore be contaminated by bacteria that are either members of the resident flora or transient colonizers of an abnormal external body surface. This includes *Staphylococcus aureus*, one of the most common causes of soft tissue infection. As there is no reliable method of distinguishing colonization from invasive soft tissue infection, *S. aureus* infection is almost certainly overdiagnosed. The best that can be done to avoid overdiagnosis is not to pick colonies of *S. aureus* for further work when growth is very scanty and there is a heavy mixture of other bacterial species. It has been suggested that the quality of specimens could be improved by first cleaning exudate from surface lesions before using a fresh, sterile swab moistened with water for injection to scour the infected surface. Alternatively, a fine needle aspirate of exudate from the leading edge of a spreading inflammatory lesion can be tried. If either method is used to improve the quality of the specimen the laboratory eventually works with, fewer Gram-negative species will be obtained. Isolates are more likely to be staphylococci, streptococci or obligate anaerobic bacteria. Ward staff should be discouraged from routine collection of swabs from the surface of venous stasis ulcers and similar lesions that show no evidence of spreading or invasive infection.

Careful recording of the results of Gram stain on inflammatory exudate specimens from soft tissue infection can be very helpful in interpreting subsequent culture results. Occasionally, alternative methods will be required for detection of unusual bacterial pathogens such as *Actinomycetes* (so-called "sulphur granules" may be evident to the naked eye), which can be stained, with a modified version of the acid-fast Ziehl-Neelsen stain.

For optimal results, well-collected specimens from hospital patients will need culture on several different types of selective and non-selective agar, and incubation under aerobic and anaerobic conditions. The addition of a metronidazole disk to the surface of the anaerobic plate will help distinguish obligate from facultative anaerobes and provide presumptive susceptibility results.

The results of processing specimens from patients with soft tissue infection depend on a multi-step process that can take several days even in some monobacterial infections. Some bacteria, for instance fastidious anaerobes, can take much longer to identify and obtain definitive susceptibility results. It is therefore standard practice to issue preliminary or interim reports as soon as clinically useful information is ready for issue. Late results can then be released in a final report when they have been fully validated.

Many clinical laboratories use a selective reporting approach to antibiotic susceptibility results. It is unhelpful to the requesting physician to simply list the results of all antibiotics tested against all bacteria identified in a clinical specimen, particularly when it has been obtained from a non-sterile site. A judgement has to be made to seleet the likely pathogen, when more than one species have been isolated. Ideally, the reported choice should be agreed in advance and comply with hospital antibiotic policy and other long-term objectives. It is usual to aim to provide an element of choice for the requesting physician and to avoid reporting more than one antibiotic from a group of closely related agents (e.g. first generation cephalosporins). Antibiotic testing and reporting guidelines are published and updated regularly by organizations such as the National Council for Clinical Laboratory Standards (NCCLS). These guidelines are based on recent experience and usually reflect technological ability and clinical priorities of the organization's home country.

(3) Respiratory Infections

Laboratory diagnosis of bacterial respiratory infections poses several notable problems for the clinical laboratory. Upper respiratory tract infection is one of the most common reasons for seeking a general practitioner con-sultation and is usually a self-limiting viral infection. A small proportion may be caused by group A streptococci (*Streptococcus pyogenes*), or *Arcanobacterium haemolyticum*. The potentially fatal diphtheria (caused by *Corynebacterium diphtheriae*) is rare in this region, but may need excluding in patients with appropriate symptoms who have traveled from an endemic area. There is therefore a very low rate of return on bacteriological examination of throat swabs, despite the frequency of submission. *C. diphtheriae* infection is so rarely diagnosed in most centers that routine screening for this pathogen is not performed in the absence of a history of contact with a known case or recent travel in a diphtheria-endemic area. While many upper respiratory tract infections are self- limiting, lower respiratory infection more often leads to serious con-sequences including death. Unfortunately, respiratory secretions are amongst the most poorly collected of all clinical specimen types. Sputum has to pass through the oropharynx before it gets into a specimen pot, and this will bring it into contact with the normal bacterial flora of the upper respiratory tract. Not only will this lead to contamination with an enormous range of commensal bacterial species that make isolation of a potential bacterial pathogen challenging, it also leads to contamination of the specimen with potential respiratory bacterial pathogens such as *Streptococcus pneumoniae* and *Haemophilus influenzae*, both of which may colonize the upper respiratory tract in normal subjects. This problem is exacerbated when patients spit the contents of their buccal cavity into the specimen container, rather than coughing secretions from the lower part of their respiratory tract directly into the container. This action needs to be taught to the patient, best by a physiotherapist or equivalent. If this proves to be impossible, the patient can still wash their mouth out with sterile water for injection to reduce the bacterial load prior to specimen production. A special effort to improve sputum specimen quality should be undertaken if evidence of contamination by oral bacteria (e.g. mixed oral species, presence of buccal epithelial cells) is present in the majority of sputum specimens submitted to the laboratory.

Normally throat swabs will only be inoculated onto 5% blood agar for isolation of *S. pyogenes* (and *C. albicans*). The use of a sterile straight wire to inoculate the secondary inoculum direct into the blood agar will help demonstrate oxygen-labile hemolysin by *S. pyogenes*. In most centers, *A. haemolyticum* can be recovered on this medium if laboratory staff is aware of the possibility. Tellurite medium should be added for isolation of *C. diphtheriae* only when indicated. Mycoplasmas and *Chlamydia pneumoniae* cannot be isolated using conventional laboratory media or methods. If investigations have been requested for these bacterial species, nucleic acid amplification tests are more appropriate.

Sputum specimens are more complicated. Given the problems associated with specimen collection noted above, many centers reject specimens that appear on macroscopic examination to be mainly saliva. After Gram stain, those that contain a luxuriant mixture of bacterial types and many epithelial cells may also be rejected. The remaining specimens should be inoculated onto a minimum of 5% blood agar and hemolyzed blood agar. When the patient has been hospitalized for three or more days, MacConkey's agar should be added for recovery of Gram-negative bacteria and a dilution on homogenized sputum performed for semi-quantitative estimation (some authorities regard $>10^7$ colony forming units per mL of a Gram-negative bacillus to indicate nosocomial pneumonia). It is common practice to add a streak of *Staphylococcus aureus* to the blood agar plate to demonstrate V factor dependent satellite growth of *H. influenzae*, and an optochin disk to the secondary inoculum on the same plate to help detect *S. pneumoniae*. *S. aureus* should also be sought in good quality specimens, particularly from patients in intensive care units who have recently been admitted following multiple trauma.

The interpretation of primary sputum culture results is a skilled task whose importance should not be underestimated. An over-enthusiasm for separating potentially pathogenic minority members of the primary culture result can lead to unnecessary identification and susceptibility determinations and in turn, misleading laboratory reports. It is common practice to comment on the presence of a heavily mixed growth or the poor quality of the specimen in order to avoid unrewarding work on poor quality bacteriological material. When the specimen has been marked "urgent" or is otherwise of greater than usual importance, the laboratory should observe the basic courtesy of informing the clinical service. This will provide an opportunity to obtain a replacement specimen quickly, and at the same time educate ward staff in the finer details of respiratory specimen collection.

Large hospitals are increasingly using flexible fibreoptic scopes to obtain lower respiratory tract specimens. These may be labeled with terms other than "sputum" such as "broncho-alveolar lavage", "BAL" or "protected brush specimen". This approach bypasses the pharynx and reduces the chance of major contamination, but the tip of the scope may still pick up large quantities of oral bacteria leading to the same contamination problem without the epithelial cells present to give the game away. Some respiratory physicians use a protected brush or BAL catheter device to overcome the contamination problem. If this has been used, any bacteria isolated are likely to be significant particularly if the patient has not yet been treated with antibiotics (a rarity). If not, interpretation of the primary culture result is largely the same as for sputum culture.

For comments about the diagnosis of pulmonary tuberculosis, see above. Many centers will not perform acid-fast stains on sputum or mycobacterial culture unless these have been requested by the patient's physician. Unfortunately, pulmonary tuberculosis does not necessarily follow the textbook descriptions, particularly in the earlier stages of infection. In communities with a high prevalence of HIV infection, the clinical features of tuberculosis are often atypical. Tuberculosis should be considered a possibility in all patients with persistent pneumonia that does not respond to conventional antibiotics, and therefore the laboratory should perform at least an acid-fast smear when the bacterial cause of a patient's pneumonia has not been determined by routine sputum bacteriology on a series of specimens submitted over a week or more. An even lower threshold of suspicion should be applied to purulent sputa that are visibly blood-stained, when the patient is a known pulmonary TB contact or in localities with a high TB or HIV prevalence. In the last of these, this is a case to be answered for screening all first sputa for acid-fast bacilli.

Specialized agar media (BCYE or ADCD) should be used on the rare occasions when a respiratory specimen has been obtained from a patient with suspected Legionnaires' disease (often a non-productive cough). Nucleic acid amplification (NAA) methods for *Legionella* nucleotide sequences are a more sensitive method of diagnosis. Although a consensus on the best means of diagnosis has not yet been reached due to substantial differences in the disease's epidemiology, the urinary antigen test is a good way to obtain a rapid diagnosis in *Legionella pneumophila* serogroup I infection. These methods are obviously much more timely than serodiagnosis which may only yield a positive result after the patient has recovered.

Mycoplasma and *Chlamydia* respiratory infections are most probably underdiagnosed. Both can be detected by nucleic acid amplification when respiratory secretions can be obtained. Unfortunately, both groups of pathogen cause a typically non-productive cough. Throat swabs can be used to obtain microbial DNA from the oropharynx for NAA-based tests. Otherwise, diagnosis is usually by serological means.

Bordetella pertussis occurs sporadically but is being recognized increasingly in older children and adults. In these groups, the full range of symptoms seen in younger children may not be present. Cough plates should be discouraged and alginate-tipped pernasal swabs used for NAA-based tests. Culture can be performed using Bordet-Gengou agar or a more modern alternative (e.g. CIN). Although this is less sensitive than NAA-based diagnosis, isolation of *B. pertussis* allows molecular epidemiological studies to be performed for disease control and vaccine design purposes.

(4) Enteric Infections

Determining the bacterial cause of an enteric infection has been likened to searching for a needle in a haystack. The problem is that most cases of enteritis result in diarrhea (frequency of defecation and/or a loose and often watery consistency of feces) irrespective of the microbial etiology. Rotavirus and other non-bacterial pathogens are common, and much diarrheal disease is non-infective. Moreover, the fecal specimen contains an abundant commensal bacterial flora. It is therefore important that the diagnostic laboratory takes a rational, planned approach to processing fecal specimens.

Microscopy plays little part in the routine handling of fecal specimens, unlike in parasitology where it is a fundamental part of the diagnostic process. It may be helpful to perform a Gram stain when either severe *Campylobacter* infection or cholera is suspected. Far more important is the macroscopic examination of fecal specimens. The consistency (liquid/ solid, fluid/formed), presence of fresh blood or inflammatory exudate, or excessive mucus should all be noted if present. These, the clinical details and any travel history on the request form will help determine which bacterial species should be sought.

The presence of the commensal fecal flora dictates that selective growth media be used to inhibit all but bacteria of likely clinical significance. There is no single selective agar suitable for this task and many of the more common enteritis-causing species do not require specific antibiotic therapy. Isolation of the etiological pathogen is therefore performed in many cases for epidemiological and public health purposes. Unfortunately, this means that only a minority of cases is pursued as far as specimen collection. Surveys suggest that this figure may only be around 10% of total diarrheal cases. It is essential to make the work performed on those specimens that are sent for workup as worthwhile and cost-effective as possible.

A minimum (in many centers the normal set of media) inoculated would be a selective enrichment broth for salmonellas (e.g. selenite broth), two different media for *Salmonella* and *Shigella* (e.g. xylose lysine decarboxylase agar and deoxycholate citrate agar) and a *Campylobacter* medium (e.g. Skirrow's agar). When indicated, a sorbitol MacConkey plate should be inoculated for *E. coli* O157:H7, a selective anaerobic plate for

Clostridium difficile, thiosulphate citrate bile salts agar (TCBS) for *Vibrio cholerae*, and a selective agar for *Yersinia enterocolitica*. The indications for use of these media should be understood and consistently observed by laboratory staff as part of standard operating procedures.

Preliminary isolation requires a series of initial, benchtop tests to rule out enteric bacteria of no clinical significance. *Proteus* species are easily confused with salmonellas, but can be distinguished by performing a rapid urease test. Positives can be discarded. Agglutinating antibody preparations are used to screen *Salmonella* and *Shigella* but positive reactions should always be compared against a negative control and should be confirmed because of the ability of selective agar to suppress second species of bacteria without totally inhibiting them. Only when a suspect colony type has been picked off and subcultured onto a non-selective agar such as 5% blood agar, can one be more certain that the single colonies are in pure growth. There are over 2,000 serotypes of *Salmonella enterica*. Even major enteric reference laboratories only carry a small fraction of the antisera required to identify these. Most clinical laboratories will only keep the most basic set of antisera required to confirm *Salmonella* and leave the detailed determinations to reference laboratories. Large teaching center labs will probably want to keep the antisera for the top ten serotypes, *S. typhi* and *paratyphi*, and leave the rest to national centers who have the turnover and skills such as phase change procedures to justify the extra work. *Shigella* seroagglutination requires a smaller collection of reagents but has a different type of complexity. Again, a smaller selection of agglutinating reagents is normally stocked.

E. coli 0157:H7 and other strains responsible for hemorrhagic colitis and hemolytic uremia syndrome need to be tested for production of verotoxin. This is usually performed by either neutralization of cytotoxic effect with antitoxin antibody or by ELISA immunoassay. It should be remembered that some verotoxin positive strains are sorbitol positive on the SMAC agar developed to detect *E. coli* O157:H7.

Confirmation of *C. difficile*-associated diarrhea requires detection of toxin in fecal specimen filtrate in addition to isolation. At present proprietary, no single immunodiagnostic method will pick up more than 90–95% cases. Reliance on any single test for *C. difficile* or its toxin may therefore result in under-diagnosis of the condition.

Preliminary and confirmatory tests for the major and other bacterial enteric pathogens are listed in Table 16.5.

Many of the bacterial enteric pathogens are important causes of community infection. They may be notifiable to the public health authority and many are either food- or water-borne infections. Typing methods are being used with increasing frequency to complement epidemiological investigations intended to identify the cause of point-source outbreaks. In the case of *Salmonella*, serotype and phage type are widely used. For other bacterial enteric pathogens, molecular typing methods such as pulsed-field gel electrophoresis or PCR-based methods are being used selectively. At present, phage typing of the commonest *Salmonella* serotypes remains the principal means of demonstrating possible links between cases. Even when molecular typing methods can provide the same level of discrimination as phage typing in a shorter time, a historical attachment to phage typing will ensure its retention as a typing method for some time to come.

(5) Genital Tract Infections

The most common bacterial infections of the genital tract are non-specific urethritis caused by *Chlamydia trachomatis* and gonorrhea. (Bacterial vaginosis is a common condition due to an overgrowth or imbalance of the bacterial flora of the female lower genital tract, but as the likely causative bacteria are difficult-to-grow anaerobic species, diagnosis is by a combination of clinical observations, non-specific tests and exclusion of other causes of vaginal discharge).

(a) Female genital tract infections

With the exception of vaginal thrush, the causes of vaginal discharge are difficult to grow in the laboratory. This is also reflected by the spread of infection mainly through intimate body contact before transmission of infection can take place. *Neisseria gonorrhoeae*, the cause of gonorrhea, is a fastidious microaerophilic Gram-negative diplococcus that requires a combination of selective agar (e.g. vancomycin-colistin-amphotericin-trimethoprim or VCAT agar) and special gaseous atmosphere for isolation. Specimens that have been collected with conventional cotton swabs or kept in transit for a long time may not allow laboratory recovery of *N. gonorrhoeae*. Chlamydial infection is even more difficult to confirm

Table 16.5. Identification of gastrointestinal bacterial pathogens.

Pathogen	Preliminary	Definitive
Campylobacter species	Motile, oxidase+, GNB Growth on selective agar at 37–42°C in 15% O_2, 5% CO_2	Species identification by cephalothin, nalidixic acid sensitivity, catalase, hippurate, H_2S and optimal growth temperature
Clostridium difficile	Spore-forming GPB Growth on selective agar at 37°C anaerobically	Assay for toxin A and B
Escherichia coli (VTEC)	Oxidase–, GNB Ferments lactose on MAC Ferments D-sorbitol	Verotoxin assay + agglutinating antisera for serotype
Salmonella enterica	GNB, oxidase–, urease– Lactose non-fermenter Agglutinates Poly-O antisera	Identification by substrate utilization agglutination of specific O and H antisera
Salmonella typhi	GNB, oxidase–, urease– Lactose non-fermenter H_2S+ on selective media Agglutinates Poly-O/Vi antisera	Identification by substrate utilization agglutination of specific O and H antisera
Shigella species	GNB, oxidase –, urease– Lactose non-fermenter	Identification by substrate utilization agglutination of specific antisera
Vibrio cholerae	Oxidase+, motile, GNB Growth on selective agar (TCBS)	Identification by substrate utilization, O/129, polymyxin sensitivity
Yersinia enterocolitica	Oxidase–, GNB Growth on selective agar	Identification by substrate utilization

since the causal agent is an obligate intracellular bacterial parasite requiring tissue culture for isolation. In both cases, the quality of specimen collection is an important determinant of successful isolation. Both these species can be detected by nucleic acid amplification tests on specimens of genital secretions, but the swabs must not have come into contact with the nutrient agar used to transport conventional cotton swabs, as this is highly inhibitory to the polymerase chain reaction. While nucleic acid amplification improves the sensitivity of *N. gonorrhoeae* and *Chlamydia* detection, it does not provide live bacterial isolates for subsequent susceptibility testing or molecular epidemiological investigations.

Other bacterial species isolated from low vaginal swabs include *Streptococcus agalactiae* (group B streptococci), *S. aureus*, *E. coli*, *H. influenzae*, *S. pneumoniae* and *N. meningitidis*. *S. agalactiae* is a minority or transient member of the commensal flora in a minority of women. It is of little consequence to the woman unless she is pregnant in which case it may require antibiotic prophylaxis if isolated shortly before delivery. *S. agalactiae* isolates should be reported in all pregnant patients, though the approach to prophylaxis will differ substantially between centers. *S. aureus* is an innocent commensal species in the majority of patients from whose lower vaginal swabs it has been isolated. Here, it probably reflects contamination of the swab by contact with the perineum where this species is quite common. *E. coli* is a common isolate in swabs from post-menopausal women, is often found concurrently with atrophic vaginitis but has no direct pathogenic role in this setting. *H. influenzae*, *S. pneumoniae* and *N. meningitidis* are occasionally noted in specimens from this location by observant laboratory staff. They are an oddity in this setting except possibly when isolated from a Bartholin's cyst.

High vaginal swabs collected with the aid of a speculum to investigate discharge are usually set up only for detection of *N. gonorrhoeae*, *Chlamydia* and yeasts. Endocervical swabs, on the other hand, are much less likely to be contaminated with the commensal vaginal flora and are therefore more suited to investigation of upper female genital tract infection. They may be used for detection of *Mycoplasma* and *Ureaplasma* infection by either culture or nucleic acid amplification. In the post-operative setting, a wider range of more conventional bacterial pathogens should be anticipated including *S. aureus* and obligate anaerobic bacteria.

Postpartum, the laboratory should be particularly vigilant for evidence of group A streptococci and clostridia.

Pelvic collections sampled during an operation or by needle aspiration should be handled as sterile site specimens, and may contain a wide range of bacteria including staphylococci, obligate anaerobes, enteric Gram-negatives, chlamydias, mycoplasmas and others.

(b) Male genital infections

The bacterial infections most commonly encountered are so-called non-specific urethritis, and gonorrhea. The former is caused by *Chlamydia* and the latter by *N. gonorrhoeae*. In view of the principal symptom being penile discharge, clinical diagnosis usually occurs faster and is more easily achieved for males than it is for females, some of whom may only be diagnosed through follow-up of their male partners. However, as in females, the success of laboratory diagnosis depends on the quality of specimen collection. The urethral swab collected in patients with non-specific urethritis should be inserted several centimeters into the distal urethral, and rotated to dislodge epithelial cells in which the bacterium resides. This should be done after a sample of discharge has been obtained for Gram stain and culture for *N. gonorrhoeae*. A Gram stain with intracellular Gram-negative diplococci is highly predictive of *N. gonorrhoeae* infection in males (much less so in females). Alternatively, a specimen of urine can be submitted for detection of both pathogens by nucleic acid amplification. Co-infection with both agents is common.

The identity of *N. gonorrhoeae* should be confirmed by a combination of oxidase reaction, superoxyl test, and either sugar utilization or co-agglutination.

Susceptibility testing of *N. gonorrhoeae* isolates is performed to detect penicillinase production and other key resistance factors that may significantly reduce the choice of therapy. Many centers publish trend data on antibiotic resistance in local *N. gonorrhoeae* isolates.

The nature of the infections caused by these bacterial species means that the results of laboratory investigations for genital tract infections should be handled with discretion and a high standard of professional confidentiality maintained at all times. Many laboratories use coding systems to ensure that these results can only be identified with a named patient by the requesting physician.

16.4. Emerging Technologies

16.4.1. *Rapid bacterial identification systems*

The requesting physician is often frustrated by the time it takes for identification procedures that rely on bacterial growth to produce a definitive result. It does not help that substrate utilization panels such as API and Microbact need an isolate in pure growth before incubation can begin. Definitive results reaching a hospital ward 48 h after specimen reception are unlikely to have much impact on acute clinical management.

The response to this problem has been varied. Some substrate identification systems have been automated and miniaturized to give a rapid, presumptive identification and antibiotic susceptibility results. These include the Vitek and Microscan systems. Results can be available on the same working day as preliminary isolation. This may appear to shorten the time to final reporting by as much as 24 h, but in reality these systems appear to have less impact on the clinical service. They must be used selectively, since definitive identification is not necessary for most bacterial isolates from non-sterile sites. Moreover, acceleration of bacterial growth can fundamentally alter microbial physiology and result in misleading results. Vitek preliminary identification of many enteric Gram-negative bacilli will often have to be confirmed by conventional substrate utilization panel, and an erroneous *"Salmonella"* result appears to be particularly common with some systems.

Another approach is to tackle the problem of identification of various bacterial species from non-sterile sites by development of specific chromogenic indicators incorporated in the primary growth medium. Chromogenic agars have yet to gain wide acceptance because of their relatively high cost but are likely to be used more widely in future.

Nucleic acid amplification has not been used as widely as originally anticipated for direct detection of bacterial pathogens. Its greatest impact has been on the diagnosis of infections caused by bacteria that are either fastidious, slow growing or easily lost in transit to the laboratory. Logistics, cost, and the problem of PCR inhibitors in particular specimen types have hindered the provision of molecular diagnostic services for out-of-hours and truly urgent specimens. Nevertheless, the new generation of thermal cycling systems such as the LightCycler and the Taqman analyser may alter this. These new systems will speed up nucleic acid amplification, reduce reliance on gel electrophoresis and enable quantitative results to be obtained. Despite these advances in laboratory technology, the complexity of genes determining antibiotic resistance and the speed with which they adapt to prevailing antibiotic use means that antibiotic susceptibility testing is likely to be determined by phenotypic methods for some time to come.

16.4.2. *Improved bacterial identification*

The identification of medically important bacteria and classification into subsidiary groups or taxa has come a long way since the original development of the Gram stain. It is often a moot point how far identification should go with a particular laboratory isolate (a suggested guide is provided in Table 16.6). Bacterial taxonomy has been assisted in recent decades by the widespread use of genomic systems with which genotypic relatedness can be explored. The use of standard names for bacteria satisfying agreed identification criteria provides a benchmark against which the accuracy of identification can be measured. Unless additional information is required for epidemiological investigations, bacteria will normally only be identified as far as species level, at best. In many cases, a genus level identification will suffice. However, species level identification of some groups of bacterial can place heavy demands on laboratory staff especially if a large range of non-standard culture-based tests are needed to follow a taxonomic key. Many laboratories will issue reports that refer to the genus or group and take the identification of these species (e.g. corynebacteria, actinomycetes or nocardias) no further. They may submit important isolates to a reference laboratory. However, there is a move to resolve some of these identification issues by selective use of newer technology. Chemotaxonomy, for instance, can be performed by analyzing fatty-acid methyl esters of the above bacteria with dedicated fine capillary column gas chromatography (MIDI System). More recent approaches are the automated ribotyping system (RiboPrinter, Dupont Qualicon) which relies on a cDNA probe directed at *Eco*RI digests of bacterial suspensions, and PCR-based gene sequencing.

Table 16.6. Guidelines on how far to extend bacterial identification procedures on bacterial isolates.

Microscopy	Colonies of potentially significant, dominant species isolated from urine, genital swabs Other non-sterile site swabs, sputa, feces Any isolates from blood cultures, CSF, other sterile site specimens
Benchtop tests	Clinically significant isolates from urine, genital swabs Other non-sterile site swabs, sputa, feces Any isolates from blood cultures, other sterile site specimens
Substrate utilization tests	Clinically significant isolates not yet identified by benchtop tests from genital swabs, other non-sterile site swabs, sputa, feces, blood cultures Other sterile site specimens
Subtyping procedures	Clinically significant isolates from feces, CSF Isolates responsible for nosocomial infections

NB: Considerable local variations will exist. More extensive identification procedures will be necessary in more serious infections e.g. urine isolates will require full characterizations to allow close comparison with blood culture isolates from a patient with urosepsis.

The use of molecular or chemotaxonomic methods for species level identification is difficult to justify on a routine basis on grounds of hardware and consumables costs. These are expected to fall as the technology comes into more widespread use. Nevertheless, specific problem isolates can be successfully identified to species level more quickly and efficiently than possible with traditional methods. There are also quality assurance issues in bacterial identification that can be resolved best by selective use of advanced microbial identification technology. The most discriminating level of bacterial identification involves the identification of reproducible subtypes of a given species. Individual subtypes may have a particular pathogenic or epidemiological significance, in which case subtype identification may be performed on a semi-routine basis. Examples of this include serotyping and phage typing of salmonellas, IS6110 typing of *M. tuberculosis*, and capsular typing of *S. pneumoniae*. The most widely used means of molecular subtyping are either PCR- of DNA macrorestriction-based (PFGE) and are generally used for epidemiological investigations, particularly in support of hospital infection control. Other methods including automated ribotyping and gene sequencing can be expected to be used with greater frequency and for a great range of applications in future. The ability to store and transmit typing data in digital form for analysis at a later date or a distant site (as with automated ribotyping) provides a particularly powerful tool for multicenter infectious disease surveillance. It is likely that outbreaks of bacterial infection will be investigated in a more timely fashion in future using molecular epidemiological methods.

As the technology settles into place, decisions will have to be taken on which methods to apply to given bacterial species in specific circumstances. In order to do this, preliminary validation of bacterial sub-typing systems must be performed, whether the system is based on phenotypic or genotypic bacterial characteristics. It is prudent to try out the system on collections of local clinical bacterial isolates as well as unrelated type culture isolates. This will give some insight into the population structure of locally-acquired bacteria, and is an essential pre-requisite to any future inferences drawn from typing studies on local isolates. The conventional parameters for comparing typing systems are typability (percentage of strains giving a subtyping result), discrimination (ability to subdivide isolates into clear-cut groups), reproducibility (ability to produce the same subtype result on consecutive occasions), and epidemiological concordance (whether clusters of indistinguishable strains within the collection are consistent with results from other subtyping systems). The term "clonal" is often used inaccurately to describe a cluster of isolates belonging to a presumed single strain after analysis with a given molecular typing system. Ribotyping examines around 1% of total bacterial genome, and DNA macrorestriction analysis around 10% of total. At present, total genome analysis is not a practical proposition for subtyping purposes. Whatever the molecular or other subtyping system used, in order to ensure a sensible use of laboratory resources, it is important that epidemiological typing addresses a properly formulated hypothesis and is performed under the same high standards of quality control as other diagnostic laboratory activities.

16.4.3. *Susceptibility testing*

Conventional antibiotic susceptibility testing methods rely on bacterial growth and therefore delay the release of critical information on clinical isolates. Susceptibility results on bacteria isolated from blood or CSF cultures typically take a further 12–24 h after primary isolation and have little impact on the outcome of severe, life-threatening sepsis. Results take even longer in the case of slow-glowing species such as *Mycobacterium*. Another problem that bedevils routine susceptibility testing is the mismatch between antibiotic resistance phenotype and genotype. Methods in common use such as breakpoint and Kirby-Bauer disk diffusion are typically reduced to "S" or "R". While many physicians will act on this kind of result as if it is an authoritative predictor of treatment outcome, the determinants of a given antibiotic resistance phenotype are obviously much more complex and only a part of the collection of factors that determine therapeutic outcome.

In some cases, the structural genes responsible for antibiotic resistance can be quickly and effectively demonstrated by nucleic acid amplification. A good example is recognition of methicillin resistance in *S. aureus* (MRSA) by probing for the *mec A* gene. This method can be used as a confirmatory test for MRSA, and in combination with rapid thermal cycling systems has the potential to significantly shorten the time to critical susceptibility test results. Unfortunately, there are far too many potential drug-bug combinations to use this approach for routine antibiotic susceptibility testing, even if the responsible gene sequences were known and stable. For the most part, genetic tests for antibiotic resistance determinants are likely to remain a research tool.

A radically different approach is to apply advanced cell biology technology for determination of antibiotic susceptibility phenotype. An example of this approach is the use of flow cytometry in susceptibility testing. Flow cytometers have become tools for the fast and accurate measurement of particles suspended in fluids. They have been used for many years in clinical haematology and can analyze data on thousands of cells in a matter of seconds. Some of the newer devices can accurately analyze particles an order of magnitude smaller than blood cells, which puts bacteria within their reach. The method that appears to provide the most effective way to measure antibiotic susceptibility is to combine viability stains (e.g. propidium iodide, which does not enter cells with a leaky cytoplasmic membrane) with cytometry and measure the viable fraction of total bacteria at a given antibiotic concentration. This method can be used to produce results on bacterial isolates in a matter of minutes. Flow cytometry has yet to be adapted for use in a routine, clinical service laboratory, but the method is promising for rapid susceptibility testing on critical bacterial isolates from blood and other sterile fluid specimens.

16.4.4. *Laboratory information systems*

The last decade has seen computerized information technology encroach on practically every corner of the diagnostic microbiology laboratory. Most larger laboratories now feed specimen request and test result data into an integrated laboratory information system, and many items of automated laboratory equipment (e.g. blood culture analyzers) are computer driven. Reliance on information technology is highlighted every time there is a central power failure, major virus alert, operating systems change and on a global scale, with the date change on 1st January, 2000. Further integration of information systems can be expected with the ultraminiaturization of nanotechnology, the fusion of computing and telecommunications, and the wider range of information portals.

In the increasingly complex world of information technology, the ability to competently navigate the laboratory network requires progressively higher skill levels. Information specialists with competence in diagnostic bacteriology will be required in ever increasing numbers. One example is in the field of molecular epidemiology (see above) where automated ribotyping can now be networked via a modem with other centers operating the same system. Data sharing provides a powerful tool for outbreak recognition and tracking, but opens up a new world of issues that will have to be resolved if analytical results are to be applied for disease control purposes. It can be anticipated that very tight confidence limits will have to be applied on typing data when it is to be shared between centers in this way. Former methods such as subjective visual comparison ("eyeballing") of electrophoretic gels will have to give way to probabilistic comparison of gel results. These in turn will have to be integrated mathematically with other typing results, some of which will be text-based, to generate measurable levels of confidence in a proposed isolate cluster.

Another development that is well within the reach of existing information technology is in the improvement of data interfaces. Graphic user interfaces and icon-driven systems have been in widespread use for well over a decade. However, most diagnostic microbiology laboratories leave their icon-driven system behind when they enter the laboratory information system. Here the complex data that has been generated is reduced into a series of alphanumeric strings, some of which convey extremely complex concepts. Little has been done to provide the end-user, the requesting physician, with a graphic result. It is no longer difficult to trap digital images and send these as file attachments. Why not use this approach with a critical Gram stain result? And if that can be done, why not present the laboratory user with a human body image map as an entry into test results. Following the logic a little further, it should be possible to link pharmacy records of antibiotics used to those tested and depict these with temperature, leucocyte count and C-reactive protein on an in-patient timeline. Perhaps the most obvious bacteriological application of a graphic-user interface is in epidemiological cluster analysis. Infection control specialists and public health physicians have historically focused on temporal clustering to produce their epidemic curves. They have largely ignored the potential of information technology as a geographical tool. But with programs such as MapInfo, the co-location of food-borne infections such as salmonellosis and campylobacterosis would be more self-evident.

The diverse information technology developments mentioned above have an important common feature. They are all a means to obtain a second tier of information from the primary data set that comprises the laboratory's main work. It is a frequent source of frustration that the laboratory's work on a given specimen finishes with the final report. Information technology gives us an opportunity to add clinical value to that work at little additional cost.

16.5. Biotechnological Advances for Rapid, Specific and Sensitive Microbial Diagnosis

Conventional methods for the diagnosis of viral infections include tissue culture, electron microscopy, and detection of viral antigens and viral antibodies using immunological techniques. While these methods have proved useful, they suffer from certain disadvantages. Tissue culture is cumbersome and time-consuming. Certain viruses are extremely difficult, if not impossible, to propagate, e.g. hepatitis B and human papilloma viruses. The propagation of highly pathogenic viruses such as human immuno-deficiency, Lassa fever, Marburg and Ebola viruses may be potentially hazardous to laboratory personnel. The interpretation of cytopathic effect is sometimes subjective and lacks type-specificity unless combined with the antibody neutralization test. Although a definitive serological diagnosis of viral infection can be achieved by observing a four-fold or greater increase in antibody titer between an acute serum sample and a convalescent sample taken about 10–14 days later, such an approach is slow. Furthermore, infected individuals (e.g. with HIV) may be negative for viral antibodies during the "window period" between viral acquisition and seroconversion.

Some viral infections may be life-threatening, e.g. chickenpox or herpes zoster in immunosuppressed patients. An increasing number of effective anti-viral agents are being introduced, which should ideally be administered early to treat certain viral infections, e.g. acyclovir for severe varicella-zoster virus infections in immunocompromised persons. In addition, the prompt recognition of viral infection allows measures to be taken to prevent further viral transmission, e.g. patient counseling, health education, vaccination of close contacts or vector control where appropriate.

Clearly there is a need for more rapid, specific and sensitive techniques to augment the conventional methods for the diagnosis of viral and other microbial infections.

Recent advances and applications of nucleic acid hybridization (NAH) with gene probes, and gene amplification by the polymerase chain reaction (PCR) have revolutionized microbial diagnosis and research in the field of infectious diseases. Being highly specific, sensitive and reproducible techniques for identifying and characterizing genes of microbial pathogens, these powerful tools are enjoying increasing applicability in the molecular diagnostics of a wide range of infectious disease agents, where large sample scale, speed and accuracy are important considerations. These improved screening methodologies allow earlier diagnosis resulting in prevention and timely intervention.

16.5.1. *Nucleic acid hybridization*

The principle of NAH is based on the remarkably similar behavior of nucleic acid molecules. Deoxyribonucleic acid (DNA) possesses a helical structure with complementary base sequences (adenine-thymine and cytosine-guanine) in the middle of the double-stranded helix, thus forming a long polymer with a net external negative charge from the phosphate groups of the sugar-phosphate backbone. Hybridization between two homologous nucleic acid strands is dependent upon the equilibrium between the attracting hydrogen bonds and the phosphate repulsions. Heat, alkali and various organic agents destabilize the attractive forces, while high salt concentrations partially neutralize the repulsive forces.

NAH involves the detection of specific target nucleic acid sequences by hybridization to a homologous gene probe. The excess probe is then washed off. In order to visualize the hybrid formation between the gene of interest and the probe, the latter must be labeled by an appropriate method.

NAH methodologies include *in situ* hybridization, filter *in situ* hybridization dot/spot/slot blot hybridization, Southern blot hybridization, sandwich hybridization and Northern blot hybridization. Depending on the size and nature of the probe, reactions for labeling probes include nick translation, end labeling, random primer labeling, RNA riboprobes, complementary DNA labeling, biotinylated bioprobes, chemiluminescence and fluorescent dyes. Non-radioactive alternatives for labeling probes are increasingly preferred in view of the hazards and relatively shorter life span of radioactive isotope labels.

16.5.2. *Polymerase chain reaction*

Despite the availability of a variety of modern amplification techniques (e.g. branched-chain DNA assay, ligase chain reaction, nucleic acid sequence-based amplification, Qβ replicase-based amplification, self-sustaining sequence replication, strand displacement amplification, transcription-based amplification), the most promising one is polymerase chain reaction (PCR) amplification which offers the important attributes of high sensitivity, specificity, speed, simplicity and sample throughput.

PCR is an *in vitro* technique for the primer-directed enzymatic amplification of specific nucleotide sequences. The exponential synthesis of millions of copies of a nucleic acid sequence of interest can be achieved in a matter of hours. Only a miniscule quantity of template, even femtogram amounts, is needed to initiate the reaction. Conventional PCR requires sequence information of the extremities flanking the target DNA segment in order that a pair of specific oligonucleotide primers can be synthesized which will hybridize to opposite strains. PCR consists of repeated thermal cycles of template denaturation, primer hybridization and primer extension. Each successive cycle thus doubles the amount of DNA synthesized in the previous cycle, theoretically resulting in the exponential accumulation of the target fragment according to the formula 2^n, where *n* equals the number of cycles. Thermostable DNA polymerases such as *Thermus aquaticus (Taq)* polymerase, are employed to catalyze DNA synthesis. Their superiority lies in their thermostability; optimum polymerization at higher temperatures with enhanced specificity; and being amenable to automation using computerized heat blocks.

16.5.3. *Molecular applications in diagnostic microbiology*

The applications of gene probes and PCR in infectious diseases have contributed immensely to a better understanding of their molecular pathology.

Rapid molecular virological assays and markers such as probes and PCR are emerging as viable diagnostic adjuncts or alternatives to slower and more cumbersome traditional methods such as tissue culture. PCR is suitable for detecting HIV during the seronegative window and perinatal periods; for latent viruses; for confirmation of herpes simplex virus, cytomegalovirus and respiratory syncytial virus infections prior to acyclovir, ganciclovir and ribavirin therapy respectively; for viral quantitation to monitor antiviral therapy; for detecting antiviral drug resistance; for hazardous laboratory pathogens e.g. Lassa fever virus; and for surveillance of insect and animal vectors. PCR is being applied to a growing list of other medically important viruses which include influenza, parainfluenza, corona, rubella, parvo B19, human papilloma, hepatitis, Epstein-Barr, human herpes 6, human T-lymphotropic, dengue, Japanese encephalitis, rabies, adeno-, entero- and rota-viruses.

Acquired immune deficiency syndrome (AIDS) represents an epidemic that threatens public health on a global scale. HIV cDNA probes and PCR have proven useful for identifying infected babies born to HIV-positive mothers, and for detecting HIV carriers who may be seronegative during the "window-period" between acquisition of infection and seroconversion.

The more rapid procedures of DNA hybridization and PCR can potentially replace cell culture and serology for the detection of potentially life-threatening pathogens such as cytomegalovirus and varicella-zoster virus in immunocompromised patients who require urgent diagnosis and treatment. PCR has already been employed for detecting herpes simplex virus DNA in cerebrospinal fluid towards the early diagnosis and therapy of herpes encephalitis.

NAH and PCR have been extensively used to detect human papillomaviruses, of which particularly genotypes 16 and 18 have been incriminated as an important co-factor in the etiology of genital cancer. Other examples of tumor viruses for which DNA probes and PCR have been employed are the Epstein-Barr virus (which is incriminated in Burkitt's lymphoma and nasopharyngeal carcinoma), and hepatitis B virus (HBV). HBV remains a global health problem with millions of chronic carriers who incur greater risks of developing chronic hepatitis, hepatic cirrhosis and primary hepatocellular carcinoma. HBV probes and PCR have been used to detect HBV DNA in serum and liver tissue specimens.

The molecular procedures of nucleic acid hybridization with gene probes and PCR technology are increasingly being applied for detecting bacterial, fungal or parasitic genomes to augment or even replace conventional cultures. PCR is ideal for identifying organisms that are dangerous, difficult or impossible to culture *in vitro*, such as *Chlamydia, Clostridium difficile, Helicobacter pylori, Legionella, Leptospira, Listeria, Mycobacterium tuberculosis, Mycobacterium leprae, Mycoplasma, Neisseria, Treponema, Yersinia, Histoplasma capsulatum, Pneumocystis carinii, Leishmania, Plasmodium, Toxoplasma* and *Trypanosoma*. Moreover, by amplifying known toxin genes, PCR can differentiate between toxigenic and nontoxigenic strains of bacteria such as *Clostridium difficile*.

Screening is essential to ensure that donor blood for transfusion purposes is free from pathogenic microbes and to prevent the spread of diseases such as cytomegalovirus, HIV infections, hepatitis B and C, yellow fever, syphilis, malaria and Chagas' disease. Thus, a promising area of diagnostic microbiology for the intensive application of gene probes and PCR is in blood bank screening. A suitable technique is multiplex PCR which involves the use of multiplex primer pairs each specific or a particular pathogen and yielding a PCR product of diagnostic size.

The high prevalence of sexually-transmitted diseases represents an enormous area of exploiting gene probe and PCR technology to potentially replace slower conventional diagnostic methods.

PCR amplification of genes mediating resistance to antimicrobial agents may serve as a useful adjunct to antimicrobial sensitivity tests and to identify antibiotic-resistant strains of microbes.

Another potential area for the use of gene probes and PCR is the detection of enteric pathogens in food and water samples as well as from suspected gastrointestinal infections. Being more sensitive than conventional cultures and immunological assays, these molecular tests may be used to detect *Campylobacter, E. coli, Salmonella, Shigella*, rotaviruses, enteroviruses, etc.

Fingerprinting of plasmid DNA isolated from pathogenic strains of *Staphylococcus* species can augment diagnosis by cultures and is also helpful for tracing the sources and spread of infection. Similarly, restriction endonuclease patterns of DNA from microbes, such as varicella-zoster virus, and *Mycobacterium tuberculosis*, can distinguish strains responsible for unrelated outbreaks and even those infecting different individuals.

Methodologies complementary or related to PCR are increasingly being exploited for improving specificity and sensitivity, as well as for molecular epidemiological and evolutionary analyses of infectious disease pathogens. Examples include RT-PCR for RNA viruses and viral transcripts; nested PCR; multiplex PCR; co-amplification with human housekeeping gene primers; *in situ* PCR; internal oligonucleotide probe hybridization;

restriction fragment length polymorphism; arbitrarily primed PCR, and direct DNA sequencing.

The absence of a diagnostic PCR product may be interpreted as a negative result or as a failure of amplification ("dropout"). To exclude the latter possibility, an additional pair of primers may be included to co-amplify a conserved human gene such as actin.

Consensus and type-specific primers are capable of detecting all four dengue virus types. Such a strategy is valuable for rapid diagnosis of dengue fever and for monitoring the presence of the virus in mosquito vectors.

Consensus or type-specific PCR primers can be employed to amplify products identified by hybridization to nucleic acid probes labeled nonisotopically e.g. using biotin/streptavidin/enzyme/substrate or chemiluminescent systems which can be adapted to a microtiter plate format.

By means of quantitative PCR, viral load (e.g. of HIV, hepatitis B or C viruses) can be assessed by determining plasma DNA or RNA levels, i.e. number of viral DNA or RNA copies per unit volume.

PCR combined with restriction fragment length polymorphism allows the fingerprinting of and differentiation between microbial strains for epidemiological purposes.

DNA cloning and/or cycle sequencing of PCR products constitute powerful tools for the analyses of hypervariable regions for studying microbial evolution and transmission pathways, and for sequencing genes encoding microbial products which are critical epitopes recognized by the host immune system, e.g. envelope proteins of viruses and the outer membrane proteins of bacteria such as *Neisseria gonorrhoeae* and *Chlamydia*.

A limitation associated with the use of PCR for diagnosis is the requirement for molecular training and instrumentation. The exquisite sensitivity of PCR is potentially also one of its greatest drawbacks in view of the danger of false positives arising from contamination with extraneous or "carryover DNA".

Notwithstanding this, contamination can be avoided by adhering to stringent pre- and post-PCR laboratory procedures and equipment to prevent cross-contamination. Suitable negative and positive controls should be valid in every batch of PCR samples for authentic results. In case of doubt, PCR of samples should be repeated. The success of routine PCR diagnosis depends on reliable internal controls, good quality assurance and DNA reference standards, and greater automation. With increasingly widespread availability and less expensive reagents, PCR and gene probe technology have found a stable niche in routine laboratory diagnosis.

16.5.4. *Other developments and advances in diagnostics*

The development of novel rapid sensitivity and specific techniques for detecting and characterizing pathogenic microorganisms has progressed with monoclonal antibody technology. Even though monoclonals are relatively less sensitive and specific than DNA methods, it is likely that both techniques will complement rather than compete with each other, in view of the generally easier and more rapid use of antibodies.

Furthermore, advances in the robust classical and IgM enzyme-linked immunosorbent assay (ELISA) as well as Western blot techniques for viral serodiagnosis are being achieved with better reagents addressing immunodominant epitopes, e.g. recombinant proteins, synthetic peptides and monoclonal antibodies. The Western or immunoblotting procedure is especially important for the confirmatory diagnosis of HIV infection.

There is currently a great demand for rapid bedside tests that are accurate and reliable in order to achieve prompt and more specific management and therapy. "Near patient" testing is ideal for obtaining almost instant diagnosis of potentially fatal or highly infectious diseases e.g. AIDS, hepatitis B, pneumococcal pneumonia, Legionnaires' disease, *H. pylori* infection, syphilis, tuberculosis, malaria which are amenable to antimicrobial therapy.

Consequently, "point-of-use" assays (e.g. latex agglutination and dipstick methods) are gaining increasing popularity for bedside and physician office testing. For example, immunochromatography involves placing the patient's specimen (e.g. urine, saliva, serum, blood) on a hand-held card, followed by the addition of a chromogenic reagent. The appearance of a single line indicates a negative result, while positivity is denoted by the presence of two lines.

With the expanding growth and importance of biotechnology in medicine and other spheres, disciplines such as computer science and biophysics can augment its development by collaborating in areas such as automation, laser and biosensor technology. Biosensors have numerous useful applications in patient care, i.e. "self-diagnostics" such as for home monitoring the concentration of chemical substances and drugs in the body (e.g. antimicrobial agents).

Diagnostic procedures are being revolutionized through miniaturization using biochip technology. Thousands or possibly millions of nucleic acid or protein/antibody probes can be embedded or printed onto chips, beads or slides made from silicon, glass or plastic, then reacted against clinical specimens, and positive signals detected by fluorescence scanners.

Biochips can perform thousands of biological reactions in seconds, thus allowing the virtually instantaneous identification of a wide range of pathogenic microbes, and facilitate prompt treatment.

Gene chips consisting of multiple DNA probes (e.g. oligonucleotides) aim to detect specific genes and even mutated genes via nucleic acid hybridization.

Protein chips are designed for recognizing specific proteins or antibodies by antigen-antibody reactions or protein-protein interactions.

Further Reading

1. Baron EJ *et al. Manual of Clinical Microbiology*. ASM, Washington, DC, 1999. (Widely regarded as the authoritative word on diagnostic microbiology. The principal emphasis is on recognition and detection of specific microbial agents of disease. Chapters are thoroughly referenced.)

2. Chow VTK. The impact of biotechnological advances in medical science: Implications for health care development. *ASEAN J. Sci. Technol. Dev.* 11: 1–10, 1994.

3. Cleland CA, White PS, Deshpande A, Wolinsky M, Song J and Nolan JP. Development of rationally designed nucleic acid signatures for microbial pathogens. *Expert Rev. Mol. Diagn.* 4: 303–315, 2004.

4. Inglis TJJ. *Clinical Microbiology Pocketbook*. Churchill-Livingstone, 1998. (A pocket-sized, problem-based compendium of useful

information for clinicians, scientists, infection control practitioners and public health physicians.)

5. Koneman E *et al. Atlas and Textbook of Diagnostic Microbiology.* Lippincott, 1997. (A comprehensive review of laboratory microbiology. that majors on the bacteriology performed in diagnostic labs. Brief accounts of clinical conditions are integrated with laboratory information. Illustrated and thoroughly referenced.)

6. Lueking A, Horn M, Eickhoff H, Bussow K, Lehrach H and Walter G. Protein microarrays for gene expression and antibody screening. *Anal. Biochem.* **270**: 103–111, 1999.

7. McFaddin JF. *Identification of Medically Important Bacteria.* Lippincott, 1999. (Widely regarded as the definitive bench-top laboratory manual on identification methods. Many flow diagrams and keys. Spiral bound for in-lab use.)

8. Searls DB. Using bioinformatics in gene and drug discovery. *Drug Discov. Today* **5**: 135–143, 2000.

9. Smith JE. *Biotechnology,* 3rd ed. Cambridge University Press, 1996.

10. Struelens MJ, Denis O and Rodriguez-Villalobos H. Microbiology of nosocomial infections: Progress and challenges. *Microbes Infect.* **6**: 1043–1048, 2004.

11. Tibayrenc M. *Encyclopedia of Infectious Diseases: Modern Methodologies.* John Wiley and Sons, 2007.

12. Topley and Wilson's Microbiology and Microbial Infections. *Vol. 1 Bacterial Diseases* and *Vol. 2 Systematic Bacteriology,* 1998. (These two volumes contain a comprehensive series of review chapters on medically-important bacteria and the diseases they cause. The text is available in computer-searchable, CD-ROM format.)

13. Versalovic J and Lupski JR. Molecular detection and genotyping of pathogens: More accurate and rapid answers. *Trends Microbiol.* **10**: S15–S21, 2002.

Websites

- All the Virology on the WWW
  http://www.virology.net

- Centers for Disease Control and Prevention, USA
  http://www.cdc.gov
- Microbiology: Neal's Microbial World
  http://www.microbesite.com
- National Institute of Allergy and Infectious Diseases, USA
  http://www.niaid.nih.gov
- Program for Monitoring Emerging Diseases (ProMED)
  http://www.fas.org/promed and www.promedmail.org
- TIGR Microbial Database
  http://www.tigr.org/tdb/mdb
- World Health Organization
  http://www.who.int

Questions for Thought...

1. Emerging and re-emerging infectious diseases represent an increasing problem worldwide. Discuss the contributory factors, and the impact of these diseases on public health. Illustrate your answer with examples.
2. Performing Gram and other staining on clinical samples may provide useful information on the etiology of an infection and may guide preliminary antibiotic therapy. Discuss the application of staining techniques on cerebrospinal fluid, sputum and blood culture samples.
3. Discuss the role of selective culture media in the laboratory diagnosis of bacterial infections.
4. Describe the methods routinely employed to measure the sensitivity of bacteria to different antibiotics in the diagnostic microbiology laboratory. How should the results be reported?
5. How would you determine the minimum inhibitory concentration of an antibiotic for a bacterium?
6. Discuss the methods used in the laboratory diagnosis of bacterial meningitis.
7. Discuss the laboratory diagnosis of community-acquired pneumonia.
8. In what ways can the microbiology laboratory assist the clinician in the diagnosis and treatment of septicemia?

9. Discuss the laboratory diagnosis of urinary tract infections.
10. Discuss the diagnostic value of midstream urine specimens and other urine samples in establishing the etiology of urinary tract infections.
11. Discuss the techniques used in the laboratory diagnosis of bacterial respiratory infections.
12. Discuss the laboratory diagnosis and epidemiology of *Salmonella* infections.
13. Describe the principles of the different laboratory methods for the diagnosis of sexually transmitted infections. Discuss their use in a routine diagnostic laboratory, including specificity, sensitivity, cost, time and medico-legal issues.
14. Discuss laboratory information systems in a diagnostic bacteriology laboratory.
15. Discuss the advantages and disadvantages of classical (non-genetic) techniques for the laboratory diagnosis of human bacterial infections.
16. Discuss the recent technical advances for rapid diagnosis and epidemiology of infectious diseases.
17. Compare and contrast enzyme-linked immunosorbent assays with gene probe techniques for microbial diagnosis. Illustrate your answer with examples.
18. Discuss the impact of new genetic and immunological techniques on the diagnosis of microbial infections.
19. Discuss the advantages and disadvantages of nucleic acid amplification for viral diagnosis.
20. "The exquisite sensitivity of the polymerase chain reaction is both its greatest strength and its greatest weakness." Discuss this statement.
21. Discuss how advances in computerization, miniaturization, automation and signal detection are significantly enhancing the specificity, sensitivity, speed, and sample throughput of microbial diagnosis.
22. Discuss the principles, techniques and applications of molecular fingerprinting of microorganisms.
23. Discuss the techniques and applications of restriction fragment length polymorphism.
24. Discuss the application of gene probes and the polymerase chain reaction in the analysis of restriction fragment length polymorphisms.

Chapter 17

Microbes: Friends or Foes?

Vincent T. K. Chow

Infectious Diseases Program
Department of Microbiology, Yong Loo Lin School of Medicine
National University Health System,
National University of Singapore, Kent Ridge, Singapore

Tim J. J. Inglis
School of Pathology and Laboratory Medicine
University of Western Australia,
Crawley WA 6009, Perth, Australia

Song Keang Peng
School of Science, Monash University
Sunway Campus, Selangor, Malaysia
Email: song.keang.peng@sci.monash.edu.my

Robert Koch was a medical doctor and a great microbiologist who first identified *Bacillus anthracis* as the causative agent of anthrax in the late 1880's. Since then, many other bacteria were identified and later shown to be responsible for different infectious diseases. Examples of some bacteria that can cause infections include *Vibrio cholerae* (causes cholera), *Pseudomonas aeruginosa* (wound infection), *Corynebacterium diphtheriae* (diphtheria), *Clostridium perfringens* (gas gangrene), *Clostridium botulinum* (botulism), *Shigella dysenteriae* (shigellosis), and *Bordetella pertussis* (whooping cough).

From the perspective of clinical microbiology, bacteria can be divided into pathogenic and non-pathogenic groups. Pathogenic bacteria are a group of organisms causing infectious diseases, and based on the molecular definition of pathogenicity, these bacteria contain the necessary genetic information to allow them to express virulence when conditions permit. Fortunately, when compared to the huge number of bacteria known to exist, the proportion represented by pathogenic bacteria is relatively very small, even though many of them are still uncharacterized. Under most instances, pathogenic bacteria are undoubtedly viewed as one of the biggest enemies of humans. However, many people always wonder what makes bacteria pathogenic? Bacteria may be small in size, but they can certainly compete and adapt for survival. Once the host on whom they depend dies, bacteria will also lose their source of food and growth environment. Therefore, just like ants or bees found in the field, normally they will only start to attack any host when their survival is threatened. This can happen when bacteria are unintentionally brought into an environment unfamiliar to their original dwelling place. For example, during an abrasion on the skin, bacteria like *C. perfringens* (that causes gas gangrene on wounds) from the soil can be introduced into the deeper tissues underneath the skin. This can also arise when their source of food is running out, which then induces the bacteria to secrete toxins and enzymes to kill host cells so that the released material from lyzed cells can be used as nutrients. The bottom line is that bacteria may become pathogenic for their own protection and survival. It may be similar to any other biological phenomena of self-defense. Therefore, most bacteria are generally harmless, but may become pathogenic when their survival is at stake.

Although bacteria, and pathogenic bacteria in particular, have caused much misery to mankind, we cannot deny the fact that some of them have also contributed directly and indirectly to the well-being of mankind. Through advances in science and technology, many species of bacteria are exploited to produce substances of importance for both medical and industrial applications. Some of the by-products of bacteria are also being used in industry. Examples of some known uses include generation of antibiotics, production of insulin, development of vaccines, and so on. As such, although some less fortunate individuals suffer from the devastating effects caused by bacterial infections, others may enjoy the benefit and convenience offered by bacteria, either directly or indirectly. So, are bacteria our friends or enemies?

17.1. Use of Bacteria as Hosts in Molecular Cloning

Diabetes is one of the biggest killers not just in developed countries, but throughout the whole world. Patients suffering from diabetes cannot make insulin, or cannot use insulin effectively. Insulin is one of the many hormones produced by the endocrine glands of the body. It is circulated in the bloodstream and is required for the regulation and maintenance of the blood sugar level in the body by aiding in the conversion of sugar to glycogen. Approximately 10% of the patients have type I diabetes, whereby they require insulin to maintain life. The rest suffer from type II diabetes, and whose condition can be managed by a combination of diet control, exercise, using agents to lower blood sugar and insulin treatment. Regardless of which type, many diabetic patients will eventually need insulin therapy, and the standard treatment is the administration of exogenous insulin.

Insulin is synthesized by the beta cells of the islets of Langerhans in the pancreas. Only a very small amount of insulin is produced each day. When one suffers from insulin-dependent diabetes, one has to obtain insulin from external sources. Many years ago, insulin was mainly derived from the pancreatic glands of animals such as sheep, cows and pigs, or even from cadavers. Not only was the amount obtained very limited, the quality of the insulin was also inconsistent. The non-human insulin was not as effective as human insulin. Moreover, some people even developed allergy when animal insulin was used. There were many side effects associated especially with the animal sources of insulin, and it was also very expensive.

Today, diabetic patients no longer have to pay a hefty sum for diabetes treatment. The insulin they receive is also genuine human insulin, thanks to the development of genetic engineering for the mass production of recombinant insulin. Human insulin consists of two polypeptides, designated as chains A and B, which are connected by disulphide bridges. During the synthesis of insulin, proinsulin is first formed prior to cleavage to generate the active form of insulin. The recombinant human insulin currently available is prepared by cloning and expressing genes coding for proinsulin in *E. coli*. After the proinsulin is extracted and purified from the bacteria to remove other contaminating substances, the connecting peptide is enzymatically cleaved to convert the proinsulin to functional insulin.

Using *E. coli* as a host to express genes for insulin provides many advantages. Bacterial cells are generally easy to grow, and the media used to culture bacteria is relatively cheap. Moreover, being a well-characterized bacterial species, manipulation of genes in *E. coli* is relatively safe, the outcomes are predictable, and any problems encountered can be easily solved. By using fermentation techniques to grow *E. coli*, large numbers of bacteria can be cultivated at any given time. This will also translate to large amounts of protein that can be extracted and purified from the bacteria. Human insulin represents the first successful commercial product resulting from genetic engineering of mammalian genes and expressing them in bacteria. It is often referred to as a "biotech drug". After the successful introduction of insulin to the market, other human hormones like growth hormone and thyroid stimulating hormones have also been developed and marketed.

17.2. Contribution of Bacterial Enzymes in Genetic Engineering

The typical size of bacteria usually ranges from 0.5 to 2 μm. This size is at least ten times smaller than that of most eukaryotic cells, which normally ranges from 2 to 25 μm. As such, many people perceive bacteria to be small and simple organisms. However, through our understanding of the genetics and physiology of bacteria, we know that this view is inaccurate because bacteria are not only complex in their internal structural organization, they are also capable of performing many complicated tasks. This is supported by the fact that over the years, despite massive efforts by scientists all around the world to study and understand bacteria, there are still many unknowns and questions remained unanswered. Although bacteria lack nuclei and membrane-bound organelles such as mitochondria, they have almost everything that an eukaryotic cell contains. This indicates that the basic machinery necessary to keep cells alive is common for both prokaryotic and eukaryotic cells. The same ongoing events such as DNA replication, RNA transcription, gene expression and cell replication are in operation for the two cell systems. One of the main players responsible for making these events possible includes proteins.

Proteins are involved in many cellular activities of bacteria, e.g. some as enzymes to catalyze biochemical reactions of the cell, some as positive regulators for the expression of genes, and others as signal molecules for the regulation of cellular functions. Among the many known proteins, enzymes are particularly important and play a very active role in keeping the cells healthy and alive. Many cellular events and processes are driven by various types of enzymes synthesized by the bacteria. Examples of some of these enzymes include DNA polymerase, RNA polymerase, restriction enzymes and DNA ligases. Some of these enzymes are so critical that bacteria lacking the ability to synthesize them may not survive. Studies using site-directed mutagenesis and deletion mutations to construct mutant bacteria that were deficient in expressing one or more of these enzymes, demonstrated that these bacteria were weak and fragile.

Following the discovery of the structure of nucleic acids, molecular biology has become the core subject in biology and life sciences curricula in colleges. Many scientists are even prepared to switch their interests to this field. The majority of scientists who are engaged in biological research are in one way or another molecular biologists. One important tool that molecular biologists use in their research is genetic engineering, and the key players in this technology are enzymes. Enzymes such as DNA polymerases and restriction enzymes have been isolated and purified from different species of bacteria. They are used in the manipulation of genes, for example cloning, sequencing, mutagenic studies, and so on. Two of the enzymes, restriction enzymes and DNA polymerases, are described below.

17.2.1. Restriction enzymes

One of the many tasks that molecular biologists perform is to "cut and paste" DNA fragments into plasmid vectors, a process known as cloning. These DNA fragments are also used as templates in genetic manipulations such as gel-shift assay, DNA foot-printing analysis, probe preparation, sequencing, RNA synthesis, and so on. There are several ways to generate DNA fragments, but one easy and fast method is through the use of restriction endonucleases to digest long nucleic acids at specific locations.

In fact, it was the discovery of restriction endonucleases that actually gave birth to, and laid the foundation for genetic engineering.

Restriction endonucleases were discovered in *E. coli* in the early 1970s. The first enzyme to be isolated was *Eco*RI by Robert N. Yoshimori of Boyer's laboratory in San Francisco. The nomenclature of restriction enzymes consists of several alphabets. The first capitalized alphabet is derived from the first letter of the genus of bacteria from which the enzyme is isolated. The next two alphabets of the enzyme are in small letters derived from the first two letters of the species of bacteria. There may be a fourth alphabet which is a capital letter derived from the strain designation of the bacteria. For example, the enzyme *Eco*RI is from the bacterium *Escherichia coli*. If the strain of microorganism produces just one restriction enzyme, the name ends with the Roman numeral I. If more than one enzyme is produced, they are numbered II, III, and so on.

Restriction endonucleases are a group of enzymes synthesized by bacterial cells. These enzymes usually recognize specific nucleic acid sequences about 4 to 6 base pairs long, and cleave the DNA at a very specific site. Their function is to protect bacterial cells from foreign DNA injected by invading bacteriophages. Since the discovery of *Eco*RI, many other restriction endonucleases have been discovered. Some of these are listed in Table 17.1.

Table 17.1. List of restriction endonucleases discovered after the discovery of *Eco*RI.

Microorganism	Enzyme	Recognition Sequence
Bacillus amyloliquefaciens H	*Bam*HI	C*GATCC
Caryophanon latum L	*Cla*I	AT*CGAT
Diplococcus pneumoniae	*Dpn*I	GA*TC
Escherichia coli RY13	*Eco*RI	G*AATTC
Escherichia coli J62	*Eco*RV	GATAT*C
Haemophilus influenzae R_d	*Hind*III	A*AGCTT
Haemophilus influenzae R_f	*Hind*FI	G*ANTC
Moraxella bovis	*Mbo*I	*GATC
Neisseria denitrificans	*Nde*I	CA*TATG
Serratia marcescens S_b	*Sma*I	CCC*GGG

One of the main *in vitro* uses of restriction endonucleases in molecular biology research is for the preparation of DNA fragments for a number of purposes. Owing to the specific activity of the restriction enzyme on the recognition and digestion of the DNA, the ends of the DNA fragment digested with the enzyme are complementary to another fragment digested with the same enzyme. These two pieces of DNA are considered to have perfect match and therefore can be joined chemically by a process of ligation involving another enzyme called DNA ligase.

DNA ligase is another enzyme used widely in genetic engineering. The enzyme isolated from *E. coli* is a polypeptide chain with a molecular mass of 75 kDa, and its activity requires NAD^+ as a cofactor. In contrast, the ligase encoded by phage T4 is 68 kDa, and requires ATP as a cofactor and an energy source. Regardless of the source, ligase works like a "super glue" in joining two DNA fragments together, or repairing two broken strands of DNA in a double-stranded DNA fragment. Ligase was also discovered at about the same time as restriction enzymes were discovered. Both enzymes which have contributed significantly to genetic engineering are produced by *E. coli*.

17.2.2. *DNA polymerases*

DNA polymerases are required for DNA replication in both prokaryotic and eukaryotic cells. The basic and primary activity of the polymerase is to synthesize DNA, i.e. duplication of DNA, using another strand of DNA as a template. Other than DNA polymerization activity, polymerases can also function as exonucleases to cleave DNA. This cleavage activity is also an important property of polymerases where it proof-reads nucleic acid sequences, and corrects any misincorporated nucleotide when necessary. This step is critical so that the replicated DNA will contain exactly the same nucleotide sequences as the parental copy. DNA mutations arise when some of the wrongly incorporated nucleotides are not noticed and repaired accordingly. The processes of polymerization and exonuclease activities are repeated many times during DNA replication, until the DNA strands are faithfully duplicated.

DNA polymerases are used in many techniques in molecular biology research. The protruding ends of DNA fragments ("sticky" ends) can be polished to become non-protruding ("blunt" ends) by using both the polymerase and exonuclease activities of the enzyme. They can also be used to incorporate radioactive labels in DNA via a process called random primer labeling, where nucleotides containing radioactive isotopes are built into DNA strands by polymerases. DNA polymerases have also been used extensively to determine nucleic acid sequences by the Sanger or Maxam-Gilbert sequencing techniques.

A major breakthrough in molecular biology was the discovery of a way to duplicate or amplify DNA in an efficient and rapid manner *in vitro*. This procedure is well known as PCR, or the abbreviated form for polymerase chain reaction. The scientist who invented this procedure, Dr. Kary Mullis, was awarded a Nobel Prize in 1993. PCR is a rapid method that can replicate DNA to at least a million-fold within a short period of time, usually in 2–3 hours. One of the components used in PCR is DNA polymerase, but this polymerase is not an ordinary enzyme. It is an enzyme that can withstand high temperatures of as high as 100°C, and because of this property, it is called a thermostable DNA polymerase. Unlike conventional DNA polymerases, which are inactivated once they are exposed to temperatures above 70°C, this DNA polymerase used in PCR attains its optimal activity instead at high temperatures. This unique enzyme is found in bacteria that live in hot springs in places such as the Yellowstone National Park in USA. An example of this type of polymerase that is available commercially is *Taq* polymerase, isolated from an organism called *Thermus aquaticus*. Its thermostability, and the availability of sophisticated apparatus for thermal cycling have allowed the PCR technique to be performed in an automated fashion.

Since the invention and automation of PCR, the methodology has been applied to other fields such as DNA sequencing and site-directed mutagenesis. The conventional way of performing DNA sequencing is very cumbersome and laborious, taking at least a few days just to prepare the necessary components needed for the reaction. If this conventional way of sequencing were to be used for sequencing bacterial genomes, it would not be cost-effective at all because much manpower and time is involved. However, by incorporating PCR methodology to sequencing, this once laborious procedure has become almost a daily routine in many laboratories. There are now even companies offering sequencing services for only a fraction of the cost. This is attributed to the main player involved, i.e. DNA polymerase, an example of how another enzyme from bacteria has contributed to the advancement of science.

17.3. Use of Bacterial Toxins in Therapeutics

Only a small group of bacteria is known to be pathogenic, but they are sufficient to cause havoc and misery to many people. Numerous occasions in history have shown how bacteria caused epidemic infections that killed millions of people. Most of these pathogenic bacteria cause diseases by producing toxins that attack and kill host cells. The level of pathogenicity of these bacteria is usually measured by the potency of the secreted toxins to kill cells. The faster these bacteria kill cells, the more pathogenic they are. Although the ultimate effect of these toxins may be the same, e.g. death of cells, the target and pathogenic mechanism of each toxin usually differ from one another. Some toxins are milder in their cytotoxic activity, but others can be very toxic by causing instant death to cells. Therefore, toxins play a very important role in the pathogenesis of bacteria. For those toxins that are protein in nature, the best way to neutralize them is via their inactivation through the use of antitoxin antibodies. These antitoxin antibodies bind to the active sites on the surface of toxins, thereby blocking access of the toxins to host cells. In fact, this is how treatment of some infectious diseases is achieved. The primary treatment for infections like botulinum poisoning and tetanus is by the administration of antitoxin antibody to neutralize the activity of toxins. Through studies on bacterial pathogenesis, scientists can better understand how pathogenic bacteria work. Based on the knowledge gained from these studies, it was realized that toxins could in fact be harnessed as molecules with beneficial functions. It was found that there are diseases which can be treated with toxins. Research on toxicology (the study of the biochemical and physiological properties of toxins) suddenly assumed a new role in the identification of potent toxins with potential therapeutic value. So far, a few products have been developed and some are available commercially for use in medical practice. Two examples will be discussed here. The first is the use of botulinal toxin for the treatment of disabling muscle spasm. The other is the development of chimeric toxins for the killing of anaplastic cells in diseases like cancer.

Botulinal toxin is secreted by a Gram-positive, spore-forming, anaerobic bacillus called *Clostridium botulinum*. This organism is the causative agent of botulism, a highly fatal food poisoning usually contracted following the ingestion of a neurotoxin produced by the organism. *C. botulinum* can be found in soil, lake sediments, and decaying vegetation. This makes the contamination of foods such as vegetables and meats relatively easy.

C. botulinum is classified into groups A to G based on the antigenic specificity of the secreted toxins. Groups A, B and E are responsible for most cases of human disease, while the rest are found most often in animals like cattle, horses, and sheep. The toxins are neurotoxic, heat-labile, immunologically distinct, and are among the most toxic compounds known. The estimated lethal dose for humans is about 1 μg, while 1 ml of culture fluid is sufficient to kill two million mice. Botulinal toxin binds to its target at the neuron and blocks the presynaptic release of acetylcholine at the neuromuscular junction, resulting in the prevention of information relay between muscle and nerve cells. This activity of the toxin leads to flaccid muscle paralysis of patients. Some of the symptoms commonly observed in patients include double vision (diplopia) and difficulty in swallowing (dysphagia). Death results from the paralysis of respiratory muscle functions.

The introduction of botulinal toxin for medical use was initiated in the early 1980s for the experimental treatment of a painful neurological disorder called dystonia. Its use as a general therapy for dystonia was finally approved in 1989 by the Food and Drug Administration (FDA) of the US. Dystonia is a group of neurologic disorders affecting more than 300,000 people in North America. It is characterized by sustained involuntary muscle contractions and spasm of the body, which results in the development of painful movements or postures in certain parts of the body. It can affect virtually any part of the body and involve any muscle. The specific cause of the disease remains unclear. Examples of such dystonias are blepharospasm (eyelids), hemifacial spasm and hand cramps. Before the use of botulinal toxin was known, the only therapy that was given to patients suffering from the disease was through surgical destruction of the nerve endings in the area affected by the spasms. This type of surgical treatment is not only expensive and dangerous, but has a low success rate. Since dystonia involves sustained contraction of muscle, it can take advantage of the neuromuscular blocking effect of botulinal toxin to alleviate muscle spasm caused by the disease. Injecting a small amount of botulinal toxin at the affected site will neutralize the spasm caused by the disease. This was indeed proven effective. Botulinal toxin therapy provides a better alternative to surgical treatment because it is cheaper and less dangerous. However, the effect provided by botulinal toxin is only temporary, and repeated treatment of the toxin has to be administered.

The botulinal toxin used currently is the group A neurotoxin, the only one approved for medical use today. It is a large protein with a molecular mass of 150 kDa. The commercial form of the toxin is a complex with a number of other proteins resulting in a molecule that is six times the size of the original neurotoxin. Attempts are currently underway to genetically engineer botulinal toxin to make it more effective at targeting neurons so that a one-shot therapy with permanent effects will be feasible.

Although botulinal toxin is a neurotoxin that can cause death in a victim during the pathological process of the infection, it does not kill the affected cells. The person dies mainly due to respiratory failure as a result of muscle paralysis. There is a group of toxins called cytotoxins that actually kill cells through certain biochemical reactions. Examples from this group of cytotoxins include *Pseudomonas* exotoxin (produced by *Pseudomonas aeruginosa*), and pertussis toxin (produced by *Bordetella pertussis*). These toxins kill cells by inhibiting the synthesis of proteins in cells. To become functional proteins, this group of toxins usually consists of three domains when they are folded properly: (a) the receptor binding domain that binds to the receptors on host cells; (b) the translocation domain that transfers the toxin to the cytoplasm of the host cells; and (c) the enzymatic domain that carries out the catalytic function of the toxin. Toxins lacking any one of these domains will lose their ability to kill cells, and thus become inactive. Generally, the degree of specificity is determined by the receptor-binding domain of the toxin, which directs the toxin specifically to only certain cell types. Once inside the host cells, the enzymatic domain of the toxin will function to inactivate its target molecule. The specificity of the enzymatic domain is usually dependent on the target molecules present in the cell, e.g. the elongation factor and ribosome responsible for translation of proteins. Therefore, if there is a strategy to introduce the enzymatic domain into the cytoplasm of certain cells such as tumor cells in a very specific manner, this portion of the toxin will be able to kill those cells selectively. This is how the idea of the chimeric toxin was conceived.

A chimeric toxin is a toxin chemically attached to a protein molecule with high specificity for certain receptors found on the surface of eukaryotic cells, so that the action of the toxin can be directed towards those targeted cells. If this protein molecule is a portion of an antibody, it is called an immunotoxin. This type of immunotoxin can be generated by attaching the enzymatic domain of the toxin to the antibody-binding (Fab) portion of a monoclonal antibody, all achieved by genetic engineering techniques. This method of toxin targeting is potentially very useful and effective for anaplastic cells like tumor cells. The conventional ways of removing tumors include either surgical excision, chemotherapy or radiotherapy, or a combination of these. The ultimate aim of the treatment is the complete elimination of all tumor cells in the body. Despite proven success in the treatment of certain tumors, these methods are not always successful, especially when the tumor growth has already progressed to the terminal stage. The major limitation of conventional methods is the inability to achieve specific killing of only tumor cells but not normal cells. With the advent of immunotoxin technology, there appears to be hope for a better and more effective way of treating cancer patients. A group of researchers from the National Institutes of Health, US, has recently demonstrated the use of immunotoxins to destroy solid tumors. This immunotoxin was created by chemically coupling the enzymatic domain of *P. aeruginosa* exotoxin to a monoclonal antibody recognizing receptors present only on tumor cells. Results from the phase I clinical trial were quite promising, although there are still both technical and biological problems to be resolved. One of the problems is the host induction of antibodies to the recombinant toxin. This research group is currently improving on the construction of the immunotoxin, including designing a smaller recombinant immunotoxin, and the use of immunotoxin in conjunction with immunosuppressants to inhibit the development of antibodies.

17.4. Use of Biotechnology to Produce Antimicrobial and other Therapeutic Agents

Microorganisms and plasmid vectors have been manipulated by genetic engineering for use in medicine, such as cloning of genes of interest to express valuable proteins in prokaryotic (e.g. *E. coli*) or eukaryotic host cells (e.g. yeast, mammalian, human, plant).

With the advent of recombinant DNA and protein engineering technology, an ever increasing number of genetically-engineered therapeutic agents are being introduced and marketed by the pharmaceutical industry, undergoing clinical trials or have potential clinical applications.

Many important examples of these novel drugs include hormones (such as growth hormone for pituitary dwarfism; insulin for diabetes mellitus; erythropoietin for anemia or renal failure), growth factors (such as insulin-like growth factor 1 for Laron dwarfism; brain-derived neurotrophic factor), biological response modifiers, immunomodulators, immuno-stimulators, cytokines (such as interferon for hairy cell leukemia, chronic myelogenous leukemia and chronic hepatitis B and C; stem cell factor; granulocyte colony stimulating factor; granulocyte-macrophage colony stimulating factor for bone marrow failure; interleukins for certain cancers; tumor necrosis factor), hematological products (such as tissue plasminogen activator for thromboembolic lysis; coagulation factor VIII for hemophilia; albumin for shock; hemoglobin as a blood substitute), alpha-1 antitrypsin for emphysema, encephalin with analgesic properties, antibacterial lactoferrin and lysozyme, microbial vaccines (such as the hepatitis B vaccine and potential recombinant AIDS vaccines currently undergoing clinical trials), and cancer vaccines. In 1995, the world market for human growth hormone alone was worth US$1.4 billion.

Furthermore, synthetic peptides are being contemplated as therapeutic agents, e.g. vaccines. Modern biotechnology is also being exploited for the biological control of pests and vectors of human disease.

Although bacteria, yeasts and cultured cells serve as the conventional hosts for the expression of commercially important proteins, "molecular biopharming" or the use of transgenic animals and plants as living pharmaceutical factories or bioreactors to produce drugs and monoclonal antibodies on a large scale appears promising. Thus, valuable proteins can be generated in quantity in milk and from plant extracts.

The knowledge acquired from the molecular biology of important disease processes such as cancer and microbial infections including AIDS, yielding sophisticated information such as X-ray crystallography and nuclear magnetic resonance data of proteins can be processed using "virtually reality" and interactive computer graphics to permit the three-dimensional modeling and structure-function analysis of complex molecules, thus facilitating a more rational, refined, precise design and molecular engineering of effective novel drugs such as agonists, antagonists, enzyme activators and inhibitors. For example, Gleevec (STI571, imatinib mesylate) is a new signal transduction inhibitor developed against the deregulated tyrosine kinase activity of the *bcr-abl* fusion protein in chronic myeloid leukemia. Molecular mimics of the substrate of a viral enzyme that differ slightly from the natural substrate can also be created to inhibit the enzyme and interrupt viral replication.

Notwithstanding the thrust of molecular genetics in medical science, pharmaceutical companies are not neglecting the rich sources of drugs and medical products such as new antibiotics from nature. In fact, there is a renewed interest in screening for these high-value products effective against global epidemics such as AIDS, particularly in the plethora of plant and microbial flora and species of the rapidly depleting tropical rainforests of the world.

Examples of the production of antimicrobial agents from microbes are cephalosporin from *Acremonium* species, chloramphenicol from *Cephalosporium* species, penicillin and griseofulvin from *Penicillium* species, rifamycin from *Nocardia* species and streptomycin from *Streptomyces* species.

Compounds with anti-tumor activity have also been derived from microbes, e.g. actinomycin D, bleomycin, daunorubicin, mitomycin C from *Streptomyces* species, and asparaginase from *Erwinia* species. Other new drugs with anti-cancer properties include enediynes, isolated from a soil bacterial strain.

Much progress has been made in the development of antimicrobial agents, and the search for new ones is intensifying in the light of the increasing problem of antibiotic resistance and the emergence of new infectious diseases. The new antimalarial drug artemisinin extracted from the leaves of the Chinese herbal plant *qinghaosu*, shows promise in controlling multi-drug resistant and cerebral malaria currently spreading in parts of South-east Asia. Tafenoquine is a new synthetic analogue of primaquine for antimalarial prophylaxis.

Available to the physician's armamentarium is an increasing number of antiviral agents effective against herpes viruses (e.g. acyclovir, famciclovir, valaciclovir, ganciclovir, foscarnet), hepatitis B virus (e.g. interferon, lamivudine), influenza A and B viruses (e.g. neuraminidase inhibitors including oseltamivir and zanamivir), and respiratory syncytial virus (e.g. ribavirin).

The threat of the AIDS pandemic stimulated the generation of new antivirals against HIV, i.e. nucleoside analogue reverse transcriptase inhibitors (e.g. azidothymidine, dideoxyinosine, dideoxycytidine), non-nucleoside reverse transcriptase inhibitors and protease inhibitors. Highly active antiretroviral therapy (HAART) for AIDS involves multiple antiretroviral drugs prescribed in combination as a cocktail to achieve synergistic effect, to suppress viral load, and to minimize drug-resistant variants.

Novel treatment modalities for infectious diseases are being devised and tested, such as immunotherapy using anti-cytokine antibodies against septic shock, e.g. antibodies against tumor necrosis factor alpha, interleukin-1 and interleukin-6.

Microbial toxins are being harnessed for beneficial uses, e.g. botulinum toxin A for the alleviation of muscle spasms and wrinkles. Modern "magic bullets" or immunotoxins in the form of custom-made, humanized monoclonal antibodies conjugated with potent toxins which specifically target cancer cells are being studied in patients with advanced malignancies. An example is the diphtheria toxin conjugated with interleukin-2 targeted against T-lymphocytic leukemia cells that express interleukin-2 receptors.

Liposomes and biodegradable polymer particles are being developed as vehicles to deliver drugs directly to target sites in the body, thus enhancing efficacy and reducing adverse or toxic effects.

Knowledge on the formation of biofilm arising from the ability of bacteria and other microbes to adhere to and colonize surfaces has even led to the development of "non-stick" coatings for medical devices such as artificial implants.

17.5. Vaccine Development

Most pathogenic forms of bacteria owe their pathogenicity to the virulence factors they produce during pathogenic events leading to damage of host cells. Some of these virulence factors are secreted into the environment to act on surrounding or remote cells (e.g. exotoxin and hydrolytic enzymes), others are produced and deposited on the surface of the bacteria (e.g. capsules and pili). The roles of these virulence factors vary from one pathogen to another, but they are usually produced either to protect the bacteria from attack by molecules like antibiotics and antibodies, so that the bacteria can spread, or to obtain nutrients for the survival of bacteria. However, in order for these virulence factors to be successful, they must first avoid or overcome the host defense system, in particular the immune responses. The human body contains several powerful defense systems against most foreign invaders. The first line of defense starts from the skin, and extends all the way to the internal structures of the body including the Peyer's patches of the intestine. One of the immune responses is the generation of antibodies. Once the primary immune response is triggered, memory cells in the body will be established, and the subsequent encounter with the same invader will bring about speedy synthesis of the specific antibodies to inactivate or neutralize the invader. This is known as the secondary immune response, which is more efficient and rapid than the primary response. Therefore, the ideal situation for an individual is to have as many of the primary responses triggered in the body, so that the body is always ready to defend against invading pathogens. This was the concept of the smallpox vaccine introduced by Dr. Edward Jenner in the eighteenth century. A vaccine is a substance that can trigger the immune response of the body without causing harm to the host. As a result of Dr. Jenner's discovery and successful mass vaccination programs worldwide, the World Health Organization officially declared the world to be free of smallpox in 1980. Currently, vaccines are already available for several infections: poliomyelitis, diphtheria, cholera, whooping cough, tetanus, mumps, measles and rubella (Table 17.2). The use of vaccines has extended the human life span by approximately 15 years.

Vaccines can be developed in many ways. In general, to make a vaccine for a pathogen, the pathogen is first deliberately inactivated so that it is no longer harmful. This inactivated form is then injected into the body of the host. Since it is rendered not pathogenic and harmless, it will not cause damage to the host. However, although the pathogen is inactivated, some of the antigenic determinants are still retained. These remaining antigenic determinants can continue to trigger the host's immune response, resulting in the generation of antibodies against the pathogen. Consequently, the

Table 17.2. Human vaccines currently used for active immunization against microbes/diseases.

Vaccines	Dead/Inactivated	Live/Attenuated
Viral	Hepatitis A	Measles
	Hepatitis B	Mumps
	Influenza	Poliomyelitis (Sabin)
	Japanese encephalitis	Rubella
	Poliomyelitis (Salk)	Smallpox
	Rabies	Varicella
		Yellow fever
Bacterial	Anthrax	Tuberculosis (BCG)
	Cholera	
	Diphtheria	
	Haemophilus influenzae b	
	Meningococcal meningitis	
	Pertussis	
	Plague	
	Pneumococcal pneumonia	
	Tetanus	
	Typhoid	

body will be ready for any subsequent invasion of the live and active pathogens because the primary immune response has been stimulated. It is interesting to note that we can exploit the pathogen itself to boost up our own defense system, e.g. the vaccine against whopping cough is prepared using killed *B. pertussis* bacteria.

Virulence factors of bacteria such as toxins can also be targeted as a source of vaccines. The majority of bacterial toxins have conserved structures, and therefore the antibodies mounted against these toxins will be effective even if the body has not encountered the toxin for years. In other words, as long as a specific pathogen is the causative agent, the toxin produced will be more or less similar structurally. Therefore, if a toxin is targeted as a vaccine, the protective effect it can provide will be quite significant. It is thus not hard to understand why toxins are one of the best candidates for vaccine development. Before any toxin is used as a vaccine, it will have to be inactivated by treatment with formalin or heat. This is necessary to ensure that no harmful effects that may damage host cells when the toxin is introduced in the body are retained. This modified "non-toxic" toxin is called a toxoid. Most people will already have been inoculated with doses of toxoid before they reach adolescence. The most common one is the DTP vaccine which is composed of inactivated toxins of three organisms: *Corynebacterium diphtheriae* (D), *Clostridium tetani* (T) and *Bordetella pertussis* (P). Injecting toxoid in humans serves as a vaccine to protect the body from harmful toxin produced by the organism from which the toxoid is made. But when the toxoid is injected into animals, it serves to generate polyclonal or monoclonal antibodies for use in scientific research.

Another well-known vaccine is that against tuberculosis, a major cause of death worldwide. This vaccine is called BCG (which stands for bacille Calmette-Guerin), and it is prepared by using a live but avirulent strain of *Mycobacterium bovis*. It should be noted that tuberculosis is caused by *M. tuberculosis* which does not produce toxin, so the option of developing toxoid as a vaccine does not exist. The BCG srain of *M. bovis*, on the other hand, shares many common structural properties with *M. tuberculosis*, but most importantly is nonpathogenic. Therefore, it is believed that the immune response triggered against *M. bovis* will be equally effective against *M. tuberculosis*. BCG is relatively cheap and safe, and has been used to immunize children for many years.

Vaccines have also been developed based on the capsules present on the surface of bacteria. One such vaccine currently available is for protection against *Streptococcus pneumoniae*, a causative agent of pneumonia and meningitis. The vaccine against *S. pneumoniae* contains at least 20 of the commonest capsular serotypes found associated with this bacterium. Individuals who are immunized with this vaccine should be protected against the commonly encountered strains of *S. pneumoniae*. However, one major limitation is the polysaccharide nature of the vaccine. This type of vaccine is generally not effective in both infants and elderly adults, which unfortunately comprise the two main high-risk groups of *S. pneumoniae* infection among the population. At present, other vaccine targets such as surface proteins of the organism are being investigated, and it is believed that given their protein nature, such vaccines will be more effective.

17.6. Bacterial Surface Display

The outer membrane of Gram-negative bacteria contains a number of proteins such as LamB, PhoE and OmpA, and lipoproteins such as Lpp, TraT and PAL. These heterologous proteins are displayed on the surface of the bacteria. Due to their strategic location, they have been employed as tools for scientific research, especially in vaccine development. One of the limitations of vaccine development is the instability of the antigen introduced in the body. To overcome this limitation, enteric bacteria that are commensals in the intestine are employed as vehicles to deliver the vaccine antigen. Enteric bacteria are not only resident in the intestine, but are also capable of replicating. Therefore, with appropriate vaccine construction using enteric bacteria, there will be a continual supply of the antigen as long as the bacteria exist as commensals. An ideal way to develop such vaccines is to genetically engineer and fuse a subunit of toxin to a surface display protein such that the toxin can also be displayed on the surface of the bacteria.

Gram-negative bacteria such as *E. coli* and *Salmonella* species have been used for surface display of several potential vaccine candidate proteins. Preliminary results from immunization of mice with live *S. dublin* expressing flagellin with a fused cholera toxin or hepatitis B virus epitope have been quite promising. The advantages offered by this type of vaccine delivery system are related to the physiological and biochemical properties of enteric bacteria, i.e. they are part of the normal intestinal bacterial flora, easy to grow, and the cloning vectors are readily available. Moreover, no toxins or enzymes need to be extracted and purified, and no chemical inactivation of the toxin is necessary. In addition, when bacteria replicate, everything including the engineered surface display protein is simultaneously replicated.

17.7. Immunization Strategies Against Human Microbial Pathogens

Despite the global eradication of smallpox and the considerable success of immunization against poliomyelitis, infectious diseases constitute the world's major killer of children and young adults, especially HIV/AIDS, measles, diarrheal diseases, pneumonia, tuberculosis and malaria in developing countries. Launched in 2001, a multi-million dollar Global Fund for Children's Vaccines will primarily be used to vaccinate children in developing nations against hepatitis B, yellow fever and *Haemophilus influenzae* type b. Prophylactic vaccines against major infectious diseases of global public health importance are urgently required. These include malaria, leishmaniasis, schistosomiasis, tuberculosis, melioidosis, sexually-transmitted diseases, AIDS, dengue, hepatitis C, enterovirus, respiratory syncytial virus and other infections for which there are currently no effective vaccines.

Conventional microbial vaccines comprise of live attenuated vaccines, inactivated vaccines and toxoids. Although they currently contribute significantly in the protection against major infectious diseases via immunization, certain conventional vaccines suffer from drawbacks. For example, one negligible but worrisome risk of live attenuated viral vaccines is the potential generation of mutants that may possess similar or even higher virulence compared to the parental wild-type viruses, especially RNA viruses with a higher propensity for genetic variability.

Thus, much research is focusing on new strategies for vaccine development, to improve the quality of current vaccines especially against diseases for which immunization is not yet routinely available, e.g. AIDS, malaria, hepatitis C, human papillomavirus infections.

Using recombinant DNA technology, selected microbial genes can be cloned into suitable vectors for the production of immunogenic vaccines. Commercially produced in yeast or mammalian host cells capable of expressing glycosylated proteins, the hepatitis B vaccine has the potential to reduce and even eliminate the morbidity and mortality associated with chronic hepatitis, cirrhosis and primary liver cancer. *Salmonella* and lactic acid bacteria are being tested as host cells for the production of vaccines to elicit mucosal immunity. Microbial antigens are also being expressed in fruits and vegetables such as bananas, tomatoes and potatoes as "edible plant vaccines". The protective efficacy of chemically synthesized peptide vaccines (e.g. against malaria and HIV) is being evaluated.

Naked DNA vaccines (NDVs) can be engineered to encode gene(s) for microbial or viral antigen(s), e.g. envelope or core proteins. The host intracellular processing of the DNA-expressed viral protein occurs via the major histocompatibility complex (MHC) class I pathway and can evoke cell-mediated immunity. In contrast, the processing of conventional vaccines is generally through the MHC class II pathway which stimulates a predominantly antibody response.

Compared with conventional microbial vaccines, DNA vaccines possess many advantages, some of which are outlined below. NDVs are capable of inducing both humoral-mediated and cellular-mediated immune responses, with a stronger T-helper response. Antigen expression is more sustained, resulting in long duration of immunity. NDVs also allow the expression of antigens that more closely resemble native viral epitopes. In addition, DNA vaccines are safer than live vaccines and can obviate problems associated with viral vectors used in gene transfer, e.g. vacciniavirus vectors. In comparison with recombinant DNA vaccines, there is no necessity to purify the final protein product. Other practical merits which make NDVs suited to developing countries include the relatively low cost of vaccine production, and the stability of DNA versus the requirement for the "cold chain" for most conventional vaccines. A further advantage is the simple concept and route of administration of DNA vaccines, i.e. via intramuscular injection or via the "needleless" gene gun, which can influence the desired outcome of immune response.

Promising initial results have been achieved in DNA immunization against specific microbial pathogens in animal models. For example, a naked DNA vaccine based on the influenza viral core nucleoprotein has been shown to confer protection against lethal doses of heterogeneous strains in mice. Experiments have also documented the successful protection of animals by DNA immunization against challenge with various other infections, including rabies, tuberculosis and malaria. Despite theoretical concerns such as the potential induction of tolerance or autoimmunity, and the possible integration of foreign DNA into the host genome culminating in insertional mutagenesis, animal studies have so far supported the safety of DNA vaccines.

DNA immunization represents the most exciting development in vaccine technology over the past few decades. Although tests have indicated the high safety and efficacy of DNA-based vaccines, current methodologies require further optimization. The outlook on direct DNA immunizations to address a variety of medical problems is indeed very promising.

The ease and rapidity of construction of DNA vaccines coupled with advances in rapid nucleotide sequencing of microbial genes offer a clear advantage in countering the threat of newly emerging, re-emerging and drug-resistant pathogens. Moreover, in the case of highly mutable microbes e.g. influenza and human immunodeficiency viruses, DNA vaccines can be easily redesigned and tailor-made according to the predominant strains prevalent in different geographic regions.

With the advent of large-scale projects to sequence entire genomes of major microbial pathogens, there are prospects for large numbers of potential target genes to be tested as candidates for inclusion in effective DNA vaccines.

It has been demonstrated in specific instances that DNA vaccines co-expressing immunostimulatory molecules are much more efficacious than those lacking certain cytokine genes, thus justifying the need to test such "hybrid" vaccines.

In the case of typhoid, gonorrhea, chlamydial and other infections in which mucosal immunity is critical in their prevention, oral delivery of DNA vaccines can be achieved via vehicles such as specific vaccine strains of *Salmonella* and *Lactobacillus*.

While prospective prophylactic DNA vaccines should be the mainstay of research and development, there appears to be potential for the use of DNA vaccines in the immunotherapy of chronic infections e.g. hepatitis B and C, HIV infections, with the aim of enhancing cell-mediated immunity to accelerate viral clearance.

Another exciting prospect is the cloning of multiple gene cassettes of different pathogens to facilitate "single shot" protection as opposed to the current practice of multiple doses given at various time intervals. Yet another potentially successful challenge is to develop DNA vaccines to stimulate immune responses against specific tumor antigens to augment killing of cancer cells *in vivo*.

17.8. Beneficial Uses of Viral Vectors for Gene Therapy

Numerous clinical trials are underway to employ gene therapy to correct genetic diseases by supplying cells with missing or defective genes with normal functional genes, often accomplished by transferring the desired genes using vectors based on viruses such as adenoviruses, adeno-associated viruses, herpesviruses and retroviruses for gene delivery.

In September 1990, the revolutionary use of human gene therapy to treat a 4-year old girl with severe combined immunodeficiency, by infusing her with cells containing the correct adenosine deaminase gene expressing the critical missing protein with subsequent encouraging clinical improvement, marked a significant milestone in medical history.

Although many clinical trials have demonstrated that viral vectors for gene therapy are non-toxic, the unfortunate death of 18-year old Jesse Gelsinger during a Phase I trial in 2000 has forced the research community into re-evaluating public risk assessment and safety issues.

The advent of gene therapy heralds the future application of this potentially powerful technology (and its associated approaches such as gene targeting and anti-sense RNA therapy) for the treatment of hitherto incurable or debilitating maladies from genetic disorders (e.g. cystic fibrosis, hypercholesterolemia, hemophilia, sickle cell anemia and Lesch-Nyhan syndrome) to AIDS and cancer. Genetically-modified cells (e.g. tumor-infiltrating lymphocytes implanted with the tumor necrosis factor gene) are being tested as immunotherapeutic autovaccines in certain cancer patients.

Anti-sense strategy aims to block messenger RNA in order to prevent the translation of genetic information for protein expression. Anti-sense molecules are classically synthetic DNA analogues whose sequences are complementary to those of specific mRNA strands, and can thus bind to inhibit the expression of specific target genes. The anti-sense approach has been shown to interrupt viral replication in infected cells (e.g. with HIV).

The world's first Institute of Human Gene Therapy established in the University of Pennsylvania, USA in 1992 is a recognition of the impact of gene therapy in the next century, and this revolution is being hailed as a potentially dominant area of medicine as important as antibiotics and vaccines in the 20th century.

To complement this important technology, the massive Human Genome Project was launched in the US in 1990 to map and sequence the three billion pairs of bases of human DNA encoding an estimated 50,000 genes, with the aid of supercomputers for data processing. First directed by the Nobel laureate and co-discoverer of the double helical structure of DNA,

Professor James Watson, this enormous task was initially projected to take 15 years at a cost of US$3 billion. Important biomedical implications and commercial spin-offs include novel candidate genes for genetic screening, gene therapy and drug discovery.

17.9. Genomics and Proteomics of Microbial Pathogenesis

The official announcement on 26 June 2000 of the completion of the first draft of the human genome sequence opened a floodgate of important genetic information with profound implications for biomedical science and practice. Molecular biologists are harnessing the newly emerging and cutting-edge technologies of genomics and proteomics, which aim to elucidate and characterize functions of genes and proteins, with downstream applications. These projects have potential applications in novel diagnostics, new drug discovery, vaccine development and clinical trials. It is envisaged that spin-offs from these new technologies may evolve and be developed for industry, e.g. pharmaceutical and biotechnology companies.

Scientists are exploiting and developing technologies that encompass or overlap the triad of genomics, proteomics and bioinformatics. These include high-throughput DNA sequencing, gene arrays and chips, subtractive and differential gene screening, protein-protein interactions, two-dimensional gel electrophoresis, mass spectrometry, protein micro-sequencing, protein chips, transgenic and gene knock-out animals, protein identification and fingerprinting.

Microbial genomic and proteomic profiling include whole genome sequencing and proteome analysis of major causative agents of infectious diseases. The complete genomes of many bacterial species and other micro-organisms have been sequenced, and many more microbial sequencing projects are in progress.

Proteomics is facilitated in part by two-dimensional polyacrylamide gel electrophoresis which can separate complex mixtures of proteins. Interesting protein spots can be further characterized by mass spectrometry.

Bioinformatics plays a key role in mining valuable genome-wide data sources, especially for the identification of virulence factors and antibiotic resistance mechanisms, as well as for anti-microbial drug discovery.

Comparative genomic analysis can identify essential genes conserved across bacterial species but not in eukaryotes, which may lead to promising broad-spectrum antibiotics.

Comparison of gene expression levels between test (infected) and reference (uninfected) cell samples can be achieved by techniques such as differential display and microarrays, which may lead to the discovery of novel genes related to pathogenesis and antimicrobial drug targets. Microarrays or chips spotted with cDNAs permit the simultaneous expression profiling of thousands of genes. Microarray technology is useful for the analysis of cellular genetic responses to microbial infections, and for screening bacterial and other microbial isolates.

The applications of pharmacogenomics and pharmacoproteomics to study remission or resistance after antimicrobial treatment involves the profiling of successful or adverse responses to drug therapy of infections, and may lead to new diagnostic and prognostic markers of therapy.

17.10. Benefits and Implications for Healthcare Development

The extraordinary power of microbial biotechnology can be harnessed to contribute effective solutions for global health problems. Scientists are seizing the opportunities for carefully exploiting the new microbial biotechnology to meet the healthcare challenges of the future. Medical biotechnology businesses are tapping huge worldwide markets worth billions of dollars. Many countries have adopted key strategic master plans and implemented long-term as well as short-term programs to develop biotechnology relevant to regional resources and requirements, with the health care aspect constituting a strong pillar.

While the exciting advances of biotechnology in medicine augur well for a better tomorrow, many economic and ethical issues remain yet unresolved. Molecular biotechnology is expensive, and new diagnostic and therapeutic methods are expected to escalate health costs, thus bearing a negative impact on health care development of developing countries that need it most. Most vital of all, moral and ethical deliberation have not matched the progress and potential of medical biotechnology. Difficult dilemmas for biomedical ethics committees to confront include confidentiality versus abuse of an individual's genetic information, and right of access to insurance. Modern microbial biotechnology should be "environment friendly", ensuring that hazardous by-products are not indiscriminately released into the environment. Finally, given its rapid pace of progress, and if we can grapple with the core issues and implications, medical biotechnology can overcome the many challenges ahead to attain better health for all humankind.

Further Reading

1. Blau HM and Springer ML. Gene therapy — A novel form of drug delivery. *New Engl. J. Med.* **333**: 1204–1207, 1995.
2. Burkardt HJ. Standardization and quality control of PCR analyses. *Clin. Chem. Lab. Med* **38**: 87–91, 2000.
3. Clark DE and Pickett SD. Computational methods for the prediction of "drug-likeness". *Drug Discov. Today* **5**: 49–58, 2000.
4. DeFilippis V, Raggo C, Moses A and Fruh K. Functional genomics in virology and antiviral drug discovery. *Trends Biotechnol.* **21**: 452–457, 2003.
5. Dertzbaugh MT. Genetically engineered vaccines: An overview. *Plasmid* **39**: 100–113, 1998.
6. Doolan DL and Hoffman SL. Multigene vaccination against malaria: A multi-stage, multi-immune response approach. *Parasitol. Today* **13**: 171–178, 1997.
7. Eisen T. Novel vaccine technologies. *Trends Biotechnol.* **15**: 483–484, 1997.
8. Ellis RW. Technologies for the design, discovery, formulation and administration of vaccines. *Vaccine* **19**: 2681–2687, 2001.
9. Galanis E, Vile R and Russell SJ. Delivery systems intended for *in vivo* gene therapy of cancer: Targeting and replication competent viral vectors. *Crit. Rev. Oncol./Hematol.* **38**: 177–192, 2001.
10. Galperin MY and Koonin EV. Searching for drug targets in microbial genomes. *Curr. Opin. Biotechnol.* **10**: 571–578, 1999.
11. Gregoriadis G. Genetic vaccines: Strategies for optimization. *Pharm. Res.* **15**: 661–670, 1998.

12. Hughes TR and Shoemaker DD. DNA microarrays for expression profiling. *Curr. Opin. Chem. Biol.* **5**: 21–25, 2001.
13. Korth MJ, Kash JC, Furlong JC and Katze MG. Virus infection and the interferon response: A global view through functional genomics. *Methods Mol Med.* **116**: 37–55, 2005.
14. Lander ES *et al.* Initial sequencing and analysis of the human genome. *Nature* **409**: 860–921, 2001.
15. Lennon GG. High-throughput gene expression analysis for drug discovery. *Drug Discov. Today* **5**: 59–66, 2000.
16. Loferer H, Jacobi A, Posch A, Gauss C, Meier-Ewert S and Seizinger B. Integrated bacterial genomics for the discovery of novel antimicrobials. *Drug Discov. Today* **5**: 107–114, 2000.
17. Montgomery DL, Ulmer JB, Donnelly JJ and Liu MA. DNA vaccines. *Pharmacol. Ther.* **74**: 195–205, 1997.
18. Ogra PL, Faden H and Welliver RC. Vaccination strategies for mucosal immune responses. *Clin. Microbiol. Rev.* **14**: 430–445, 2001.
19. Pastan J and FitzGerald D. Recombinant toxins for cancer treatment. *Science* **254**: 1173–1177, 1991.
20. Patrinos G and Ansorge W. *Molecular Diagnostics*, 2nd ed. Elsevier Science and Technology, 2009.
21. Ramsay AJ, Ramshaw IA and Ada GL. DNA immunization. *Immunol. Cell Biol.* **75**: 360–363, 1997.
22. Rappuoli R and Covacci A. Reverse vaccinology and genomics. *Science* **302**: 602, 2003.
23. Shearer GM and Clerici M. Vaccine strategies: Selective elicitation of cellular or humoral immunity? *Trends Biotechnol.* **15**: 106–109, 1997.
24. Smith HA and Klinman DM. The regulation of DNA vaccines. *Curr. Opin. Biotechnol.* **12**: 299–303, 2001.
25. Somia N and Verma IM. Gene therapy: Trials and tribulations. *Nat. Rev. Genet.* **1**: 91–99, 2000.
26. Stahl S and Uhlen M. Bacterial surface display: Trends and progress. *Trends Biotechnol.* **15**: 184–192, 1997.
27. Tang H, Peng T and Wong-Staal F. Novel technologies for studying virus-host interaction and discovering new drug targets for HCV and HIV. *Curr. Opin. Pharmacol.* **2**: 541–547, 2002.

28. Whelen AC and Persing DH. The role of nucleic acid amplification and detection in the clinical microbiology laboratory. *Ann. Rev. Microbiol.* **50**: 349–373, 1996.

29. Young RA. Biomedical discovery with DNA arrays. *Cell* **102**: 9–15, 2000.

Websites

- DNA Vaccines
  http://www.dnavaccine.com
- Human Genome Project Information
  http://www.ornl.gov/hgmis

Questions for Thought...

1. Using insulin as an example, discuss the advantages and disadvantages of using bacterial hosts to manufacture products from eukaryotic genes.
2. Discuss the contributions of bacterial enzymes to biotechnology.
3. With the aid of examples, discuss the role of toxins in the pathogenesis of bacterial diseases.
4. Explain the term "immunotoxin". Discuss the potential limitations of an immunotoxin when it is used as a therapeutic agent.
5. Discuss the advantages and disadvantages of using chimeric toxins to treat cancer.
6. Discuss how the growth of microorganisms can be useful to humans.
7. When preparing a toxoid vaccine, what precautions must be taken to ensure that the toxoid will be effective?
8. What are the requirements for an ideal microbial vaccine? With examples, discuss how conventional and molecular-based vaccines can fulfill these requirements.
9. Discuss the contributions of molecular biology to the development of new and improved microbial vaccines.
10. Discuss the merits of recombinant DNA vaccines for immunization against infectious diseases.

11. Discuss the molecular approaches for the development of novel vaccines against pathogenic RNA viruses. Illustrate your answer with relevant examples.
12. Discuss the use of viral vectors for human gene therapy.
13. What are the criteria for selecting an ideal inherited disease candidate for gene therapy?
14. Explain whether gene therapy can be used to cure diabetes mellitus.
15. "The tools of molecular biology have facilitated the elucidation of certain mechanisms of microbial pathogenesis." Discuss this statement, illustrating your answer with examples.
16. "Studies on microbial genes, genomics and proteomics can contribute towards a better understanding of infectious diseases." Discuss.

Part V

Microbes in Environmental Biotechnology

By the turn of the last millenium, the rapid development in various areas such as microbiology, genetics, biochemistry and chemical engineering has led to tremendous advances in the knowledge and applications in biotechnology. Previous chapters of this book describe the applications and development of biotechnology in industry, agriculture and medicine. In Part V of this book, the development of biotechnology in environmental applications will be introduced. Because there are numerous applications in this area, only selected topics will be discussed in detail with emphasis on the applications of microorganisms in the biotechnological processes for solving environmental problems.

The unique properties of microorganisms, which are not present in plants and animals, are fast growth rate, small cell size and diversified metabolic pathways. The first two properties allow microbial applications to operate in a fast mode and in a small reactor, while the third characteristic enables microorganisms to degrade or transform a large number of compounds that are harmful to our environment.

In addition, advances in genetic engineering provide further improvement and facilitate the applications of microorganisms in solving environmental problems.

Several classes of microorganisms, namely viruses, bacteria, fungi and algae are commonly used in environmental biotechnology. Examples will be introduced to illustrate the importance of these microorganisms in environmental biotechnology. Many of these microbes work in concert to form biofilm for the processing of complex substrates in wastewater and for protection to adverse environmental conditions. It is therefore important to understand the mechanism of biofilm formation and its role in wastewater management.

The specific application of biotechnology to environmental management and pollution control is generally referred to as environmental biotechnology. It is used in waste and wastewater treatment as well as environmental cleanup and bioremediation encompassing toxic waste decontamination and the biodegradation of oils, pesticides and other organics. Microorganisms are in general more effective than multicellular organisms in environmental treatment, although higher plants are frequently used in phytoremediation and hydrophyte treatment systems. Municipal systems are usually mixed substrate, mixed culture continuous operations. Most processes are aerobic, but few are anaerobic. Some processes such as composting and anaerobic digestion are multi-stage and involve both mesophilic (15–40°C) and thermophilic (45–70°C) organisms under different operational conditions. The majority of treatment processes makes use of the indigenous populations in the wastes and provides favorable conditions for the growth and metabolic activity of these microorganisms. Industrial treatment systems often use pure strain (monoculture) of artificially screened microorganisms, while the degradation of xenobiotics may employ genetically engineered microorganisms in laboratory trials or pilot scale systems.

Chapter 18

Municipal Wastewater Treatment

Chu Lee Man and Wong Po-Keung
*Department of Biology, The Chinese University of Hong Kong
Shatin, NT, Hong Kong SAR, China*

Human activities consume water, producing wastewater which, when discharged into watercourses, will have serious impact on public sanitation and the well-being of the ecosystem. Used waters are characterized by high loading in gross organic matter as assessed by parameters such as COD and BOD, nutrients including nitrogen and phosphorus, toxic organics like hydrocarbons and inorganics such as heavy metals, as well as microorganisms such as enterobacteria (Table 18.1). The level of these parameters depends upon the source of wastewater, composition of water supply, pattern of water consumption, hydraulic flow, local habits and industrial contamination. As a consequence of urbanization and industrialization, the proper discharge of wastewater has become a problem of nearly all municipalities. Treatment is essential. Biochemical processes are usually involved in the secondary stage to remove suspended solids and organic matter after preliminary screening which removes coarse particles, and primary sedimentation which allows suspended solids to settle. The secondary treatment steps are generally considered environmental biotechnologies that make use of bioreactors for the biodegradation of organic matter and bioconversion of soluble nutrients in the wastewater. Industrial wastewater may have atypical composition and in-house treatment works may be needed prior to discharge to watercourses or further treatment in municipal treatment plants.

The discharge of municipal wastewater to open water will inevitably lead to water pollution, the degree of which depends highly on the self-purification

Table 18.1.   Typical composition of municipal wastewater.

	Concentration
pH	6.5–8.0 (no unit)
Total solids	300–1,200
Suspended solids	100–350
COD	250–1,000
BOD	100–500
Total nitrogen	15–80
Ammoniacal nitrogen	10–50
Total phosphorus	5–20
Inorganic phosphorus	4–15
Total coliforms	$1 \times 10^6 - 1 \times 10^8$ (MPN/100 mL)

Units in mg/L unless specified.

capacity of the receiving water. This can be typically exemplified by changes in the environmental quality (e.g. oxygen sag) and biological communities in a river in response to wastewater discharge from an outfall upstream. For systems with high assimilative ability, as determined by flow rate, dilution factor, degree of aeration and water temperature, the ecosystem damage will be minimal and recovery will be rapid. The rate of degradation is also governed by the concentration of microorganisms that are either suspended in the water column or attached to plant or rock surfaces. The low biomass concentration in the natural environment is usually the limiting factor for effective purification. This is enhanced by the intensification of biochemical processes through technical and engineering designs in bioreactors or treatment works for wastewater, which harness the self-purification capacity, thus creating an artificially controlled environment for optimal microbial activity and subsequent biodegradation of organic matter (Fig. 18.1).

Municipal wastewater contains a mixed population of aerobic and anaerobic microorganisms that are mostly heterotrophs. The microbial composition of wastewater fluctuates widely, consisting of either pathogenic or nonpathogenic organisms, which are common microflora of the intestine and soil. In biochemical treatment systems, the most adapted and effective microbes are selected. The selective pressure varies with the combination and intensity of various factors that can be categorized into two types. Primary selection factors, which are based on microbial physiological properties, determine

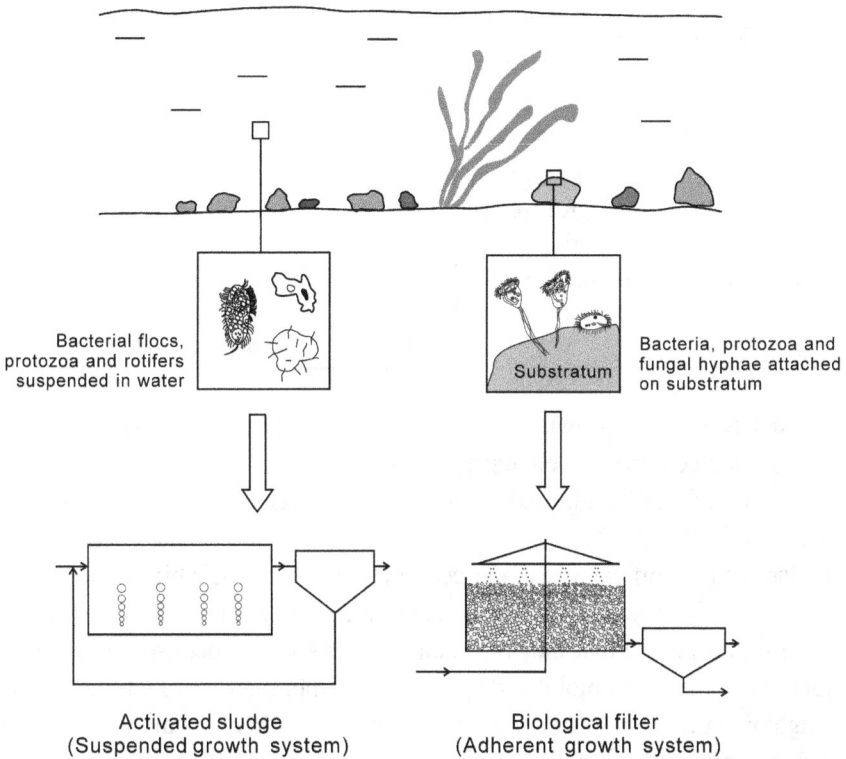

Fig. 18.1. Wastewater treatment plants are in fact artificial systems that intensify natural biological process in a defined and controlled environment so as to enhance the self-assimilative capacity of our environment (redrawn from Mudrack and Kunst, 1986).

whether autotrophs or heterotrophs will be selected, and include energy and carbon sources as well as oxygen availability. The majority of microorganisms involved in the biodegradation of organic matter in solid wastes and wastewaters are chemoorganotrophic (Table 18.2). A small number are chemolithotrophic, such as all nitrifying bacteria and some denitrifying bacteria, which are responsible for biological nitrogen removal. In addition, photolithotrophs such as green algae are active in shallow pond systems and surface of biological filters. Secondary factors, which are based on microbial physical properties, exert pressure on continuous flow treatment system. The electrical charge of cells determines initial attachment in fixed film reactors, and cell mass or cell density selects population based on the settling properties of microbes in the secondary sedimentation tank. These selection factors create unique microbial consortia in the various biological treatment

Table 18.2.   All living organisms can be categorized into four major groups according to their nutritional or metabolic diversity.

Energy source	Carbon source	
	Inorganic	Organic
Solar	Photolithotrophs (autotrophs) e.g. green plants, green bacteria	Photoorganotrophs e.g. purple non-sulphur bacteria
Chemical	Chemolithotrophs e.g. nitrifiers, thiobacilli	Chemoorganotrophs (heterotrophs) e.g. animals, fungi, most bacteria

systems. Suspended growth systems are dominated by free living organisms in dispersed colonies or cell aggregates while attached growth systems consist mainly of surface attached organisms in stationary biofilms. Such microbial associations display different physicochemical properties, which are of fundamental significance to their ecological role and their biotechnological function in wastewater treatment. However, the two growth forms share some similarities in that they are stable synergistic microconsortia in which microbial cells are immobilized by an exopolymer matrix of high molecular weight in a three-dimensional gel-like network, enforced by divalent cations such as Ca^{2+}.

18.1. Aerobic Treatment System

Aerobic systems use dissolved oxygen (DO) in the wastewater for carbonaceous and nitrogenous oxidation. Oxygen solubility is dependent upon gas pressure, water temperature and salinity, and oxygen is usually supplied by aeration to avoid depletion by oxygen consuming processes. The transfer of oxygen at the air-water interface and its subsequent diffusion to the cellular components is affected by agitation, flow pattern, viscosity, oxygen deficit and concentration gradient. Aerobic processes are commonly used as the systems are more reliable, stable and efficient.

18.1.1. Activated sludge process

Activated sludge process (Fig. 18.2), which was first developed in the United Kingdom in the early 1900s is the most commonly used aerobic

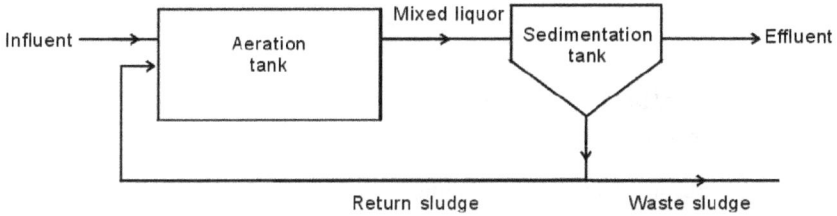

Fig. 18.2. Activated sludge system consists of a long aeration tank where the bio-oxidation of organic matter in the wastewater takes place, and a final sedimentation tank where the microbial biomass and solids are separated from the mixed liquor to produce a clarified effluent. A certain proportion of the sludge is recycled to the aeration tank as an inoculum.

Table 18.3. Bacterial population in wastewater samples from different stages in plants treating mainly domestic wastewater.

	Bacterial count (number/mL)	
	Total	Viable
Primary settled wastewater	6.8×10^8	1.4×10^7
Activated sludge mixed liquor	6.6×10^9	5.6×10^7
Secondary effluent	5.2×10^7	5.7×10^5
Tertiary effluent	3.4×10^7	4.1×10^4

(Adapted from Pike and Curds, 1971)

process by wastewater treatment plants worldwide, treating both municipal and industrial effluents, due to its operational flexibility and reliability. Activated sludge process is a suspended growth, mixed population and continuous culture system in which free living organisms are exploited to breakdown the biochemically oxygen demanding organic matter in the wastewater. During this process, wastewater is treated aerobically by a microbial consortium dominated by heterotrophic bacteria that are flocculated in the mixed liquor under a good supply of oxygen in an aeration tank 3–5 m deep to form discrete clumps of microorganisms called flocs. Protozoa and rotifers that enhance floc formation and effluent clarification are also present in significant numbers. The biomass is allowed to settle out of the suspension with other flocculent solids in a secondary clarifier to form sludge, a proportion of which is returned to the aeration tank as microbial inoculum to maintain a high microbial population for biodegradation (Table 18.3). The sludge contains a high concentration of active

Box 18.1

Why are Mixtures of Protozoa and Bacteria in Aerobic Treatment Systems very Efficient?

In aerobic treatment systems it is common to find a wide range of bacteria and protozoa. What effect does this combination of microorganisms have on the breakdown of the organic waste? As a general rule bacteria are important primary degraders and destroyers of solid organic matter. However, experiments showed that when bacteria and protozoa were exposed to detritus separately and in combination, the detritus disappeared much more quickly when the two kinds of organisms were grown together.

In the natural environment this phenomenon has been noticed and named the "decomposition–facilitation paradox." It can be stated as "the rate or the extent of organic matter decomposition often increases in the presence of bacterivorous protists that substantially reduce bacterial abundances." However the reason for this seeming paradox has been under discussion for some time. One hypothesis is to do with 'nutrient recycling'. It is suggested that protists excrete mineral nutrients they obtain from the ingested bacteria that ultimately stimulate the metabolism of bacteria so that they use even more of the carbon in organic matter.

Nitrogen and phosphorus are often limiting in the natural environment. In the case of bacteria, the cells are rich in these elements by comparison with other living things. When the protozoa consume the bacteria they release these limiting nutrients into the environment where bacteria can use them to grow so they are very active metabolically.

Based on modeling results it has been shown that the nutrient recycling hypothesis can explain the decomposition–facilitation paradox. The grazers release nutrients, even if there are a small number of them, thus making it possible for the remaining bacteria to be free from nutrient limitation so they can grow unrestrained. As a consequence, the protistan grazers "have strong positive effects on organic matter decomposition in a nutrient-limiting environment" while having "no effect on organic matter decomposition in a carbon-limiting environment."

Box 18.1   *(Cont'd)*

The results in the Table illustrate this point.

The effect of a mixed microbial microcosm of bacteria and bacterivorous grazing protozoa on degradation of organic matter in nutrient poor and nutrient rich environments. Adopted from Table 3 of Wang *et al.* (2009).

Experimental microcosm	Source of organic matter	Duration of experiment (days)	% Loss of organic matter Without grazers	With grazers	Bacterivorous grazers
Nutrient poor water environment					
Sea water	Barley hay	40	22	82	Flagellates, ciliates
Lake water	Lyophilized *Peridinium* cells	14	0	70	Microflagellates
Stream water	Leaf litter	120	1	11	Flagellate (*Spumella* sp.), ciliate (*Dexiostoma campyla*)
Carolina biological supply company	Wheat seeds	28	33	67	Ciliate (*Colpidium striatum*)
Pond water	Wheat seeds	42	15	23	Ciliate (*Tetrahymena*)
Nutrient rich water environment					
Saltmarsh sediment	Cordgrass litter	75	65	63	Nematodes (*D. meyli, D. dievengatensis, D. oschei, P. paetzoldi*)
Eutrophic lake water	Macrophyte leachate	27	41	43	Heterotrophic nanoflagellates (*Bodosaltans, Spumella* sp.)

mixed micro-flora and fauna, hence the name "activated sludge". A conventional activated sludge system has a hydraulic retention time (HRT) of 3–8 hours but solid retention time (SRT) is prolonged (4–15 days) because of sludge recycle.

18.1.1.1. *Floc characteristics*

Floc or sludge biomass is the basic operational unit in the mixed liquor in the aeration tank of the activated sludge system. Under the light microscope, flocs appear as discrete clumps of flocculated slurry with irregular boundaries containing organisms that include bacteria, protozoa and metazoa in a non-living gelatinous mesh-like matrix. The cells form small aggregates to large flocs commonly of diameter ranging from 50–300 μm with a surface area of 40–150 m^2/g, the size of which is determined by cohesive strength and the degree of shear in the aeration tank. Floc size is the resultant balance between floc formation and floc breakage, which is a dynamic process of floc restructuring and reform. Floc formation is enhanced by collision and adhesion to result in aggregation which is aided by extracellular polymeric substances. Recent size distribution studies by electron microscopy and image processing show that the aggregation process begins with primary particles (2.5 μm), which aggregate to form secondary particles (13 μm) which in turn form tertiary particles (125 μm). On the other hand, hydrodynamic forces and turbulence in the aeration tank disintegrate flocs at their weakest points and cause destabilization. This results in smaller flocs of similar or different sizes. Flocs can vary between 1–1000 μm; biomass in the mixed liquor is made up of flocs of size 65–260 μm. Spherical and compact microfloc is formed when the particle diameter is less than 75 μm and specific gravity greater than 1.0. When filamentous microorganisms are predominant, the floc will be in the form of extended mat, loose (specific gravity <1.0) and irregular in shape. This macrofloc is not inducive to settlement and will result in settling problems in the secondary clarifier. Chemically, floc consists of biomass, organic matter which is mainly flocculating polymers including polysaccharides, amino polysaccharides, proteins and lipids, and inorganic materials such as carbonate, phosphate, nitrate and metal hydroxides. Floc organisms usually

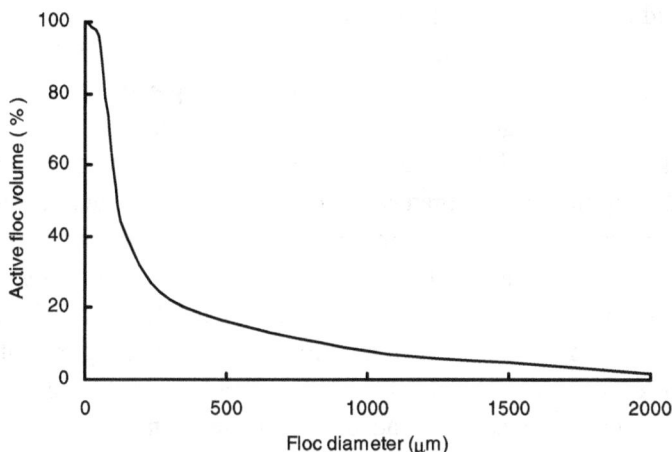

Fig. 18.3. The active floc volume (volume fraction of flocs containing oxygen) that contributes to aerobic degradation as a function of the floc diameter (Pasveer, 1959).

concentrate in the slimy matrix in the peripheral region while the inner core is composed of inorganic material and inert organic matter. Cells are flocculated with polymeric matrix or polyermic bridges linked by mechanical entanglement or by electrostatic bridging. The active floc volume that contributes to aerobic degradation is a function of the floc particle diameter (Fig. 18.3). Loose irregular floc has high bulk volume but low active fraction.

Regarding biological composition, bacteria are the dominant group and over 300 strains have been isolated from activated sludge systems. The bacteria are mostly chemoorganotrophic, Gram-negative motile rods, many of which originated from human fecal materials. Common genera are *Pseudomonas, Flavobacterium, Zoogloea, Arthrobacter, Acinetobacter, Achromobacter, Alcaligenes, Cytophaga, Bacillus, Nitrosomonas* and *Nitrobacter*. Most of these are floc-forming bacteria, which either aggregate in sludge flocs or are freely suspended in the mixed liquor. Holozoic or saprobic protozoa also play an important role in wastewater purification. Over 200 species have been identified and can be found to occur at levels as high as 50,000 per mL. Seventy percent are ciliates, such as *Vorticella* spp., *Aspidisca* spp., *Opercularia* spp. and *Epistylis* spp. They are sedentary, sessile or stalked, attaching to individual floc, crawling over the floc surface or free-swimming in the mixed liquor. They feed on bacteria and tiny suspended particles thereby regulating bacterial density and clarifying effluent. Rotifers are also present in substantial number; they encourage new floc formation by breaking up flocs into smaller particles, contribute to floc formation by producing undigested materials and mucus, and reduce turbidity by removing dispersed bacteria and filtering suspended particles. Fungi rarely dominate in mixed liquor except in acidic wastewater. Such a biological composition is the result of microbial selection under aerobic condition and the shear stress of aeration. Nonflocculating microbes that are unable to form settleable flocs are washed out in the final effluent. Substrates in wastewater are adsorbed to the surface and transported from the aerobic peripheral region of a floc, where carbon oxidation and nitrification take place, to the inner region. In floc of substantial size, the inner zone may be anaerobic which allows denitrification to occur. In the aeration tank, biodegradation of organic matter by C-oxidizers is the primary goal, and the dynamic ecosystem functioning depends very much on the bacterial-protozoal relationship (Fig. 18.4). Nevertheless plant designers still regard activated sludge system essentially a

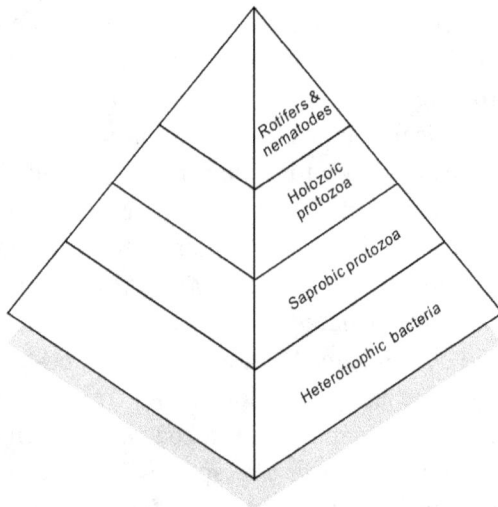

Fig. 18.4.   The relatively simple bacterial-protozoal relationship in activated sludge system.

bacterial process. New analytical tools such as molecular techniques allow microorganisms in the activated sludge environment to be characterized, but knowledge in microbiology and biochemistry has not yet been fully integrated into the technology and the engineering of plant design and operation.

18.1.1.2. Biological nitrogen removal

The biological oxidation of ammonium to nitrate by nitrifiers takes place as long as the DO content is above 2 mg/L. This is a 2-step process: (a) ammonium oxidation in which *Nitrosomonas* spp. converts ammonium to nitrite, and (b) nitrite oxidation in which *Nitrobacter* spp. converts nitrite to nitrate. Nitrifiers are chemolithotrophic and strictly aerobic, consuming 4.2 g oxygen for each gram of ammonium-nitrogen oxidized. They have lower specific growth rates than C-oxidizers and can therefore only establish and grow in old flocs. A sludge age of 3–5 days is required, which is longer than the solid retention time in a conventional activated sludge system. A two-stage activated sludge process has been developed, the first aeration tank for BOD removal and the second one for nitrification. Nitrification alone cannot remove nitrogen since nitrate is formed stochiochemically. It must be followed by denitrification in which nitrate in the nitrified effluent is reduced to molecular nitrogen and removed as gas. Denitrifiers are facultative anaerobes, which use nitrate as an electron acceptor. They are mostly heterotrophic bacteria such as *Pseudomonas* spp., *Alcaligenes* spp. and *Achromobacter* spp., which require organic substrates. However, some are autotrophic which do not require organic substrates, such as *Micrococcus* spp. and *Thiobacillus* spp. Both types grow well at pH above 5 and DO level lower than 1 mg/L. Because of this, nitrogen removal in activated sludge process used to involve two systems (Fig. 18.5). The first system is a conventional activated sludge process with an aerobic tank for carbonaceous oxidation and nitrification and a clarifier for sedimentation. The second consists of an anoxic tank which has no aeration but only mixing to encourage denitrification and keep the flocs in suspension, and a clarifier for sludge settlement. As most organic matter is removed in the initial oxidation step, the concentration of organic substrate is insufficient to meet the requirement of heterotrophic

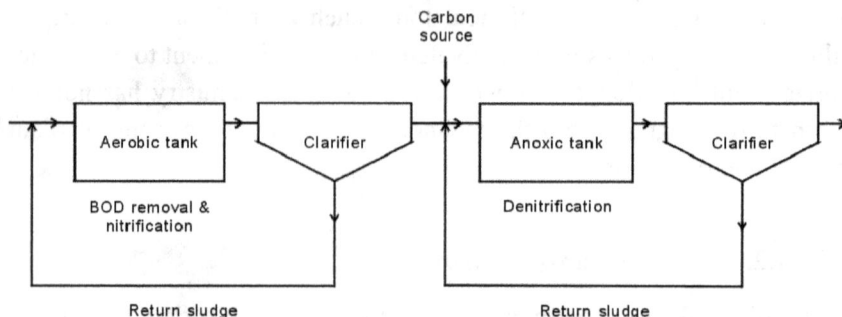

Fig. 18.5.   Two-clarifier system in which the aerobic tank is for BOD oxidation and nitrification while the anoxic tank is for denitrification. An external carbon source is needed for denitrification.

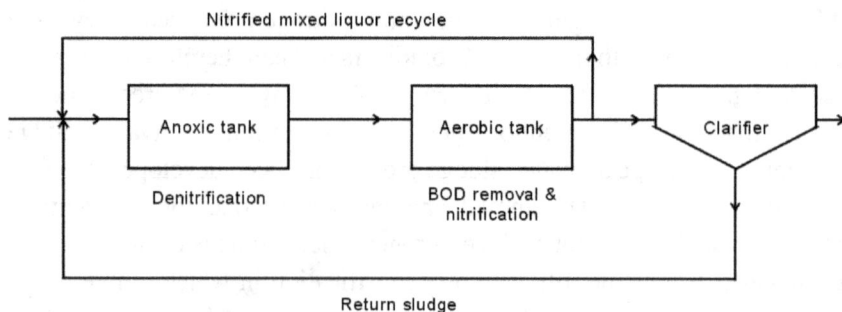

Fig. 18.6.   One-clarifier system with an anoxic tank for denitrification and an aerobic tank for BOD oxidation and nitrification. The nitrified mixed liquor is recycled back to the anoxic tank for denitrification.

denitrifiers. In such a system, an external carbon source in the form of methanol or acetate is usually added. To remove the requirement for carbon source addition, a one-clarifier system (also called one-sludge system) is used (Fig. 18.6). Effluent from primary treatment enters an anoxic tank where denitrification and some BOD removal take place. The mixed liquor then flows to a second aeration tank where nitrification and BOD removal are carried out. The nitrified mixed liquor will recycle back to the anoxic tank so that nitrate is converted to nitrogen by denitrification. Biological nitrogen removal is usually incorporated in modern activated sludge systems, which are no longer solely aerobic processes. There are a number of versions for such anoxic-oxic (A/O) systems. In the modified

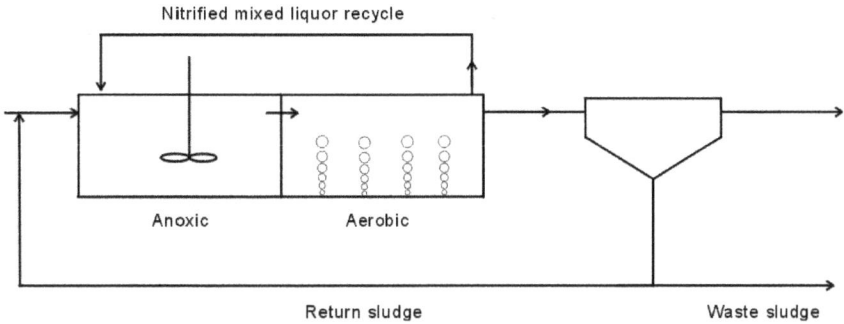

Fig. 18.7.   Modified Ludzak-Ettinger process with an anoxic unit and an aerobic unit.

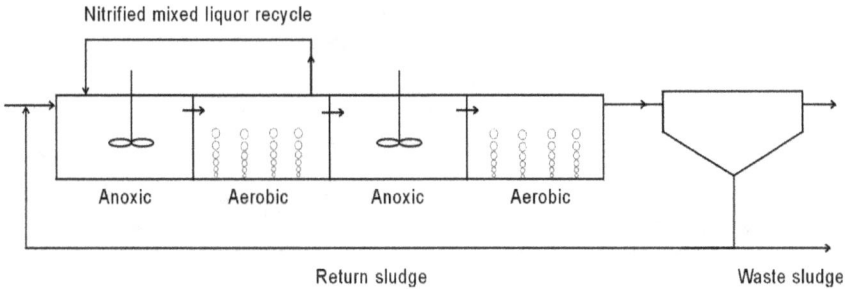

Fig. 18.8.   The four-stage Bardenpho process with two anoxic units for denitrification and a second aerobic unit for the removal of nitrogen gas from denitrified mixed liquor.

Ludzak-Ettinger process developed in South Africa, an anoxic unit, in which denitrification takes place, precedes an aerobic unit (Fig. 18.7). In this process, effluent formed has a total nitrogen content of slightly greater than 5 mg/L. More complicated processes include the 4-stage Bardenpho process which is developed in the US. A second anoxic unit increases the total denitirification efficiency and a second aerobic unit removes nitrogen gas from denitrified mixed liquor. Total nitrogen content in the polished effluent is below 3 mg/L (Fig. 18.8).

18.1.1.3. Biological phosphorus removal

Phosphorus in wastewater, which cannot be removed by a conventional activated sludge system, is commonly removed by chemical precipitation.

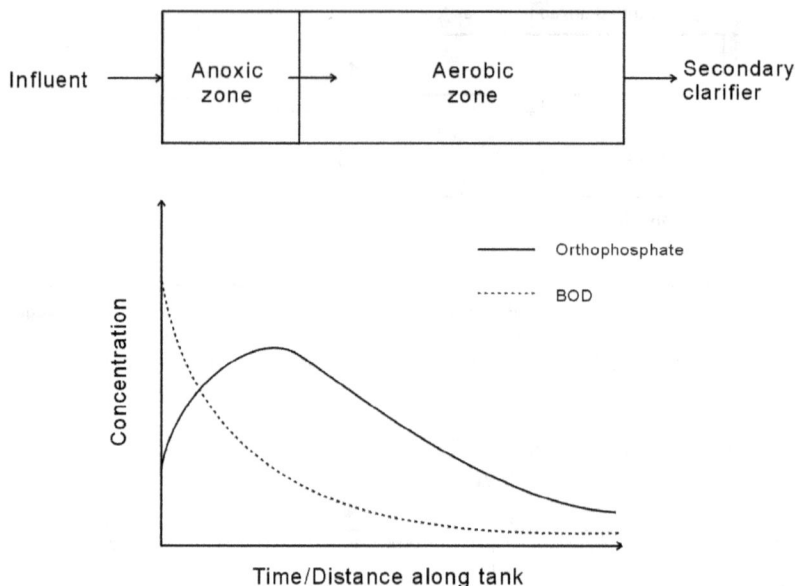

Fig. 18.9.   Changes of BOD and orthophosphate in the anoxic and aerobic units of the activated sludge system.

During the past two decades, the biological removal of phosphorus has become more widely used. Enhanced biological phosphorus removal (EBPR) is the result of luxury uptake by microbial cells, which involves enhanced uptake of phosphorus during the aerobic stage and its release during the anaerobic stage. This is evident by an increase in orthophosphate concentration in the anoxic zone and a substantial reduction in the oxic zone of the activated sludge system (Fig. 18.9). Using the energy liberated from BOD oxidation, phosphorus is stored as polyphosphates in bacterial cells under aerobic conditions (Fig. 18.10). The accumulated polyphosphates are hydrolyzed and energy is released for assimilation under anaerobic conditions. Although the detailed processes are not yet entirely understood, it is postulated that the intracellular polyphosphate storage is carried out by phosphate accumulating or removing microorganisms (PAM or PRM) or polyphosphate microorganisms. There is large number of nonspecific bacteria, the majority of which is *Acinetobacter calcoaceticus* var. *lwoffi*. *Aeromonas* spp., *Achromobacter* spp., *Arthrobacter* spp., *Klebsiella* spp. and *Pseudomonas* spp. also contribute

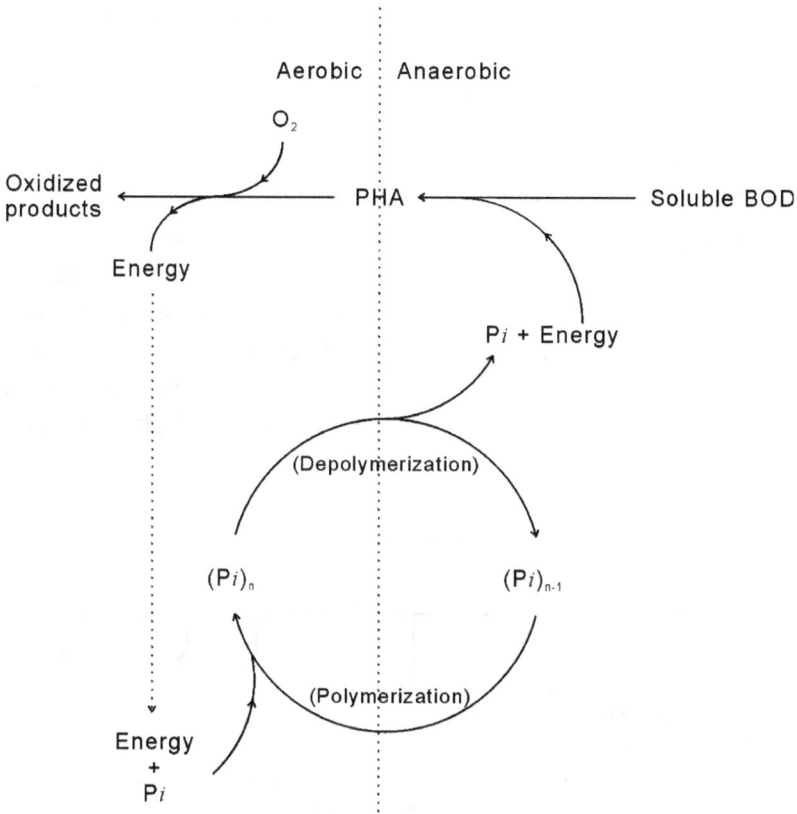

Fig. 18.10. The mechanism of biological phosphorus removal which involves the synthesis and degradation of polyphosphates and a storage product (poly-β-hydroxyalkanoates or PHA) under aerobic and anaerobic conditions.

to EBPR. In the anaerobic stage, facultative bacteria convert soluble BOD to volatile fatty acids like acetate, which are then stored as poly-β-hydroxyalkanoates (PHA) such as poly-β-hydroxybutyrate (PHB), using the energy released from the depolymerization of polyphosphate; phosphorus is released from the bacterial cells. In the aerobic stage, PHB is oxidized to provide energy for growth and polyphosphate accumulation. An enhanced phosphorus uptake occurs when conditions change from anaerobic to aerobic, and phosphorus deficient cells replenish lost polyphosphate by rapid accumulation (oversurplus compensation). Growth causes an uptake of phosphorus higher than the amount released in the anaerobic stage to result in a luxury uptake, and the excess phosphorus is stored as polyphosphates.

There are a number of commercial systems for phosphorus removal. The Phoredox process is a two-stage (anoxic-aerobic) A/O system with short solid retention time to minimize nitrification and hence denitrification. The A²/O process (Fig. 18.11) and the Virginia Initiative Plant (VIP) process (Fig. 18.12) are three-stage (anaerobic-anoxic-aerobic) activated sludge systems. The five-stage modified Bardenpho process developed in 1978 is the three-stage process with a second anoxic stage to enhance denitrification and a second aerobic stage to ensure aerobic conditions in the final clarifier (Fig. 18.13). With the exception of the Phoredox process, all are mainstream processes and are efficient for both nitrogen and phosphorus removal.

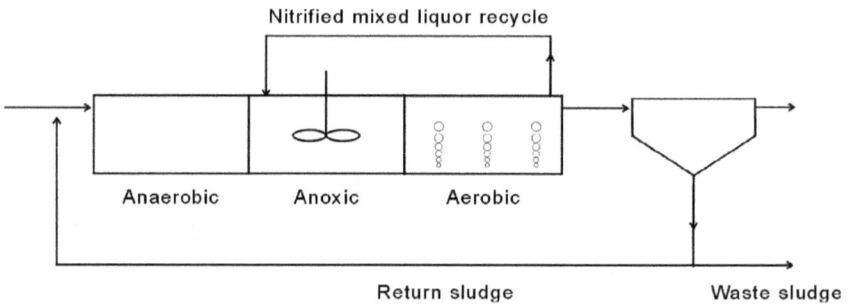

Nitrified mixed liquor recycle

Anaerobic   Anoxic   Aerobic

Return sludge   Waste sludge

Fig. 18.11.   A²/O process consists of anaerobic, anoxic and aerobic tanks, with the recirculation of nitrified mixed liquor from the aerobic to the anoxic tank.

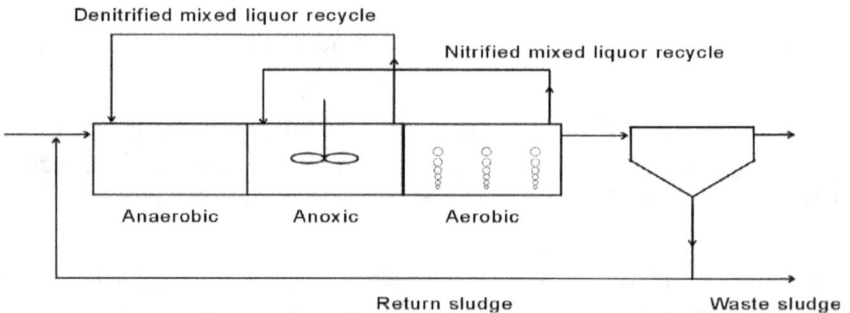

Denitrified mixed liquor recycle

Nitrified mixed liquor recycle

Anaerobic   Anoxic   Aerobic

Return sludge   Waste sludge

Fig. 18.12.   The VIP process which is similar to the A²/O process but with the addition of denitrified mixed liquor recycle from the anoxic to the anaerobic tank.

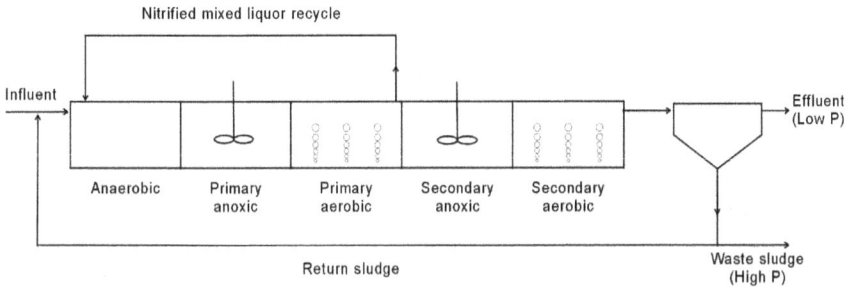

Fig. 18.13. The five-stage Bardenpho process with a second anoxic and aerobic tanks to enhance denitrification and nitrogen gas removal.

18.1.1.4. Treatment efficiency

The purification efficiency of activated sludge process depends not only on the biodegradability of the organic matter and the removal of nutrients in the wastewater, but also on the separability of the suspended solids or flocs in the sedimentation process. Floc settlement separates solids from the treated wastewater to result in a clarified effluent, and allows sufficient amounts of sludge to be returned to the aeration tank in order to maintain mixed liquor suspended solids (MLSS) which is usually used as a crude measure of the biomass for routine operational control. If solids end up in the effluent, its quality will deteriorate. Floc separability, which is affected by the settling characteristics, is governed by the physical nature of flocs. This is usually measured in terms of sludge density or sludge volume and expressed as the sludge density index (SDI) or sludge volume index (SVI) respectively. Tests are performed by determining the sludge volume of a 1 L mixed liquor sample after being settled for 30 minutes and taking into account the MLSS of the sample. The values of the two indices for sludge of good and poor settling properties are presented in Table 18.4.

18.1.1.5. Sludge settleability

Effluent quality depends not only on the treatment process in the aeration tank, but also on the separation of the flocs from the treated effluent at the final sedimentation tank. Settling problems result in the loss of biomass and

Table 18.4.   Sludge settling characteristics. The settleability of sludge is assessed by determining the sludge volume of a 1 L mixed liquor sample after being settled for 30 min, and the SVI or SDI is calculated, taking into account the MLSS of the sample.

Sludge settling quality	Sludge volume index (SVI)[a]	Sludge density index (SDI)
Excellent	<80	>1.25
Moderate	80–150	0.66–1.25
Poor	>150	<0.66

[a] $SVI = \dfrac{Sludge\ volume\ 30\ min\ (mL/L) \times 1000}{MLSS\ (mg/L)}$

$SVI \times SDI = 100$

poor quality effluent, and there are three major categories of settling problems. Floc formation problems can be resulted which are attributed to (i) dispersed growth under high BOD or high food to microorganism (f:m) ratio; (ii) formation of pin-point flocs which is due to the disruption of flocs as a result of insufficient food (low f: m ratio), particularly prevalent for old flocs under extended aeration; or (iii) deflocculation which normally takes place under conditions such as low pH, low DO, the presence of toxic pollutants or severe turbulence. These result in turbid effluent. Poor settling can also be caused by density problems in the settling tank which has no aeration. Anaerobic gases such as CO_2, CH_4 and H_2S form bubbles that are trapped in the sludge and raise the sludge to surface. Nitrogen gas from denitrification under anoxic conditions can also be entrained in flocs which float. Commonly, poor settling is due to the formation of bulking sludge which interferes with the performance of many treatment works both locally and overseas. Bulking sludge has massive filamentous growth and the length of extended filament can exceed 107 μm/mL (Fig. 18.14). Filamentous bulking results in the formation of loose flocs with SVI>150, and interferes with compaction and sludge settling. It is caused by filamentous microorganisms, predominantly filamentous bacteria such as *Sphaerotilus* spp., *Microthrix* spp. and *Streptothrix* spp., and actinomycetes such as *Beggiatoa* spp. and *Nocardia* spp. Occasionally, fungi such as *Geotrichum* spp. and *Leptomitus* spp. are found especially under acidic conditions. About 30 types of filamentous microorganisms have been identified in bulking sludge so far and 10 species have been accounted to cause 90% of bulking. Of these ten species, *Sphaerotilus natans*, originally believed to be the only

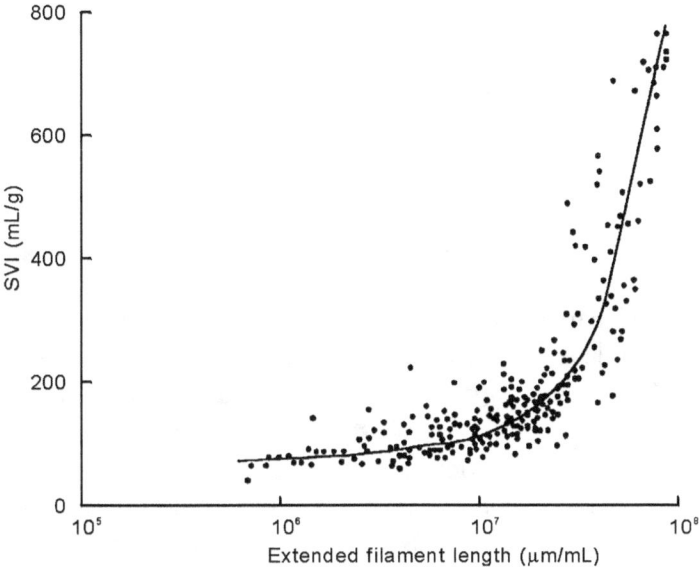

Fig. 18.14. Effect of total length of extended filament on sludge volume index (Palm *et al.*, 1980).

filamentous bacteria that cause bulking, and *Microthrix parvicella* are the most common species. The inventory of bulking microorganisms is not completed; some filamentous species have no strict taxonomic identification but are tagged by a numerical code, e.g. Types 0041, 0092, 1701, 1851, 021N, because their properties are not yet fully known. Recently phylogenetic analyses based on 16S rDNA sequencing have helped in better identification, e.g. Type 021N is named *Thiothrix eikelboomii*. When settling is poor and the settling zone is enlarged, little clear supernatant is obtained. Bulking can be caused by a variety of conditions, including high BOD, salts, H_2S and fatty acids in the influent, and low pH, DO and soluble nutrients (nitrogen and phosphorus). Filamentous microorganisms have a higher surface to volume ratio than cell aggregates, and this allows them to be more competitive under low concentrations of soluble substrates, e.g. nitrogen, phosphorus and DO, or low f: m ratio. Also, it has been empirically proven that H^+, S^{2-} and soluble organics such as phenols, organic acids and aldehydes are more toxic to flocculating cells than filamentous microorganisms. This selective toxicity explains why filamentous growth is predominant at reduced pH and high concentration of H_2S.

Apart from sludge bulking, bacterial foaming is the second most common operational problems encountered in activated sludge systems. Bacterial foaming is formed when thick brown scum occurs in the aeration tank, which is caused by excessive growth of filamentous actinomycetes such as *Nocardia amarae*, *N. pinensis*, *N. rhodochrous* and *Microthrix parvicella*. The most common species found in foams is *Nocardia amarae*, and the major genus of foam-causing agents is *Nocardia*, hence the name *Norcadia* foam. *Norcadia* is highly branched filament, which traps air bubbles and floats to the surface of aeration tank, appearing as chocolate mousse in color and texture. Foams are usually associated with high fatty material content, warm weather (>18°C) and long sludge age (>9 days). Foams trap MLSS which are carried out of the aeration tank when overflowed, hence deteriorates effluent quality. In some cases, foam overflows onto footpaths and access ways in activated sludge treatment plants, or is dispersed by wind causing health hazards to workers in the treatment plant.

To increase treatment reliability regardless of the performance of final sedimentation, membrane-coupled activated sludge processes incorporating microfiltration have been used with success. These processes are particularly effective in case of high effluent requirement. However, the process is still not widely used due to cost considerations.

18.1.1.6. *Operational design*

Aeration is one of the key operational requirements, which ensures aerobic activities in the mixed liquor and contributes to the efficient degradation of organic matter (BOD removal). This is achieved by bottom air blowers such as porous diffusers, or surface aerators such as cone aerators and rotating brushes. Aeration serves dual purpose: oxygenation satisfies the oxygen demand of microbes, and mixing keeps the flocs in suspension and maximizes the contact between the microbial cells and the substrates in the wastewater. Carbon oxidation takes place at an oxygen level greater than 0.5 mg O_2/L while nitrification does not occur at below 2.0 mg O_2/L. Other operational parameters which affect process efficiency include mixed liquor suspended solids, organic loading, food to microorganisms (f:m) ratio, sludge residence time and sludge recycle ratio. Sludge is

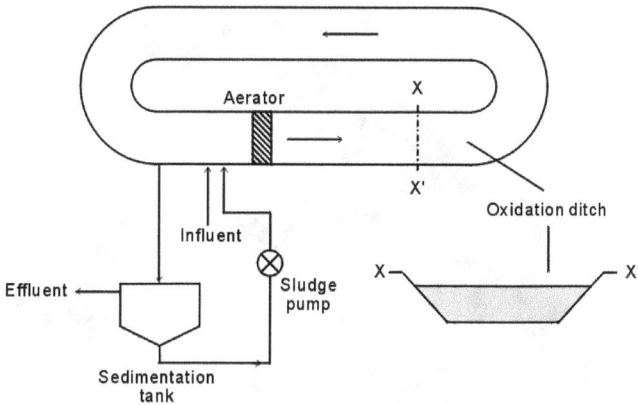

Fig. 18.15.   Typical layout of an oxidation ditch.

recycled to maintain microbial biomass in the aeration tank. Normally 50–90% is returned and surplus sludge is wasted. This is done in order to attain the desired sludge age. Insufficient sludge results in low MLSS, high f:m and young sludge age (<3 days) while excessive sludge will lead to high MLSS, low f:m ratio and old sludge age (>4 days). Young sludge age is usually poorly flocculated while old sludge age will result in pin-point floc, which are not conducive to good settling. The rate of sludge recycle ultimately depends upon MLSS, organic loading, HRT (a function of flow rate and reactor volume) and effluent quality.

There are a number of versions of the activated sludge process, such as the oxidation ditches and the vertical shaft bioreactor. An oxidation ditch is a continuous shallow oval circuit 1–3 m deep and rectangular or trap-ezoidal in cross section (Fig. 18.15). It is usually equipped with one or two surface aerators, commonly in the form of rotating brushes or discs (Fig. 18.16), which circulate and aerate the mixed liquor in the channel. There are few systems in operation, which include the mammoth system, split channel system and carrousel system (Fig. 18.17). The vertical shaft bio-reactor, also called deep shaft system, is a high-rate activated sludge pro-cess which employs a subsurface well 30–200 m deep to provide the aerobic treatment (Fig. 18.18). It was developed to increase the rate of oxygen transfer from the gaseous phase to the liquid phase by increasing the hydrostatic pressure in the circulating wastewater through the vertical

Fig. 18.16.   Rotating dics of an oxidation ditch in operation.

shaft, either in the form of a simple U-tube or a central tube configuration (a central downflow pipe and an outer upflow section). Such system can achieve very good BOD removal at a very short HRT (1–2 hours). Aeration using oxygen has been used in high purity oxygen activated sludge systems for treating high strength industrial wastewaters.

18.1.2. Aerobic fixed film reactors

Another type of commonly used biological treatment process is the aerobic fixed film or biofilm reactor which utilizes surface attached organisms growing as a biological film on stationary or moving media for wastewater purification. Reactors with stationary media are exemplified by conventional biological filters, which are also called trickling or

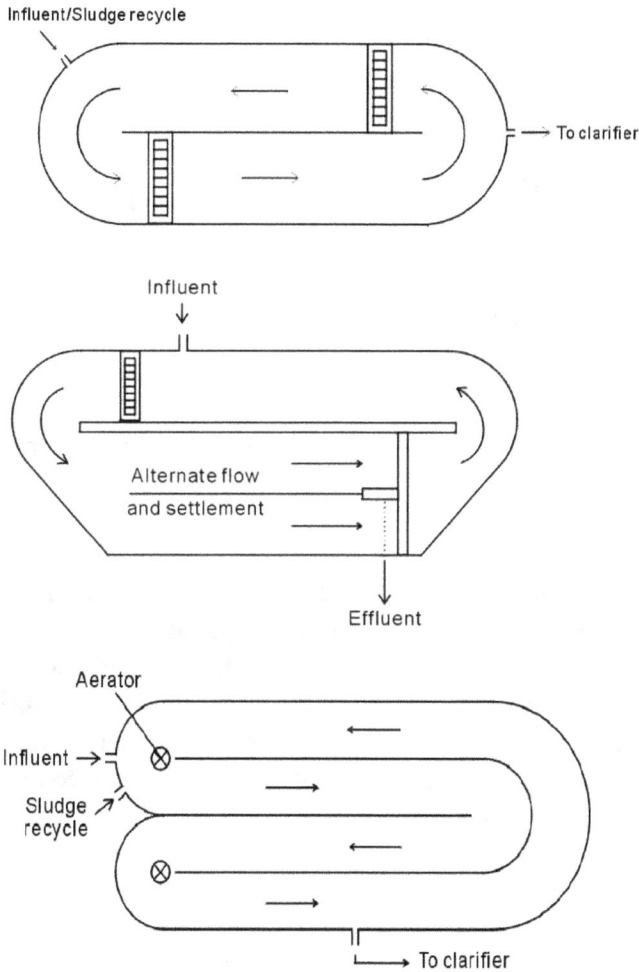

Fig. 18.17.   Different oxidation ditch systems: mammoth system (top), split channel system (middle) and carrousel system (bottom).

percolating filters. The first tricking filter was built in 1892 in Salford, England and is one of the earliest biological treatment systems for wastewater. This system consists of a circular or rectangular filtering tank containing bed of permeable filter media. Wastewater is sprayed from the top onto the filter bed via a rotary or reciprocating distributor for the circular or rectangular filter respectively, and trickles through the filter media which provide an extensive growth surface to support the

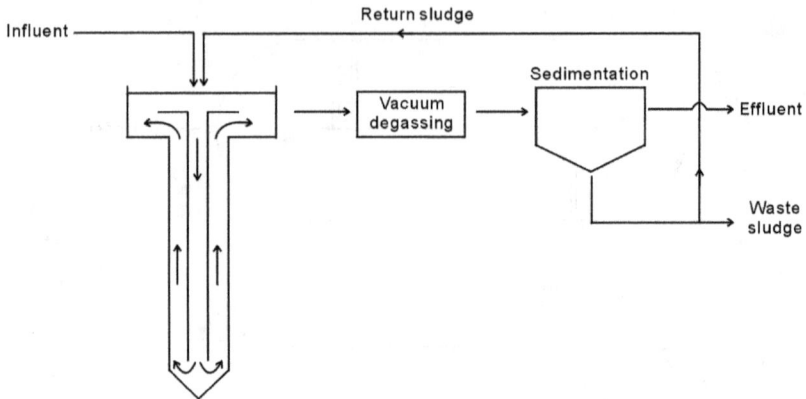

Fig. 18.18.   The vertical shaft bioreactor consists of a simple U-tube or central tube with outer upflow pipe as shown. No primary sedimentation is required owing to its high treatment efficiency.

microbial film and associated fauna necessary for wastewater purification. Most conventional filters are of the depth of 1.5–2.5 m and HRT of 0.5–1 hour. The tank is furnished with natural or forced draft ventilation with a pipe connecting the interior of the filter to the atmosphere or to a ventilation fan to provide free air movement. An underdrain collects treated wastewater from the bottom, which is then pumped to a clarifier for solid separation.

18.1.2.1. *Filter media*

The filter is filled with natural material such as mineral or synthetic material such as plastic to provide a surface for growth of organisms. Natural media are usually made up of gravel or rock of 25–100 mm in diameter which is good for conventional low-rate loading (<0.1 kg BOD/m^3/day). Plastic media which were introduced in the 70s are composed of PVC, polystyrene or polypropylene, and have a high surface area to volume ratio and high voidage. Plastic media are lightweight and can be packed into deeper beds (8 m) for stronger influent. Plastic media can be of random pack of individual moulds of specific shape, usually in the form of ring or short tubes with external and/or internal fins and ribs to increase surface area. The other type of plastic media is modular packing either in

Table 18.5.   Properties of mineral and plastic media used in trickling filters.

	Surface area (m^2/m^3)	Void volume (%)
Mineral aggregates	50–100	50
Plastic media	100–300	>90

the form of zigzag channels between vertical sheets or block of closely assembled vertical tubes with concentric tubules. Plastic media increase the surface area per unit volume by two to six times and the void volume for at least 80% (Table 18.5).

18.1.2.2. Biofilm

The biological filter is not a physical filter but has a solid support medium on which the biofilm develops. The surface of the biofilm is not smooth but convoluted and corrugated. The biofilm is a complex community of bacteria, fungi, protozoa and other meso- and macro-invertebrates in a slimy coating on the supporting surface of the filter medium. Wastewater, as the aqueous phase, trickles through the filter bed and flows freely over and through the biofilm, which brings about purification. The biofilm is a semi-aquatic habitat with diverse life forms, and it is the basic unit of purification in biological filters (Fig. 18.19). The primary feeders are major decomposers and are responsible for the extracellular enzymatic solubilization of solids and oxidation of carbonaceous matter. They are bacterial cells and fungal mycelia growing in a microbial slime of polysaccharide matrix which forms the bulk of the biofilm. Bacteria are the largest in population, both in number and biomass and the major genera include *Zoogloea*, *Pseudomonas*, *Flavobacterium*, *Beggiatoa*, *Alcaligenes*, *Sphaerotilus*, *Nitrosomonas* and *Nitrobacter*. Fungi are also common in the aerobic regions and can flourish under low pH or high organic content. Common fungal genera are *Fusarium*, *Geotrichum*, *Penicillium*, *Mucor*, *Ascoidea*, *Subbaromyces* and *Sepedonium*. Algae that are present in the portion of biofilm exposed to light (usually on the upper 5 cm) contribute relatively little to wastewater treatment. Microbial composition varies seasonally with wastewater composition and filter depth. The microbial film and cells are

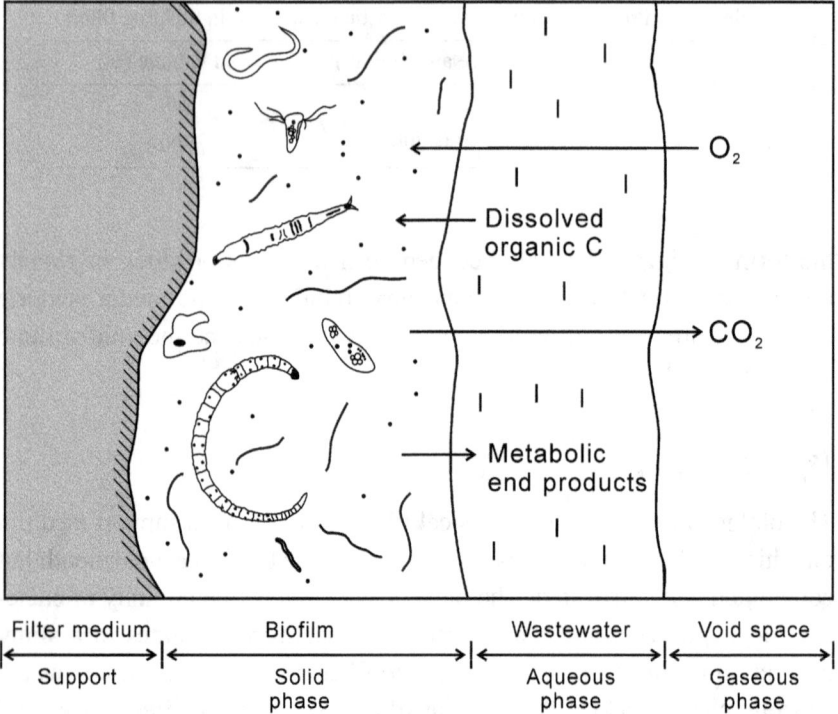

Filter medium	Biofilm	Wastewater	Void space
Support	Solid phase	Aqueous phase	Gaseous phase

Fig. 18.19.   The biological filter works with biofilm on filter media as the basic operational unit. Wastewater percolates over and through the biofilm matrix with the trasfer of substrate and end products in and out of the biofilm. The microbial communities growing in the biofilm are not shown, while the secondary feeders drawn are not to the same scale.

grazed by secondary feeders which are mostly protozoa, nematodes, rotifers, annelids and insect larvae. There are about 2,000–100,000 protozoa per mL sludge. Over 200 species of protozoa have been isolated and over 50% are ciliates such as *Paramecium* spp., *Vorticella* spp. and *Opercularia* spp. Amoebae such as *Amoeba* spp. and zooflagellates such as *Trigonomonas* spp. are also present. They feed on dispersed bacterial cells and help clarify the effluent. The burrowing activities of nematodes, annelids such as epigic earthworms and larvae of insects such as chironomid encourage nutrient and oxygen flow within biofilm. The biological filter has a food chain pyramid containing several trophic levels, and is more complex than that of the activated sludge process (Fig. 18.20). The metazoa, though not important

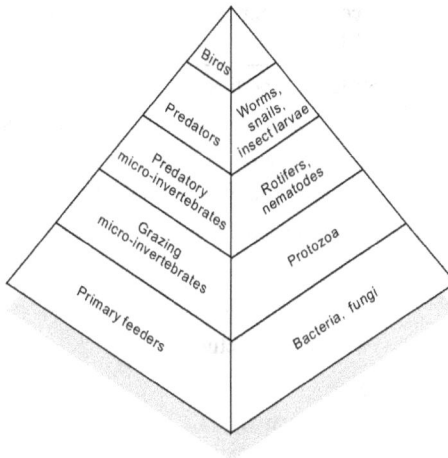

Fig. 18.20.   The biological filter has a pyramid of food chain which contains several trophic levels, and is more complex than that of the activated sludge.

for design purposes, also play an important role in the ecological balance needed for efficient operation.

The biological composition of biofilm is selected according to attachability as governed by the electrostatic surface adhesion of cells, degree of hydrophilicity between cells and the medium surface, surface roughness and the formation of a slimy matrix. Biofilm growth depends on wastewater strength and biofilm thickness, which affects substrate transfer and aerobic activity. Growth is unidirectional, moving out from the supporting medium to the aqueous and gaseous phases. The major factors that control biofilm composition and structure and development have been summarized in Table 18.6.

Biofilm formation is affected by factors such as hydrodynamics of the bulk fluid, nature of the substratum, species composition and nutrient availability. The traditional concept views biofilm as a virtually dual-layered structure with a base film which consists of a structured layer and a surface film which has irregular boundaries and is a transition between the base film and the bulk liquid (Fig. 18.21). However, recent experimental data suggest that biofilm is not structurally uniform and homogeneous as generally assumed, but a dynamic layer of microbial clumps adhering to and detaching from the support surface.

Table 18.6. Major factors that control composition, structure and development of biofilm (Wingender and Flemming, 1991).

Substratum	Surface properties (e.g. roughness, hydrophobicity)
Bulk liquid (wastewater)	Physicochemical conditions (e.g. temperature, pH, salinity, ions, organic matter)
	Concentration of available organic substrates
	Hydraulic conditions (e.g. flow rate, shear stress)
Microorganisms	Morphology
	Surface properties
	Physiological activity
	Lysis
Micro- and meso-fauna	Grazing activity
Biofilm	Age

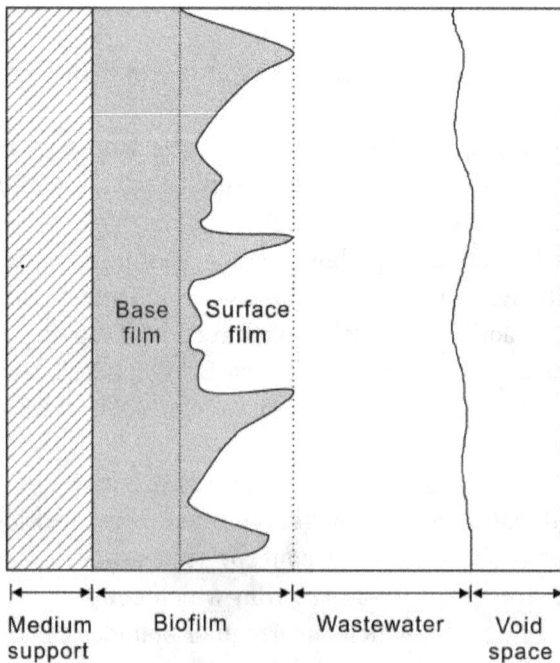

Fig. 18.21.   Traditional concept of a biofilm which consists of a basal layer and a rough surface layer in intimate contact with the flowing wastewater.

It is a spatially and temporally heterogeneous layer of variable thickness and density, which contains microstructures such as cell clusters and cell-free pores cemented by extracellular polymeric material. Spaces exist between cell clusters to form vertical voids (pores) and horizontal voids (channels). Mass transport of substrates and end-products is facilitated in the biofilm voids and channels at high flow velocities. Various biofilm conceptual models have been put forward. Among them, the dense confluent biofilm model (1991) views biofilm as a confluent layer with non-uniform biomass distribution but no water channels or porous structures. The penetrated water-channel mushroom-like biofilm model (1994) views biofilm as a non-uniform structure consisting of discrete cell clusters (microbial aggregates) cemented by extracellular polymeric substances; it is a porous network with voids which allow wastewater to flow through (Fig. 18.22). In the mosaic biofilm model (1995), the biofilm consists of a thin basal layer of microorganisms (about 5 μm) with vertically aligned stacks of microcolonies (up to 100 μm) outgrow from the surface (Fig. 18.23). Fluxes are multidirectional instead of unidirectional. In addition, mass transfer in biofilm is not just diffusional because convectional transport has also been demonstrated.

18.1.2.3. *Purification efficiency*

The transport of substrates and end products take place by diffusion in the pores and channels in the biofilm. Organic substrates, nutrients and metabolic end products move between the wastewater (aqueous phase) and the biofilm (solid phase). Gaseous transfer occurs through the air-waste-water interface, but oxygen concentration in the biofilm is limited by the DO concentration in the bulk liquid phase. Oxygen is first dissolved in the wastewater and diffused through the biofilm along a concentration gradient. The total amount of oxygen transferred is governed by the surface area of the media exposed; therefore smaller gravel is better in terms of ventilation because of the greater surface area per unit volume. However, the choice of gravel size is not ultimately aimed at maximum surface area as the void space is reduced when smaller gravel is used. The thickness of aerobic layer depends on the extent of oxygen diffusion within the biofilm. The critical

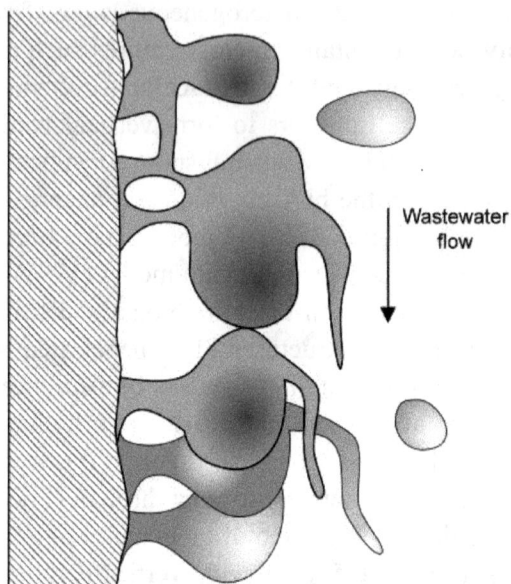

Fig. 18.22.   The penetrated water-channel mushroom-like biofilm model views biofilm as a heterogeneous and dynamic layer of microbial aggregates (microcolonies) adhering to and detaching from the support surface. The discrete clumps are surrounded by a network of interstitial voids filled with water (redrawn from Costerton *et al.,* 1994; Lewandowski and Beyenal, 2003).

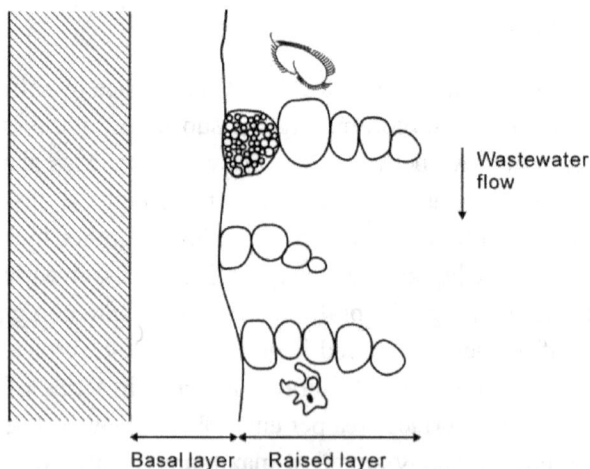

Fig. 18.23. Biofilm in the heterogeneous mosaic biofilm model consists of a basal layer and a raised layer with chains of microcolonies rising from the basal layer (redrawn from Walker *et al.,* 1995).

depth, i.e. the depth of oxygen penetration, varies from 0.05–2.00 μm, which is dependent upon the composition and density of the biofilm. The inner layer of the biofilm may be anaerobic and substrate limited. Thick film tends to have a thicker anaerobic layer than a thin film.

The biological activity of the biofilm is a function of the organic (BOD) loading, oxygen availability and temperature. BOD is removed by the active portion of the biofilm with purification efficiency being highest at optimal biofilm thickness. When the film is too thin, biodegradation is limited by the low microbial biomass. For thick film, interstitial volume, which is essential to sewage flow, ventilation and microbial growth, is reduced and ventilation is hampered. In some cases, the film can be so thick that sewage flow is prevented and the filter becomes clogged, and in severe cases, ponding is caused.

The growth of the biofilm is a function of the BOD loading in the wastewater. However, gaseous and nutrient diffusion is affected when the film is thicker than 0.5 mm. When the inner layer is depleted of nutrients and becomes anaerobic, cells will undergo endogenous respiration and the film will slough off. The thickness of the biofilm is also regulated by the feeding activity of the grazing detritivores, the flushing action of the wastewater and the stripping action of detached film from the upper region. All these processes act to remove biofilm and keep the thickness at a dynamic equilibrium of 2 mm.

18.1.2.4. *Vertical stratification*

Differential growth occurs along the depth of the biological filter. At the top, higher BOD loading and a faster cycle of biofilm growth and sloughing off select microbes with short generation time (e.g. C oxidizers). Heterotrophs outgrow nitrifiers when BOD is available. Oxygen concentration in the wastewater is also limited in the upper region and this inhibits the growth of nitrifying bacteria which are strict aerobes. At the bottom, microbes of longer generation time (e.g. nitrifying bacteria) establish, with *Nitrosomonas* spp. becoming established before *Nitrobacter* spp. as ammonium is present in higher concentration in the upper zone. Nitrifying activity increases with depth and nitrification is enhanced at a greater depth when BOD < 20 mg/L. A vertical gradient is established as the pollutants are removed when the wastewater trickles downward with microbial density and bioconversion

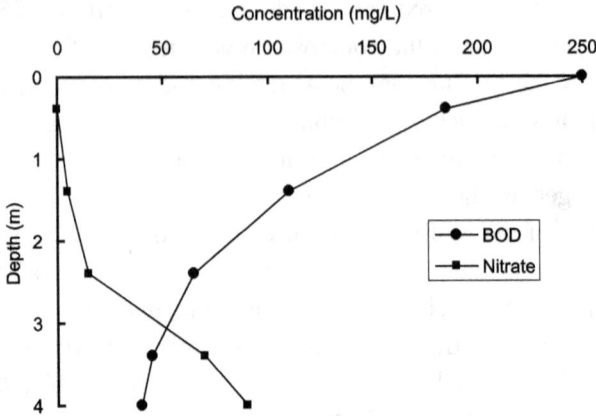

Fig. 18.24.   The changes in BOD and nitrate concentration with filter depth in a trickling filter of 4 m deep. Carbon oxidation takes place in the upper region while nitrification only commences at greater depth (Mudrack and Kunst, 1986).

rate being highest at the top. However, nitrifiers build up only at the bottom of the tank. Chemically, BOD declines while nitrate concentration increases with depth (Fig. 18.24).

18.1.2.5. Recirculation

At low hydraulic and organic loadings, the conventional low-rate single-pass filtration can produce a final effluent of very high quality after solid settling. For stronger industrial wastewater, filtration towers with plastic packing are preferred. Recirculation of effluent helps, as recirculating treated effluent to the filter dilutes strong wastewater and increases the flushing action, which effectively restricts biofilm accumulation and prevents clogging of the filter. Recirculation at low flow periods can even out diurnal fluctuations in volume, and can enhance nitrification as the recirculated effluent contains nitrate and DO at low BOD. The amount of recirculation depends on the organic strength (BOD) of the influent and effluent, and the recirculation ratio (recirculated flow to influent flow) is typically 0.1–2.0.

18.1.2.6. Filter systems

In the traditional single-stage process, wastewater is treated by a single filter or by filters working in parallel. This is used for a conventionally

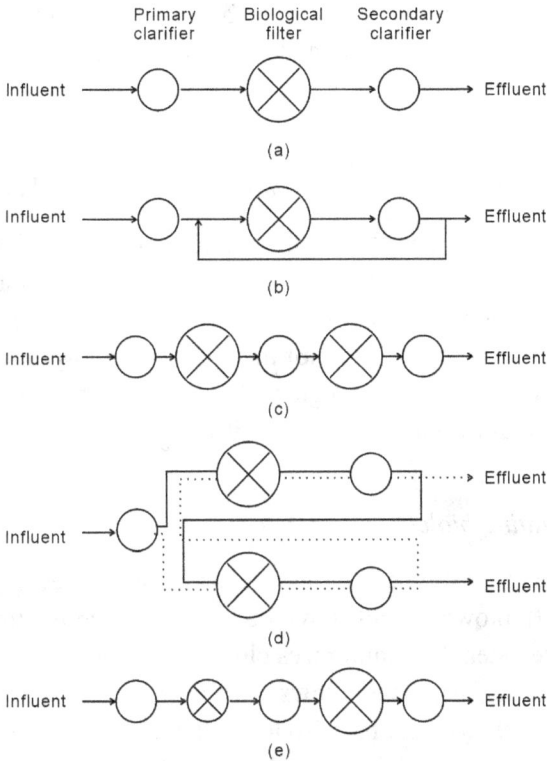

Fig. 18.25. (a) Filters can be arranged in single-pass system in which wastewater is treated by a single filter or a number of filters in parallel; (b) recirculation can take place for strong wastewater. Multi-pass systems can be used in which wastewater is treated by more than one filters in series (c)–(e); (c) double filtration consists of two sets of filters in series; (d) alternating double filtration changes sequence of filters periodically with flow direction reverses before the primary filter shows sign of ponding; (e) two-stage filtration uses two different filters in series, usually a primary high-rate filter.

low organic loading rate of 0.1–0.5 kg BOD/m³/day. For higher rates (>0.5 kg BOD/m³/day), multi-pass mode is employed in which wastewater is treated progressively by filters operating in a series (Fig. 18.25). Two sets of similar filter beds can operate in series as in double filtration. For alternating double filtration, the sequence of filters used is periodically reversed to relieve overloading of the conventional filter. Partially treated effluent is applied from one filter, but the direction of flow reverses before the primary filter shows signs of ponding (7–14 days).

In these situations, nitrifiers never establish and nitrification is poor. In two-stage filtration, two filters operate in series; the first one usually acts as a primary high rate filter or the secondary filter is used for nitrification.

18.1.2.7. Nitrification and denitrification in biological filters

Nitrification usually takes place at the lower regions of the filter with rates dependent upon temperature and the concentration of ammonium ion. Denitrification, however, takes places in the upper filter area where organic matter is abundant and biofilm is thick to allow anaerobic conditions to develop in the inner region of the film. Nitrate is supplied via recirculation to allow simultaneous denitrification.

18.1.2.8. Rotating biological contactor

The invention of plastic packing media has led to the development of tall tower filters (biotowers) with forced aeration for the treatment of high-strength wastewater. This minimizes clogging in gravel media to produce a higher quality effluent, and allows nitrification to occur.

The use of plastic media has also led to the development of the rotating biological contactor which is an aerobic fixed film reactor with horizontal flow and moving media (Fig. 18.26). It was developed in the late 1920s and was commercially available in 1965. This system is popular for small communities and industries. The moving media are a series of circular discs made of plastic, PVC or expanded polystyrene in groups, mechanically driven around a horizontal axis. The discs are 10–20 mm thick, 2–3 m in diameter and spaced 20 mm apart. They rotate slowly (0.5–10 rpm) at right angle to the horizontal wastewater flow with 40% of the area immersed. The whole system can be covered to protect from bad weather and low temperature, and to control fly and odour nuisance. Biofilm grows on the disc surface which is alternately immersed as the discs are rotated and ventilation is achieved when the discs are exposed to the air. There is a succession along the reactor, with heterotrophic bacteria and fungi in the first few discs, protozoa on successive discs and nitrifiers towards the end of the system. Sloughing will be more frequent at the front portion of the system. Sewage flow and disc rotation ensure efficient

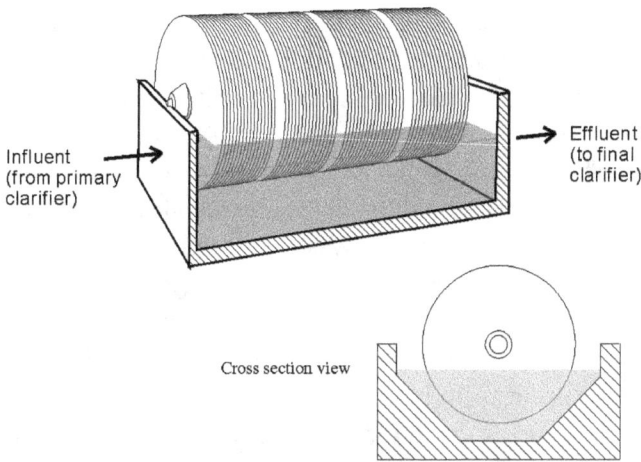

Fig. 18.26. Rotating biological contactor consists of circular discs on which biofilm develops. The discs rotate slowly in right angle to the wastewater flow.

mass transfer from liquid to film and control excessive film accumulation. This system has several advantages over percolating filters. The media are completely wetted and there is no clogging even in the case of excessive film accumulation, ventilation is good as the discs are regularly exposed to air during rotation, less space is needed (which is about 10% of that by low-rate filtration), and nuisance can be reduced when the reactor is covered. However, the system is sensitive to overloading and is more expensive to run especially in motor drive and bearing maintenance when compared to other small biological treatment plants.

18.1.2.9. Biological aerated filters

Though not as common as those conventionally used in municipalities, there are modified versions of the biological filter, some of which have been used for over twenty years. The activated biofilter integrates activated sludge and biological filter to provide a combined suspended and attached growth process. However, there are only a small number of full-scale operations in use due to declining interest. Technological advances led to the development of different ranges of biological aerated filters (BAF) or submerged attached growth systems. The aerated packed bed reactors use attached biomass, as in trickling filters, but the microbial film grows on smaller granular carriers (0.2–3 mm) made of mineral, glass or plastic, which are immersed in the wastewater and aerated by injection of air or pure oxygen. The small particle size provides a large specific surface area for biofilm attachment. Wastewater enters the filter either from the top (downflow filter) or the bottom (upflow filter). In the upflow system, the bed of carriers (e.g. silica sand with a diameter of 0.2–0.7 mm) can be fluidized by the wastewater flow (aerated fluidized bed reactor). An alternative is the airlift bioreactor in which the bed, which has thicker biofilm, is fluidized by the introduction of air at the bottom of the reactor. The free floating media move with the wastewater circulation in the reactor. High specific surface area ($1,000–4,000\ m^2/m^3$) improves contact with substrate in the wastewater. High biomass density allows shorter hydraulic retention time, and clogging or ponding is no longer a problem. Microfauna with the normal species diversity and rotifers were found. These moving bed biofilm reactors are considered very efficient for treating high strength wastewater such as food industry effluents. Immobilization of nitrifying bacteria on carriers allows the development of older biofilm which is essential for nitrification. However, nitrate can only be denitrified if anaerobic condition is developed. This can be achieved using either a separated anoxic filter for denitrification before an aerobic filter (pre-denitrification), which is similar to the A/O system for activated sludge, or after BOD removal and nitrification with carbon addition (post-denitrification). Biological phosphorus removal has also been demonstrated in some of these filter systems.

18.2. Anaerobic Wastewater Treatment

Aerobic treatment processes require aeration which is energy consuming and expensive. In this regard, anaerobic plants would be an advantage and can treat strong wastewater (BOD > 500 mg/L) which is otherwise difficult to be degraded aerobically. Anaerobic systems produce biogas as an endproduct, which can be used as a fuel. Sludge handling is also easier as about 10% of the substrate is converted to microbial biomass compared with about 50% for aerobic processes. The anaerobically digested effluent can be further processed by aerobic treatment.

Anaerobic digestion of wastewater is more or less the same as that of solid wastes or slurries in that the biological material is broken down microbially under anaerobic conditions with the generation of biogas. However, as wastewater treatment plants usually work under a high hydraulic flow rate with low (1–4%) solid content, a longer retention time is required when using conventional stirred tank reactors or very large reactors have to be built. With the development of modern digesters, anaerobic treatment is a feasible and practically better option for wastewaters of high organic loading.

18.2.1. *Microbiology of anaerobic digestion*

Anaerobic reaction is solely a bacterial process in which organic matter is degraded by a succession of mixed bacterial populations. It was previously thought to be a two-stage process consisting of fermentation and methanogensis, but it is now clear that it has four stages, namely hydrolytic, acidogenic, acetogenic and methanogenic phases (Fig. 18.27). In the hydrolytic phase, the organic substrate is hydrolyzed enzymatically by facultative anaerobic bacteria. Macromolecules of high molecular weight such as carbohydrates, proteins and lipids are broken down extracellularly into soluble monomers. Cellulolytic bacteria such as *Clostridium* spp. secrete cellulolytic enzymes which convert cellulose into cellobiose and glucose; proteolytic bacteria such as *Bacteroides* spp., *Peptococcus* spp. and *Peptostreptococcus* spp. break down protein into amino acid. In the acidogenic phase, the hydrolytic products such as sugars and amino acids are acted on by acid-forming bacteria (e.g. *Lactobacillus* spp., *Bifidobacterium* spp. and *Butyrivibrio* spp.), converted to pyruvate and subsequently to acetate and carbon dioxide or alcohol and hydrogen. The fate of metabolites depends on the type of bacteria and culture conditions (e.g. redox potential). At low partial pressure of hydrogen ($<10^{-4}$ atmosphere), acetate and/or hydrogen are formed; at high partial pressure of hydrogen ($>10^{-4}$ atmosphere), ethanol or organic acid is produced. Short chain fatty acids (e.g. acetate, propionate and butyrate) are the main products, which lower the pH, and hence the name acidification phase as a

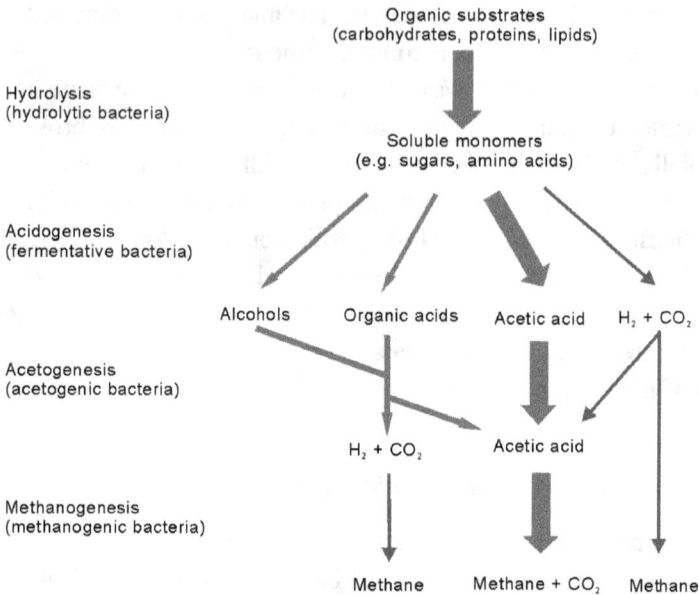

Fig. 18.27.   Anaerobic digestion is a four-stage bacterial degradative process which converts polymeric substrates to methane and carbon dioxide.

synonym to acidogenic phase. Acetate is the major product of carbohydrate fermentation.

The third phase is the acetogenic phase. Alcohols and short chain fatty acids are utilized by hydrogen-producing heteroacetogens (e.g. *Enterobacter* spp., *Citrobacter* spp., *Serratia* spp. and *Syntrobacter* spp.), which produce acetate and/or hydrogen and carbon dioxide through acetogenic dehydrogenation. Hydrogen-consuming homoacetogens (e.g. *Clostridium* spp. and *Acetobacterium* spp.) use hydrogen and carbon dioxide as substrates to form acetate via acetogenic hydrogenation. Acetate and carbon dioxide formed from Phases 1 to 3 are methanogenic substrates. Acetate-removing methanogens utilize acetate and form methane and carbon dioxide through acetate decarboxylation, and this accounts for 70% of the methane formed. The remaining 30% is produced by reductive methane formation with hydrogen-removing methanogens which convert hydrogen and carbon dioxide to methane and water.

Methanogens which are strict anaerobes belong to Archaebacteria. They are morphologically diverse group which are restricted to a small number

Table 18.7.   Morphological characteristics and growth substrates of different methanogenic bacteria.

Genus	Shape	Gram stain	Substrates
Acetate-removing methanogens			
Methanosarcina	Cocci in clusters	Positive	Acetate, methanol, methylamines
Methanothrix	Long rods to filaments	Negative	Acetate
Methanosaeta	Long rods to filaments	Negative	Acetate
Methanococcoides	Cocci	Negative	Methanol, methylamines
Methanolobus	Cocci in clusters	Negative	Methanol, methylamines
Hydrogen-removing methanogens			
Methanobacterium	Long rods	Positive	Hydrogen, formate
Methanomicrobium	Short rods	Negative	Hydrogen, formate
Methanococcus	Cocci	Negative	Hydrogen, formate
Methanogenium	Cocci	Negative	Hydrogen, formate
Methanospirillum	Curved rods	Negative	Hydrogen, formate

of simple substrates such as acetate, formate, methanol, methylamines, carbon dioxide and hydrogen (Table 18.7). They are the terminal organisms in biogasification which produce methane as the metabolic endproduct. There are about 50 species. Acetate-removing methanogens include *Methanosarcina* spp., *Methanothrix* spp. and *Methanosaeta* spp. which produce methane from acetate, and *Methanococcoides* spp. and *Methanolobus* spp. which utilize methanol or methylamines. Hydrogen-removing methanogens such as *Methanomicrobium* spp., *Methanobacterium* spp., *Methanobrevibacter* spp., *Methanococcus* spp., *Methanogenium* spp. and *Methanospirillum* spp. utilize hydrogen and formate in addition to acetate.

Methanogens are more demanding in environmental requirements and are sensitive to shock conditions. Oxygen, as low as 0.01 mg/L, is toxic to methanogens. Methanogens are highly sensitive to oxidized materials such as nitrate and sulphate, with an optimal redox potential at below –300 mV. Methanogens are also sensitive to pH and grow best in the narrow range of 6.5–8.0. A pH less than 6.6 is inhibitory and acidification will reduce the growth of methanogens, which further acidifies the digester content owing to an accumulation of volatile fatty acids, as the acidogenic bacteria are less sensitive to low pH. Methanogens are slow-growing with a generation time of 2–3 days at 35°C as compared with 2–3 hours for acid-formers. Methanogenesis is therefore the rate-limiting step of anaerobic decomposition.

The bacterial consortium is metabolically dependent on each other for survival and activity, and the acid-formers will be inhibited by their own endproducts if these are not catabolized by the methanogens. Mutualistic effects occur between different metabolic groups for the maintenance of pH and redox potential. Hydrogen is usually recognized as the controlling intermediate and an interspecies hydrogen transfer is involved. Low hydrogen concentration is maintained not just by diffusion alone, but also by intimate cell contact between acetogenic and methanogenic bacteria.

Most anaerobic digestions operate at a mesophilic temperature range of 30–38°C. As anaerobic reaction produces little free energy as heat, the digester needs energy input for heating up the content to above 35°C. Thermophilic digestion gives higher yields, allows higher loading and greater pathogen destruction, but greater heating costs and the need for insulation to reduce excessive heat loss outweigh the advantages.

18.2.2. *Anaerobic reactors*

All reactions in anaerobic digestion can take place in one reactor in single-stage system, though the use of two-stage system in which fermentation and methanogenesis are separated is not uncommon for solid digestion. Anaerobic processes can take place in open lagoons or in closed digesters. Anaerobic ponds in the stabilization pond systems are usually used as pretreatment steps for strong wastewater. In a pond of 2–5 m deep with no aeration, anaerobic conditions prevail. These are followed by facultative and maturation ponds, in series, for complete treatment. Septic tanks, which are the underground individual household systems for wastewater disposal in unsewered municipalities, are in fact low volume anaerobic systems in which solid settlement and anaerobic digestion of sludge take place in one tank.

For closed reactors, stirred tank digesters, which are developed for the digestion of high-solid feedstocks, mixing performed by mechanical mixers such as stirrers and screw pumps or by methane recirculation. For

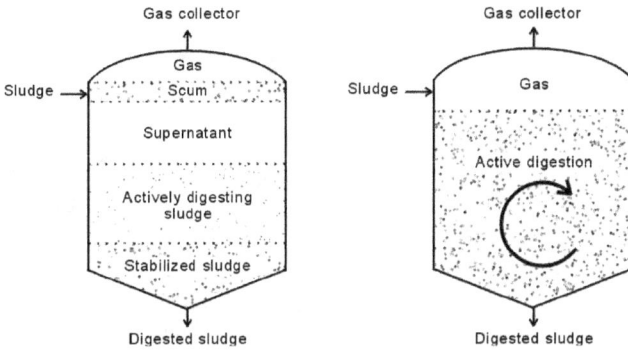

Fig. 18.28.   Conventional digestion has stratified content while high-rate digestion has active digestion throughout the digester content.

conventional digestion (0.5–1.5 kg volatile solids/m³/day) of slurry in the absence of agitation, the content is mixed solely by the movement of gas up through the digesting sludge. Well defined strata are formed with an upper relatively inactive scum layer, central layer of tank liquor, lower active layer of digesting solids and a bottom layer of sludge (Fig. 18.28), with a retention time is 30–60 days. In high rate digestion (1.5–5.0 kg volatile solids/m³/day), the content is homogenous due to mixing, and active digestion takes place throughout the digester content. Mixing improves the process tremendously by reducing the required retention time shorter than 15 days. High rate anaerobic digesters are mostly used in the brewery and beverage industry, distillery and fermentation industry, food industry as well as paper and pulp industry.

Digesters can be of different configurations with different height to diameter ratios (Fig. 18.29). A conical bottom retains grit and sludge for periodic removal. Egg-shaped digesters have the benefits of better mixing properties, lower energy demand for heating and smaller surface which minimizes the size of a scum blanket, and have become very popular in Asian countries such as Japan for wastewater treatment. However, all these configurations when operated as flow-through system, suffer from washout of the bacterial biomass and impractically long retention times for wastewater treatment.

Stirred tank digesters are not suitable for treating low strength waste-water since the hydraulic flow is great and the gas yield is very low, which

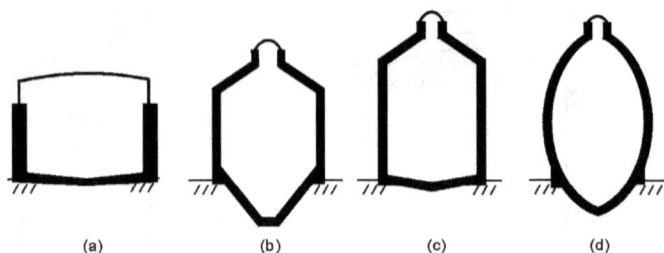

Fig. 18.29.   Major shapes of anaerobic digesters most commonly used in different parts of the world. (a) Anglo-American digester; (b) classical European digester; (c) European digester; and (d) pear- or egg-shaped digester (Hobson and Wheatley, 1993; Bode and Klauwer, 1999).

would otherwise require a very large digester. The separation of solids from the effluent is another problem in that the biogas produced tends to cause the sludge to float and affects settleability. With the development of biomass-retained systems, the anaerobic treatment of wastewater becomes feasible. Contact systems are mostly used with the mean cell residence time (MCRT) far exceeding the HRT, as in aerated biological filters for aerobic treatment (Fig. 18.30). Biomass is preserved in the reactor by immobilization with the cells being trapped on a solid support in the digester or by an upward flow of the wastewater in which the biomass is suspended and prevented from being washed out. In fact, immobilized biomass containing the proper balanced anaerobic bacterial consortia is the key for success in anaerobic digestion of wastewater. MCRT is usually greater than 20 days and cells are acclimatized to the influent. Anaerobic fixed film digester in the form of anaerobic static media biofilter or anaerobic rotating disc contactor can be used. The former is actually a submerged trickling filter with packed bed of gravel, or beads of PVC or polyester of 6.0–60 mm in diameter and wastewater either flows upward or downward. Anaerobic microbial cells develop as attached film immobilized on the immersed media and as dispersed biomass in the interstitial space. Treatment performance, therefore, does not depend highly on the size and specific surface area of the media as in aerobic filters. However, when small filter media are used, there is the potential risk of clogging by solids in the wastewater. The latter is similar to the aerobic form of RBC but with the discs totally submerged.

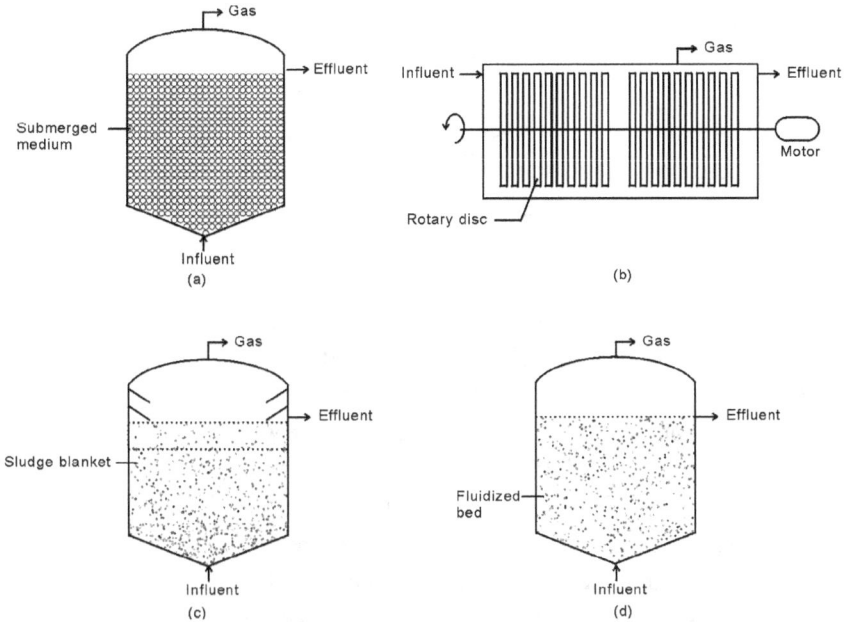

Fig. 18.30. The various anaerobic reactors commonly used for wastewater digestion. (a) Anaerobic static media biofilter; (b) anaerobic rotating disc contactor; (c) upflow anaerobic sludge blanket reactor; and (d) anaerobic fluidized bed reactor.

The upflow anaerobic sludge blanket (UASB) reactor is a floc-based anaerobic system. Wastewater flows upwards at velocities of 0.5– 1.0 m/h through the bottom sludge bed and an upper sludge blanket which consists of lighter solid fraction (flocculated sludge granules) kept in suspension by the wastewater flow. Bacterial organization and arrangement govern the granulation of sludge, and the bacterial composition, size and microstructure of the granules were dependent upon the nature of the substrate. It is generally accepted that the UASB granules, which are of 0.5–4.5 mm in size, are stratified structures. The granule core was almost exclusively composed of *Microthrix* spp., which was a key structural component in granulation. Acetoclastic methanogens and hydrogen-producing acetogens develop in the middle layer while acidogenic (fermentative) bacteria and hydrogenotrophic methanogens occupy the outer layer. Finer sludge settles back to the blanket in the upper zone while heavier sludge settles to the bottom of the digester. Mixing is done as the influent flows upward through the blanket and gas bubbles released from the digester content. The digestion tank has inclined louvres or baffles at the top, which improve sludge settleability and reduce biomass washout. Advantages of UASB granules over sludge flocs include dense compact biomass, high mechanical strength, high settleability (30–80 m/h) (which prevents washing out), balanced microbial community, closely associated syntrophic partners, high methanogenic activity, and resistance to toxic shock. There are more than 300 full-scale plants all over the world for treating wastewaters mainly from food and beverage industries. A variant of the UASB reactor is the expanded granular sludge bed (EGSB) reactor or simply anaerobic expanded bed (AEB) reactor. It is a tall version of upflow sludge bed reactors which has faster rate of upward influent velocity (5–10 m/h) to suspend the granular biomass and result in high hydraulic mixing, expanded granular sludge bed and better wastewater-biomass contact. It is effective in treating low strength wastewaters (< 1 g soluble COD/L). UASB and EGSB reactors make up the majority of the full-scale plants in operation worldwide and the development is being considered as a sustainable integrated low cost wastewater biotechnology.

The anaerobic fluidized bed (AFB) reactor is a hybrid of fixed film and floc-based reactor in which the fine grained inert carriers, made of quartz sand, diatomaceous earth, glass beads, PVC particles or carbon granules, of size ranging 0.2–2 mm are coated with bacterial biomass. The small-sized particles provide a very large surface area and better mixing of the substrates and the bacterial cells. The upflow wastewater movement fluidizes the biomass on carrier particles (bioparticles) and the shear forces of fluidization create a thin dense biofilm with active biomass. The upflow velocity is high enough to expand the bed by typically 20–25% of the resting volume. The MCRT can be as long as 300 days at HRT of about 2 hours. This biomass-on-carrier technology has the advantages of good contact between bacterial cells and wastewater, high biomass concentration, biomass thickness under control, no potential blockage, and reduced reactor volume. It is a highly engineered system with high energy requirement for fluidization. Practical problem in controlling the attachment of biofilm to the carrier particles also limits its use. The technique is not widely applied yet with full-scale plants in the US, the UK, the Netherlands and Australia.

References

1. Bode H and Klauwer E. Advantages and disadvantages of different shapes in digester design. *Water Qual. Int.* **March/April**: 35–40, 1999.

2. Costerton JW, Lewandowski Z, DeBeer D, Caldwell D, Korber D and James G. Biofilms, the customized microniche. *J. Bacteriol.* **176**: 2137–2142, 1994.

3. Hobson PN and Wheatley AD. *Anaerobic Digestion. Modern Theory and Practice.* Elsevier Applied Science, London, 1993.

4. Lewandowski Z and Beyenal H. Mass transport in heterogeneous biofilms. In: *Biofilms in Wastewater Treatment*, eds. S Wuertz, PL Bishop and PA Wilderer. International Water Association, London, 2003, pp. 147–177.

5. Mudrack K and Kunst S. *Biology of Sewerage Treatment and Water Pollution Control.* Ellis Horwood, Chichester, 1986.

6. Palm JC, Jenkins D and Parker DS. Relationship between organic loading, dissolved oxygen concentration and sludge settleability in the completely-mixed activated sludge process. *J. Water Pollut. Control Fed.* **52**: 2484–2506, 1980.

7. Pasveer A. A contribution of the development of activated sludge treatment. *J. Proc. Inst. Sewage Purif.* **4**: 436–465, 1959.

8. Pike EB and Curds CR. Microbial ecology of the activated sludge process. In: *Microbial Aspects of Pollution*, eds. G Sykes and FA Skinner. Academic Press, New York, 1971, pp. 123–147.

9. Walker JT, Mackerness CW, Rogers J and Keevil CW. Heterogeneous mosaic biofilm — a haven for waterborne pathogens. In: *Microbial Biofilm*, eds. HM Lappin-Scott and JW Costerton. Cambridge University Press, Cambridge, 1995, pp. 196–204.

10. Wang H, Lin Jiang L and Weitz JS. Bacterivorous grazers facilitate organic matter decomposition: a stoichiometric modeling approach. *FEMS Microbial Ecol* **69**: 170–179, 2009.

11. Wingender J and Flemming H-C. Autoaggregation of microorganisms: Flocs and biofilms. In: *Biotechnology Vol. 11a*, eds. H-J Rehm and G Reed. Wiley-VCH, Weinheim, 1991, pp. 65–83.

Further Reading

1. Bitton G. *Wastewater Microbiology*. Wiley-Liss, New York, 1994.
2. Forster CF. *Biotechnology and Wastewater Treatment*. Cambridge University Press, Cambridge, 1985.
3. Forster CF and Wase DA. *Environmental Biotechnology*. Ellis Horwood. Chichester, 1987.
4. Grady CPL Jr, Daigger GT and Lim HC. *Biological Wastewater Treatment*, 2nd ed. Marcel Dekker, New York, 1999.
5. Henze M, van Loosdrecth MCM, Ekama GA and Brdjanovic D. *Biological Wastewater Treatment. Principles, Modelling and Design*. IWA Publishing, London, 2008.
6. Hobson PN and Wheatley AD. *Anaerobic Digestion. Modern Theory and Practice*. Elsevier Applied Science, London, 1993.
7. Scragg A. *Environmental Biotechnology*. Longman, London, 1999.
8. Wuertz S, Bishop PL and Wilderer PA. *Biofilms in Wastewater Treatment*. International Water Association, London, 2003.

Questions for Thought...

1. The aeration tank in a modern activated sludge treatment plant is not entirely aerobic. Discuss the purpose of this design.
2. What is the role of the filter media in a percolating filter? Why are plastic media superior to mineral ones in the treatment of high strength sewage?
3. Compare and contrast conventional biological filters and submerged biofilters as wastewater treatment options.
4. Discuss briefly the advantages of using activated sludge system over percolating filter for the treatment of municipal sewage.
5. Describe the operational features of an anaerobic fluidized bed reactor. What are its advantages over the other types of anaerobic reactor in high rate digestion?
6. What is meant by mesophilic anaerobic digestion? How does it compare with the thermophilic digestion in anaerobic wastewater treatment?
7. Discuss the advantages of anaerobic processes over aerobic ones for the treatment of high-strength wastewater.

Chapter 19

Industrial Wastewater Treatment

Wong Po-Keung and Chu Lee Man
Department of Biology, The Chinese University of Hong Kong
Shatin, NT, Hong Kong SAR, China

Industrial wastewater treatment differs from municipal wastewater treatment in that there is far less variation in the composition of the wastewater because the water is being discharged from a large single purpose process. Municipal waste contains a far greater range of wastewater from a greater range of sources. The advantage of industrial wastewater is that any treatment may be more targeted because the major pollutant may be of one class of material and not the mixed classes of materials seen in municipal wastewater industrial.

19.1. Removal of Inorganic Pollutants

Organic waste is composed mainly of carbon compounds that can be easily transformed and broken down while being incorporated into other living things. Inorganic pollutants are not so easily dealt with. As the amount of the inorganic ions are often far greater than any living thing can incorporate into their living cells there are special approaches to their removal. By far the heavy metals are the most important problem because their ingestion can result in toxic consequences.

19.1.1. Removal of metal ions in industrial wastewater

One of the most common definitions of "heavy metal" is a metal with a density of greater than 5 g/cm^3. Under this definition, 69 elements belong to this group of which 16 are synthetic and these metal ions are actually of very diverse chemical properties. Those that commonly cause serious pollution include antimony, arsenic, cadmium, chromium, cobalt, copper, iron, lead, mercury, nickel, silver, thallium, tin, vanadium and zinc. Recently, a new scheme has been provided to categorize "heavy metals". They are recognized as three classes of metals based on the electro-negativity, charge and ionic radius of metal ions and equilibrium constants that describe the formation of metal ion/ligand complexes. These three categories are:

(1) Class A metals show preferences for ligands containing oxygen,
(2) Class B metals show preferences for ligands with nitrogen or sulfur, and
(3) Metals with characters between Classes A and B, which also reflect their Lewis acid properties.

Although some metal ions are essential to life (e.g. copper, iron, zinc and chromium), excessive intake of these metal ions is toxic to living organisms. The toxicity of metal ions to living organisms is well documented. Metal ions can be transported along the food chain and finally enter the human body. Toxic actions of metal ions to other living organisms also cause perturbation of ecological processes in the biogeochemical cycles. When the metal ions pass through the environment which has no natural assimilative capacity for them, the metal ion concentration in the environment will be elevated to a level that poses risk to ecosystem and human health.

Metal ions are transported to air, water and soil from various sources through diverse routes. These include natural sources released from geochemical materials and human activities such as mining, agricultural activities, industrial processes and miscellaneous activities in urban life. As a result, these metal ions are transported to the environment as mine wastes, pesticides, fertilizers, preservatives, sewage sludge and various forms of wastes produced by both metalliferous and non-metalliferous industries.

19.1.2. *Metal pollution in Hong Kong*

In Hong Kong, metal ions that are released into the environment come mainly from the effluent of the metal finishing industries such as electroplating industry (Fig. 19.1). The chemical composition of electroplating effluent has been determined. Table 19.1 shows the metal

Fig. 19.1. A plating tank in an electroplating factory in Hong Kong. Depending on the work-piece to be plated, the plating bath contains Cu^{2+}, Ni^{2+} or Zn^{2+}.

Table 19.1. Content of metal ions and cyanide, and pH of electroplating effluent discharged from 16 electroplating factories in Hong Kong (Environmental Management Division, 1986).

Parameters	Range
Chromium (Cr^{6+})	1–40 mg/L
Copper (Cu^{2+})	1–30 mg/L
Nickel (Ni^{2+})	3–365 mg/L
Zinc (Zn^{2+})	4–250 mg/L
Aluminum (Al^{3+})	10–230 mg/L
Silver (Ag^{2+})	2–3 mg/L
Cyanide (CN^{-})	1–6 mg/L
pH	1.7–8.2

ion content and pH of plating wastewater discharged from sixteen electroplating factories in Hong Kong. The effluent is highly acidic or alkaline and contains toxic substances such as cyanide and metal ions. As these chemicals possess a certain degree of toxicity to human health and aquatic life, the electroplating effluent should be pretreated to minimize the impact on the receiving water bodies. Moreover, the production of metal-containing wastes is expected to increase in the future. Copper-laden waste-water production is the highest among metals now and in the future.

In Hong Kong, metal ion-containing effluent is either transported to the conventional wastewater treatment plant without pretreatment or discharged directly into the environment. The degree of pollution can be revealed by the elevated metal contents in the marine sediment and in shellfish that are found in the region. There is no specific process for the removal of metal ions in the wastewater treatment in Hong Kong, but large amounts of metal ions accumulate in the "activated sludge" — the major agent in biological wastewater treatment. The elevated level of metals is harmful to the microorganisms in the "activated sludge". Thus low concentrations of metal ion in untreated wastewater can significantly reduce the efficiency of biological wastewater treatment by inhibiting microorganisms in "activated sludge" process and anaerobic digestion of sludge. Cu^{2+} concentrations in excess of 1.9 mg/L inhibited anaerobic digestion in wastewater treatment plants. The threshold copper level in wastewater, defined as the concentration at which the treatment efficiency is reduced by 5%, was 1.9 mg/L. The rank of toxicity of metal ions to the "activated sludge" process and anaerobic digestion of sludge is: Ni^{2+} > Cu^{2+} > Cr^{6+} > Zn^{2+}.

19.1.3. *Toxicity of metal ions in the environment*

(1) Copper

Copper is a transition metal with three common oxidation states: Cu^0 (copper metal), Cu^{+1} (cuprous ion) and Cu^{2+} (cupric ion). Cu^{2+} is the most commonly occurring species that readily forms free hydrated ion in water. It prevails in the environment and is the most toxic form to living organisms among the three forms. Cu^{2+} starts to precipitate at above pH 5.5. However, in the presence of organic or inorganic ligands, Cu^{2+}

forms complexes with the ligands and remains free in solution. It can be strongly complexed by electron donor groups (O-, N- and S-containing groups) in organic compounds.

Copper is an essential trace element, but it is toxic to various sensitive members of the food chains (e.g. plants, algae and human) at moderate levels and the toxicity of Cu^{2+} to microorganisms is well documented. Cu^{2+} is generally one of the most toxic metal ions to heterotrophic bacteria in the aquatic environments. The order of sensitivity to metal ions of heterotrophic microorganisms is: $Ag^+ \geq Cu^{2+} = Ni^{2+} > Ba^{2+} = Cr^{6+} = Hg^{2+} > Zn^{2+} = Na^+ = Cd^{2+}$. The toxic actions of Cu^{2+} on a living system are essentially exerted on enzymes, especially enzymes whose activities depend on sulfhydryl and amino groups, as metal ions have high affinity for ligands containing nitrogen and sulfur donors. The binding of metal ions to enzyme molecules, thus masking the catalytic sites, altering the conformations of the biomolecules, or competing with and displacing the enzyme-activating metal ions may inhibit enzymatic activities. Besides, nucleic acids and metabolites may also be targets of metal toxic actions. Cupric salts increased the frequency of non-complementary nucleotides in the synthesized DNA-double helix.

(2) Nickel

Nickel is highly toxic because of its role as an antagonist of essential metals. Nickel has two oxidation states: Ni^0 (nickel metal) and Ni^{2+} (nickel ion). Toxic effects of Ni^{2+} will exert on various kinds of organisms along the food chain. The toxicity of Ni^{2+} on microorganisms such as bacteria, algae and fungi is well documented. Besides, the toxic effects of Ni^{2+} on plants and animals have also been reported. In addition, Ni^{2+} was found to induce chromosomal aberration in mammalian cells and to suppress the immunological responses of animal cells.

(3) Zinc

Zinc, which has a molecular weight of 65.38 g/mol and a density of 7.14 g/cm^3, is a Group IIB transition metal. Zinc has only two oxidation states: Zn^0 (zinc metal) and Zn^{2+} (zinc ion). Being an amphoteric metal, Zn^{2+} exhibits a "U" shaped solubility curve with the theoretically minimum solubility at pH 10.5 and with increasing solubility at both ends. The toxicity of Zn^{2+} is considered to be mild comparing to other metal ions.

This is due to the importance of Zn^{2+} to organisms in the nature where a greater need usually translates to a greater tolerance. In the terrestrial food chains, Zn^{2+} is more toxic to the plants than to the animals eating them because crop yields tends to drop before Zn^{2+} content become hazardous to the next trophic level. In aquatic food chains, the level of Zn^{2+} tends to decrease rather than increase in higher trophic levels. For example, it is shown that fish tends to die or cease reproduction before tissue accumulation of Zn^{2+} reach to a level to be toxic to humans. A generalized statement can be made: Zn^{2+} does not accumulate along the terrestrial or the aquatic food chains. It is because that certain important members of the food chains are particularly susceptible to the toxicity of Zn^{2+} and upward transport of Zn^{2+} halts as those susceptible members deceased.

The adverse effects of Zn^{2+} toxicity on microorganisms are well documented. Growth inhibition, impaired adhesion, morphological changes and altered biochemical activities have been reported in several microorganisms.

19.1.4. *Methods of removal of metal ions from industrial effluent*

Conventional methods used for the removal of metal ions from industrial effluent can be divided into two categories. They are physico-chemical methods and biological methods.

19.1.4.1. *Physico-chemical methods*

The classical physico-chemical methods are chemical precipitation, chemical oxidation and reduction, ion exchange, electrochemical reduction, reverse osmosis, freeze crystallization, electrodialysis, cementation and starch xanthate adsorption. These methods, owing to the high operational cost or the difficulty to treat solid wastes subsequently generated, are generally expensive and ineffective especially when the metal ions are dissolved in large volumes of solutions at relatively low concentrations (around 1–100 mg/L).

Reverse osmosis (RO) is based on the application of high pressure to overcome osmotic pressure, forcing metal ion-laden wastewater to pass through the RO membrane that enclosed in vessels. Metal ion-laden wastewater will be separated by selective permeation into purified, permeate stream and concentrated stream of metal solution. The concentrated metal ion solution can be recycled into plating bath while water with sufficient purity can be reused for rinsing.

Precipitation is the most common and widely used method in wastewater treatment. Undesired metal ions in the effluent can be precipitated as insoluble metal hydroxides by adding calcium hydroxide or sodium hydroxide. They can also be precipitated as the more insoluble metal sulfides by adding sodium sulfide, sodium hydrosulfide or ferrous sulfide. The insoluble precipitates formed are then allowed to settle in the sedimentation tank by addition of coagulating agents. Finally, the sludge can be removed and the clarified effluent can be discharged to the sewerage system.

Ion exchange is a stoichiometric and reversible chemical reaction in which a metal ion (e.g. Ni^{2+}) from solution is exchanged for a similarly charged ion (e.g. H^+) attached to an immobile solid particle. The immobilization matrix can be naturally existing inorganic zeolites, synthetic resin materials or activated carbon, which are contained in a series of columns in the ion exchange system. Through the regeneration of the columns by adding an acidic solution, metal ions in the eluent of the cation exchange column may be recovered by evaporation.

19.1.4.2. Biological methods

(1) Microbial resistance to metal ions
Since the beginning of industrialization 200 years ago, there have been huge changes in the distribution of metal ions on the surface of the earth due to anthropogenic activities. Bacteria adapted to these changes and have evolved strategies to maintain low intracellular toxic metal ion concentration. Firstly, the adaptation may have innate abilities to organisms living in extreme environment. Secondly, resistance may have been easily acquired from plasmids. These metal ions resistance strategies can be acquired by a number of mechanisms:

(a) The development of active transport systems that maintain non-toxic level of toxicants within the cell. Examples of active transport systems are the *cad A* determinant from *Staphalococcus aureus* plasmid and the secondary role of arsenate detoxification of *pst* (phosphate) transport system found in various bacteria.

(b) Enzymatic mediated oxidation or reduction that converts toxicants to less toxic forms. Examples are the oxidation of As^{6+} to As^{5+} and the reduction of Hg^{2+} to Hg^0 (or Cr^{6+} to Cr^{3+}).

(c) Biosynthesis of intracellular or extracellular biological compounds that serve as traps for the removal of metal ion from solution. Siderophores refer collectively to a group of low molecular-weight iron-binding compounds secreted by various organisms. Such extra-cellular biological compounds have the ability to sequester iron and other metal ions in solution. Intracellular sulfhydryl-containing protein such as metallothionen has also been identified as intracellular trap for cadmium and copper.

(d) The binding of metal ions onto bacterial cell surfaces.

(e) The precipitation of insoluble metal salts on the bacterial cell surfaces through the actions of membrane associated sulfate reductase or through the biosynthesis of oxidizing agents like oxygen and hydrogen peroxide.

(f) Biomethylation and transport through cell membranes by diffusion-controlled processes.

Usually, metal resistant strains of the microorganisms are being used for the removal of metal ions since microorganism-based processes are more economic. For life forms, bacteria have the greatest tolerance to unusually high saline conditions and can cope with saturated solutions of salt. Moreover, they have a greater capacity to adsorb (Mechanism d) and precipitate (Mechanism e) metal ions from solution since adsorption depends mostly on diffusion and the cell surface area to volume ratio. The structural and functional simplicity and short generation time of bacteria facilitate the elucidation metal ions removal mechanism and manipulation of the genetic structure for the enhancement of metal ion accumulation ability. Bacterial cells of *Zoogloea ramigera* removed Cd^{2+} up to 40% of the microbial biomass dry weight when it was grown in medium containing about 0.345 g/L of Cd^{2+}. For instance, *Klebsiella aerogens* are capable of producing polysaccharide capsules that are known to be more tolerant to 10 mg/L Cu^{2+} than the non-capsulated strain. The details of metal ion removal by microorganisms will be discussed in the following sections.

(2) Cell wall of gram-negative bacteria

Cu^{2+}-resistant bacteria cells of *Pseudomonas putida* 5-X that can adsorb a relatively large amount of Cu^{2+} was isolated from metal contaminated

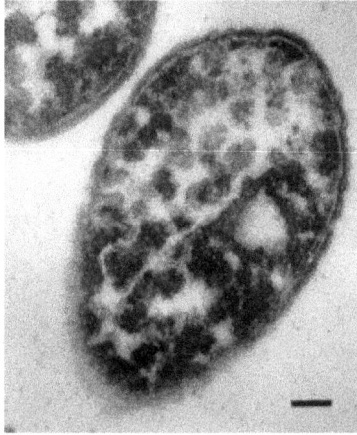

Fig. 19.2. Transmission electron microscopic (TEM) image of cells of *Pseudomonas putida* 5-X stained with Cu^{2+}. The cells were suspended in 50 mg/L of Cu^{2+} solution for 1 hour before proceeding TEM staining. The bar is 1 μm. (Lu and Wong, unpublished results)

sludge (Fig. 19.2). It is a Gram-negative bacterium, the cell envelope of which consists of two membrane bilayers (the outer and plasma membranes) that are chemically distinct from each other and that sandwich a thin petidoglycan (PG) layer between them. The arrangement and exposure of the various functional groups (e.g. carboxylates, phosphorylates, hydroxylates and other negatively charges sites) in cell wall produce a net electronegative charge, which allow the cell wall to interact with metal cations within their immediate surroundings.

The outer membrane of *Escherichia coli* has been studied most extensively (Fig. 19.3). It consists of an asymmetric arrangement of lipid (the outer surface of the bilayer contains mostly lipopolysaccharide (LPS), whereas the inner leaflet is mostly phosphatidylethanolamine) combined together with at least four major outer membrane proteins (OmpF, OmpC, OmpA and lipoprotein).

All are intrinsic proteins, except for OmpA protein which is peripherally embedded in the outer leaflet. The lipoprotein forms a chemical union between the outer membrane and the underlying PG and effectively cements the two wall layers together. The OmpF and OmpC intrinsic proteins act as porin proteins that closely associated with lipoprotein and LPS and form small, hydrophilic channels that span the bilayer. Because of the preponderance of

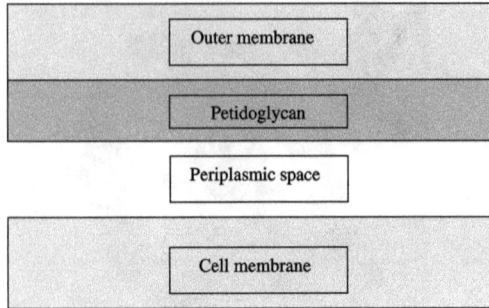

Fig. 19.3. Diagrammatic location of the outer membrane of a Gram-negative bacterium, *Escherichia coli.*

electronegative amino acids at the channel mouth, small metal ions would be encouraged to pass. Thus, it functions as a sieve of metal ions. LPS is anionic and consists of phosphate groups. Phosphatidylethanolamine has a single polar head group and its constituent phosphate group is anionic. The same holds true with phosphatidylglycerol and cardiolipin which are found in smaller amounts within the outer membrane.

The outer membrane orients its lipid molecules in a way that the hydrophilic polar head groups align themselves along each of the bilayer's faces, while their hydrophobic acyl groups point toward the interior of the bilayer. Consequently, each hydrophilic face has reactive chemical groups available to interact with metal ions. The prime candidates for this interaction are the phosphate groups that are resident within the polar head groups of the LPS and phospholipid. Mg^{2+} and Ca^{2+} form an integral component of the outer membrane and are required for the correct packing order of the lipid constituents. The lipid components of phospholipid or LPS constantly exhibit rotational and lateral movements, and the acyl chains of LPS that inhabit the outer leaflet are more rigid and closely packed than their phospholipid counterparts. Consequently, the packing order (and fluidity) of LPS is more dependent on the polar head groups, which requires stabilization by divalent metal ions such as Mg^{2+} and Ca^{2+}. Moreover, LPS possesses more phosphoryl groups than phospholipid, these metal ions are especially important for its retention in the outer membrane.

The PG layer is most probably a monolayer in *E. coli*. It interacts more strongly with metal ions than the outer membrane, presumably because of available carboxylate groups. This chemical reactivity of cell walls is important since they can no longer be thought of as inert, stagnant structures. They are able to interact actively with the environment. It is possible that their reactions with the counter ions of the external milieu could send a perceptible signal to the protoplast to make it aware of its surroundings. Thus, the cell wall can be considered a shroud which acts as a chemical buffer at the periphery of the cell to either collect essential ions (e.g. Mg^{2+} and Ca^{2+}) or to immobilize toxic substances (e.g. Cu^{2+} and Cd^{2+}).

(3) Metal ions and ligands

Metal ions are relatively unattracted to hard (non-polarizable) ligands compared to the attraction experienced by the metal ions, and do not form ionic bonds readily. Binding to hard ligands tends to be covalent where the metal ion and the counter ion each contribute an electron to the bond. The metal ions tend to bind to the "softer" (polarizable) ligands, such as sulfur and nitrogen. For example, Cu^{2+}, Ni^{2+} and Co^{2+} bind to ligands in the order $RS^- > RNH^- > OH^- > H_2O$. Moreover, there is variation of binding affinity within the metal ion group. For example, Cu^{2+} is attracted to the nitrogens of histidine, while Zn^{2+} has an affinity for the sulfhydryl groups of cysteine. The order of softness of metal ions is: $Cu^{2+} > Ni^{2+} > Zn^{2+} > Co^{2+}$. Biomass contains many of these softer binding sites, and therefore has greater binding affinity for metal ions. This in turn means that biomass can selectively accumulate metal ions, with a reduced loss of these ions than that of traditional ion exchangers. Ca^{2+} and Mn^{2+}, for instance, compete effectively with metal ions for electrostatic binding sites on ion exchangers.

(4) Kinetics of metal ion uptake

Metal ion taken up by microorganisms frequently exhibits two-stage kinetics (Fig. 19.4). The first stage is a very rapid, passive, reversible and metabolism-independent reaction called physical adsorption or ion exchange at the cell surface. During this rapid stage, microorganisms can adsorb metal ions onto their surfaces by many processes. This kind of adsorption is called "biosorption". The second step is slower, active and metabolism-dependent, called chemisorption with intracellular metal ion accumulation. Dead cells accumulate metal ions to the same or greater extent than living cells.

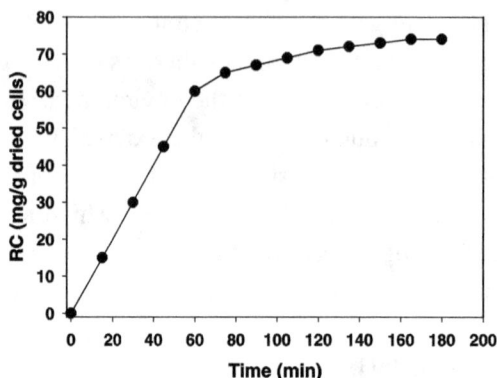

Fig. 19.4. Kinetics of metal ion removal by microbial biomass. RC is the metal ion removal capacity of the microbial cells. From 0–60 minutes, a rapid metal ion uptake phase, mainly by adsorption by microbial biomass is observed. Then from 60–120 min, slow metal ion uptake phase by intracellular accumulation is proceeded.

Biosorption is largely attributed to the presence of metal ion-coordinating ligands such as phosphoryl, sulhydryl, carboxyl, hydroxyl and amino groups on the cell surface and in membrane proteins. In addition, biosorption involved various processes such as adsorption, coordination, complexation, chelation, ion exchange, precipitation or crystallization of metal ions. These processes can occur either at the microbial cell surface or near the cells, and certain insoluble metal ions may become physically entrapped in extracellular polymers or in cavities within the cell wall. Moreover, analytic observations show those surface metal ion concentrations far exceed the expected stoichiometry of metal ion bound per reactive chemical site within the cell wall. Electron microscopy clearly establishes that metal ion precipitates within and on the cell walls are common (Fig. 19.2), and a two-step mechanism to account for this observation was invoked. Firstly, metal ions interact in a stoichiometric manner with accessible reactive cell wall groups. Then, these sites nucleate the growth of a metal ion precipitate.

Metabolism-dependent intracellular metal ion accumulation generally occurs via monovalent or divalent cation transport system and in response to a transmembrane electrochemical potential generated by plasma membrane-bound H^+-ATPases. Metabolism-dependent intracellular metal ion accumulation may result in higher levels of metal ion uptake than those attributable to adsorption processes alone. It is also sensitive to inhibition by low temperature, metabolic inhibitors, and the absence of an energy source and low light intensities in photosynthetic organisms. Moreover, this process is irreversible in many cases and requires destructive treatments for metal ion recovery. The intracellular uptake of toxic metal ions is primarily by pathways existing for the uptake of metabolically necessary metal ions.

(5) Adsorption isotherm

The adsorption of metal ions on microorganisms is a reversible phenomenon and can be represented by two adsorption isotherms — Langmuir isotherm and Freundlich isotherm.

(a) Langmuir isotherm

The Langmuir isotherm, which is a widely accepted monolayer adsorption model, is often used to describe the adsorption phenomenon of various biosorbents. This model is based on the following assumptions: (1) all the sites on the adsorbent surface are identical; (2) the energy of adsorption is constant (homogeneous); (3) there is no migration of adsorbate molecules in the adsorbent surface; (4) there is no interaction between the adsorbed molecules; (5) adsorption is only limitd to a monolayer. The Langmuir model can be described as:

$$q_e = \frac{bC_e q_{max}}{(1 + bC_e)} \qquad (19.1)$$

in which q_e is the amount of solute adsorbed per unit dry weight of adsorbent at concentration C_e (mg metal ion/g biosorbent or mmol metal ion/g biosorbent); C_e represents the equilibrium concentration of metal ion in bulk aqueous phase after adsorption (mg/L or mmol/L); q_{max} is the theoretical maximum uptake of solute per unit dry weight of adsorbent (mg metal ion/g biosorbent or mmol metal ion/g biosorbent); b is the adsorption affinity constant related to energy of adsorption (L/mg or L/mmol) Twoconvenient linear forms of the Langmuir equation are:

$$\frac{C_e}{q_e} = \frac{1}{bq_{max}} + \frac{C_e}{q_{max}} \qquad (19.2)$$

or

$$\frac{1}{q_e} = \frac{1}{bq_{max}} + \frac{1}{bC_e \, q_{max}}. \qquad (19.3)$$

A plot of C_e/q_e against C_e (or $1/q_e$ against $1/C_e$) gives a straight line with a slope of $1/q_{max}$ and an intercept of $1/bq_{max}$ (or with slope of $1/bq_{max}$ and intercept of $1/q_{max}$). Hence the adsorption parameters, q_{max} and b, can be evaluated and used to compare the adsorption behavior quantitatively in different adsorbate-adsorbent systems.

It is worth noting that when the amount of adsorption is small such that $bC_e \ll 1$, the Langmuir equation would give a linear adsorption relationship as

$$q_e = q_{max} \, bC_e. \qquad (19.4)$$

On the other hand, when the amount of adsorption is large such that $bC_e \gg 1$, the Langmuir equation would become

$$q_e = q_{max}. \qquad (19.5)$$

(b) Freundlich isotherm

The Freundlich isotherm is another widely accepted model for describing the monolayer adsorption. The empirical Freundlich equation assumes that the surfaces of adsorbents are heterogeneous and therefore the adsorbents are expected to have heterogeneous energis for adsorbing adsorbates. This equation can be expressed as

$$q_e = kC_e^{1/n} \qquad (19.6)$$

in which q_e is the amount of solute adsorbed per unit dry weight of adsorbent at concentration C_e (mg metal ion/g biosorbent or mmol metal ion/g biosorbent); C_e represents the equilibrium concentration of metal ion in bulk aqueous phase after adsorption (mg/L or mmol/L); k and n are empirical constants and indicative of adsorption capacity and intensity, respectively. To simplify the derivation of k and n, the

Freundlich equation is often linearized by taking the natural logarithm on both sides of the equation as

$$\ln q_e = \ln k + 1/n \ln C_e. \qquad (19.7)$$

A plot of $\ln q_e$ against C_e would yield a straight line with a slope of $1/n$ and an intercept equal to the value of $\ln k$ for $C_e = 1$ ($\ln C_e = 0$). The magnitude of n is an indicator of system suitability, with values of $n > 1$ (or $1 < 1/n < 0$) representing favorable adsorption conditions, while the intercept $\ln k$ is related to the capacity of the adsorbent for the adsorbate.

The Freundlich equation generally agrees quite well with the Langmuir equation and experimental data over moderate ranges of concentration, C_e. However, unlike Langmuir equation, it does not become linear particularly at low concentrations but remains convex to the concentration axis.

19.1.4.3. Methods of recovery of metal ion from metal-loaded biosorbent

Metal ions loaded on biosorbent may be recovered by a destructive method in which the biosorbent may either be destroyed to release the adsorbed metal ions, or a non-destructive method in which the biosorbent can be regenerated for future use by treating mildly with chemicals to desorb the metal ions. The choice of methods is principally determined by the relative costs of the two recovery methods. Moreover, the metal ion uptake mechanism of the biosorbent also limits the choices. For an economically efficient biosorption operation, it is usually mandatory that the biosorbent can be conserved and regenerated. A pre-requisite for biosorbent regeneration is the non-destructive removal of metal ion from loaded biosorbent.

Many studies try to recover loaded metal ions from biosorbents. The choice of eluents for metal ion recovery and regeneration of biosorbent should be determined mainly by (1) the metal ion recovery efficiency of the eluents, (2) the cost of the eluents, and (3) the effect of the eluents on metal ion removal capacity of the biosorbent regenerated. Recovery of the loaded metal ion could be achieved by the use of an appropriate elution solution that can effectively strip the loaded metal ions from the biomass. It is desirable that the elution is complete or the irreversibly loaded metal ion could be kept to a minimum. In addition, it is desirable that the least possible damage occurs to the adsorption properties of the biomass so as to allow the reuse of the biomass in subsequent adsorption-desorption cycles.

Various kinds of eluents are used for the desorption of adsorbed metal ions from cell surfaces. Examples include eluting Cs^{2+}-laden fungal pellets with 1 M HNO_3 or 1 M $Na_2CO_3/NaHCO_3$ solutions, Au^{3+}-laden algal biomass with mercaptoethanol in 0.05 M sodium acetate (pH = 6), U-laden *Penicillium* biomass with 0.1 M EDTA (pH = 6 and 4) and sodium carbonate/sodium bicarbonate (15/5 and 30/20 g/L), Cu^{2+}-laden fungi (*Ganoderma lucidum*) with EDTA and 0.1N HCl solutions, Au^{3+}-laden immobilized cells of *Persimmon tannin* with 1 M thiourea (pH = 3) and Ni^{2+}-laden immobilized bacterial cells *Enterobacter* sp. 4-2 with 0.5 M citrate buffer (pH = 4).

19.1.4.4. *Immobilization of metal-adsorbing microbial cells*

For more rigorous industrial applications, the use of immobilized microbial cells is preferable. Immobilized microbial cells have many advantages over freely suspended cells. These include (1) better capability of repeatable reuse of the biomass, (2) easy separation of cells from the reaction mixture, (3) high biomass loading, (4) minimal clogging in continuous flow systems, (5) particle size can be controlled and high flow rates achieved with or without recirculation, (6) prevent cell wash and allow a high cell density to be maintained in a continuous reactor at any flow rate, (7) improved catalytic stability of immobilized cells, (8) provide resistance to shear for shear-sensitive cells, (9) microorganisms more inert to microbial contamination and degradation, and (10) operational stability is generally high. Moreover, immobilized microbial cells usually had good mechanical stability. As a result, reactor with immobilized microbial cells can reach a high metal loading capacity and achieve a high efficiency for metal removal.

In order to evaluate the suitability of a biosorbent (e.g. immobilized cells) for metal ion removal, it is better to select the matrix that can fulfill many of the following criteria: (1) the active immobilized biosorbent should be produced at a low cost and should be reusable; (2) particles size, shape and mechanical properties should be suitable for use in a continuous-flow system in completely mixed, packed or fluidized-bed-reactor configurations; (3) uptake and release of the metal ion should be efficient and rapid; (4) separation of the biosorbent from solution should be cheap, efficient and rapid; (5) the sorbent should be metal ion-selective; and (6) regeneration of metal ions from the sorbent should preferably be metal ion-selective and economically feasible, and the sorbent should be in a physical state that can be used. Since the use of magnetite-immobilized microbial cells have many advantages over the traditional technologies such as freely suspended cells, and possess the above criteria as a biosorbent on metal ions removal, it is an attractive method for sewage treatment.

Many studies have shown that using microorganisms (especially bacteria) that adsorbed onto magnetites (Fe_3O_4) can be applied for removal and recovery of useful substances from wastewater. Cells of *Rhodopseudomonas sphaeroides* were immobilized on magnetites to remove pesticides such as lindane, 2,4-dichchlorophenoxyacetic acid (2,4-D) and 2,4,5-trichlorophenoxyacetic acid (2,4,5-T) from wastewater. The removal of chlorinated hydrocarbons such as p,p'-DDT, aldin, heptachlor, dieldrin, α-hexachlorocyclohexane and α-hexachloro-cyclohexane by immobilized bacterial cells have been reported. Cells of a Mn^{2+}-oxidizing bacterial cells immobilized onto magnetites removed Mn^{2+} from water. Although there are numerous reports on the removal of metal ions by microbial biomass, mainly by biosorption, the application of biosorption using microbial biomass to remove metal ion from aqueous environment has been hindered by the difficulty of recovering metal ion laden microbial cells due to the minute size. Centrifugation and filtration are impractical due to the high cost of the operation. Immobilization of microbial cells onto appropriate matrix is the most adopted method to solve the above-mentioned separation problem. Matrices such as alginate, carrageen and even acrylamide have been used to immobilize microbial cells for adsorption of metal ions. Their drawbacks are that these immobilized cells have to be used in a packed bed reactor and the hydraulic problem reduces the applicability of the reactor to treat large volume and high flow rate solution containing metal ions. In order to solve the problem, cells of *Enterobacter* sp. 4-2 were immobilized onto magnetites and used in a fluidized bed reactor to remove Ni^{2+}. There was no hydraulic problem

Table 19.2. The removal capacity of Cu^{2+} in concentrated metal ion solution by magnetites, free and magnetite-immobilized bacterial cells of *Pseudomonas putida* 5-x (So 1992).

Adsorbent	Removal capacity (mg/g)[a]
Free bacterial cells	46.43 ± 0.72
Magnetite-immobilized cells	48.70 ± 0.25
Magnetites	0.45 ± 0.09

[a]The experimental conditions: initial concentration of Cu^{2+} before precipitation = 30 mg/L, pH for precipitation = 7.5, equilibrium time for precipitation = 40 min, acid dissolution of metal sludge = 5 mM H_2SO_4, biosorbent concentration = 50 mg/50 mL, pH for adsorption = 5.5, retention time for adsorption = 30 min and temperature = 23 ± 2°C. Data represent means ± standard deviations from triplicates.

due to the design of the reactor and the removal efficiency of the reactor was very high (Table 19.2) and suggested this technology can be used to remove metal ions from aqueous medium at a fast flow rate.

19.2. Removal of Organic Pollutants from Industrial Wastewater

Among all the wastes produced, industrial wastes demand the largest attention because they have been long regarded as being of the greatest diversity and toxicity, and are mostly recalcitrant. For many countries where the textile industry is prosperous, the industry presents a major water pollution problem since this industry not only exhausts a large quantity of clean water for processing, but also contributes a significant amount of wastewater. Figure 19.5 shows the discolored area near the wastewater outlet of a textile factory in Hong Kong. Synthetic dyes are toxic to aquatic organisms. The increasing potential hazard of synthetic dyes in textile wastewater has made them a major pollution problem that draws serious concerns.

In 1992, approximately 10,000 different synthetic dyes are used industrially, and over 700,000 tonnes of these dyes are produced annually worldwide. The loss of dyes in the effluent during the dyeing process

Fig. 19.5. A discolored area next to a textile factory in Hong Kong. The area was heavily polluted with wastewater discharged from the factory.

was about 10 to 15%. In other words, 84,000 tonnes or more of dyes were discharged into our environment in 1992. These hundred thousand tonnes of dye invade our environment via the air, water and soil media as synthetic dyes are recalcitrant to biodegradation and persistent in the environment. For example, Solvent Red 1, an azo dye, and Solvent Yellow 33, an quinoline dye, were transformed with half-lives of a few days and months, respectively under anaerobic condition which is favorable for dye reduction.

Obviously, the first impact of such great dye contamination is the coloring of our aquatic and terrestrial environments which are aesthetically unacceptable. However, the main concerns regarding these pollutants are the influence and toxicity of the dye-containing materials to the ecosystem and, even more importantly, to the living organisms. As one of the properties of dyes is their high color value, the treated wastewater containing non-biodegraded dyes causes the receiving water-courses to be highly colored, which would normally observable at about 1 mg/L level which is unacceptable on aesthetic grounds. The color of dye not only affects the aquatic environment aesthetically but also the growth and physiological activities of aquatic flora and fauna by reducing light penetration into the water. It has been reported that some commonly used synthetic dyes and their metabolites or derivatives were tested to be mutagenic using *Salmonella*-mammalian microsome mutagenicity test. The large number of synthetic dyes and their high potential for toxicity have made this class of compounds one of the most extensively reviewed and regulated under the Toxic Substances Control Act (TSCA) new chemical programs. The pollution problems arising from dye wastes therefore deserve more efforts to handle and deal with these wastes being discharged before into the aquatic environment.

19.2.1. *Azo dyes*

Among the many classes of synthetic dyes used in various industries, azo dyes are the largest and the most versatile, and play a prominent role in almost every type of application. Azo dyes are a group of compounds characterized by the presence of one or more azo groups (-N=N-) in association with one or more aromatic systems. Azo dyes are cost effective because of their ease and versatility of synthesis, and relatively higher tinctorial strength compare to other synthetic dyes. The versatile applications of azo dyes have initiated many studies on their adverse effects on human and the environment as azo dye-containing effluents are discharged without suitable treatment. The increase in the occurrence of intestinal cancer in highly industrialized society has been related to the increasing use of azo dyes. Some European countries have banned the application of azo dyes in the late 1990s.

19.2.2.   Treatment of industrial effluent containing azo dyes

(1) Physico-chemical methods

The contamination of the environment with azo dyes during the 20th century has overwhelmed the natural cleaning capacities of micro-organisms. Scientists had put in much effort in finding effective methods to solve the problem. More effective though less sensitive methods were developed and tested for dye removal. The feasibility of several methods was evaluated, including chlorination, ozonation, adsorption, flocculation, coagulation and Fenton's oxidation.

(2) Biological methods

(a) Activated sludge process

In the past, municipal wastewater treatment systems were mainly used for the purification of textile mill wastewater. These systems, however, depended mainly on biological activity and were mostly found to be inefficient in the removal of the more resistant, recalcitrant synthetic dyes. In the synthetic dye manufacturing industry, synthetic dyes produced have to be more and more resistant to sunlight, washing and microbial action in order to be successful in the commercial market. It is not surprising that most studies on the biodegradation of synthetic dyes produce negative result when synthetic dyes are designed to resist this type of treatment. In a study of the fate of 18 azo dyes in the "activated sludge" processes, 11 dyes were found to pass through the "activated sludge" process substantially untreated, 4 were significantly absorbed on the surplus "activated sludge" while only 3 were apparently degraded.

Some azo dyes do show a tendency to be adsorbed onto "activated sludge", but acid and reactive dyes were only slightly adsorbed. However, these studies were done on "activated sludge" plants that were not regularly treating dyeing effluent. It seems likely that in plants that continuously or regularly treat dye, the adsorption sites on microbial cells will be rapidly occupied if the dye is strongly adsorbed. Renewal of adsorption sites will be by creation of new sludge flocs, desorption of the dye or biodegradation of the dye. With normal sludge ages, the creation of new flocs is too slow to remove dyes adequately, and desorption leads to resolubilization of color. Biodegradation of dyes under the aerobic conditions in an "activated sludge" mode wastewater treatment plant is slow. Long adaptation periods are required and the enzymes produced appear to be highly specific. The net result of increased dye loads and poor dye removal on conventional wastewater treatment is an increased level of color in the receiving watercourse.

(b) Biodegradation by isolated microorganisms

Rat gut bacterial suspensions (with starved *Proteus* sp. suspension or its cell-free extract) decolorized azo dyes such as *m*-Methyl Orange, Neoprontosil, Tartrazine, Yellow 2G, Amaranth, Proceau SX, Fast Yellow, Naphthalene Fast Orange 2G and Sunset Yellow FCF. Thin-layer chromatographic study indicated that these azo dyes had been reduced

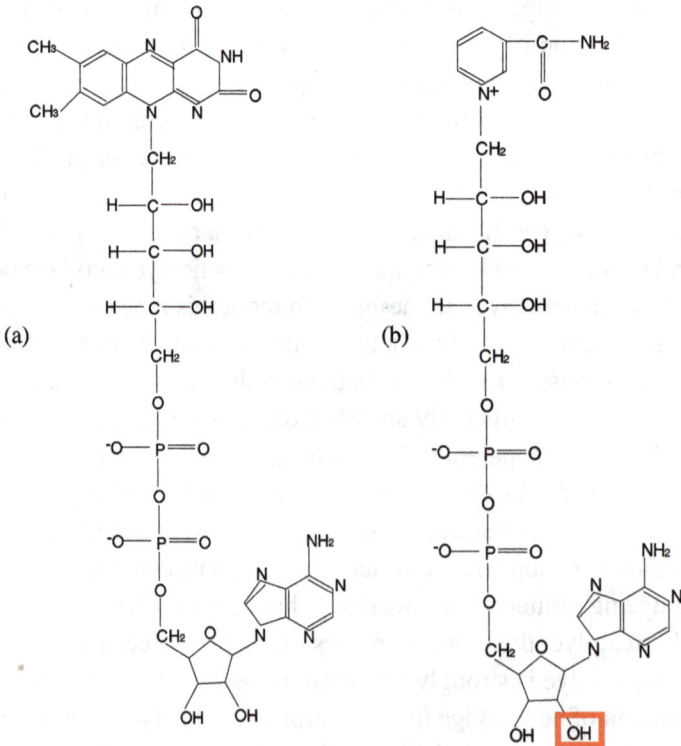

Fig. 19.6. Structures of (a) flavin adenine dinucleotide (FAD); and (b) nicotinamide adenine dinucleotide (NAD(P)). The NADP is the marked hydroxyl group (OH), is esterified with phosphate (PO_4^{3-}).

to amines. The enzyme of *Proteus* sp. that reduced Tartrazine was a flavoprotein requiring reduced forms FAD (flavin adenine dinucleotide, Fig. 19.6(a)) and NAD(P) (nicotinamide adenine dinucleotide phosphate, Fig. 19.6(b)) for reduction to take place.

A study was conducted using the established NADH-generating system incorporating in the standard assay procedure to study the azo-reductase activity of cell-free extracts of *Streptococcus faecalis* under optimal conditions. It was found that flavins (e.g. FAD) added to the preparations were initially reduced enzymatically by the crude enzymes in the presence of NADH-generating system. They rapidly and non-enzymatically reduced azo dye, Red 2G, into the primary amines. The reduced flavins were then oxidized in the process. Without $FADH_2$, NADH reduced Red 2G very slowly if at all. This azo reduction process would more likely be a double two-electron transfer via the hydrazo intermediate (R-NH-HN-R') under anaerobic condition:

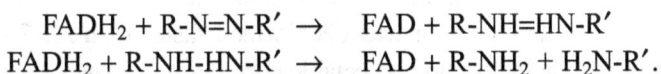

$$FADH_2 + R\text{-}N\text{=}N\text{-}R' \rightarrow FAD + R\text{-}NH\text{=}HN\text{-}R'$$
$$FADH_2 + R\text{-}NH\text{-}HN\text{-}R' \rightarrow FAD + R\text{-}NH_2 + H_2N\text{-}R'.$$

Since there was no enzyme-substrate complex formed between the bacterial azo reductase and Red 2G, it showed that FAD was the true substrate of the bacterial azo reductase, being reduced by flavoproteins, and then reduced the dyes non-enzymatically. This indirect mechanism of azo reduction would explain why both bacterial and mammalian hepatic preparations could reduce a wide range of azo dyes. They also found that under aerobic conditions, the azo dye acts as another electron shuttle for reduced flavin.

(i) Bacteria
Studies that concern the fate of azo dyes in the "activated sludge" process showed that most azo dyes were mainly and significantly removed by adsorption on the "activated sludge" flocs and only a very few appeared to be biodegraded. As early as 1937, decolorization of azo dyes in spoiled dairy products was attributed to predominant lactic acid bacteria, which turned the azo dyes into amines. Later, more studies reported that reduction of a large variety of azo dyes occurred extensively among intestinal microflora. Appreciating the versatile abilities of microorganisms

in degrading and mineralizing nearly all organic matters in the nature, more studies were encouraged to determine the feasibility of using azo dye-degrading microorganisms to treat dye wastes. *Bacillus subtilis*, which was selected for having faster and stronger decolorization rate of *p*-aminoazobenzene (PAAB), an azo dye hardly degraded by activated sludge and even decreases the activity of the activated sludge, converted PAAB into aniline and *p*-phenylenediamine. *Pseudomonas cepacia* 13NA, isolated from soil of draining trenches at a dyeing factory, also degraded PAAB. The bacterium metabolized the dye better under poor nutrient conditions. *P. cepacia* 13NA also decolorized a number of synthetic dyes and their decolorization abilities were in the order Acid Orange 20 (AO20) > *p*-aminoazobenzene (PAAB) > Acid Red 88 (AR88) > Acid Orange 7 (AO7) > Acid Orange 12 (AO12) > Direct Yellow 4 (DY4) > Direct Red 28 (DR28). It appeared that when hydroxyl group was in the *ortho*-position, such as that in AO7, compared with AO20 that has the hydroxyl group at the *para*-position, decolorization was facilitated. Also, AR88 that has a naphthalene ring was decolorized faster than AO7 that has a benzene ring. There was no appreciable correlation between decolorization ability of synthetic dyes and their octane-water partition coefficient. This may indicate that the decolorization ability of synthetic dyes is not related to their permeability through bacterial cell wall, but to their molecular weight of synthetic dyes.

Growing cells, intact non-growing cells, cell-free extract and purified enzymes of *Pseudomonas* sp. S-42, which was isolated from "activated sludge", could decolorize azo dyes Diamira Brilliant Orange RR, Direct Brown M and Eriochrome Brown R under the conditions of pH 7.0 and temperature of 37°C. Oxygen inhibited the decolorizing activities of cell-free extract and purified enzymes. Studying the structure-activity relationship was difficult for intact cells due to the structure-dependent permeability of azo dyes through cell membranes. Cell-free extract of *Pseudomonas stutzeri* degraded several azo dyes and the results of the study indicated the redox potential and hydrophobicity of dyes were important to understanding the relationship between the ease of degradation of dyes and their structures. Intact cells and cell-free extract of *P. stutzeri* IAM 12097, *P. cepacia* 13NA and *Bacillus subtilis* IFO 13719 decolorized several sulfonated and non-sulfonated azo dyes. These bacteria showed very different decolorization abilities towards the dyes due to different specificity of their azo reductases, different permeabilities of their cell membranes and different structures of the dyes. Azo dye reduction is microorganism-specific and dye-specific.

Pseudomonas sp. K24 and *Pseudomonas* sp. KF46 could specifically utilize azo dyes Orange I and Orange II respectively through adaptation in chemostat culture. The azo reductases of the two bacterial strains, Orange I azo reductase and Orange II azo reductase, were found to evolve from a common precursor protein of the bacterium from which the above two bacterial strains derived. Wong and Yuen (1993) studied degradation (decolorization) of an azo dye, Methyl Red (MR), by *Klebsiella pneumoniae* RS-13 isolated from activated sludge (Fig. 19.7). The initial degradation of Methyl Red by *K. pneumoniae* RS-13 was a reductive cleavage of the azo bond and produced an aromatic acid (2-amino-benzoic acid = ABA) and a comparative toxic aromatic amine (N,N′-dimethyl-*p*-diphenylene-diamine = DMPD) (Fig. 19.8). DMPD was proven to be even more toxic than the parent dye. However with longer incubation in medium containing glucose, *K. pneumoniae* RS-13 could degrade DMPD into non-toxic intermediate, and the intermediate, like ABA, would be used by the cells for other biomolecule synthesis.

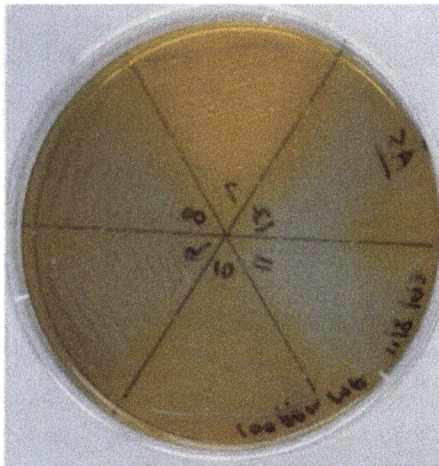

Fig. 19.7. Decolorization of Methyl Red by *Klebsiella pneumoniae* RS-13 under aerobic condition.

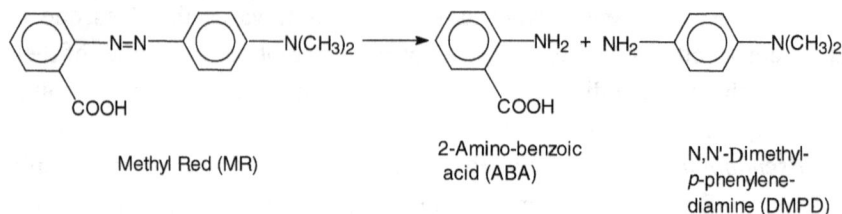

Fig. 19.8. Initial degradation of Methyl Red by *Klebsiella pneumoniae* RS-13 under aerobic condition.

(ii) Fungi

The feasibility of using fungal system to degrade synthetic compounds has been extensively studied. White-rot basidiomycetes, especially *Phanerochaete chrysosporium*, received most of the attention due to their versatile and rather non-specific aerobic lignin-degrading ability. The lignin-depending ability is initiated during secondary metabolic (idiophasic) growth, and could degrade synthetic organic compounds with structures similar to that of lignin. Lignin is an optically inactive, random phenylpropanoid polymer, which is relatively recalcitrant to biodegradation in the nature.

Many *Streptomyces* spp. (ligninocellulolytic) and *P. chrysosporium* BKM-F-1767 (ATCC 24725) (ligninolytic) degraded commercial azo dyes Acid Yellow 9, sulfanilic acid, synthesized azo dyes 1 and 2. The latter two were synthesized by conjugating guaiacol molecule (2-methoxyphenol) onto Acid Yellow 9 and sulfanilic acid via azo-linkage. Five of six *Streptomyces* spp., which were capable of assimilating vanillic acid — a compound having the same ring substitution pattern (4-hydroxy-3-methoxy) as guaiacol, could significantly decolorize the two synthesized azo dyes. These *Streptomyces* spp. could not decolorize sulfanilic acid and Acid Yellow 9. Moreover, *P. chrysosporium* could almost completely decolorized synthesized azo dyes 1 and 2 whereas it only could decolorize sulfanilic acid and Acid Yellow 9 to a limited extent. These results indicated that the degradation of sulfanilic acid and Acid Yellow 9 by *Streptomyces* spp. were rendered feasible by the bacteria which start the assimilation by attacking the guaiacol substituent first, while the linkage of guaiacol molecule to these two compounds increased their susceptibility to the degradation by *P. chrysosporium*. The newly synthesized compounds are rendered more easily biodegradable by selective linkage with readily degradable substituents into the compounds.

By comparing the degrading ability of *P. chrysosporium* BKM-1667 (ATCC24725) with that of *Streptomyces chromofuscus* A11 (ATCC55184) towards sulfanilic acid, sulfonated and non-sulfonated azo dyes, these dyes were coupled with guaiacol molecules, aromatic ring or naphthol group. *P. chrysosporium* BKM-1667 demonstrated a greater ability than *S. chromofuscus* A11 in mineralizing azo dyes. *P. chrysosporium* BKM-1667 could mineralize (completely degrade) all the sulfonated azo dyes, and the substitution pattern did not significantly influence the susceptibility of the synthetic dyes to degradation. In contrast, *S. chromofuscus* A11 was unable to mineralize aromatics with sulfo groups except those coupled with guaiacol molecule.

It seems that the lignin-like guaiacol structure attached to sulfonated azo dye was a must to initiate degradation of *S. chromofuscus* A11 and suggested that this bacterial enzymatic systems responsible for degradation may be somehow more selective to lignin-like structures than that of *P. chrysosporium* BKM-1667. A white-rot fungus, *Geotrichum candidum* CU-1, was isolated and found to be able to degrade a number of water soluble azo dyes including Procion Red MX-5B and Methyl Red under aerobic condition (Fig. 19.9). The mechanism of degradation of these azo dyes seems to be an initial oxidative cleavage of the aromatic rings of the dyes followed by further degradation of the intermediates into very "small and simple" compound(s) that could not be resolved by high performance liquid chromatography (HPLC) and gas chromatography-mass spectrometry (GC-MS) analyses. In addition, the "small and simple compounds" had no detectable toxicity towards Microtox® test (a bacterial enzymatic fluorescence test developed by Microbics Corporation).

(iii) Algae

Chlorella pyrenoidosa, *C. vulgaris* and *Oscillateria tenuis* degraded 31 azo dyes at 20 mg/L over 96-hour incubation. *C. vulgaris* decolorized 29 out of 31 azo dyes with the extent varying from 5 to 100%. The extent of decolorization of dyes seemed to be related to their molecular structures.

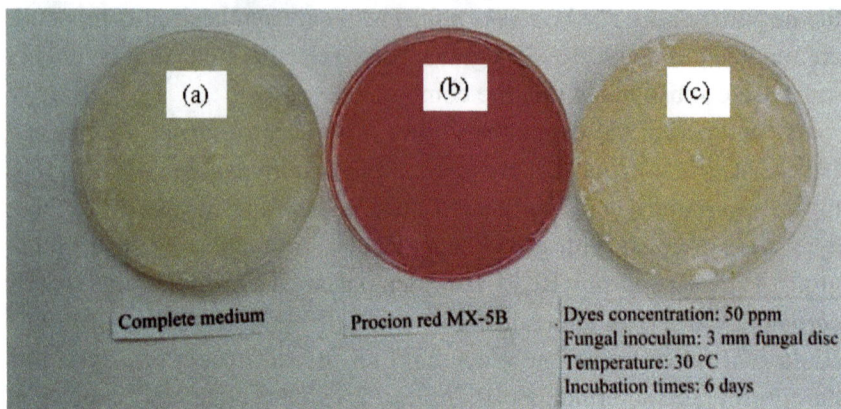

Fig. 19.9. Degradation of Procion Red MX-5B by *Geotrichum candidum* CU-1 under aerobic condition. (a) Complete medium with *G. candidum* CU-1; (b) complete medium with Procion Red MX-5B (50 mg/L); and (c) complete medium with Procion Red MX-5B (50 mg/L) and *G. candidum* CU-1 (part of the fungal mycelia were removed after full growth of the fungus to show the decolorization of the azo dye).

Dyes with an amino or a hydroxyl group could be readily decolorized. These groups also could counteract the inhibitory effect of sulfonyl group on reduction. Substitution with methyl, methyoxyl, nitro or sulfo group made dyes hard to decolorized. All three algae were found to be able to effectively degrade aniline products into simple inorganic materials. The absence of inorganic carbon and nitrogen in the medium facilitated the algae to utilize Erichrome Blue SE and Black T, azo dyes, as sole sources of carbon and nitrogen. The addition of NADH or NADPH increased the algal azo dye reduction. These observations demonstrated that NADH or NADPH could act as electron donors for the algal azo reductases. Moreover, algae that had been cultured with 10 mg/L of dye displayed higher reduction activity of that dye. This result suggested that azo reductase is inducible.

It is believed that algae serve as an oxygen source for the aerobic action of bacteria, and have a direct effect on the degradation of azo dyes.

(c) Biosorption

Almost all biological materials can be used as a biosorbent to adsorb different materials: peat, wood-shaving, biomass, starch, moss, bagasse, red mud, rice husk, biogas residue, orange peel, activated sludge, wood charcoal, coconut husk, and banana pith. It is because nearly all cell structures are basically made up of protein, fatty acid and mono/polysaccharide. Protein constructed from a set of 20 different amino acids. Each amino acid contains at least one amino (NH_2) and one acidic group (COOH). Fatty acid composes of an acid group (COOH) and a saturated or unsaturated branched or unbranched hydrocarbon chain. Monosaccharide consists of aldo (CHO) or keto (CO) group and hydroxyl group (OH) on the carbon skeleton. Thus, biological compounds contain many potential charged binding sites for different adsorbates. Hence, the success of biosorption is dependent on the isolation and selection of the appropriate biological material.

As mentioned previously, synthetic dyes are poorly degraded in biological waste treatment systems because of their recalcitrance. The plentiful discharge of dye-containing industrial effluents into the wastewater treatment systems has initiated many studies on the fate of these dyes and the treatment efficiency of activated sludge process on these substances.

It has been reported that thirty-one water soluble and water slightly soluble azo dyes at 50 mg/L were degraded by 1,000 mg/L of sludge in 100 mL culture medium under shaking condition. The biodegradation was limited to azo reduction and the decolorization was readily with hydroxyl or amino group, not readily with methyl group and hardly with hydrogen, chloro, methoxyl and nitro group. In addition, it was found that carboxyl group itself, similar to sulfonyl group, decreased reduction because of its hydrophobicity. The substitution of electron-withdrawing groups (SO_3H and SO_2NH_2) did not increase the reduction.

The low BOD:COD ratio (<0.1) of the majority of textile effluents leads to low biodegradability of azo dyes that are partially removed in biological treatment by adsorption onto the activated sludge flocs. This adsorption includes exchange adsorption, van der Waals forces and hydrogen bonds. The adsorption of basic dyes increased with pH and the opposite effect is observed for acidic dyes. Dye adsorption is mainly dependent on type, chemical composition, structure of dyes, structure of adsorbent surface, pH and salinity. Bacteria and fungi are the microorganisms that are always selected as biosorbents because they grow very rapidly and are easy to grow. Moreover, the surface area to volume ratio of these microorganisms is relatively higher than the other living organisms.

(i) Bacteria

Bacteria such as *Aeromonas* sp., *Rhodococcus* sp. and actinomycetes such as *Streptomyces* sp. have been reported as good biosorbents for textile dyes. The adsorption mechanisms of bacteria and actinomycetes are mainly attributed to the microbial cell wall. Many bacterial cells exert adsorption during the stationary phase in which exopolysaccharides are produced. Exopolysaccharides are primarily composed of carbohydrates such as D-glucose, D-galactose and D-mannose which are produced in the stationary phase. Moreover, exopolysaccharides are polyanionic in nature. It may explain why textile dyes, especially reactive dyes (cationic dyes), can be adsorbed by the bacterial cells. There are some exceptions, in which the bacterial cells perform better reactive dye adsorption in the earlier growth phase as in *Aeromonas* sp. These may indicate different adsorption sites and binding mechanisms between different dyes and bacterial strains. Lai (1997) selected a bacterial strain, *Pseudomonas* sp. K-1, which was isolated from a sludge sample generated by a textile factory. The bacterium adsorbed a large amount of an azo dye, Procion Red MX-5B (Fig. 19.10) and no decolorization of the dye was observed even after incubating the dye with the bacterium on agar plate for about a week (Fig. 19.11). This result suggests that the bacterium only adsorbed the azo dye but had no degradation ability for this azo dye.

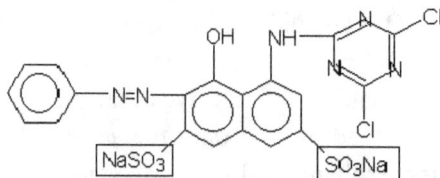

Fig. 19.10. Structure of a sulfonated water-soluble monazo dye, Procion Red MX-5B. The sulfonate groups are marked.

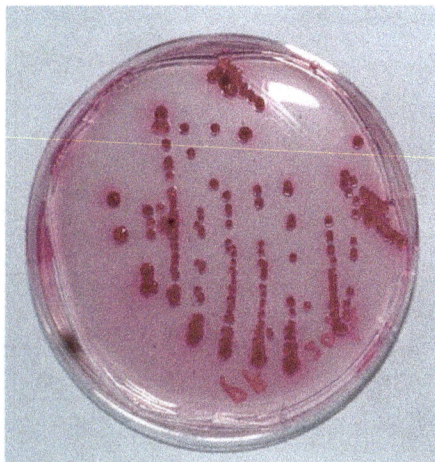

Fig. 19.11. Colonies of *Pseudomonas* sp. K-1 stained by Procion Red MX-5B on a minimal medium (MM) agar plate.

(ii) Fungi

Fungus such as *Myrothecium verrucaria* was isolated from soil and reported as an excellent dye-binding microorganism. Fungal biomass was as much as 100-fold more effective than "activated sludge" in the removal of dyes. The main binding materials on fungal cell wall are chitin and chitosan. Chitin is one of the most abundant organic materials in the nature. It occurs in fungal cell wall as the principal fibrillar polymer and in animal, particularly in mollusks, crustaceans and insects, as the constituent of the exoskeleton. The structure poly[β-(1-4)-2-acetamido-2-deoxy-D-glucopyranose] of chitin (Fig. 19.12(a)) is similar to cellulose (Fig. 19.13), except that the C2-hydroxyl group of cellulose is replaced by an acetamido group. In addition, chitin structure is also similar to murein in that it is the main constituent of bacterial cell wall. The principal derivative of chitin is chitosan (Fig. 19.12(b)). Chitosan is produced by alkaline deacetylation of chitin. It also occurs naturally in some fungi but the occurrence is much less widespread than that of chitin. However, in some fungal species, such as *Mucor rouxii*, the composition of chitosan is more than that of chitin.

Chitin polymer present in fungal cell wall is polycatioinc in nature due to the amino (NH) and carbonyl (CO) groups substituted to the hydroxyl

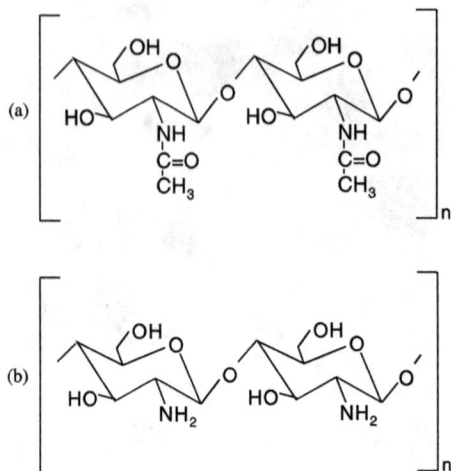

Fig. 19.12. Structure of (a) chitin and (b) chitosan.

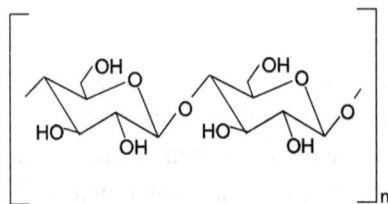

Fig. 19.13. Structure of cellulose.

(OH$^-$) groups. Because of the cationic nature of chitin polymer rather than anionic nature of cellulose or exopolysaccharides in bacterial cell wall, their adsorption mechanisms are different. Using a reactive dye as an example, the adsorption sites are the electrophilic reactive group and the nucelophilic sulfate (SO$_3^-$) group. Under alkaline conditions, the dye molecules expose the electrophilic reactive group (positively charged group). Therefore, cellulose that exposes the nucleophilic OH$^-$ group (negatively charged group) can adsorb the dye molecules and form covalent bond, while the chitin polymer (positively charged) will repel the dye molecules. On the contrary, under acidic conditions, since the dye molecules cannot expose the electrophilic reactive group, adsorption on cellulose surface decreases, while the chitin polymer (the amount of protonated amine (NH_3^+) groups increases under acidic condition) has more sites for sulfate groups (SO_3^-) binding. The corresponding reaction is illustrated below:

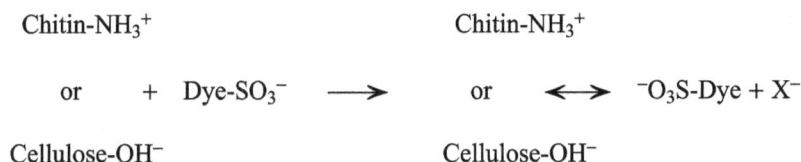

Chitin-NH_3^+ Chitin-NH_3^+

or   +   Dye-SO_3^- $\longrightarrow$ or $\longleftrightarrow$ ^-O_3S-Dye + X^-

Cellulose-OH^- Cellulose-OH^-

Therefore, chitin polymer adsorbs dye molecules more effectively under acidic condition. Chitosan is more effective in dye adsorption than chitin. Because of the high dye adsorption capacity of chitin/chitosan, they are extensively studied as a dye adsorbent. When azo dyes had high sulfonic acid substitution, poor or no adsorption of the dye by the microbial cell or cell by products would occur. This also limits the chance of aerobic biodegradation.

References

1. Lai KM. *Integration of Adsorption and Biodegradation of Azo Dyes*. M. Phil. thesis, The Chinese University of Hong Kong, Hong Kong, 1997.
2. So CM. *Removal of Copper Ion (Cu^{2+}) from Industrial Effluent by Immobilized Microbial Cells*. M. Phil. thesis, The Chinese University of Hong Kong, Hong Kong, 1991.
3. Wong PK and Yuen PY. *Microbial Degradation of Methyl Red and Its Cleavage Products*. M. Phil. thesis, The Chinese University of Hong Kong, Hong Kong, 1993.

Questions for Thought

1. What is the definition of a heavy metal?
2. List all toxic metal ions.
3. What are the major source(s) of metal ions in aquatic environment?

4. What is the priority of treatment of toxic metal ions?
5. What are the problems associated with physical and chemical methods in removing metal ions from industrial wastewater?
6. Describe the unique features of the kinetics of biosorption of metal ion (e.g. Cu^{2+}).
7. Describe the similarities and differences between Langmuir and Fruendlich adsorption isotherms.
8. Describe the methods of immobilization of bacterial/microbial cells for metal ion removal by biosorption.
9. What are the common steps of the pathway for anaerobic degradation of azo dyes.
10. Discuss the unique features of bacteria and fungi that degrade azo dyes.
11. Define biosorption.
12. Compare the metal ion adsorption ability of chitin with that of chitosan.

Chapter 20

Municipal and Industrial Solid Waste Treatment

Wong Po-Keung and Chu Lee Man
*Department of Biology, The Chinese University of Hong Kong
Shatin, NT, Hong Kong SAR, China*

20.1. Composting

Traditionally, plant residues and animal excreta are piled up to decompose and then returned to the soil for crop production. This is composting which is the accelerated microbial decomposition of organic materials of biological origin under controlled conditions of ventilation, moisture and temperature by organisms in the wastes to form a stabilized organic product. Composting is a mixed substrate, mixed culture operation (Table 20.1). Although the process can be anaerobic, it is preferentially or predominantly aerobic in most operations, involving the oxidative breakdown of part of the carbon in the wastes. Intermediate products of carbon oxidation, secondary metabolites, microbial biomass and recalcitrant compounds form humus-like substances as the product. As aerobic decomposition is exothermic, heat is released which raises the temperature of the composting mass. Consequently, conventional composting is thermophilic and during the process, there is typical temperature change that divides composting into five stages (Fig. 20.1). There is a short latent stage, during which soil microorganisms colonize the composting mass. Microbial breakdown of the readily accessible polysaccharides, proteins and simple carbon substrates raises the temperature to

25–40°C in the mesophilic stage. As temperature further increases to the thermophilic stage, heat kills most mesophiles, and thermophilic bacteria and fungi take over at about 45–55°C. At temperature above 55°C, thermophilic actinomycetes dominate, which attack hemicellulose and cellulose. At temperature greater than 60°C, spore-forming microorganisms remain viable. As readily decomposable materials are exhausted, microbial activity decreases, and the mass cools down as heat loss is much greater than heat production. Mesophilic fungi (mainly basidiomycetes) and actinomycetes re-invade and attack hemicellulose, cellulose and lignin in this cooling stage. The composting mass returns to ambient temperature, and humification and nitrification occur in the maturation stage, which may last for several weeks. Such temperature change is more apparent in

Table 20.1.   Biological composition (microbial genera only) of a composting heap.

Groups		Organisms	
Decomposers	Bacteria	Fungi	Actinomycetes
Mesophiles	*Bacillus*	*Rhizopus*	*Nocardia*
	Pseudomonas	*Coprinus*	
	Enterobacter	*Polyporus*	
	Clostridium	*Pleurotus*	
		Penicillium	
		Aspergillus	
		Geotrichum	
		Trichoderma	
		Cephalosporium	
Thermophiles	*Bacillus*	*Mucor*	*Stretomyces*
	Clostridium	*Thermoascus*	*Thermoactinomyces*
	Lactobacillus	*Chaetomium*	*Thermomonospora*
		Penicillium	
		Aspergillus	
		Sporotrichum	
		Humicola	
Consumers	Protozoa	Microinvertebrates	Macroinvertebrates
		Nematodes	Millipedes
		Mites	Centipedes
			Ants
			Earthworms

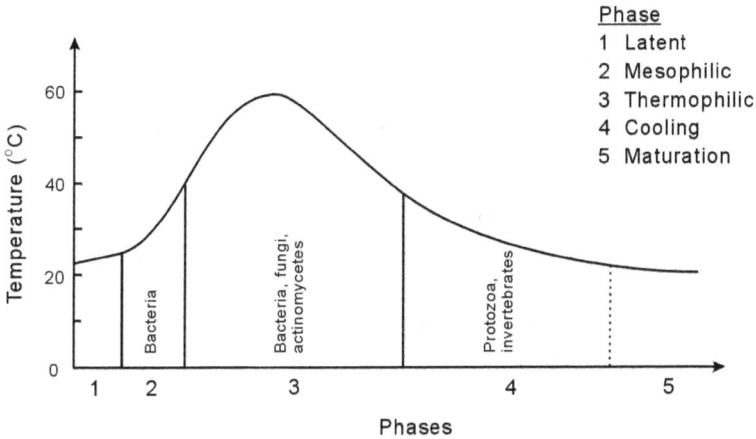

Fig. 20.1. Temperature change and biological succession in compost piles (modified from Polprasert 1996).

windrow composting (in open rows), but high temperatures (60–70°C) may not be reached in mechanical composting due to continuous agitation and thus excessive heat loss. A succession of microbial population parallels the temperature and biochemical changes, and several metabolic stages that represent different microbial populations can be observed. A dynamic community selection is acted on by a continuing alteration of environmental conditions such as temperature and substrate availability. Although composting is largely a microbial process, mesofauna such as detritivorous invertebrates (e.g. nematodes, millipedes and earthworms) and predatory invertebrates (e.g. ants, centipedes and beetles) invade the composting mass (Table 20.2), but only in the maturation stage when temperature is favorable. Color and particle size also change as a result of humification and abrasion by soil fauna.

Factors affecting decomposition will inevitably influence the rate of composting, and the physical and chemical nature of the substrate to be composted are key factors in determining the rate of the process. Particle size, which affects the surface area for microbial enzyme attack, determines the biodegradation rate. However, if the particles are too small, the structural strength and interstitial space for air circulation will be affected. Chemically, substrate assimilability as reflected by the C:N ratio is also vital in assessing the suitability of the waste as composting

Table 20.2.   Organisms involved and their populations in a composting heap (Gray *et al.* 1971).

Groups	Organisms	Number/g compost
Microflora	Bacteria	10^8–10^9
	Actinomycetes	10^5–10^8
	Fungi	10^4–10^6
	Algae	$<10^4$
Macroflora	Fungi	—
Microfauna	Protozoa	10^4–10^5
Mesofauna	Detritivorous, herbivorous and predatory invertebrates	—

substrate; a C:N ratio of 20–35 is optimal for composting. For environmental conditions during composting, the optimal moisture content is 45–70%. If moisture content is too high, this will reduce aeration as water and air share the same interstices. Oxygen is essential for modern aerobic composting to avoid objectionable odors, allow compost to reach high temperature and achieve rapid decomposition. Additives can be used to improve the properties of the material to be composted. Organic or inorganic amendment can be added to adjust the C:N ratio of the wastes, sludge for high C:N wastes and sawdust for low C:N wastes. Bulking agent such as sawdust, woodchips, bark and straw can be mixed with wet and weak materials to overcome poor structural strength, lack of airspace and excessive moisture. Oxygen is usually provided by air in the interstices, which is provided by either periodic turning in windrow composting or forced aeration in mechanical composting, but too much aeration may result in excessive heat loss, which is not beneficial in terms of pathogen destruction, whereas inadequate aeration will lead to anaerobiosis and malodor.

Oxygen availability is the key factor in system design and operation (Table 20.3), and is provided by agitation or aeration. In windrow composting, agitation by turning introduce air to the interior of composting mass, expose unattached material to promote uniform decomposition and expose pathogens to heat. Periodic agitation is usually performed, either manually or by

Table 20.3. Different types of composting classified by the mode of aeration (modified from Haug, 1980).

Types		Modes of agitation and oxygenation
Non-reactor composting (open or roofed)		
Pile or heap	Agitated solid bed	Windrow Oxygen provided by natural diffusion and periodic turning
	Static solid bed	Aerated static pile No agitation or turning; oxygen provided by aeration system at bottom of pile
Reactor composting		
Vertical flow reactor	Moving agitated bed	Waste agitated during movement down multi-story digester; oxygen provided by forced aeration
	Moving packed bed	Waste moves downward in digester with no compartmentation; oxygen provided by forced aeration
Horizontal flow reactor	Tumbling solid bed	Rotating drum Tumbling provides dispersion; oxygen provided by forced aeration
	Agitated solid bed	Bin reactor Waste agitated; oxygen provided by forced aeration
No flow reactor	Aerated non-agitated solid bed	Container or box composting Waste usually not agitated; oxygen provided by forced aeration

composting machines (Fig. 20.2), the frequency of which should be adjusted to maintain good ventilation but avoid excessive heat and water loss at the surface. Turning can be more frequent in the first few days when decomposition is rapid (once or twice per day) but reduces to once every 3–7 days, depending on the rate of oxygen consumption. In aerated static pile, there is no mixing and ventilation is achieved by aeration through perforated pipes at the bottom of the pile (Fig. 20.3). Aeration can be in the form of continuous blowing (positive pressure aeration) or suction (vacuum-induced or negative pressure aeration), which differ in their pattern of heat distribution (Fig. 20.4). To obtain a more uniform heating and decomposition, alternate blowing and suction can be utilized. Composting in an open system whether roofed or not is inevitably

Fig. 20.2. Composting machine is commonly used to turn the composting piles on open ground or in roofed facility.

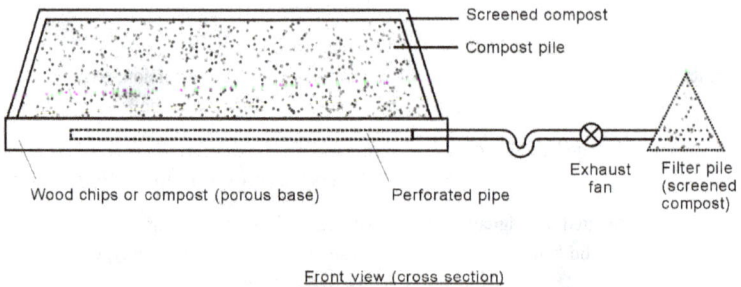

Front view (cross section)

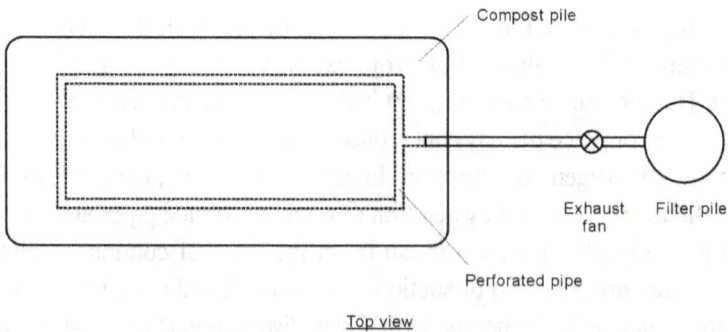

Top view

Fig. 20.3. Static solid bed composting with aeration by suction or blowing through perforated pipe at the bottom of the pile.

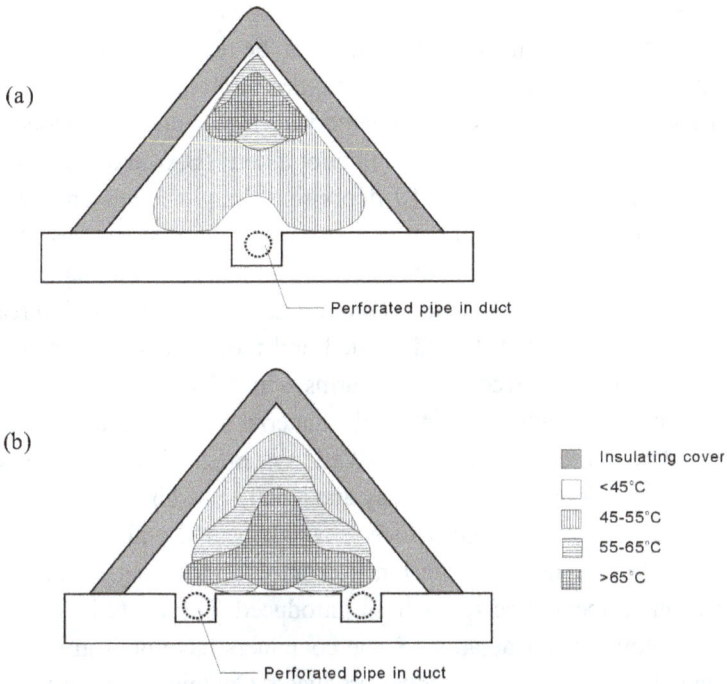

Fig. 20.4. Temperature distribution in an aerated static pile with (a) blowing aeration; and (b) suction aeration (redrawn from Stentiford *et al.* 1985).

Fig. 20.5. Rotating drum is one of the most popular reactors for municipal refuse composting.

affected by ambient conditions especially temperature. To be independent of the changing environment, composting can be done in a closed system in specially constructed reactors that are designed to provide adequate aeration, temperature and moisture control. Cylindrical or rectangular reactors, either in the form of silo towers without interior floors (BAV and Triga system) or multi-story digesters with 8–10 floors (e.g. Earp-Thomas and Thermax processes) are forced aerated to provide oxygen as the waste moves down the reactor. Reactors can also be in the form of a horizontally rotating drum with or without compartments as in the Dano process (Fig. 20.5). For bin reactors which are usually roofed, forced aerated and mechanically agitated, waste being composted is moved by radial arms which have a series of rotating perforated augers which force air into the material with some mixing as in the Fairfield-Hardy process. In the Metro-Waste system, a rail-mounted agiloader passes through a long bin once per day and agitates the bin content. Agitation provides the oxygen required, which is supplemented with forced aeration through perforated floors in the bin. Recently, composting in large boxes or containers (biocontainers) has been introduced. The waste is not agitated and air is blown in for aeration. Some containers have built-in exhaust air treatment system to treat malodours produced. Container composting saves space as containers can be stacked.

Composted nightsoil, sludge and even municipal refuse are usually utilized as a slow released fertilizer or soil conditioner. For composting to be satisfactory, the product (compost) should be stabilized and cured so that its utilization poses no harms to crops and no risks to public health. The compost should be low in pathogenic microorganisms and the extent of pathogen destruction is entirely dependent on the temperature reached and the duration of the thermophilic stage. This is important if composting is considered as a viable waste treatment option. Of course, the quality of compost (e.g. concentration of toxic elements) will be greatly hinged to the agronomic value of the mature product, which is ultimately dependent on the chemical composition of the waste before composting.

20.2. Bioremediation of Contaminated Surface and Subsurface Sites

Bioremediation is a collective term to express the application of biological agents (either whole cells or their components) to remove or treat toxic inorganic or organic compounds in aquatic or terrestrial environments. One of the most well studied bioremediation is the final step in the clean up of spilled oil by microbial cells with nutrient enrichment. Although there is a stringent international regulation to control the disposal of oil and petroleum into sea, accidental oil spills and leaks from tankers carrying crude oil or refined petroleum products are unavoidable. Among the bioremediation of spilled oil by microbes, the most striking example is the clean up of oil spill of Exxon Valdez at the coastal areas of Alaska in 1989. About 42,000 m^3 of oil was released from the super oil tanker and spread into the Prince William Sound off the coast of Alaska and produced a slick to cover over 250 km^2 of sea. After the accident, a large portion of spilled oil was first removed by physical means such as suction, and then the remaining oil droplets were emulsified with oil dispersant(s) in order to increase the surface area of the oil droplets that facilitate the degradation chemically or biologically in water. The spilled oil was oxidized by marine bacteria; however, the sticky, recalcitrant components of oil were oxidized slowly and the residual oil spread to a very large area along the coastline. When oil spread onto the coastal areas, biological remediation became the major force to remove the spilled oil since physical and chemical treatments seemed to be impractical to remove the oil from contaminated rocks or sand on the coastal areas. Although spraying with seawater and even with chemical foam was used to wash away the oil slick from the coastal rocks and sand, most oil slick remained in the solid matrices. Enrichment by addition of nutrients was used to increase the biological activity of endogenous organisms, mainly microorganisms, to degrade the oil slick in rocks and sand in coastal area.

Other examples of cleaning up oil leakage by bioremediation were reported in 1980s. About 570,000 m^3 of oil was accidentally released from oil well into the Gulf of Mexico in 1980 and substantial amount of oil leaked from oil well into the Persian Gulf during the Iraq-Iran war in the mid-1980s. Disasters of these accidents seriously contaminated the aquatic environment of the nearby water bodies and posed extremely dangerous threat to the living organisms in these areas. Thousands of sea birds and marine mammals were killed, fishery and crustacean industries in these gulfs were seriously damaged and other ecosystems were strongly disturbed. In these occasions, microbial remediation was once again recruited to solve the pollution problem caused by oil slick. Due to the slow action of microbial degradation of oil slick, the leaked oil have to be retained by physical mean: floating booms were used to contain the oil slick and sawdust was used to help the oil sink and to give micro-organisms a good surface from which oil slick can be degraded. In addition, dispersant(s) was also used to clear the coastal areas and to help disperse floating oil droplets. However, since some dispersants are toxic to oil-degrading microorganisms, the use of dispersant was under stringent monitoring and control. Even with an excellent management of oil spill clean up strategy, the harmful effects of spilled oil can exert on the contaminated areas for several years, especially for the temperate and cold regions where microbial activity that degrades residual oil slick is low.

In order to improve the efficiency of microbial remediation of spilled oil, several aspects were investigated. As mentioned above, the addition of nitrogen and phosphate appeared to greatly enhance the oil degradation by endogenous microorganisms. On the other hand, selection of appropriate microbial strain(s) or genetic manipulation of microbial strains to have more efficient oil degradation ability have provided a significant improvement in oil degradation by microorganisms. For instance, strains of *Pseudomonas* spp. have been isolated from natural environment or genetically engineered to degrade hydrocarbon components in oil slick. A set of genes that reside on plasmid, an extra-chromosomal element, was identified in these bacteria to produce required enzymes for the degradation of hydrocarbons in oil. The trait can be easily transferred to other microorganisms since this set of genes is located in a plasmid. Genetic manipulation of this set of genes on the plasmids, enable the bacterium to degrade hydrocarbons in oil faster and more completely. Further development on the construction of a bacterial strain containing plasmids that encode genes for rapid degradation of all components in oil slick has been a main topic in this area of environmental biotechnology.

Other than the cleaning up of spilled oil, the use of microorganisms to reduce the level of toxic organic compounds such as polycyclic aromatic hydrocarbons (PAHs), polychlorinated biphenyls (PCBs) and even pesticides in the underground of abandoned agricultural land or military bases attract a great effort. There are many impressive studies on the use of natural or genetically engineered microorganisms to remove (or treat) unwanted toxic materials from the contaminated natural environment. The use of microorganisms, either as pure culture or as mixed populations, to remediate the incidents of industrial or military pollution becomes the potential key technology to deal with environmental clean up.

20.3. Microbes in Mining

During mining activity, there is a large amount of metal ions remaining in the ore. The dumping of untreated ore in the mining site will result in the leaching of metal ions into the groundwater or nearby aquatic environment, causing serious pollution problems. However, the complete extraction of remaining metal ions from the ore is impractical due to low cost effectiveness. In order to minimize the metal ion contamination problem and to recover valuable metal ions from the ore, microbial leaching has become a very attractive operation in the mining industry. Bacteria such as sulfate oxidizers can produce acidic solution for the leaching of metal ions from the low grade ore. The solubilized metal ions can be concentrated and recovered as useful metal salt or solution for commercial applications.

Most of the valuable metal ion in low grade ore are in water insoluble sulfide forms. In the presence of sulfate oxidizer, sulfide can be oxidized into sulfate and the metal sulfate becomes water soluble. Sulfate oxidizer such as *Thiobacillus ferrooxidans* can obtain energy by oxidizing either ferrous or sulfide, and produce ferric sulfate according to the following equations:

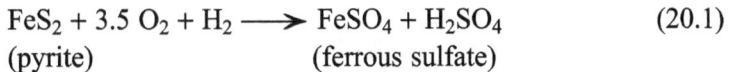

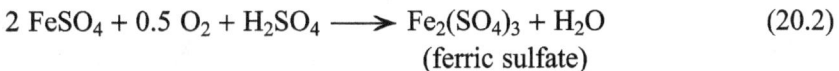

$$FeS_2 + 3.5\ O_2 + H_2 \longrightarrow FeSO_4 + H_2SO_4 \qquad (20.1)$$
$$\text{(pyrite)} \qquad\qquad\qquad \text{(ferrous sulfate)}$$

$$2\ FeSO_4 + 0.5\ O_2 + H_2SO_4 \longrightarrow Fe_2(SO_4)_3 + H_2O \qquad (20.2)$$
$$\text{(ferric sulfate)}$$

Then ferric sulfate will be available to solublize Cu^{2+} and Fe^{3+} from the CuFeS (Eq. 18.3):

$$CuFeS + 2\ Fe_2(SO_4)_3 \longrightarrow CuSO_4 + 5\ FeSO_4 + 2\ S^0 \qquad (20.3)$$
$$\text{(chalcopyrite)}$$

Ferric sulfate is a strong oxidant and is able to dissolve other metal sulfides. Leaching by ferric sulfate is an indirect reaction because it is independent of the presence of oxygen and microbial action. The rate of extraction of Eq. (18.3) depends on the concentration of ferric sulfate.

On the other hand, *T. ferrooxidans* can attach onto the mineral particles in the ore and enzymes associated with the cell membrane catalyze the oxidative attack on the crystal lattice of the metal sulfide. Oxidation of the metal sulfide such as copper sulfide proceeds as the following equations:

$$CuS + 0.5\ O_2 + 2H^+ \longrightarrow Cu^{2+} + S^0 + H_2O \qquad (20.4)$$

$$S^0 + 1.5\ O_2 + H_2O \longrightarrow H_2SO_4 \qquad (20.5)$$

Among the microorganisms that can oxidize metal ore, most are acidophilic chemolithotrophic bacteria, for instance, *Thiobacillus ferrooxidans*, *Leptospirillum ferroxidans*, and species belonging to *Sulfolobus*. *T. ferrooxidans* is a small rod, Gram-negative bacterium. It can grow in solution with pH as low as 1.5–2.5. The optimal growth temperature is 10–30°C with an upper limit of 37°C. As a chemolitotroph, *T. ferrooxidans* obtains its energy from oxidation of ferrous ion to ferric ion, and utilizes CO_2 to generate its carbon-containing macromolecules.

There are other *Thiobacillus* species such as *T. thiooxidans*, *T. acidophilus* and *T. organoparus*. These acidophilic bacteria oxidize sulfide instead of ferrous ion to obtain their energy. Usually these *Thiobacillus* species are able to co-exist with *T. ferrooxidans* and have the role of facilitating mineral leaching by *T. ferrooxidans*. *L. ferroxidans* can grow at a pH about 1.2 and more acidophilic than *T. ferrooxidans* and grow at a higher temperature (40°C). The bacterium is a highly motile, curved rod and obtains energy by oxidizing ferrous ion to ferric ion, but it is unable to oxidize sulfur compounds. *Sulfolobus*, an archaebacterium, is also involved in the mineral leaching process. The bacterium grows autotrophically at pH 1.0–3.0 and temperature of 50–90°C. It can derive energy from oxidizing either ferrous ion or sulfur compounds.

Copper ore is a smelting residual containing less than 0.5% of copper. The low grade copper ore is piled for dump leaching. Water is sprayed onto the pile and leached water is sprayed repeatedly onto the ore. After a pre-determined period, pyrite (copper sulfide) is solubilized with resultant acidic ferric sulfate solution. Continuous recirculation of the acidic ferric sulfate solution into the copper ore solubilizes the copper ion. Finally, the copper ion-rich solution is pumped into a container and precipitated by adding iron scraps. The precipitation reaction is:

$$Cu^{2+} + F^0 \longrightarrow Cu^0 + Fe^{2+} \qquad (20.6)$$

The Fe^{2+}-rich solution remaining after copper precipitation is returned to the oxidation pond where *T. ferrooxidans* rapidly oxidizes Fe^{2+} to Fe^{3+} and forms additional H_2SO_4 through the oxidation of sulfur compound. Fe^{3+} is precipitated with NaOH to $Fe(OH)_3$. The supernatant acidic ferric sulfate solution is pumped back to the top of the ore pile.

20.4. Microbial Desulfurization of Coal

Coal is one of the major energy sources for the production of electricity. High concentration of sulfur in coal will lead to the production of SO_2 in coal combustion. As SO_2 reacts with water in mist, corrosive sulfuric acid will be formed and this leads to acid rain. The damages to our environment caused by acid rain have been well documented. Acid rain not only causes problems in the aquatic environment, but also leads to property damage in urban environment. Thus the use of coal faces increasingly stringent environmental control in industrial countries. Low-sulfur coal has been recommended in order to reduce the level of SO_2 emitted. However, the supply of such low-sulfur coal is limited. Most of the sulfur in coal is in the form of heterocyclic organic compounds. Although there are physical and chemical methods to reduce SO_2 emission before, during and after coal combustion, the reduction of sulfur content in coal is an environmental friendly approach. The operation cost of these physical and chemical methods is relatively expensive and hinders their use to reduce sulfur content in coal. Recently, microbial technology has been proven to be a good alternative for coal desulfurization.

Biodesulfurization of coal is a process that uses microorganisms to reduce the sulfur content in coal. The sulfur found in coal is either part of the molecular structure of coal (organically bound sulfur), for instance, DBT (dibenzothiophene) in coal can be used as the sole sulfur source by microorganisms; or in mineral (e.g. pyrite in inorganic form — FeS_2).

One of the simplest reactions is the oxidation of coalmine with sulfur such as $CuFeS$ and FeS_2. Some microorganisms only remove sulfur from organic components of coal (i.e. desulfurization of DBT by *Pseudomonas* spp.), some remove sulfur from inorganic component of coal (i.e. desulfurization of pyrite by *Moraxella* spp.), while remove sulfur from both organic and inorganic components of coal (e.g. *Xanthomonas maltophila*).

For the removal of sulfur from organic components of coal, one of the most typical examples is the cleavage of DBT into inorganic sulfur and hydroxybiphenyl. Some species of *Pseudomonas* attack the sulfur in DBT without altering the carbon skeleton or affecting other valuable components of coal. Biodesulfurization mechanisms, which convert sulfur-containing components into water soluble oxidation products, could cause loss of the hydrocarbons associated with the sulfur atom in coal. These mechanisms should not be used since they will reduce the usage of coal by reducing the energy stored in the hydrocarbon-core. Recently the genes involved in the sulfur-specific DBT desulfruization were identified and the corresponding enzymes have been investigated. From the practical point of view, it has been proven that the microbial desulfurization process for DBT and its analogs is economically feasible.

A group of bacteria called sulfate bacteria can oxidize sulfur to sulfate. The acid solution can be used in mine leaching (the importance of micro-organisms in mine leaching has been discussed in previous section). The bacteria can remove sulfur atom in inorganic components of coal such as pyrite. This process can minimize the problem caused by the high level of sulfur in coal and also provides an acidic solution to recover mineral resource from the ore. In addition, the use of microorganisms offers an economical and energy-saving method of desulfurization of coal.

Beside the bacteria mentioned above, many S-oxidizing bacteria oxidize the sulfur-containing compounds in coal. *Thiobacillus ferrooxidans*, *T. thiooxidans*, *Leptospirillum ferroxidan* and *Sulfulobus* spp. have been reported to involve in the sulfur oxidation. The common drawback of using these native bacteria in biodesulfurization is their slow growth rate (i.e. slow reaction rate). In addition, most of these bacteria are oligotrophic organisms; addition of extra nutrient (including carbon and nitrogen) is required to stimulate the desulfurization rate of these bacteria in laboratory and field studies.

The idea of using acidophilic pyrite-oxidizers to clean up sulfur in coal is not well developed before 1980. The experimental and field trials were conducted in Europe and the US in 1970s–1980s. However, the large-scale application of biodesulfurization did not appear even in the large coal production countries. In early 1990, more studies were initiated on the feasibility of biodesulfurization of coal. The advantages of biodesulfurization are as listed below.

(1) Selectivity

As mentioned above, acidophilic sulfur-oxidizers attack only inorganic sulfur in coal mine, while other bacteria such as *Rhodococcus*, *Bacillus*, *Corynebacterium* and *Arthrobacter* species attack only the sulfur atom of sulfur-containing organic compounds such as dibenzothiopene (DBT) in coal mine. These bacteria do not attack the carbon core of coal. Thus, unlike physical and chemical methods, there is no significant loss of carbon or calorific content of coal.

(2) Metal removal

The production of sulfate or sulfur leads to the production of sulfuric acid. The acidic solution thus produced is able to leach out the remaining high level of metals in the coal mine. Consequently, the resultant biodesulfurized coal is low in both sulfur and metal content.

(3) Ash removal

Owing to the production of acidic solution during biodesulfurization, most metal ions in a coal mine are dissolved and mobilized. This lowers the ash content of biodesulfurized coal.

However, there are some disadvantages such as the slow reaction rate and the production of waste (including biomass and acidic metal/mineral solution) which requires further treatment after biodesulfurization. Heat production during the process is another problem that has to be dealt with. However, with appropriate design and setup, the metal ions in the acidic solution can be recovered and provide a valuable source of metals, while the bacterial biomass can be harvested and recycled back into the process to speed up biodesulfurization. The heat produced can be used directly as energy or to generate another form of useful energy — electricity. With these modifications, biodesulfurization of coal is an excellent choice to minimize pollution caused by the sulfur content in coal with the added advantage of the production of useful by-products — metal and energy during the process.

20.5. Biodegradable Plastics

Materials that are synthesized from petrochemical such as polyethylene, polyvinylchloride and polystyrene are commonly used as plastic wares in our daily life. Owing to the synthetic and recalcitrant properties of petroleum-derived plastics, the disposal of these solid wastes becomes a major problem in waste management in many modern cities. The production of photodegradable plastics becomes meaningless since many cities use landfill as the means for disposal of solid waste. Once photodegradable plastics are disposed of in a landfill and covered with soil, the photodegradable plastics cannot be exposed to sunlight and thus remain intact in landfill sites.

The alternative degradable plastics are biodegradable plastics. The first biodegradable plastic was plastic polymer blended with a biologically degradable polymeric compound such as starch. Once the biodegradable plastic was disposed of, microorganisms in the natural environment can degrade the starch molecules in the copolymer and result in the disintegration of the plastic polymer. However, the physical and chemical properties of the starch-substituted biodegradable plastics are not suitable for practical usage. In addition, a large part of non-degradable plastic polymers remains as residues that can accumulate in our environment and cause pollution problems.

The production of another types of biodegradable plastics that uses polyhydroxyalkanoates (PHAs, Fig. 20.6), which has the physical and chemical properties very similar to those of polypropylene, has been studied extensively. PHAs are microbial storage polyesters. In the presence of excessive carbon sources, many microorganisms form PHAs granules intracellularly. There are many companies engaged in the large-scale production of PHA biodegradable plastics. One of the examples is "Biopol^{TM}", which is a copolymer of P(3HB-Co-HV) [poly(3-hydroxoybutyrate-co-hydroxyvalerate)], produced by Zeneca Bio Products

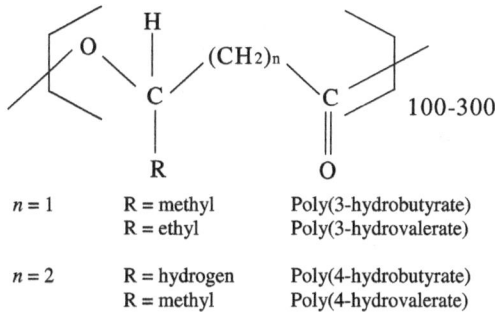

$n = 1$	R = methyl	Poly(3-hydrobutyrate)
	R = ethyl	Poly(3-hydrovalerate)
$n = 2$	R = hydrogen	Poly(4-hydrobutyrate)
	R = methyl	Poly(4-hydrovalerate)

Fig. 20.6. Structures of PHAs (polyhydroxyalkanoates).

(formerly Imperial Chemicals Industries) in the UK. Later, Monsanto, a leading agricultural chemical manufacturer in the US, applied genetically engineered oilseed rape plant to produce PHA biodegradable plastic polymer. However, the properties of the PHAs produced by the plant have physical and chemical properties different from those produced by microorganisms. After that, another large chemical manufacturer in the US, Union Carbide, also produced microbial biodegradable PHA plastics called "Tone Polymer", while a chemical company in Japan, Showa Denko, produced "Bionolle" which is a chemically synthesized biodegradable PHA plastic. Among these biodegradable plastics, poly-β-hydroxybutyrate (PHB) and its copolymer, P(3HB-Co-HV) [poly(3-hydroxoybutyrate-co-hydroxyvalerate)] are the most widespread and thoroughly characterized PHAs. A number of bacteria, including *Alcaligenes* spp., *Pseudomonas* spp. and a number of filamentous genera such as *Nocardia* spp. produce these PHAs when they encounter unfavorable growing conditions.

In general, a specific microorganism, which occurs widely in nature, is utilized in a novel transgenic fermentation process to convert carbohydrates into the PHA polymers. After the fermentation, the microorganisms have accumulated 80% of their dry weight as PHA polymer. The polymer is purified by breaking open the cells and then harvesting the PHA using an aqueous based extraction method. Owing to their amorphous character, these bioploymers can be ideally handled as lattices. The bioploymer can be extracted with organic solvents and then purified to the usage form. A nonsolvent based process for recovery of PHAs from bacteria was reported. The process is to solubilize the bacterial biomass by sequential treatments of heating, protease digestion and detergent, and leave the bacterial peptidoglycan in cell wall intact for the ease of separation. Then the PHA granules will be purified by filtration and the final product is in a latex form. The estimated cost of production and extraction of PHA by this route will be US$5/kg, which is still higher than that of synthetic plastics.

PHA polymer can be compounded with safe, biodegradable additives to tailor processing and mechanical properties before being converted into plastic articles using standard techniques. Although stable in normal use, the PHAs can be biologically degraded when deposited in an active microbial environment such as compost. Certain microorganisms can metabolize and consume PHA polymers as their nutrients. Biodegradation of the PHAs under aerobic conditions yields carbon dioxide and water in the same amounts originally used during photosynthesis, completing the PHA biocycle. In order to test the biodegradability of the PHA-plastics, a series of test under laboratory and field conditions should be conducted. There are standard methods published by the American Society for Testing and Materials (ASTM) for the degradation of synthetic and biodegradable plastics under laboratory conditions. A series of microorganisms, including bacteria and fungi, has been characterized as being able to degrade PHA-plastics. Most of these microorganisms were isolated from aerobic environment. Over 10 different extracellular PHA deploymerases have been purified and characterized.

The critical factor limiting the use of PHA plastics is their high production cost. Alternatively, one can produce novel PHAs with improved physical and chemical properties that suit the specific needs of certain usages. In Hong Kong, there are several studies on the production of biodegradable PHA-plastics by microorganisms. Studies have been conducted to investigate the use of synthetic media with various amount of organic acid (e.g. butyric acid) as the sole carbon source to produce PHA with improved physical and chemical properties by selected microorganisms such as *Alcaligenes eutrophus* and *Klebsiella pneumoniae*. By varying the concentrations of nutrient supplied, the bacteria produced PHA-plastics with the ratio of HB:HV (hydroxbutyrate: hydroxvalerate) were 97:3 and 79:21. These PHA-plastics have different thermal stabilities.

These results were due to the amount and ratio of various organic acids added to control the polymeric composition of PHAs, that in turn change the physical and chemical properties of the PHA-plastics. Another approach is to produce PHA-plastics with a more economical substrate such as wastewaters from food industry and chemical industry with xenobiotic organics. There are studies in which soya wastes from a soya milk dairy and malt waste from a beer brewery were successfully used as feedstock to produce PHAs from microorganisms in activated sludge. Microorganisms, mainly bacteria, in the activated sludge fed with various amount of wastewater from food and chemical industries produced copolymer (P(3HB-Co-HV)) of PHA-plastics. Various types of PHAs were produced when the C:N ratio of the wastewater varied, with the amount of PHAs increased as the C:N ratio increased. The result was due to the wide variety of bacteria in the activated sludge that produce various types of PHA. Different bacteria have specific PHA synthetases with different substrate specificities and produce different types of PHAs. Sugars such as maltose, sucrose and fructose are present in the industrial wastewater as carbon sources for the bacteria to produce PHAs. The use of activated sludge to convert carbon substrates into PHAs not only can produce biodegradable plastics, but also serves as a wastewater treatment in which the problem of disposal of activated sludge after the wastewater treatment is solved.

Recent advances in understanding the metabolism, molecular biology and genetics of the PHA-producing bacteria and cloning of more than 20 different PHA biosynthesis genes allow the construction of various recombinant strains that are able to produce a much larger amount of biodegradable plastics with different types of PHA copolymers. These advances lead to the production biodegradable plastics with costs that are comparable to or lower than that of synthetic plastics and with greatly improved physico-chemical properties.

20.6. Detoxification of Metal Ions

20.6.1. *Demethylation and reduction of mercury*

Several hundred million tonnes of toxic mercury-containing wastes are discharged into our environment. The ionic forms of mercury such as Hg^+

772 Microbial Biotechnology: Principles and Applications

and Hg^{2+} and organic mercury (R-Hg^+) such as methyl mercury and phenyl mercury are extremely toxic and even mutagenic to living organisms. Many microorganisms are resistant to the toxic ionic forms of mercury. Mercury resistance found in bacteria can be categorized into two classes, narrow-spectrum resistance (resistant to inorganic mercury only) and broad-spectrum resistance (resistance to inorganic and organic mercury). Among these microorganisms, the mercury resistance of Gram-positive strains such as *Staphylococcus* spp. and *Bacillus* spp.; and Gram-negative bacteria such as *Pseudomonas* spp. and *Shigella* spp. are extensively studied. Organic mercury has been reported to be more toxic than its inorganic forms since the penetration of organic mercury into microbial cells is comparatively easier and more rapid than inorganic mercury. However, most of the discharged inorganic mercury can be rapidly and easily transformed biologically into organic forms in the environment. Therefore, the broad spectrum mercury resistant microorganisms are important to detoxify mercury-contaminated sites in our environment.

What is the difference between mercury narrow-spectrum resistance and broad-spectrum resistance in microorganisms (Fig. 20.7)? Many studies on the biochemistry and mechanisms of the mercury resistance of Gram-negative and Gram-positive bacteria reveal that the narrow-spectrum resistant bacterium has only an enzyme called mercury reductase that is

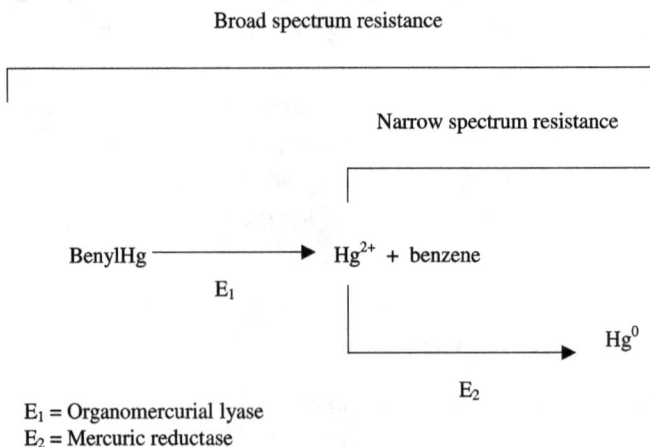

Broad spectrum resistance

Narrow spectrum resistance

BenylHg $\longrightarrow$ Hg^{2+} + benzene

E_1

Hg^0

E_2

E_1 = Organomercurial lyase
E_2 = Mercuric reductase

Fig. 20.7. Narrow spectrum and broad spectrum mercury resistant bacteria.

used to catalyze the reduction of Hg^{2+} to Hg^0. Further studies on the mercury resistance genes of *Pseudomonas stutzeri* indicate that a plasmid pPB confers broad-spectrum mercury resistance. The plasmid is self-transmissible. Two pPB regions, separated by 25–30 Kb and sharing homology with the transposon, Tn501 *mer* (Hg detoxification) genes. Two regions were cloned separately and each was shown to carry a cluster of functional and independently regulated *mer* genes. One cluster conferred resistance only to inorganic mercury (with at least *merA* and other mercury transporting genes such as *merT* for narrow-spectrum mercury resistance) and other gene cluster, upstream from *merA* has a novel *merB* gene encoding for organomercurial lyase.

The microorganisms with broad-spectrum mercury resistance have, in addition to the mercury reductase, an organomercurial lyase which remove the organic residue(s) of organic mercury. The function of organomercurial lyase is to remove the methyl group of methyl mercury and to produce free methyl group (CH_3) and inorganic mercury (Hg^{2+}). The inorganic mercury can be further transformed into Hg^0 by the reduction mediated by the mercury reductase. Thus the broad spectrum mercury resistant microorganisms can grow in the presence of organic and/or inorganic mercury.

In addition, the elementary mercury (Hg^0) is volatile and most of them leave the aquatic environment and enter the atmospheric environment. Although the total mercury content of the aquatic and atmospheric environments remains unchanged, the availability of mercury to organisms in aquatic environment, mainly in ionic form, will be greatly reduced. Thus the reduction of Hg^{2+} to Hg^0 (also called mercury volatilization) is one of the most effective methods to detoxify mercury in aquatic environment.

Although many microorganisms contain mercury reductases and organomercurial lyases, only few mercury reductases, and quite a few organomercurial lyases have been purified from bacteria. Most of the properties of these enzymes are based on the purified enzymes from Gram-negative bacteria (e.g. *Pseudomonas* spp.).

20.6.1.1. *Mercury reductase*

Mercury reductases purified from the cellular membrane of Gram-negative bacteria are either dimeric or trimeric with monomers of 58,700 daltons.

The native enzymes require FAD (flavin) for reduction function and preferentially use NAD(P)H as electron donor. A reducible active-site disulfide (cys-135, cys-14) and a C-terminal pair of cysteines (cys-558, cys-559) were identified. All 4 cysteines are required for efficient Hg^{2+} reduction. These cysteine residues at its active site and the electrons donated by NAD(P)H are transferred via FAD to reduce the disulfur bond (-S-S-) of two adjacent cysteine residues at the active site; and converted them into two cysteine residues (with -SH groups). The electrons captured by the enzyme are then used to reduce Hg^{2+} to Hg^0.

20.6.1.2. *Organomercurial lyase*

The enzyme is inducible by the presence of organic mercury. It has been reported to be located in the cytoplasm of the bacterial cell. The enzyme is 19,000–20,000 daltons (19–20 KD). Its function is the cleavage of the C-Hg bond of organic mercury by protonolysis. For instance, methylmercury is detoxified by demethylation to mercuric ion by bacterial enzyme.

Other than mercury reductase and organomercurial lyase, the enzymes involved in the transportation of Hg^{2+} such as *merT* and *merP* are also important to the mercury resistance of bacteria. *MerT* encodes a transmembrane protein. The molecular biology of *merT* and *merP* proteins of Tn501 isolated from *Escherichia coli* carrying the pPB plasmid was studied. Similar to the biochemistry of *merA* protein, an active site with cys-24 and cys-25 in the first transmembrane region of *merT* protein was essential for transport of Hg^{2+} through the cytoplasmic membrane. A cys-33-ser mutation in *merP* appears to block transport of Hg^{2+} by *merT* protein, while deletion of whole *merP* only slightly reduced Hg^{2+} resistance of the bacterium and suggests that a functional *merT* protein is sufficient for Hg^{2+} transport across the cytoplasmic membrane.

20.6.2. *Reduction of chromium*

Hexavalent chromium ion (Cr^{6+}) is a common and toxic pollutant in soils and waters. Trivalent chromium ion (Cr^{3+}) is less toxic as compared with its oxidized form (i.e. Cr^{6+}). Chemical reduction of Cr^{6+} is feasible and some operations have been practised. However, the addition of reduced compounds such as NAD(P)H that are potential toxicant for living organisms render the large-scale application of chemical method to detoxify Cr^{6+}. Biological detoxification of Cr^{6+} becomes a promising approach to solve the pollution problem of toxic chromium ion. In some cases, chromate (i.e. Cr^{6+}) is used as terminal electron acceptor during anaerobic respiration by microorganisms. Once Cr^{6+} in chromate is reduced to Cr^{3+} Eq. (18.4), the toxicity of the chromium is greatly reduced. In addition, since Cr^{3+} compounds are insoluble and tend to precipitate out from the aqueous phase, the reduced chromium ion (Cr^{2+}) will not be available for uptake by microorganisms as well as other organisms in the contaminated environments.

$$r^{6+}{}_2O_7{}^{2-} \longrightarrow Cr^{3+}O_4{}^{2-} \tag{20.7}$$

There are many cases using bioreactors with bacteria to detoxify Cr^{6+} in aqueous solution. Many bacteria such as *Bacillus* spp., *Pseudomonas* spp. and *Thiobacillus ferrooxidans* are able to reduce Cr^{6+}. Both monoculture of Cr^{6+}-reducing bacteria or bacterial consortium have been used in detoxification of Cr^{6+} in aqueous solution. Bacterial cells immobilized by appropriate matrices (e.g. DuPont Bio-Sep beads) in either packed-bed or in fixed-film (biofilm) reactors efficiently reduced Cr^{6+} in the presence of reducing compounds such as sulfite or thiosulfate. Under selected conditions, up to 200 mg/L of Cr^{6+} can be completely reduced to Cr^{3+} within 24 hours. Further study indicates that a chromate reductase is associated with the Cr^{6+}-reducing bacteria such as *Pseudomonas putida* PRS2000. Both cell suspension and cell free extract can reduce Cr^{6+}. The enzyme is a soluble protein and the crude enzyme activity is heat labile and with a Km of 40 μM of chromate and is NAD(P)H dependent. Neither sulfate nor nitrate affects chromate (Cr^{6+}) reduction either *in vitro* or with intact cells. In the reduction of Cr^{6+} by *Pseudomonas ambigua* G-1, an intermediate, Cr^{5+}, was identified by electron spin resonance during the enzymatic reduction of Cr^{6+}; this suggested that the chromate reductase in this bacterium reduced Cr^{6+} to Cr^{3+} with at least two reaction steps via Cr^{5+} as an intermediate.

Further Reading

1. Blazquez ML, Balklester A, Gonzales F and Mier JL. Coal bio-desulfurization: A review. *Biorecovery* **2**: 155–177, 1993.
2. Demain AL and Davies JE. *Manual of Industrial Microbiology and Biotechnology*. American Society for Microbiology, Washington, DC, 1999.
3. Evans GM and Furlong JC. *Environmental Biotechnology: Theory and Application*. John Wiley & Sons, Chichester, 2003.
4. Gray KR, Sherman K and Biddlestone AJ. A review of composting. Part I. *Process Biochem.* **6**: 32–36, 1971.
5. Haug RT. *Compost Engineering: Principles and Practice*. Technomic Publication, Lancaster, 1980.
6. Olguin EJ, Sanchez G and Hernandez E. *Environmental Biotechnology and Cleaner Bioprocesses*. Taylor & Francis, New York, 2000.
7. Polprasert C. *Organic Waste Recycling*, 2nd ed. John Wiley and Sons, Chichester, 1996.
8. Stentiford EI, Taylor PL, Leton TG and Mara DD. Forced aeration composting of domestic refuse and sewage sludge. *Water Pollut. Control* **84**: 23–32, 1985.

Questions for Thought...

1. What could be done to control odor in the composting of organic wastes?
2. Discuss the unique features of microorganisms that are used in bioleaching.
3. What are the differences between the biosulfurization of inorganic and organic components of coal?
4. Discuss the new developments in the production of economically feasible PHA-plastics.
5. Dicuss the difference between the narrow spectrum and board spectrum mercury resistance of microorganisms.

Chapter 21

Bacterial Biofilm: Molecular Characterization and Impacts on Water Management

Teng Wee Lin,[a] Gao Pingping and Chang Siao Yun
Technology & Water Quality Office, Public Utilities Board (PUB)
82 Toh Guan Road East, WaterHub, Singapore 608576
[a]Email: sarah_teng@pub.gov.sg

Biofilm is composed of viable and dead cells, organic matters, inorganic deposits and particulates that are held together in the extracellular polymeric substance (EPS) matrix. EPS comprises a complex mixture of biopolymers secreted by the embedded cells and consists mainly of polysaccharides, proteins and DNA. Bacteria can attach to and form biofilm on both biotic and abiotic surfaces found in living tissues, medical and industrial devices and, diverse soil and aquatic environments. In the aquatic environment, the free-living bacterial cells are constantly challenged by limited nutrients and other unpredictable conditions in the local environment which are dependent on weather, human activities, dosing of antimicrobial agents and invasion of predator species. The EPS matrix of a biofilm meanwhile helps to sequester and concentrate the diluted nutrients found in the water phase, mediates active interactions between species for expression and transfer of advantageous genetic material and protects embedded cells from antimicrobial agents and predators. Biofilm formation is thus advantageous and is analogous to smart-living for many bacterial cells found in the aquatic environment.

21.1. Biofilm Formation and Development

Many microorganisms can form biofilm and bacterial biofilm is perhaps the best studied. Complex mechanism underlies the biofilm formation process even though it can be simplified into three major steps as illustrated in Fig. 21.1. The free-living or planktonic cells found in the water phase will first adhere to a solid surface usually pre-adsorbed by conditioning organic molecules. Once attached irreversibly, active cell growth and division will take place resulting in formation of micro-colonies frequently separated by fluid-filled channels creating a heterogenous biofilm surface. At maturation, some cells will start to detach and slough off from the biofilm. However, little is known on what triggers the cell dispersion process, possibly due to lack of nutrients or EPS.

EPS is a major component of bacterial biofilm. Extracellular polymeric substances close to the cell surface assist in the adherence of cells to

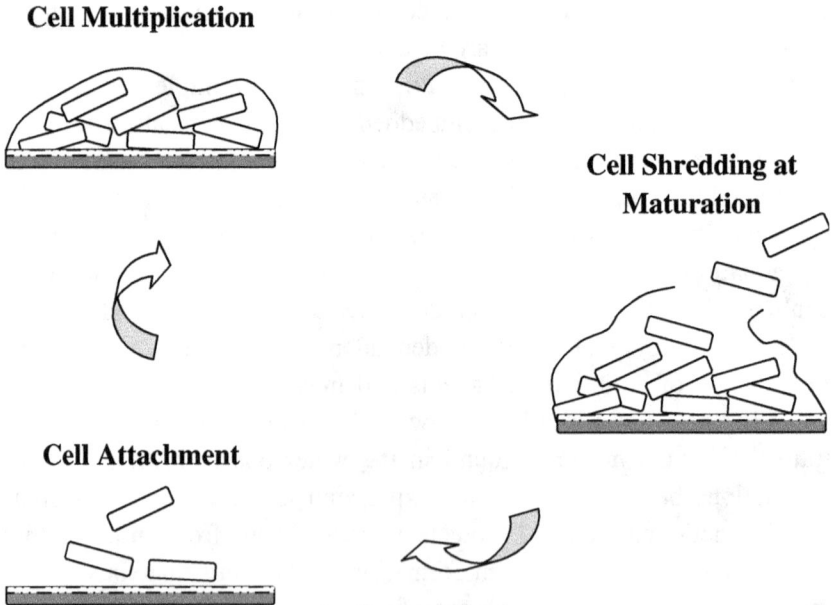

Cell Multiplication

Cell Shredding at Maturation

Cell Attachment

Fig. 21.1.   Schematic presentation depicting the major steps in biofim formation.

surfaces and other cells in the biofilm, while other secreted biopolymers polymerize with adsorbed inorganic ions to form and strengthen the three-dimensional structure of the biofilm matrix. The high sorption property of EPS can be attributed to the charged and hydrophobic polysaccharides and protein components. As a result, the EPS matrix can bind, retain and stabilize effectively various extracellular enzymes that are advantageous for bacterial cell growth in the biofilm. In addition, the EPS components tend to retain water very well forming a highly hydrated matrix that protects the embedded cells against desiccation and grazing by predators such as protozoa.

Bacteria modulate and change their gene expression system in order to adapt to the transition from planktonic to sessile stage in a biofilm. There is often up-regulation of genes associated with biofilm-related phenotypes such as spread of mobile genetic elements, resistance to UV and other antimicrobial agents, enhancement of metabolic capability and, production and secretion of EPS. The latter phenotype is said to be regulated by sensing increased cell density, a type of quorum sensing control (Box 21.1).

Box 21.1

The Concept of Quorum Sensing in Biofilm Development

Quorum sensing (QS) is a type of regulatory mechanism used by bacterial cells to modulate gene expression in response to increasing cell density. QS-dependent response typically follows three major steps: i) secretion of QS signal molecule that increases with cell density ii) coupling of QS signaling molecule with intracellular signal transduction that usually comprises a sensor and response regulator (RR) and, iii) eliciting the appropriate QS-dependent response by regulating the expression of the target gene(s). N-acyl-homoserine lactones (AHL) secreted by gram-negative bacteria is perhaps the most well-studied QS signaling molecule while auto-inducing peptides or otherwise known as peptide pheromones are typically confined to Gram-positive bacteria. QS signaling based on auto-inducer-2 (AI-2) is shared by both Gram-positive and Gram–negative bacteria.

Box 21.1   *(Cont'd)*

Production and secretion of virulence factors, bioluminescence, motility and biofilm development are some examples of bacterial activities that have been shown to be under the control of QS. The coupling of AHL with LuxI/LuxR in *Vibrio fischeri* represents the first described QS system in bacteria. LuxI produces the QS signaling molecule AHL while the Lux operon (*luxICD-ABE*) is responsible for bioluminescence. The uptake and coupling of AHL with LuxR, which increases with cell density, promotes the transcription of Lux operon resulting in light production.

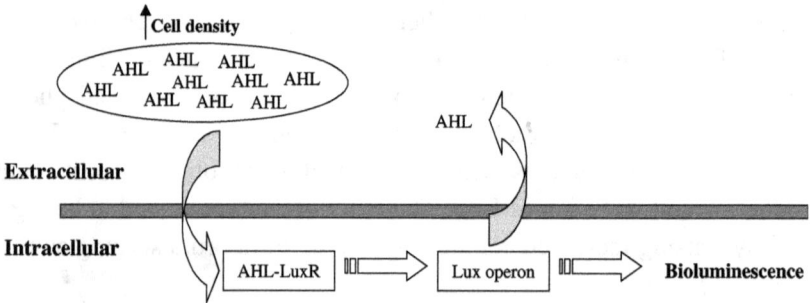

In biofilm development, QS is frequently linked with EPS production. For example, in *Pseudomonas aeruginosa*, mutation in the *lasI* gene which encodes for the diffusible extracellular cell-to-cell signaling molecule N-(3-oxododecanoyl)-L-homoserine lactone ($3OC_{12}$-HSL) results in abnormal biofilm formation that is generally thinner and more uniform compares to wild-type biofilm (Davies *et al.*, 1998). Supplementation with $3OC_{12}$-HSL in the culture medium enables the *lasI* mutant to form biofilm that differentiates into clusters of relatively loosely packed cells with considerable intervening space between bacteria, just like the wild-type. Also, the biofilm belonging to the *lasI* mutant could be rapidly dispersed by the addition of biocide made up of 0.2% sodium dodecyl sulfate (SDS). As such, quenching the QS signal may be a promising approach to eradicate or to overcome biofilm development on both medically and industrially important surfaces.

21.2. Molecular Tools to Dissect Bacterial Community in Biofilm

The bacterial population embedded in the EPS matrix of a biofilm is highly diverse with respect to phylogenetic and physiology. The population is dynamic, constantly shifts and evolves with changing environmental conditions. To understand the spatiotemporal variations in biofilm community from different environments or to investigate the impact of a defined factor on biofilm development, it is imperative to analyze the composition of the EPS matrix and the embedded bacterial population. The EPS matrix has a defined three-dimensional structure that comprises an assortment of biopolymers secreted by the bacterial cells. Due to the complex mixture of the biopolymers and their conjugates, it is technically challenging to isolate individual EPS constituent from environmental biofilm at high level of purity and concentration suitable for downstream analysis. However, feasibility to study EPS production using pure bacterial isolates and readily available fluorescent dyes specific to polysaccharide, protein and some of their derivatives or conjugates help to facilitate *in situ* staining and visualization of the EPS matrix in a biofilm thus advancing our knowledge on the biofilm spatial structure.

Traditional cultivation technique or more recently, molecular DNA fingerprinting tools can be applied to assess and monitor population diversity and dynamics in biofilm. The former has the distinct advantage in isolating pure cultures for further manipulation and study. However, bacterial population identified by culture does not reflect the original community structure since artificial culture medium selectively promotes growth of some bacterial species which is also dependent on cultivation conditions. The faster growing species also tend to outcompete others and the nutrient composition such as the main carbon and nitrogen source may not necessarily simulates actual environmental condition. Most importantly, significant environmental isolates are refractory to culture.

Metagenomic is defined as the genomic analysis of microorganisms by direct extraction and cloning of DNA from an assemblage of microorganisms (Handelsman, 2004). The feasibility of molecular tools to perform the genomic analysis without the need to culture has increased significantly our understanding on the complexity and diversity of bacterial community found in biofilm. In the following sections, we will review the basic principles, advantages and limitations belonging to some of these molecular techniques which are mainly coupled with 16S rRNA gene analysis (Box 21.2).

Box 21.2

16S rRNA gene as phylogenetic marker for bacterial community analysis

Ribosome is a ribonucleoprotein complex that catalyses protein synthesis in living cells. It is composed of two different sized subunits; the small subunit helps to translate the genetic code in mRNA by mediating the interaction between mRNA and tRNA whereas the large subunit facilitates peptide bond formation in the growing polypeptide chain. Construction of high resolution crystal structures of the large subunit about a decade ago revealed important contributions of rRNA molecules in protein synthesis. In prokaryotes, 16S rRNA is found in the small subunit which sediments at 30S and, 23S and 5S rRNA molecules are associated with the larger subunit (50S). These rRNA molecules are highly conserved in bacteria with regards to size, nucleotide sequence and secondary structure. 16S rRNA is perhaps the most conserved amongst the three rRNA species.

The primary diagnostic value of 16S rRNA gene as phylogenetic marker lies in the presence of conserved regions that enable design of universal primers to amplify target genes from a wide variety of species and variable region specific to species. A total of nine variable regions (V1–V9) are identified in the 16S rRNA gene as shown in the diagram below.

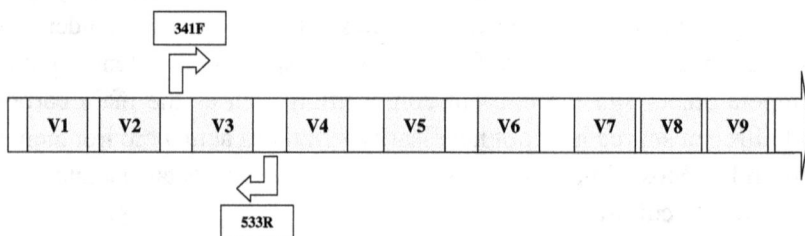

Although V1 is most varied, PCR amplicons derived from V3 provide the best discrimination between bacterial species in the DNA fingerprinting technique DGGE (X.2.1) (Yu and Morrison, 2004). The universal forward (F) and reverse (R) primers priming the V3 region is denoted in the diagram and numbering of the primers corresponds to the nucleotide position in the *Escherichia coli* 16S rRNA gene. Besides well-characterized universal primers, availability of robust database for 16S rRNA gene sequences further enhanced the application of 16S rRNA gene as a phylogenetic tool. Ribosomal Database Project (RDP) contains more than 600,000 16S rRNA gene

Box 21.2   (*Cont'd*)

sequences, with at least 80% derived from environmental samples (Cole *et al.*, 2009). Small subunit rRNA gene alignment and analysis tools including a recent pyrosequencing pipeline for the analysis of ultra high-throughput rRNA sequencing data are accessible at http://rdp.cme.msu.edu/. Therefore, sequence analysis of 16S rRNA gene is a powerful tool for typing bacterial population.

21.2.1. *Denaturing gradient gel electrophoresis (DGGE)*

Denaturing gradient gel electrophoresis (DGGE) separates double-stranded DNA molecules in a polyacrylamide gel containing a mixture of DNA denaturants made up of urea and formamide in a linear concentration gradient. In the closely related system of temperature gradient gel electrophoresis (TGGE), a linear gradient of temperature is used instead of the denaturants.

Historically, DGGE has been widely used to detect mutations associated with human diseases attributed by single to few nucleotides changes within specific regions of genomic DNA. In 1993, Muyzer and co-workers first applied DGGE in combination with 16S rDNA-PCR to profile a complex mixture of uncharacterized microorganisms. Since then, this DNA fingerprinting technique has been routinely applied to study the diversity and dynamics of microbial populations present in natural habitats such as soil, sea and hot springs, gastrointestinal tracts of animals and more recently, bioreactors used for wastewater treatment and food fermentation.

The combined approach of 16S rDNA-PCR and DGGE is now widely used to elucidate the diversity and succession of bacterial population in biofilm. Total genomic DNA extracted from the biofilm is used as template in PCR amplification and the resultant DNA amplicons are then subjected to DGGE analysis using a pre-determined and optimized concentration gradient of denaturants. A 40-base pair GC-rich sequence is commonly incorporated into the PCR amplified products to further improve the resolution of all possible nucleotide sequence variations (Sheffield *et al.*, 1989). The migration and retardation of the DNA amplicons in the gel is based on the melting properties of the double-stranded DNA molecules; as such, separation is dependent on nucleotide sequence and not the size of the DNA fragment. As the helical DNA amplicons migrate into an increasing gradient of denaturants, localized disruption of the double-stranded DNA molecule in the so-called "melting domain" will occur to the extent that its mobility will get retarded at a characteristic depth specific to its nucleotide sequence. The DNA molecules with identical nucleotide sequence appear as a single band on the DGGE gel when stained with DNA-binding dyes.

A bacterial 16S rDNA-DGGE profile belonging to an aquatic biofilm examined at three different time points is shown in Fig. 21.2 (Teng *et al.*, unpublished data). It is assumed that each DNA band in a DGGE gel represents at least one bacterial species or variant of a functional gene. The occurrence of many distinct DGGE bands in Fig. 21.2 therefore indicates the presence of a complex and diverse bacterial population in the aquatic bioflm. The dynamics of the population could be projected by the intensity of specific DNA bands and their distribution across multiple samples representing different treatment conditions or time of study. As shown in

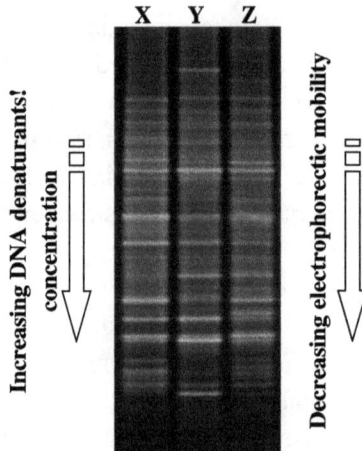

Fig. 21.2.   Bacterial 16S rDNA-DGGE profile belonging to an aquatic biofilm. 16S rDNA-PCR amplicons derived from an aquatic biofilm harvested at time points arbitrarily labeled as X, Y and Z are separated in a parallel DGGE gel consisting of 35–65% denaturants.

Fig. 21.2, the banding patterns varied between the different time points thus suggesting a dynamic bacterial community. The individual DNA band in the DGGE profile could be further identified by either hybridizing using a genus- or cluster-specific oligonucleotide probe, or cloning and sequencing of the excised DNA band. The latter tends to be more laborious and time-consuming.

The 16S rDNA-DGGE profile depicting the bacterial community structure is subjected to errors and biases inherent in any PCR amplification technique. Efficiency of DNA extraction protocols including removal of PCR inhibitors from the extracted genomic DNA may result in preferential or differential amplification of rRNA genes. Unwanted annealing of two different PCR products resulting in heteroduplex DNA molecules could potentially falsify the actual diversity of the bacterial community studied. Besides the limitations associated with PCR technique, handling and storage of environmental samples could also affect bacterial species composition if the conditions deviate significantly from the original source. Lastly, limited phylogenetic-related studies could be inferred from DGGE and TGGE profiles since both techniques resolve well only relatively small DNA fragments, notwithstanding that the resolution is also highly dependent on the primer sequence used and various gel electrophoresis conditions. Nevertheless, PCR in combination with either denaturant or temperature gradient gel electrophoresis is a powerful and rapid tool to analyze the complexity and dynamics of bacterial population across different biofilm samples simultaneously.

21.2.2. *Single-strand-conformation polymorphism (SSCP)*

Single-Strand-Conformation Polymorphism (SSCP), as the name implies, separates similar length single-stranded DNA molecules based on sequence-specific secondary conformations. Similarly with DGGE, SSCP was initially used to detect polymorphisms and mutations in human genes. SSCP analysis coupled with 16S rDNA-PCR was then adapted by Lee and co-workers in 1996 for studying the structure and diversity of bacterial community in the aquatic ecosystem.

In SSCP-PCR analysis, the PCR amplicons are first denatured and the resulting single-stranded DNA molecules separated in a non-denaturing acrylamide gel. The electrophoretic mobility of these similar length single-stranded DNA molecules depends on its secondary conformation which is dictated by the DNA sequence. The nucleotide composition directs intra-molecular chain interactions producing specific localized folding pattern resulting in an overall unique conformation with distinct electrophoretic mobility. Similarly with DGGE, the DNA band of interest can be excised, sub-cloned and sequenced for identification at species level. However, no extended GC-rich sequences are required for addition to the PCR amplicons.

The characteristic SSCP band patterning can be applied for comparative analysis between bacterial communities derived from different biofilms or the same biofilm subjected to different treatments or conditions of study. However, the caveat of this technique is that the single-stranded DNA molecules tend to re-anneal in the gel thereby increasing the complexity of the SSCP band patterning. The formation of duplex DNA bands however could be prevented by using alkali denaturing buffer and glycerol-free gel albeit somewhat compromised SSCP pattern resolution.

21.2.3. *Terminal restriction fragment length polymorphism (T-RFLP)*

Terminal Restriction Fragment Length Polymorphism (T-RFLP) is another robust DNA fingerprinting tool that can be used in combination with either 16S rRNA genes or functional genes to dissect the complex bacterial community in biofilm (Liu *et al.*, 1997). This technique is an extended version of Restriction Fragment Length Polymorphism (RFLP) which is also known as Amplified rDNA Restriction Analysis (ARDRA). Both techniques share the initial steps of DNA isolation, PCR amplification and restriction digest with the exception in the mode of analysis of the restriction fragments.

In ARDRA, 16S rRNA genes are amplified from the total genomic DNA belonging to a bacterial community and digested using one or more restriction enzymes which typically have four base-pair (bp) recognition sites. The varying length restriction fragments are then separated in an agarose gel. The RFLP patterns derived from a complex bacterial community are generally very complicated hence not easily resolved by the relatively low resolution agarose gels. Polyacrylamide gels may help to better resolve the large number of restriction fragments, especially the smaller fragments. But, due to the limited staining sensitivity of DNA binding dyes, visualization of less abundant DNA bands may still be suppressed hence not reflecting the actual diversity of the bacterial community.

Alternatively, the amplified target genes could be sub-cloned and re-amplified from individual clones before being digested similarly using either one or more 4-bp restriction enzymes. Each specific restriction pattern could be clustered separately as individual operational taxonomic unit (OTU) (Moyer et al., 1994). The occurrence and frequency of these discrete OTUs are reflective of the diversity and structure of the bacterial community studied. In most cases, the almost full-length 16S rRNA genes are further sequenced for the purpose of species identification and phylogenetic-relatedness analysis. The construction and screening of clone libraries are however laborious and time-consuming, hence ARDRA is more suitable for studying relatively simple bacterial community.

In contrast with ARDRA, T-RFLP provides rapid resolution and quantification of terminally-labeled restriction fragments using an automated DNA sequencer. 16S rRNA genes are amplified using at least one primer that is fluorescently labeled; the resulting amplicons are most frequently labeled at 5' end. Simultaneous application of two differently labeled primers, resulting in two DNA fingerprinting profiles per community, is also common and helps to increase the level of confidence for the prediction of diversity in the bacterial community studied. FAM, HEX, VIC, NED and PET are some of the commonly used fluorescent dyes applied in T-RFLP. It is important to note that different dyes may affect the migration of the fragments possibly resulting in some discrepancy with regards to the size of the labeled fragments. The PCR amplicons are then subjected to digestion using restriction enzymes possessing four base-pair recognition sites and the varying length terminally-labeled restriction fragments resolved using an automated DNA sequencer.

A T-RFLP profile belonging to an aquatic biofilm (Teng et al., unpublished data) is shown in Fig. 21.3. At least 10 terminal fragments with the size up to 400 bp and varying fluorescent intensity could be represented in this example. Each peak in the electropherogram represents a specific sized terminal-labeled restriction fragment belonging to one

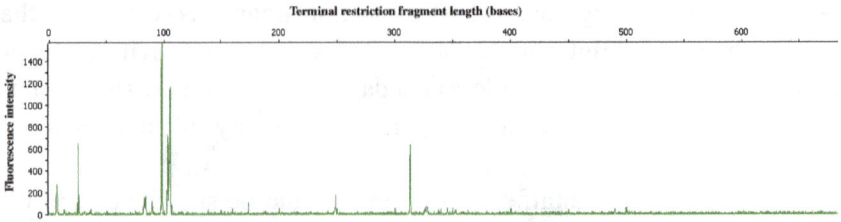

Fig. 21.3.   Bacterial 16S rDNA t-RFLP profile belonging to an aquatic biofilm. Bacterial 16S rRNA genes were amplified using primers binding at *Escherichia coli* 16S rRNA gene position 1055–1392 bp. The 5′ HEX-labeled amplicons were digested using MspI enzyme and the labeled fragments sorted in an automated DNA sequencer.

closely-related phylogenetic group. The intensity of the fluorescent signal for each fragment reflects its relative abundance. This distinct peak or fragment patterning, otherwise called ribotypes, could be analyzed against the 16S rDNA database in RDP in order to assign best-matched phylogenetic groups present in the community studied. Thus, the gain or loss of specific fragments and their relative distributions helps to indicate any temporal or spatial changes occurring in the bacterial community.

The main advantage of T-RFLP is its high-throughput nature and resolution capability of capillary electrophoresis or sequencing gel compares with the electrophoresis in DGGE/TGGE, SSCP or ARDRA. T-RFLP, however, also shares biases inherent in any PCR-based tool and lacks definite identification of bacterial species which could be achieved in the other tools simply by performing cloning and sequencing of the separated DNA species. Nevertheless, the application of multiple restriction enzymes in combination with optimal choice of primers enables rapid construction of highly discriminating restriction patterns to predict the structure and diversity of a bacterial community in biofilm.

21.2.4. *Automated ribosomal intergenic spacer analysis (ARISA)*

Automated ribosomal intergenic spacer analysis (ARISA) is a microbial diversity analysis tool that relies on the length and sequence heterogeneity found in the intergenic transcribed spacer (ITS) region localized between the 16S and 23S ribosomal genes (Fisher and Triplett, 1999). The ITS region belonging to different bacterial species is amplified using fluorescent labeled universal primers and the varying length ITS amplicons are detected using an automated sequencer. The main advantage of ARISA is that it could be coupled with RFLP analysis to sort the varying length ITS amplicons. The setback, however, is that the technique is PCR biased with regards to preferential amplification of shorter fragments and the possibility of more than one ITS length variants arising from a single species itself.

21.2.5. *Emerging molecular tools*

A search in the PUBMED database showed that DGGE is the most applied technique for studying bacterial biofilm amongst the molecular tools discussed above (Sec. 21.2.1 to Sec. 21.2.4). The relative distribution of reported investigations that arise from these tools in bacterial biofilm studies is summarized in a pie chart in Fig. 21.4. This observation could be explained in terms of the relatively ease-of-use and low cost features in DGGE for bacterial profile analysis and importantly, it can be coupled with DNA sequencing for identification of separated DNA amplicons. However, there are a limited number of clones that can be sampled and sequenced in DGGE for identification and typing of the bacterial population at species level.

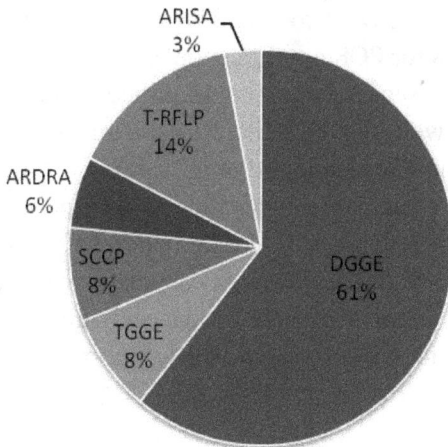

Fig. 21.4.   Relative distribution of molecular tools applied for bacterial biofilm studies. The percentage represents relative frequency of reports related to the respective molecular tools in bacterial biofilm studies found in Pubmed database surveyed until December 31st, 2010.

The recently developed ultra-high throughput sequencing technologies known collectively as the next-generation sequencing (NGS) are able to produce massive output of sequencing data rapidly and inexpensively in comparison to the conventional Sanger/capillary-based sequencing. Roche/454 FLX Pyrosequencer, Illumina/Solexa Genome Analyzer and Life/APG SOLiD System are three relatively widely used NGS platforms, pyrosequencer being the first platform made commercially available in 2005. NGS belonging to Roche/454, Illumina/Solexa and Life/APG SOLiD system will be briefly described in this section. Interested readers are directed to a review article by Metzker (2010) for more detailed description of these platforms including other recent NGS technologies such as Helicos BioSciences, Polonator and Pacific Biosciences.

The NGS platforms utilize unique chemistry for sequencing but share the same features in template preparation which is not dependent on bacterial cloning. The template DNA is first randomly fragmented and the fragment ends made blunt before being annealed to platform-specific adapters for amplification either by emulsion PCR or solid-phase amplification. In Roche/454 Pyrosequencer, the adapter-ligated single-stranded DNA fragments are captured on beads containing surface-immobilized oligonulceotides that are complementary to the adapter sequence. Each bead that is associated with a single DNA fragment is placed in a water-in-oil emulsion containing reagents for PCR amplification to make sufficient copies of the DNA fragment for sequencing. The amplified bead is then distributed within individual well of a PicoTiterPlate for sequencing. The wells are supplied with a single nucleotide solution at any one time and a pyrophosphate is released whenever a nucleotide is incorporated into the DNA strand by the DNA polymerase. The pyrophosphate is converted to ATP which is coupled with the downstream enzymatic reaction of luciferase to produce light; the amount of light produced is thus proportional to the number of nucleotides incorporated. All unincorporated nucleotides will be degraded to allow repetitive addition of other nucleotides.

In Illumina/Solexa NGS system, the fragmented template DNA is amplified by solid-phase amplification and sequencing is performed using nucleotides that carry a base-specific fluorescent label with the 3'-OH group reversibly blocked to ensure that incorporation of each nucleotide by DNA polymerase is imaged as a unique event. The chemically blocked

3'-OH group is freed for next nucleotide incorporation only after the imaging step. In contrast, Life/APG SOLiD System's sequencing approach is based on DNA ligation. Template DNA is first amplified using emulsion PCR. After amplification, a sequencing primer is hybridized to the complementary adaptor sequence linked to the fragmented DNA. Fluorescently-labeled 8-mers oligonucleotide probe that contains two matching bases with the template strand is then ligated by DNA ligase to the hybridized sequencing primer. The fluorescent signal is captured for each ligation before the fluorescent group is cleaved off including bases from the probe to prepare for new ligation event in the extending sequencing primer. The whole process is repeated from beginning using n-1, n-2, n-3 etc. primers in order to probe the complete template strand.

To date, Roche FLX Pyrosequencer (Titanium series) provides an average of 400 bases read length, whereas Illumina/Solexa and Life/APG SOLiD systems provide comparable read length that is less than 100 bp. As such, pyrosequencing platform is more advantageous for the application in typing complex bacterial community in biofilm using 16S rRNA genes as phylogenetic marker. Nevertheless, the NGS technology is continually evolving to provide higher read length and throughput with innovative sequencing chemistry. NGS system from Pacific Biosciences, for example, is able to detect in real-time the addition of a nucleotide to a growing strand of DNA by the polymerase enzyme without amplification of template DNA thus avoiding amplification bias inherent in PCR approach. The resulting read length is also significantly higher, potentially more than 1 kb. NGS is therefore paving the way for high throughput typing of diverse and complex microbial communities.

Besides bacterial population analysis, the detection of specific bacterial species is informative in determining the spatiotemporal role contributed by individual species in biofilm formation. Detection and *in situ* visualization of specific bacterial species in biofilm could be performed using nucleic acid-based hybridization approaches such as fluorescence *in situ* hybridization (FISH). This technique allows direct hybridization of nuclei acids belonging to specific bacterial species found in a biofilm using fluorescent probes without the application of PCR amplification. Microscopic examination could be performed using Confocal Laser Scanning Microscopy (CLSM). Coupling of FISH with CLSM enables high-resolution visualization of occurring bacterial species at different depths of the biofilm matrix.

Clearly, molecular tools are becoming increasingly important to complement engineering tools in monitoring and controlling biofilm development. However, the molecular tools are still limiting with respect to quantification and viability determination of the occurring bacterial species. Therefore, it remains to be seen how new molecular tools will evolve to provide more information on the relationship between genetic diversity and cell physiology in the bacterial community found in biofilm in order to understand the dynamic and complex phenomenon of biofilm development.

21.3. Impacts of Bacterial Biofilm on Water Management

Biofilm can be found at different stages in water systems, ranging from the source water to treatment and distribution of treated water. Bacteria can form biofilm on surfaces such as those found in the water storage tanks, membrane filtration systems used in water treatment, cooling systems and pipings in potable water distribution system. The occurrence of bacterial biofilm is significant, posing potential public health hazard and aesthetic concerns. In the water distribution system, bacterial biofilm can contribute to pipe corrosion and importantly, compromise water quality by secreting off-flavor compounds and sloughing off bacterial cells which could include opportunistic pathogens into the water phase. The diverse bacterial species living in close proximity within the biofilm may also promote spreading of antimicrobial resistance genes. Interestingly, the unwanted occurrence of these biofilm in water distribution system could be partially attributed to processes such as ozonation and dosing of chloramines which may inevitably increase the biological instability of water. In ozonated water, an increased amount of smaller and biodegradable organic products could be released from the partially oxidized natural organic matters (Hammes et al., 2006) while the decay of chloramines produces NH_4^+ which may act as the inorganic electron donor thus further enhances the biological instability of water (Regan et al., 2002).

In water treatment, membrane-based technologies such as micro- and ultra filtration and reverse osmosis processes are gaining significance due to small footprint, decreasing cost of membranes and rapid advances in developing high-performance membranes. The membranes, however, are susceptible to fouling which is primarily caused by particulate matter, organic and inorganic compounds, and biological growth, with the latter being the most difficult to eliminate. Excessive formation of biofilm due to microbial growth on the membrane surfaces otherwise known as bio-fouling has adverse effects on the membrane's function, as it reduces the filtration efficiency of membrane, resulting in higher operating cost of membrane processes attributed to increased chemical usage for membrane cleaning and increased frequency of membrane replacement. As such, better understanding of the science underlying biofilm development on different surfaces and factors that promote biofilm formation need to be gained in order to implement pragmatic measures to control and monitor biofilm development in the water industry.

Lastly, the occurrence of bacterial biofilm in the water sector may not necessarily be undesirable. The complex bacterial community present in biofilm containing diverse metabolic capability is of immeasurable help to degrade nutrients of organic material in domestic and industrial waste-waters (Table 21.1). The design of membrane bioreactors (MBR) which

Table 21.1. Biofilm-based applications in drinking water and wastewater treatments.

Applications	Outcomes
Aerobic biofilm process in the pretreatment of source water for drinking water production	• Reduction of Biochemical Oxygen Demand (BOD)
• Biofilms formed on a porous sand bed (e.g. Slow sand filter)	• Trapping of pathogens
Aerobic biofilm process for wastewater treatment	• Reduction of BOD
• Biofilms formed on either a bed of irregularly shaped rocks (e.g. Trickling filter) or corrugated plastic sheets with large pores (e.g. Biological tower)	• Reduction of inorganic electron donors such as NH_4^+ through nitrification
• Biofilms formed on small-granule filters that are packed either in a fixed bed (e.g. Biological aerated filter) or moving-bed reactors (fluidized or circulated)	• Reduction of electron acceptors such as NO_3^- through denitrification
Anaerobe biofilm process for wastewater treatment	• Digestion of organic solids
• Biofilms formed on the granulated sludge (e.g. upflow anaerobic sludge-blanket reactor)	• Generation of methane gas, a valuable energy resource

combines some of these biofilm processes with membrane filtration for solid separation is one of the key new developments in wastewater treatment. Besides wastewater treatment, biofilm could also be applied *in situ* to remove hazardous contaminants such as heavy metals from contaminated groundwater. Clearly, the beneficial applications of biofilms are extensive and will remain as one of the key catalysts for the advancement of water treatment technologies.

References

1. Cole JR, Wang Q, Cardenas E, Fish J, Chai B, Farris RJ, Kulam-Syed-Mohideen AS, McGarrell DM, Marsh T, Garrity GM and Tiedje JM. The Ribosomal Database Project: improved alignments and new tools for rRNA analysis. *Nucleic Acids Res.* **37**(Database issue): D141–D145; doi: 10.1093/nar/gkn879, 2009.
2. Davies DG, Parsek MR, Pearson JP, Iglewski BH, Costerton JW and Greenberg EP. The involvement of cell-to-cell signals in the development of a bacterial biofilm. *Science* **280**(5361): 295–298, 1998.
3. Fisher MM and Triplett EW. Automated approach for ribosomal intergenic spacer analysis of microbial diversity and its application to freshwater bacterial communities. *Appl. Environ. Microbiol.* **65**(10): 4630–4636, 1999.
4. Hammes F, Salhi E, Köster O, Kaiser HP, Egli T and von Gunten U. Mechanistic and kinetic evaluation of organic disinfection by-product and assimilable organic carbon (AOC) formation during the ozonation of drinking water. *Water Res.* **40**(12): 2275–2286, 2006.
5. Handelsman J. Metagenomics: application of genomics to uncultured microorganisms. *Microbiol. Mol. Biol. Rev.* **68**(4): 669–685, 2004.
6. Lee DH, Zo YG, Kim SJ. Nonradioactive method to study genetic profiles of natural bacterial communities by PCR-single-strand-conformation polymorphism. *Appl. Environ. Microbiol.* **62**(9): 3112–3120, 1996.
7. Liu WT, Marsh TL, Cheng H and Forney LJ. Characterization of microbial diversity by determining terminal restriction fragment length polymorphisms of genes encoding 16S rRNA. *Appl. Environ. Microbiol.* **63**(11): 4516–4522, 1997.

8. Metzker ML. Sequencing technologies — The next generation. *Nat. Rev. Genet.* **11**: 31–46, 2010.
9. Moyer CL, Dobbs FC and Karl DM. Estimation of diversity and community structure through restriction fragment length polymorphism distribution analysis of bacterial 16S rRNA genes from a microbial mat at an active, hydrothermal vent system, Loihi Seamount, Hawaii. *Appl. Environ. Microbiol.* **60**(3): 871–879, 1994.
10. Muyzer G, de Waal EC and Uitterlinden AG. Profiling of complex microbial populations by denaturing gradient gel electrophoresis analysis of polymerase chain reaction-amplified genes coding for 16S rRNA. *Appl. Environ. Microbiol.* **59**(3): 695–700, 1993.
11. Regan JM, Harrington GW and Noguera DR. Ammonia- and nitrite-oxidizing bacterial communities in a pilot-scale chloraminated drinking water distribution system. *Appl. Environ. Microbiol.* **68**(1): 73–81, 2002.
12. Sheffield VC, Cox DR, Lerman LS and Myers RM. Attachment of a 40-base-pair G+C-rich sequence (GC-clamp) to genomic DNA fragments by the polymerase chain reaction results in improved detection of single-base changes. *Proc. Natl. Acad. Sci. USA* **86**: 232–236, 1989.
13. Teng WL, Tan GLX and Gao PP. Elucidation of microbial species in desalination membranes and microbial water quality monitoring using DNA-based techniques. Technical Report, Technology and Water Quality Office, PUB, Singapore.
14. Yu Z and Morrison M. Comparisons of different hypervariable regions of rrs genes for use in fingerprinting of microbial communities by PCR-denaturing gradient gel electrophoresis. *Appl. Environ. Microbiol.* **70**(8): 4800–4806, 2004.

Questions for Thought...

1. Describe a laboratory device suitable for high-throughput study of biofilm formation and development.
2. Discuss all potential hurdles to overcome in designing a real-time sensor for monitoring biofilm in drinking water distribution system.

Annex

Regulatory Issues on Application of Natural and Genetically Engineered Microbes in Environmental Biotechnology

Wong Po-Keung and Chu Lee Man

The rapid development of biotechnology in various areas including their applications in environment has attracted numerous legal, safety and public policy issues. The use of "unnatural" microorganisms, especially the genetically engineered microorganisms, in solving environmental problems such as waste treatment has lead to the intensive debate on the safety issue. The fate and risk of releasing these "unnatural" microorganisms into the environment are largely unknown. Although many studies preliminarily indicate that the problem caused by these microorganisms can be minimized by introduced a "suicide" (i.e. self-destruct) trait into the microorganisms used in environmental biotechnology. Once these microorganisms accomplished their expected roles, they will be eliminated from the environment by a specific trait specially designed and embedded in their genetic materials. An example for this approach is that the microbial strain released into the environment losses its ability in propagation into the natural environment, the released microorganisms will be eliminated until a new batch of the microorganisms are introduced into the environment. Another approach is to make these "unnatural" microorganisms such that they are removed easily from the environment by physical or chemical means. One of the common examples is the flocculant strain of microorganisms. During the operation, aeration is provided to maintain the suspension of the microbial cells in aqueous solution. At the end, microbial cells will settle to the bottom of the reactor once the aeration is discontinued and the settled microbial cells can be easily removed by physical means.

The safety of using natural and genetically engineered microbial strains in environmental biotechnological application has been debated since 1970. The public fears that the uncontrolled release of these microorganisms into our environment will disturb the balance of ecosystems in the natural environment. The imbalance will in turn cause more problems to our environment than the pollution by wastes/toxicants. There are comparatively comprehensive and stringent regulatory guidelines for the use of recombinant DNA technology and genetically engineered organisms in biotechnological applications; there are quite a few policies on the use of natural microorganisms in environmental biotechnology. In order to prepare a regulatory guideline for the use of "natural" microorganisms in environmental biotechnology, the regulatory guidelines for the release of genetically engineered microorganisms will be reviewed in order to lay the foundation for formulating the guidelines for the "natural" microorganisms.

There are several government agencies in the US that deal with the regulatory issues for the genetically engineered microorganisms. They are the National Institutes of Health (NIH), Environmental Protection Agency (EPA), Food and Drug Administration (FDA) and the United States Department of Agriculture (USDA). Some of the representative acts that have been prepared by these government agencies for the regulatory issue of genetically engineered microorganisms are Federal Insecticide, Fungicide, and Rodenticide Act (FIFRA); Toxic Substances Control Act (TSCA); National Environmental Policy Act (NEPA); Federal Plant Pest Act (FPPA) and Plant Quarantine Act (PQA).

The recombinant DNA Molecule Program Advisory Committee (RAC) of NIH established voluntary guidelines for researches involving recombinant DNA in 1976. However, the NIH did not have the statutory authority to regulate research and the guidelines, and had real impact only on the recipients of federal grants. Until the late 1970s, NIH, USDA, FDA and NSF (National Science Foundation) coordinately implemented the guidelines of NIH. However, the coordination became more difficult as the recombinant DNA technology reached the stage of commercial products. With large-scale commercial production and the potential for deliberate release into the environment, additional regulatory agencies were required. The EPA was established in 1970 to execute the authority over the activities that have potential to pollute the environment (air, water and land) in the US. At the same time, although USDA had regulations on the use of biological pesticides, each microbial pesticide was registered on an *ad hoc* basis since no consolidated policy had been issued until 1979.

In 1986, the EPA issued the FIFRA and applied TSCA to cover the genetically engineered microorganisms for a variety of uses such as environmental cleanup and industrial uses. The FIFRA also regulates microbial pesticides including viruses, bacteria, fungi and algae and derivatives of any of these microorganisms. The Act prohibits the distribution, sale and use of microbial pesticides that have not been registered with the EPA. The EPA reviews all data submitted for microbial pesticides and determines whether registration is appropriate. For registration, applicants must cite data on product composition, human health effects, environmental fate, and effects on non-target organisms. The EPA assesses this information when making decisions about safety. An experimental use permit (EUP) is required to conduct initial tests, even field studies. FIFRA has generally required an EUP even for small-scale testing of an area less than ten acres if genetically engineered microorganisms are used for the test.

The TSCA gives EPA the authority to regulate chemicals that may pose a threat to human health and the environment in research and commercial products development. The Act requires EPA to review test data within a specific period of time and determine whether the product is safe or pose a risk. TSCA was enacted to regulate organic and inorganic substances or a combination of substances, and applied to all micro-organisms produced for industrial, consumer and environmental uses with the exception of their manufacture, processing, or distribution as foods, food additives, cosmetics, medicines, or pesticides that are regulated by other acts.

The NEPA requires that all federal agencies prepare an environmental impact statement or an environmental assessment of any major federal action that might adversely affect the human environment. The impact statement must describe the predicted effect of the action, any adverse impacts, and alternatives for the proposed action. The NEPA is not a regulatory statute and it simply ensures that government agencies assess risks and environmental impact of genetically engineered microorganisms.

The FPPA and PQA provide the authority for regulating the movement of genetically engineered microorganisms that are potential plant pests into or within the US. Field tests of genetically engineered crops and plant pests require the approval of USDA under the FPPA and PQA.

Part VI

Microbes in Alternative Energy and in Mining

Photosynthetic microorganisms, such as micro-algae and cyanobacteria are able to harness low-intensity solar energy and store it as latent chemical energy in the biomass, which can then be released via biochemical conversion. Various photobioreactors have been developed to allow maximum uptake and storage of solar energy. The structural and storage carbohydrates in biomass have low energy content and it is necessary to concentrate the energy content further for fuel application. Anaerobic microbial fermentation is an efficient and widely used method for such conversion processes. Useful renewable fuels produced by microorganisms include hydrocarbons, ethanol, methane and hydrogen. Biofuel cells that are able to release energy in fuel chemicals to generate electrical energy at ambient temperatures have been developed.

The unique ability of microbes in the recovery of metals from low grade ores formed the science and biotechnology of biomining. Together with the microbial CO_2 remediation and biofuel production, they contribute toward a sustainable future for the human kind.

Chapter 22

Microorganisms and Production of Alternative Energy

Lee Yuan Kun

Department of Microbiology, Yong Loo Lin School of Medicine
National University of Singapore
5 Science Drive 2, Singapore 117597
Email: micleeyk@nus.edu.sg

The 1973 oil embargo and accompanying soaring oil prices together with the recognition that natural fuel reserves are being rapidly depleted, led to a worldwide interest in the development of alternative renewable fuels. Microbial production of fuels has the potential for helping to meet world energy demands. Living organisms assimilate and concentrate energy in their biomass and products. Hence, biomass in its various forms is an attractive alternative source of energy. Through photosynthesis, biomass collects and stores low-intensity solar energy, which can then be harvested and released via biochemical conversion. Biological processes are involved in both the harnessing of solar energy and upgrading of low energy feed stocks to biomass fuels. Useful fuels produced by microorganisms include biodiesel (produced from algal oils), ethanol, methane and hydrogen.

22.1. Harnessing Solar Energy

The sum of solar energy receives by earth per year is estimated to be 3×10^{24} J, which is about 100 times that of the proven total reserves of non-renewable energy (2.5×10^{22} J or 8×10^{11} ton coal equivalent).

This means that solar collectors covering 0.1% of earth's surface with 10% efficiency will meet all current world annual needs for energy (3×10^{20} J). Besides, the energy source is almost unlimited and the energy is renewable.

However, there are technical difficulties involved when harnessing solar energy as industrial fuel. These include the following: (i) the energy supply is variable and intermittent (solar irradiance changes from dawn to noon to dust, and total darkness in the night); (ii) the energy supply is diffuse (the highest irradiance measured at noon in the tropic and during summer in the temperate is about 4000 μE/m^2/s). Thus, it requires some form of storage system and a large area of collectors.

22.1.1. *Photosynthetic Production of Biomass*

Photosynthetic organisms (including cyanobacteria, algae, plants) use solar irradiance between 400 and 700 nm (known as the photosynthetically available radiance, PAR), this comprises about 50% of the total solar energy.

The overall photosynthetic process is represented by the following equation:

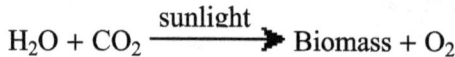

$$H_2O + CO_2 \xrightarrow{\text{sunlight}} \text{Biomass} + O_2$$

In this case, solar energy is captured by the photosynthetic apparatus, and stored as latent energy in the chemical structure of the biomass produced. The biomass can then be converted to renewable fuels, replacing fossil fuels.

To improve the productivity of cultivation, various **photobioreactors** have been developed to allow maximum uptake and storage of solar energy in photosynthetic microalgae and cyanobacteria. In general, besides having large illuminated surface area to volume ratio, these photobioreactors possess the following features:

(i) High turbulent streaming
High turbulence is required to move cells rapidly across the light path, which is perpendicular to the surface of the bioreactor (Fig. 22.1). This is to prevent the photosynthetic cells from staying too long in low light

(bottom of the culture) resulting in energy starvation, or to prevent their surface from being expose to inhibitory level of high sun light intensity. High turbulence is also necessary to facilitate mass transfer of gases (CO_2 and O_2), substrates and waste products at the surface of cell, and at gas-liquid interface.

Degree of turbulence in a flowing liquid along a pipe is proportional to the linear flow rate of the liquid, diameter of the pipe, and inversely proportional to the viscosity of the liquid. It is usually expressed as Reynolds Number, a partial turbulence is attended at a Reynolds Number

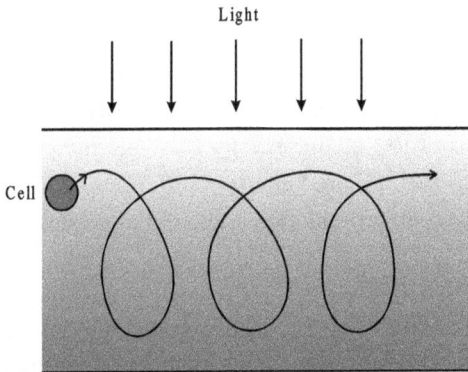

Fig. 22.1. Movement of a photosynthetic cell across the light intensity gradient in a culture illuminated from one direction.

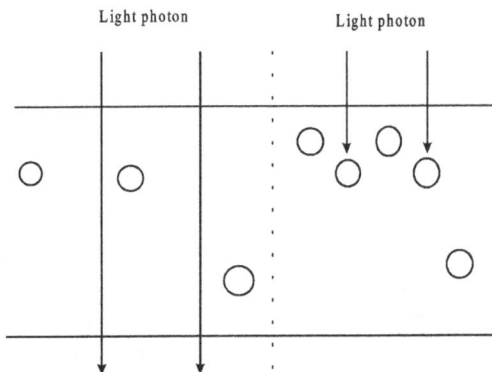

Fig. 22.2. Light penetration in a photosynthetic culture.

of 2000, and complete turbulence is achieved at a Reynolds Number of 4000.

(ii) High cell density
This is necessary to capture all light photons that impinge on the culture. A light photon would pass through the bioreactor if it is not captured by a cell while traveling through the culture (Fig. 22.2).

(iii) Light absorption
The density of photosynthetic pigments and light absorption cross section determine the absorption of light photons by a single cell in the light path. A cell of low pigment density and small light absorption cross section would allow more light to pass through and reach cells at the lower level of the light path. A cell that captures excessive quantity of light energy would lead to either dissipation of the energy (for the photosynthetic system is not able to process the captured energy sufficiently fast), or destruction of the photosynthetic components and reduce photosynthetic activity.

(iv) Shallow culture
High cell density would result in shading of cells by other cells in the light path. Discontinuous supply of light energy due to the light gradient in the

---

Box 22.1

An Enclosed α-Type Tubular Photobioreactor Developed in Singapore

The photosynthetic culture is lifted up in air-riser tubes to a receiver tank 5 m above ground, and then flows down the tubular photobioreactor to reach the opposite set of air-riser tubes. The culture was then lifted up 5 m to another receiving tank, flowing down to parallel photobioreactor tubes connected to the base of the first set of riser tubes. Because the flow of culture in the tubular bioreactor does not change direction except when lifted up in the air-lift system, a high liquid flow rate and Reynolds number, and short residence time are achieved at relatively low air supply rates. As the α-photobioreactor is placed at an angle with the horizon, a fairly constant solar irradiance is received by the culture throughout the day, allowing high biomass productivity at high culture density.

Box 22.1   (*Cont'd*)

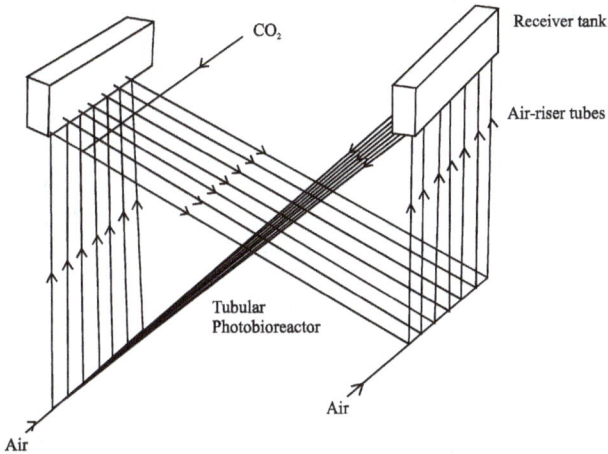

Box 22.2

Effect of Photobioreactor Inclination on the Profile of Sunlight Measured at the Surface of the Reactor

The photon flux density incidented on the upper surface of a bioreactor panel facing the sun is $I \cdot \cos(\delta - \alpha)$, where I = the intensity of sun-light incident on a surface perpendicular to the sun rays, δ = the angle between the direction of sun and the vertical, α = the angle of inclination between the

Box 22.2 *(Cont'd)*

bioreactor and the horizontal. Thus, the photon flux density measured at the upper surface of the bioreactor panel facing the sun increased with increasing α, and reached the maximum at $\delta = \alpha$. At a fixed-angle α, the intensity of direct light measured at the bioreactor surface decreased with decreasing δ, as shown in the figure below.

Along the equatorial (like in Singapore), for a bioreactor inclined at an angle of about 45°, the photosynthetic radiance measured at the surface remains almost constant throughout the day.

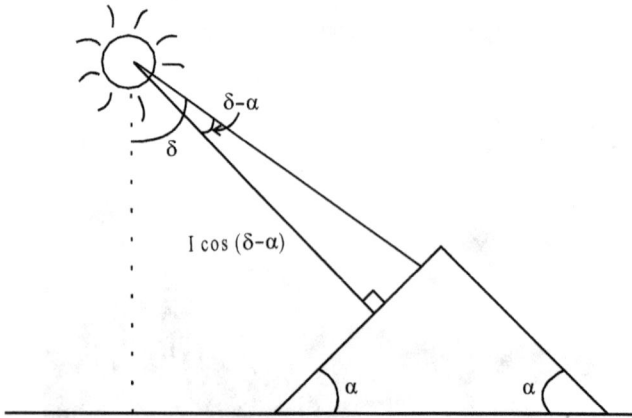

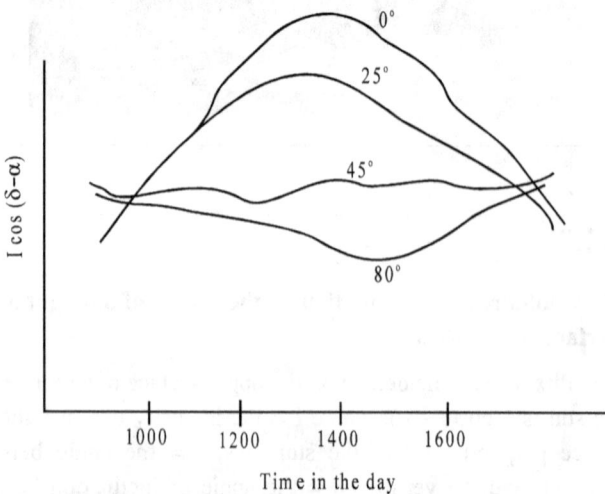

culture, would lead to enhance maintenance energy requirement and lower growth yield (see Chap. 2, Sec. 2.1.4). Shallow culture and high turbulence would allow frequent short exposure of cells to light at the surface of the bioreactor.

(v) Inclined at an appropriate angle with the horizon
Inclining a photobioreactor at an angle allows the bioreactor to receive more irradiance in the morning and afternoon, and less at noon. As such, the intensity of irradiance receives by the photosynthetic culture during the day is even out. It had been reported that higher photosynthetic conversion efficiency was achieved in this kind of set up.

In the field, photobioreactors using suitably selected microalgae or cyanobacteria achieved nearly 10% of photosynthetic conversion efficiency. This compare favorably with higher plant, which could convert <1% irradiance absorbed by its leaves to biomass energy.

22.1.2. *Photo-biological hydrogen production*

Photosynthetic apparatus, chloroplast, of some photosynthetic micro-organisms such as the green alga *Chlorella* in the presence of suitable electron acceptors is capable of producing H_2 and O_2 through direct photolysis of water.

In the system, the substrate (electron donor) is water, sunlight as the energy source is unlimited, and the product (hydrogen) can be stored

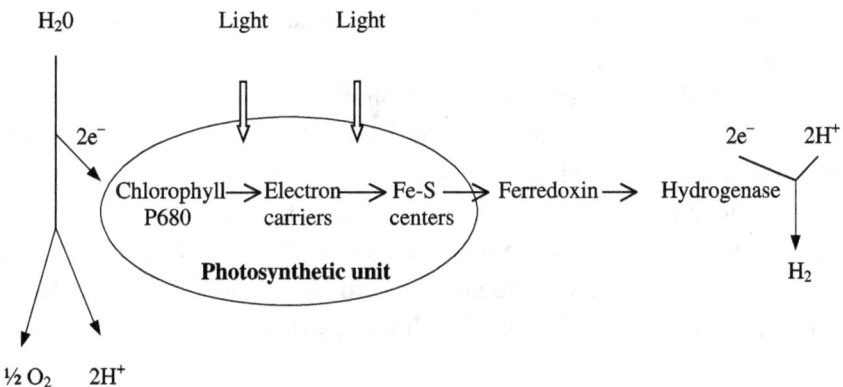

and is non-polluting. Moreover, the process is renewable, because when the energy is consumed, the substrate (water) is regenerated.

However, the process currently has no practical value, as the system is not stable (chloroplast is structurally unstable), and could function for only a few hours outside the plant cells. Artificial membrane mimicking biological photosynthetic unit is being investigated.

22.2. Conversion of Biomass Energy to Fuels

Structural and storage carbohydrates in biomass have low energy content and could not be used as fuel directly. It is necessary to concentrate the energy content further for fuel applications. The use of microorganisms to produce commercially valuable fuels depends on getting the right micro-organisms that are able to produce the desired fuel efficiently and in having an inexpensive supply of substrates available for the fermentation process. It is obviously imperative that the production of synthetic fuels does not consume more natural fuels resources than are produced. Anaerobic microbial fermentation is an efficient and widely used route for such conversion processes.

22.2.1. *Alcohol (ethanol) fermentation*

The microbial production of ethanol has become an important source of a valuable fuel, particularly in regions of the world that have abundant supplies of plant residues. Brazil produces and uses large amounts of ethanol as an automotive fuel. Mixing gasoline and ethanol in a 9:1 ratio is popular in some part of US. Ethanol combustion is less polluting than gasoline combustion.

Fermentation production of fuel alcohol can be accomplished through microbial conversion of low cost agricultural substrates high in starch and sugar content such as molasses (by-product of sugar industry), sugar-cane juice, maize starch, cassava starch and other fermentable carbohydrates. Though some microalgae have been reported to produce ethanol as one of the fermentation products, the production from microalgae as a finite biomass, is still being examined.

Numerous microorganisms are capable of producing ethanol, but not all are suitable for industrial processes. Yeast cultures, particularly *Saccharomyces*, have been most extensively examined because they are very efficient in converting sugars into ethanol, i.e. cost competitive, and are not as strongly inhibited by high ethanol concentrations as are other microbes.

The following equation illustrates the basic biochemical mechanism the ethanol is produced by the fermentation procss.

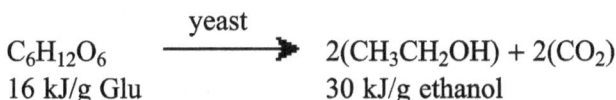

$$C_6H_{12}O_6 \xrightarrow{\text{yeast}} 2(CH_3CH_2OH) + 2(CO_2)$$

$$\text{16 kJ/g Glu} \qquad\qquad \text{30 kJ/g ethanol}$$

Theoretically, two moles of ethanol can be produced from one mole of glucose. On actual calculation:

1 kg invert sugar can produce 0.484 kg ethanol (or 0.61 L)
1 kg sucrose can produce 0.510 kg ethanol (or 0.65 L)
1 kg starch can produce 0.530 kg ethanol (or 0.68 L)

The yeasts commonly used in industrial alcohol production include *Saccharomyces cerevisiae* (ferment glucose, fructose, maltose, maltoriose), *S. uvarum* (carlsbergensis), *S. diataticus* (ferment dextrins), *Kluyveromycesfragilis* and *K. lactus* (ferment lactose). The ethanol productivity ranges between 1–2 g/h/g cells.

Selected bacterial cultures were examined for use in ethanol production processes because of their higher temperature tolerance. However, their yield of ethanol was not as high as in yeasts fermentations. Recently, the bacterium *Zymomonas mobilis* has been selected to achieve a high productivity of 2.5–3.8 g ethanol/h/g cells.

Various processes can be employed for the fermentation production of ethanol as a fuel. The development of the fermentation processes through the years reveals the strength and weaknesses of the various culture methods as discussed in Chap. 2, Sec. 2.4.

(i) For batch fermentation, the practicable level of carbohydrate used is 16–25% (w/v), giving a final ethanol concentration of 6–12% (v/v), with an average output rate of 1.4 g ethanol/L/h. Higher carbohydrate concentrations would result in substrate inhibition or product inhibition of cell growth, as high sugar concentrations inhibit metabolism due to the increased osmotic pressure. Despite of the limitation, the advantages of batch process is that it is resistant to process abuse and can be operated by less skilled personnel.

(ii) Ethanol is relatively toxic to microorganisms, and therefore only limited concentrations of ethanol can accumulate in a fermentation process. To overcome the effect of product (ethanol) inhibition in order to achieve higher ethanol output rate, batch culture with cell recycle or continuous chemostat culture systems have been used and productivity of >10 g ethanol/L/h has been achieved. Such systems required higher operator skills, capital costs and more sophisticated control devices.

In the **batch recycle system** (Fig. 22.3), ethanol is recovered and cells are removed from the brew at the end of a fermentation cycle. As ethanol inhibition reduces the specific rate of ethanol production with resultant lower output of ethanol, higher cell density is therefore desired to maintain high output rate, and this can be achieved by recycling the cells removed from the effluent back into the fermenter in the following batch. This process allows the fermentation time to be shortened.

In a chemostat system, substrate is continuously being added to the culture system and the culture (including the product ethanol) is being harvested at the same rate, so that the volume of the culture

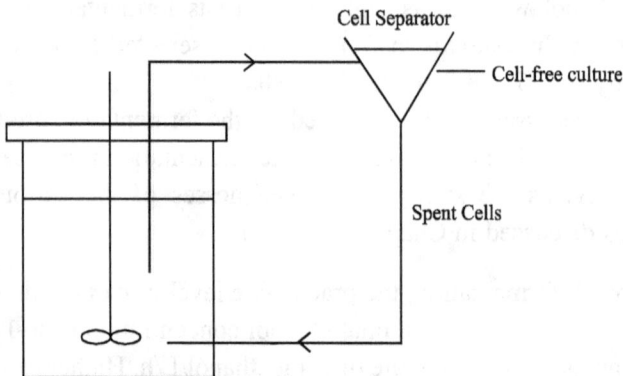

Fig. 22.3. A batch culture system with biomass recycled.

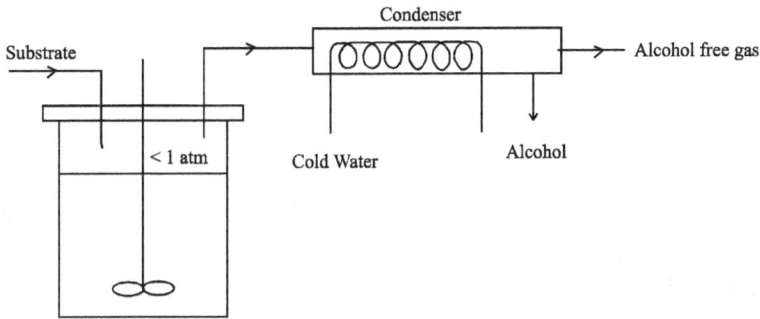

Fig. 22.4. Vacuum fermentation system for alcohol production.

remains constant and inhibitory level of ethanol does not occur to cause decreased growth rates. If the microbial population grows slower than the dilution rate due to ethanol inhibition, the culture will be diluted and "washed out" of the fermenter. High product output rate is achieved at high dilution rate (the feeding rate of medium).

(iii) Alternatively, the vacuum fermentation system has been proposed (Fig. 22.4). In this case, the fermentation is performed under reduced pressure. The ethanol is distilled off at the fermentation temperature, while concentrated substrate is fed continuously, thus the inhibitory effect of ethanol on cell growth is reduced. However, the energy balance in such a system is less than 1 (i.e. consumed more energy than being recovered in alcohol).

Alcohol fermentation is a well-established technology. However, the energy input necessary to process the feed material and product is about the same as the energy output in the form of ethanol as distillation to recover ethanol requires energy input. Besides, human food, which is relatively expensive, is often used as the feedstock; hence the cost of ethanol production by fermentation is high.

22.2.2. Methane production by anaerobic digestion

Methane (CH_4) is an energy-rich fuel that can be used for the generation of mechanical, electrical, and heat energy. Large amounts of methane can be produced by anaerobic decomposition of waste materials. Efficient generation of methane can be achieved by using algal biomass grown in pond cultures, sewage sludge, municipal waste, plant residue and animal waste.

In microbial production of methane, naturally occurring mixed anaerobic bacteria population is always used, and cells are retained within the digester. During the fermentation process, a large amount of organic matter is degraded, with a low yield of microbial cells, while about 90% of the energy available in the substrate is retained in the easily purified gaseous products CH_4. The end product is a mixture of methane gas and CO_2 (also called biogas).

The degradation process involves three major metabolic groups of interacting bacteria, most of which do not directly produce CH_4. Fermentative bacteria hydrolyze the degradable primary substrate polymers such as proteins, lipids, and polysaccharides and decompose to smaller molecules with the production of acetate and other saturated fatty acids, CO_2, and H_2 as major end products. The second group is the obligate H_2-producing acetogenic bacteria, which metabolize low molecule organic acids (end products of the first group) to H_2 and acetate (and sometimes CO_2). Finally, the methanogenic bacteria catabolize mainly acetate, CO_2 and H_2 to the terminal products. Methanogenic bacteria found in such mixed population include species of *Methanobacterium*, *Methanobrevibacter*, *Methanoccus*, *Methanogenum*, *Methanospirillum* and *Methanosarcina*.

The stages of the fermentation process are as follows:

(i) Stage 1: Acid-forming stage
Polymeric materials such as polysaccharides, proteins, and lipids were broken-down to form volatile organic acids (in particular acetic acid), H_2 and CO_2.

(ii) Stage 2: Methanogenesis (methane generation)
Methanogens can utilize only a small number of simple compounds as energy sources. Some utilize the H_2 produced by the other bacteria to reduce CO_2 to CH_4, and some cleave acetate to CO_2 and CH_4. The methanogens are the only organisms that are able to catabolize acetate and

hydrogen to gaseous products in the absence of light energy or exogenous electron acceptors such as oxygen, sulfate, or nitrate.

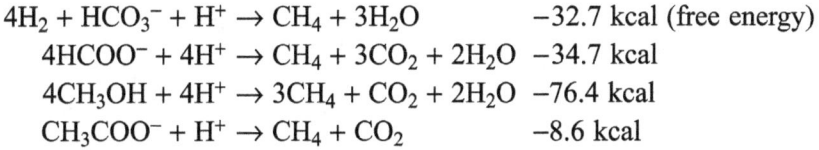

$$4H_2 + HCO_3^- + H^+ \rightarrow CH_4 + 3H_2O \qquad -32.7 \text{ kcal (free energy)}$$
$$4HCOO^- + 4H^+ \rightarrow CH_4 + 3CO_2 + 2H_2O \ -34.7 \text{ kcal}$$
$$4CH_3OH + 4H^+ \rightarrow 3CH_4 + CO_2 + 2H_2O \ -76.4 \text{ kcal}$$
$$CH_3COO^- + H^+ \rightarrow CH_4 + CO_2 \qquad -8.6 \text{ kcal}$$

The advantage of anaerobic production of fuel is that the energy conversion efficiency is high. A yield value ($Y_{methane/glucose}$) of 0.27 g/g is often observed, which corresponds to an energy yield (Y_{energy}) of 90%.

The disadvantage is that the digestion process is slow. Typically, for a gas consists of 60–70% methane, the production rate is 0.5 m^3/kg dry volatile solids over a resident time of 15 days.

22.3. Electricity from Biofuel Cells

Biofuel cells could release energy in fuel chemicals to generate electrical energy at ambient temperature. Fuel cells convert energy more efficiently than conventional power engines such as the internal combustion engine and produce almost no pollution. The basic set-up of the fuel cell is two electrodes placed in an electricity conducting electrolyte, separated by an ion exchange membrane. The arrangement allows the electrochemical equivalent of combustion to occur.

Example 1: To generate electricity from hyrogen gas (Fig. 22.5):

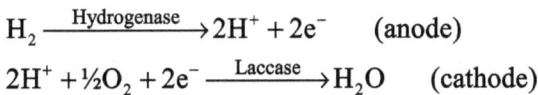

$$H_2 \xrightarrow{\text{Hydrogenase}} 2H^+ + 2e^- \quad \text{(anode)}$$
$$2H^+ + \tfrac{1}{2}O_2 + 2e^- \xrightarrow{\text{Laccase}} H_2O \quad \text{(cathode)}$$

Hydrogen gas as a fuel enters at one electrode, and oxidant, usually oxygen from the air, at the other. The anode is coated with the enzyme hydrogenase, while the cathode is coated with laccase. On the anode, hydrogen molecules split into their constituent protons and electrons. The enzymes catalyze a reduction-oxidation reaction across the membrane,

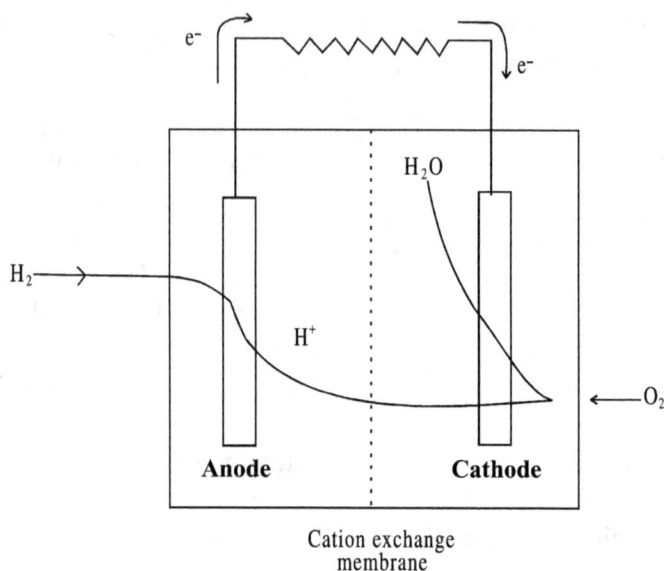

Fig. 22.5. A biofuel cell using hydrogen gas as the electron donor.

releasing energy, which pushes electrons round an external circuit. At the cathode, the protons and electrons combine with oxygen to form water. The fuel does not actually burn, therefore fuel cells do not produce pollutants associated with combustion, such as carbon oxides and oxides of nitrogen. Cells that use hydrogen generate only water as compare to fossil fuel, which produce water and carbon dioxide as waste product. This type of fuel cells is called proton exchange membrane cell (PEM).

Example 2: To generate electricity fro methanol:

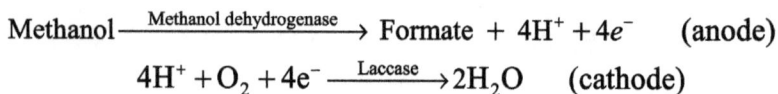

$$Methanol \xrightarrow{\text{Methanol dehydrogenase}} Formate + 4H^+ + 4e^- \quad \text{(anode)}$$
$$4H^+ + O_2 + 4e^- \xrightarrow{\text{Laccase}} 2H_2O \quad \text{(cathode)}$$

In this case, the fuel is methanol. The direct methanol fuel cell (DMFC) runs on a diluted mixture of about 2% methanol in water. The methanol is converted into formate on the anode. The proton then reacts with oxygen as in a PEM cell.

The metabolically active microorganisms, such as *Proteus vulgaricus* and *Anabaena variabilis* immobilized in a biofuel cell could convert energy in their substrate (glucose for the former and light for the later) into electricity. A biofuel cell in which bacteria *Proteus vulgaircus* and *Escherichia coli* were used as sulfate reduction catalysts was in operation for 5 years, thus demonstrated its long-term stability.

The disadvantage of biofuel cell is that the power output is low (1 kW at 40 mA/cm^2). Thus, it is used for specific purposes, such as small medical and military apparatuses used in the field and in space missions.

Further Reading

1. Allen RM and Bennetto HP. Microbial fuel-cells. Electricity production from carbohydrates. *Appl. Biochem. Biotechnol.* **39/40**: 27–40, 1993.
2. Dellweg H, ed. *Biotechnology Vol. 3*. Chapter 3a: Ethanol fermentation, pp. 257–386; Chapter 4: Energy from renewable resources, pp. 595–626. Weinheim, Verlag Chemie, 1983.
3. Habermann W and Pommer EH. Biological fuel cells with sulphide storage capacity. *Appl. Microbiol. Biotechnol.* **35**: 128–133, 1991.
4. Hall DO, Coombs J and Higgins IJ. Energy and biotechnology. In: *Biotechnology, Principles and Application*, Chapter 2, eds. IJ Higgins, DJ Best and J Jones. Oxford, Blackwell Scientific, 1985, pp. 24–72.
5. Laminie J and Dicks A. *Fuel Cell Systems Explained*. Wiley, New York, 2000.
6. Lee YK and Richmond A. Bioreactor technology for mass cultivation of photosynthetic microalgae. In: *Marine Biotechnology Vol. 2*, eds. M Fingerman, R Nagabhushanam and M-F Thompson. Oxford and TBH Publishing, New Delhi, 1998, pp. 271–288.
7. Tanaka K, Kashiwagi N and Ogawa T. Effects if light on the electrical output of bioelectrochemical fuel-cells containing *Anabaena variabilis* M-2: Mechanism of the post-illumination burst. *J. Chem. Technol. Biotechnol.* **42**: 235–240, 1988.

Questions for Thought...

1. Using alcohol production as an example, illustrate how the principles of microbial fermentation could be applied in the development of processes for upgrading of agricultural products to liquid fuel.
2. Design a self-contained space station that is made up of space travelers and microorganisms.

Chapter 23

Microbial Biomining

Yew Wen Shan
Department of Biochemistry, Yong Loo Lin School of Medicine
National University of Singapore
MD 7, 8 Medical Drive, Singapore 117597

23.1. Microbial Biomining — An Introduction from History and a Glimpse of the Future

Microbial biomining involves the use of microorganisms in the recovery of metals from traditional sources such as ores. This process, also known as bioleaching, has been applied by the mineral industry since the 15th century. Paracelsus (1495–1541), a Swiss Renaissance alchemist, described how the rustics in Hungary were able to leach copper from marcasite, an iron sulfide (FeS_2)-containing mineral, apparently relying on microbial processes. In Spain, Sweden, Germany, China and other countries, the biooxidation of sulfide ores for copper recovery has been practiced for centuries, although the roles of the microbial agents were not immediately acknowledged. Biomining has several distinctive advantages over the traditional mining procedures, namely that it does not require huge amounts of energy (generated from fossil fuels) during roasting and smelting, and does not generate pollution in the form of harmful gaseous emissions such as sulfur dioxide, and environmentally pollutive acids and metals associated with acid mine drainage. Microbial biomining is also preferred in the recovery of metals from low-grade ores which cannot be economically processed with chemical methods.

Commercial copper extraction from mine waste through biomining was documented with the recognition of the acidophilic *Thiobacillus ferrooxidans* in 1958; this microbe maintained iron in the oxidized ferric form to serve as the oxidant for sulfide copper minerals so that copper can be solubilized. At the Kennecott Bingham Mine near Salt Lake City, Utah, USA, the run-of-mine material of low-grade copper, stacked in waste dumps to depths of over 100 meters, was leached using an acidic ferric iron solution for economic recovery of copper. Although the role of the iron-oxiding microbe was recognized, the mine dump, as a bioreactor, was not designed to promote the activity of the bacteria, and thus limited recovery. In 1980, bioreactors were designed to facilitate the activity of the microorganisms in the bioleaching of copper from heaps by providing nutrients and aeration; this led to a series of successful and extensive commissioning of numerous copper biomining operations.

Apart from copper, commercial biomining has also been applied to the recovery of uranium, and other metals such as nickel, lead, and zinc, have been processed from low-grade ores by microbial leaching. At the Denison Mine, Ontario, Canada, a bioreactor was designed to promote bacterial growth of *Acidithiobacillus ferrooxidans* by providing nutrients and aeration, resulting in the commercial extraction of uranium from underground low-grade ore. Another successful commercial biomining example is the biooxidation pretreatment of refractory sulfidic gold ores. Microorganisms are used to oxidize pyrites to allow the subsequent extraction and recovery of gold by conventional hydrometallurgical processes, such as cyanide leaching and recovery on carbon or precipitation on zinc. The biooxidation pretreatment plant with the longest history of operation is Goldfield's BIOX process at the Fairview Mine in South Africa; this plant, operating since 1986, uses biooxidative pretreatment of ores to allow downstream gold extraction through conventional milling.

This brief historical introduction on microbial biomining illustrates how this microbial biotechnology has been an integral and economically sound part of humanity's history. However, as our planet's reserves begin to be depleted, and rapid urbanization spreads, we have to look for alternatives to replace the traditional sources for microbial biomining. As microbial biotechnology advances, can we adapt our current practices toward our evolving landscape? This chapter serves to define the current principles of microbial biomining and provides a primer for its future application towards sustainable cities and environments.

23.2. Microbiology of Biomining

Microorganisms play a critical role in bioleaching processes. In 1951, the first microbe to be isolated and described from an acid mine drainage was the acidophilic iron- and sulfur-oxidizing bacterium, *Thiobacillus ferrooxidans*. Subsequently, it was recognized that most of the microbes in biomining environments thrive under acid pH conditions (acidophilic). Generally, these acidophilic microbes can be grouped according to their temperature optima: they are either mesophilic microbes or thermophilic microbes. Mesophilic microbes grow best in moderate temperature, ranging from 25°C to 40°C; whilst the thermophilic microbes thrive in high temperatures, ranging from 45°C up to temperatures exceeding 120°C!

(1) Mesophilic and Thermophilic Microbes in Biomining

The main mesophilic microbes involved in biomining are *Thiobacillus ferrooxidans, Thiobacillus thiooxidans*, and *Leptospirillum ferooxidans* (Olson *et al.*, 2003). In 2000, *Thiobacillus ferrooxidans* and *Thiobacillus thiooxidans* were reassigned to the new genus *Acidithiobacillus*. *Acidithiobacillus ferrooxidans* is an auxotrophic Fe^{2+}/S-oxidizer and Fe^{3+}-reducer. Most of the auxotrophic microbes in biomining processes are chemolithoautotrophs; these microbes are capable of using inorganic chemical compounds to generate organic compounds such as carbohydrates, lipids and proteins from inorganic carbon dioxide. *Acidithiobacillus ferrooxidans* obtains its energy from the oxidation of ferrous iron and elemental sulfur, aerobically transferring the electrons from ferrous iron to oxygen in an electron pathway analogous to our electron transport chain in our mitochondria.

Leptospirillum ferooxidans, the other major mesophilic microbe in biomining, is also an auxotroph; however, it is a major Fe^{2+}-oxidizer, and is a better ferrous iron oxidizer than *Acidithiobacillus ferrooxidans*. In fact, *Leptospirillum ferooxidans* has a higher affinity to ferrous iron (K_M of 0.25 mM compared to 1.34 mM for *A. ferrooxidans*) and a lower sensitivity to inhibition by ferric ion (K_i of 42.8 mM compared to 3.10 mM for *A. ferrooxidans*). Hence, this mesophilic microbe is often found to be the predominant iron oxidizer in industrial continuous flow biooxidation tanks, where the high ferric-to-ferrous iron ratio affects the bioleaching abilities of all species except *L. ferrooxidans*.

The bioleaching process is an exothermic process, producing significant amount of heating regardless of the reactor or heap design. As such, moderately thermophilic microbes such as *Ferroplasma acidiphilum* and *Acidithiobacillus caldus* are increasingly used in bioleaching processes. The use of these microbes, as well as thermophilic archaea such as *Sulfolobus* species, *Acidianus brierleyi* and *Metallosphaera sedula*, improves bioleaching in two ways: 1) with exothermic bioleaching, reaction rates increased with increasing temperature; 2) an increase in temperatures enhanced the extent of metal recovery from the solid mineral ores. The current practice of using moderately thermophilic microbes in the bioleaching of copper from copper ores such as chalcopyrite ($CuFeS_2$) illustrates the utility of thermophiles in biomining: chalcopyrite is one of the most important ore of copper, and copper biomining in the past has been unsuccessful with mesophilic microbes due to the generation of heat in the bioreactors; there are currently bioreactors containing chalcopyrites undergoing bioleaching with extreme thermophiles at pilot scales, with the expectation of increased copper extraction and decrease in process-cooling requirements.

(2) Contact and Non-Contact Bacterial Bioleaching

In microbial biomining, there are two general processes for bioleaching (Sand *et al.*, 2001), direct or contact bioleaching, and indirect or non-contact bioleaching (Fig. 23.1). In direct bacterial bioleaching, a physical contact exists between the bacteria and the metal, and bioleaching occurs through bacterial enzyme-catalyzed reactions. This form of bioleaching is dependent on the electrochemical processes that result in the dissolution of metals, and occurs at the interface between the bacterial cell wall and the metallic surface. This contact bioleaching process is dependent on the attachment of bacteria to the metal ore surface. The primary mechanism of bacterial attachment is normally electrostatic in nature, although other mechanisms of cellular attachment occur, depending on the nature of the ore: in the case of *Acidithiobacillus ferrooxidans*, bacterial exopolymers on the cell surface contain Fe^{3+}-complexed uronic acid residues,

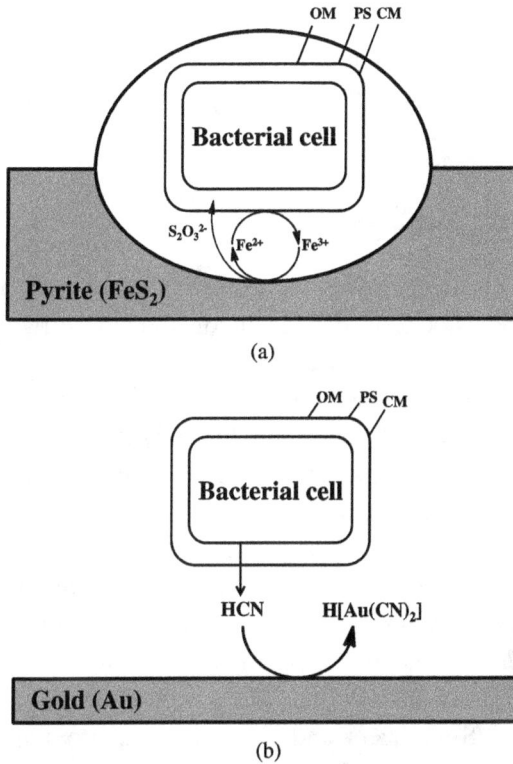

Fig. 23.1. General processes for bioleaching. (a) Direct bacterial bioleaching, showing direct physical contact between the bacterial cell and the metal feedstock. (b) Indirect bacterial bioleaching, showing a physical separation between the bacterial cell and the metal feedstock, and the production of leaching agents such as hydrogen cyanide. OM: outer membrane; PS: periplasmic space; CM: cytoplasmic membrane.

allowing the attachment of the positively-charged bacterial surfaces to the negatively-charged pyrite ores.

In indirect or non-contact bioleaching, the bacteria are not in direct contact with the metal ores, but leaching agents or lixiviants (such as sulfuric acid or hydrogen cyanide) are produced by the bacteria to oxidize or leach the metallic ores. The differentiation between non-contact and contact bacterial bioleaching lies on the planktonic and non-planktonic nature of the bacterial species, respectively. In both contact and non-contact mechanisms, the chemical reaction of metal bioleaching occurs outside the cells; in contact bioleaching, the chemistry occurs within the microenvironment between the microbe and the metallic surface, whilst in non-contact bioleaching, the chemistry is not limited by physical proximity, but by the nature and type of lixiviant produced.

(3) Microbial Consortia in Biomining

The success and efficiency of a bioreactor in the recovery of metals from ores or other feedstocks is dependent on the type of microorganisms present within the bioreactor (Rawlings and Johnson, 2007). As such, an important consideration in the establishment of a bioreactor for biomining would involve the use of the most suitable and efficient microbial consortia (involving the most effective strains and species). An example of a moderately thermophilic biomining consortium would include the following microbes: *Leptospirillum ferriphilum* (strain MT6), *Acidimicrobium ferrooxidans* (strain TH3), *Ferroplasma acidiphilum* (strain MT17), *Actinobacterium* isolate Y005, *Sulfobacillus acidophilus* (strain YTF1), and *Sulfobacillus* isolate BRGM2. In this consortium, there are auxotrophic Fe^{2+}-oxidizers, heterotrophic Fe^{2+}-oxidisers and Fe^{3+}-reducers, and mixotrophic Fe^{2+}/S-oxidizers and Fe^{3+}-reducers, providing a balanced, stable and effective bioleaching consortium for polymetallic mineral biomining of copper, zinc and iron sulfides.

There are two distinct approaches to the assembly of a microbial consortium in biomining, a "top-down" or reductive approach, and a "bottom-up" or additive approach. In the "top-down" approach, a mixture of microorganisms is used to inoculate the mineral feedstock, and the assumption is made that a limited number of these microbes will emerge as a stable and effective bioleaching consortium. In this "survival of the fittest" approach, the aim is to have sufficient microbial diversity (both physiological and phylogenetic) available in the starting inoculums from which to adapt and select an efficient and stable consortium for biomining. An advantage of this approach is that the large biodiversity present increases the robustness of the bioleaching system, enabling it to better adapt to (given its large gene pool) and recover from sudden operational changes such as pH and temperature fluctuations due to interruption in environmental control of the bioreactor.

In the "bottom-up" approach, the logic lies within the rational selection and inclusion of superior and efficient species and strains of microbes. This approach necessitates a labor-intensive, if not protracted, search for an optimum bioleaching consortium, to bioleach a particular feedstock or ore or concentrate, on the basis of the operational conditions of the bioreactor. In both approaches, there will be a domination of microbes most suited for the bioreactor, whilst other microbes may persist as minor members of the community. Ultimately, there is an objective that is common to both approaches, in that the metallic mineral has to be efficiently recovered; this consideration supersedes all others, including the choice of microorganisms to be used. The use of a microbial consortium in biomining also relies on the assumption that a sufficiently varied and adaptable community of microbes is assembled so that after adaptation, those in the final consortium would be as efficient in metal recovery as any other adapted consortium is expected to be under similar conditions.

A case study: Is there only one ideal combination of microorganisms?

The question of whether only one ideal combination of microbes exists for the efficient recovery of metal from mineral ores was explored when the biooxidation rates (and the metal recovery rates) of two arsenopyrite biomining plants, the Fairview plant (South Africa) and Tamboraque plant (Peru), respectively, were compared. The evidence suggested that a change in microbial consortium can be made without affecting the rates of biomining. At the Fairview biomining plant, the microbial consortium inoculum was made from one that had been established to adapt to grow on arsenopyrite in the stirred tanks of the Fairview mine. The Fairview microbial consortium is dominated by the iron-oxidizer *Leptospirillum ferriphilum* and the sulfur-oxidizer *Acidithiobacillus caldus*. However, when the Tamboraque biomining plant was started, a microbial consortium, that had been adapted from acid drainage in the Coricancha Mine in Peru, was used. After several years of operation, the consortium was found to be dominated by the iron-oxidizer *Leptospirillum ferooxidans*, rather than *Leptospirillum ferriphilum*. As such, there may be alternative combinations of microbes that can produce comparable recovery efficiencies in the biomining of metals from mineral ores.

23.3. Biochemistry of Biomining

There are a number of biochemical reactions that occur during the biomining process; these reactions are dependent on the chemical nature of the feedstock or ore, as well as the inherent metabolic pathways within the associated microorganisms. The classical biochemical processes in biomining, involving the dissolution of feedstocks or ores containing metal sulfides, include the thiosulfate pathway and the polysulfide pathway (Schippers and Sand, 1999, and Rohwerder *et. al.*, 2003). Recent non-classical biochemical processes in biomining, involving the secondary metabolic production of lixiviants, include the hydrogen cyanide biosynthetic pathway.

(1) Classical Biochemical Biomining

The thiosulfate pathway (Fig. 23.2) allows the recovery of metal sulfides from acid-insoluble substrates, such as the metal sulfides pyrite,

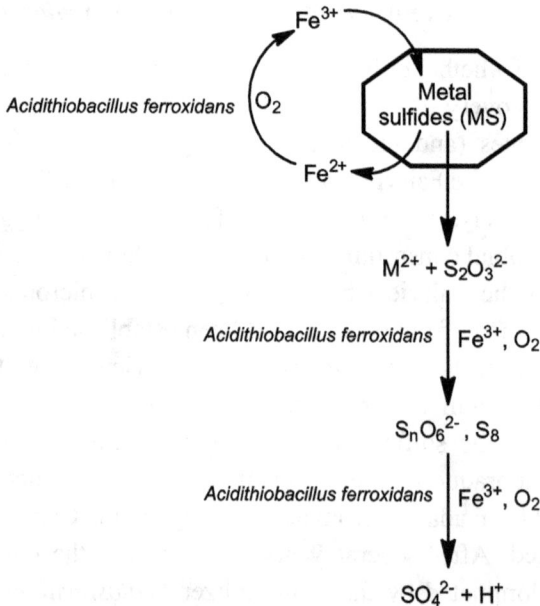

Fig. 23.2. The thiosulfate biochemical pathway for metal sulfide bioleaching. Metal ions (M^{2+}) are obtained from corresponding metal sulfides (MS) through the oxidative bioleaching actions of bacteria such as *Acidithiobacillus ferroxidans*.

molybdenite, and tungstenite (FeS_2, MoS_2, and WS_2, respectively). Within this group of metal sulfides, metallic dissolution or metal liberation (i.e. the breaking of chemical bonds between the metal and sulfur) occurs after oxidation *via* electron transfer to Fe^{3+} ions. The pathway is named after the production of thiosulfate ($S_2O_3^{2-}$) that is liberated upon the electron transfer from the solid metal sulfides to Fe^{3+}; subsequent oxidation of thiosulfate results in the production of sulfate (SO_4^{2-}). From the thiosulfate pathway, it is evident that only Fe^{2+}-oxidising microbes will be able to bioleach acid-insoluble metal sulfides; these microbes are able to carry out the bioleaching process as they can regenerate the Fe^{3+} ions required for the initial oxidation process that results in metal recovery.

The polysulfide pathway (Fig. 23.3) allows the recovery of metal sulfides from acid-soluble substrates, such as sphalerite, galena, arsenopyrite, chalcopyrite, and hauerite (ZnS, PbS, FeAsS, $CuFeS_2$, and MnS_2, respectively). Within this group of metal sulfides, metal dissolution

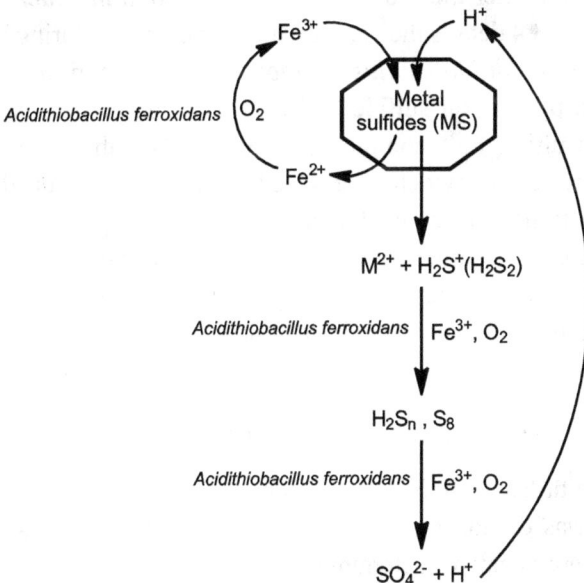

Fig. 23.3. The polysulfide biochemical pathway for metal sulfide bioleaching. Metal ions (M^{2+}) are obtained from corresponding metal sulfides (MS) through the combined actions of oxidative bioleaching and acidic proton attack by bacteria such as *Acidithiobacillus ferroxidans*.

(i.e. the breaking of chemical bonds between the metal and sulfur) occurs after protonation and subsequent oxidation *via* electron transfer to Fe^{3+} ions. In this pathway, a sulfide cation, H_2S^+, is generated upon metal dissolution, and subsequent oxidation processes result in the production of polysulfides, and eventually to sulfate (SO_4^{2-}). As a result of this chain of chemical events, the mechanism is known as the polysulfide pathway. Since the oxidizing activity of Fe^{3+} ions is not strictly required in the poly-sulfide pathway, acid-soluble metal sulfides can also be bioleached by sulfur-oxidizing microbes. In the absence of Fe^{3+} ions, the microbes can oxidize hydrogen sulfide (H_2S) resulting from protonation on the metal sulfide, thereby regenerating the protons previously consumed by the metal sulfide dissolution.

During the course of bioleaching, electron transfer from Fe^{3+} ions to molecular oxygen occurs. The electron transfer pathways are expected to differ between different microbes, especially in the details of redox components; however, a general electron transfer pathway can be proposed, modeled after the prototypical microbe *Acidithiobacillus ferrooxidans*. Figure 23.4 details the electron transfer pathway during bioleaching of pyrite by *Acidithiobacillus ferrooxidans*. Electron transfer occurs between the pyrite iron sulfide and Fe^{3+} ions complexed by glucuronic acid; the resulting Fe^{2+} complex is reoxidized at the outer membrane of the microbe by cytochrome Cyc2. Electron transfer then occurs through the transport chain of rusticyanin and periplasmic cytochrome Cyc1, to the cytoplasmic membrane-bound aa_3-type cytochrome oxidase. Through the catalytic action of aa_3-type cytochrome oxidase, the electrons are terminally transferred to molecular oxygen, resulting in the reduction to water.

(2) Non-Classical Biochemical Biomining

In 2001, the utility of lixiviants as secondary metabolites by microbes in biomining was demonstrated in the laboratory (Campbell *et al.*, 2001): using the Gram-negative bacterium *Chromobacterium violaceum* (Fig. 23.5), Campbell and coworkers showed that all of the gold on glass test slides was solubilized by the bacterium; in addition, the group also demonstrated that the microbe could mobilize gold from a sulfidic ore concentrate. The biomining or recovery of solid gold metal is brought about by the lixiviant

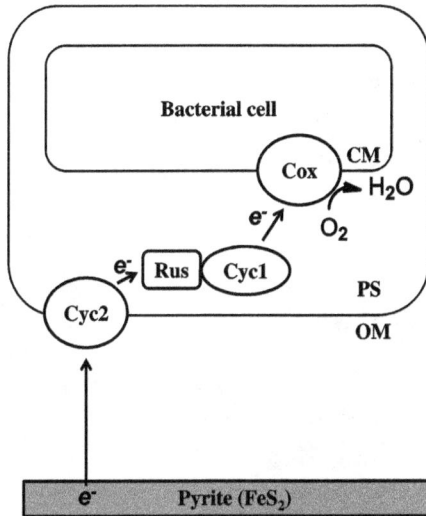

Fig. 23.4. The electron transfer pathway during metal bioleaching. Electrons are transferred from pyrite iron sulfide to cytochrome Cyc2 (Cyc2) located at the outer membrane (OM) of the microbe; the electron transfer then proceeds through the transport chain proteins of rusticyanin (Rus) and cytochrome Cyc1 (Cyc1), located within the periplasmic space (PS), and to the cytoplasmic membrane(CM)-bound aa$_3$-type cytochrome oxidase (Cox). The transfer pathway terminates with the transfer to molecular oxygen, resulting in the reduction of molecular oxygen to water.

hydrogen cyanide. Hydrogen cyanide is a secondary metabolite of a number of microorganisms, including *Chromobacterium violaceum* and *Pseudomonas aeruginosa*.

Hydrogen cyanide, or its reactive cyanide anion (CN$^-$), is a potent ligand for many transition metals, as well as for the precious metals gold and silver. In the gold cyanidation process (mining of gold from ores using cyanide), the precious metal is complexed by cyanide anions to form soluble aurate derivatives, and recovered *via* electrolysis. A typical chemical gold cyanidation reaction can be represented as follows:

$$2Au + 4HCN + \tfrac{1}{2}O_2 \rightarrow 2H[Au(CN)]_2 + H_2O$$

Hydrogen cyanide biosynthesis occurs in *Chromobacterium violaceum* through the catalytic action of the enzyme, hydrogen cyanide synthase.

Fig. 23.5. A petri dish culture of gram-negative bacteria *Chromobacterium violaceum*. The purple coloration is due to the presence of violacein, an antibiotic produced by the bacteria.

This enzyme uses glycine as a substrate, and through an oxidative decarboxylative reaction, produces the lixiviant hydrogen cyanide. Since hydrogen cyanide is a potent inhibitor of respiratory cytochrome oxidase and several other metalloenzymes, its intracellular toxicity is negated by cyanolytic processes in the bacterium. A major cyanolytic process in *Chromobacterium violaceum* involves the *cyn* operon within the bacterium's genome, encoding for detoxifying enzymes such as cyanase.

Cyanide has been used in the gold mining industry since the end of the 19th century. However, the possibility of cyanide leaks in the mines, resulting in spillage of cyanide into the environment, has led to the employment of microorganisms such as *Chromobacterium violaceum* in the bioleaching of gold in the gold mining industry. Biomining of gold has the advantage of reducing operational costs (chemical cyanide production is comparatively costly), as well as the reduction of environmental contamination associated with gold recovery.

23.4. Commercial Biomining

Commercial biomining, or the bioleaching of metals for commercial gains, has now become a profitable and sustainable practice in the industry. The common use of microorganisms in the recovery of copper has led to the establishment of a number of copper bioleaching plants throughout the world. Table 23.1 lists some copper biomining plants that recover copper from the mineral chalcocite (CuS). Within these plants, copper is leached from chalcocite by ferric iron formed from microbial oxidation of ferrous iron.

Commercial biomining has also been extended to the biooxidative pretreatment of precious metal ores to enhance gold recovery. Table 23.2 lists some commercial biomining plants for pretreatment of gold concentrates located throughout the world. In these biomining plants, biooxidation pretreatment of sulfidic gold concentrates enable sulfides in pyrite and arsenianpyrite to be occluded from gold. Following biooxidation, the oxidized ore is removed from the biomining plant and processed by milling and gold cyanidation (cyanide recovery of gold). In these plants, a mix of microorganisms including mesophilic *A. ferroxidans, L. ferrooxidans,* moderately thermophilic *Sulfobacillus* species and thermophilic archaea *Acidianus* and *Metallospheara* species are used. The mix or consortium of

Table 23.1.   Commercial copper biomining plants.

Name of plant and location	Processing size of plant (tones of ore per day)	Years in operation
gunpowder's Mammoth Mine, Australia	1.2 million	1991 till present
Cerro Colorado, Chile	16,000	1993 till present
Ivan-Zar, Chile	1,500	1994 till present
Quebrada Blanca, Chile	17,300	1994 till present
Andacollo, Chile	10,000	1996 till present
Dos Amigos, Chile	3,000	1996 till present
Cerro Verde, Peru	32,000	1996 till present
Zaldivar, Chile	20,000	1998 till present
S&K Copper, Myanmar	18,000	1998 till present
Equatorial Tonopah, USA	24,500	2000 to 2001

Table 23.2.   Commercial biomining plants for pretreatment of gold concentrates.

Name of plant and location	Processing size of plant (tones of concentrate per day)	Technology	Years in operation
fairview, South Africa	40	BIOX	1986 till present
Sao Bento, Brazil	150	BIOX	1990 till present
Wiluna, Australia	158	BIOX	1994 till present
Sansu, Ghana	960	BIOX	1994 till present
Tamboraque, Peru	60	BIOX	1990 till present
Beaconsfield, Australia	70	BacTech	2000 till present
Laizhou, China	100	BacTech	2001 till present

microbes is necessary as pyrite oxidation increase areas within the bioreactor to temperatures as high as 81°C, thus requiring the inclusion of thermophilic archaea as well as mesophilic bacteria.

23.5. Biomining and Bioremediation

In our increasingly populated environment, and for that matter, in any densely populated city in the world, it is important to ensure that our society progresses in a sustainable manner. To achieve environmental sustainability, it is important that we are able to recycle and slow down resource depletion. Although industrial biomining of metals has allowed access to a larger spectrum of natural resources, the fact remains that these resources are being depleted in an astonishing and unsustainable rate; in the absence of new technologies or global political efforts, we will soon run out of natural resources to maintain our infrastructure and ensure that our civilization survives the next few decades. In a world that is increasingly dependent on electronics and technology, the generation of a torrent of electronic waste is an unfortunate and inevitable reality. However, the ability to recycle and reuse materials would mediate the effects of growth and progress, ensuring a green and sustainable environment for future generations. In this respect, the bioremediation of electronic waste could possibly provide a constant feedstock, replacing the traditional limited sources of metallic ores with an abundant and sustainable source.

Current conventional treatment methods to recover precious and useful metals from electronic waste involve the use of harsh acids such as concentrated sulphuric and nitric acids; these methods pose considerable environmental risks as the run-offs from such treatment are extremely pollutive in nature. To counter the threat of environmental pollution, bioremediation efforts have been made in using microbes to provide a sustainable platform to recover precious metals from electronic waste. In such microbes, the lixiviant or bioleaching agent involved in bioremediation and recovery of metals is hydrogen cyanide. Although possible hydrogen cyanide leakage poses a considerable threat to the environment, the use of microbes in the biomining industry limits and minimizes such concerns as the microbes in context are both cyanogenic (capable of generating cyanide equivalents) and cyanolytic (capable of detoxifying cyanide equivalents), virtually ensuring that there will be no bulk release of cyanide into the environment during the bioleaching process.

References

1. Campbell SC, Olson GJ, Clark TR and McFeters G. Biogenic production of cyanide and its application to gold recovery. *J. Indust. Microbiol. Biotechnol.* **26**: 134–139, 2001.

2. Olson GJ, Brierley JA and Brierley CL. Bioleaching review part B. Progress in bioleaching: Applications of microbial processes by the mineral industries. *Appl. Microbiol. Biotechnol.* **63**: 249–257, 2003.

3. Rawlings DE and Johnson BD. The microbiology of biomining: Development and optimization of mineral-oxidizing microbial consortia. *Microbiology* **153**: 315–324, 2007.

4. Rohwerder T, Gehrke T, Kinzler K and Sand W. Bioleaching review part A. Progress in bioleaching: Fundamentals and mechanisms of bacterial metal sulfide oxidation. *Appl. Microbiol. Biotechnol.* **63**: 239–248, 2003.

5.   Sand W, Gehrke T, Jozsa PG and Schippers A. (Bio)chemistry of bacterial leaching-direct vs. indirect bioleaching. *Hydrometallurgy* **59**: 159–175, 2001.

6.   Schippers A and Sand W. Bacterial leaching of metal sulfide proceeds by two indirect mechanisms via thiosulfate or via polysulfides and sulfur. *Appl. Microbiol. Biotechnol.* **65**: 319–321, 1999.

Further Reading

1.   Barr DW, Ingledew WJ and Norris PR. Respiratory chain components of iron-oxidizing, acidophilic bacteria. *FEMS Microbiol. Lett.* **70**: 85–90, 1990.

2.   Cardenas JP, Valdes J, Quatrini R, Duarte F and Holmes DS. Lessons from the genomes of extremely acidophilic bacteria and archaea with special emphasis on bioleaching microorganisms. *Appl. Microbiol. Biotechnol.* **88**: 605–620, 2010.

3.   Michaels R and Corpe WA. Cyanide formation by *Chromobacterium violaceum. J. Bacteriol.* **89**: 106–112, 1965.

4.   Rawlings DE. Heavy metal mining using microbes. *Annu. Rev. Microbiol.* **56**: 65–91, 2002.

Questions for Thought...

1.   Certain types of microorganisms are especially suited for use in the biomining of metals.

   a)   Name two microbes that are commonly used in biomining.

   b)   List two important properties of such microbes used in the biomining process.

   c)   Mesophilic and thermophilic microbes are normally encountered in biomining. Explain why tolerance to moderately high temperatures is an advantage for microbes involved in biomining.

2.   You are tasked to set up a new microbial gold biomining facility. This facility will use arsenopyrite (FeAsS) as the mineral feedstock for the extraction of gold. Select up to three appropriate microorganisms from the table below for your microbial consortium, and explain the rationale for your selections.

Microorganism	Physiological Traits
Leptospirillum ferooxidans	Mesophile, Fe^{2+}-oxidizer
Acidithiobacillus ferrooxidans	Mesophile, Fe^{2+}/S-oxidizer, Fe^{3+}-reducer
Acidithiobacillus caldus	Moderate thermophile, S-oxidizer
Chromobacterium violaceum	Cyanogenic strain (HCN producer)
Metallosphaera sedula	Thermophile

3. The recovery of metal sulfides from arsenopyrite (FeAsS) by microbes can occur by the polysulfide biochemical pathway. Explain what is meant by the polysulfide pathway of metal recovery.

4. Electronic waste, such as computers, office electronic equipment and mobile phones, have become a major urban waste stream. The bioremediation of electronic waste can potentially provide a constant feedstock for metal recovery and contribute towards environmental sustainability. Can you name five metals that are commonly found in electronic waste?

Part VII

Patenting Microbial Biotechnology

Most governments see at least part of national future prosperity in the expansion of life sciences research, either by exploiting natural resources such as indigenous medicinal plants and thermophilic bacteria, or by supporting a workforce skilled enough to compete in the *de novo* design of new drugs, as well as genetically modified animals and crops. So much money will be spent on development that those who finance it will insist that wherever possible its fruits will be protected from exploitation by others, that is, patented. All scientists, including microbiologists, would do well to be aware of the prerequisites and procedures involved.[1] Put briefly, the invention must be novel (not already identifiable in practice or in the literature), it must involve an inventive step (something not obvious to those already in the field) and it must have some industrial application. In return for meeting these criteria, the invention will be protected against exploitation by others for twenty years. In regard to exactly what may be patented however, there are certain exclusions, such as devices, which offend public morality, as well as theories and discoveries. Methods of medical treatment were previously non-patentable but are now of debatable status.[2]

It is often said that scientists are the worst judges of the patentability of their work and this is generally true. They will often look upon their own advances as obvious or trivial when in fact they may be readily commercially exploitable — they may be *inventions* in the legal sense, and although the scientist may not look upon himself in that light, he is an *inventor*. Although some patents are taken out by private citizens with good ideas, inventions in the life sciences are likely to be achieved within large governmental, educational or commercial organizations. Whereas these organizations will readily recognize the scientist as an inventor, they will almost certainly be the legal owners of any patent awarded to him.

Chapter 24

Patenting Inventions in Microbiology

John Candlish
Faculty of Medicine and Health Sciences, University of Malaysia Sarawak, Kuching, East Malaysia
Email: jkcandlish@yahoo.com.sg

24.1. Territorial Nature of Patents

By law, almost universally, a resident in any country must file for a patent in that country first. The rationale is to allow the fruits of the invention to be enjoyed by the society which engenders it, and — more importantly — allow the relevant government, if it so desires, to acquire a compulsory license to exploit it. For example, if a microbiology patent is apposite to national defense — perhaps a way of halting an advancing army by subjecting them to a debilitating diarrheal illness — a government will not be slow to acquire it.

A patent granted in one country gives no protection in any other, so it is important to make applications abroad under the Patent Cooperation Treaty (PCT) as detailed below. The PCT as administered by the World Intellectual Property Organization (WIPO) is gradually making patent law more uniform worldwide. For example, the length of patent protection in the US, Australia and China has recently been lengthened to 20 years in conformity with the rest of the world.

Groupings to harmonize patent regulations within specific geographical areas are also arising. The European Patent Office (EPO) has some 20 countries as subscribers. This is not to be confused with the attempts of the European Union, with some 15 members, to identify common features within its own, strictly economically based, community. The Eurasian Patent Office deals with applications from the Russian Federation and a number of former republics of the USSR. The Association of South East Asian Nations (ASEAN) has declared itself to be desirous of a regional intellectual property framework, but this has not yet come to fruition.

The patent laws of most countries state that "any person" may make an application. That is, anybody can be an inventor. Usually the inventor cannot be a corporation or an organization. The inventor can go through the filing process on his own, or appoint another person to do so. Thus, in that sense, anybody can be a patent agent. However most countries have practicing patent attorneys whose qualifications are recognized by law and who alone are permitted to describe themselves as such. In some countries, the ordinary members of the legal profession may so describe themselves. Large universities, research institutes, and corporations often have officers who in effect act as patent agents for their employees.

24.2. Ownership and Royalties

In most, if not all, jurisdictions, an employee's inventions are regarded as property of the employer, but of course, the initiative must come from the inventor. As owner of any patent granted, the royalties from any licensing agreement accrue to the organization. However, the employer is likely to be generous in any share out, for an employee who can produce patentable inventions is obviously worth retaining. Where there are two or more inventors, they are regarded as joint owners, with "undivided shares". As a legal term, this means that they cannot claim specific proportions of the invention, say one third each if there are three of them. Rather, if one of them mortgages his share (which is possible) the lender also has an interest in the whole, not in one third.

In universities, where so much work is done by postdoctoral fellows, they are presumably to be regarded as employees, as has been indicated in a recent federal case involving the University of Chicago, in which the reasoning would probably be persuasive elsewhere.[3] However, inventions arising from graduate student projects may give rise to problems. As indicated above, there must be no public disclosure (oral or written) of the invention prior to filing (the "novelty" requirement) and this includes theses deposited in the library. Most universities then will delay emplacement in the library until the patent is filed, for it is then about to become public knowledge anyway. If the reports of projects used for the assessment of postgraduates are retained in the relevant department, one would imagine, on a reasonable interpretation, that this does not put them in the public domain. It might be as well, however, to put anyone reading commercially sensitive reports who is not already implicitly under a bar of confidentiality (such as members of the same research team) into that category. This prevents disclosure to any third party who on gaining the information innocently and publishing it, would not be liable. Thus, anybody put under a bar of confidence who discloses information rendering a patent invalid for lack of novelty could theoretically be sued, in tort, for damages; whether this would be worthwhile would depend on the circumstances.

24.3. Novelty, Disclosure and Confidentiality

It is a general proposition of the novelty requirement that a patent is not obtainable if the nature of the invention has been available in any written, oral or electronic form, anywhere in the world, before filing. (The term often used is "in the public domain"). The principle could conceivably be overturned if publication is in a really obscure form or locale. This was the burden of the decision in a keynote US case, *Gayler*[4] and also in an important UK case, *General Tire & Rubber Co* v *Firestone Tire & Rubber Co*[5] wherein the judge stated that novelty is negated only if the information had been accessible to the public as of right, with or without fee, anywhere in the country.

Recently, a patent for the use of Taxol in treating cancer was held to be invalid because there had been an oral disclosure at a scientific meeting attended by 500 persons.[6] This illustrates a potential pitfall for scientists, who instinctively want their work to become known as quickly as possible via learned journals, books and conferences. This can well destroy novelty. This problem is obviated to some extent in the US by way of its "first to invent" system. If there is publication between the invention itself and application (which must be not more than 12 months) the novelty requirement is not violated. Canada operates similarly. Most jurisdictions have some sort of a qualification to the strictness of the rule. In UK if the disclosure results from a display at an "international exhibition" and the applicant files notification thereof, novelty is not destroyed. However, the definition of an international exhibition is very restrictive. In Japan, there is a six month window between disclosure and filing. In Singapore, to take an example from a smaller country, if publication occurs no more than 12 months before the date of filing then it will not invalidate the patent if it was made at "an international exhibition" or in paper to a "learned society". The former seems to have the same meaning in the UK Act (which to some extent acted as a model) but the latter is broadly framed, covering a club or association anywhere in the world whose main object is the promotion of any branch of science. Even this might well become subject to judicial interpretation, however, and it might be as well to take advice before making disclosure in this manner. One of the purposes of the rather narrow 12 month window, by the way, is to prevent inventors sitting passively on potentially valuable inventions which competitors (perhaps in competing economies) might pick up.

Generally, disclosure to colleagues in the same organization will not constitute putting an invention into the public domain; they are under an implied obligation of confidentiality.

24.4. Inventive Step

All patent legislation mandates that the invention be not obvious, rather it must incorporate an "inventive step". Here the various jurisdictions introduce a legal fiction, the "hypothetical skilled artisan" or "person skilled in the art". This is a person who knows all about the technology surrounding the invention, but does not have the imagination to conceive of the inventive step in the invention claimed. If he *can* imagine it, the invention will be void for obviousness. The patent examiners, and indeed the tribunals which have to adjudicate if there is a dispute about validity, must assume the mantle of this individual.

When patent applications for biotechnological inventions fail, it is usually on the grounds of obviousness. Obviously each dispute will turn on the facts, and judges have a hard time making decisions. In the UK case of

Biogen v *Madeva*[7] (which is famous as the first biotechnology case to go all the way to the House of Lords, the supreme court of the UK) the Biogen invention was accepted as non-obvious although the technique was well established at the time; the lack of obviousness lay in attempting a shotgun approach which disregarded the probability that introns might disrupt the expression of hepatitis B antigen in *E. coli*. Therefore, if this judicial approach is persuasive elsewhere, it means that if you try an approach to a problem which seems extremely unlikely to work, but use established techniques, and if you are lucky, then you will get a patent. Here the person skilled in the art,[8] would not have tried the shotgun technique because he did not have the imagination to believe that he might be lucky.

To sum up, you might fail to apply for a patent because being heavily involved in your field your technical advance seemed obvious to you; what the patent examiners are interested in, however, is whether it was obvious to anybody else.

24.5. Industrial Application

Some of the earlier legislation in English (the Statute of Monopolies, 1623) spoke of a patent as necessarily being a "manner of new manufacture", emphasizing the utility requirement. There is seldom any difficulty in establishing the utility of a microbial process, but the concept has caused problems in other areas of biotechnology. Many scientists and members of the public have objected to patenting genes or nucleotide sequences on the grounds that they are discoveries rather than inventions. The response has been that patents are not issued for genes in their natural state but rather on DNA sequences after they have been isolated and (usually) amplified, and even then only if the requirements of utility and novelty are met.[9]

24.6. Procedure and Form

24.6.1. *Search*

Initially, of course there has to be the realization that an invention may have been created and so a search has to be made of existing patents worldwide, to ensure that it has not been protected previously. The patent office, in the person of the "examiner" will do this check after filing, of course,[10] but it is obviously a complete waste of time and money to file when the invention already exists. The authorities, recognizing the extreme diversity of possible inventions, have devised classification systems, so that if one is a virologist, he can begin by looking at previous patents relating to viruses rather than to broom handles. There is an International Patent Classification (IPC) system,[11] devised by the WIPO based on one of 8 initial letter codes plus Arabic numerals and running to over 67,000 subdivisions. Any one invention, as might be expected, can have several different classifications. Most microbiology inventions will fall under category C. For example, the patent entitled "a vaccine-induced hepatitis B viral strain and uses thereof"[12] has six classifications beginning with C12N 15/51. The EPO generally uses this system but the United States Patent Office (USPTO) has an alternative system based on Arabic numerals, which will gather most microbiology patents under category 435. Thus "dengue virus peptides and methods"[13] has 13 categories; the first of which is 435/69.3. Where known, a published patent will list numbers under both systems.

A patent attorney will of course do the search, but at an appropriate fee. The task is becoming much easier with modern methods of information technology. Websites with search engines, which take keywords, especially those operating Boolean logic, make the task quite feasible from one's own office. A list of sites is given in Table 24.1.

24.6.2. Drawing up the application

The document does not have to adhere to an unalterably rigid format, but must include a title, a description/specification, claims and abstract. It can also include drawings where relevant and preferred embodiments, that is forms, which the invention may eventually take. The US authorities prefer the following sequence:

Title → Abstract → Inventors → References/related patents → Claims → Description → Examples

Table 24.1. Useful websites.

Site	Comments
http://patents.uspto.gov	Front pages of all US patents from 1976
http://www.european-patent-office.org	Search of EPO, PCT and Japanese patents in facsimile
http://www.surfip.gov.sg	Singapore initiative giving access to many databases (EPO, UKPTO, WIPO, USPTO)
http://www.patent.gov.uk/ dpatents/ how_prep.html	Instructions on how to frame a patent application in UK
http://qpat1.qpat.com	Front page information of all US patents since 1974
http://patent.womplex.ibm.com	IBM site — complete US patent information for last 23 years

All of these sites are free but some may need registration. Some sites offer large numbers of links to search sites, see for example: http://www-sul.stanford.edu/depts/swain/patent/patdbases.html (Stanford University) and http://www.ipmall.fplc.edu.psa/ (Franklin Pierce Law Center)

24.6.3. *Specifications*

This section of the application is intended to be read by a person having skills related to the invention. Such an addressee may be one person, but if an invention encompasses two or more disciplines, this role can be split. For example, a microbiology application may combine a genetically engineered bacterium with a chemical purification process for its product. It will state the prior art and then go on to describe the invention in terms, which allow the person skilled in the art to make it and use it. This is called "enablement".

It is often the case that a skilled person cannot perform the processes described in a microbiology invention because the microorganism is not available to him, or cannot be described, however fully, in manner which will allow reproduction of the technique unless the microorganism is cultured. There is thus a scheme, under an international convention, The Budapest Treaty of 1977, for the deposition of microorganisms in a registered depository. In Europe, this occurs at the date of publication of the application, that is, before the grant of the patent. In the US, it is at the time of the grant.[14] It is mandated by many authorities, for example the EPO.

24.6.4. *Claims*

These can be multiple and overlapping. For example, a patent for "cDNA sequence of Dengue virus serotype 1"[15] has 21 claims, the first for the isolated virus, leading on to a kit for its detection and various antibodies and antigens. The claims are often set down first when drafting a patent, and then they lead the draft of the specifications. It is therefore important that they be clear and succinct.

24.6.5. *Embodiments*

Often an invention may be conceived in several different forms, called embodiments. For example, the kit to detect the Dengue virus strain referred to above could be envisaged as using liquid reagents, freeze dried reagents, or even as the dipstick type of test. "Preferred" embodiments are often noted. The idea here is to show that the invention can be reduced to a useful, working form. If it is left at the stage of an idea or discovery, it will probably not be patentable.

24.6.6. *Filing*

This is done on prescribed forms, with the fee and the patent application attached. The costs levied by the various patent offices are now almost invariably accessible on the Internet (Table 24.1). Indeed filing on line is rapidly coming into use and its status in the selected patent office can be checked by going into these same sites.

24.6.7. *Priority*

Most countries operate a "first to file" system for priority, that is the patent is deemed to be owned from the date of the application to the relevant national patent office. Again, the US system is different in that it operates a "first to invent system" for the priority date. First to invent must be interpreted as first to reduce to practice, or in other words, show that it works. This is why properly maintained laboratory notebooks are so important in the interests of obtaining protection in the US. Should any dispute arise over priority (called "interference" in the US) the record in the notebook would be decisive.[16] Notably, an applicant from another country who files in the US has the same advantage in respect of being able to prove priority by being first to invent.

24.6.8. *International application*

The grant of a patent in your own or host country is important, but patents are highly territorial and will not prevent somebody from exploiting an invention elsewhere. Protection in the US, where your invention could be exploited by others in the face of the huge and affluent market, is obviously important, but might be equally so in Europe, possibly also Australia and Japan depending on its nature. For protection in the US — whose patent granting regime, by the way, is acknowledged by all commentators to be extremely liberal[17] — and any other country included in the PCT an international application is filed in the home patent office. There are now 109 contracting states and any number can be chosen. The local patent office will act as a post office in transmitting the application to an International Searching Authority (in effect one of the high quality national patent offices) who will prepare a list of document relevant to the prior art and send this to each of the countries included in the application. There is also the option of an international preliminary examination report, which will obviously save costs should it indicate that filing is unlikely to be successful. Only a single set of fees need be to paid in respect of an application under the PCT.

The final set of patents will however be granted separately by the countries cited in the application, if it is decided to proceed in them. This again emphasizes the enduring territorial nature of patent law.

24.6.9. *Publication*

In most countries, applications are published automatically without examination, 18 months after filing. This is also the case with PCT applications. Until recently, in the US publication only occurred at the time of the grant of the patent itself. However its Intellectual Property and Community Omnibus Reform Act (1999) harmonizes US practice with the rest of the world. Publication is important for it is only thereafter that action for infringement can be mounted. Infringement refers to the use of the invention by a person other than the inventor or the person to whom he has sold, mortgaged, or licensed it. Should this happen the patent holder has the right to challenge the infringement in court and seek damages, or alternatively an "account of profits". The law in this area is complex and replete with case precedents, so it will not be considered further here.

24.6.10. *Grant*

The actual conferment of the patent is called the "grant". If all goes well and there is no opposition, this will be between three and four years after the priority date. There are usually renewal fees payable after a certain number of years, up to 20 years of course. Some pharmaceutical patents including, for example new antibiotics from bacteria or fungi, cannot be marketed until there is extensive safety testing by way of clinical trials. There is thus provision, in some jurisdictions, for an extension of the protection period for this type of patent. An attempt at an elementary flowchart for the whole process is made in Fig. 24.1.

24.7. Recent Trends in Biotechnology and Microbiology Patents

What can be patented in the life sciences? It has been a longstanding principle that theories and discoveries cannot be patented. Thus, Crick and Watson's discovery-cum-theory of the structure of DNA, useful though it has been, was not patentable. If you discover a new thermophilic bacterium as a product of nature that will not in itself enable a patent; if however you find a way of culturing it in the laboratory the situation is entirely different. See for example "*Aspergillus* culture claim".[18] Products, processes, methods and devices, all attract protection under the patent system.

24.7.1. *Microbial products*

There is no difficulty about patenting anything produced by micro-organisms as long as the triad of novelty, inventive step and utility are met. Proteins produced spontaneously or by recombinant DNA techniques,

Identify invention

↓

Search for prior art

↓

Prepare application

↓

File and get
priority date

↓

Formalities
Examination

International
formalities examination

↓

International search

↓

Patent Cooperation
Treaty (PCT) publication

Decide on PCT
application within 12 months

Examination/search

↓ ↓

Publication

↓ ↓

Grant

↓ ↓

Renewal

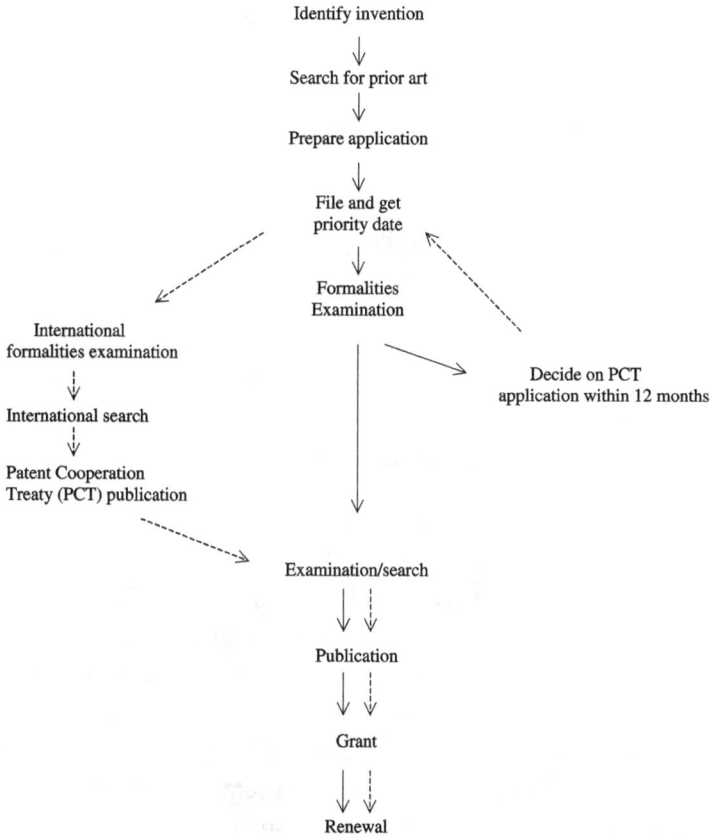

Fig. 24.1. Procedural flow diagram. This is intended as a very general scheme applicable worldwide; from filing to grant will usually take 4–5 years. PTC: Patent Cooperation Treaty.

antibiotics, flavors, and the myriad of other chemicals produced by the marvelous synthetic capacity of bacteria are all eminently patentable. See for example "The fermentation production of eicosapentaenoic acid containing oil by cultivation of heterotrophic diatoms".[19]

24.7.2. Base and amino acid sequences

Most people will be aware of the debate in the media over the "patenting of genes". One of the latest statements came jointly from Sir Aaron Klug

(Royal Society of London) and Bruce Alperts (National Institutes of Health) in *Nature*[20] when they stated that "The intention ... to patent DNA sequences themselves, thereby staking a claim to large numbers of human genes without necessarily having a full understanding of their functioning, strikes us as contrary to the essence of patent law ..." However it is clear that genes of known expressivity, if isolated, can be patented in most, probably all, jurisdictions. The European Patent Convention (EPC) articulates the enablement simply thus:

> *"Biotechnological inventions shall also be patentable if they concern (a) biological material which is isolated from its natural environment or (b) produced by means of a technical process even if it previously occurred in nature."*[21]

This obviously covers nucleic acids, proteins and peptides, if they can be shown to be useful. Particularly contentious has been the status of expressed sequence tags (ESTs). Celera Genomics, Dr Craig Venter's company in the US appears initially to have been successful if the case in protecting them if it could maintain that they could be used as probes to elucidate the sequence and function of a whole gene, if only some time in the future.[22] The European Patent Office (EPO) is however opposed to the patenting of ESTs and the USPTO has recently tightened up slightly.[23] But the guidelines reiterate that no patent can be filed for a natural product occurring in nature. So a genome in itself, human, mammalian bacterial, or other cannot be protected. However, if the gene is cloned, it is then patentable like any ordinary chemical, as long as it can be demonstrated that it has "value in the real world". Notwithstanding that however, if one were to isolate a hitherto unknown microorganism from soil or a protein of unknown function from a tissue, one could claim that a use would be found for it one day.

24.7.3. *Life forms per se*

Historically there has been no difficulty in patenting lowly life forms. Louis Pasteur received a patent for a yeast strain in 1873[24] and in the landmark US case *Diamond* v *Chakraborty*,[25] an engineered bacterium was successfully protected by its inventor. Only recently, with the possibility of cloning or engineering higher organisms has a new debate erupted. The Harvard "oncomouse" was patented in US although the application was refused in several other jurisdictions, and it is still subject to opposition proceedings in Europe. Objections are partially based on the width of the claims. Thus, the patents (there are a series) seek to cover the technique of inserting the gene in any animal leading to the specter of oncocats and oncogiraffes. In Canada, the patent office refused it but on the appeal, the Federal Court allowed the grant. The Canadian Government is in the course of appealing to the Supreme Court, which has just refused it.

Human beings of course are a special case. The patent system in the US is notoriously favorable to the inventor but the 13th Amendment to the constitution forbids slavery and involuntary servitude of any human beings and so one presumes that a cloned human being would be refused as an invention although there is apparently nothing on the US patent legislation specifically blocking such applications. In any case, an application to patent any form of human being is implicitly forbidden by Article 4 of the UN Universal Declaration of Human Rights, and probably any modified germline would be refused since this could lead to a usable human form. The European Patent Convention's (EPC) exclusions policy is cogently expressed:

(1) *Processes for reproductive cloning of human beings.*
(2) *Processes for modifying the germline genetic identity of human beings.*
(3) *Use of human embryos for industrial or commercial purposes.*
(4) *Processes for modifying the genetic identity of animals ... likely to cause them suffering without substantial medical benefit to man....*[26]

It is difficult to believe that in any jurisdiction with a humane basis, this will not be followed. Nevertheless, there should be minimal problems in respect of microorganisms as long as they are not environmentally dangerous — indeed they are specifically exempted from any protection in several instruments, notably Article 53 (b) of the EPC — they are too lowly to have any rights.

24.7.4. *Algorithms, methods and processes*

Historically, discoveries, theories, artistic works, methods for doing business and presentation of information (to summarize) have been excluded. Business methods are now patented in the US — (Amazon.com has received a US patent for its so called "one click" marketing system, but this is now under challenge) and in microbiology or molecular biology it is difficult to think of any new process or method of doing things, perhaps cutting down the steps or time involved in the isolation of a fermentation product, which if it works, is not useful in one way or another.

However, you may think of a method of making a bacterium produce a protein, say, in a manner distinct from that in a pre-existing patent for producing the same protein, and the owner of this might challenge. Such is the situation in the currently unresolved dispute in the US over the production of erythropoietin.[27] Amgen markets it after having cloned the gene and inserted it into Chinese hamster ovary cells; the other company Transkaryotic Therapies persuades human cells to produce it by means of a promoter.

A difficulty also arises when the process is known already but either a new starting material or new product is claimed. For example, a patented bacterium could be used to convert compound A into compound B, and either A or B is novel; obviously, A could be a component of the culture medium and B a protein. Until recently, in the US (via what is known as the *Durden* case[28]) such a maneuver was not patentable, but the law was changed.[29]

24.7.5. *Mosaicism*

Good ideas can come in many forms — one might realize that elements of a number of known processes or inventions can be abstracted, combined in some novel way, and made useful in some sense. This is known as "mosaicism" and is attacked by patent lawyers on the grounds of obviousness; however, the rejoinder must be that if assembly has been obvious then somebody else would have constructed it. Therefore, it may be that if one can put together a variety of techniques in a novel way, then the result will be patentable. In microbiology, one could think of using an unusual but established culture medium for a known bacterium, along with a known but hitherto (in that system) unused method for purifying its product. A patent application might succeed.

24.8. Potential Licensees

Even if an invention is patentable, the cost thereof may be prohibitive unless a clear commercial advantage is likely; for that reason an inventor will be at an advantage if he can identify a company or organization likely to take up a license for exploitation of the invention. It is as well to keep an eye on this when developing the invention. Before discussion with an outside party, one should get a non-disclosure agreement signed. In rare cases, especially in universities, an invention will be regarded so highly that a spin-off company might be set up with the employee taking an executive role.

Governments actively encourage inventors to take out patents. Inventions put technology into the public domain and stimulate business and further inventions. In most countries then, apart from a patent being taken over for the use of the government for perhaps defense purposes, there is a system of compulsory licensing. If an invention, having been patented, is not being developed after say three years, any person can apply for a compulsory license. Even this lag period can be waived if the invention is related to food, medicine, or treatment of disease.

Further Reading

1. As might be expected, the literature on patents in biology, including microbiology, is huge. Since the law is highly territorial, in theory there could be a separate book to explain the system in each country of the world. The scientific journals *Nature* and *Science* take a keen interest in scientific patents and keep the reader up to date. The specialist journal *Biotechnology Law Report* is especially valuable. For a bird's eye view worldwide, see Gutterman AS and Anderson BJ, *Intellectual Property in Global Markets*, Kluwer Law International, London, 1997.

2. There has been a long standing ban on patents for medical treatment in most if not all countries, in the US from 1862. Of course, the scope of "medical treatment" has to be evaluated. For a discussion, see Garris JL, The case for patenting medical procedures, *Am. J. Law Med.* **XII**: 86, 1996.

3. The case involved the claim by a postdoctoral fellow, Jenny Chou, for a share in a patent for a herpes simplex vaccine in which her former supervisor was cited as the sole inventor. The judge held that she had no standing since by law an employee's inventions are the property of the employer. See *Science*, 31 March 2000, p. 2399. For a comprehensive discussion, see Puri K, Ownership of employee's inventions: A comparative study, *12 Intellect. Prop. J.*, December 1997, p. 34.

4. The *Gayler* case is discussed in Halpern S, Nard CA and Port KL, *Fundamentals of United States Intellectual Property Law*, Kluwer Law International, The Hague, 1999, p. 201.

5. *General Tire and Rubber Co v Firestone Tire and Rubber Co, 1 Weekly Law Reports* 799, 1972.

6. *Bristol Myers Squibb v Baker Norton.* See discussion in *CIPA J.*, June 2000, p. 293.

7. *Biogen v Madeva* (1996) has been extensively discussed and analyzed. See for example Cornish WR, *Intellectual Property*, Sweet and Maxwell, London, 1999, p. 150. Although it has no binding force outside England, due to the closeness of the reasoning it is likely to be persuasive in most common law jurisdictions. In fact, the Biogen patent was revoked by the House of Lords for other reasons, including over-broadness of claim. Despite that, it is said that Sir Keith Murray, who did the work in Edinburgh in the late seventies (the first of his patents being granted in 1978), became a multimillionaire. *Biogen* also appeared in the Singapore courts in 1994 in connection with the infringement of its rights there, the patent having been registered in 1987; see *Biogen Inc v Scitech Medical Products* reported in 1 Asia Intellectual Property Reports, 1996.

8. This archaic term persists unchanged to date, being used freely for example in the *Utility Examination Guidelines for Biotechnology*, Federal Register, US Vol. 66, No. 4, 5 January 2001.

9. Federal Register, citation 8, p. 1093.
10. In some countries there is a self-assessment system, whereby, all other requirements being met, the patent will be issued without formal enquiry into novelty; it is then open to interested parties to challenge the application after publication.
11. For international classification see http://www.wipo.org/eng/general/ipip/strasbourg.htm and for US see http://www.uspto.gov/web/offices/ac/ido/oeip/taf/def/index.htm.
12. WO (World Patent) 9966047.
13. US Patent 5,824,5060.
14. For a discussion of the microbial deposit system see Cornish, citation 7, p. 234.
15. US Patent 6,017,535.
16. The US patent attorneys Ladas and Parry suggest *inter alia* that lab books be (1) permanently bound (not loose leaf); (2) gels and chromatographs be permanently mounted; (3) regularly scrutinized by a third party and signed. This last is obviously the most difficult. See http://www.ladas.com.
17. For discussions of the more liberal US system see Mahendra B, Novelty and method in drug treatment, *New Law J.*, 3 November, 2000, p. 1502; also in Cook T, Doyle C and Jabbari D, *Biophar-maceuticals, Biotechnology and The Law*, Stockton Press, New York, 1991, p. 120.
18. US Patent 4,593,005.
19. US Patent 5,567,732.
20. See *Nature* **404**: 325, 2000.
21. Article 3 of 98/44/EC: Legal protection of biotechnological inventions, 1998.
22. For discussion, see Sheraton H, Gene genius — patent rights and the human genome project, *Biotechnology*, November 1999, p. 9.
23. Federal Register, citation 8, p. 1094.
24. US Patent 141,072.
25. *Diamond v Chakrabarty* 447 US 303, 1980.
26. Art 6 of 98/44/EC.
27. See *Nature* **404**: 532, 2000 [Note added: to the surprise of many, the courts appear to have decided that Transkarystuis process infringes that of Amgen].

28. In *re Durden* 763 F.2d 1406, 226 USPQ 359 (Fed. Cir. 1985).
29. Training materials for treatment of product and process claims in the light of in *re Brouer* and in *re Ochai* and 35 U.S.C 103 (b), see http://www.uspto.gov/web/offices/pac/dapp/oppd/ppclms.htm.

Index